Three Faces of Sun Tzu

Sun Tzu's *Art of War* is widely regarded as the most influential military and strategic classic of all time. Through reverse engineering of the text structured around fourteen Sun Tzu themes, this rigorous analysis furnishes a thorough picture of what the text actually says, drawing on Chinese-language analyses, historical, philological, and archaeological sources, traditional commentaries, computational ideas, and strategic and logistics perspectives. Building on this anchoring, the book provides a unique roadmap of Sun Tzu's military and intelligence insights and their applications to strategic competitions in many times and places worldwide, from Warring States China to contemporary US/China strategic competition and other twenty-first-century competitions involving cyber warfare, computing, other high-tech conflict, espionage, and more. Simultaneously, the analysis offers a window into Sun Tzu's limitations and blind spots relevant to managing twenty-first-century strategic competitions with Sun Tzu–inspired adversaries or rivals.

SCOTT A. BOORMAN is Professor of Sociology at Yale University.

The Masterpiece: A Thousand Peaks and Myriad Ravines by Gong Xian 龔賢 (c. 1617–1689)
Image courtesy of Museum Rietberg in Zürich

Three Faces of Sun Tzu

Analyzing Sun Tzu's Art of War, A Manual on Strategy

SCOTT A. BOORMAN
Yale University

With the collaboration of Sun Jianyuan

CAMBRIDGE
UNIVERSITY PRESS

Shaftesbury Road, Cambridge CB2 8EA, United Kingdom

One Liberty Plaza, 20th Floor, New York, NY 10006, USA

477 Williamstown Road, Port Melbourne, VIC 3207, Australia

314–321, 3rd Floor, Plot 3, Splendor Forum, Jasola District Centre, New Delhi – 110025, India

103 Penang Road, #05–06/07, Visioncrest Commercial, Singapore 238467

Cambridge University Press is part of Cambridge University Press & Assessment,
a department of the University of Cambridge.

We share the University's mission to contribute to society through the pursuit of
education, learning and research at the highest international levels of excellence.

www.cambridge.org
Information on this title: www.cambridge.org/9781108471039

DOI: 10.1017/9781108686877

First published 2024

A catalogue record for this publication is available from the British Library

A Cataloging-in-Publication data record for this book is available from the Library of Congress

ISBN 978-1-108-47103-9 Hardback
ISBN 978-1-108-45698-2 Paperback

Contents

Figures

Tables

Acknowledgments

First and foremost, thanks are owing to the Brady-Johnson Program in Grand Strategy at Yale University and to its founders – John Gaddis, Charles Hill, and Paul Kennedy – for providing an intellectually stimulating environment when the present study was in its formative stages. Thanks are also owing to the Whitney and Betty MacMillan Center for International and Area Studies at Yale, which provided financial support.

Special thanks are owing to Katsiaryna Zinavenka, alumna of the Boston University School of Law Graduate Tax Program, who provided strategic insights pertaining to the last section of the Conclusion, "A Sun Tzu for the Twenty-First Century."

Assistance on many parts of the Sun Tzu project was provided by the Yale Library and its dedicated staff, including Suzanne Mirabile and Michael Meng, Head of East Asia Library and Librarian for Chinese Studies at Yale. As one of the world's great research libraries, the Yale Library was an invaluable resource at every stage of the Three Faces of Sun Tzu project. Additional library and archival support came from the Eccles Library of the United States Naval War College; from the Naval Historical Collection there; and from the Liddell Hart Centre for Military Archives, King's College London. Professors Grace Kao and Philip Gorski, as Chairs of the Yale Sociology Department in a time of pandemic, provided a steadily supportive Departmental environment while this book was approaching fruition.

Among scholars with early China specialties, thanks for assistance on specific issues are owing to Roger Ames, Victor Mair, Keith McMahon, and John Minford.

Martin Shubik (Yale and Cowles Foundation for Research in Economics) provided ideas relevant to each of the three faces of Sun Tzu from Shubik's deep background spanning economics, finance, game theory, and defense analysis. Ideas of William J. McGuire (Yale) on the role of creative heuristics have influenced the present line of research on Sun Tzu. I am indebted to Natasha Sood for discussion of ideas of Carl Jung that have influenced Theme #8's treatment of coincidences. Thanks are also owing to Rene Almeling (Yale), Antony Best (London School of Economics), Janet Harris (Cardiff), James Hevia (Chicago), David Howarth (Cambridge), Jacob Howarth (MSt Oxon), Peter Kornicki (Cambridge), Gagandeep S. Sood (London School of Economics), and Yuri Tschinkel (Courant Institute of Mathematical Sciences, New York University) for insights into specific issues relating to the present work. Franklin Antonio (Qualcomm), Paul Horowitz (Harvard), Sharanya Majumdar (University of Miami), Inna Polichtchouk (European Centre for

Medium-Range Weather Forecasts), Mary Power (University of California, Berkeley), and Thomas Schoener (University of California, Davis) provided scientific background for evaluation of Sun Tzu substance having physical or biological science aspects.

Thanks are also owing to Sihini Trinidad (Tec de Monterrey alumna) for her support cohosting the Yale International Alliance common room events entitled "A Sun Tzu for our time."

A further intellectual debt is owing to three anonymous referees who reviewed the concept for the present book for Cambridge University Press. Their constructive suggestions have contributed significantly to the substance and architecture of the present book.

Research assistance for the project, including data science aspects, was provided by Pei Hua Chen (Yale College '21). For Russian-language translation assistance thanks are also owing to Irina Gavrilova (Yale College '17). H. Michael Bailey provided invaluable project logistics support throughout, including but not limited to computing aspects.

Special thanks are owing to Robert Dreesen, Senior Editor at Cambridge University Press, for his steadfast encouragement and moral support while the present book was in preparation.

Expert assistance in book production was provided by Ken Moxham, Chloe Quinn, Erika Walsh, and Hannah Weber at Cambridge University Press.

Any errors of fact or interpretation in the present study are Scott Boorman's alone.

Author's Note on Chinese-Language Romanization

The goal is to navigate Chinese-language romanization and related issues as unobtrusively as possible, allowing for the diversity of romanization systems (none of them ideal). Doing so calls for exercise of judgment and common sense. The pinyin romanization system, now the most widely used, is treated as a default choice. Reflecting the fact that the name is now almost as widely known in the West and elsewhere around the world as in China, an exception is made for Sun Tzu himself, where reference will be to Sun Tzu, not to the pinyin romanization Sunzi.

Book, journal, and book chapter titles containing romanized Chinese-language words or proper names will be cited without change regardless of the romanization used (in particular to avoid complicating database searches).

Author names of Chinese heritage appearing in published English-language sources are similarly left unchanged, again regardless of romanization used.

Non-pinyin romanizations present in quotations (e.g., when quoting from various translations of Sun Tzu or works about Sun Tzu) are converted to pinyin. One purpose of doing so is to avoid potential reader confusion when the same words used in the present book's surrounding discussion appear in pinyin.

Because the equivalence

Sunzi (in pinyin romanization) = Sun Tzu (in Wade–Giles romanization),

should be clear from context, no romanization substitution will be made in quotations containing a reference to "Sunzi" or "Sun Zi."

Author's Note on Chinese Characters

As a general rule, Chinese characters associated with pinyin romanizations of words, phrases, proper names, etc. are assigned to the Glossary and related lists located at the start of the online annex.

Exceptions:

- modern Chinese-language scholarly materials, where citations are accompanied by Chinese characters;
- titles of pre-modern Chinese works of historical type when first mentioned on pp. xxiii–xxiv;
- Chinese history information in the Chronology section (pp. xviii–xx);
- Chinese characters embedded in titles of Western-language scholarly works.

Occasional further exceptions are made for purposes of conceptual clarity:

- Chinese characters may be included in the main text or its footnotes for reasons of disambiguation of concepts.

  Example: The concept of *wu xing* 無形 ("formlessness") central to Theme #8 has a pinyin romanization coinciding with that of the distinct concept of *wu xing* 五行 (the so-called five elements or five phases) that appears widely in military, medical, and other early texts, including in Sun Tzu verse VI.31 (see Passages #5.16, #8.1, etc.).[1]

  Example: The traditional Chinese concept of *qi* 氣 ("spirit"), whose meanings are complex but in a military context include fighting spirit or morale, has a pinyin romanization coinciding with the distinct concept of *qi* 奇, Griffith's "extraordinary force" also variously translated as "crafty," "unorthodox," "surprise," etc. (see Theme #14 analysis).[2]

Chinese characters appear more extensively in the online appendices, where assorted technical issues pertaining to the Chinese text of Sun Tzu are addressed.

---

[1] For further background on *wu xing* 五行 see #5 footnote 103 and #9 footnote 13.
[2] For further background on *qi* 氣 see #3 footnote 80.

Author's Note on Online Annex

In order to limit the bulk of the physical book, Appendices 1–17 are located in an online annex to the present volume maintained by Cambridge University Press.

The online material starts with a Glossary and other lists (of terrain types, Chinese texts, etc.) supplying Chinese characters to accompany pinyin romanizations.

Appendices 1–14 (corresponding to Themes #1–#14, respectively) serve two basic purposes (see also pp. A-30ff. in Scholarly Controls section in online annex).

First, Appendices 1–14 repeat the main text passages with further comments or footnotes. Many of those address Chinese-language textual, interpretive, or translation issues, thereby serving to backstop and amplify analysis in the main text without interrupting its flow. While not exhaustive of all Sun Tzu issues (a tall order given the lengthy history of the Sun Tzu text as well as the highly developed state of Sun Tzu studies in China, the West, and elsewhere), this material positions the present study of Sun Tzu vis-à-vis various ambiguities or uncertainties in the text. Some of those involve live scholarly debates, commonly technical, sometimes contentious. Where the present study takes a stand on those debates, the goal is to make that stand and its rationale as transparent as possible. Where there is no need to take a position, the reader is simply alerted to what the debate is about.

In order to avoid duplication, content of footnotes accompanying the main text appearance of a particular Sun Tzu passage is generally not repeated when the same passage reappears in an appendix, and vice versa. A reader interested in engaging with all aspects of the present study's treatment of a given Sun Tzu passage, including discussion of Chinese-language issues, should therefore consult both of that passage's appearances, one in the main text, one in the pertinent appendix in the online annex. Most Chinese-language issues are addressed in appendix footnotes located in the online annex.

Second, Appendices 1–14 serve as repositories of supplementary Sun Tzu passages, rounding up Sun Tzu content that contributes to development of a given theme, yet less so than passages assigned to the main text (e.g., because obscure, burdened by textual issues, or simply less relevant to developing that theme's focus). Capitalizing on the flexibilities allowed by cyberspace, inclusion of these "also-ran" passages helps the analysis achieve a level of thoroughness in coverage of the Sun Tzu text that would be lacking were they to be ignored. At the same time, this scholarly device helps avoid complicated, possibly heavily qualified, additions to the main text (which, of course, would also add to the physical bulk of the present book).

As an aid to substantively oriented readers, comments at the start of each of Appendices 1–14 offer an estimate of where its supplementary passages deliver most value-added.

Also assigned to the online annex (Appendix 15) are analyses of two additional military-technical topics (command, control, and communications; special weapons) where the Sun Tzu text makes significant comments but where its coverage is not extensive enough to warrant adding further themes to the present set of fourteen.

Appendix 16 is a concordance of Sun Tzu verses and passages used in the present study.

Appendix 17 is Griffith's Sun Tzu translation (Oxford University Press, 1963), which is the translation on which the present study primarily builds (THE ART OF WAR by Sun Tzu, translated by Griffith (1963) 9,600w from pp. 63–149 © 1963 by Oxford University Press, Inc. By permission of Oxford University Press, USA).

The online annex can be accessed via this link: www.cambridge.org/Boorman

Chronology of Chinese History

Dynasties	Dates[1]	Sun Tzu–related events
Xia 夏 (whether, and in what sense, such a dynasty existed is disputed)	c. 2070–1600 BC	
Shang 商	c. 1600–1046 BC	
Zhou 周	1046–256 BC	
Western Zhou 西周	1046–771 BC	
Eastern Zhou 東周	770–256 BC	
Spring and Autumn period 春秋	722–476 BC	Sun Tzu text early date, c. 500 BC[4]
Warring States 戰國[2]	475–221 BC[3]	Sun Tzu text, a later date, c. 453–403;[5] or c. 345–272 (another later date)[6]

[1] Dates are based on Wilkinson, §A.2, pp. 4–5.

[2] For detailed chronology of the Warring States period, the period most relevant to the present study, see Yang Kuan 楊寬, *Zhanguo shiliao biannian jizheng* 战国史料编年辑证 (*Excerpted Evidence for the Chronology of the Warring States Period*) (Shanghai: Shanghai renmin, 2001), coordinating multiple types of sources (including archaeological information) broken down by year, 476–221 BC. For a short description of this Yang Kuan work, whose source material includes historical, strategic, and philosophical texts, both traditionally transmitted and excavated, see Wilkinson §58.6.3.2, p. 777.

[3] There is little consensus on a start date to assign to the Warring States period. Sima Qian's *Shiji* (see pp. xxiii–xxiv below) gives a 475 BC start date; for context, 479 BC is Confucius's traditional date of death. A common alternative periodization in Chinese historiographical tradition would align the start of the Warring States period on the partition of the state of Jin into the three states of Han, Wei, and Zhao. Although the date of that partition is also debated, by many accounts it would start the Warring States period in 403 BC, leaving an unnamed interval of some seventy years following the end of the Spring and Autumn period that preceded it. For further background on these periodization issues see Wilkinson, §58.1, p. 767.

[4] Date based on Sun Tzu's *Shiji* 65 biography, containing the famous story of his drilling the ruler's concubines. That biography is regarded by some modern scholarly authorities as a stylized pseudo-biography and Sun Tzu the person as fictive. See pp. 12–13 below.

[5] See p. 218 of 1988 Yates article cited in Introduction footnote 28, also cautioning that not all the Sun Tzu text is from the same period.

[6] See Brooks, pp. 70–73 chronology (for summary see Table 12A on p. A-35 in online annex). Brooks assigns a particularly late date (c. 272 BC) to Sun Tzu's espionage chapter, a human generation or more after the rest of the thirteen-chapter Sun Tzu text. Further perspectives come from Griffith, p. 11 (supported by Ames, p. 24), placing the Sun Tzu text as c. 400–320 BC; Wilkinson §28.4.1, pp. 348–50 entry on Sun Tzu, assigning it a date "probably from the fourth to the early third century BCE."

(cont.)

Dynasties	Dates	Sun Tzu–related events
Qin 秦 (unification of China)	221–206 BC	
Han 漢	202 BC– 220 AD	
Former Han 前漢 (also known as Western Han)	202 BC–8 AD	Archaeologically recovered Sun Tzu text, dated to *c.* 140–118 BC
Later Han 後漢 (also known as Eastern Han)	27–220	Cao Cao 曹操 famous general and Sun Tzu commentator, 155–220
Wei, Jin, Nan-Bei Chao 魏晉南北朝	220–589	
Three Kingdoms or Sanguo 三國	220–80	
Wei 魏	220–65	
Han 漢	221–63	
Wu 吳	222–80	
Jin 晋	265–420	
Nan-Bei Chao 南北朝 (Northern and Southern Dynasties)	420–589	
Sui 隋	581–618	
Tang 唐	618–907	
Wudai Shiguo 五代十國 (The Five Dynasties and Ten Kingdoms)	907–79	
Song 宋	960–1279	
Northern Song 北宋	960–1127	*Seven Military Classics*
Southern Song 南宋	1127–1279	*(Wujing qishu* 武經七書, *c.* 1080)[7]
Liao 遼 (Qidan 契丹, Khitan)	916–1125	
Jin 金 (Nüzhen 女真, Jurchen)	1115–1234	
Xixia 西夏 (Dangxiang 黨項, Tangut)	1038–1227	Tangut translation of Sun Tzu *c.* 1100–50
Yuan 元 (Menggu 蒙古, Mongol)	1271–1368	

[7] In part galvanized by ongoing struggles with the Tangut as well as with the Liao states, emperor Shenzong of Song 宋神宗 (r. 1067–85) authorized compilation, for purposes of the revitalized Military Examination, of the newly established "Seven Books" military canon (*Seven Military Classics*) comprising seven military works including the Sun Tzu text. This was the military counterpart to the Civil Service Examination at whose core stood mastery of the Confucian canon. For further background on the *Seven Military Classics* see Wilkinson, §24.8.3, pp. 350–51.

(cont.)

Dynasties	Dates	Sun Tzu–related events
Ming 明	1368–1644	
Qing 清 (Manzhou 滿洲, Manchu)	1636–1912	
Post-dynastic China:		
Republic of China 中華民國	1912–49	
People's Republic of China (PRC) 中華人民共和國	1949–	

List of Abbreviations for Frequently Cited Works

To streamline citations, a reference in the present book to one of the names highlighted in bold in (1)–(3) below, followed by a page number – but without further citation details – is a reference to the work cited below. Example: "Griffith, p. 100."

(1) Abbreviations pertaining to translations of Sun Tzu into English

Ames: Sun-tzu, The Art of Warfare: the first English translation incorporating the recently discovered Yin-ch'üeh-shan texts (translated, with an introduction and commentary, by Roger T. Ames; New York: Ballantine Books, 1993).

Denma: Sun Tzu, The Art of War: the Denma translation (translation, essays and commentary by The Denma Translation Group; Boston, MA/London: Shambhala, 2002).

(Note should also be made of a website created by Kidder Smith, co-leader of The Denma Translation Group, which provides a word-for-word correspondence between each Chinese character in the Sun Tzu text, its pinyin romanization, and a suggested English counterpart word. See http://learn.bowdoin.edu/suntzu/index.html [visited October 6, 2021]. This website is an important tool for Sun Tzu scholarship, with the caveat that the critical edition text of Sun Tzu to which its word-for-word correspondence is pegged differs in places from that in the presently preferred edition cited on p. 40 below.)

Giles: Sun Tzŭ on the Art of War, the oldest military treatise in the world, translated from the Chinese with introduction and critical notes by Lionel Giles (London: Luzac & Co., 1910).

Griffith: Sun Tzu: The Art of War (translated and with an introduction by Samuel B. Griffith, with a foreword by B. H. Liddell Hart; Oxford: Clarendon Press, 1963).

Mair: The Art of War: Sun Zi's Military Methods (translated by Victor H. Mair; New York: Columbia University Press, 2007).

Mair's translation draws on the same Sun Tzu Chinese text as does the present study. Mention should also be made of a monograph-length companion work by Mair providing further background on his Sun Tzu translation, along with observations on the Manchu translation of Sun Tzu (translation by the same Qing official entrusted by the Qing emperor to conclude a peace treaty with Britain after the First Opium War, leading to the Treaty of Nanking). See:

Victor H. Mair, "*Soldierly Methods*: vade mecum for an iconoclastic translation of the *Sun Zi bingfa* (with a complete transcription and word-for-word glosses of the Manchu translation, by H. T. Toh)," Sino-Platonic Papers series (Philadelphia, PA: University of Pennsylvania, Department of East Asian Languages and Civilizations), No. 178 (February, 2008), 195 pp.

Minford: *Sun-tzu: The Art of War* (translated, with an introduction and commentary, by John Minford; New York: Viking, 2002).

Sawyer: *Sun-tzu: The Art of War* (translated, with introduction and commentary, by Ralph D. Sawyer, with the collaboration of Mei-chün Lee Sawyer; New York: Basic Books, 1994).

(2) Abbreviations pertaining to other basic Sun Tzu–related sources

Brooks: E. Bruce Brooks, "Review article: the present state and future prospects of pre-Hàn text studies," pp. 1–74 in *Sino-Platonic Papers*, No. 46 (July, 1994).

Lau, 1965 article: D. C. Lau, "Some notes on the Sun Tzu 孫子," *Bulletin of the School of Oriental and African Studies* (University of London), 1965, 28(2), 319–35.

Lau & Ames: *Sun Bin: The Art of Warfare, a translation of the classic Chinese work of philosophy and strategy* (translated, with an introduction and commentary, by D. C. Lau and Roger T. Ames; Albany, NY: State University of New York Press, 2003). (Translation of an archaeologically recovered early Chinese military text that is later than the Sun Tzu text.)

Needham & Yates: Joseph Needham and Robin D. S. Yates (with the collaboration of Krzysztof Gawlikowski, Edward McEwen, and Wang Ling), *Science and Civilisation in China*, Vol. 5: *Chemistry and Chemical Technology*, Part VI: *Military Technology: missiles and sieges* (Cambridge: Cambridge University Press, 1994).

Page ix and Author's Note, p. xxvii identify Gawlikowski as having had a lead role in drafting coverage particularly relevant to Sun Tzu, including the subsection on Chinese literature on the art of war (Needham & Yates, pp. 10–66).

Sawyer, *Seven Military Classics*: *The Seven Military Classics of Ancient China* (translation and commentary by Ralph D. Sawyer with Mei-chün Sawyer; Boulder, CO: Westview Press, 1993). (Includes Sun Tzu plus six other military works.)

(3) Abbreviations pertaining to basic reference works

Early Chinese Texts: Michael Loewe (ed.), *Early Chinese Texts: a bibliographical guide* (Early China Special Monograph Series No. 2; Berkeley, CA: The Society for the Study of Early China and the Institute of East Asian Studies, University of California, Berkeley, 1993).

Wilkinson: Endymion Wilkinson, *Chinese History: a new manual* (5th ed.; Cambridge, MA: Harvard University Asia Center, 2018).

(Citations to Wilkinson's manual in the present study will include both section and page numbers.)

Among its many other uses as a fundamental reference resource, Wilkinson, §58.6.2–58.6.4, pp. 775–78 provides a very useful roundup of traditionally transmitted pre-Han texts and translations.)

(4) Abbreviations pertaining to early Chinese historical works

Zuozhuan 左傳: Historical compilation of uncertain date, estimated as being largely complete by end of the fourth century BC, whose focus is the Spring and Autumn period (771–c. 476 BC).

General background on the Zuozhuan (Wade–Giles romanization: Tso Chuan): See Wilkinson, §48.1.1, pp. 679–80. See also "Introduction" to The Tso Chuan: selections from China's oldest narrative history (translated by Burton Watson; New York: Columbia University Press, 1989).

A complete translation of the Zuozhuan is Zuo Tradition (Zuozhuan): commentary on the "Spring and Autumn Annals" (translated and introduced by Stephen Durrant, Wai-yee Li, and David Schaberg; Seattle: University of Washington Press, 2016), 3 vols.

Shiji 史記: Historical work attributed to Sima Qian (Wade–Giles romanization: Ssu-ma Ch'ien), c. 145–86 BC, traditionally regarded as the father of Chinese historiography (contains biography of the traditionally ascribed author of the Sun Tzu text, translated by Griffith, pp. 57–59).

General background on the Shiji: See entry on the Shiji in Early Chinese Texts, pp. 405–14; Wilkinson, §59.1, pp. 787–93. References to the Shiji will be accompanied by chapter numbers cited in an abbreviated way, for example Shiji 92 for Shiji chapter 92, etc.

Translations: The present study uses translations by Burton Watson:

Records of the Grand Historian: Han Dynasty I (rev. ed.; Hong Kong/New York: Renditions–Columbia University Press, 1993).

Records of the Grand Historian: Han Dynasty II (rev. ed.; Hong Kong/New York: Renditions–Columbia University Press, 1993).

Records of the Grand Historian: Qin Dynasty (rev. ed.; Hong Kong/New York: Renditions–Columbia University Press, 1993).

Watson's translations do not cover the entire Shiji. They will be supplemented as needed by references to the multi-volume Shiji translation initiative of William H. Nienhauser, Jr. and co-workers, entitled The Grand Scribe's Records (Bloomington/Indianapolis: Indiana University Press, 1994–), which also contains extensive annotations and anchoring in further scholarly sources.

See in particular:

William H. Nienhauser, Jr. (ed.), The Grand Scribe's Records. Vol. V.1. The Hereditary Houses of Pre-Han China, Part I by Ssu-ma Ch'ien (Bloomington/Indianapolis, IN: Indiana University Press, 2006).

William H. Nienhauser, Jr. (ed.), The Grand Scribe's Records. Vol. VII. The Memoirs of Pre-Han China by Ssu-ma Ch'ien (Bloomington/Indianapolis: Indiana University Press, 1994).

Nienhauser's Volume VII has also been published in a 2021 revised edition (Bloomington/Nanjing: Indiana University Press/Nanjing University Press, 2021).

With an eye on facilitating scholarly access, citations to Volume VII will be to both 1994 and 2021 editions.

(5) **Sanguo yanyi** (三國演義): Often known as the *Sanguo*, this is a novel attributed to Luo Guanzhong, *fl.* late Yuan–early Ming dynasties, about the Three Kingdoms era of Chinese history, 220–80 AD, see Chronology, p. xix above).

General background on the *Sanguo*: See Wilkinson, §31.1.6, p. 451.

A complete translation is *Three Kingdoms: a historical novel, attributed to Luo Guanzhong* (translated from the Chinese with afterword and notes by Moss Roberts; Berkeley/Beijing: University of California Press/Foreign Languages Press, 1991).

Albeit influential in modern times – including as an influence on the strategic thought of Mao Zedong, who may in fact have absorbed much of Sun Tzu's counsel through this conduit – this is a novel and that fact needs to be kept in mind.

To reinforce this point, mentions of the *Sanguo* in the present study will refer to it by the *Romance of the Three Kingdoms* title of the earlier translation by C. H. Brewitt-Taylor (Shanghai, 1925). That title also appears in relevant analytical literature, e.g., C. T. Hsia, *The Classic Chinese Novel: a critical introduction* (rev. ed.; Hong Kong: The Chinese University of Hong Kong, 2016), chapter II, "The Romance of the Three Kingdoms," pp. 33–70.

In citing the *Romance*, however, page references will always be to the annotated Moss Roberts edition cited above. For further orientation to the *Romance* and its role in the present study of Sun Tzu see pp. 45–46 below.

Introduction

The touchstone of the present work is a cluster of fundamental substantive questions pertaining to the *Sunzi bingfa* ("Military Methods of Master Sun," commonly rendered as "The Art of War"), an ancient Chinese text widely regarded in both China and the West as one of the most insightful strategy writings of all time.

Sun Tzu had many timeless insights about strategy and warfare, providing incisive angles on how to achieve success in conflict, especially if one has no scruples. Yet he did not have all the smart ideas. Which ones did he have? Which ones not? Do the ideas he had fit together in a coherent larger pattern? If so, what is that pattern? In what kinds of conflict does Sun Tzu's way of war work especially well? In what other kinds might it fall short?

In the background of these questions stand two further ones having a distinctly practical aspect. In fifteen words, the first is: *"What did Master Sun know that we still don't (or have yet to absorb adequately)?"* – knowledge that, if put into practice by an alert, aggressive foe (perhaps a state actor, perhaps not), holds seeds of adverse outcomes for the United States and its allies, if not on the battlefield then in a strategic competition outside of any shooting war. Such a question invites particular twenty-first-century scrutiny because Sun Tzu can stake a credible claim to be the world's first information warfare theorist.

Complementing this first question is a second one, equally important if one ever faces a Sun Tzu–inspired adversary: *"What are Sun Tzu's limitations or blind spots?"*

Shedding useful light on the twin questions just posed calls for a carefully crafted analytical approach. The Sun Tzu text has long enjoyed iconic status as one of a small canon of classics of strategy that have emerged from several civilizations, worldwide – in fact, in the eyes of many knowledgeable observers, the leading exemplar of that canon. That towering reputation shows no sign of diminishing. Indeed it may be gathering momentum. Sun Tzu's reputation has benefits for the advancement of Sun Tzu studies, ensuring a continuing flow of attention from influential thinkers and practitioners spanning many professions. Yet it is also a double-edged sword. Attention garnered for reputational reasons is commonly superficial. Sun Tzu's treatise is all too often reduced to capsule summaries, belying the more nuanced treatment that Sun Tzu's ideas merit. Against this backdrop, the present study is the product of a careful reading of the Sun Tzu text, developing an approach anchored in "three faces of Sun Tzu" (or "three Sun Tzus," for short):

Sun Tzu (1), the Warring States Chinese military text, likely the work of multiple hands, geared to achieving success in the warfare of that era;

Sun Tzu (2), the theorist of the military art of war in many times and places;

Sun Tzu (3), the far-reaching strategist and conflict theorist whose insights span grand strategy, cyber conflict, other high-tech conflict, and more.

This tripartition into three Sun Tzus provides much-needed basic orientation for tackling the twin questions posed above. Such a structured approach – which is also a tool for demystifying Sun Tzu – is still in strikingly scarce supply given Sun Tzu's intellectual reputation and influence. It is the type of reading that a major classic of applied philosophy should get.

To that end, the present analysis aims to find and navigate a middle way between text and ideas in the Sun Tzu work. The text is central; equally, so too is its intellectual level.

Without adequate textual grounding, there is a real risk that analyses inspired by Sun Tzu will drift away like helium balloons – conceivably revealing intriguing vistas but not ones that can lay valid claim to Sun Tzu's intellectual mantle and its insights tested in the cauldron of the bitter existential warfare of Warring States China and later over some twenty-five centuries of East Asian history. Without adequate attention to Sun Tzu's substantive content – an idea level of strategic thought comparable to that found in the writings of Clausewitz or Thomas Schelling or Andrew Marshall (to cite some leading Western examples of the genre) – there is a risk that study of the Sun Tzu text will become enmired in a multitude of specific, highly technical, and possibly insoluble philological and historical puzzles.

To navigate these competing imperatives, a systematic, notably straightforward approach to a thorough reading of Sun Tzu will be implemented here. An important point that does not seem to have received its due in the Sun Tzu literature (possibly because the utility of distinguishing among the three Sun Tzus has itself not been clarified and developed) is that many bones of contention among translators and commentators, while certainly significant on a Sun Tzu (1) level, actually have little impact on our understanding of the enduring military or strategic substance of the Sun Tzu text – i.e., on Sun Tzu (2) or (3) levels. That observation makes it possible to encapsulate, albeit not eliminate, many Sun Tzu textual and interpretive puzzles, thereby sidestepping a wide range of longstanding, refractory scholarly debates and muddles.

The approach of the present study centers on identifying and clarifying, always with textual anchoring, a set of fourteen basic Sun Tzu substantive themes around which, as Alfred Thayer Mahan put it so well, "considerations of detail group themselves."[1] Fourteen such themes are identified. Analysis developing them is a way of rising, in Sun Tzu context, to one of the most fundamental and recurring challenges in all military and strategic theory: overcoming the tendency of strategy, certainly outside of game theory areas (which have their own major limitations born of extreme simplification), to be an amorphous domain of thought.

As pioneering cognitive scientist Jerome Bruner warned: "Knowledge one has acquired without sufficient structure to tie it together is knowledge that is likely to be forgotten."[2]

---

[1] Alfred Thayer Mahan, *Naval Strategy* (Boston, MA: Little, Brown, 1911), p. 118. In more extended form, the quotation from Mahan reads: "The search for and establishment of leading principles – always few – around which considerations of detail group themselves, will tend to reduce confusion of impression to simplicity and directness of thought, with consequent facility of comprehension."

[2] Jerome S. Bruner, *The Process of Education* (Cambridge, MA: Harvard University Press, 1960), p. 31.

Although strategy is the primary focus in the present study, many of these same fourteen themes also operate, commonly in simpler ways, on a tactical level.[3] Importantly, while each of the fourteen themes has its own analytical identity and textual anchoring, they interweave with one another in a way that creates a coherent larger intellectual tapestry. Modern applications of that tapestry can and do vary greatly in how, and with what tools and technology, they combine the fourteen elements of the package, highlighting some, possibly muting others. But for achieving greatest traction on the practical applications of Sun Tzu's thinking, including twenty-first-century ones, there is great value to be had in engaging with this package of fourteen themes as a whole, if only to decide which parts of it to play up or play down in building on Sun Tzu's ideas.[4]

Three Faces of Sun Tzu

It is clear from even an initial encounter with the Sun Tzu text that extracting much, perhaps most, of its value-added calls for an active yet disciplined analogical imagination. The immediate focus of the text is overwhelmingly geared toward conventional warfare between state actors (and land warfare specifically; there is not even a hint of blue water naval warfare in it), with important further attention to what is now often called grand strategy and, additionally, to espionage. The specifics of the conventional warfare known to Sun Tzu are, of course, antiquarian today. That conventional warfare focus does not directly reach the many other contexts (often largely or entirely non-military, certainly in any traditional sense) where potential for applications of Sun Tzu's thinking has elicited keen twentieth- and twenty-first-century interest, worldwide. Figure 1 offers a bird's-eye view of some of the possibilities.

Analogical thinking, indeed often of more than one type, takes center stage here. By virtue of the text's emphasis on warfare's information level, its important use of water and other imageries, and overall abstractness, such thinking is integral to Sun Tzu in a way that has few parallels in most other military theory, ancient or modern. Analogies come in many shapes and sizes and, like trademarks in the law, are by no means all of equal strength. All have their limits that need to be spotted, navigated, and in some cases exploited for additional insight. With an eye on developing the rich potential of Sun Tzu's thinking for being applied in a creative analogical spirit – which is one of its most basic and appealing characteristics – the distinction among three Sun Tzus is now set forth more systematically:

> Sun Tzu (1) or "Sun Tzu proper," focusing as rigorously as possible on what "Sun Tzu said" (a stylized phrase with which each of the text's thirteen chapters begins) in a context of war as Sun Tzu knew it (i.e., situations faced by Warring States Chinese generals and rulers).

---

[3] Notably, the Sun Tzu text itself does not embrace a strategy/tactics distinction (see pp. 47–48 below), though such a distinction, as applied by a modern analyst, remains useful in clarifying different levels of application of Sun Tzu's thinking.

[4] Drawing on a terminology used by Harrison White in a 1970 paper on the small-world phenomenon, the fourteen themes can also be envisioned as fourteen "probes" into Sun Tzu's thought. For the concept of probes as a tool of strategic analysis see p. 99 of Scott A. Boorman, "Fundamentals of strategy: the legacy of Henry Eccles," Naval War College Review, Spring 2009, 62(2), 91–115.

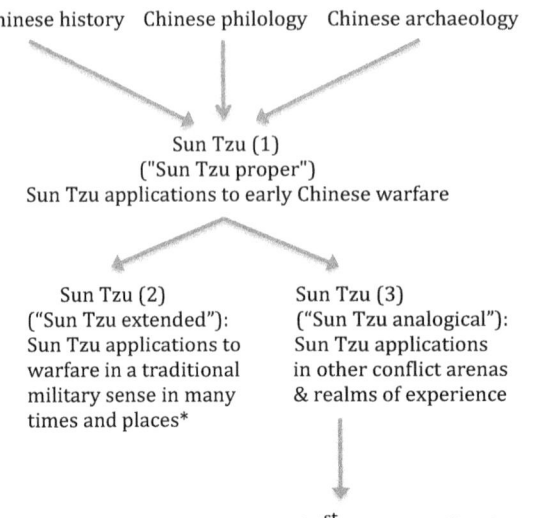

Chinese history   Chinese philology   Chinese archaeology

Sun Tzu (1)
("Sun Tzu proper")
Sun Tzu applications to early Chinese warfare

Sun Tzu (2)
("Sun Tzu extended"):
Sun Tzu applications to
warfare in a traditional
military sense in many
times and places*

Sun Tzu (3)
("Sun Tzu analogical"):
Sun Tzu applications
in other conflict arenas
& realms of experience

21stcentury applications of Sun Tzu
building on Sun Tzu as pioneer of information warfare, broadly conceived in
a way that emphasizes human factors

Without being exhaustive of Sun Tzu's
applications, four complex environments,
each centered on a particular type of structure, create – singly and in combination –
many opportunities for Sun Tzu's ideas
to flourish in the 21st century:
• digital structures and devices, including
  the algorithms on which they build
• social networks
• complex organizations & bureaucracy
• complex statutes.

For further development see Figure 7
(p. 514).

Figure 1. Overview of Sun Tzu (1), Sun Tzu (2), and Sun Tzu (3) perspectives.

*In keeping with the focus of the Sun Tzu text, a natural Sun Tzu (2) emphasis is on land
warfare, though Sun Tzu's ideas also have relevance to naval or air warfare cases (as regards
the latter, Sun Tzu himself in Griffith verse V.14 – see p. 300 and Passage #9.3 on
pp. 314–15 – makes a relevant observation, using the image of a raptor and its prey).

While Sun Tzu (1) is typically the least analogically demanding of Sun
Tzu's three "personae," even here an important niche for analogical
thinking exists. For example, some Sun Tzu passages have a natural
battlefield focus but also have broader implications that repay exploration
on each of (a) strategic, (b) operational (in a military sense), and
(c) tactical levels of war.

Sun Tzu (2) or "Sun Tzu *extended*," applying Sun Tzu's thinking to contexts
involving warfare in a traditional military sense in times and places
other than Sun Tzu's own – among them, post–Warring States eras of

Chinese history down to modern times, where the political–military conditions have often differed greatly from those prevalent in Sun Tzu's world (not least because asymmetric warfare commonly took center stage in clashes between Chinese and Central or Inner Asian military forces).

At the root of many Sun Tzu (2) analytical opportunities is an underlying similarity of problems faced by land warfare commanders in all ages of war, many emanating from the crucial dependence of all conventional armies on their logistics. At the same time, in exploring applications of Sun Tzu outside of a specifically Warring States context, an element of analogy is always present.

Sun Tzu (3) or "Sun Tzu analogical," extrapolating Sun Tzu's thinking in ways that call for more venturesome acts of analogical imagination, e.g., applications to purely political warfare in times other than Warring States China; or to conflicts unfolding in other realms, possibly involving non-state actors; or ones rooted in contemporary technological milieux such as cyber environments or twenty-first-century biological capabilities.[5]

More than with Sun Tzu (1) and Sun Tzu (2), Sun Tzu (3) insight is often holistic, not readily or fully anchored in any single specific passage in the text. What matters here is credible continuity with Sun Tzu's principles and ways of thinking, albeit sometimes dressed in unfamiliar garb (e.g., cyber environments). In helping to evaluate when such continuity exists, the intellectual discipline of the fourteen themes analytical framework provides a basic resource.

Sun Tzu (3) explorations should be noted as by no means solely a creature of modern engagements with the text. Already in the Warring States period there is reason to believe that Sun Tzu's treatise was widely known among literate civilians.[6] Sun Tzu's thought has certainly been a source of inspiration to countless thinkers and doers over the long course of Chinese history, many of them having interests far afield from the military.[7] Some of Sun Tzu's influential traditional commentators were civilians, poets among them.[8] Nor was that influence limited to China alone. For example, though beyond the scope of the present study, Sun Tzu's ideas have also had a lengthy and vibrant impact in Japan spanning many centuries.[9]

---

[5] For example, see Table 3 on pp. 151–52 below.

[6] See #4 footnote 6 for comments of Han Feizi; separately, of Warring States businessman Bai Gui.

[7] For context, it should be noted that Sun Tzu's text had a somewhat offline, even faintly "underground" status in Confucian China. As Paul Goldin has observed, "The intellectual elite that fixed Confucianesque orthodoxy for future generations did not incorporate such texts as *Stratagems of the Warring States, Laozi, Sunzi,* or *Han Feizi* into their canon, even if they commonly read and enjoyed these works in private." See p. 18 of Paul R. Goldin, "The theme of the primacy of the situation in classical Chinese philosophy and rhetoric," *Asia Major,* Third Series, 2005, 18(2), 1–25.

[8] See Yan Shengguo 閻盛國, "Songdai shiren bixia de sunwu yu Sunzi bingfa" 宋代詩人筆下的孫武與孫子兵法 (Sun Wu and the *Art of War* as seen through works of Song dynasty poets), Junshi lishi yanjiu 軍事歷史研究 (*Military History Research*), September 2011(3), pp. 191–97 (examining twenty poets, some of whom were also eminent statesmen, who mention Sun Tzu in their poetic oeuvres).

[9] For a survey of how Sun Tzu has been studied and received in Japanese circles, non-military as well as military, see Satō Kenji 佐藤堅司, *Sonshi no shisōshiteki kenkyū; shu to shite Nihon no tachiba kara* 孫子の思想史的研究：主として日本の立場から (*A Study of the History of Sun Tzu's Thought, Mainly from the Standpoint of Japan*) (Tokyo: Kazama Shōbo, Shōwa 37, 1962).

Each of these three Sun Tzus presents its own distinctive issues and challenges. Within its own realm each can offer important insights. But since quite different modes of interpretive work are called for, those applications should not be freely or indiscriminately commingled, zigzagging back and forth from one Sun Tzu to another. That is too commonly the case in modern literature on Sun Tzu. In particular, a useful byproduct of distinguishing the three Sun Tzus is to help alleviate self-imposed pressure, found in some Sun Tzu studies, to intermingle Sun Tzu (1) content specific to early China, sometimes tinged with exoticism, with Sun Tzu (2) or (3) observations often geared to establishing Sun Tzu's contemporary relevance, when separable writeups might better bring out the essential analytical points.

For work in Sun Tzu (1) mode, the basic toolkit is methods of Chinese history, philology, and archaeology, applied to shed light on the nature of war in Sun Tzu's time and its influence on Sun Tzu's thought.

Work in Sun Tzu (2) mode – which will always be of basic importance, for all the reasons that the art of war on land is a bedrock military focus and skill-set that no high tech will ever supplant entirely[10] – calls for specifically military knowledge pertaining to military strategy, logistics, and tactics and the interplay of all of these with technology and institutions. Sun Tzu (2) work is best approached against a backdrop of what, drawing inspiration from Wayne Hughes's analysis of naval warfare, might be labeled the "great constants."[11] In a land warfare context those constants would include the enduring relevance of human nature and emotions, morale, time and space, terrain features, logistics, fog of war, and a few other comparably fundamental factors.

The historiography of land warfare – involving the kind of expertise and judgment for which Michael Howard's 1961 classic The Franco-Prussian War sets a standard – is a basic tool for Sun Tzu (2) development, providing an immensely rich fund of situations, cases, and examples for bringing Sun Tzu's ideas to life on a world stage. Among pioneering Western students of Sun Tzu, Lionel Giles showed particular initiative in fleshing out Sun Tzu's ideas with examples from the annals of world military history, thereby attracting kudos from Lord Roberts, one the most successful British commanders of the latter nineteenth century (Roberts's experience had included both the Indian Rebellion of 1857 and the Second Anglo-Afghan War).[12]

For related reasons, Sun Tzu (2) work has the makings of being an important teaching tool, encouraging thinking about Sun Tzu's ideas in settings that are more familiar than early China (and where available historical information is frequently

---

[10] This point has been made, with a modernist tilt (albeit reflecting an era prior to the rise of drones!), by Rear Admiral J. C. Wylie, Jr., USN: "the ultimate tool of control in war is the man on the scene with a gun" (quoted by Boorman, Introduction footnote 4, see p. 110 note 33 there).

[11] See Captain Wayne P. Hughes, Jr., USN (Retired), Fleet Tactics: theory and practice (Annapolis, MD: Naval Institute Press, 1986), chapter 6 ("The great trends") and chapter 7 ("The great constants"), highlighting maneuver, firepower, counter-force, scouting (p. 182 treatment leads off with Sun Tzu) and anti-scouting, and C² (command and control) and C²CM (command and control counter-measures).

[12] Presented with pre-publication proofs of the 1910 Giles translation, Lord Roberts commented that "Many of Sun Wu's maxims are perfectly applicable to the present day." See Giles, p. xlii footnote 6. Kind words indeed; but penned less than five years before the guns of August 1914, heralding a most un-Sun-Tzu-esque kind of war on the Western Front.

much richer), yet which remain tethered by the "great constants" to the military roots of Sun Tzu's thinking.

Sun Tzu (3) adds to the mix a need for careful judgment in crafting more ambitious leaps, at times of a sustained or extended analogy type.[13] In exploring Sun Tzu (3) it is helpful for an analyst to have an omnivorous interest in phenomena of human conflict, since Sun Tzu's insights can frequently assist in cross-fertilizing realms commonly regarded as distinct. By contrast to Sun Tzu (2), in some major Sun Tzu (3) areas – e.g., contemporary high-tech settings like some cyber warfare – usable history is scant, at times virtually non-existent, and analysis may therefore require supplementation by other tools (e.g., formal modeling, computing, or gaming). Doing so puts in play analytical skill-sets far removed from Sinology and traditional military history alike.

Work on a Sun Tzu (3) level is more difficult than is commonly recognized, since it necessitates intellectual bridge-building between Sun Tzu's terse text and often remote and at times highly technical areas of human endeavor, many of which did not exist in Sun Tzu's time. Such forays call for acts of extrapolation and pattern matching across diverse technological and human contexts where criteria of judgment for evaluating analytical success are frequently not well worked out and it is easy to overreach or otherwise strike a false note. Unsurprisingly, avoiding such missteps best starts with thorough familiarity with Sun Tzu (1) – text, ideas, historical context.

It is also important to recognize that strategic or tactical steps that on one level may appear to profit by Sun Tzu's advice (say, by giving free play to his famous advocacy for deception) may create vulnerabilities on a different level that a Sun Tzu–inspired adversary could exploit (say, by undermining integrity of command or social or organizational trust). To draw on cybernetics imagery, when eyeing Sun Tzu (3) applications it is essential to keep in mind all the feedback loops likely to be relevant, not just a favorite (but included) subset of those feedback loops. The latter is a common pitfall among modern students of Sun Tzu who find themselves drawn, often shortsightedly, to Sun Tzu's incarnation as a bad boy.

In developing the three Sun Tzus perspective, the entire text of Sun Tzu – which, as a 1970s archaeological find of a partial text corroborates, had already taken substantially its present form by the second half of the second century BC – will be analyzed. That textual analysis has two basic components. The first, organized theme-by-theme, focuses on the most important Sun Tzu content, providing anchoring for main text thematic analyses that begin with Part A (p. 54) and Theme #1 (pp. 55–87) below. The second involves "also-ran" Sun Tzu content (e.g., because it is less clear or less on point). Likewise presented theme-by-theme, the latter material is assigned to Appendices 1–14 in the online annex. An addendum

---

[13] Illustrating extended analogies, each developed at book length, are the following two analyses:
   (1) Scott A. Boorman, *The Protracted Game: a wei-ch'i interpretation of Maoist revolutionary strategy* (New York: Oxford University Press, 1969);
   (2) David Howarth, *Law As Engineering: thinking about what lawyers do* (Cheltenham, UK: Edward Elgar, 2013).
   Sun Tzu–relevant domains that provide foci for these books are, respectively, (1) Chinese Communist revolutionary warfare, 1927–49; (2) the law (a profession having its own major strategic aspects).

to this coverage involves specialized Sun Tzu content having a technological slant. It is assigned to Appendix 15.

A basic payoff of this division of labor between main text and appendices is to take pressure off students of Sun Tzu to try to treat everything found in Sun Tzu as comparably insightful or of larger lasting relevance. Attempting to do so carries a real risk of obscuring or eclipsing the most significant and enduring Sun Tzu content by commingling it with the rest.

Three features of the Sun Tzu text make it especially well suited to implementing and deriving value-added from the approach to the text just sketched.

Brevity. At around 6,000 characters (count varies with edition used) the Sun Tzu text is short – far shorter, for example, than is Clausewitz's *On War* or countless modern strategic writings.[14] Brevity makes it feasible to hold the tally of basic Sun Tzu themes to a moderate number.[15] Furthermore (and in welcome contrast to many of those modern writings!), Sun Tzu does not waste words, harboring little redundancy. That sparseness also makes it feasible to anchor each theme in a tractable number of Sun Tzu passages (usually around twenty, mostly short). That is one basic reason why the present passage selection can be kept user-friendly.

Disaggregated structure. D. C. Lau, an important modern Sun Tzu authority, has observed that the Sun Tzu text conforms to a widespread tendency found in early Chinese texts, whereby the text tends to fractionate into short passages which are only loosely (if at all) articulated with one another.[16] In the case of Sun Tzu, as with many other early Chinese texts, a factor contributing to this fractured quality is a penchant for short, often numbered, heuristic lists (of factors, elements, problems, patterns, situations, rules, etc.). The upshot is that the Sun Tzu text exhibits, overall, a kind of loose-knit, granular character – very different from the sustained logic chains found in parts of Clausewitz and far more so in modern game theory.

Undercurrents of intellectual unity, but scattered through the text. Each Sun Tzu theme from the set of fourteen has multiple aspects or facets, often diverse and requiring active effort to ferret out.[17] Varying expressions of a given theme commonly surface in different parts of the Sun Tzu text. Pulling the relevant content together stands at the core of the present Sun Tzu analysis. Theme #1 (calculation) illustrates the exercise, since "calculation" is an activity with which Sun Tzu engages from several interconnected but clearly distinguishable angles.

---

[14] By way of comparison, the US Constitution (including amendments) is around 7,600 words. See p. 399 of Stephen Gardbaum, "The myth and the reality of American constitutional exceptionalism," *Michigan Law Review*, 2008, 107(3), 391–466.

[15] Even fourteen could, of course, call to mind a remark attributed to Georges Clemenceau, referring to Woodrow Wilson's Fourteen Points: "Fourteen? The good Lord had only ten." But a more apt comparison benchmark would be Clausewitz, where a scholarly concordance covers thirty-five-plus subjects each equipped with an associated list of pertinent passages from *On War* – just a list, not those passages themselves – spanning over fifty journal pages. See Jon Sumida, "A concordance of selected subjects in Carl von Clausewitz's *On War*," *Journal of Military History*, 2014, 78(1), 271–331.

[16] Lau, 1965 article, p. 322.

[17] Cf. Michael Handel, *Masters of War: classical strategic thought* (3rd rvd. and expanded ed.; London/Portland, OR: Frank Cass, 2001), p. 21: "Sun Tzu's *The Art of War* may seem easier [than Clausewitz] on first reading, but it is actually more difficult to understand in depth."

By thus establishing priorities in reading the text, and drawing clarifying benefits from delineating and grouping passages, the present approach creates a platform for a disciplined exegesis of Sun Tzu's ideas and their applications. The anchoring in Sun Tzu passages is crucial here. It lets Sun Tzu speak with his own voice and preferred emphases (which were, of course, shaped by the military realities of his era).

This approach also holds seeds of a somewhat different further payoff. Traditional Chinese education was very much a culture of memorization. Given the prodigious feats of memory required by the traditional Chinese examination system in its mature form, it would have been no great further feat for many educated Chinese over the centuries to have similarly memorized the Sun Tzu text, short as it is.[18] Doing so would in turn have created cohorts of individuals who would *really* know their Sun Tzu on a level of detail and specificity that few Westerners (and probably also few twenty-first-century East Asians) can match. As a stand-in for the chains of association and analogy that such memorization (if coupled with reflection on the substance) would encourage, the present Sun Tzu analysis offers a degree of insight into how Sun Tzu's thought might have been internalized by a traditionally educated Chinese steeped in a culture of textual memorization – insight not readily available to modern audiences unaccustomed to that way of absorbing basic texts.

Overview of the Present Approach

The Sun Tzu translation on which primary reliance will be placed is the 1963 one by Brigadier General Samuel B. Griffith, USMC (Ret.), based on Griffith's DPhil thesis in Chinese history at Oxford.[19] This translation draws on, and benefits from, Griffith's knowledge base and intuitions as a professional military officer. Griffith's career background included a distinguished record of World War II combat service, receiving the Navy Cross for action on Guadalcanal in 1942. Importantly (since there are many kinds of military service), Griffith's background involved a type of military

---

[18] As late as the 1890s, memorization of Sun Tzu (in fact of the full *Seven Military Classics* canon, see footnote 7 on p. xix) was required of candidates taking the official written Military Examination. In practice, things did not work quite that way given widespread illiteracy among examination takers, leading to practices of candidates' seeking third-party help. See Le P. Etienne Zi (Siu), SJ, *Pratique des examens militaires en Chine* (Chang-hai: Imprimerie de la Mission Catholique, 1896), p. 21. The more basic point, however, is that memorization of the Sun Tzu text would have been an easy task for traditionally educated Chinese civilians, who might have done so for many reasons including simply personal interest. By way of comparison involving the Confucian canon (Wilkinson, §28.4, pp. 400–403), Miyazaki Ichisada has estimated that candidates for Tang-Song civil examinations had to memorize texts having a Chinese-language counterpart to word count of *c.* 570,000 words – a volume of material exceeding Sun Tzu by some two orders of magnitude. See Benjamin A. Elman, *A Cultural History of Civil Examinations in Late Imperial China* (Berkeley: University of California Press, 2000), p. 267. See also Wilkinson, §22.1.6, pp. 319–20 ("How many texts were memorized?").

[19] Supervised by Wu Shichang (then teaching at Oxford). A companion to the Griffith translation is Lau's 1965 article cited on p. xxii. In that article, which mobilizes extensive Sinological expertise, Lau is very critical of Griffith. At times Lau states his criticisms too affirmatively, since crisp "right answers" in scholarship on early China are often few and far between. The upshot, however, is a constructive one, yielding many suggestions for critical thinking about particulars of Griffith's work.

experience that aligns well on many features of the military situations emphasized by Sun Tzu: land warfare between two opposing conventional armies; operations in complex and demanding terrain; operations where both sides mount all-out efforts and where margins of combat success are often razor-thin (with further lessons for the contributions of sound strategy, logistics, and tactics to being on the winning side of those thin margins).[20] Also importantly, Griffith's post-war active duty military career included three further years spent at the US Naval War College in Newport, RI – one as a student, two on the faculty – in that institution's highly creative early post–World War II era.[21] Griffith's translation did not have the benefit of a 1970s archaeological find of which more recent Sun Tzu translations take cognizance.[22] However, it is easier to compensate for that deficit by drawing on more recent Sun Tzu scholarship than it is to replace the intangible but genuine edge found in Griffith's translation, emanating from Griffith's visceral grasp of core military realities (among them, the fundamental part played by logistics in hard-fought combat operations).[23]

The present approach to Sun Tzu draws on Griffith's subdivision of the Sun Tzu text into short segments or "verses" whose brevity and (often) single-military-issue focus lays a foundation for creating passages illustrating the fourteen themes.[24] In effect, Griffith's verses serve for present purposes as "molecules" of the Sun Tzu text. For present analytical needs, Griffith's verse-based structuring pays its way handsomely, giving just the right degree of flexibility needed to implement passage identification with a minimum grinding of scholarly gears. Although improvements on Griffith's specific versification scheme could be suggested, as a practical matter any such gains seem minor. The well-established status of Griffith's translation

---

[20] For a lens on Griffith's experience in intensive, sustained land warfare combat see his book The Battle for Guadalcanal (Philadelphia, PA/New York: Lippincott, 1963). That work, which cites Sun Tzu on certain military analysis points, is a further natural companion to Griffith's Sun Tzu translation.

For historical context, especially important as World War II recedes in collective memory, it should be noted that Guadalcanal was America's first major offensive action against the Japanese Empire following the Pearl Harbor attack and a succession of humiliating US defeats – Guam, Wake, Bataan, Corregidor – and as such held a special place in the American psyche at a crucial stage of US mobilization for World War II. It was a bitterly fought 1942–43 campaign where US success long hung in the balance. Griffith was executive officer and later commander of the First Marine Raider Battalion during the campaign. For his part in this campaign, in which he was wounded, Griffith was awarded the Navy Cross for "extreme heroism and courageous devotion to duty."

[21] For background on the Naval War College in that period see Boorman, Introduction footnote 4.

[22] This find included the "Han strips" text of Sun Tzu described on pp. 41–42 below.

[23] Griffith, p. xi gives Sun Tzu kudos as one who "appreciated the decisive influence of supply on the conduct of operations." A wry aside showing Griffith's own alertness to logistics issues is found in a late 1950s letter from Griffith to his friend B. H. Liddell Hart: "I think you will be amused at [Tang dynasty Sun Tzu commentator] Du Mu's comment ... that of a company of one hundred, there were twenty-five administrative people to serve the seventy-five combat troops and care for the horses and oxen. Things don't seem to get any better in this connection, do they?" This letter is filed in LH 1/333/2 in the Liddell Hart Papers, Liddell Hart Centre for Military Archives, King's College London.

[24] For further specifics see pp. 52–53 below and "Scholarly Controls" in the online annex.

makes efforts at tinkering with his division of the text into verses likely to create more complications than they would be worth.

Leading off with the deservedly famous Passage #1.1 on net assessment (pp. 74–76 below), the present study organizes Sun Tzu content into numbered passages each comprising a sequence of Griffith verses. In all, just over 400 such passages have been used to develop the fourteen themes. Some two-thirds of those passages – the ones deemed most important – appear in the main text of the present study. The remaining passages are assigned to appendices in the online annex.[25]

A basic feature of the present approach is that specific Sun Tzu verses (and longer passages too) commonly appear under more than one theme. Such versatile verses and passages should be regarded as normal. They facilitate capture of different facets of the same material, frequently pointing to alternative paths of generalization or application. The common sense of such multiple thematic assignments is that choice of theme to which to assign a given verse or set of verses frequently boils down to where one places the emphasis. To get the most out of Sun Tzu it is wise not to be too rigid about which emphasis is seen as the "correct" one. As an important case in point, some verses assigned to other themes by reason of their military content lend themselves to additional civilian interpretations ("dual use"), hence also receive assignment to Theme #4 (winning by non-military ways and means).

The present initiative might be described as a type of intellectual "reverse engineering" of Sun Tzu content.[26] Any exercise of this type inescapably involves many judgment calls. In some cases, reasonable readers might dispute assignments of particular passages to particular themes. In certain instances too, the "adhesion" between a given passage and a theme to which it is assigned is not as strong as in other cases (scarcely an unknown problem in concept-matching exercises!). However (subject, of course, to ever-present classical Chinese-language issues), the transparency of the present approach should be emphasized. In a world of digital editing capabilities, a critically inclined reader can try her hand at moving material from one theme to another or between main text and appendices.[27] Insight often comes from the exercise of implementing passage assignments, as when upon rereading a particular passage suddenly leaps into a fresh light; or when, on successive readings, its assignment shifts from "above the line" (main text) to "below the line" (appendices), and perhaps back again, as considerations pro and con are weighed and reweighed. Such possibility for dynamic extension of the

---

[25] In spirit (though the present goals differ from Lau's), this rounding up of passages for comparison and analysis rises to a challenge thrown down by Lau, who proposes that

> we should treat with impartiality all [Sun Tzu] passages which deal with a common topic (in the case under discussion, the classification of terrain), and there is much to be said for placing all such passages side by side, since when they are read together it is possible that they may serve to illuminate one another, in no matter how small a way.
>
> (Lau, 1965 article, pp. 328–29)

[26] It is worth noting affinities between the present approach and certain types of legal analysis. Manifestations of the same underlying legal idea or principle commonly surface in non-contiguous locations in a statute or other legal text (or set of related texts), often in disparate and at times non-obvious ways. A basic challenge for legal analysis (and law teaching as well) is to reveal the underlying principle and to place its variegated manifestations in clear perspective and relationship.

[27] Or, more ambitiously, proposing a new theme and populating it with Sun Tzu passages.

present approach merits attention for larger reasons too. Notwithstanding highly developed traditions of wargaming and other gaming applications, strategy education – more so than its counterparts in tactics – remains much in need of further tools, concrete or abstract, that can facilitate active learning about strategic ideas (e.g., much as problem sets facilitate such learning in mathematics). Viewed in that way, the present approach, with its possibilities for fostering disciplined but creative engagement with the text, opens a door for active learning about Sun Tzu.

No textually informed approach can escape the fact that the Sun Tzu text is an ancient one, with accompanying obscurities and textual challenges. The present task is to navigate those minefields in order to clarify Sun Tzu (1), Sun Tzu (2), and – perhaps most importantly for our own time – Sun Tzu (3) substance. The reward from doing so is that the fourteen themes all contain shards of deep strategic insight capable of sustaining far-reaching development, in both theory and applications.

Broader Perspectives on Reading Sun Tzu

By one modern estimate, much of the Sun Tzu text can be assigned a date in the second half of the fifth century BC, close to the start of the Warring States period (for chronological background see p. xviii above).[28] Other estimates, in keeping with dates assigned to Sun-Tzu-the-person by Chinese historiographical tradition, point to an earlier date (e.g., late sixth or early fifth century BC, at the tail end of the Spring and Autumn era of Chinese history). There is also an important body of modern scholarly opinion that would place the Sun Tzu text or major parts of it (notably including the chapter on espionage) considerably later in the Warring States era, say, late fourth century to early third.[29] We would certainly like to know more about dating of the text than we do, especially given that these centuries were a time when Chinese society, including its military institutions and technology, was in great flux.

There has been a longstanding controversy, originating in pre-modern Chinese scholarship, as to whether Sun Tzu was in fact a historical person.[30] Certainly what

---

[28] See Robin D. S. Yates, "New light on ancient Chinese military texts: notes on their nature and evolution, and the development of military specialization in Warring States China," T'oung Pao, 1988, 74(4/5), 211–48, noting (p. 218) that "we may therefore tentatively date the Sun-tzu to this period [453 BC – 403 BC], while recognizing that not all the sections derive from the same period." (That dating's rationale has not escaped controversy: see Petersen's work cited in Introduction footnote 32.)

[29] For a strong position in favor of such more recent dating see "Précis" in Mair, (unnumbered) p. li. Taking a compatible stance is Wilkinson, §24.8.1, p. 349. Griffith, pp. 6–11 also rounds up a range of historical and linguistic evidence that bolsters identification of Sun Tzu as a Warring States text.

[30] Evidence that Sun Tzu the person is fictive, not historical, is presented in detail by Mair, pp. 9–23. Weighing against Sun Tzu's historical existence is the fact that the Zuozhuan, regarded as China's earliest narrative history (see p. xxiii above), fails to mention Sun Tzu despite extensive coverage of his purported historical time and place. Giles, pp. xxv–xxx makes a tentative case that Sun-Tzu-the-person did exist (fl. c. 500 BC) but was at best a minor figure on the historical canvas of his time (perhaps a little like Clausewitz in modern times). Sawyer, p. 84 sums up our ignorance: "Sun Wu [i.e., Sun Tzu] remains an enigma not only because of the absence of historical data in the so-called authentic texts of the period, but also because his life never generated the anecdotes and illustrative stories frequently found about famous figures in the works of succeeding periods."

we know about Sun Tzu's life is extremely limited. Virtually all of it comes from his biography in the work of the pioneering historian Sima Qian, the "grand historian of China" who lived c. 145 BC–86 BC – i.e., centuries after any historical Sun Tzu. That biography is dominated by just one item: the colorful and oft-told tale of the drilling of the ruler's concubines by Sun Wu (traditionally equated to the Sun Tzu of the Sun Tzu text), with outcomes fatal to two of them for disobeying Sun Tzu's orders.[31] The authenticity of Sima Qian's biography of Sun Tzu has long been called into question and modern scholarship has intensified those challenges.[32] Tellingly for a military figure, there is little we are sure of regarding a putative historical Sun Tzu's military experience (e.g., what, if any, were his consequential command decisions?).[33] There is even uncertainty regarding which of several different warring states should be treated as the locale where the Sun Tzu text's perspective on war and geopolitics took shape. Sima Qian's biography places Sun Tzu as a native of the state of Qi (whose capital lay in north China, in modern Shandong province), though his famous anecdote about the concubines has Sun Tzu advising the ruler of Wu, a state centered several hundred miles to the south and whose territory included the Yangzi Delta where modern Shanghai and Nanjing would be built. Some modern scholarship finds intellectual roots of the Sun Tzu text in ambitions and geopolitical problems of the state of Lu, a small state (best remembered for having been home to Confucius, who died in 479 BC) situated between greater powers to the north and south but having aspirations (as Brooks, p. 59, puts it) "to play in the big leagues."

Two basic insights come from asking about origins of the Sun Tzu text.

[31] For translations of this Shiji 65 biography (which Mair labels a pseudo-biography), see Griffith, pp. 57–59; Ames, pp. 32–34; Mair, pp. 133–35. Translations with further scholarly apparatus are Nienhauser, Vol. VII (1994), pp. 37–38; Vol. VII (2021), pp. 69–71 (both editions are cited on pp. xxiii–xxiv above). A somewhat different, fragmentary version of the concubines story was recovered from the same archaeological find that recovered the "Han strips" text of Sun Tzu. For that different version see #10 footnote 9 below.

[32] An ingenious theory that this personage was created as a kind of "double out of nothing" (Petersen's evocative phrase) – a fictive collaborator of Wu Zixu (d. 484 BC), a prominent historical figure (naval as well as military) of the late Spring and Autumn era – has been propounded by Danish Sinologist Jens Østergård Petersen. See pp. 14–15 of his "What's in a name? On the sources concerning Sun Wu," Asia Major, Third Series, 1992, 5(1), 1–31 (also cited by Wilkinson, §24.8.1, p. 349).

[33] The battle of Boju (506 BC) between the states of Chu and Wu, in which Wu prevailed and went on to take the Chu capital of Ying, is the only battle we know of in which Sun Tzu may have participated. Our source, the Shiji 65 biography (see Introduction footnote 31 above), tells us little more than that he was on the winning side. The Zuozhuan (p. xxiii above) gives more detail on Boju, but notably without any mention of Sun Tzu. (Read on a level of military ideas, that Zuozhuan account does tell a story that evokes several signature Sun Tzu concepts, including death ground. See p. 323 below.)

Shiji 66 does note two cases of Sun Tzu's advising the ruler of Wu on specific high-level military strategy matters. On at least one of those occasions – the other probably too, though the Shiji account is slightly less clear – the advice is described as emanating jointly from Wu Zixu, one of the distinguished generals of his time, and from Sun Tzu, which leaves Sun Tzu's contribution to the conversations uncertain at best. For both episodes, see Wu Zixu's Shiji 66 biography, translated in Nienhauser, Vol. VII (1994), p. 53; Nienhauser, Vol. VII (2021), pp. 96–97 (both cited on pp. xxiii–xxiv above).

A similar account appears in Shiji 31, translated in Nienhauser, Vol. V.1 (cited on p. xxiii above), p. 17.

First, as noted earlier, archaeological evidence supports the notion that Sun Tzu's thirteen-chapter text had attained close-to-modern form by the second century BC.[34] This was the time of the Western Han dynasty (202 BC–8 AD), successor to the short-lived Qin dynasty (221–206 BC) of terracotta soldiers fame that brought the Warring States era to a close by unifying China. That dating gives grounds to believe that, for the past two millennia and more, a textually well-defined body of military theory and doctrine has existed under the Sun Tzu rubric. In that sense the Sun Tzu "brand" has been essentially stable for over 2,000 years. Quibbles with this statement are possible – but the basic evidence is archaeologically sound and the quibbles do not seem major.

Second, there is considerable, if still inferential, reason to treat the Sun Tzu text as a compilation – the work of more than one hand, quite possibly spanning decades in the making. Reflecting on the eddies of scholarly opinion swirling around who Sun Tzu may have been, Roger Ames, a contemporary Sun Tzu scholar (and student of D. C. Lau), has observed that a "quest for a single text authored by one person" harbors a "real danger . . . of pursuing the wrong questions and, in so doing, losing sight of what might be more important insights."[35] He goes on to suggest that "works such as the Sun-tzu might have emerged more as a process than as a single event, and those involved in its authorship might well have been several persons over several generations."

Carrying this vein of thinking a step or two further, a boldly specific account of how Sun Tzu's thirteen chapters may have taken shape through successive accretions – sequentially adding chapters over a period spanning from the mid-fourth century until the process reached closure around 270 BC – has been advanced by E. Bruce Brooks.[36] While parts of the story he tells seem inescapably conjectural, there is no need to accept all of the details in order to derive plausible and useful insights from it. His story points to an early cluster of Sun Tzu chapters where terrain issues loom large; a later cluster (coming after the first by perhaps half a human

[34] Citing the reconstruction of the archaeologically recovered Sun Tzu text by the Yinqueshan Committee (its name reflects the geographic location in modern Shandong province where that text was found), Ames, p. 36 notes that these "remnants of the thirteen-chapter edition (over 2,700 characters)" contain "representative text from all of the chapters except Chapter 10 [i.e., Sun Tzu X]." For background on the Yinqueshan find – which recovered much more textual material, some Sun Tzu–related, some not, than this partial text of Sun Tzu's thirteen chapters – see Wilkinson, §59.6.2.2, p. 806.

[35] Both this and the following quotation are from Ames, p. 21.

[36] This theory of the genesis of the Sun Tzu text is put forward in Brooks's insightful, if rather telegraphic, mini-essay on Sun Tzu embedded in a much larger review article by Brooks (cited on p. xxii) addressing a wide range of early Chinese texts. For highlights of Brooks's argument, which is informed by analysis of Sun Tzu's military content, see Brooks, pp. 59–62. His proposed accretion sequence appears in Table 12A on p. A-35 in the online annex. With the proviso that some of Brooks's conjectures (e.g., proposing definite calendar years for particular Sun Tzu chapters) seem destined to outrun the evidence, this terse essay, published in a specialized forum, deserves to be widely known to the large community of analysts of many backgrounds and persuasions interested in Sun Tzu's substance. For background on relevant pioneering work of E. Bruce Brooks & A. Taeko Brooks see Mair, pp. 67–68 note 56. See also E. Bruce Brooks & A. Taeko Brooks, *The Emergence of China: from Confucius to the empire* (Warring States Project, University of Massachusetts at Amherst, 2015), p. 238, presenting a bar chart indicating, for a range of early Chinese texts (Sun Tzu among them), estimates of approximate time intervals in which particular texts took shape by an inferred process of accretion.

generation) comprising all but one of the remaining chapters; and one much later chapter (the espionage one) conjectured as having been added around a human generation later still. While the present Sun Tzu thematic analysis does not rest on the validity of Brooks's proposed ordering of Sun Tzu chapters, his analysis frequently helps sharpen thinking about substantive issues and will be often be used as a heuristic tool for that purpose.

These observations suggest an important perspective on the Sun Tzu text and the military and strategic ideas found in it. Certain types of efforts at a close reading of Sun Tzu may in fact be overreaching in light of the "intergenerational group product" that the text may well represent. It needs to be recognized and accepted that there are limits on the extent to which weight can be placed on the Sun Tzu text as a polished – or even fully internally consistent – repository of military or strategic thought, either textual or intellectual. The lines of thinking exemplified by Ames and Brooks should sound a warning about efforts to deduce too elaborate a logical edifice from delicately titrated exegesis of different parts of the text, after the fashion of reading modern statutes or works of analytical philosophy.

At the same time (and echoing the spirit of Lau's 1965 article) this caveat should definitely not deter efforts to connect and coordinate the thrust of ideas found in different parts of the Sun Tzu text. Importantly, approaching Sun Tzu as a coherent package of ideas need not require taking a stand on the presence of coherent overall organization of the text. Strategic thought is always an act of synthesis, and high-quality strategic writings are notorious for their commonly mongrel roots (as well as for often not being very well organized!). As the writings of Mao in modern times attest – representing as they do an amalgam culled from many sources, among them, parts of the Marxist-Leninist tradition, Mao's own direct experience and observation of Chinese society and politics in his time, Chinese historical romances, and, along the way, some strategy of the game of *weiqi* plus some Sun Tzu – strategic ideas of very different provenance and vintage can at times serendipitously blend to create an original and awesomely powerful punch. Much turns on how those sources are chosen and even more on how adroitly they are conceptually integrated. The Sun Tzu text may indeed be a compilation decades-long or even longer in the making, the work of multiple hands and in many ways loose-knit in its makeup, but it can still express an intellectually coherent strategic style.

That is the fundamental perspective on the text adopted in the present analysis. With a dash of poetic license, it might be labeled the "Axiom of Conceptual Unity for Sun Tzu" (or perhaps more exactly, drawing on the measured language of the law, a "rebuttable presumption" of such unity).[37] It is an approach that works for

---

[37] Also stressing the consistency of Sun Tzu's concepts and principles is Ralph D. Sawyer, "Military writings," in David A. Graff & Robin Higham (eds.), A Military History of China (Lexington: The University Press of Kentucky, 2012), pp. 97–114 (see p. 100 there). For a kindred "conceptual unity" perspective on the Laozi Daoist classic see Benjamin Schwartz, "The thought of the Tao-de-ching," pp. 189–210 in Livia Kohn & Michael LaFargue (eds.), Lao-tzu and the Tao-te-ching (Albany: State University of New York Press, 1998). His p. 209 is an acute sketch of the challenges faced in identifying conceptual unity underlying seeming intellectual sprawl.

For background on the Laozi text (also known as the Daodejing), which is often regarded as having intellectual affinities with Sun Tzu, see Early Chinese Texts (cited on p. xxii above), pp. 269–92; Wilkinson, §29.3.2, pp. 414–15.

Sun Tzu, though definitely not for all early Chinese texts. In the chapters that follow, support for this axiom's credibility is built, one step at a time, by a careful reading of the text that turns up many interconnections and harmonies, some readily apparent, others less so.[38] Support of a different sort comes from within Chinese tradition itself, where Sun Tzu's thirteen chapters have long been read as whole, enjoying one reputation and being treated as one source of intellectual inspiration, indivisible.[39]

Of course, accepting an Axiom of Conceptual Unity for Sun Tzu does not eliminate a battery of textual and interpretive issues that impinge on how we read the text and engage with particular strands of Sun Tzu's thinking. Many of the footnotes to Sun Tzu passages below, starting with Passage #1.1, represent an effort to clarify those issues and, where it seems appropriate, to take a stand on them.

Standing back from the myriad details and disputes, certain approaches to reading Sun Tzu stand out as more productive than others. Much as in many applied mathematics settings, all approaches involve approximations of some kind. The key to analytical success is identifying a serviceable approximation, taking into account the pertinent goals and constraints. Again as in much applied mathematics work, there is "no one size fits all" approach that is best for all purposes. Different approximations may harbor inconsistencies with one another, yet each may have its uses.

Two forks in the road in reading Sun Tzu merit specific comment here.

First, there is a basic divide between the present fourteen themes approach, which pivots on disassembling the Sun Tzu text to anchor and illustrate those themes, and other approaches that seek insights from identifying overall structure in Sun Tzu's thirteen chapters or major parts of the text (in particular at the chapter level). Traces of larger structure may exist even if the text is a compilation (indeed Brooks, p. 61 expands on his accretionist theory by what may be read as a sketch of such a structure, which in his analysis is broadly evolutionary in nature). However, a quest for larger structure in the Sun Tzu text is left here for other Sun Tzu scholarship to address.[40]

There is a second, even more important, divide between approaches that emphasize Sun Tzu's Chinese cultural roots and content and other approaches that play up its place in the universalistic analysis of warfare and strategy. The former focus is

---

[38] A case in point involves harmonies between Sun Tzu's military thinking about "formlessness" – which one modern scholar has characterized as the "ultimate counterintelligence" (see p. 269 below) – and human intelligence emphases in Sun Tzu's espionage chapter (Sun Tzu XIII).

[39] Of the Sun Tzu book's longevity, a review essay on Griffith's translation observes that "Few books have had a better record." See p. 129 of Scott A. Boorman & Howard L. Boorman, "Mao Tse-tung and the Art of War," *Journal of Asian Studies*, 1964, 24(1), 129–37.

[40] A relevant line of analysis has been suggested by David Robert Howell. It is premised on the concept that some (though not all) of Sun Tzu's thirteen chapters lend themselves to being read as organized, coherent essays rather than as loose-knit, unorganized assemblages of short passages. Howell has developed his standpoint in application to Sun Tzu's first chapter via a customized course handout prepared by him for January 2014 Sun Tzu–focused class meetings at the start of that year's Yale University Brady-Johnson Program in Grand Strategy seminar.

illustrated by the Ames translation.[41] It is also exemplified by work of others who find kinship between Sun Tzu and Daoist strands of Chinese thought.[42]

With some qualifications, the present study hews to the latter – universalistic – emphasis.[43] More specifically, the present perspective on Sun Tzu builds on a premise that there exists a body of fundamental military and strategic knowledge of a universal type – whose delineation is an intellectual ambition shared with Clausewitz and game theory alike – into some of whose deeper recesses Sun Tzu's thinking offers major insights.[44] Generalizing an earlier observation about warfare in its traditional military sense (p. 6 above), universal knowledge grows out of an underlying commonality of many of the problems faced by strategists, transcending particulars of time, place, and technology. An alternative, more nuanced, statement of a universalist position is that there exists a universal military theory with

---

[41] See Ames, pp. 39ff. Ames, pp. 6–7 very effectively summarizes his culturally oriented standpoint:

> Most accounts of the Sun-tzu have tended to be historical; mine is cultural. In the Introduction that precedes the translations, I have attempted to identify those cultural presuppositions that must be consciously entertained if we are to place the text within its own world view. In our encounter with a text from a tradition as different from ours as is classical China's, we must exercise our minds and our imaginations to locate it within its own ways of thinking and living. Otherwise we cannot help but see only our own reflection appearing on the surface of Chinese culture when we give prominence to what is culturally familiar and important to us, while inadvertently ignoring precisely those more exotic elements that are essential to an appreciation of China's differences. By contrasting our assumptions with those of the classical Chinese world view, I have tried to secure and lift to the surface those peculiar features of classical Chinese thought which are in danger of receding in our interpretation of the text.

That cultural standpoint makes Ames's work on Sun Tzu an excellent foil for the present study, precisely because it represents a coherent, sustained effort to develop and advocate for a way of reading Sun Tzu that contrasts with the universalistic standpoint emphasized in the present study.

[42] On Daoistic aspects of Sun Tzu see Mair, pp. 2, 47–49. Of course, such affinities do not spell identity and it is also important to recognize ambiguities in the concept of Daoism, "a term that covers diverse strains of ancient Chinese thought." See p. 237 of Benjamin I. Schwartz's review of Lisa Raphals, *Knowing Words: wisdom and cunning in the classical traditions of China and Greece* (Ithaca, NY: Cornell University Press, 1992), published in *Harvard Journal of Asiatic Studies*, 1996, 56(1), 227–44.

[43] This is largely also Griffith's standpoint. See, e.g., Passage #1.1 discussion on pp. 74–76 below.

[44] This intellectual position is forcefully stated in the introduction to Brigadier General Vincent J. Esposito and Colonel John R. Elting, *A Military History and Atlas of the Napoleonic Wars* (rvd. ed.; London/Mechanicsburg, PA: Greenhill/Stackpole, 1999): "the conduct of war is an art based on ageless fundamental concepts that have remained valid irrespective of the prevailing means and methods of warfare." See also Handel, Introduction footnote 17, p. xvii ("the universal logic of war still exists whether or not it is codified").

Perhaps more compelling than are any general affirmations about universal theory is comparing Sun Tzu with observations from *The Good Soldier* (London: Macmillan, 1948) by Field-Marshall Archibald Wavell, one of the distinguished British commanders of World War II and a major advocate of military deception. The similarities of various key ideas and emphases, separated in time by over two millennia, fairly leap off the page. Compare, for example, Wavell's "Manoeuvre and stratagem" (his pp. 157–61) and Sun Tzu Passage #7.9, entitled "Sun Tzu's 'Second Symphony': deception, calculation, and operational style."

stylistic variations in its development and application.[45] Adopting that perspective, Sun Tzu represents one style within such a larger family of styles, some of Chinese origin, some not.

Especially by encouraging attention to logistics and related structural constraints that all strategists face in one way or another, a universalistic emphasis gives analytical traction, providing a powerful and versatile tool for shedding light on interpretive issues arising in the Sun Tzu text. Wielding it involves continually posing the question: If I was a general (or strategist), what *usable* insight might I derive from this part of Sun Tzu? Responses to that question unlock many doors. It is a tool that lacks a close counterpart in reading many other early Chinese texts dealing with broad social, political, or philosophical subjects where pertinent context and constraints are commonly far less clear.

Such "usable insight" analysis should be exploited to the fullest possible extent.

Universalistic and culturally oriented approaches to Sun Tzu are by no means mortal enemies of one another, and many possibilities for cross-fertilization exist.[46] Indeed the more seriously a universalistic type of reading is pursued, the more avenues for combining it with a culturally oriented reading start to surface naturally. But efforts to unify a universalistic with a culturally oriented reading prematurely – before work along each path has been developed far enough – can easily lead to confused results, serving neither goal well.

In keeping with the perspective just set forth, the present study of Sun Tzu aims to clarify what Sun Tzu's universal insights are and to point directions for their further development and application, including in our own time in contexts remote from early Chinese warfare. Along the way, Chinese cultural content of various of Sun Tzu's military concepts and principles will be noted, but (with a few important specific exceptions) will not be given center stage. One such exception arises in connection with Theme #5, where the fundamental Chinese strategic concept of *shi* (commonly rendered "strategic advantage," though *shi* has other aspects too) is pivotal and should not be sidestepped. Another involves Theme #14, whose focus is the contrast between *qi* and *zheng* approaches in warfare (which Lau's 1965 article renders as "crafty" and "straightforward," respectively, but which lack truly satisfactory English translations). Even these exceptions will be approached in a way that seeks to identify general insights not limited to Chinese warfare or cultural context. The present study's focus on a universalistic reading of Sun Tzu aspires to capture a

---

[45] The versatile and useful concept of strategic style owes much to work of Nathan Leites. See, e.g., his *Soviet Style in War* (rvd. ed.; Santa Monica, CA: RAND, 1992; first published by Crane, Russak, New York, 1982). Boorman's *weiqi* book, cited in Introduction footnote 13, is in many ways a study of a specific strategic style expressed by Mao's revolutionary warfare writings, one that has considerable common ground with Sun Tzu's style of warfare but that also retains its own identity.

[46] Choices between a universalistic lens and a culturally specific one also arise in history of mathematics, an activity that has had a symbiotic relationship with warfare throughout recorded human history. Useful perspective may come from that quarter. It has been written of one of the great historians of mathematics, Otto Neugebauer (whose mathematical roots lay in Göttingen, where he served as Richard Courant's assistant), that "even through years of allowing that mathematics was grounded in culture, he [Neugebauer] really believed that in a more profound sense it was not." See Noel M. Swerdlow, "Otto E. Neugebauer (26 May 1899 – 19 February 1990)," *Proceedings of the American Philosophical Society*, March 1993, 137(1), 138–65 (quote is from p. 160). Kindred intellectual issues should be kept in mind when engaging with Sun Tzu.

remarkable quality of timelessness of Sun Tzu's thought of which B. H. Liddell Hart wrote in his "Foreword" to Griffith's Sun Tzu translation (p. v):

> Among all the military thinkers of the past, only Clausewitz is comparable, and even he is more "dated" than Sun Tzu, and in part antiquated, although he was writing more than two thousand years later. Sun Tzu has clearer vision, more profound insight, and eternal freshness.

A task of distilling universal strategic insights from Sun Tzu is in some important ways more tractable than the culturally oriented exegesis to which some Sun Tzu scholarship aspires. Far from being the politically and culturally consolidated entity that China would later become – at least to a very significant extent – with Confucianism as its predominant cultural force, China in the Warring States era was in a "hundred flowers" period. Intellectually, it brimmed with contending schools of applied philosophy – Confucian, Legalist, Daoist, Mohist, Military Experts (bingjia), and many more – with highly variegated menus of analytical, ethical, and policy offerings, the Sun Tzu text being but one.[47] The diversity of schools of thought flourishing in Warring States China, with their many turbulent eddies and cross-currents, contributed to making that era one of the major transitions in world history.[48] It gave rise to an intellectual ferment that would certainly have affected the reading and reception of Sun Tzu's text (which is decidedly not of Confucian persuasion), and in all likelihood the formation of the text itself as well.

That context means that any effort at a cultural interpretation of Sun Tzu's teaching, certainly a unified one that also remains faithful to its historical milieu, needs to be handled with utmost care to avoid overreaching, retrospectively imputing more cultural order than existed when the text took shape. Threading the needle here may at best be only partially feasible with available sources. Nor do the challenges and pitfalls facing culturally oriented readings of Sun Tzu end with the Warring States period. Later Chinese strategic tradition, far from being a monolith, exhibited major internal diversity.[49] In particular, scholarship on the military strand of Chinese strategic thought should not to be allowed to crowd out attention to its comparably longstanding and highly developed civilian bureaucratic strand.[50] The priorities and

---

[47] For a snapshot of various of the "hundred schools" of China's axial age see Wilkinson, §58.6.1, p. 775. The bingjia school goes by various English-language names, e.g., Militarists, School of the Military, School of Strategy, etc. Referring to it as "Military Experts" follows Wilkinson, §24.8, p. 348, and avoids some extraneous connotations as well as possible confusions with other schools.

[48] See Brooks & Brooks, footnote 36 above, p. 14. In this and other work they convey to modern readers the spirit of give-and-take, of advocacy and counter-advocacy, that suffused Warring States intellectual life. As they put it (p. 74), "the theorists responded to each other's work, copying the good ideas or opposing the erroneous ones. Such was the interactive nature of the [fourth-century BC] Golden Age of Chinese Thought." At the same time, modern audiences should be alert to a "general tendency within Chinese philosophical works for rival schools to use the same vocabulary to advance very different ideas" (Raphals, Introduction footnote 42, p. 7) – at times amounting to a type of semantic warfare!

[49] As a case in point see Peter C. Perdue, "Culture, history, and imperial Chinese strategy: legacies of the Qing conquests," pp. 252–87 in Hans van de Ven (ed.), Warfare in Chinese History (Leiden: Brill, 2000), in particular his pp. 266–72 regarding "Differences between Ming and Qing strategic thinking."

[50] Roots of the Chinese bureaucratic phenomenon are very old. For a vantage on Chinese bureaucracy from a time long before Sun Tzu see Li Feng, Bureaucracy and the State in Early China: governing the Western Zhou (Cambridge, UK/New York: Cambridge University Press, 2008).

cognitive patterns of the bureaucrats were certainly of enormous significance in shaping Chinese grand strategy and sometimes also military strategy down through the dynasties, yet in many ways were very different from those of the military tradition (certainly in context and emphases, starting with choice of tools of conflict).[51] As a further overlay of complexity, it is also essential to take cognizance of the influence on Chinese warfare of Inner Asian strains of military and strategic thought and practice. These are prominently represented, inter alia, by the several conquest dynasties (Liao, Jin, Yuan, and Qing) that ruled China for much of the past thousand years (see Chronology, pp. xix–xx above). There are traditions here that have their own identities distinguishable from Chinese ones (not least because of the formidable logistics challenges facing Inner Asian military operations, as well as the central role therein of mounted troops).[52]

Adding to these complications, efforts to extrapolate a coherent Chinese strategic tradition to our own time runs into the further difficulty that China and the United States are now on the "other side of the river" from one another, to use Edgar Snow's haunting metaphor – a relationship increasingly far too close and fraught with paths of reciprocal influence to be conducive to crisp propositions about cultural and psychological divergences, albeit that some of those persist.

For all these reasons, one should be very wary of generalizations that purport to sum up an overall Chinese "strategic culture" as a neat package, and along the way uncritically to merge readings of Sun Tzu with other major strands of it.[53] Especially

---

[51] See James T. C. Liu, "Eleventh-century Chinese bureaucrats: some historical classifications and behavioral types," *Administrative Science Quarterly*, 1959, 4(2), 207–26. One type of bureaucrat Liu labels "manipulative," profiling its characteristics on his p. 224. Such bureaucrats could well have profited from studying Sun Tzu (which many surely did!), but often in ways involving spheres of action remote from military combat. Cross-influences also existed. As Yates observes,

> While it may be difficult to determine in any one instance whether a given military officer was literate or learned enough to modify his behavior on campaign in the light of previous historical examples quoted by these [Sun Tzu commentators] or by the rules of action enunciated by the Sunzi, yet it is likely that this mode of commentary did influence the dispassionate analysis of officers' actions by their superior civilian officials and by their civilian contemporaries.

See p. 75 of Robin Yates, "Early modes on interpretation of the military canons: the case of the Sunzi bingfa," pp. 65–79 in Ching-I Tu (ed.), *Interpretation and Intellectual Change: Chinese hermeneutics in historical perspective* (New Brunswick, NJ/London: Transaction Publishers, 2005).

[52] For overview see Nicola Di Cosmo, "Introduction: Inner Asian ways of warfare in historical perspective," pp. 1–29 in Nicola Di Cosmo (ed.), *Warfare in Inner Asian History (500–1800)* (Leiden/Boston, MA: Brill, 2002). On the Manchu case, with extensive attention to logistics challenges, see Peter C. Perdue, *China Marches West: the Qing conquest of central Eurasia* (Cambridge, MA/London: The Belknap Press of Harvard University Press, 2005). For overview of Mongol warfare in the time of Chinggis Khan see Timothy May, *The Mongol Art of War: Chinggis Khan and the Mongol military system* (Barnsley, England: Pen & Sword Military, 2007). For broader analysis of a steppe empire's non-bureaucratic nature, worlds apart from the Chinese bureaucratic phenomenon, see also Joseph F. Fletcher, Jr., "The Mongols: ecological and social perspectives," *Harvard Journal of Asiatic Studies*, 1986, 46(1), 11–50 (reprinted in Fletcher's posthumously published *Studies on Chinese and Islamic Inner Asia* [Beatrice Forbes Manz, ed.; Aldershot, UK/Brookfield, VT: Variorum, 1995]).

[53] A relevant question, too often neglected, centers on means of transmission of Chinese military and strategic thought. Such means would certainly have important bearing on what was transmitted (and along the way, variances, deletions, elaborations, embellishments, or garbles). In Sun Tzu's time (and in later times too) much of that transmission would have taken place via oral tradition – a pattern by no means unknown in how military ideas and doctrine are transmitted

when one starts to move away from informed Sinological scholarship to parts of the non-specialist modern literature invoking Sun Tzu, this is a very common pitfall. A testament to the many missteps possible here is the proliferation of stereotyped, even largely imaginary, ancient Chinese worlds that populate parts of the contemporary Sun Tzu literature and discourse.

A particular risk arises in attempting to align Sun Tzu (2) or (3) work too closely on culturally specific interpretations of Sun Tzu, especially where contemporary issues are the focus and the goal of the analysis is practical. In cultural analysis settings, the old fallacies of essentialism and reification are difficult to eradicate entirely and often persist in subtle, low-visibility forms. These sources of analytical bias have significant potential for inadvertent encouragement of cultural stereotypes, including ones pertaining to "strategic culture" or behavior patterns regarded as associated with it.[54] In many scholarly settings, the adverse effects of tendencies like these are manageable, if regrettable. But in practically oriented strategic analysis, such intellectual traps can all too easily provide openings for an astute foe to play along with some received and oft-recited cultural stereotype – and then to break from that stereotype, without warning, at a singularly inconvenient moment.[55] Such a stratagem would indeed be vintage Sun Tzu!

Weighing these considerations, the present position is that it would be counterproductive to become so immersed in dissecting the nuances of cultural interpretations of Sun Tzu (and perhaps also so wary of getting them wrong) that the universal military and strategic content of Sun Tzu's thought received only secondary or afterthought attention. That would be a major mistake, unnecessarily hobbling our ability to identify Sun Tzu insights pertinent to our own time and thereby to gain traction on the twin questions with which the present analysis began:

"What did Master Sun know that we still don't (or have yet to absorb adequately)?"[56]

"What are Sun Tzu's limitations or blind spots?"

Ideas and leads sparked by the first question point in diverse directions. The present recommendation is that they definitely not be approached in too literal-minded a way or as constituting a unified "system." To attempt to do so would

---

today. Relatedly, rhyming passages identified in part of the Sun Tzu text may have assisted its memorization by the marginally literate. See "Overview" section of Kidder Smith website, citing sources (p. xxi above). See also Yates, Introduction footnote 51, p. 67.

[54] Military or strategic doctrinal pronouncements commonly tend to be somewhat simplified, at times very much so. Such simplification is understandable in light of military doctrine's core role as a type of communications system, enabling a degree of coordinated action to be achieved even when timely, secure, direct communication is difficult or impossible (a common circumstance in warfare). But doctrinal simplification, however valid its purpose, can also easily feed cultural stereotypes.

[55] Such a break with tradition is a known move in some pre-modern Chinese military writings, e.g., the military manual Huqian jing of Xu Dong (fl. 1000 AD). See Ralph D. Sawyer (with the collaboration of Mei-chün Lee Sawyer), The Tao of Deception: unorthodox warfare in historic and modern China (New York: Basic Books, 2007), pp. 260–73, in particular p. 260: "[Xu Dong] not only warns against slavishly following ancient methods, but also advocates deliberately contravening them in order to shed inimical constraints and avoid predictability" (emphasis supplied).

[56] Failures to absorb importantly include "failures to apply what we already know," with resulting challenges for strategy and logistics education. See Boorman, Introduction footnote 4, p. 112 note 45.

undercut one of the Sun Tzu text's most appealing and valuable features: namely, its ability to catalyze fresh applications in novel situations (some of which Sun Tzu could not have imagined). In the main body of the present study, candidates for "What did Master Sun know ..." insights will be suggested from time to time, usually in the Sun Tzu (2) and (3) "frontiers" discussions that accompany each of the fourteen themes.[57] Creative adaptation of Sun Tzu's ideas takes center stage here; specifics vary greatly. This book ends with two further batches of ideas also relevant to the "What did Master Sun know ..." question, which appear in the last two sections of the Conclusion. The first batch (pp. 504–13) analyzes Sun Tzu as a pioneering information warfare theorist. A second batch (pp. 513–21) profiles "A Sun Tzu for the Twenty-first Century."

The second question (Sun Tzu's limitations) will also be addressed in more than one way, providing an element of critique much needed in modern treatments of early Chinese warfare and strategy which too often fall into a cheerleading trap. First, each of the thematic chapters developing Themes #1–#14 contains a section entitled "Roads Not Taken by Sun Tzu," which is allocated to an overview of analytical directions that Sun Tzu might plausibly have pursued in context of his time, yet did *not* pursue (see p. 51 below). A broader vantage on Sun Tzu's limitations comes from applying to Sun Tzu the Eccles–Rosinski strategy-as-control framework analyzed in Boorman's 2009 *Naval War College Review* strategy article (see Theme #5 "frontiers," pp. 187–94 below).[58] The Conclusion chapter then casts a still wider net, profiling limitations on Sun Tzu's thinking in the broadest terms (see "Reality Check: Sun Tzu's Limitations," pp. 497–504). Table 8 (pp. 498–99) provides an overview.

Themes #1–#14 Chapters As Self-contained Units

Building on basic orientation provided by the Background and Preliminaries chapters, each thematic chapter is designed to be read in a self-contained way. Main text footnotes help indicate connections between themes. Textual support for each theme is anchored by the Sun Tzu passage selection with which every thematic chapter begins.

★★★★★★

---

[57] Table 9 in the Conclusion (p. 506 below) provides a roundup and finding aid for the present study's harvest of insights of "What did Master Sun know ..." type. Many of these contribute ideas from Sun Tzu pertaining to the role of information in strategy and conflict.

[58] This framework's seven basic elements (Boorman, Introduction footnote 4, p. 103) are set forth on p. 187 below.

Background: Historical and Textual

The present chapter provides historical and textual background for the present study. Theme-by-theme Sun Tzu analysis, beginning on p. 54 below with a short introduction to Themes #1 and #2, will draw on this background material.

As noted in the Introduction, there is a spectrum of informed modern scholarly opinion pertaining to dating and authorship of the Sun Tzu text. Faced with that diversity of views – and rather than taking on the unrewarding task of constantly trying to juggle them – the present study makes some choices.[1] Without committing to a detailed theory of the genesis of the text (not necessary for present analytical purposes), the present study will treat the Sun Tzu text as largely a product of the Warring States period (a dating consistent with a sizable body of modern scholarship).[2] It also accepts an accretionist stance on the Sun Tzu work, acknowledging that particular parts of the text are likely to hail from different times.[3] That in turn means that the Sun Tzu text we have is unlikely to represent the work of one hand.[4] However, on a pragmatic note (with a goal of avoiding awkward or pedantic phraseology), this book will routinely refer to Sun Tzu as if Sun Tzu were an individual person and author, on an understanding that this is only a modern-day language of convenience and that Sun Tzu's historicity as a person is not essential.

---

[1] An observation of the late Professor Charles L. Black, Jr. of the Yale Law School applies well here. "The hardest thing in the world," he said, "is to be clear about confusion."

[2] For a contrarian position (which is not a mainstream one), that the Sun Tzu text is a sole-authored work finished by 512 BC (i.e., in the late Spring and Autumn period), see He Bingdi 何炳棣, *You guan "Sunzi" "Laozi" de san pian kao zheng* 有關《孫子》《老子》的三篇考證 (*Three Textual Studies on Sunzi and Laozi*) (Taibei: Institute of Modern History, Academia Sinica, 2002), pp. 37–70.

[3] This perspective accords with widely held modern views on the development of many pre-Qin transmitted texts. See Wilkinson §58.6.2, pp. 775–76; also §28.1.1, p. 397. For both accretionist perspectives and their critics, focusing on analysis of the Confucian *Analects* (for many reasons, a very different text from Sun Tzu), see Michael Hunter & Martin Kern (eds.), *Confucius and the Analects Revisited: new perspectives on composition, dating, and authorship* (Leiden/Boston, MA: Brill, 2018).

[4] It should be mentioned that Griffith himself (Griffith, p. 12), operating on his intuition regarding "the originality, the consistent style, and the thematic development" of Sun Tzu's thirteen chapters, favored a position that the work "is not a compilation, but was written by a singularly imaginative individual who had considerable practical experience in war."

These choices point to focusing the historical discussion on conditions prevalent in Warring States China as contrasted with the prior Spring and Autumn period (to which Sun Tzu's traditional biography assigns Sun Tzu the person). Because an accretionist perspective also implies a text that took shape gradually, possibly over a period of many decades, adopting such a perspective also brings with it a need to be circumspect in pointing to any particular historical event or series of related events to shed light on Sun Tzu.

Military and Other Historical Context

Given the goals of the present study, which center on clarifying Sun Tzu's military and strategic ideas, the following discussion is limited to basic orientation that a reader should have regarding military conditions in Warring States China. That information will be supplemented later, as needed, in the course of developing Themes #1–#14.

Orientation should start on a note of chronology. Not only did multiple polities (states) already exist during the Spring and Autumn era (722–476 BC) preceding the Warring States period, but also many of the trends in warfare that emerged in full-fledged form in the Warring States were already well in evidence in that earlier time.[5] The non-specialist reader should therefore resist a natural urge to read some sort of brightline into the periodization.[6] Meanwhile, any Sun Tzu (1) proposition needs to be scrutinized in light of the joint effect of what may be labeled (with a bow toward mnemonic practices in Chinese culture) the Three Uncertainties:

- uncertainty about timing of composition of the text itself (or specific parts of it);
- uncertainty about timing of the advent, rise to prominence, or decline of specific military tools, institutions, or practices (chariots, crossbows, cavalry, etc.);
- uncertainty about timing of broader changes reshaping warfare and society in early China.

For example, depending on relative timing, a particular archaeologically recovered artifact may be highly relevant to understanding Sun Tzu – or not at all.[7] An additional

---

[5] Aspects of a case for continuity are set forth in Edward L. Shaughnessy, "The military histories of early China: a review article," *Early China*, 1996, 21, 159–82 (see, e.g., pp. 178–79, suggesting that it is possible to over-emphasize the extent to which a focus on "honor" set warfare in the Spring and Autumn era apart from warfare in the Warring States period that came after). Likewise underscoring continuities, Sawyer, pp. 267–68 note 89 observes: "'Be deceptive' was a dictum clearly in the minds of commanders in this [Spring and Autumn] era, a century or two before Sun-tzu's *Art of War*."

[6] That periodization derives, not from later scholarship but from a Confucian classic, the *Spring and Autumn Annals* (*Chunqiu*), a chronicle of the state of Lu that lends its name to the period it covers (722–481 BC). For background on the *Chunqiu* see Wilkinson, §48.1.1, p. 679.

[7] Uncertainty increases once the subject matter starts to touch on information questions ("who knew what, when, how," etc.). For example, we simply do not have direct evidence, textual or archaeological, to prove that reliable household registers existed in the time(s) and place(s) where the Sun Tzu text took shape. For a pertinent warning that the origins of an institution sometimes occurred centuries before it came to represent systematic practice, see p. 611 of Mark Edward Lewis, "Warring States political history," pp. 587–650 in Michael Loewe and Edward L. Shaughnessy (eds.), *Cambridge History of Ancient China: from the origins of civilization to 221 B.C.* (Cambridge, UK/New York: Cambridge University Press, 1999).

layer of complication arises because these uncertainties are themselves moving targets, as new archaeological finds occur and their analysis progresses, sometimes calling into question or falsifying seemingly solid bodies of scholarly opinion.

The Warring States period has been characterized as one of the major transition eras in world history. Its events have certainly exerted influence on the entire later course of Chinese history down to the present day. On a political–military landscape it was a time of great political disunity and unending warfare, as the *ancien régime* of the Zhou dynasty, already sagging for centuries, crumbled into oblivion.[8] As struggles among successor centers of political power unfolded, smaller states were snapped up by more powerful ones. Seven principal Warring States came to dominate the geopolitical landscape, though lesser ones persisted.[9] This agglomeration process culminated in 221 BC with the unification of China under the sole surviving state, that of Qin, whose dynasty – albeit short-lived – launched the imperial era in China that would last, with intervening periods of disunity, to 1912.

The transition that occurred in the Warring States period cut far deeper than its political (including its military) manifestations.[10] Despite challenges raised by revisionist historians regarding the extent of the transformation, both traditionalist and revisionist historians agree that it was a transformation that touched all strands of Chinese society, politics, economy, culture, and intellectual life.[11] As one strand of this transformation, warfare itself (as well as the state machinery supporting war) was

---

[8] The extent of warfare in early China is illustrated by lists of wars appearing in Victoria Tin-Bor Hui, *War and State Formation in Ancient China and Early Modern Europe* (Cambridge, UK: Cambridge University Press, 2005). Her appendix I (pp. 239–41) is "List of wars involving great powers in early modern Europe (1495–1815)." Appendix II (pp. 242–48) is "List of wars involving great powers in ancient China (656–221 BC)." The Western tally of wars is 89; the Chinese one is 256.

    Commenting on counts of Chinese battles, Wilkinson, §24.1.1, p. 339 observes: "whatever total [of battles] is arrived at, one thing certain is that the importance of Chinese warfare in world history is still underestimated by military historians in general and by historians of China."

[9] For profiles of the seven states see Lewis, Background footnote 7, pp. 593–97. Disappearance of some states, great and small, and incorporation of their territory into other states, is roughly tracked by maps 1–4 (pp. 186–88, 192) – spanning from middle/late Spring and Autumn period to *c.* 257 BC – in Victoria Tin-bor Hui, "Toward a dynamic theory of international politics: insights from comparing ancient China and early modern Europe," *International Organization*, 2004, 58(1), 175–205.

    What these maps don't tell us is the influence of specific states' physical and human geography on their military establishments, strategies, and doctrine. For example, the geographic situation of the state of Wu, which embraced the mouth of the Yangzi, seems a likely factor in Wu Zixu – a major historical military figure traditionally regarded as a contemporary of Sun Tzu, and who likewise served the state of Wu – producing a naval treatise, now lost. Further background on the strategic topography of China is provided by Wilkinson §24.4, pp. 343–44.

[10] For a terse but powerful vision of the challenge thereby presented to historians see Brooks, p. 74.

[11] Lewis, *Sanctioned Violence in Early China* (Albany: State University of New York Press, 1990), pp. 5–6 offers a panoramic sketch:

    The Warring States period in China, dated from various years in the fifth century to 221 B.C., is generally regarded as a transitional period between the "feudalism" of the high Zhou state and the creation of a unified Chinese empire by the Qin state. It was a time of unprecedented changes in government, society, economy, and culture. The economic changes included the rise of private ownership of land, the development of water control and iron tools leading to expanding agricultural output, a probable increase in population, a massive upsurge in the scale and diversity of urban centers, the burgeoning of handicraft industries and trade, the

concurrently itself undergoing rapid evolution. As the Zhou dynasty broke down, and rival centers of political power sprung up all over China, fighting became bitter as conflicts between states took on an existential quality (an aspect of warfare that Sun Tzu repeatedly notes). Professionalism increased; armies grew in size; weapons improved; social organization underpinning warfare, including merit-based personnel assignments as well as ways of mobilizing the civilian workforce for war, evolved alongside those developments. Meanwhile combat intensified; considerations of honor and an ethos privileging physical heroism fell by the wayside; casualty lists grew long. Perhaps most importantly from a Sun Tzu vantage, those developments were accompanied, and in no small part were fueled, by the rise of a new set of professional military commanders, edging aside the nobility who had dominated military command roles in an earlier era. Unsurprisingly, in ways that should strike a familiar chord from twentieth- and twenty-first-century experience, one upshot was a surge of interest in ways of fighting more effectively, of fighting smarter. That included keen interest in military and strategic theory and doctrine questions (Sun Tzu's forte).[12] In a pattern not paralleled in most non-Chinese armies prior to modern times, the new breed of generals owed much of its distinctive claim to professional competence to their mastery of substantive ideas found in military texts.[13]

---

increased use of money, and the introduction of hired labor. The social order was transformed by the disappearance of the old aristocracy, the rise of a large class of land-owning peasants, and a substantial increase in the urban classes devoted to trade and handicraft production. Political developments included the creation of a government staffed entirely by appointed, salaried officials, the assertion of direct control over local governments by the court, the promulgation of legal codes, and the comprehensive registration of the population. Among the military innovations were the introduction of cavalry, the invention of the crossbow, the increased use of iron weapons, the rise of mass, infantry armies, and the appearance of commanders who held office through their mastery of the arts of generalship. In the cultural sphere the period is most noted for its competing philosophical schools, which for the first time in China articulated theories of the state, the social order, language, and logic, but it also witnessed significant developments in literary genres, the visual arts, and such diverse sciences as medicine and astronomy.

Lewis goes on (*Sanctioned Violence*, p. 6) to make the telling (though not surprising) observation that

while scholars have agreed on the importance and the revolutionary nature of these collective changes, there is little agreement as to the inner coherence, if any, of the total ensemble. How the various phenomena were related to one another, which were primary and which secondary, and how to characterize the entire transition remain open questions.

[12] While the Sun Tzu text emerged as the pre-eminent text of the military methods (*bingfa*) genre, it was far from being the only text of that kind. Some other military texts of Warring States vintage have survived, including ones recovered by modern archaeology. Many more did not. The centrality of warfare in the dynamics of the Warring States also meant that many texts of other philosophical schools had significant military content. For example, teachings of the Mohist school – which flourished during the Warring States period though did not long survive it, at least officially – paid major attention to the arts of fortification and city defense. See *The Mozi: a complete translation* (translated and annotated by Ian Johnston; New York: Columbia University Press, 2010), pp. 731–921.

[13] Lewis, Background footnote 11, "The Commander and Texts," pp. 98–103. See also Sawyer, p. 54 noting appearance of an "extensive body of military theory."

Moving from texts to battlefields, the fact that the warfare of Sun Tzu's era was itself a moving target – actively evolving at the same time on levels of material technology, of "soft technology" (e.g., organizational concepts), and of ideas – creates hurdles for the modern analyst trying to pin down how prevailing military conditions may have shaped Sun Tzu's thinking. That is especially the case, of course, if the text itself took shape over a drawn-out period. Only a few basic observations, aiming to be as robust as possible in the face of major holes in our knowledge, will be offered here.[14]

First, sources point to a large, possibly even order of magnitude, increase in Warring States army sizes by comparison to those in the precursor Spring and Autumn period.[15] The Sun Tzu text itself cites army sizes in the vicinity of 100,000; in some cases still larger Warring States armies are reported and, especially towards the end of the period when truly all-out efforts were being mounted, such numbers may be credible.

These Warring States armies should be envisioned as fundamentally mass-infantry armies, replacing the chariot-dominated armies of the Spring and Autumn period.[16] A key military corollary – one that goes far to explaining Sun Tzu's preoccupation with terrain types and their implications (Theme #2) – was that the armies he knew were able to operate in far more diverse and often rugged terrain than chariot armies, which require an expanse of level, uncomplicated ground. Conjoined with larger armies, that expanded capability would also reinforce a new-found interest in military theory and doctrine, since options available to commanders would now be far greater, opening many new possibilities not confined to a tactical (battlefield) level but also reaching into the operational and strategic levels of war. The Sun Tzu text reflects such broader horizons.

A major factor in the combat effectiveness of Warring States armies, at least by the fourth century BC, possibly earlier, was the crossbow (mentioned three times in

[14] For a condensed (hence inevitably somewhat simplified) mainstream account of topics pertaining to Warring States armies, see Zhongguo junshi shi bianxie zu中國軍事史編寫組 (Committee for the Writing of Chinese Military History), Zhongguo lidai junshi zhidu 中國歷代軍事制度 (Military Organization of Successive Chinese Dynasties) (Beijing: Jiefangjun chubanshe, 2006), pp. 60–79.

[15] Lewis, Background footnote 11, pp. 60–61. Focus here should be on general trends, not purported specific head counts. Wilkinson, §24.5.1, p. 344 cautions that "The size of enemy armies and the number of enemy troops captured or killed are often greatly inflated in ancient Chinese sources," especially warning that pre–Qin dynasty statistics are "almost always suspect."

[16] Reflecting terrain factors, some states, notably Wu and Yue (both mentioned in the Sun Tzu text), appear to have avoided reliance on chariots, investing instead in infantry and naval forces as well as in weapons technology (e.g., swords of distinctively high quality). See p. 223 of Edward L. Shaughnessy, "Historical perspectives on the introduction of the chariot into China," Harvard Journal of Asiatic Studies, 1988, 48(1), 189–237; Sawyer, p. 66. Wilkinson §24.1.3, p. 342 points to the southern kingdoms of Wu and Chu as "the first naval powers in Chinese history." For further background see also Lan Yongwei 藍永蔚, Chunqiu shiqi de bubing 春秋時期的步兵 (Infantry in the Spring and Autumn Period) (Beijing: Zhonghua Shuju, 1979) as well as Raimund Theodor Kolb's pathbreaking study Die Infanterie im Alten China: ein Beitrag zur Militärgeschichte der Vor-Zhan-Guo-Zeit (Mainz am Rhein: P. von Zabern, 1991).

the Sun Tzu text).[17] In some ways it was the wonder weapon of its time, enabling relatively unskilled and unfit common soldiers to exert deadly force at a safe distance from hand-wielded weapons.[18] By contrast, something of the level of skill required by chariot drivers is conveyed by pondering the combat implications of the fact that, riding in a vehicle without shock absorbers, "they were forced to control two or more horses of less than identical physical capability and personality."[19] There is also fascinating archaeological evidence of the existence of a *repeating* crossbow of c. 400 BC vintage (from Tomb 47 at Qinjiazui in modern Hubei province, a site within the territory of the ancient state of Chu, the most southerly of the seven major Warring States).[20] Such a repeating capability might conjure images of a Warring States counterpart to the New Haven Arms Company's Henry repeating rifle that made its debut in the American Civil War, mainly on the Union side (reportedly prompting a Confederate to refer to it as that "damn Yankee rifle that can be loaded on Sunday and fired all week").[21] Unfortunately we know far too little about availability and operational characteristics of Warring States repeating crossbows (e.g., range; accuracy; maintenance and repair issues; etc.) for much weight to be placed on the mere existence of the instrument. A modern experiment suggests an effective range for the archaeologically recovered artifact of just twenty to thirty

[17] An archaeologically recovered crossbow trigger mechanism dated to 600 BC is described by Kuo-Hung Hsiao and Hong-Sen Yan, *Mechanisms in Ancient Chinese Books with Illustrations* (Cham: Springer International, 2014), p. 220. However, as with any technological advance, it is necessary to distinguish between invention per se and its widespread adoption. In Mair's estimate (Mair, p. 28) the crossbow "was not common in East Asia until the fourth c. B.C." Sawyer, p. 270 note 97 points to a pattern of cumulative technological advances: "Although early Chinese compound bows were extremely powerful, crossbows provided dramatically more formidable firepower; their strength and effective killing range generally increased over the centuries as their mechanisms were perfected."

[18] See Stephen Selby, *Chinese Archery* (Hong Kong: Hong Kong University Press, 2000; study cited by Wilkinson, §24.11, p. 354), p. 154:

> The reason that the crossbow found favour on the battlefield was not simple superiority over the bow and arrow. It was recognized from early on that the bow and arrow were superior in many situations. But by allowing a relatively unfit and untrained recruit to pull a heavy draw-weight (by using two hands with the middle of the bow held under the feet), many more soldiers could exploit the range and power of the bow without needing a large investment in training and strength-building.

[19] See Ralph D. Sawyer, with the bibliographic collaboration of Mei-chün Lee Sawyer, *Ancient Chinese Warfare* (New York: Basic Books, 2011), p. 349. Sawyer, p. 66 makes a telling observation: "Imagine racing across a corn field without shock absorbers and attempting to fire a bow or strike a moving, equally unpredictable opponent with a shock weapon at the last instant." Quite apart from any hostile action, chariots (somewhat like early motor vehicles) were also prone to breakdowns: "broken axles, becoming mired, getting tangled in branches, and falling into unseen gullies" (Sawyer, p. 67).

[20] For structural description of this repeating crossbow (including detailed analysis of its trigger mechanism), see Hsiao & Yan, Background footnote 17, pp. 221, 230–34. See also Kuo-hung Hsiao and Hong-sen Yan, "Structural synthesis of ancient Chinese Chu State repeating crossbow," pp. 749–58 in Jian S. Dai, Matteo Zoppi, and Xianwen Kong (eds.), *Advances in Reconfigurable Mechanisms and Robots I* (London: Springer, 2014).

[21] Laura Trevelyan, *The Winchester: the gun that built an American dynasty* (New Haven, CT/London: Yale University Press, 2016), p. 28. See also Robert B. Gordon, review of Joseph G. Bilby, *A Revolution in Arms: a history of the first repeating rifles* (Yardley, PA: Westholm Publishing, 2006), published in *Technology and Culture*, 2007, 48(2), 457–58.

meters, making it suited, at best, to be only a defensive weapon.[22] But the more telling point is that frontiers of weapons technology were being pushed that far that early.[23]

Available information suggests a relatively uncomplicated Warring States tactical military organization, rooted in a nested sequence of progressively larger units beginning with the five-man squad.[24] Sun Tzu's repeated affirmations attest to the aspiration of Warring States military organization to create military forces fused into a single instrument ("as if it was one man"), totally responsive to the commander's will.[25] In keeping with such a goal, it is plausible that military training, at the squad level and higher, privileged perfect discipline over individual combat skills (a pattern familiar from some eras of Western military history, for example the Prussian armies of Frederick the Great).[26]

Bracketing siegecraft (definitely not Sun Tzu's forte or focus) and, of course, chariots, we unfortunately know next to nothing about the nature and use of special-purpose or special-capability units that might have augmented a Warring States army's infantry core, possibly in interesting ways.[27] Perhaps most importantly, we

[22] As a conjecture, a primary purpose might have been hunting (possibly for birds). The experiment (and importantly, also its limitations, since the string of the original crossbow cannot be reproduced) is described on p. 37 of Lin Yun 林沄, "Nu de lishi" 弩的歷史 (History of the crossbow), Zhongguo dianji yu wenhua 中國典籍與文化 (Chinese Classics & Culture), 1993(4), pp. 33–37. For further background see Zhong Shaoyi 種少異, Zhongguo gudai junshi gongcheng jishu shi – shanggu zhi wudai 中國古代軍事工程技術史 – 上古至五代 (History of Military Engineering Technologies in Ancient China – From Ancient Times to the Five Dynasties) (Taiyuan: Shanxi Education Publishing House, 2008), pp. 236–46 (also cited in Hsiao & Yan, Background footnote 17, p. 242).

[23] Illustrating another development pushing the state of the art, there were also "linked crossbows" which "could be fired mechanically in unison, sometimes mounted on carriages" (Lewis, Background footnote 7, p. 623). Noting an aiming mechanism and mechanical way of retrieving (some) arrows, the Mozi text, Background footnote 12, pp. 781–83 proclaims "no limit to the number of crossbows" that could be thus linked. A crew of ten is mentioned (p. 783).

[24] Griffith, p. 126 footnote 1. For overview of early Chinese army organization dating back to times before Sun Tzu see Sawyer, pp. 73–76; Wilkinson §24.10, pp. 352–53. Mair, pp. 25 and 142 note 3 cautions that Warring States military organization varied by state and across time. A basic source on organizational ideal types (which, as in many eras of warfare worldwide, may not have aligned closely on military practice) is the Rites of Zhou (Zhouli) which traces out successively higher levels of organization from the five-man squad up to the "army" (jun) of 12,500 men. Apart from one case where it is four, the span of control is five throughout the structure. For background on the Zhouli – one of the Confucian canon's thirteen classics – see Early Chinese Texts (cited on p. xxii above), pp. 24–32; Wilkinson, §28.4.3, pp. 402–3 and §58.6.3.1, p. 777.

[25] See, for example, Sun Tzu verses XI.41a (Passage #13.24) and XI.54 (Passage #13.32).

[26] An observer of Frederick's Prussian army described it as "a vast and regular machine," a description that evokes the start of Sun Tzu's verse XI.41 (Passage #13.24), "To cultivate a uniform level of valour is the object of military administration." See Frederick the Great on the Art of War (Jay Luvaas, ed. and trans.; New York: Free Press, 1966), p. 12.

[27] The Wuzi military treatise, another of the Seven Military Classics, mentions several criteria used to form special units (e.g., in one case based on exceptional physical abilities; in another, comprising personnel seeking to redeem earlier conduct deemed disgraceful). See Griffith, pp. 153–54; Sawyer, Seven Military Classics, p. 209. Wilkinson, §28.4.3, p. 350 assigns the Wuzi a late Warring States date. On the Wuzi text see also Sawyer, Seven Military Classics, pp. 191–205. For description of Warring States elite forces notable for their extraordinary mobility see Lewis, Background footnote 7, p. 621.

are uncertain as to the historical point when cavalry appeared as a significant force on early Chinese battlefields. Cavalry's addition to Chinese military force structure is traditionally assigned a relatively late date, around 307 BC, associated with military reforms of King Wuling of Zhao who was feeling pressure from nomadic peoples to the north and decided to emulate their archer-cavalry mounted methods.[28] Even accepting Brooks's relatively late date for the Sun Tzu work (see Chronology, p. xviii), a 307 BC military innovation seems too late to have influenced the text, which notably does not mention cavalry at all (by contrast to its mentions of chariots and crossbows). However, the Sun Tzu text is so short and abstract that one cannot exclude the possibility that a role for cavalry was understood though not mentioned.

There is also evidence of a version of artillery (trebuchet, a type of catapult), possibly available by late Spring and Autumn times.[29] However, we know little of its mobility or range of tactical uses beyond an unsurprising association with siege warfare. Barring an obscure part of verse XII.1 (Passage #9.7), which might point to incendiary missiles, artillery-like capabilities do not figure in the Sun Tzu text.

A vitamin usually missing from attempts to reconstruct the armies of an ancient time is a visceral sense of what their actual battles were like.[30] That is a question where our sources from Sun Tzu's time largely fail us.[31] It is a question different in kind from recreating performance characteristics of individual weapons or items of equipment, and also different from static estimates of army size, force structure, etc.

From just after Warring States times a serendipitous lens on battle dynamics – albeit more cerebral than visceral – comes from the *Shiji*'s account of a successful battle waged by Han Xin, a famous early Han general who played a major part in the founding of the Han dynasty.[32] The story is richly imbued with many of the moves in

---

[28] See Wilkinson, §21.11, p. 354; §11.2.1, p. 191. There are grounds to believe that some Warring States army commanders rode horses in the field a century before 307 BC, but such equestrian capabilities of individual senior officers should be distinguished from creation of actual cavalry and were considerably less important as a military innovation. See note 16 on pp. 220–21 of Chauncey S. Goodrich, "Ssu-ma Ch'ien's biography of Wu Ch'i 吳起," *Monumenta Serica*, 1981–83, 35, 197–233. Even after the advent of cavalry there is evidence that cavalry remained a modest if important part of the force structure of Warring States armies. See pp. 103–104, 156–57 of Xiang Wan, "The horse in pre-imperial China," Ph.D. dissertation, University of Pennsylvania, 2013, available at https://repository.upenn.edu/edissertations/720 (visited August 18, 2021).

[29] See Needham & Yates, p. 206 (from section acknowledging contribution of Wang Ling).

[30] E.g., the sort of detail found in Philip Sabin, "The mechanics of battle in the Second Punic War," *Bulletin of the Institute of Classical Studies*. Supplement, No. 67, The Second Punic War: A Reappraisal (1996), 59–79 (acknowledging insights from modern wargamers).

[31] Brooks & Brooks, Introduction footnote 36, p. 101 observe that the *Zuozhuan* and early Chinese literature in general "do not abound in battle descriptions. Those who approach this literature with the Iliad ringing in their ears are courting disappointment." Utilizing an analytical outline format to clarify the phases of a Spring and Autumn period battle, Kierman has undertaken a reconstruction of the 632 BC battle of Chengpu, whose description in the *Zuozhuan* has been characterized as the "longest and best" description of some fifty battles chronicled there (Wilkinson, §24.1.2, p. 340). Yet making sense of the actual sequence of events in that battle requires numerous additional inferences or assumptions. See pp. 30–31, 47–56 of Kierman's "Phases and modes of combat in early China," in Frank A. Kierman, Jr. and John K. Fairbank (eds.), *Chinese Ways in Warfare* (Cambridge, MA: Harvard University Press, 1974), pp. 27–66.

[32] For *Shiji* 92 account see Watson translation (cited on p. xxiii above), *Han Dynasty I*, pp. 168–71. A modern analysis of the battle, with diagrams, is given by Kierman, previous footnote, pp. 56–62.

Sun Tzu's playbook, at least on one side of the engagement – indeed, a veritable roadmap of that playbook in practice.

The battle account starts with an offensive thrust by Han Xin's troops, said to number 20,000–30,000, through a lengthy gorge in the direction of the enemy Zhao army's fortified camp.[33] As an important piece of context, Han Xin's army had been depleted of its best troops, allocated by the Han ruler to another mission.

Learning that the Han forces were moving to attack, an adviser to Zhao's top command proposed to launch a force of "surprise troops" (qi bing) by a secret route to cut off Han Xin's supply train bringing up the rear of his army.[34] If successful, the effect would have been to trap Han Xin's army in the gorge between a fortified enemy position in front and another strong enemy force in the rear, unable either to advance or retreat and also unable by reasons of enemy presence and topography to plunder the surrounding countryside for food.[35]

This bright idea was rejected by Zhao's commanding general, a Confucianist who had no time for tricky schemes. The Shiji account has the Confucian also justifying his position by reciting military principles favoring a no-tricks direct attack, at least under certain circumstances.[36] However, key assumptions of that analysis were faulty, starting with underestimation of the size of Han Xin's force (and, perhaps yet more importantly as matters turned out, also its combat effectiveness).[37]

Han Xin learned of this upshot through spies sent to learn the enemy's thinking.[38] Armed with this inside knowledge, he then proceeded to lead his troops through the gorge. As he drew near its mouth, but had not yet left it, he stopped, made camp, and dispatched a 2,000-man force of light cavalry – his own version of surprise troops – to proceed along a secret route to a location where they could observe the Zhao army while themselves remaining unseen.[39] A crucial part of this ruse-in-the-making was that each man in the detached force should carry a red flag. The key tactical idea now appears in Han Xin's instructions to the detachment:

> When the Zhao forces see me marching out [of the gorge], they are sure to abandon their fortifications and come in pursuit. Then you must enter their walls with all speed, tear down the Zhao flags, and set up the red flags of Han in their place.
>
> (Watson translation, Han Dynasty I, p. 169.)

Han Xin's plan then came to life. First, he sent 10,000 men to march out of the gorge and draw up their ranks with their backs to the river that ran through it.

---

[33] On ravines (xian) in Sun Tzu see Theme #2 (pp. 95, 98 below) and Appendix 2 anchor footnote 31.

[34] The concept of "surprise troops" (qi bing) is a facet of the qi/zheng concept pair basic to Sun Tzu's thinking. For development see Theme #14 below. Ability to avail this force of a "secret route" suggests local terrain knowledge, perhaps gleaned from local guides (see verse XI.51 in Passage #11.13).

[35] Plundering for supplies is a basic Sun Tzu prescription. See Passages #3.6, #13.6, etc.

[36] Interestingly, one of those ideas ("when ten to the enemy's one, surround him") is explicit in Sun Tzu (Passage #1.32A's verse III.12). Arguably, another is also Sun Tzu's (see Appendix 1 footnote 47).

[37] The estimate made no allowance for Han Xin's creation and use of a death ground situation.

[38] Here espionage is seen playing a key role in a battle plan (see Theme #11, "Know the Enemy").

[39] Cf. verse VI.13 (Passage #8.1), "If I am able to determine the enemy's dispositions while at the same time I conceal my own [wu xing] . . ." As noted earlier (see p. 30) cavalry seems not to have been part of force structure in Sun Tzu's time, so the specifics of this part of Han Xin's battle plan reflect a post–Sun Tzu stage in the evolution of Chinese warfare.

The Zhao army interpreted this as a big blunder and is said to have roared with laughter.

Next, as dawn broke, Han Xin raised his flags as commanding general, thereby signaling his location (and serving up additional bait to the Zhao army), and marched out of the gorge with further troops. At this point the Zhao army took the bait and poured out of their fortifications to attack. The battle began. After a lengthy period of combat, Han Xin made a deceptive controlled retreat to the forces arrayed along the river, abandoning flags and drums.[40] The battle raged, but the army arrayed along the river, having nowhere to retreat to (and possibly also encouraged by Han Xin's presence), fought ferociously and could not be defeated. That tenacity of troops in extreme circumstances ("death ground") is a core Sun Tzu principle.[41] Han Xin was plainly counting on it for his larger scheme to work.[42]

Pressing their apparent advantage, the Zhao forces denuded their camp of troops. At that crucial juncture, the surprise force of 2,000 cavalry breached the Zhao fortifications. Tearing down the Zhao flags there they set up 2,000 red flags of Han.

This was the tipping point. Seeing the red flags of Han in their erstwhile home base, the Zhao troops concluded (erroneously) that the Han army had already captured their generals. It proceeded to disintegrate as a fighting force.[43] Although the Zhao commanders resorted to slaying some of their own men who were attempting to flee, they could not stop the rout. Han Xin emerged the decisive victor. On his orders, Han Xin's troops captured the smart Zhao adviser alive and, in a pre-modern twist to the story, Han Xin proceeded symbolically to pay him honor "with the respect due a teacher." He then also proceeded to pump him for bright strategic ideas, eliciting advice – sound as matters turned out – that the next stage of the campaign should center on psychological and political warfare, not combat.[44] (The gentlemanly Confucian was summarily executed.)

In an after-action analysis, conferring with his subordinate generals (who had been skeptical of the plan before it succeeded, and were still confused by the reason for its success), the Shiji reports Han Xin as observing:[45]

> Does it not say in *The Art of War*, "Drive them into a fatal position and they will come out alive; place them in a hopeless spot and they will survive"? Moreover, I did not have at my disposal troops that I had trained and led from past times, but was forced, as the saying goes, to round up men from the market place and use them to fight with. Under such circumstances, if I had not placed them in a desperate situation where each man was obliged to fight for his own life, but had allowed them to remain in a safe place, they would all have run away. Then what good would they have been to me?[46]

---

[40] Verse VII.27 (Passage #7.13) alludes to feigned retreats, a basic ruse in antiquity.

[41] For more complete Sun Tzu statements see verses XI.33 and XI.55 in Passages #9.14, #10.13, etc.

[42] Theme #1's calculation emphasis – including faith in the general's ability to do accurate calculation regarding effects of emotional forces – is on display here. See pp. 60–61 and Passage #1.12 below.

[43] Note the critical role played here by psychological dynamics of the sort envisioned by verse VII.20 (Passage #9.10), "Now an army may be robbed of its spirit . . ."

[44] See Theme #4 (winning without fighting by non-military ways and means).

[45] Although there were other works of this title (bingfa, lit. "military methods," often translated as *The Art of War*), Han Xin may here be pointing to the Sun Tzu text, in particular verse XI.55 (Passage #10.14).

[46] Theme #10 (no scruples) is on display here. In the background of Han Xin's analysis stands the Han ruler's decision to detach his best troops for service elsewhere, leaving Han Xin with leftovers.

A basic hole in our knowledge of Sun Tzu's armies involves the logistics support on which they all depended.[47] As van Creveld's masterly account of supply constraints facing early modern European armies has documented, the task of feeding, on a sustained basis, tens of thousands of physically active young males presents a truly dramatic challenge – one that, if honored in the breach, can swiftly degrade combat effectiveness and indeed unravel the cohesion of the force itself.[48] There is no reason to believe that Warring States Chinese armies were any less constrained by basic logistics limitations than the early modern European armies van Creveld chronicles, where a "campaign" was at times a misnomer for blundering around a countryside in quest of food for an army to survive on. As van Creveld's European examples suggest, Sun Tzu's somewhat facile advice to an invading army to survive by plundering the surrounding countryside elides much – including a need for such an army to keep constantly moving, for reasons quite distinct from any specific military goal vis-à-vis the foe, as the resources of the area where that army is currently located rapidly dwindle. There is certainly evidence, both from within the Sun Tzu text and from other early Chinese sources, that the basic caloric needs of Warring States armies were by no means always met.[49]

How did Warring States Chinese armies rise to the challenge thus presented? Primary sources tell us that the challenge was recognized. Sun Tzu himself says as much in his net assessment passage's verse I.8, which in Griffith's translation reads:

I.8.
By doctrine I mean organization, control, assignment of appropriate ranks to officers, regulation of supply routes, and the provision of principal items used by the army.

(from Passage #1.1)

---

[47] Positively noting work of Kolb (cited in Background footnote 16), Shaughnessy, Background footnote 5, p. 163 note 7 observes that "the topic of logistics is almost unmentioned in pre-Han military treatises," and with scattered exceptions remains likewise "almost unmentioned in [modern Western] studies of early Chinese warfare." A short account of Chinese military logistics from earliest times is Sawyer, Background footnote 19, chapter 24, "Ancient logistics," pp. 390–401. Focusing on a much later period, but also providing insight into supply and transport challenges, is David A. Graff, "Dou Jiande's dilemma: logistics, strategy, and state formation in seventh-century China," pp. 77–105 in van de Ven (ed.), Introduction footnote 49. See also Raimund Th. Kolb, "Excursions in Chinese military history," *Monumenta Serica*, 2006, 54, 435–64 (emphasizing logistics factors).

[48] Martin van Creveld, *Supplying War: logistics from Wallenstein to Patton* (2nd ed.; Cambridge, UK: Cambridge University Press, 2004), chapter 1, "The background of two centuries," pp. 5–39. Regarding military provisioning issues arising over the course of Chinese history see Wilkinson, §24.13, p. 357. He puts the issue bluntly (p. 357): "Once the initial grain rations were finished, provisioning a campaigning army became a major logistical problem." Those rations, based on what an individual soldier could carry, were good for about ten days. Looked at another way, if an ancient army could travel, say, nine kilometers per day (see p. A-70), that allotment would run out after about fifty to sixty miles.

[49] Sun Tzu verse IX.32 (Passage #9.16) notes famished soldiers leaning on their weapons. Another Warring States text makes reference to starving troops as being easily wounded (a point whose force is easy to overlook in an age of lightweight assault rifles!). See p. 31 of Olivia Milburn, "The book of the Young Master of Accountancy: an ancient Chinese economics text," *Journal of the Economic and Social History of the Orient*, 2007, 50(1), 19–40 (text tentatively dated to third or fourth century BC; see p. 30 of Milburn's article).

Another early Chinese military work, the Liutao, goes further, identifying, inter alia, a role for four supply officers on the staff of the commanding general. Other staff roles on the Liutao list also have duties with recognizable logistics aspects.[50]

Different types and styles of warfare differ enormously in the nature and extent of the logistics burdens they place on the combatants. To wage the specific type of warfare Sun Tzu envisioned – conventional offensive warfare, with large armies relying on well-coordinated maneuvers and operations deep in hostile territory – places vast demands on the capabilities and flexibilities of a general's logistics arrangements. Among the most difficult challenges for any logistics system support-ing a large army is the challenge of supporting operations that constantly and seamlessly reshuffle tactical and operational goals and priorities, which is precisely what warfare according to Sun Tzu demands (e.g., Passage #8.9).[51] A comprehen-sive concept of logistics – the concept of logistics for which in modern times Admiral Eccles strongly advocated – is the right yardstick by which to measure logistics achievement here.[52] For example, demands on an early Chinese army's logistics system would have included not only food and fodder, arrows and crossbow bolts, but also provision of diverse other resources, ranging from supply wagons and horses to lacquer supplies (for preserving and strengthening weapons and chariot parts) to siege equipment to implements for stratagems.[53] Services too would be part of that logistics package – among them, medical and veterinary care (the latter likely to be of exceptional importance for transport logistics reasons).[54]

Supporting the armies of Sun Tzu's time there must have been major, in some respects sophisticated, logistics organizations and planning capabilities. The contri-butions of modern archaeology, invaluable as they are, are weak in precisely this core aspect of logistics history – namely, how generals planned, organized, and

[50] Regarding those staff roles, see pp. 60–62 of the Liutao translation by Sawyer, *Seven Military Classics*. The date of the Liutao – which from the Song dynasty on was bundled with Sun Tzu as one of the *Seven Military Classics* – has been a subject of controversy (summarized in Sawyer, *Seven Military Classics*, pp. 35–37). Wilkinson §28.4.3, p. 351 assigns it a late Warring States date.

[51] From modern warfare, a vivid sense of logistics dynamics – and drastic constraints often resulting – during a major offensive campaign is provided by Roland G. Ruppenthal, *Logistical Support of the Armies: European Theater of Operations* (Washington, DC: Office of the Chief of Military History, Department of the Army), Vol. I (1953) and Vol. II (1959). For highlights see Admiral Eccles's review of vol. I in US Naval Institute Proceedings, July 1954, 80(7), 813–14; of vol. II, in US Naval Institute Proceedings, May 1960, 86(5), 108–11 and in Naval Research Logistics Quarterly, March 1960, 7(1), 95–98.

[52] See Henry E. Eccles (Rear Admiral, USN, Ret.), *Logistics in the National Defense* (Harrisburg, PA: Stackpole, 1959) (a book that has seen numerous republications and translations). A snapshot of Eccles's comprehensive concept of logistics – the view that "logistics covers the creation and sustained support of combat forces," i.e., everything to do with the means of war – contrasting it with narrower concepts of logistics – is Eccles, "A note on management and logistics," *Naval Research Logistics Quarterly*, March 1967, 14(1), 131. See also Boorman, Introduction footnote 4, pp. 99ff.

[53] Regarding military uses of lacquer see p. 304 of Robin D. S. Yates, "The Qin slips and boards from Well No. 1, Liye, Hunan: a brief introduction to the Qin Qianling county archives," *Early China*, 2012–13, 35–36, 291–329. Noting implements for stratagems see Kierman, Background footnote 31, p. 30.

[54] For background on warfare's toll on an army's animals (cf. verse II.14 in Passage #1.17) see Gervase Phillips, "Writing horses into American Civil War history," *War in History*, 2013, 20(2), 160–81.

implemented command control of their logistics. That is a level of heavily intangible, very specific, detail commonly not preserved in tombs or other archaeological sources.[55] Meanwhile, much as extant Greco-Roman classical sources provide scant information about the specifics of the command control of logistics that supported Alexander the Great's campaigns, so too available early Chinese textual sources likewise shed scant light on those crucial intangibles.[56] We are left with a keen appreciation of the extent of the challenge, with relatively little insight as to how it was met and its impact on conduct of early Chinese military campaigns.[57]

From what few sources are available – conjoined with the attitude of immense seriousness about all warfare for which Sun Tzu stands – one can infer that much brainpower and effort was invested by early Chinese generals (probably with staff support) in rising to these challenges, quite possibly more than that invested by many European generals of much more recent times.[58] Support for such a conjecture comes from the archaeologically recovered Mawangdui military map.[59] Focusing on an area in the vicinity of modern Changsha in Hunan province, that map is some fifty years after the end of the Warring States period (estimated date is c. 181–168 BC), but would have been within living memory of it.[60] Besides topographic features (e.g., hydrology, mountains) it depicts some ninety-nine civil installations and twenty-five

---

[55] For a sense of what archaeology – including archaeologically recovered government records – can tell us, and by implication what it is unlikely to tell us, see Yates, Background footnote 53, pp. 303–4, "Military logistics."

[56] Regarding deficit of sources on Alexander's logistics see Donald W. Engels, *Alexander the Great and the Logistics of the Macedonian Army* (Berkeley: University of California Press, 1978), p. 119.

[57] For a contrast case, see Jonathan P. Roth, *The Logistics of the Roman Army at War (264 B.C. – A.D. 235)* (Leiden/Boston, MA: Brill, 1999), observing (p. 280) that "given the general neglect of logistics in military history, our [Roman] sources mention such [logistics] planning remarkably often." His chapter 6, "The administration of logistics," pp. 244–78, provides key intangible details, such as field commander control – throughout the period of the Republic – of operational bases, "the link between supply lines and the area of operations" (p. 277). For further remarks on a "systems level" of Roman logistics see also Arthur M. Eckstein, review of Roth, *Journal of Military History*, 2000, 64(1), 182–84.

[58] Cf. Robin McNeal, *Conquer and Govern: early Chinese military texts from the Yi zhou shu* (Honolulu: University of Hawai'i Press, 2012), pp. 33–34:

Highly specialized technical knowledge about weaponry, terrain, battle formations, training methods, supplies, and the like must have become a prized commodity in the war-torn multistate world of the Eastern Zhou. We know little about the reproduction and transmission of such expertise, but there must have been fairly sophisticated institutions within the larger states charged with training military leaders and all the technical specialists they would have needed.

[59] For background on this "garrison map" (on silk) see Wilkinson, §14.1.2, p. 214. Further analysis of it using geographic information system (GIS) tools is Hsin-Mei Agnes Hsu and Anne Martin-Montgomery, "An emic perspective on the mapmaker's art in Western Han China," *Journal of the Royal Asiatic Society*, Third Series, 2007, 17(4), 443–57. See also Hsu's "Structured perceptions of real and imagined landscapes in early China," pp. 43–63 in Kurt A. Raaflaub and Richard J. A. Talbert (eds.), *Geography and Ethnography: perceptions of the world in pre-modern societies* (Chichester, UK/Malden, MA: Wiley-Blackwell, 2010).

[60] Martin Shubik (personal communication) has defined "within living memory" as a context "where someone who is extant knew someone now dead who transmitted an observation that he had witnessed or had obtained directly from someone who had witnessed it (2 degrees of fact attenuation)."

military facilities (e.g., command control nodes, garrison camps, weapons storage locations, observation towers, roads, defensive obstructions, etc.). In some ways even more striking than that level of detail, the same map has a temporal dimension, depicting three distinct phases of military planning. It also has basic data on civilian population movements induced by the fighting. Most importantly from the present standpoint, the map captures to a remarkable degree the spirit of information integration at the heart of logistics command control, requiring analytical integration of diverse types of logistics information pertaining not only to supply but also to transportation, weapon systems logistics, personnel, bases, and more.[61] The map corroborates existence of a type of information support for practice of logistics command control – admittedly in a setting simpler than the fast-moving needs of a major army on campaign, but involving many of the same skill-sets.

To this point the discussion has centered on what is sometimes called the consumer phase of logistics, focusing on logistics of a Warring States army in the field (the pertinent "consumer"), as contrasted with the "producer" phase of logistics pertaining to the economic sinews of military power and its mobilization for war (e.g., activities of weapons makers).[62] Some relevant insight into the latter can be squeezed out of the Sun Tzu text itself (though by the same token cannot serve as an external source of corroboration of the accuracy of Sun Tzu's observations).[63] For present purposes, a very brief summary can suffice.[64]

As the image of the "Warring State" suggests, the state and its people were mobilized for war on a continuing basis, subject only to limits of state capacity of that time (limits which, as in many other places in the pre-modern world, would have been substantial).[65] Although there were professional military personnel at the upper

---

[61] Underscoring the complexity of that information integration task – which is, of course, yet one more facet of the need for a comprehensive view of the logistics process – see Shaughnessy, Background footnote 5, pp. 163–64, noting a battery of logistics factors impinging on early Chinese warfare.

[62] For brief description of the producer/consumer distinction in logistics see Eccles, Background footnote 52, p. 64. Further structural aspects are developed in Eccles's *Military Power in a Free Society* (Newport, RI: Naval War College Press, 1979), pp. 63–66 and fig. 2, "The phases of logistics" (p. 65).

[63] See Yan Shengguo 閻盛國, "Sunzi bingfa suo jian zhanshi jingji wenti tanxi"《孫子兵法》所見戰時經濟問題探析 (A study of wartime economics based on evidence from Sun Tzu), *Sunzi Yanjiu* 孫子研究 (Sunzi Studies), 2018(2), pp. 29–36.

[64] For general factual background, a basic source still widely cited is Zhanguo shi 戰國史 (*A History of the Warring States Period*) by Yang Kuan 楊寬 (1914–2005), first published in 1955.

[65] A useful perspective flows from juxtaposing two observations about the Warring States period, each of which is basic to understanding the environment in which Sun Tzu's thinking took shape.

The first is from Lewis, Background footnote 11, starkly noting that "In these states warfare was no longer the means by which an aristocracy defined its authority, but rather *the primary institution used by the rulers of states to organize, rank, and control their subjects*" (p. 67, emphasis supplied).

The second – an important qualification – is from McNeal, Background footnote 58, p. 110: "This shift toward mobilizing entire populations was no doubt piecemeal, always easier on paper than on the ground, and probably easier in newly annexed territories than in areas where long-entrenched nobility would surely have resisted the change."

For analysis focusing on the state of Yue, foe of the state of Wu that Sun Tzu is said to have served, see Chen Zhang 陳章 and Zou Xuguang 鄒旭光, "Bing nong he yi, quan min jie bing: yueguo junshi zhidu chutan" 兵農合一, 全民皆兵：越國軍事制度初探 (Soldiers and farmers

echelons and a standing army component, personnel needs of a Warring State were largely met by military conscription.[66] Equipment needs were met by weapons factories or workshops;[67] financial needs by a functioning tax system.[68] Illustrating the sophistication of some of the pertinent capabilities is a case where evidence points not only to provision for medical leave from military duties but also to anti-fraud measures pertaining thereto.[69] Meanwhile, the Warring States period saw a shift toward sharp iron weapons whose production was scalable to high volumes, facilitating larger armies.[70] As of the fourth century BC there is also tantalizing evidence of some role for market competition forces associated with weapons

are one, all people are soldiers: a probe into the state of Yue's military institutions), *Journal of Nanjing Agricultural University (Social Sciences Edition)*, March 2007, pp. 93–98.

[66] For brief overview see Sawyer, p. 76; Lewis, Background footnote 11, pp. 91–92. For one of the seven principal Warring States – that of Chu, against which the putative historical Sun Tzu is said to have fought – archaeologically recovered legal documents of late fourth-century vintage provide an invaluable, vivid "insider perspective" on a complex society with far-reaching social control aspirations, one where the state took a keen interest in up-to-date, accurate population and worker registers.

See Susan Roosevelt Weld, "Chu law in action: legal documents from Tomb 2 at Baoshan," pp. 77–97 in Constance A. Cook and John S. Major (eds.), *Defining Chu: image and reality in ancient China* (Honolulu: University of Hawai'i Press, 1999). Regarding the enormous importance attached to household registration by early Chinese governments see Wilkinson §21.1–2, pp. 305ff. However, a weather eye needs to be kept on double-barreled uncertainty regarding (a) dating of the Sun Tzu text and (b) implementation of such registration, an ambitious logistics task whose degree of success surely varied across states and time periods, for reasons that modern tax administrators might attest to.

[67] See Brooks & Brooks, Introduction footnote 36, p. 42 pointing to *c.* fifth-century BC weapons factories. From shortly after the Warring States period, there is also archaeological evidence of batch or cellular production of weapons. See Marcos Martinón-Torres, Xiuzhen Janice Li, et al., "Forty thousand arms for a single emperor: from chemical data to the labor organization behind the bronze arrows of the Terracotta Army," *Journal of Archaeological Method and Theory*, 2014, 21(3), 534–62. Noting that "multi-skilled cells could be more easily moved with a real army to repair and produce other weapons as needed," p. 553 suggests that cellular production organization in this mortuary setting may have been similar to that involved in ordinary weapons' manufacture for battlefield use.

[68] Sawyer, p. 308 note 31 gives a thumbnail sketch of the military economics of the Warring States. Regarding land taxation and services see also Wilkinson §21.8, p. 311. Both the tax system and the military conscription system built on household registration.

[69] For specifics see #12 footnote 57.

[70] Mair, p. 28; Wilkinson §24.11, p. 354. Archaeological evidence confirms that steel weapons had appeared in China as early as 600 BC, though here again it is necessary to distinguish between existence of the technology and its widespread implementation and use. In that vein, Yang Hong, *Weapons in Ancient China* (Zhang Lijing, trans.; New York: Science Press, 1992), p. 174 notes of an archaeologically recovered steel artifact: "While this small steel sword may be said to be symbolic of the beginning of iron weapons, down to the early Warring States period, armies were not yet actually equipped with iron weapons." Timing of large-scale adoption of iron weapons by early Chinese armies remains debated, with some scholars arguing that true ascendancy of iron weapons only occurred in Han times. For overview of some of the debates see Sawyer, p. 271 note 100.

procurement practices.[71] By that time (and probably earlier) these many and sprawling producer logistics activities created a need for bureaus and administrators.

The Warring States also saw extensive development of transport and other infrastructure, much of it of dual use type (i.e., serving both civilian and military purposes). An impressionistic comparison suffices to convey a flavor.[72] The late Shang state (which fell to the Zhou dynasty around 1046 BC) has been depicted as

> a series of nodal points, with the important ones frequently appearing at cross-roads on the king's various journeys ... a thin network of pathways and encampments; the king and information and resources traveled along the pathways, but the network was laid in a hinterland that rarely saw or felt the king's presence and authority.[73]

Moving forward in time to the Warring States era, a very different picture appears:

> the feudal lords undertook the expanded defense of borders, constructing "great walls," ramparts, forts, and guard towers throughout the countryside to defend the entire territory against incursion. ... Fortified cities, previously military and administrative centers, grew enormously in significance as industry, trade, and population all flourished, and they became focal points in the road network.[74]

Developments like these would have given rise to a multitude of strategic decisions involving indivisibilities, set-up costs, lead times, etc. – all issues very familiar from modern military planning and operations research. The result would have been to make effective planning for war and peace more challenging, and along the way to enlarge the intellectual niche of works of the *bingfa* (military methods) genre.[75]

---

[71] See Brooks & Brooks, Introduction footnote 36, p. 49, quoting from a *Guanzi* passage they identify as late fourth-century BC, regarding weapons procurement and product inspection practices:

> Therefore, he assembles the world's finest products; he examines the sharp weapons of the various artisans. In spring and autumn, there are competitions to make a selection. The best and sharpest are given a superior rating. Until they have been inspected, the finished weapons are not used. Until they have been tested, they are not stored away.

As they remark, "The best is the best – and the best will be identified by the market process." For background on the *Guanzi*, a text with substantial military content bundled with non-military material, see *Early Chinese Texts* (cited on p. xxii above), pp. 244–51; Wilkinson, §58.6.3.1, p. 777.

[72] For further background see the pioneering study of Bai Shouyi 白壽彝, *Zhongguo jiaotong shi* 中國交通史 (*History of Transportation in China*) (Shanghai: Shangwu yinshuguan, 1937). Regarding road and transportation issues in the Warring States period specifically see also Shi Nianhai 史念海, "Zhanguo shiqi de jiaotong daolu" 戰國時期的交通道路 (Transportation and roads in the Warring States period), *Zhongguo lishi dili luncong* 中國歷史地理論叢 (*Journal of Chinese Historical Geography*), 1991(1), pp. 19–57. See also Xiang Wan, Background footnote 28, pp. 154ff.

[73] See p. 548 of David N. Keightley, "The late Shang state: when, where, and what?," in *The Origins of Chinese Civilization* (David N. Keightley, ed.; Berkeley: University of California Press, 1983), pp. 523–64. Map 17.1 on his p. 538 (from work of a Japanese scholar, Shima Kunio) shows a relevant network.

[74] Sawyer, p. 55. "Great walls" refers to initiatives by several of the warring states to construct defensive walls sometimes hundreds of miles in length. For background, dating the earliest walls to a pre-third-century BC time period, see Wilkinson, §24.12.1, pp. 355–56; Lewis, Background footnote 7, pp. 629–30 (walls to protect against northern nomads distinguished from walls erected along frontiers between Chinese states, the latter being more significant in the Warring States era). See also Yuri Pines, "The earliest 'Great Wall'? The long wall of Qi revisited," *Journal of the American Oriental Society*, 2018, 138(4), 743–62.

[75] For modern perspective on the complexities see Martin Shubik and J. Hoult Verkerke, "Open questions in defense economics and economic warfare," *Journal of Conflict Resolution*, 1989, 33(1), 480–99.

Anchoring in Chinese Text of Sun Tzu

As is the case with many early texts, Chinese and non-Chinese alike, there is no such entity as "the" original Sun Tzu text, nor should one be expected.[76] Militating against any such belief is the likelihood, discussed earlier, that the text we have is the upshot of a sustained process of accretion, guided by no single hand. Amplifying possibilities for textual variants is the inevitability of copyist mistakes (a fact of life compounded, especially in earliest times, by lack of standardization of the early Chinese language itself).[77] For a gaggle of reasons (an axe to grind; correcting an earlier problem, real or perceived; aspiration to leave a personal imprint on a famous text; etc.) somebody, indeed several different somebodys, might easily have slipped a change or two into the Sun Tzu text in the course of its transmission, creating downstream variance across versions if that edit was not universally known and accepted. Then too, there is the ever-present possibility of jumbled bamboo strips suggested by D. C. Lau, whereby the text might become disarranged at some point in its early transmission, creating still further possibilities for textual problems.[78] There are also lingering issues of a boundary definition type involving uncertainty, unlikely ever to be resolved crisply, as to whether in early times the Sun Tzu text (or Sun Tzu corpus) might have encompassed more content than its present thirteen chapters.[79] For all these reasons, it should come as no surprise that live scholarly debates about Sun Tzu textual issues are unlikely to disappear any time soon.

Yet none of these considerations should be allowed to obscure the more fundamental point that the Sun Tzu text we have – vouchsafed to us via both traditional transmission and archaeological means – is actually in pretty good shape for a text of its antiquity (much better shape, in fact, than many other important early Chinese

---

[76] Relatedly, there is a Chinese tradition that tends (by contrast to much of the Western philological tradition from Greco-Roman times forward) to exhibit a high degree of sangfroid with respect to differences across copies of classical texts, viewing the variations not as prima facie evidence of error but simply as alternative versions.

[77] On copyist errors see Wilkinson §38.13.1, pp. 539–40. A frequently neglected point of considerable importance is that Warring States authors or copyists also did not have access to dictionaries such as those we take for granted today. Meanwhile, larger editing or textual harmonization tasks would have been impeded by the sheer bulkiness and weight of bamboo books, which meant (as Wilkinson, §69.1.3, p. 1026 puts it wryly) that even "Reading from bamboo slips was a weighty matter." He further observes that "Because of the weight of the bamboo slips, texts did not circulate widely, and, if they did, under normal circumstances, it must have been in separate rolls or bundles," not as complete works. The latter fact alone seems conducive to emergence of textual variations.

[78] Lau, 1965 article, p. 325. The possibility of rotted, broken, etc. strings is plainly a real one (for a non–Sun Tzu example see Wilkinson §60.3, p. 820). However, any such disorder comes in degrees. The present analysis of Sun Tzu, while not ruling out possibilities for some disorganization born of jumbled or broken bamboo strips, suggests little reason to regard the Sun Tzu text as any sort of outlier or extreme case. Indeed a close reading of Sun Tzu turns up cases where adjoining passages, which might initially appear disjointed, on further scrutiny convey related military or strategic ideas.

[79] For evidence, which is inconclusive, see pp. 447–49 of Krzysztof Gawlikowski and Michael Loewe, "Sun tzu ping fa 孫子兵法" in Early Chinese Texts (cited on p. xxii above), pp. 446–55. Drawing on work of Li Ling, Yates, Introduction footnote 51 gives a perspective on what is known. See also Mair, p. 11.

texts).[80] Indeed the point could be put more strongly: Sun Tzu is actually one of the *least* problematic of the traditionally transmitted early Chinese texts in terms of textual difficulties. For military and strategic analysis purposes it is therefore accurate – indeed desirable, to avoid needless confusion – to envision a well-defined Sun Tzu text in thirteen chapters and thus to speak, sans a raft of qualifiers, about what "Sun Tzu said."[81] That text is the focus of the present study.[82]

Of course, some specific textual issues do persist. Notable among these is a smattering of cases (some perhaps attributable to "ancient typos" of one sort or another) where received versions of the Sun Tzu text differ, sometimes by a single Chinese character, in a way that creates a significant ambiguity as to the meaning. There are also a handful of instances suggestive of more extensive textual defects, making understanding of some parts of Sun Tzu perforce tentative.[83] Yet these larger difficulties remain confined to specific segments of the text and their effects are containable.

Preferred copy. The Sun Tzu text used to anchor the present study is the Song dynasty copy of Sun Tzu with eleven commentators ("preferred copy"), of which the presently preferred edition is the People's Liberation Army one.[84] It should be noted

---

[80] For example, A. C. Graham, an authority on the *Mozi* text (see Background footnote 12) bluntly observes that the "text of *Mozi* is notoriously corrupt." See *Mozi* entry by him in *Early Chinese Texts* (cited on p. xxii), p. 339.

[81] The stylized phrase with which each of Sun Tzu's thirteen chapters begins.

[82] The same archaeological find that recovered the Han strips copy of much of Sun Tzu's thirteen chapters content (see pp. 41–42 below) also turned up a mass of other Sun Tzu–related material that Ames translates as Part II of his Sun Tzu edition (Ames, pp. 173–96) and describes (Ames, pp. 36–37) as bearing a "commentarial" relation to Sun Tzu's thirteen-chapter "core." Unlike the work of Sun Tzu's eleven traditional commentators (see pp. 42–44 below) this archaeologically recovered material is unsigned, leaving us in the dark as to its authorship. Other material attributed to Sun Tzu, much of it fragmentary, culled from a variety of later Chinese encyclopedic works and other sources (such as a 1978 archaeological find in western China bordering Tibet), is collected and translated in Ames's Part III (Ames, pp. 197–259). However, comparing the military and strategic content of "Sun Tzu Part II" translated by Ames with that of Sun Tzu's thirteen-chapter core, loss of substance from de-emphasizing Part II material seems minor. With the exception of some content from Du You's *Tongdian* (Ames, pp. 199–223), an encyclopedic work that contributes to clarifying Sun Tzu's "nine terrains" (see Passage #2.1), a similar evaluation applies to Part III.

[83] Six problematic parts of the Sun Tzu text are collected in Table 13A on p. A-37.

[84] **Note on editions and sources:** The presently preferred copy is the *Shiyi jia zhu Sunzi* 十一家註孫子 (Sun Tzu with eleven commentators). The preferred edition of that copy used in the present study is:
  Zhongguo Renmin Jiefang Jun Junshi Kexue Yuan, Zhanzheng Lilun Yanjiu Bu Sunzi Zhushi Xiaozu 中國人民解放軍軍事科學院、戰爭理論研究部《孫子》註釋小組 (Group for the Annotation and Explanation of the Sun Tzu in the Institute for Military Theory of the Chinese People's Liberation Army Academy of Military Science), *Sunzi bingfa xinzhu* 孫子兵法新註 (Sun Tzu's Military Methods, Newly Annotated) (Beijing: Zhonghua Shuju, 2014).
  Also consulted (in addition to other work by Li Ling, a leading Sun Tzu authority) was Li Ling's modern commentary on Sun Tzu along with Yang Bing'an's collection of textual variants and emendations:
  Li Ling 李零, *Sunzi yi zhu* 孫子譯註 (Sun Tzu, Translated and Annotated) (Beijing: Zhonghua Shuju, 2014);
  Yang Bing'an 楊炳安, *Shiyi jia zhu Sunzi jiaoli* 十一家註孫子校理 (An Edited Emendation of Sun Tzu with Eleven Commentators) (Beijing: Zhonghua Shuju, 2016).

that Griffith's translation relied on a different copy of the text, one that also incorporated emendations by Sun Xingyan, a Qing scholar (1753–1818).[85] However, the divergences between the presently preferred text and that used by Griffith only very occasionally rise to a level affecting key substance.[86]

Han strips text. An outlier among extant copies of the Sun Tzu text is the archaeologically recovered partial text of Sun Tzu (mentioned on p. 7 above) dating from the second century BC and found in 1972 at a site in Shandong province.[87] The present study will standardly refer to that partial text as the "Han strips," thereby underscoring the Han dynasty era provenance of the bamboo strips on which it is recorded.[88] This archaeologically recovered text largely agrees with the traditionally transmitted ones (though there are enough differences to keep scholars

For a short overview of the genealogy of the Sun Tzu text see also Hattori Chiharu 服部千春 and Liu Chunsheng 劉春生, "Sunzi bingfa de banben he jiaokan yanjiu" 《孫子兵法》的版本和校勘研究 (Studies of Sun Tzu's copies and emendations), Tianjin Nankai Xuebao 天津南開學報 (Tianjin Nankai University Bulletin), 1998(6), pp. 57–61. The five copies identified there as the most important are:

Copy 1 (not complete): archaeologically recovered "Han strips" partial text found in 1972;

Copy 2 (fragmentary): found at Dunhuang site in China's northwest, a major Silk Road hub, and published in 1918 by Luo Zhenyu, a modern scholar (copy possibly of Jin dynasty vintage [265–420 AD, see Chronology, p. xix above]; it has helped to correct certain errors in later copies);

Copy 3: Sun Tzu with eleven commentators (the presently preferred copy, see above);

Copy 4: Song dynasty copy with annotations by Cao Cao (see next footnote);

Copy 5: further Song copy contained in the *Seven Military Classics* compilation (see p. xix above). For a somewhat different but complementary overview of early editions and versions of the Sun Tzu text see also Gawlikowski & Loewe, Background footnote 79, pp. 449–51.

[85] The Sun Xingyan edition is based on, while also emending, Copy 4 (see previous footnote). This is *Wei Wu Di zhu Sunzi* 魏武帝註孫子 (*Wei Wu Di's annotation of Sun Tzu*), a Song dynasty copy of Sun Tzu with commentary by Cao Cao (155–220 AD) who is the same person as the title's Wei Wu Di. For further background on Cao Cao see p. 43 below. Raw material for a textual comparison between the preferred edition (based on Copy 3) and the Sun Xingyan edition (based on Copy 4) can be found in Li Ling 李零, "Sunzi" guben yanjiu 孫子古本研究 (*A Study of the Ancient Copies of Sun Tzu*) (Beijing: Beijing University Press, 1995), pp. 80–106.

For biographical background on Sun Xingyan, an important scholar of his time, see "Sun Hsing-yen" entry in Arthur W. Hummel (ed.), *Eminent Chinese of the Ch'ing Period (1644–1912)* (Washington, DC: United States Government Printing Office, 1944), Vol. II, pp. 675–77.

[86] In his 1965 article's lengthy footnote 9 (see pp. 320–21 there) Lau takes Griffith to task for relying on the Sun Xingyan text, illustrating the point with verse VI.5 where Sun Tzu's military advice is drastically altered depending on whether Sun Xingyan's edition or the preferred one is used. Specific issues presented by verse VI.5 are analyzed in Appendix 6 anchor footnote 12, coming down in favor of Griffith (and Sun Xingyan). However, this textual disparity is unusual, not the norm.

[87] For a reconstructed text of the Han strips see Wu Jiulong 吳九龍, *Yinqueshan Hanjian shiwen* 銀雀山漢簡釋文 (*Transcriptions of the Han bamboo slips from Mt. Silver Sparrow*) (Beijing: Wenwu, 1985) (cited in Wilkinson, §59.6.2.2, p. 806). Wilkinson characterizes the archaeological find that included this Han strips partial text of Sun Tzu as one of "the most important discoveries [of bamboo slips] in the twentieth century."

[88] Confusingly, this text goes by quite a few names in Sun Tzu literature, among them: "bamboo slips," "tomb writings," "the Yinqueshan manuscript" (after the location in Shandong province where the text was found), Silver Sparrow Mountain slips (a translation of that Chinese place name), and more.

on their toes!). As with other Sun Tzu textual issues, variations between the Han strips and the preferred text will be noted and addressed as needed, primarily in footnotes.

Reinforcing the point that there is no single "pristine" or "original" Sun Tzu text, it is worth weighing in against sometimes-uncritical modern acceptance of the Han strips as representing such a text.[89] For example, in a possible case of later interpolation, the Han strips contains a reference to a historical figure that the traditionally transmitted texts of Sun Tzu omit.[90] Again suggesting a text not yet quite solidified, the Han strips archaeological find also turned up two variants of Sun Tzu IV (though the differences between them are minor).[91] More importantly from a military vantage, the Han strips offers advice for several types of terrain that significantly differs from advice found in the traditionally transmitted text, thereby creating a specialized puzzle for modern students of this part of Sun Tzu's substance to ponder.[92]

Perspective on Sun Tzu's Commentators (and *Romance of the Three Kingdoms*)

Over the course of its long history, the Sun Tzu text has attracted the efforts of many commentators.[93] Eleven traditional commentators – often treated, with some considerable arbitrariness, as a canonical set – are represented in the presently preferred Song dynasty copy of Sun Tzu.[94] Highlights of their biographies, when known, are profiled by Griffith, who also translated what he regarded – drawing on his

---

[89] For a corrective standpoint, advocating for a balanced view of the Han strips, see Sawyer, p. 303.

[90] Specifically, the Han strips text corresponding to Sun Tzu verse XIII.22 (Passage #5.19) includes a reference to Warring States strategist Su Qin (380–284 BC), a far later figure than the two personages mentioned in the traditionally transmitted text of this verse (whose purported exploits relate to eras many centuries before Sun Tzu). See Ames, pp. 22–24 (noting a body of scholarly opinion that this Su Qin reference is a later interpolation).

[91] See Li Ling, Background footnote 85, pp. 9–10 for the two textual variants (which Ames, p. 286 note 140 labels "A" and "B"). The reason for inclusion of both versions in the tomb is unclear.

[92] For further specifics see Appendix 2 footnote 3.

[93] For orientation, see Yates, Introduction footnote 51.

[94] The present study reserves the phrase "traditional commentator" for the eleven commentators represented in Shiyi jia zhu Sunzi (cited in Background footnote 84). Other pre-modern commentators whose commentary survives certainly also exist. Among the latter, the present study draws in particular on Shi Zimei (Southern Song; earliest commentator on the *Seven Military Classics* compilation, which includes Sun Tzu), Liu Yin (early Ming), and Zhao Benxue (1478–1544).

For housekeeping purposes, the eleven traditional commentators, in estimated chronological order (following Mair, p. 3), are: (1) Cao Cao (155–220), (2) Meng shi (lit. "Mr. Meng," Liang dynasty [502–557]), (3) Li Quan (*fl.* 750), (4) Jia Lin (*fl. c.* 775–800), (5) Du You (735–812), (6) Du Mu (803–852), (7) Chen Hao (*fl.* late Tang dynasty [618–907]), (8) Mei Yaochen (1002–1060), (9) Wang Xi (*fl.* 1082), (10) He Yanxi (after mid-eleventh century), and (11) Zhang Yu (Southern Song dynasty [1127–1279]). By Tang dynasty times a commentator network of a sort had begun to develop, with Du You citing work of Li Quan (Mair, p. 5) and Du Mu, grandson of Du You, criticizing Li Quan as well as Cao Cao for their apparent approval of a military role for divination (Yates, Introduction footnote 51, p. 74).

Mentions of "ten commentaries" refer to the above list excluding Du You, whose comments are extracted from Du You's encyclopedic work, the *Tongdian*. See Mair, p. 57 note 8; Griffith, pp. 17–18.

professional military judgment – as the best parts of their commentary (which are interspersed between segments of Griffith's Sun Tzu translation).[95] There are also a large number of further commentaries or commentary-equivalents, both pre-modern and modern, emanating from many quarters, now worldwide.

Among the eleven commentators particular notice should be taken of Cao Cao (155–220 AD), the earliest Sun Tzu commentator whose signed commentary has survived. One of the most famous soldier-strategists in Chinese history (he was also a poet), Cao Cao flourished in the waning years of the Han dynasty, centuries after Sun Tzu's time. His stature as one of the greatest generals of Chinese (indeed world) military history lends his commentary unique prima facie stature that few if any civilians in any era could rival.[96]

Certain of Cao Cao's observations point in interesting directions. For example, commenting on verse XII.2 in Sun Tzu's fire attack material (Passage #9.7), Cao Cao contributes a thought of his own (Griffith, p. 141), to the effect that one should "rely on traitors among the enemy" to start fires. Although he clearly ranges beyond the four corners of the Sun Tzu text to propose this point, this Cao Cao idea could be a stimulus for creative Sun Tzu (3) thinking about what a well-placed traitor might be able to accomplish using *other* special weapons (say, twenty-first century cyber or biological capabilities, or political warfare ones).[97] Yet too much focus on Cao Cao would move the present inquiry a step away from a focus on Sun Tzu (or possibly two steps if clarification of Cao Cao's ideas necessitates entering other debates, such as evolution of Chinese warfare from Sun Tzu's time to Cao Cao's, centuries later).[98]

Most of the remaining commentators in the traditional set of eleven are bunched in two separate periods – one in the Tang dynasty, one in the Song

---

[95] For brief biographical sketches, see Griffith, pp. 184–86; Giles, pp. xxxiv–xlii; Mair, pp. 3–5. See also Li Ling, Background footnote 85, pp. 263–74. Several of the traditional commentators have been subjects of extended English-language studies (see previous footnote for identifying numbers):
- (1) Cao Cao: see Rafe de Crespigny, *Imperial Warlord: a biography of Cao Cao 155–220 A.D.* (Leiden/Boston, MA: Brill, 2010);
- (3) Li Quan: see Christopher C. Rand, "Li Ch'üan and Chinese military thought," *Harvard Journal of Asiatic Studies*, 1979, 39(1), 107–37;
- (6) Du Mu: see Stephen Owen, *The Late Tang: Chinese poetry of the mid-ninth century (827–860)* (Cambridge, MA: Harvard University Asia Center, distributed by Harvard University Press, 2006), chapter 8, "Du Mu," pp. 255–314 (focus is on Du Mu as a poet, but relevant career information and analysis is interspersed);
- (8) Mei Yaochen: see Jonathan Chaves, *Mei Yao-ch'en and the Development of Early Sung Poetry* (New York: Columbia University Press, 1976) (see pp. 17–18 on Mei's Sun Tzu commentary).

[96] Cao Cao's relation to the Sun Tzu text is discussed in de Crespigny, previous footnote, pp. 319–32.

[97] Cao Cao's remark pertaining to traitors who set fires also illustrates why tangling with the commentators can be frustrating. For example, alluding to Cao Cao's comment, Chen Hao remarks that "one does not only rely on traitors" (Griffith, p. 141 footnote 2). Fair enough; but Chen's fussy addendum tends to detract from Cao Cao's sharp-edged (albeit not textually anchored) insight.

[98] Cao Cao's commentary is famous for its terseness. With some justice, Giles, pp. xxxv–xxxvi (citing a supporting observation from a Chinese source) complains that, owing to their "extreme compression," Cao Cao's notes on Sun Tzu are "scarcely intelligible and stand no less in need of a commentary than the text itself."

dynasty – each spanning around a hundred years. Their author's knowledge of the warfare of Sun Tzu's time (or indeed of warfare at all, certainly in its practical aspects including logistics ones) is in many cases unclear and may well have been highly limited.

While commentators of many stripes, including modern-day ones, can at times shed light on Sun Tzu, they may have their own axes to grind (sometimes visible, sometimes less so). Commentators often volunteer insights that are of uneven quality and relevance (and in some instances are, candidly, poorly coherent or simply banal).[99] Not infrequently, their observations are also at odds with one another.

These caveats apply with redoubled force to the ever more vast and chaotic modern secondary literature relating to Sun Tzu that ranges beyond the commentary genre. Contributions to that literature sometimes seem to pay more attention to each other than they do to the Sun Tzu text, which can give rise to a kind of echo chamber whereby an idea or hypothesis about Sun Tzu reverberates many times over, possibly lending it an unmerited appearance of credibility or gravitas. Many of those who write about Sun Tzu, Asian as well as Western, are military people who, understandably, have limited textual exegesis skill-sets (or indeed interest in that type of task). General Griffith, with his DPhil from Oxford in Chinese military history, remains an outlier here. Meanwhile many civilians, attracted to the Sun Tzu text by a perception that Sun Tzu is a kindred spirit, have limited interest in (indeed no more than glancing awareness of) core military topics like logistics or command and control or indeed any serious interest at all in the art of war on land.[100] The resulting Sun Tzu–inspired writings can (and often do) wind up conjuring more or less imaginary Sun Tzus. Other problems born of the winds of ideology and regime change (and spawning yet more imaginary Sun Tzus!) crop up in modern literature on Sun Tzu emanating from the Maoist era in China or the Soviet period in Russia.

For these reasons, the present choice is to focus as single-mindedly as possible on the Sun Tzu text itself. Apposite commentator insights or suggestions, rarely definitive, should be taken for what they are worth. But there also seems little compelling reason to take on the gargantuan task of trying to round up and coordinate all the interpretations and observations, defensible or otherwise, that have emanated from assorted commentators and other students of Sun Tzu ancient and

---

[99] Griffith was keenly aware of this pitfall. While his Sun Tzu translation was in preparation, Griffith's correspondence with Liddell Hart notably included a request for the latter's advice as to which of the commentary passages Griffith had translated should be omitted in the published version of Griffith's work – e.g., because "boring, silly, if not down right non sensical." See Griffith's informal notes headed "Chapter 1" filed (along with accompanying letter to Liddell Hart) in LH 1/333/2 in the Liddell Hart Papers, Liddell Hart Centre for Military Archives, King's College London.

[100] This is not a new phenomenon. Noting that traditional commentator Du Mu "was not a soldier and can hardly have had any practical experience of fighting," Giles, p. 101 casts a quizzical eye on Du Mu's commentary on verse X.2 (Passage #2.4A) which "speaks of protecting the [army's] line of communications by a wall, or enclosing it by embankments on each side." Another example involves verse VII.29 (Passage #1.10) on not swallowing enemy baits. Giles, p. 68 observes, "Li Quan and Du Mu, with extraordinary inability to see a metaphor, take these words quite literally of food and drink that have been poisoned by the enemy." A Song dynasty source tells us that "although [Du Mu] had no practical experience of war, he was extremely fond of discussing the subject." See Giles, p. xxxvii.

modern. Certainly none of the Sun Tzu commentary literature – starting with commentators in the traditional set of eleven – merits being elevated to a status of "secondary classics."[101]

★★★★★★

One other pre-modern Chinese work merits special mention in a Sun Tzu context. A work of fiction, in a literal sense it is not part of the Sun Tzu tradition at all. Still sometimes known to English-language readers by the title of its first English translation, the *Romance of the Three Kingdoms*, the *Sanguo yanyi* – commonly identified for short as the *Sanguo* ("Three Kingdoms") – is a great Chinese historical novel (estimated to be of early Ming dynasty vintage) about the unsettled period leading up to the fall of the Han dynasty and subsequent Three Kingdoms era of Chinese history (220–280 AD).[102] A widely quoted aphorism of Qing scholar Zhang Xuecheng (1738–1801) is that the *Sanguo* is "seven parts history and three parts fiction."

Famous for its vivid accounts of stratagems galore and other military and leadership vignettes – "a dramatic record of personal combat, raids, surprises, offensives, sieges, pitched battles, and protracted campaigns," as translator Moss Roberts puts it – the *Sanguo* is indelibly linked with Sun Tzu through exploits of one of its central characters, none other than the prominent general and Sun Tzu commentator Cao Cao. Although the *Sanguo* does not mention Sun Tzu's treatise as such, candidates for content from nine of Sun Tzu's thirteen chapters have been identified as present in it.

In modern times the massive *Sanguo*, containing more words and more human characters than Tolstoy's *War and Peace*, stirred the imagination of the young Mao Zedong.[103] Its beloved status and influence on audiences from many walks of life in

---

[101] A Sun Tzu–related practice, worth mentioning because it is sometimes found in otherwise high-quality modern literature, takes a large step further. That is to allow statements from the traditional commentators to be conflated with propositions found in the Sun Tzu text itself – in effect, merging the commentators into Sun Tzu. That move needs to be recognized as truly problematic, for many reasons starting with the fact that quality of military insight found in the commentators is decidedly uneven (see previous footnote). There is a real risk here of leading astray Sun Tzu audiences who are not acquainted with the complexities of the commentary genre in Chinese tradition and the need to approach many commentaries gingerly, in a caveat emptor spirit.

Such a conflation of Sun Tzu text and commentary should be resoundingly rejected.

[102] For brief orientation to the *Sankuo* see Raphals, Introduction footnote 42, pp. 134–36. The *Sanguo*'s historical setting, authorship, dating, use of sources, and related issues (including relationship to Han nationalism in two times of dynastic transition, Yuan/Ming and Ming/Qing, both involving an Inner Asian conquest of China) are addressed in "Afterword: about *Three Kingdoms*," pp. 937–79 in the Moss Roberts translation (cited on p. xxiv above).

The *Sanguo*, a work of fiction, importantly needs to be distinguished from the *Sanguo zhi* (*Records of the Three Kingdoms*), a similarly titled non-fiction historical work of third-century AD vintage on which the novel loosely builds and with which it can be confused. For basic background on the *Sanguo zhi* – compiled more than a thousand years prior to the *Sanguo* – see Wilkinson, §60.1.1, p. 816; on the *Sanguo yanyi*, see Wilkinson, §31.1.6, p. 451.

[103] See Mao's autobiographical reminiscences as quoted in Edgar Snow, *Red Star over China* (revised and enlarged edition, with introduction by John K. Fairbank; London: Victor Gollancz, 1973), p. 133.

late imperial and modern China and elsewhere in East Asia has sometimes been compared with that of Shakespeare in the West.[104] From a Sun Tzu standpoint, the *Sanguo* is a trove of concrete, albeit fictionalized, examples of many of Sun Tzu's precepts in action, along with other ideas that Sun Tzu never articulated as such but where many educated East Asian readers would instantly spot resonance with "thinking like Sun Tzu." The *Sanguo* is a basic lens through which those audiences have long approached and appreciated the idea level of Sun Tzu. For these reasons – while avoiding reliance on the *Sanguo* as a historical source – the main body of the present work will sometimes draw on the *Sanguo* to help make Sun Tzu's abstractly phrased concepts and principles come to life in a Chinese context.

★★★★★

---

[104] It has been estimated that scenes from the *Romance of the Three Kingdoms* "account for 70 per cent of the popular theatre repertoire at the turn of this century." See p. 1224 of Arthur Waldron, review of Needham & Yates, *The China Quarterly*, No. 144, December 1995, pp. 1222–25.

Preliminaries

The three Sun Tzus approach brings with it an expository challenge: how to accommodate the fact that exploring the three Sun Tzus unavoidably puts in play not just a single body of background information but at least three different bodies of background, one for Sun Tzu (1), another for Sun Tzu (2), and yet another for Sun Tzu (3). In the latter case more than one type of background is likely to be needed.

The present response to this challenge makes use of two basic tools.

The first is a suite of basic military/strategic concepts, timeless in nature and versatile in their applications, which in their many permutations and combinations generate a lingua franca that can be used to talk about each of the three Sun Tzus.

The second is a structured approach to developing the fourteen Sun Tzu themes at the heart of the present analysis. It involves the same structure for each. That approach, which splits off Sun Tzu (2) and (3) treatment from Sun Tzu (1) coverage, is introduced in the second section of Preliminaries (see pp. 50–53 below). Further details are assigned to the "Scholarly Controls" section in the online annex.

To facilitate anchoring analysis in the Sun Tzu text, p. 53 below also introduces simple notational conventions used to identify specific Griffith verses and passages.

Basic Military Terms and Concepts: Strategy, Tactics, Logistics, Operations

Despite its many razor-sharp insights, along with some forays of a recognizably definitional type, the Sun Tzu text does not take cognizance of a distinction between strategy and tactics, surely one of the most insightful conceptual distinctions military theory offers.[1] Possible glimmerings of such a distinction can be spotted in one or two places in the text, but on closer scrutiny fall short. A case in point is verse XII.15 (Passage #5.18):

> XII.15.
> Now to win battles and take your objectives, but to fail to exploit these achievements is ominous and may be described as "wasteful delay".

The idea thus expressed is suggestive of seeking strategic exploitation of tactical success, one that could bring about a qualitative change in the situation. However,

---

[1] Clausewitz in his own time was able to treat the strategy/tactics distinction as well known, albeit (as he noted) not always well analyzed. See Carl von Clausewitz, *On War* (ed. and trans. by Michael Howard and Peter Paret; rvd. ed.; Princeton, NJ: Princeton University Press, 1984), p. 128. In our day, the favorite formalisms of game theory largely elide the fundamental distinction between strategy and tactics, designating all choices as "strategic" ones.

Efforts by Sun Tzu at defining some of his own basic concepts are noted in #5 footnote 4 below.

verse XII.15 does not allude to achieving, or even attempting to achieve, the sort of qualitative change that the modern phrase "strategic exploitation" suggests.[2]

It is even conceivable that Sun Tzu would have rejected the strategy/tactics distinction outright, since in more than one place in the text Sun Tzu plays up a fundamental similarity of different scales of military action.[3]

These caveats notwithstanding, the present study will feel at liberty to employ the strategy/tactics distinction as a modern analyst's tool. Many parts of Sun Tzu can in fact be given both a strategic and a tactical reading.[4] Of the two, a strategic reading is often the more productive because less tethered to antiquarian forms of warfare (or indeed to conventional warfare at all).

Taken from a modern Western source, the following coordinated descriptions of strategy and tactics serve to anchor these interlocking concepts:[5]

> strategy = the comprehensive direction of power to control situations and areas to attain broad objectives;
>
> tactics = the immediate employment of specific forces and weapons to attain strategic objectives.

---

[2] For verse XII.15 interpretation issues see Appendix 5 anchor footnote 21.

A further candidate for Sun Tzu material suggesting a strategy/tactics distinction is verse I.16, discussed in Theme #1 context on p. 59 below. Here, however, a "strategic level" appears via a net assessment calculation carried out before a war (or possibly a campaign) starts (Passage #1.1). Equating that calculation to a concept of "strategy" risks reducing "strategy" to a one-time choice and, what's more, one having binary overtones (attack/do not attack). Important as it is as a lens on Sun Tzu's concepts of calculation, verse I.16 therefore seems too narrow to be parlayed into a nascent Sun Tzu distinction between "strategy" and "tactics."

[3] Verses V.1–V.2 (Passage #1.13) point to Sun Tzu's belief in a similarity of command-and-control problems across multiple scales (see Appendix 15 for further analysis). Verse III.2 (Passage #3.3) likewise has a scale-invariant tonality.

There may be a pattern here that ranges beyond Sun Tzu. As Sawyer observes: "In general, unlike much Western military doctrine, the principles and concepts of Chinese military science are not rigidly constrained nor limited to one level or sphere of application, such as tactical rather than operational or strategic." See Sawyer, Introduction footnote 55, p. 411 note 16.

[4] Griffith, p. 43 picks up on this latter possibility when he observes that qi and zheng operations (see Theme #14) may be either tactical or strategic. Another example involves verse VII.21 (Passage #5.15) which traditional commentator Mei Yaochen suggests can be read expansively as applying to phases of a long campaign. See Griffith, p. 108 footnote 2. Mair, p. 149 note 6 formulates verse VII.21's idea in its natural generality: "The falling off of an army's spirit is described here [in verse VII.21] as occurring in the course of a day, but this is only a metaphor for a process that takes place over time."

In reading Sun Tzu it is usually advisable not to conceptualize timescales too rigidly. For example, there is reason to believe that the "death ground" response of a trapped army that is an important focus of Sun Tzu's thinking may unfold over a lengthier period of time than that suggested by an image of trapped troops going berserk or exhibiting comparably desperate behavior (as implied by traditional commentator Zhang Yu [Giles, p. 125]). See #1 footnote 92 (example from Cao Cao's military career).

[5] See Boorman, Introduction footnote 4, p. 93 (strategy); p. 108 note 11 (tactics). Both descriptions, along with that of logistics on p. 49 below (see also Boorman, p. 100), are drawn from Admiral Eccles's writings. See in particular Henry E. Eccles, *Military Concepts and Philosophy* (New Brunswick, NJ: Rutgers University Press, 1965), p. 69 (Eccles graphic entitled "The Fundamental Relations from the Perspective of Command," also incorporating roles for intelligence and communications).

While in no sense a comprehensive treatment of the subject, indeed in some ways a fragmentary one, the Sun Tzu text takes significant cognizance of logistics issues (e.g., Passages #5.7 and #5.8). A description of logistics, developed in coordination with the preceding descriptions of strategy and tactics, is therefore provided here:

> logistics = the creation and sustained support of weapons and forces to be tactically employed to attain strategic objectives.

"Operations" in a military sense may be described as a level of the direction of war that interpolates between a strategic level and a tactical level.[6] Operations are a blend of tactics and logistics.[7] In the warfare of Sun Tzu's era, as in many other times and places in world military history, tactics closely aligns on a battlefield level of war, while the operational level involves a strong component of maneuver outside of, and in some cases preparatory to, battle. Reflecting Sun Tzu's affinity for thinking at what would now be called an operational level of war, important Sun Tzu content aligns well on what is now called operational art (including a pronounced distaste for battles having a less-than-certain likelihood of success).[8] Sun Tzu's interest in the operational level of war shows with clarity in Sun Tzu VI, VII, and XI chapters.[9]

For expository purposes – to avoid continual pedantic allusions to the distinction between operations and tactics – the term "tactics" will often be used as a shorthand term covering both military operations (in the past sometimes called "grand tactics") and tactics. Context should make the intended meaning clear.

Filling a further terminological need, particularly arising in Sun Tzu (3) discussions, the term "strategic" will sometimes be employed as a generic label, describing intendedly rational, goal-directed activities anywhere on the spectrum of conflict, including its non-military parts. Context should again make clear when "strategic" is being used in this generic sense, which is close to the use of the term "strategic" in what Martin Shubik has dubbed conversational game theory.[10]

---

[6] As Liddell Hart notes, "The term 'operations' has a more specific meaning in German military language than in English – covering the intermediate sphere between strategy and tactics, and being applied to generalship in the handling of forces in the field [emphasis in original]." See The Rommel Papers (ed. B. H. Liddell Hart; trans. Paul Findlay; New York: Harcourt, Brace and Company, 1953), p. 329 footnote 1, also making the further important point that focus on an operational level risks siphoning attention away from strategy.

[7] Eccles, Preliminaries footnote 5, p. 49. "Operations" in an intelligence sense is a distinct concept. For Sun Tzu's thinking relevant to the latter see Theme #11 below, particularly p. 361. See also Passage #7.10, entitled "Sun Tzu's 'Third Symphony' or 'Symphony of Spies.'"

[8] For more on the concept of operational art, with grounding in modern military history, see Operational Level of War – Its Art (A Special Text for the Department of Military Strategy, Planning and Operations, United States Army War College, Carlisle Barracks, PA 17013-5050, 1985–1986). The late Cold War vintage of that text, falling at a historical juncture when the US Army was actively planning for theater-level operations in Europe against a presumed Soviet adversary, makes its coverage of operational thinking particularly compelling.

[9] Of Sun Tzu XI, Sun Tzu's lengthiest chapter, Brooks, p. 61 observes that it "introduces long-range campaigns, showing a technical and organizational advance over previous chapters." The reference to "previous chapters" here is to Brooks's proposed Sun Tzu chapter accretion sequence, which starts with Sun Tzu IX and continues with Sun Tzu X, VIII, and XI in that order. See Table 12A on p. A-35.

[10] See pp. 1515–17 of Martin Shubik, "What is an application and when is theory a waste of time?," Management Science, 1987, 33(12), 1511–22 (proposing a distinction among three strands

Format Used to Structure Development of the Fourteen Sun Tzu Themes

Each theme is developed in the same way, in five sections.

First section: To give the reader a quick orientation to what's coming, development begins with an inventory of the most relevant Sun Tzu passages, each with a passage caption (i.e, a short parenthetical indicating its contents).[11] The passages in the inventory are informally clustered, bringing together passages (often from different Sun Tzu chapters) deemed to have similar or related foci. Reflecting the Sun Tzu text's brevity, these passage lists are around a page or less in length (the Theme #13 list is slightly longer), thereby facilitating a coup d'oeil of Sun Tzu's coverage and emphases.

Criteria for passage identification are discussed in Scholarly Controls in the online annex (see pp. A–25ff.)

Second section: Next comes an analysis of Sun Tzu's substantive ideas pertaining to the given theme. In some cases (e.g., in Themes #2 and #4) that overview is combined with methodological thoughts on how to engage with relevant Sun Tzu content.[12] A comment by a Russian scholar (Ksenia Kepping), describing the efforts of an unknown Inner Asian translator of Sun Tzu into the Tangut language a thousand years ago, transposes well to describe the present goal: "his main aim was to convey adequately the contents of the military treatise."[13] One sign of progress in that endeavor is when multiple Sun Tzu verses leap into a previously unnoticed but coherent analytical relationship.

Although the analysis is usually couched in general terms – thereby facilitating later generalizations – the basic anchoring here is Sun Tzu (1), geared to the broad parameters of war in Sun Tzu's time (e.g., mass-infantry armies; very limited and awkward written communication capabilities; etc.). That emphasis encourages keeping focus on Sun Tzu's simple, concrete military advice, avoiding a possible temptation to make Sun Tzu's thoughts on war out to be more complex or modern than they actually are. At all times it is important to keep in clear view that the text is a Warring States China military advice manual. Recognizing that there will inevitably be differences of opinion pertaining to the text and its interpretation, a question always worth asking is: If a certain line of criticism pointing to an alternative understanding of the text is accepted, what difference in Sun Tzu's substantive military advice would it make?

---

of game theory, which Shubik labels, respectively, high church, low church, and conversational game theory, categories that broadly align on the motivation for the analysis and the use to which it is put).

[11] Adjusting for the fact that Clausewitz's *On War* is massive while the Sun Tzu text is short, there is a family resemblance between the present Sun Tzu passage inventory and the approach of Sumida's Clausewitz concordance cited in Introduction footnote 15.

[12] Relatedly, this second section of a given theme's development will sometimes draw on formal concepts (mathematical, computational, etc.) to assist in clarifying Sun Tzu's ideas and emphases. Examples include rule-based expert systems (Theme #2) and hypergames (Theme #5).

[13] See p. 15 of Ksenia Kepping and Gong Hwang-cherng, "Zhuge Liang's 'The General's Garden' in the Mi-nia [Tangut] translation," pp. 12–23 in Ksenia Kepping, *Last Works and Documents* (posthumously published tribute to the memory of Dr. Ksenia Kepping, St. Petersburg, Russia, 2003).

Third section: Next comes a counter-factual exercise, likewise anchored in Sun Tzu (1), exploring facets of the given theme that the conditions of war and of politics in Sun Tzu's time suggest that *Sun Tzu might plausibly have discussed, yet did not.*[14] By definition, this "roads not taken" exercise excludes cases involving technology or institutions plainly foreign to Sun Tzu's historical context; also excluded are instances where Sun Tzu simply fails to supply more details, e.g., regarding weapons, equipment, force structure, scenario specifics, etc. This section of the analysis is important for insights it can offer into the second of two central questions posed at the start of the Introduction: *"What are Sun Tzu's limitations or blind spots?"*

Fourth section: Under a rubric of "Sun Tzu (2) and (3) frontiers" attention now shifts from Sun Tzu (1) to Sun Tzu (2) ("Sun Tzu extended") and Sun Tzu (3) ("Sun Tzu analogical"). The mix of emphases varies somewhat by theme. In part reflecting the fact that Sun Tzu (3) analysis often requires introducing additional context, Sun Tzu (2) topics are generally discussed first.

While there is no royal road to productive generalizations of Sun Tzu, the most useful lines of thought identify larger general principles to which Sun Tzu's thinking points, then also identify applications contexts, possibly remote from Sun Tzu's historical setting, where those principles provide insights.[15] Of basic importance here is a pragmatic recognition of limits of Sun Tzu's generalizability – e.g., when a proposed application is the proverbial bridge too far. A hallmark of successful generalization is when Sun Tzu's ideas bring something to the table – a challenge of assumptions, a direction for development, a way of unifying seemingly disparate problems, etc. – commonly missed or marginalized in other approaches to strategic topics.[16]

Sun Tzu's ideas are versatile, so that the "frontiers" coverage is perforce highly selective; the animating spirit is that a few compelling broader applications of Sun Tzu's ideas beat a raft of lesser ones. To hold topic sprawl and length in check, an effort is made to introduce pertinent Sun Tzu (2) and (3) applications succinctly, sketching just enough of each application's direction to serve as a catalyst to the imaginations of readers with relevant backgrounds and interests. Reflecting contemporary trends that are influencing many areas of science and engineering, strategy, and policy, attention will be paid to identifying connections between Sun Tzu's ideas

---

[14] This framing is inspired in part by Joseph Needham with the collaboration of Wang Ling, *Science and Civilisation in China*, Vol. 3: *Mathematics and the Sciences of the Heavens and the Earth* (2nd ed.; Cambridge, UK: Cambridge University Press, 1970), p. 167, proposing topics that traditional Chinese science "should have been interested in," yet was not (one example they cite is mechanics for ships; another is hydrostatics for a canal system).

[15] An observation of General George Catlett Marshall, Jr. (from a speech to the Army Ordnance Association, October 11, 1939) captures a spirit that transposes well to getting the most out of Sun Tzu: "[W]henever changes are proposed, modern theories advanced, or surprising developments are brought to my attention, I automatically search for the fundamental principle involved in the particular matter at hand."

See Larry I. Bland, Sharon R. Ritenour, and Clarence E. Wunderlin, Jr. (eds.), *The Papers of George Catlett Marshall*, Vol. 2: "We Cannot Delay" (July 1, 1939–December 6, 1941) (Baltimore, MD and London: The Johns Hopkins University Press, 1986), p. 83.

[16] E.g., Sun Tzu's contribution to showcasing the continuing need for investment in human intelligence (HUMINT) capabilities, even in (perhaps especially in!) a high-tech-dominated world.

and calculational/computational thinking, broadly understood.[17] A Sun Tzu textual anchor for such attention (though not the only one) is verse I.28 (Passage #1.2).[18]

Fifth section: The spotlight now swings full circle back to the Sun Tzu text itself, reproducing excerpts from the actual text (in Griffith's translation), not just summaries of those excerpts. Such textual anchoring for analysis of Sun Tzu's ideas is crucial: it is our best defense against clothing Sun Tzu in garments he never wore.[19] As that textual focus suggests, emphasis is again on Sun Tzu (1), though a consistent goal is to spot ideas having potential to deliver value-added in Sun Tzu (2) or (3) contexts. One basic reason for positioning the textual material at the end of a given theme's development is that its content lends itself better to browsing than to linear reading (a feature shared with many other ancient texts).

Coverage involves a set of numbered Sun Tzu passages, the same set as in the passage inventory located at the start of a given theme's development (and presented in the same order, equipped with the same captions). Each passage comprises a set of Griffith verses.[20] As an analytical device to help avoid swamping readers with technical scholarly detail (which is tremendously easy to do in analyzing an ancient text like Sun Tzu) each passage appears *twice*:

- a *first* time at the end of a given theme's main text development (where the supporting comments focus on Sun Tzu's most important insights);
- a *second* time in the relevant appendix in the online annex (where most discussion of Chinese language–related textual or interpretive issues is located, generally in accompanying footnotes).[21]

---

[17] Cf. Yuhang Liu, Xian-he Sun, Yang Wang, & Yungang Bao, "HCDA: From computational thinking to a generalized thinking paradigm," *Communications of the ACM* (Association for Computing Machinery), 2021, 84(5), 66–75. Their acronym HCDA stands for "historical thinking, computational thinking, data-centric thinking, and architectural thinking" (p. 68). On p. 70 they note the Warring States computational tool discussed in #1 footnote 82. If "architectural" is understood in a generalized sense, decoupled from modern information technology, this article's "architectural thinking" focus also lends support to the present study's highly structured approach to analyzing Sun Tzu.

[18] See verse I.28 analysis, pp. 57 and 76 below together with p. A-41ff. and Appendix 1 anchor footnotes 6 and 8.

[19] Image inspired by p. viii of Jacques Gernet's "Foreword" to Jean-Claude Martzloff, *A History of Chinese Mathematics* (revised and expanded version of book published in Paris in 1987; Stephen S. Wilson, trans.; Berlin/New York: Springer-Verlag, 1997), warning of pitfalls in writing history in that area.

[20] In almost all cases Griffith's translation is used. For a list of the exceptions see Scholarly Controls footnote 3 (pp. A-24f. in online annex). In other cases where Griffith's translation might be criticized or improved the issues are handled in footnotes. Such an approach seems preferable to breaking up the coherent style of Griffith's translation by repeatedly substituting other translations.

[21] Since the same verse may appear under more than one theme (a basic feature of the present analytical approach; see p. 11 above amplified by pp. A-28f.), judgment is needed in choosing where to position explanatory footnotes, juggling competing goals of user-friendliness and redundancy or clutter avoidance. A pragmatic approach to such choices is taken, reflecting an evaluation of where an explanation might be most helpful and using cross-references to guide the interested reader there.

This device of duplication makes use of the vastness and flexibility of cyberspace as a venue for tackling sometimes thorny textual, interpretive, and translation issues whose detailed coverage in main text would risk making it unduly intricate.

Notational Conventions

Two simple notations, both of a decimal type, are employed throughout this study:

(a) To identify *specific Griffith verses*, "$x.y$" refers to Griffith verse y assigned to Sun Tzu Chapter x. Following Griffith, Sun Tzu's thirteen chapters are identified by Roman numerals, i.e., as Sun Tzu I, Sun Tzu II, ..., Sun Tzu XIII.

Example: Sun Tzu I's first verse – which Griffith translates as "War is a matter of vital importance to the State; the province of life or death; the road to survival or ruin. It is mandatory that it be thoroughly studied." – is verse I.1.

Example: Griffith verse 14 in Sun Tzu XIII – "Delicate indeed! Truly delicate! There is no place where espionage is not used." – is verse XIII.14.

(b) To identify *specific Sun Tzu passages*, Passage #$w.z$ is the zth main text passage assigned to Theme #w, $w = 1, 2, \ldots, 14$.

Supplementary passages appearing *only* in an appendix are flagged using a suffix "A" (stands for "Appendix"), thus #$w.z$A.[22]

Example: The first passage assigned to Theme #1 (setting forth Sun Tzu's treatment of net assessment) is Passage #1.1.

Example: The first Theme #1 supplementary passage (i.e., passage assigned only to Appendix 1) is Passage #1.25A, "Steps to take at the outset of a war."

★★★★★★

---

[22] Extending the notation, Appendix 15 passages are designated #15.1A, #15.2A, ..., #15.7A, #16.1A.

A

Strategist Should Be Calculating

Starting off substantive engagement with Sun Tzu with a focus on calculation serves a positive purpose. It is a way of emphasizing to contemporary audiences that there is more to Sun Tzu than being tricky or unorthodox – the strands of his way of war that readers, at least Western ones, widely note and often lionize. In present usage, the umbrella term "calculation" is intended to serve as a flexible rubric capable of covering intendedly rational judgments of more than one kind, many intuitive, others more formally structured.[1] Importantly, it also seeks to sidestep the modernist, often bureaucratic, connotations of the word "planning" or the technological ones of the word "computation."

Calculation is a ubiquitous motif in Sun Tzu and Sun Tzu deserves kudos for pioneering recognition that military or strategic calculations come in different types. One type, addressed at the start of the Sun Tzu text (and a candidate for being primus inter pares because notably systematic), centers on the activity commonly known nowadays as net assessment.[2] Other exemplars of the calculation genre found in Sun Tzu are more short-term in focus. While many of those are qualitative – Sun Tzu's "calculations" pertaining to human emotions should be mentioned here – Sun Tzu also exhibits aspirations, at times prescient, toward quantitative thinking.

Yet a further species of the calculation genre, one that is both central enough and distinctive enough in the Sun Tzu text to warrant separate development as Theme #2, involves a form of rules-based calculation.[3] That further species centers on a kitbag of prescriptive rules designed to be invoked in a given terrain situation to counsel actions to follow or to avoid there. For reasons developed under Theme #2, Sun Tzu's investment in such a knowledge base of prescriptive rules has aspects of an early Chinese groping toward modern computing's rule-based expert systems.

A crucial, if specialized, offshoot of Sun Tzu's affinity for calculation exercises (as well as a concrete basis for putting them into practice) involves the judgments associated with creating and running an intelligence network. Because Sun Tzu's

---

[1] The idea that a military or strategic calculation may possess a "structure" (illustrated by the net assessment Passage #1.1) is itself a significant Sun Tzu contribution, possibly original with Sun Tzu.

[2] See Paul Bracken, "Net assessment: a practical guide," *Parameters*, Spring 2006, pp. 90–100.

[3] If one accepts Brooks's accretionist theory of the genesis of the Sun Tzu text (Table 12A on p. A-35), Sun Tzu's terrain rules congregate in the earliest chapters (Sun Tzu VIII–XI), whereas key quantitative material (e.g., verse I.28; assorted quantitative planning factors in Sun Tzu II, III, and VII) appears in chapters assigned a substantially later date. In an era when basic Chinese military concepts were still taking shape (Yates, Introduction footnote 28, p. 220; Lau & Ames, p. 64), here is a possible sign of evolving emphases pertaining to the role and nature of calculation in warfare, shifting from an early focus on prescriptive rules to a later focus on quantitative types of calculation.

espionage chapter gets its own thematic heading (Theme #11, "Know the Enemy"), primary coverage of that additional form of calculation, which in Sun Tzu's treatment of it has major psychological aspects, will be deferred to Theme #11.

THEME #1 NET ASSESSMENT AND OTHER CALCULATIONAL IMPERATIVES

first species of calculation emphasized by Sun Tzu – net assessment:
Passage #1.1 (verses I.1–I.14) (analysis of comparative war potential or net assessment)
Passage #1.2 (verse I.28) (quantitative analysis of war potential)

second species of calculation – expedient assessment:
Passage #1.3 (verse VII.15) (expedient assessment)
Passage #1.4 (verses V.14–V.16) (possibility of precise military calculation)

integrating different species of calculation; calculation and deception in cross-fertilizing relationship:
Passage #1.5 (verses I.16–I.17) (integrating different species of calculation; calculation and deception)
Passage #1.6 (verse VII.12) (more about calculation and deception)
Passage #1.7 (verses IV.16–IV.19) (cumulative nature of military and logistics calculations)

calculation and emotions:
Passage #1.8 (verses XII.17–XII.19) (failure to calculate (1): avoiding irreversible major decisions taken for emotional reasons, sans calculation)
Passage #1.9 (verse III.9) (failure to calculate (2): general's emotions as source of tactical error)
Passage #1.10 (verses VII.27–VII.29) (failure to calculate (3): rising to the bait)
Passage #1.11 (verses XIII.12–XII.14 + XIII.19) (delicate psychological calculations pertaining to use of secret agents)
Passage #1.12 (verses XI.23 + XI.33–XI.34 + XI.49 + XI.55) (emotional engineering: calculated use of emotional triggers to elicit supranormal effort from troops on brink of disaster)

quantitative planning factors:
Passage #1.13 (verses V.1–V.2 + V.19) (quantitative planning factors in military organization)
Passage #1.14 (verse II.18) (quantitative planning factor pertaining to giving out rewards)
Passage #1.15 (verses VII.2–VII.9) (mobility planning factors)
Passage #1.16 (verse III.8) (planning factors involving lead times and their concatenation)
Passage #1.17 (verses II.1–II.2 + II.9 + II.12–II.15) (further quantitative planning factors, some suggestive of information aggregation on a country-wide level)
Passage #1.18 (verse IX.45) (pitfalls of plausible but overly narrow planning factors)

limits of calculation:

Passage #1.19 (verses XIII.3–XIII.4) (limits of calculation (1): appropriate empirical data as crucial)

Passage #1.20 (verse I.27) (limits of calculation (2): warfare's indeterminate aspect)

Passage #1.21 (verses V.6–V.12) (limits of calculation (3): combinatorics overwhelming capacity to calculate, by the same token affording a bottomless quiver of opportunities)

roles and positions of human decision makers who do military or strategic calculations:

Passage #1.22 (verses III.18–III.23) (general's natural sphere as differing from that of ruler)

Passage #1.23 (verse XII.16) (division of labor between general and ruler)

Passage #1.24 (verses IV.8–IV.12) (widespread underestimation of successful strategists)

✶✶✶✶✶✶

The Sun Tzu material assembled here starts with calculation on a plane of strategy or grand strategy (net assessment) and is then augmented by further passages bringing in military operations, logistics, and tactics.

Lest Sun Tzu's calculational content be too swiftly or uncritically assimilated to a rubric of modern military or national security planning, it is well to bear in mind just how early the Sun Tzu text is. It antedates by several centuries the traditional date for the invention of papermaking and the revolution in document-generating and disseminating that such a capability – once fully developed – unleashes.[1] There is textual evidence indicating that a version of a military general staff apparently existed in Warring States China.[2] However, it is prudent to make no assumption that armies

---

[1] The traditional Chinese date for the invention of papermaking is 105 AD, though modern archaeology has pushed the date back as far as an early Western Han map (second century BC) made from hemp fibers from rags (Wilkinson §69.3, pp. 1029–30 and §14.1.2, p. 214). By way of larger context it is important to distinguish among:
    (a) the earliest origins of a communications technology or medium;
    (b) its widespread practical availability and use;
    (c) its practical availability for *use under operational military or espionage conditions.*
The thorny challenges arising in (c) have few counterparts in ordinary civilian life (including routine bureaucratic activities, where the tempo of activity is often lower and urgency of steps taken far less).
    Although written from a civilian perspective that elides (c), a basic study with additional information about paper specimens of Western Han vintage (202 BC–8 AD) is Tsien Tsuen-hsuin, *Written on Bamboo & Silk: the beginnings of Chinese books & inscriptions* (2nd ed.; Chicago/London: University of Chicago Press, 2004). See in particular pp. 146–47 there; p. 148 makes an important logistics point that materials used for papermaking purposes prior to c. 105 AD were in limited supply.
[2] An elaborate version of a staff apparatus appears in the *Liutao*, another early Chinese military work (see p. 34 above). For the *Liutao*'s delineation of seventy-two staff positions agglomerated into eighteen functions, with counts of how many positions should be assigned to each, see Sawyer, *Seven Military Classics*, pp. 60–62. One may reasonably question how far that elaborate apparatus

in Sun Tzu's time had any close counterpart to the type of institutionalized analytical or planning capabilities that modern general staffs make feasible. Literacy limitations aside, it is hard to see how anything like a modern staff could function in the absence of readily available paper (or its more recent digital counterparts).[3] Relatedly, the Sun Tzu text contains no reference to staff-generated contingency planning, one of the core duties of a modern general staff. Also notably absent from Sun Tzu is any reference to military cartography, a capability of basic importance for higher direction of the mass-infantry armies that typified war in Sun Tzu's age and which in modern times has been closely associated with the rise of general staffs from the nineteenth century onward.[4]

Viewed from these standpoints, the Sun Tzu text shows its early vintage.

The heart of calculation in Sun Tzu's Theme #1 sense is therefore best envisioned in relatively modest terms, aligning more closely on common sense notions of "thinking the situation through carefully" than on any more elaborate supporting machinery, human or mechanical.[5] Sun Tzu's one verse that might be an allusion to mechanical planning aids – probably the counting rods central to early Chinese mathematics – is verse I.28 (Passage #1.2).[6] It points to a quantitative approach to military planning (possibly involving working out arithmetic aggregates like troop counts or quantities of supplies, weapons, or other military assets).[7]

That the calculation advocated by Sun Tzu is in many ways elementary should not be confused with its being easy. Far from it – in Sun Tzu's time as today – for reasons never better captured than by a dictum in Clausewitz's *On War* introducing his famous idea of friction: "Everything in war is very simple, but the simplest thing is difficult." In keeping with this Clausewitzian fundamental, some of Sun Tzu's most insightful Theme #1 content aligns, not on any details of calculation but on overcoming emotional hurdles to doing any calculation at all (see Passages #1.8–#1.10).

was reflected in practice. Military staff organizations throughout history have tended to conform to what is congenial to the concepts of the commander, often highly personalized.

[3] Immense unwieldiness of bamboo strip "books" has already been noted (see Background footnote 77). Wooden boards were one alternative, but capacity was limited to about 100 characters. Silk was an alternative writing medium, in many ways preferred because "more easily consulted, stored, read, or transported" (Wilkinson, §69.1.2, p. 1025), but also expensive and non-reusable (thereby undercutting its utility for recording and disseminating military plans, which are prone to change).

Human memory would be another alternative but has its own major limitations (especially under field conditions where death, injury, or capture of a messenger or other key actor might lead to disaster). Special problems of espionage communications are discussed by Ralph D. Sawyer, with the collaboration of Mei-chün Lee Sawyer, *The Tao of Spycraft: intelligence theory and practice in traditional China* (Boulder, CO: Westview Press, 1998), "Time and communications," pp. 28–30.

[4] For overview of known early Chinese maps from Zhou to Han see Wilkinson §14.1.2, pp. 214–15 (noting that the first reference to maps in Chinese sources c. 1000 BC appears in a military context).

[5] Language in quotation marks derives from Ames, p. 166, translating verse XII.16 (Passage #1.23). Verses XII.17–XII.19 (Passage #1.8) develop the thought in greater detail.

[6] On counting rods see Minford, p. 118 (translating comments of Li Ling); Martzloff, Preliminaries footnote 19, pp. 210–11, 217ff.

[7] For an alternative understanding of the verse I.28 calculations, one that better accommodates Sun Tzu's human factors emphases, see Appendix 1 anchor footnote 8.

Two basic types of calculation are here singled out as anchors for Sun Tzu Theme #1 development. One is off-battlefield and precedes the initiation of a war (or possibly a major campaign in it). The second centers on what Mark Lewis has called "expedient assessment," a label that especially fits command decisions taken in the heat of battle but that may also apply more broadly at the tactical or operational level of war.[8] Both types are essential to effective exercise of military command.[9]

Sun Tzu's roadmap of off-battlefield calculation (Passage #1.1) – planning for the war as a whole – is remarkable for its rigor and the extent to which it foreshadows what are often viewed as sophisticated modern developments. Highlights include:

- thinking *comparatively* about war potential;
- thinking *comprehensively* about war potential, synthesizing diverse factors;
  (That comprehensiveness, of course, needs to be placed in a context of war in Sun Tzu's time, when complex economic and technological issues were far less salient than they are in modern war.)
- thinking *quantitatively*, where feasible (quantification remains implicit in Passage #1.1 but surfaces explicitly in Passage #1.2 from the end of Sun Tzu I).[10]

Sun Tzu's expedient assessment (*quan*) strand is less elaborated – unsurprisingly so given the commonly unscripted character of that activity. Two common meanings of the Chinese term *quan* are "weighing" and "leverage."[11] *Quan* appears in verse VII.15 (Passage #1.3) whose context is mobile warfare:[12]

VII.15.
Weigh [*quan*] the situation, then move.

As Mark Lewis has suggested, citing other early Chinese military writings, *quan* often has a narrower meaning than expedient assessment in general, centering on the

---

[8] Lewis, Background footnote 11, p. 118.

[9] These two kinds of calculation distilled from Sun Tzu should be treated as ideal types. They are also not exhaustive. For example, verses II.13 and II.14 (Passage #1.17), if treated as pointing to descriptive statistics (i.e., not as merely figurative or rhetorical examples), suggest a third species of calculation for administrative purposes. Further candidates for Sun Tzu attention to administrative calculation include verses I.8 (Passage #1.1), V.1 (Passage #1.13), and III.21 (Passage #1.22).

[10] Verses I.1–I.14 (Passage #1.1) and I.28 (Passage #1.2) in many ways form a natural package. See McNeal, Background footnote 58, pp. 118–19; Mark E. Lewis, *Writing and Authority in Early China* (Albany, NY: State University of New York Press, 1999), p. 282. However, there is a complication. In Sun Tzu's time, as in far more recent times, available quantification tools had distinctly limited ability to capture complex human factors (which is what Passages #1.1 and #1.2, taken together, appear to call for). Building in part on McNeal's work on a different early text, one way that calculators in Sun Tzu's time might have sought to surmount such limitations is suggested in Appendix 1 anchor footnote 8.

[11] Tracing roots of *quan* based on both paleographic and transmitted evidence, see Boqun Zhou, "Mechanical metaphors in early Chinese thought," Ph.D. dissertation, Department of Far Eastern Languages and Civilizations, University of Chicago, August 2019, pp. 61–74. See also Griet Vankeerberghen, "Choosing balance: weighing (*quan* 權) as a metaphor for action in early Chinese texts," *Early China*, 2005–2006, 30, pp. 47–89, especially pp. 67ff. (on Sun Tzu see p. 69 footnote 80 there).

[12] Verse VII.15 is discussed by Lau & Ames, p. 65 as part of a larger discussion of the lever scales concept in early Chinese military writing. For further lever scales background see Joseph Needham (with Wang Ling and Kenneth Girdwood Robinson), *Science and Civilisation in China*, Vol. 4: *Physics and Physical Technology*, Part I: *Physics* (Cambridge, UK: Cambridge University Press, 1962), pp. 19–27.

weighing of just "two opposed factors, such as 'fear and shame' or 'many and few,' that would affect the conduct of battle."[13] Such a narrowing of focus rings true as a practical matter, given constraints on a commander's ability to do calculations in the field. Sun Tzu alludes several times to weighing two opposed factors.[14]

How net assessment and expedient assessment fit together as a package is clarified by verse I.16 (Passage #1.5). Its interlocking ideas warrant its repetition here:

I.16.

> Having paid heed to the advantages of my plans [ji], the general must create situations [shi] which will contribute to their accomplishment. By "situations" [shi] I mean that he should act expediently in accordance with what is advantageous [yin li] and so control the balance [quan].

Although verse I.16 has given difficulty to commentators and translators, the present position – following Mark Lewis – is that it makes a basic military calculation point, in two parts.[15] First comes net assessment before the war begins. Then – once field operations have commenced – further sources of advantage, not reflected in that initial assessment, must be cultivated, evaluated, and exploited.

While those further sources of advantage might arise for other reasons too (e.g., enemy mistake or newfound enemy vulnerability), verse I.16's back-to-back juxtaposition with verse I.17's far-reaching statement that "All warfare is based on deception" suggests a special place for deception in creating them. More broadly, the fact that calculation and deception ideas repeatedly rub shoulders in Sun Tzu (a second example is Passage #1.6's verse VII.12; candidate for a third is verse XI.56 in Passage #7.18A) invites further exploration of the relationship between these two strands of Sun Tzu's thinking. Theme #7 analysis develops that line of thought (see pp. 245–47 below) as, in Sun Tzu (2) and (3) contexts, so also do Themes #1 and #6 "frontiers" discussions (pp. 72–74 and pp. 220–23, respectively).

Passage #1.7, reinforced by Passage #1.4 (possibility of precise military calculation) puts in play a further basic idea. Military calculations – certainly in warfare involving organizations as complex as Sun Tzu's armies – commonly build on one another. They cumulate to form a kind of pyramid, with results of calculations at lower levels informing inputs to ones higher up.[16] An elementary example would be arithmetic aggregation of subtotals. By the same token, Passage #1.7 also serves as a salutary reminder of just how badly an error in the basement of a multi-tiered

---

[13] See Lewis, Background footnote 11, p. 118 and pp. 290–91 note 75 (citing pertinent material from other *Seven Military Classics* texts).

[14] See, e.g., Passages #1.34A–#1.36A. In each of those passages Sun Tzu refers to half of an army being in one situation, half in another, but that seemingly quantitative formulation is plainly only a way of pointing to a one-variable decision problem involving relative proportions, whose optimum solution falls somewhere in a middle range.

[15] Among contemporary scholars of Sun Tzu, Mark Lewis has led the way in spotting the integrative role of verse I.16 which, as he notes, "is particularly valuable because it links 'calculations' [ji], 'powers of circumstance' [shi], and 'expedient assessments' [quan]" – thereby shedding light on a division of labor among those several concepts and how they fit together. See Lewis, Background footnote 11, p. 119. By way of clarification, Lewis's "calculations" are Griffith's "plans" (ji); Lewis's "powers of circumstance" are Griffith's "situations" (shi, the focus of Theme #5); and Lewis's "expedient assessments" are Griffith's "control the balance" (quan).

[16] Unfortunately for appliers of Sun Tzu's ideas, Passage #1.7 is obscurely expressed and specific understandings vary greatly across commentators and translators. However, the idea of some sort of cumulative calculation seems robust. Such thinking also harmonizes with algorithmic emphases found in Chinese mathematical tradition. See Martzloff, Preliminaries footnote 19, pp. 58–60.

calculation can distort actions relying on it.[17] By easy implication, calculational pyramids are also a natural target for enemy deception efforts.

Taken together, Passages #1.4–#1.7 and others in Appendix 1 document Sun Tzu's continual restless probings in numerous calculational directions. Of course, he often runs up against a lack of analytical tools to clinch promising intuitive insights.

Shifting focus to a different strand of Theme #1, it is clear that Sun Tzu is acutely conscious of the role of human frailties and foibles in decision making. Unsurprisingly, therefore, he gives much attention to the role of emotions.[18] Emotions in the commanding general or ruler are, for Sun Tzu, a source of danger, undermining ability to think and calculate clearly with the ever-present possibility of catastrophic consequences (Passages #1.8–#1.10). Mutatis mutandis, when present in rivals or foes, some types of emotions can be sources of great opportunity and leverage. Passages #1.11 and #1.12 suggest further targets for emotional manipulations involving spies or a general's own troops. The extreme (death ground) situation profiled in Passage #1.12 places extraordinary demands on a commander to do a careful estimate of the titanic emotional forces unleashed by his choices, calculating if, how, and when they might be harnessed to his advantage. For Sun Tzu, ability to do shrewd emotional calculations of that sort is basic equipment for a "compleat strategyst."[19]

Unfortunately Sun Tzu gives no pointers on how a commander is expected to carry out emotion-focused calculations (a marked contrast with Passage #1.1's detailed guidance on how to do a net assessment).[20] It seems clear that the challenges thus presented require an acute sense of nonlinear dynamics, as illustrated by a qualitative change ("rupture") in the behavior of troops (or of spies) subjected to certain kinds of

---

[17] For a modern military history example see Lee B. Kennett, *Sherman: a soldier's life* (New York: HarperCollins, 2001), p. 213, describing November 1863 operations near Chattanooga: "Frank Blair had reported all of his command at Stevenson, Alabama, before they had arrived, and this led Sherman to make erroneous calculations." Cf. also Wilkinson, §38.13.2, p. 540 (problems of false reporting of military or administrative data from assorted eras of Chinese history); Wilkinson, §38.13.1, p. 539 (numerical errors from careless aggregation of subtotals, citing Han dynasty examples).

[18] It should be noted that the demarcation between rational-cognitive and emotional domains was by no means nearly so clearcut in traditional Chinese thought as it is in much modern Western thinking, especially its rational choice strands (though cognitive neuroscience may be propelling us in a traditional Chinese direction!). See Passage #6.7 (concept of *xin* or "heart-mind").

[19] A well-known story from the *Shiji* tells of the defending general in a besieged city using a secret agent to plant in the besiegers' heads the notion that they might usefully harass the defending city dwellers by despoiling graves of the latter's ancestors. This the industrious besiegers proceeded to do, thereby triggering a rage response whose upshot was a smashing victory for the defenders (though notably with a time-lag, giving time for the defending general to put further stratagems in play). See Shiji 82, translated in Nienhauser, Vol. VII (1994), pp. 276–77; Nienhauser, Vol. VII (2021), pp. 506–509 (both cited on pp. xxiii–xxiv above). Although the story does not mention Sun Tzu, its relevance was not lost on traditional comentators Du Mu and He Yanxi, both of whom allude to it (Griffith, p. 75; Minford, p. 128). Hard on the heels of this story in the *Shiji* come remarks that closely track ideas in verses V.6 and V.12 (Passage #1.21) and XI.61 (Passage #7.11).

[20] In Sun Tzu's worldview, the perceived feasibility of such calculations may have been buttressed by a view of common soldiers, common in Warring States texts, as having minimal if any individual agency, lacking powers of calculation or deliberation. Symbolic of such a view would be the sheep of verse XI.49 (Passage #1.12, also Passage #9.14); yet more so, the rolling logs and stones of verses V.24 and V.25 (Passage #9.3). For further discussion see #9 footnote 7.

extreme stress. What's more, the calculation needs to be able to spot prospects for advantageous behavioral discontinuity – say, a ferocious fighting response animated by the courage of desperation (as contrasted with a very different sort of rupture, where the commander's troops surrender or, worse, go over to the enemy, possibilities that Passages #3.5 and #1.43A appear to contemplate). Such calculations seem perforce heavily intuitive, probably requiring long experience to master. Verse XI.24 (Passage #13.28) intimates as much, calling for the general to examine "principles of human nature" with greatest care.

Yet another strand of Sun Tzu's multifaceted Theme #1 elaborates on Passage #1.2's quantitative thrust. The sheer volume of quantitatively oriented content in Sun Tzu is noteworthy, certainly given the exceedingly limited quantitative tools at his disposal. It goes a sizable way toward establishing a bedrock place for quantitative thinking in Sun Tzu's way of war.[21]

There is, to be sure, a need to distinguish quantitative planning factors in Sun Tzu from any foreshadowing by Sun Tzu of modern scientific method, with its signature use of empirical data to test falsifiable hypotheses to reveal empirically supported principles. There is no basis to believe that Sun Tzu thought in that way. It is also essential to avoid being too quick to treat all quantitative references in Sun Tzu as guides to practical military decision making. Some are clearly that.[22] Others may be merely figures of speech (notional, rhetorical, etc.).

Passages #1.13–#1.17 are a sampler of the issues. Passage #1.13 establishes a basic place for quantitative thinking, pertaining to quantitative regularities in army organization. In Passage #1.14 there is no reason not to take at face value the stated planning factor, pertaining to handing out a reward when the number of captured chariots exceeds ten. The forced march planning factors stated in Passage #1.15 and the lead time estimates in Passage #1.16 are of a different type, since (if taken at face value) they amount to empirical rules of thumb, not (as with Passage #1.14's chariot capture rewards) rules based on administrative fiat. These further steps toward quantification seem just the sort of rules that experienced soldiers might cogitate out of long experience, direct or vicarious. For example, blistered feet and effects of heat or cold, adding up to produce attrition of a military force, would naturally be of keen interest to commanders weighing their maneuver options – and

---

[21] Sun Tzu does not stand alone or even necessarily pre-eminent in this regard. Quantitative planning factors inspired by warfare are also found in other early Chinese texts. For example, a quantitative military personnel problem appears in an early Chinese mathematical treatise. See pp. 26–27 of Lam Lay Yong, "Jiu zhang suanshu 九章算術 (Nine Chapters on the Mathematical Art): an overview," *Archive for History of Exact Sciences*, 1994, 47(1), 1–51. The *Mozi* text (see Background footnote 12) serves up a plethora of quantitative planning factors in connection with defensive fortifications for a city – far more such factors than in all of the Sun Tzu text. However, it should be noted that these *Mozi* planning factors are geared to a much more static form of warfare than that emphasized by Sun Tzu, where useful quantitative planning factors are harder to come by.

[22] Sun Tzu's interest in non-figurative quantitative planning factors is on prominent display in verses III.12–III.17 (Passage #1.32A), setting forth a battery of action recommendations based on force size ratios (own force/enemy force) to guide tactical (or possibly operational-level) military decisions.

Logistics planning factors are noted by Wilkinson, §24.13, p. 357: "Since antiquity it was common to use a standard formula to calculate the amount of grain required for a military campaign."

PERSONNEL ATTRITION BY LENGTH OF FORCED MARCH

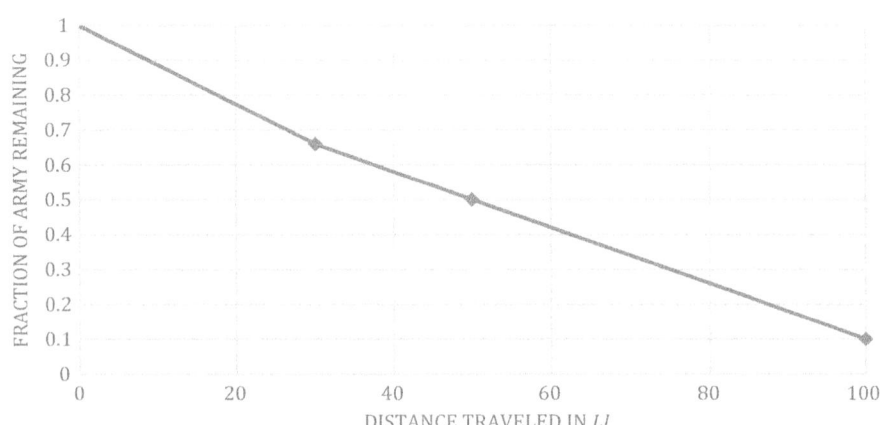

Figure 2. "Sun Tzu the quant": attrition of personnel by length of forced march, estimates from Sun Tzu formatted as survival curve.
Quantities plotted here are from Sun Tzu verses VII.7 and VII.8 (Passage #1.15).*

*Lest too much precision be imputed to Sun Tzu, li as a traditional Chinese distance measure is famously imprecise. It is often taken to be about 1/3 mile (Wilkinson, §42.3, p. 612, Box 81; p. 611, table 100).

they would have a strong motivation to get estimates of that attrition as accurate as possible.[23] The quantities in Passage #1.15 hence merit acceptance as practical planning factors, rather than being sidelined as figurative only. They prefigure the modern analytical framework known as survival analysis (see Figure 2).[24]

By contrast to the planning factors just discussed, the percentages cited in Passage #1.17's verses II.13 and II.14 involve country-level aggregates. Taken literally, verse II.13's 70 percent figure appears to point to a type of organized country-level statistical capability. Given what we know of Warring States China, it is certainly not impossible that a relevant capability existed in some times and places there.[25] Verse II.14's 60 percent figure is consistent with obtaining country-level usage data for the named military categories (broken-down chariots, worn-out horses, etc.). However, these

---

[23] Something of the importance attached to mobility issues may be inferred from the case of King Helü of Wu, the ruler whom the putative historical Sun Tzu served, who had a "special corps of 3,000 runners who could continue 300 li without respite." See Lewis, Background footnote 7, p. 621.

[24] For modern survival analysis methods see E. L. Kaplan and Paul Meier, "Nonparametric estimation from incomplete observations," *Journal of the American Statistical Association*, 1958, 53(282), 457–81.

[25] A statement credited to Legalist thinker Shang Yang (c. 390–338 BC) conveys a relevant tonality: A strong country knows thirteen kinds of statistical [information]: the number of granaries and treasury within its borders, the number of able-bodied men and women, the number of old and of weak people, the number of officials and great officers, the number of those making a livelihood by talking, the number of useful people, the number of horses and of oxen, and the quantity of fodder and of straw.

percentages Sun Tzu cites might also be no more than a way of painting a quantitative word picture of a state in distress, hence may be figurative only.

A basic military truth is that a mass-infantry army requires vast supplies of food and other supplies on an ongoing basis. For that reason, quantitative logistics planning factors – a subcategory of the broader category of operational planning factors – are a "pro's pro" part of military knowledge and a core ingredient of warfighting.[26] As an illustration of the genre, verse II.15 in Passage #1.17 stands out. The striking 20-to-1 disparity Sun Tzu cites – "One bushel of the enemy's provisions is equivalent to twenty of [one's own]," etc. – suggests awareness of basic problems in transport of supplies across desolate or unfriendly territory. Transport animals crossing such territory need to carry their own supplies, thereby piling burdens on a supply line possibly already stretched.[27] Other realities (theft; spoilage in transit; wastage of transport animals, etc.) also enter to make pertinent planning factors yet more unfavorable.[28] Even in modern times the implications of such possibilities, which become all the more formidable when supply bases along the supply route need to be set up, maintained, and defended, have at times been missed by military planners – among them, some British Indian Army higher-ups in the days of the Raj, preoccupied with risk of war with Russia.[29] Verse II.15 suggests Sun Tzu knew better. Of course, it is an example only, geared to alerting generals to the issue.

Rounding out the present overview of Sun Tzu's calculational thinking, four further passages (Passages #1.18–#1.21) document Sun Tzu's awareness of limits of military calculation, in fact no fewer than three distinct varieties of such limits:

- data-driven (Passage #1.19);
- situational-dynamics-driven (Passage #1.20, which Passage #1.18 reinforces);
- combinatorial-complexity-driven (Passage #1.21).

See The Book of Lord Shang: a classic of the Chinese School of Law (translated with introduction and notes by J. J. L. Duyvendak; Chicago: University of Chicago Press, 1928/1963), p. 205. For further background on early Chinese statistical activities see also Andrea Eberhard-Bréard, "Where shall the history of statistics in China begin?," in Gudrun Wolfschmidt (ed.), "Es gibt für Könige keinen besonderen Weg zur Geometrie" – Festschrift für Karin Reich (Augsburg: E. Rauner, 2007), pp. 93–100.

[26] See Henry E. Eccles, "Logistics planning factors," Journal of the American Society of Naval Engineers, 1954, 66(2), 421–31.

[27] Traditional commentator Zhang Yu (Griffith, p. 75) makes the basic transportation point (Griffith, p. 75). Quantitatively illustrating the phenomenon for mules and, separately, for camels (which are more economical in regular terrain), see G. J. Younghusband, Indian Frontier Warfare (London: Kegan Paul, Trench, Trübner & Co., 1898), p. 229. Based on his numbers, a mule transporting 160 lb. of grain for a distance of 200 miles (and allowing for a return trip) will consume 128 lb., delivering 32 lb. (or only 20 percent of the original load) to the military consumer. A slightly longer journey would further reduce the grain delivered down to just 5 percent of the original load (thereby aligning on the 1-for-20 ratio Sun Tzu cites). A similar concept appears in the Hanshu (translated by Minford, p. 122), recounting how a first-century AD Chinese military adviser invoked it to underscore the logistics infeasibility of a scenario requiring supply of Chinese troops fighting an Inner Asian foe in desert terrain. (On the Hanshu – the dynastic history of the Western Han – see Early Chinese Texts [cited on p. xxii above], pp. 129–36; Wilkinson, §59.2, pp. 793–96.)

[28] See James L. Hevia, Animal Labor and Colonial Warfare (Chicago: University of Chicago Press, 2018), chapter 1, "'Slayers of camels': the Second Afghan War and pack animal 'wastage.'"

[29] Hevia sums up the planning deficit succinctly: "Why, in other words, couldn't Simla do the math?" See his The Imperial Security State: colonial knowledge and empire-building in Asia (Cambridge, UK: Cambridge University Press, 2012), p. 169 (his p. 168 footnote 22 cites Younghusband's analysis).

Taken together, these passages – showing Sun Tzu bumping up against limits of calculation – provide some of the strongest evidence of Sun Tzu's level of seriousness in thinking through war's countless calculational issues.

In a shift of register, Passages #1.22–#1.24 conclude Theme #1 coverage with a separate constellation of issues, now centering not on calculation itself but on the human decision maker who does it. This strand meshes with a personalized emphasis running throughout Sun Tzu's thinking – one that foregrounds the general's or strategist's temperament and personality (and implicitly also the surrounding social structure) in ways that many modern treatments of decision-making problems, often reflecting game theory formalist influences, fail to do.

Passages #1.22 and #1.23 underscore that military calculations affecting an army's operations in the field and military-administrative calculations are tasks for the general, *not* the ruler (whose information, certainly given communications conditions of Sun Tzu's time, is likely to be woefully out of date or simply inadequate).

Passage #1.24 proffers the further cautionary observation that the "engineer of victory" should not expect to receive kudos; if by some fluke it is accorded, it is unlikely to be accorded for the right reasons. Simply put, deep strategic calculation is beyond most people's ken. Even successful strategic moves based on truly insightful calculations may appear to outsiders as little more than a lucky throw of the dice. Part of that reputational deficit stems, of course, from the role of secret intelligence in achieving victory (see Theme #11, analyzing Sun Tzu's espionage chapter). Part, however, may also stem from a winning strategist's superior insight into how a complex situation will unfold, a judgment that often defies communication to a broad non-specialist audience in terms that are both understandable and convincing.[30] A trove of examples illustrating the latter point comes from the game of *weiqi*, which sources suggest was already extant in some form in Sun Tzu's time.[31]

Roads Not Taken by Sun Tzu in Developing Theme #1

First (and perhaps most saliently for modern audiences steeped in game and decision theory), Sun Tzu proffers little if any advice to guide judgment under conditions of uncertainty, a basic challenge for warfare in any age. On the contrary, the Sun Tzu text shows a marked tendency for definitive affirmations of conclusions or advice, apparently amounting to a radical denial of the Clausewitzian dictum that war is the province of chance. Lau & Ames, p. 59 put this tendency bluntly: "For the Sun lineage [meaning Sun Tzu and his reputed descendant, Sun Bin], 'chance' is not in their vocabulary."[32] Some of that tendency is, of course, merely rhetorical. A contributing factor could be the working conditions of an itinerant advice-peddler,

---

[30] Verse VI.25 (Passages #5.11, #8.1) conveys a similar idea. Giles, p. 52 sums it up: "everybody can see superficially how a battle was won; what they cannot see is the long series of plans and combinations which preceded the battle." Many of those plans and combinations would involve logistics.

[31] See Wilkinson, §24.14.2, p. 360.

[32] Lau & Ames's surrounding discussion associates this stance with a "positive attitude toward the indeterminate element in warfare." For additional perspective see also Mark Elvin, "Personal luck: why premodern China – probably – did not develop probabilistic thinking," pp. 400–68 in Hans

which the putative historical Sun Tzu (like many others of his time) may have been, whose recommendations must show boundless self-confidence if they are to run a gauntlet of questions from a skeptical ruler and his still more skeptical entourage.

Yet judgment under uncertainty remains inescapable in warfare, and not all approaches to addressing it are equivalent or equally efficacious. One class of such approaches might, for example, build on links between gambling and divination "whose psychology, terminology, procedures, and even the implements" have often overlapped in Chinese and many other world cultures.[33] But Sun Tzu appears to reject divination out of hand (Passage #1.19), thereby leaving something of an analytical void even with the concepts and reasoning tools of his own time.[34]

It is entirely possible, of course, that Sun Tzu – having taken a look at available tools for tackling uncertainty in the situations likely to face his general – decided that none of them were particularly useful at the existing state of the art.[35] Most of what Sun Tzu feels able to do falls under one of two headings:

(a) flagging certain situations where (to use one modern terminology) some version of taking a calculated risk is needed, even if only guided by intuition;[36]

Ulrich Vogel & Günter Dux (eds.), with an overview and introduction by Mark Elvin, *Concepts of Nature: a Chinese–European cross-cultural perspective* (Leiden/Boston, MA: Brill, 2010).

Sun Tzu's failure to engage with probabilistic thinking should not be obscured by translations, including Griffith's, that infuse some Sun Tzu statements with a probabilistic twist. See, e.g., verse I.6 in Passage #1.1 (Griffith's "the chances of life or death"). See also verse III.32 (Passage #11.1), whose ending Griffith translates as ". . . your chances of winning or losing are equal." The Chinese text of verse III.32. lliterally says "one victory, one defeat," a stylized statement.

[33] Mark E. Lewis, *Dicing and Divination in Early China*, Sino-Platonic Papers, No. 121 (July, 2002), p. 1. That paper's p. 3 brings in Sun Tzu, via a reference to verse I.28 (Passage #1.2). (For bibliographic details relating to the Sino-Platonic Papers series see p. xxii above.)

[34] A case could be made that the *Laozi* text (see Introduction footnote 37) goes further than Sun Tzu in describing uncertainty, even doing so in quantitative terms: "When going one way means life and going the other way means death, three in ten will be comrades of life, three in ten will be comrades of death, and there are those who value life and as a result move into the realm of death, and these also number three in ten." See *Lao Tzu. Tao de Ching* (translated with an introduction by D. C. Lau; Harmondsworth, UK/Baltimore, MD: Penguin Books, 1963), p. 111. In a footnote (also on his p. 111) Lau observes that "'three in ten' is a rough way of saying 'one third'."

[35] Not only is there no quantification of uncertainty in Sun Tzu, but Sun Tzu also makes scant effort at even qualitative descriptions of degrees of uncertainty. (The difference between outcomes depicted in verses III.32 and III.33 [Passage #11.1] might be read as a very limited move in that direction.)

For a snapshot of modern practices of that type see Michael Herman, *Intelligence Power in Peace and War* (Cambridge, UK/New York: Cambridge University Press, 1996), p. 104, observing that "[c]onveying probabilities is one of intelligence's art forms, usually through stylized verbal codes (such as 'likely,' 'probable' and 'possible')." The language of the law provides a rich vein of further examples.

[36] Examples would include Passages #1.34A–#1.36A (involving what Passage #1.34A discussion calls Sun Tzu's "1/2 and 1/2" thinking). To be clear, in Passage #1.36A the calculated risk is being taken by the *enemy* general. For more on Sun Tzu's stance vis-à-vis calculated risk see pp. A-58f. Another situation where Sun Tzu may be read as calling for a calculated risk to be taken is verse XII.8 (Passage #1.38A) relating to an ongoing conflagration.

More commonly, need for judgment under uncertainty is not visible in Sun Tzu's advice.

(b) identifying factors (in particular information conditions) whose presence can convert a high uncertainty situation into one where a favorable outcome is effectively assured.[37]

Beyond that, Sun Tzu simply leaves his general to do the best he can using his military intuition applied to the information at hand.

Second, Sun Tzu also offers little guidance – strikingly so given the geopolitical milieu in which the text took shape – regarding the sorts of calculation needed when conflict situations involve more than two significant actors, with all the possibilities for tacit or special-purpose or shifting alliances that such multiplicity creates.[38]

Sun Tzu does place substantial emphasis on creating disharmony in the enemy's camp (see Theme #4 passage list on p. 149) and there are also suggestions to "look to your alliances" in parts of Sun Tzu's terrain advice (see Passage #2.1). But what Sun Tzu has to say on point goes only a little way. There is nothing in Sun Tzu remotely resembling the level of analytical development found in the elegant, refined (and utterly cold-blooded!) classification of types of allies found in Kauṭilya's Arthaśāstra, the early Indian strategy classic.[39]

Sun Tzu's thinking also stays almost entirely focused on zero-sum conflict situations (a point particularly germane to Theme #4 development involving civilian extensions and applications of Sun Tzu; see p. 152 there). It is unalloyed by attention to spells of strategic cooperation between sometime foes, however limited, with the obvious place for nuanced calculations that such possibilities bring to the fore.[40]

A third broad omission in Sun Tzu's Theme #1 pertains to lack of guidance as to how far ahead a strategist in Sun Tzu's mold can realistically (or usefully) be expected to plan and what such long-term planning should entail. Holding aside Sun Tzu's prudent advice to keep one's forces intact as sequential strategic competitions unfold – see verse III.11 (Passage #3.3) – we know nothing of any thoughts Sun Tzu

---

[37] See Passages #11.1 and #11.2.

[38] The political climate of Warring States China created a never-ending need for calculations of that ilk. Mair, p. 24 profiles the deep political–military instability of the era:

> A key feature of the politics of the Warring States period was the ambivalent relationship between a ruler [of a state] and the feudal lords associated with him. Though the feudal lords may have sworn fealty to the ruler, they were often on the verge of revolting and were constantly trying to assume the dominant position he occupied. The situation was by nature highly unstable ...

Writing of the precursor Spring and Autumn period, Kierman (Background footnote 31, p. 34) observes: "Alliances were constantly shifting and were to be relied upon only with the utmost wariness." A particularly striking case of alliance alignments toggling back and forth many times over, reflecting the state of Zheng's unstable relationship with the states of Jin and Chu, is noted in #12 footnote 2.

[39] See Patrick Olivelle, King, Governance, and Law in Ancient India: Kauṭilya's Arthaśāstra, a new annotated translation (New York: Oxford University Press, 2013), pp. 302–305.

[40] Verse XI.39 (Passage #9.8) is a potential exception where Sun Tzu does recognize possibilities for win–win cooperation between foes. But lest it be given more than its due, the surrounding Sun Tzu discussion is plainly focused on factors affecting cooperation among troops on one's own side.

may have had about planning for another war after the present one (surely a never-ending need in Warring States China!),[41] let alone about thinking strategically in intergenerational terms.[42] One major class of issues of the latter type involves a ruler's succession planning.[43] There can be very tricky strategic issues here, ones difficult enough so that even Cao Cao, one of the most celebrated strategists in Chinese history and the earliest of Sun Tzu's eleven traditional commentators, has been faulted for showing feet of clay in tackling them.[44] Yet there can be cases of success in strategic planning of an intergenerational type, often underwritten by bureaucracies or legal systems and, in a more tangible vein, by water projects as well as by urban planning and construction initiatives. The latter are ancient activities whose consequences can certainly have major intergenerational strategic implications (e.g., effects of decisions regarding location and development of a ruler's capital and influence wielded from that place).[45] True to a warfighting focus, Sun Tzu has nothing to contribute here.

---

[41] See Background footnote 8 (count of wars). Truly long-term planning was certainly not foreign to early Chinese war and politics. A prescient observation pointing to planning on a demographic timescale by the state of Wu's archenemy Yue is attributed to Wu Zixu (died 484 BC), a distinguished general and colleague of the putative historical Sun Tzu: "For ten years, Yue will produce and raise [children], and for ten years they will educate and train them, and then after twenty years, Wu will be destroyed!" (Translation from the Zuozhuan [see p. xxiii above] by Olivia Milburn, The Glory of Yue: an annotated translation of the Yuejue Shu [Leiden/Boston, MA: Brill, 2010], p. 22.)

[42] A borderline case involves comments attributed to Sun Wu (traditionally equated to Sun Tzu of the Sun Tzu text; see p. 13 above), predicting clans' political longevities on the basis of their tax policies. The relevant archaeologically recovered text is translated by Ames, pp. 175–76 and by Sawyer, pp. 246–48. It seems far enough afield, in both topic and reasoning, from Sun Tzu's analysis of warfare (or of espionage) to raise questions about the intellectual kinship to Sun Tzu's thirteen chapters. For analysis of Sun Wu's thinking as reflected in this recovered text – noting some apparent holes in its logic – see Petersen, Introduction footnote 32, pp. 12–18, "Sun Wu foresees the future."

[43] Wilkinson, §17.1.1, p. 270 observes that "Long before the empire was founded, ancient Chinese thinkers identified a ruler's choice of successor as one of the main criteria of his ability."

[44] Cao Cao's shortcomings in this regard are described in a famous essay by Tang Taizong, the second emperor of the Tang dynasty (r. 626–49 AD), as translated by Denis Twitchett, "How to be an emperor: T'ang T'ai-tsung's vision of his role," Asia Major, Third Series, 1996, 9(1/2), pp. 1–102:

> When Emperor Wu of Wei (Cao Cao 曹操; d. 220 AD) set up his dynasty, he took no regard of long-term policy. Not one of his sons or younger brothers was enfeoffed: the members of the royal clan did not have enough land on which to stand up an awl. Outside there were no [brothers to act as a] wall to protect himself. Within there was no great base stone to act as the foundation for the state. Consequently, the royal authority was preserved by others; and the Altars of Soil and Grain [of his dynasty] fell to members of a family with a different surname. The ancients had a saying, "If the flow of a stream is exhausted, the spring has dried up; if the branches fall, the trunk is rotten," and it applies to this situation."
>
> (p. 61; words supplied in parentheses or brackets are Twitchett's)

[45] Wilkinson, §36.9, p. 493 notes: "The siting of Chinese capitals from the Shang and Zhou to the later dynasties is usually rightly discussed in terms of their strategic location or proximity to water and food or to river transportation (for grain shipments)" – i.e., strategic-logistics factors.

Sun Tzu (2) ("Sun Tzu Extended") and Sun Tzu (3) ("Sun Tzu Analogical") Frontiers of Theme #1

Focusing first on Sun Tzu (2), note that careful calculation – Theme #1's core ingredient – has an exceptionally natural place in Central Asian warfare, for reasons reflecting both mobility needs and logistics constraints there.[46] The need for exquisite titration of what supplies and equipment to lug along vs. leave behind (see Passage #1.15) applies even more to Central Asian warfare in desolate, sparsely populated country than to the warfare Sun Tzu knew. A leading pre-modern example of the spirit of Theme #1 put into practice would be the meticulous operational planning and coordination that underpinned Mongol successes on a continent-wide scale in the time of Chinggis Khan (r. 1206–27).[47] Beyond its world-historical importance, this example is instructive as an illustration of how non-Chinese actors have very effectively waged "war according to Sun Tzu," sometimes better than the Chinese.

For twenty-first-century consumers of Sun Tzu – many of whom must deal with situations Sun Tzu never imagined – the single most basic nugget of Theme #1 advice concerns the need for ceaseless strategic calculation (here understanding strategy in its broad sense of intendedly rational action anywhere on the spectrum of conflict, civilian as well as military). For a strategist truly in Sun Tzu's mold, the calculating impulse is never switched off, no matter how pressing or frustrating the circumstances – or, at an opposite pole, however much the situational ambiance is one of "being among friends"[48] (perhaps sharing experience of exile).[49] Of course, from many modern perspectives – from Freud to cognitive neuroscience – Sun Tzu's concepts of calculation may appear resoundingly naive, and there are definitely

---

[46] For some of those constraints see Alexander Morrison, "Camels and colonial armies: the logistics of warfare in Central Asia in the early 19th century," *Journal of the Economic and Social History of the Orient*, 2014, 57(4), 443–85 (making the important point that a need to round up baggage animals served as a barrier to launching spontaneous or unauthorized campaigns).

[47] Denis Sinor has argued that at the core of this planning was "a very rigid timetable to which all Mongol commanders were expected to adhere strictly." See p. 240 of Sinor's "On Mongol strategy," pp. 238–47 in *Proceedings of the Fourth East Asian Altaistic Conference, December 26–31, 1971, Taipei, China* (Ch'en Chieh-hsien ed.; Tainan, Taiwan: Dept. of History, National Ch'engkung University, 1975), a paper cited in May, Introduction footnote 52, p. 203. See also May, Introduction footnote 52, p. 65 ("Mongol armies operated according to predetermined time schedules").

For ways in which Chinggis Khan's Mongol army addressed logistics issues see May, Introduction footnote 52, pp. 62–63 ("Organization of logistics"). See also May's "The training of an Inner Asian nomad army in the pre-modern period," *Journal of Military History*, 2006, 70(3), 617–35.

[48] A World War II anecdote tells of an Office of Strategic Services (OSS) school where the trainees were not supposed to know each other by their real names, only by OSS nicknames. At the course's end "all trainees were invited to a banquet where they were given unlimited drinks, with toasts of congratulation from their instructors. In the good fellowship nearly half told their names to their classmates. They all flunked." See Paul Frillman & Graham Peck, *China: the remembered life* (introduction by John K. Fairbank; Boston, MA: Houghton Mifflin, 1968), p. 249.

It is not known if the course's designers had read Sun Tzu (though there were OSS personnel who had), but Sun Tzu of Passage #1.5 (calculation and deception) would surely have approved this upshot.

[49] See Emilie L. Mortensen, "Being care-ful among friends: the ambiguities of friendship in exile," *Etnofoor*, 2019, 31(1), 29–46 (focus is how "social and political precarity of exile shape friendship relations among Syrian men in Amman, Jordan").

depths they fail to reach. Yet there is a ring of timeless realism to the Sun Tzu–inspired notion that important and avoidable strategic errors frequently stem, not from failure to get a calculation just right, but from failure to do any calculation at all.[50] For example, read in a way that ranges beyond warfighting, Passages #1.3, #1.10, and #1.15 convey warnings never to shelve the calculating impulse under operational conditions of high tempo and momentum – to cite modern examples, as when initiating a crash R&D project or a "normalization of deviance" course of action of the sort found at times in the history of manned space exploration.[51] Those are the sort of Scarlett O'Hara moments where all instincts (often abetted by social pressure and social proof) say to throw calculation to the winds and "I'll think about that tomorrow." Sun Tzu says: "Not so." Here is grist for *"What did Master Sun know that we still don't (or have yet to absorb adequately)?"*[52]

A further fundamental Sun Tzu insight, anchored in verse I.16, is that the competencies required to realize Sun Tzu's calculational imperatives to their fullest – thereby achieving the strategic advantage that such mastery allows – are of more than one type. Here is a further candidate for *"What did Master Sun know . . .?"* Lest this idea seem self-evident, it should be noted that at the core of the McNamara era at the Pentagon (1961–68, his years as Secretary of Defense), and the bitter intramural conflicts for which it is famous, stood efforts to shoehorn a vast variety of military calculations into a uniform calculational framework rooted in concepts transposed from civilian economics.[53] That initiative elicited visceral resistance from many military professionals who believed that much valuable context was thereby being lost, and that their judgments arrived at by very different, often intuitive, paths were sounder. Of course, those same professionals were commonly oblivious to the real merits of economic calculi, thereby themselves *also* violating Sun Tzu's ecumenical spirit recognizing contributions of a variety of types of calculating.

Looking toward a living legacy for Sun Tzu–inspired calculational thinking, three Sun Tzu (3) directions stand out, each grounded in the text.

The first is, of course, the thrust of Passage #1.1 setting forth a net assessment perspective and agenda.[54] Passage #1.1's specifics are particularly important for the way in which they stand clear of the forces-and-weapons "bean counting" that in

---

[50] Cf. Leites, *Soviet Style in War* (cited in Introduction footnote 45), section VI.1, "The disinclination to calculate," pp. 299–312. An early twenty-first-century example might involve unguarded remarks in a social setting where digitally enabled virtual assistants may be dutifully preserving them for posterity.

[51] See Diane Vaughan, *The Challenger Launch Decision: risky technology, culture, and deviance at NASA* (Chicago: University of Chicago Press, 1996).

[52] In a world of phishing emails, compelling examples come from the uneven success of many big organizations in inculcating in their large workforce a "think before you click on that link" mindset.

[53] A classic statement of that framework is Charles J. Hitch and Roland N. McKean, with contributions by others, *The Economics of Defense in the Nuclear Age* (RAND Corporation Research Study; Cambridge, MA: Harvard University Press, 1960).

[54] Once one moves beyond its naturally comparative focus, net assessment is famously hard to define, certainly in a way that captures the essence of the activity when done well. Here it is worth noting Bracken's characterization of net assessment (article cited in Part A footnote 2, see p. 54 above) as being more like a practice than either an art or a science. For further perspectives on net assessment see Stephen Peter Rosen, "Net assessment as an analytical concept," pp. 283–301 in

modern times has at times dominated this type of endeavor, instead playing up the central importance of intangibles rooted in human factors (also relevant here is Passage #1.18's verse IX.45). In Sun Tzu's formulation the latter factors are of several types, including what Griffith renders as "moral influence" (for which an approximate modern counterpart might be "conceptual unity" on a level of a polity and its people); quality of high command; quality of implementation of regulations and instructions; quality of training (nowadays also quality of education more broadly); and quality of incentives. For our own time this battery of Sun Tzu emphases challenges modern net assessment to engage as well as Sun Tzu did in his era with a full range of factors affecting the "healthiness" (or lack thereof) of pertinent social structures (again, "What did Master Sun know that we still don't . . .?"). For example, one could imagine a modern Sun Tzu urging attention to extremes of socioeconomic inequality as a core source of societal weakness and strategic vulnerability – possibly propelling a country whose inequality becomes extreme in the direction of clientelistic social structures like those of Mexico or Russia, with consequences conducive to the flourishing of high-level corruption and organized crime.[55]

While Sun Tzu gives less help here, since asymmetric warfare is not his bailiwick, a further tantalizing and very important direction for expanding on Passage #1.1 ideas centers on net assessment for asymmetric conflicts, often involving non-state actors. The analytical challenges thus posed can be difficult – often more so, in fact, than those addressed by Passage #1.1 with its implicit assumption of symmetric conflict between state actors. Here Sun Tzu's terrain thinking (Theme #2) may be a resource if "terrain" is generalized in social network directions (see pp. 98ff.).

Asymmetric warfare extensions of Passage #1.1 crucially depend on credible data (Passage #1.19). This need in turn underscores that the espionage capabilities basic to Sun Tzu's thinking are just as important in conflicts involving non-state actors as in ones involving states alone. Indeed they may be even more important there.[56]

A second direction for further developing Theme #1 pivots on the fact that Sun Tzu is at home in emotional territory – in fact, like some modern trial lawyers, might be said to thrive on exploiting emotional dynamics for strategic and tactical gain. In this regard Sun Tzu presents a welcome intellectual contrast to the emotionally

Andrew W. Marshall, J. J. Martin, and Henry S. Rowen (eds.), *On Not Confusing Ourselves: essays on national security strategy in honor of Albert and Roberta Wohlstetter* (Boulder, CO: Westview Press, 1991).

[55] In a basically positive report card on pre–World War II British net assessment Paul Kennedy identifies strategic assessment errors in a key document as "all to some extent involv[ing] a misreading of the economic element in the proposed Allied grand strategy." See p. 52 of his "British 'net assessment' and the coming of the Second World War," in Williamson Murray & Allan R. Millett (eds.), *Calculations: net assessment and the coming of World War II* (New York: The Free Press, 1992), pp. 19–59. This book's first chapter leads off by quoting verse I.28 (Passage #1.2) in the Griffith translation.

In the twenty-first century, a comparably important source of error in net assessments could arise from misreading – including by simple omission – the role played by social structure and process factors.

[56] See Stathis N. Kalyvas, *The Logic of Violence in Civil War* (Cambridge, UK/New York: Cambridge University Press, 2006), pp. 174–76 ("Information"), which begins with a proposition evocative of Sun Tzu: "Information is a key resource in irregular war . . .; it is the link connecting one side's strength with the other side's weakness." Compare Sun Tzu's *xu/shi* principle (p. 113 below) – "the principle of striking with 'fullness' (*shi*) against 'emptiness' (*xu*)" (Needham & Yates, p. 35).

desiccated treatments of strategy endemic in the rational choice literature. That affinity could be explored in several Sun Tzu (2) or (3) directions, of which one – a contribution from Sun Tzu to sociology of strategy – is singled out here. It centers on Passage #1.24, particularly its verse IV.11:[57]

IV.11.

And therefore the victories won by a master of war gain him neither reputation for wisdom nor merit for valour.

This Sun Tzu observation sounds a career warning for strategists (and for those who employ them!) that the opacity of successful strategic calculation can undercut the job satisfaction of strategists, many of whom crave precisely the sort of kudos that verse IV.11 says will be denied to them. The resulting dynamics are both emotional and political and their effects can be significant. One possible effect is fragility of the strategist's power base (which, paradoxically, may become more fragile the more subtly successful is the strategy proposed, making attribution of that success more difficult). Another involves compensatory steps by strategists (or their fans and social support networks) to build up the strategist in a role of wizard, in effect attempting to squeeze prestige and social status from a situation where the intrinsic nature of the activity works to deny it – an ego defense that may, unsurprisingly, lead to bad advice born of sheer frustration.[58] A starting point for an antidote is to proselytize for the concept that the profession of strategist – like that of spymaster and a few other elite walks of life – is inherently a niche for *inconspicuous* elites and needs to be recognized and accepted as such by all concerned.[59] That concept of inconspicuous elites – who they are, how they operate, what their sources of satisfaction and frustration are, etc. – merits attention and analysis.[60]

A third peg for creative Theme #1 development builds on the prescient vein of quantitative thinking found in Sun Tzu. One launching point for Sun Tzu (3) development of "Sun Tzu the quant" is Passages #1.2 and #1.7, the latter of which Mark Lewis has characterized as expressing a "belief that victory could be

---

[57] As traditional commentator Du Mu observes (Minford, p. 155): "His victories are based on seeing the 'subtler origins,' so the world at large knows nothing of them, and they bring no fame for wisdom."

[58] A pertinent warning comes from Admiral Eccles, writing in the US Naval Institute *Proceedings*, July 1954, 80(7), p. 809: "Frustration frequently leads men of high spirit to commit acts of reckless irresponsibility." (Also quoted in Boorman, Introduction footnote 4, p. 114 note 58.)

[59] There is resonance here with Daoist tradition, which places great emphasis on the positive value of being inconspicuous. See Arthur Waley, *The Way and Its Power: a study of the Tao tê ching and its place in Chinese thought* (London: Allen and Unwin, 1934), p. 168 quoting *Laozi* XX: "The world is full of people that shine; I alone am dark."

A translator's comment on a passage in the *Zhuangzi* – also a basic text of Daoist tradition – refers to a master of the Way as being displeased "because his worth has been discovered, whereas the true sage remains hidden and unrecognized." See p. 248 footnote 1 of *The Complete Works of Chuang Tzu* (Burton Watson, trans.; New York/London: Columbia University Press, 1968). For *Zhuangzi* background see *Early Chinese Texts* (cited on p. xxii above), pp. 56–66; Wilkinson, §58.6.3.2, p. 777.

[60] A major historical case study of inconspicuous non-Western elites in pre-modern Islamicate Eurasia is Gagan D. S. Sood, *India and the Islamic Heartlands: an eighteenth-century world of circulation and exchange* (Cambridge, UK: Cambridge University Press, 2016). For overview see pp. 17, 25 there.

mathematically calculated."[61] Although we do not reliably know the specifics, Passage #1.2's verse I.28 – especially when set alongside Passage #1.1's verses I.11–I.14 – calls to mind R. A. Fisher's discriminant function (DIF) methodology that has drawn attention from modern soldiers, tax administrators, and analysts of many types.[62] Some of the traditional commentary broadly supports that line of thought.[63] It is also tempting to try to find in the Sun Tzu text anticipations of other specific problems of modern-day operations research. To a limited extent that is possible – here Figure 2 should command attention – albeit with a crucial caveat that Sun Tzu or whoever compiled the text would surely not have known what "research" in a modern sense is.[64]

A further starting point for exploring updates to "Sun Tzu the quant" comes from Passage #1.19's endorsement of what might nowadays be regarded as an evidence-based perspective.[65] That concept aligns well on one of the healthiest contributions to the study of war to emerge from empirically minded operations research: namely, to encourage skepticism regarding efforts to extract action implications from analytical exercises that operate on far too little pertinent data regarding the problem at hand. The divination and other "misguided" bases for inference that Sun Tzu so resoundingly rejects in Passage #1.19's verse XIII.4 are worth a second look here, with an eye on spotting modern missteps having a similar flavor (perhaps simulation exercises that amount to modern-day counterparts to verse XIII.4's spirit world?).[66]

The Sun Tzu text itself does not directly address innovation issues (as is scarcely surprising for a military tract of its vintage). That deficit should, however, be no bar to exploring a rich Sun Tzu (3) frontier involving applications of Sun Tzu's Theme #1 ideas – in particular the close connection between calculation and deception – to twenty-first-century intersects among *innovation*, *calculation*, and *sabotage* (the latter

---

[61] Lewis, Background footnote 11, p. 115.

[62] For machine learning updates see Leo Breiman, "Statistical modeling: the two cultures," *Statistical Science*, August 2001, 16(3), 199–215 (pp. 216–31 there contain comments from D. R. Cox and others).

[63] See, e.g., Zhang Yu (Minford, p. 110): "If all seven of these factors [listed in verses I.11–I.13] are favorable, then victory is certain before battle has been engaged; if they are unfavorable, then defeat is equally certain." However, discerning a precursor to DIF in verse I.28 (Passage #1.2), with its rather opaque reference to reasoning probably involving counting rods, remains an intellectual leap. Verse I.28 might also be understood along less analytically ambitious lines, as simply referring to the many long problems in simple arithmetic that large-scale warfare continually churns up.

[64] Illustrating a case where Sun Tzu's insights might be developed in operations research directions, Sun Tzu's death ground (Passage #1.12) shares with mass disasters the fact that both situations may allow few behavioral options (certainly for ordinary participants), by contrast to more "normal" situations where (at least in modern times) behavioral options are legion. In mass disaster scenarios those few behavioral options may make credible agent-based modeling feasible, as it otherwise might not be, thus bringing new vitality to Sun Tzu's belief in the essential simplicity, hence calculability, of the behavior of common soldiers (see #9 footnotes 7 and 8). For more background see p. 147 of M. Mitchell Waldrop, "Free agents," *Science*, April 13, 2018, 360(6385), 144–47 (focus is National Planning Scenario 1, involving a hypothetical nuclear terrorist attack in downtown Washington, DC), noting that "in crisis situations modeling behavior becomes easier because it tends to be primal."

[65] See P. M. S. Blackett, *Studies of War: nuclear and conventional* (New York: Hill and Wang, 1962).

[66] For a framework for developing analysis along those lines see Table 5 on pp. 405–406 below.

best understood in a broad sense that has affinities with Soviet-era Russian theory and practice, focusing on degrading a foe's ability to calculate).[67]

Start with the fundamental observation that computationally demanding models of the sort that are now popping up everywhere with the rise of "big data" are rarely, if ever, entirely built on "hard" data about the applied problems that motivate them. Furthermore, scientific and technical problems that animate the twenty-first century's ubiquitous computer models are almost always capable of alternative modeling formulations. For both reasons there is much scope for backstage exercise of human judgment in developing and applying model-based findings. With that scope comes room for strategic manipulations. Here a twenty-first-century "world according to Sun Tzu" comes into its own.[68] The complex, sprawling, often murky environment where "three empires meet" – science, administration, and politics – affords major latitude for a modern Sun Tzu to try for leverage in shaping to her own ends the landscape of applied computational models:

- which specific models are developed (reflecting tradeoffs among model scope, complexity, etc., as well as choice of specific substantive assumptions);[69]
- which models see light of day (are published, otherwise disseminated, etc.);
- which models are actually deployed for operational decision-making purposes;
- how those models are used to generate numbers and predictions;
- how the resulting numbers or patterns are interpreted;
- how those interpretations are used to inform strategic or policy decisions; and, cross-cutting all of the above issues,
- by what means are the computational workings of relevant models verified.[70]

Effective strategic interventions on this landscape need not themselves be particularly high-tech. An observation of Thomas Schelling, "One of the lamentable principles of human productivity is that it is easier to destroy than to create," has vitality here.[71] Some interventions may, for example, center on steps geared toward destabilizing an activity – namely, pure or applied research – that, by its nature, is

---

[67] See Leites, *Soviet Style in War* (cited in Introduction footnote 45), chapter VI, "Enhancing one's capacity to calculate and degrading the enemy's," pp. 299–355. See especially his pp. 312–14, "The vulnerability from within of the capacity to calculate," which leads off with a comment reminiscent of Passage #1.8: "According to the Authorities, reason in human beings is incessantly threatened by mood" (a comment that also applies excellently to research work, both pure and applied!).

[68] Theme #1 here overlaps with Theme #6 (attacks on the adversary's strategy, now meaning stymieing or retarding that adversary's capacity to innovate). In an R&D context, a focus of such attacks might be a talented researcher or research administrator (or, drawing inspiration from Passage #6.1's verse III.5 regarding attacks on alliances, a personal or institutional research collaboration).

[69] See Richard Levins, "The strategy of model building in population biology," *American Scientist*, 1966, 54(4), 421–31 (comparative framework whose application is much broader than Levins's title).

[70] For relevant perspective see Victoria Stodden, Marcia McNutt, et al., "Enhancing reproducibility for computational methods," *Science*, December 9, 2016, 354(6317), 1240–41. Illustrating some of the challenges, a physical science paper warns that "specific quantitative predictions ... [from the models being analyzed in it] should be viewed with caution" and that "we do not expect the results ... to be *exactly* reproducible when a different compiler is used and/or on different platforms." See pp. 355, 373 of I. Polichtchouk, J. Y-K. Cho, et al., *Icarus*, 2014, 229, 355–77 (emphasis in original).

[71] Thomas C. Schelling, *Arms and Influence* (New Haven, CT/London: Yale University Press, 1966), p. v.

frequently not very stable under the best of circumstances.[72] Successful innovation is as delicate a process as dealing with the psychology of spies (verses XIII.13–XIII.14 in Passage #1.11), and perhaps even more easily thrown out of kilter! One can see the spirit of Sun Tzu at work when, for example, a research group is "favored" by insertion of a disruptive new member. Opportunities for manipulation multiply when only a very limited number of individuals have mastered certain critical but largely unwritten skills (say, a specific computing approach or familiarity with a particular database or lab technique), creating the human equivalent of a logistics choke point.[73] Further opportunities arise when the research activity, whatever it is, must operate under weird constraints, technical or political.[74] With the continuing rise of big data, and with ever more decisions depending on it, room for strategic manipulations of this type can only be expected to grow. There is ample reason to suspect that their potential remains underestimated, when it is recognized at all, certainly by the standards of wariness that Sun Tzu's thinking demands. Here is a major twenty-first-century incarnation of *"What did Master Sun know that we still don't . . .?"*

✶✶✶✶✶✶

Set of principal passages used to illustrate Sun Tzu Theme #1 (calculation)

Passage #1.1. First Species of Calculation Emphasized by Sun Tzu: Analysis of Comparative War Potential or Net Assessment

(A striking feature of the Sun Tzu text is that it leads off with the net assessment task. If done right, that sophisticated and tricky exercise has a three-in-one quality, calling for three questions to be addressed:

- what are the enemy's strengths and weaknesses?
- what are my own side's strengths and weaknesses?
- what is the relationship between them?

---

[72] Shubik, Preliminaries footnote 10, p. 1520 notes a range of early post–World War II research organizations (mentioning RAND Corporation, General Electric, and the Office of Naval Research of the 1950s), commenting: "They were not stable. Whatever stability they had was dynamic."

[73] Bottlenecks of this type are common in science and technological innovation. The artificial neural networks area brims with "tricks of the trade" requiring an experienced "clinician's eye" for successful use in applications. A further relevant concept is that of intellectual capital in the sense of Lynne G. Zucker, Michael R. Darby, and Marilynn B. Brewer, "Intellectual human capital and the birth of U.S. biotechnology enterprises," *American Economic Review*, 1998, 88(1), 290–306 (published version of National Bureau of Economic Research Working Paper No. 4653, February 1994). See in particular p. 6 of the NBER paper: "Intellectual capital is by nature a transient property of disequilibria. It arises because few people have mastered certain valuable techniques relative to the number that will master those techniques and earn normal returns on that mastery when full equilibrium is achieved."

[74] An insightful appreciation of just how distorting such constraints can be, born of his World War II research experience, is conveyed by the distinguished insect physiologist, V. B. Wigglesworth, "The contribution of pure science to applied biology," *Annals of Applied Biology*, 1955, 42(1), 34–44, observing (p. 34) that applied biology (under World War II conditions experienced by many biologists) "is a totally different subject requiring a totally different attitude of mind."

The first is an intelligence task; the second, a self-assessment; the third, a comparative analysis exercise, ever probing for ways of pitting one's own strengths against adversary weaknesses.[75]

A basic issue facing any translator of Sun Tzu is how far to imbue Passage #1.1 with Chinese cultural content. Griffith's choice is to hold that content to a minimum. Thus verses I.3 and I.11 invoke the Chinese concept of *dao*, which Griffith renders "moral influence." In verse I.3 the terms *tian* and *di*, widely rendered as "heaven" and "earth," Griffith translates respectively as "weather" and "terrain".[76])

I.1. War is a matter of vital importance to the State; the province of life or death; the road to survival or ruin. It is mandatory that it be thoroughly studied.

I.2. Therefore, appraise it[77] in terms of the five fundamental factors and make comparisons of the seven elements later named. So you may assess its essentials.

I.3. The first of these factors is moral influence [*dao*]; the second, weather; the third, terrain; the fourth, command; and the fifth, doctrine.

I.4. By moral influence I mean that which causes the people to be in harmony with their leaders, so that they will accompany them in life and unto death without fear of mortal peril.

I.5. By weather I mean the interaction of natural forces [*yin/yang*]; the effects of winter's cold and summer's heat and the conduct of military operations in accordance with the seasons.

I.6. By terrain I mean distances, whether the ground is traversed with ease or difficulty, whether it is open or constricted, and the chances of life or death.[78]

I.7. By command I mean the general's qualities of wisdom, sincerity, humanity, courage, and strictness.

I.8. By doctrine [*fa*] I mean organization, control, assignment of appropriate ranks to officers, regulation of supply routes, and the provision of principal items used by the army.[79]

---

[75] Cf. *xu/shi* (lit. "emptiness/fullness") concept pair which is a bridge to Theme #3 (see p. 113 below).

[76] For discussion of the cultural context of translation see Appendix 1 anchor footnote 2.

[77] Sawyer's translation as "structure it" (Sawyer, p. 167) captures something useful here, placing emphasis on how military theory and analysis can help to structure amorphous domains like war and strategy, thereby rising to Jerome Bruner's challenge set forth in the Introduction (see p. 2 above).

[78] No probabilistic thinking on Sun Tzu's part should be inferred from "chances" in Griffith's verse I.6. Mair, p. 77 avoids probabilistic overtones by translating "... positions conducive to death or life."

[79] For analysis of *fa* (Griffith's "doctrine") see Appendix 1 anchor footnote 5 as well as Passage #13.30 (where verse I.8 also appears). Cao Cao's commentary on the middle segment of verse I.8 points to "proper distinction between the [different spheres of responsibility of] various officers" (Minford, p. 107). That would include logistics officers, a category that would have been well-known to Cao Cao through his own military exploits (though Cao Cao's Sun Tzu commentary is reported to be an early career product [de Crespigny, Background footnote 95, p. 319]). Over the history of warfare, logistics organizations, certainly for large armies, have tended to be more complex than tactical ones. Where and how to insert logistics functions into a chain of command,

I.9. There is no general who has not heard of these five matters. Those who master them win; those who do not are defeated.

I.10. Therefore in laying plans compare the following elements, appraising them with the utmost care.

I.11. If you say which ruler possesses moral influence [dao], which commander is the more able, which army obtains the advantages of nature and the terrain, in which regulations and instructions are better carried out, which troops are the stronger;

I.12. Which has the better trained officers and men;

I.13. And which administers rewards and punishments in a more enlightened manner;

I.14. I will be able to forecast which side will be victorious and which defeated.

Passage #1.2. Quantitative Analysis of War Potential

(This verse lays a foundation for a robust connection between Sun Tzu and computational thinking writ large.[80] At the same time, no mathematical level beyond basic arithmetic should be envisioned in what Sun Tzu says here.

Against this backdrop, a basic difficulty is how to reconcile verse I.28's quantitative thrust with the questions presented by Passage #1.1's net assessment task. These pivot on human factors where available data – certainly in Sun Tzu's time – seems inescapably qualitative. For a potential way out see Appendix 1.[81])

I.28. Now if the estimates made in the temple before hostilities indicate victory it is because calculations show one's strength to be superior to that of his enemy; if they indicate defeat, it is because calculations show that one is inferior. With many calculations, one can win; with few one cannot. How much less chance of victory has one who makes none at all! By this means I examine the situation and the outcome will be clearly apparent.[82]

---

and how to craft incentives for logistics personnel, has continued to give trouble to militaries into modern times.

[80] Mair, pp. 140–41 note 25 observes:

> This temple is not a religious institution, but the inner sanctum within the palace precincts where the ruler made the most important decisions of state in consultation with his most trusted advisers. . . . "Temple calculations" refers to taking stock before engaging in battle, the temple being comparable to what today might be called a war room.

Note that a war room characterization need not preclude coexistence of divination practices with planning of a "rationalist" type more congenial to a modern audience. See Appendix 1 footnote 7.

[81] See Appendix 1 anchor footnote 8 (exploring summated rating scale interpretation of verse I.28).

[82] Insight into both the extent and the limits of early Chinese computing capabilities comes from the discovery of a Warring States artifact from around 305 BC, roughly contemporaneous with the latter part of Brooks's estimated dating of the Sun Tzu text's compilation (see Table 12A on p. A-35). That item, now held by Tsinghua University, is believed to have been used by officials for tax administration calculations. See Jane Qiu, "Ancient times table hidden in Chinese bamboo strips: the 2,300-year-old matrix is the world's oldest decimal multiplication table" (*Nature*, 7 January 2014), available at www.nature.com/news/ancient-times-table-hidden-in-chinese-bamboo-strips-1.14482 (visited October 6, 2021). See also Wilkinson, §38.16.1, p. 542. This table is

Passage #1.3. Second Species of Calculation: Expedient Assessment

(The expedient assessment [quan] concept appears here.[83] The Sun Tzu VII context is mobile warfare, possibly of high tempo. As that context suggests, expedient assessment is often rapid assessment, though that is not always the case.)

VII.15. Weigh [quan] the situation, then move.

Passage #1.4. Possibility of Precise Military Calculation

(Sun Tzu's verse V.14 image of a raptor striking its prey is an excellent example of expedient assessment in a fast-moving three-dimensional tactical setting.)

V.14. When the strike of a hawk breaks the body of its prey, it is because of timing.

V.15. Thus the momentum [shi] of one skilled in war is overwhelming, and his attack precisely regulated.

V.16. His potential [shi] is that of a fully drawn crossbow; his timing, the release of the trigger.

Passage #1.5. Integrating Different Species of Calculation; Calculation and Deception

I.16. Having paid heed to the advantages of my plans [ji], the general must create situations [shi] which will contribute to their accomplishment. By "situations" [shi] I mean that he should act expediently in accordance with what is advantageous [yin li] and so control the balance [quan].

I.17. All warfare is based on deception.

Passage #1.6. More about Calculation and Deception

VII.12. Now war is based on deception. Move when it is advantageous and create changes in the situation by dispersal and concentration of forces.[84]

Passage #1.7. Cumulative Nature of Military and Logistics Calculations

(A calculational instinct shines here. Unfortunately, it is also telegraphically expressed, so that this passage has long given commentators and translators

---

effectively an ancient calculator, importantly showing familiarity with the mathematical concept of a matrix structure. But juxtaposing a calculator of this type with the calculational needs likely to arise in day-to-day management of an army of 100,000 men and its logistics and other planning requirements, the rudimentary state of quantitative tools available to Sun Tzu is apparent.

[83] Other species of calculation, notably ones of administrative types, are found in Sun Tzu too, though attention to them is scattered across several chapters with scant details. See #1 footnote 9.

[84] Verse VII.12's first part is the title of a major modern study of Sun Tzu's thought by Li Ling 李零, Bing yi zha li 兵以詐立 (War is Based on Deception) (Beijing: Zhonghua shuju, 2006). Ames, p. 130 plays up a calculational aspect of verse VII.12, translating: "Therefore, in warfare rely on deceptive maneuvers to establish your ground, calculate advantages in deciding your movements …"

trouble. Nonetheless, the idea of a pyramid of quantitative constructs – each level building on results of calculations one level down – seems robust.[85])

IV.16. Now the elements of the art of war are first, measurement of space; second, estimation of quantities; third, calculations; fourth, comparisons; and fifth, chances of victory.[86]

IV.17. Measurements of space are derived from the ground.[87]

IV.18. Quantities derive from measurement, figures [shu] from quantities, comparisons from figures [shu], and victory from comparisons.

IV.19. Thus a victorious army is as a hundredweight balanced against a grain; a defeated army as a grain balanced against a hundredweight.

Passage #1.8. Failure to Calculate (1): Avoiding Irreversible Major Decisions Taken for Emotional Reasons, sans Calculation

(Passage #1.8 is a natural companion to Passage #1.1.[88] Verse XII.18 has particular relevance to a propensity, not unknown among those in high places, to "personalize" – e.g., to find personal affronts in – situations better analyzed dispassionately, weighing structural issues and options.

Sun Tzu places emphasis here on an existential threat to a strategic actor – a "cost" of an entirely different order from other costs, and one that commonly defies compelling quantification.[89] Verse XII.18 alludes to existential outcomes for both individuals and polities, tying back to the stark imagery of verse I.1 in Passage #1.1. Existential threats might also include existential threats to a people.)

XII.17. If not in the interests of the state, do not act. If you cannot succeed, do not use troops. If you are not in danger, do not fight.

XII.18. A sovereign cannot raise an army because he is enraged, nor can a general fight because he is resentful. For while an angered man may again be happy, and a resentful man again be pleased, a state that has perished cannot be restored, nor can the dead be brought back to life.

---

[85] For further analysis see pp. A-43f. and Appendix 1 anchor footnote 10 (sampler of interpretations).

[86] No probabilistic thinking on Sun Tzu's part should be inferred from Griffith's verse IV.16 "chances."

[87] Terrain issues have, of course, long been a catalyst to measurement in many eras of land warfare, including work of surveyors and military mapmakers. The Mawangdui military map (est. 181–168 BC, discussed on pp. 35–36 above) was based on surveying (Wilkinson, §14.1.2, pp. 214–15).

[88] If the espionage chapter is excluded as a late addition to the text (see Brooks, p. 61), Passage #1.1 starts the Sun Tzu text and Passage #1.8 ends it, a symmetric positioning of Sun Tzu's most detailed statement about calculation and a warning of catastrophic implications of failure to calculate.

[89] Sun Tzu here engages with issues that have proved not easy to fit into what is widely regarded as the dominant modern paradigm for strategic analysis, namely, formal modeling frameworks of game theory and economics. Those frameworks commonly sidestep head-on engagement with existential issues of the type that verse XII.18 brings to the fore. For a pioneering exception see analysis in Tjalling C. Koopmans, *Three Essays on the State of Economic Science* (New York: McGraw-Hill, 1957), pp. 62–63.

XII.19. Therefore, the enlightened ruler is prudent and the good general is warned against rash action. Thus the state is kept secure and the army preserved.

Passage #1.9. Failure to Calculate (2): General's Emotions As Source of Tactical Error

(Passage #1.9 underscores by example that some of the most difficult calculational challenges facing a commander pivot on self-control, retaining a capacity to calculate calmly and effectively even under conditions of fantastic stress.[90])

III.9. If the general is unable to control his impatience and orders his troops to swarm up the wall like ants, one-third of them will be killed without taking the [walled] city. Such is the calamity of these attacks.

Passage #1.10. Failure to Calculate (3): Rising to the Bait

VII.27. When he pretends to flee, do not pursue.
VII.28. Do not attack his *élite* troops.
VII.29. Do not gobble proffered baits [*er bing*, lit. "bait troops"]

Passage #1.11. Delicate Psychological Calculations Pertaining to Use of Secret Agents

(These verses from Sun Tzu's espionage chapter convey a tonality of espionage calculations that Griffith's "Delicate indeed! Truly delicate!" captures well. Note that verse XIII.19 requires weighing, not just one mentality but at least three different mentalities – the double agent's, the expendable spy's, and the enemy's.)

XIII.12. Of all those in the army close to the commander none is more intimate than the secret agent; of all rewards none more liberal than those given to secret agents; of all matters none is more confidential than those relating to secret operations.
XIII.13. He who is not sage and wise, humane and just, cannot use secret agents. And he who is not delicate and subtle cannot get the truth out of them.
XIII.14. Delicate indeed! Truly delicate! There is no place where espionage is not used.

...

XIII.19. And it is by this means [the double agent] that the expendable agent, armed with false information, can be sent to convey it to the enemy.

---

[90] In a marker of emphasis, Sun Tzu gives similar advice – never to allow one's emotions to dictate command decisions – directed to three different authority levels: ruler (Passage #1.8); general (Passages #1.8 and #1.9); senior subordinate officers (verse X.13 in Passage #1.26A).

Passage #1.12. Emotional Engineering: Calculated Use of Emotional Triggers to Elicit Supranormal Effort from Troops on Brink of Disaster

(Sun Tzu here takes note of the possibility that extreme situations arise in war where established behavioral norms no longer apply, because one side or the other in an existential conflict feels itself starting to lose and therefore starts to behave in extraordinary ways whose extreme dangerousness needs to be grasped by the adversary even though that behavior is certainly unsustainable in the longer run.[91] A modern terminology might label this a "nonlinear" response, connoting occurrence of a type of behavioral rupture. Possibility of such rupture presents a need to retool normal practices of cost–benefit evaluation by the side that, by any ordinary criterion of judgment, seems destined to win.

"Death ground" situations, and the advantages and pitfalls they harbor, are an important focus in Sun Tzu and Sun Tzu deserves full credit for focusing on the phenomenon.[92] What Sun Tzu does *not* do, however, is to provide any criterion of judgment to help distinguish situations where death ground reflexes may yet snatch victory from the jaws of defeat from other truly hopeless cases.)

XI.23. In death ground I could make it evident that there is no chance of survival. For it is in the nature of soldiers to resist when surrounded; to fight to the death when there is no alternative, and when desperate to follow commands implicitly.

. . .

[91] The idea of desperation as an emotion capable of triggering "an intense and general struggle which is often advantageous" – and which makes enemies fear to induce it for just that reason – is by no means limited to early Chinese armies or warfare. See Hiram M. Stanley, *Studies in the Evolutionary Psychology of Feeling* (London/New York: Swan Sonnenschein/Macmillan, 1905), p. 122. For relevant modern neuroscience background see also R. Douglas Fields, *Why We Snap: understanding the rage circuit in your brain* (New York: Dutton, 2015). Related ideas appear in non-Chinese military contexts and in Soviet-era Russian Communist thinking. See Table 2 note i (p. 98 below).

[92] More may be involved here than hard-pressed troops fighting ferociously. One of Sun Tzu's traditional commentators, He Yanxi, quotes an adage: "He who is in dire straits and can find no stratagem [mou], is lost" (Minford, p. 275). An example is suggested by the *Sanguo zhi* (historical work not to be confused with the fictional *Romance of the Three Kingdoms*; see Background footnote 102). In Giles's telling (Giles, pp. 68–69), the great general (and Sun Tzu commentator) Cao Cao, trapped between two of his enemies while maneuvering through a narrow valley, knows he is in a desperate situation. He proceeds to some clever military engineering; lays an ambush based on it; uses his baggage train as bait; and proceeds to deal a resounding defeat to the enemy army that was moving in to finish him off – a sequence of steps hardly reducible to simple berserker behavior. The story also appears in de Crespigny, Background footnote 95, pp. 107–108 and his footnote 23; and, with embellishments, in *Romance of the Three Kingdoms* (see pp. 45–46 above), chapter 18, pp. 141–42.

For further comments in a Theme #9 context see also Passage #9.14 discussion (pp. 321–22 below).

XI.33. Throw the troops into a position from which there is no escape and even when faced with death they will not flee. For if prepared to die, what can they not achieve? Then officers and men together put forth their utmost efforts. In a desperate situation they fear nothing; when there is no way out they stand firm. Deep in a hostile land they are bound together, and there, where there is no alternative, they will engage the enemy in hand to hand combat.

XI.34. Thus, such troops need no encouragement to be vigilant. Without extorting their support the general obtains it; without inviting their affection he gains it; without demanding their trust he wins it.

. . .

XI.49. He burns his boats and smashes his cooking pots; *he urges the army on as if driving a flock of sheep*, now in one direction, now in another, and none knows where he is going. [Emphasis supplied.]

. . .

XI.55. Set the troops to their tasks without imparting your designs; use them to gain advantage without revealing the dangers involved. Throw them into a perilous situation and they survive; put them in death ground and they will live. For when the army is placed in such a situation it can snatch victory from defeat.

Passage #1.13. Quantitative Planning Factors in Military Organization

(Army organization in many times and places favors regularity of components.[93] From there is it a short step, abetted by ways in which regularity greatly facilitates logistics planning, to approaching military organization from a quantitative standpoint, typically a very simple one.[94] A literal meaning of the term *shu* appearing in Passage #1.13 is "number." *Fen shu*, which Griffith translates as "organization," Mair renders as "division and enumeration," noting that "'division' [*fen*] refers to dispensing tasks and responsibilities while 'enumeration' [*shu*] signifies fixing the amounts of personnel for various units."[95]

Holding aside their value as evidence of a quantitative mindset on Sun Tzu's part, verses V.1 and V.2 make exorbitantly ambitious claims that need to be taken with a grain of salt. For further analysis see the command, control, and communications section of Appendix 15.)

---

[93] For early Chinese examples see Background footnote 24 (citing the *Zhouli* text).

[94] For an example from Roman army logistics see Roth, Background footnote 57, pp. 21–22.

[95] Mair, p. 91 and p. 146 note 2. Ranging beyond the Chinese text of Sun Tzu, though in a direction compatible with how modern professional soldiers think, Ames, p. 120 translates verse V.19's occurrence of *shu* as "logistics." Here, however – echoing #1 footnote 79 above – it should again be noted that logistics organizations, certainly in practice, tend to be more complex and also less tidy than tactical ones (a pattern that continues in modern times). What Sun Tzu asserts in verse V.1 might therefore be read as privileging tactical over logistics aspects of war – thereby revealing a tacit bias extremely common in military thinking throughout the history of organized warfare. Regarding division of labor between verses V.1 and V.2 see Appendix 1 anchor footnote 15.

V.1. Generally, management of many is the same as management of few. It is a matter of *organization* [*fen shu*, lit. "divide and count"].

V.2. And to control many is the same as to control few. This is a matter of formations and signals.

. . .

V.19. *Order or disorder depends on organization* [shu]; *courage or cowardice on circumstances* [shi]; *strength or weakness on dispositions* [xing]. [Emphases supplied.]

Passage #1.14. Quantitative Planning Factor Pertaining to Giving out Rewards

(The ten-chariot planning factor stated here is plainly intended to be established by administrative fiat. There is nothing figurative about it.)

II.18. Therefore, when in chariot fighting more than ten chariots are captured, reward those who take the first. Replace the enemy's flags and banners with your own, mix the captured chariots with yours, and mount them.

Passage #1.15. Mobility Planning Factors

(Logistics issues raised here, pertaining to how much of an army's impedimenta should be lugged along vs. left behind, illustrate a multidimensional type of expedient assessment not reducible to a general's weighing of just two opposed factors, as in some other cases of expedient assessment noted on pp. 58–59 above.[96])

VII.2. Nothing is more difficult than the art of manoeuvre. What is difficult about manoeuvre is to make the devious route the most direct and to turn misfortune to advantage.

VII.3. Thus, march by an indirect route and divert the enemy by enticing him with a bait. So doing, you may set out after he does and arrive before him. One able to do this understands the strategy of the direct and the indirect.

VII.4. Now both advantage and danger are inherent in manoeuvre.

VII.5. One who sets the entire army in motion to chase an advantage will not attain it.

VII.6. If he abandons the camp to contend for advantage the stores will be lost.

VII.7. It follows that when one rolls up the armour and sets out speedily, stopping neither day nor night and marching at double time for a hundred li, the three commanders will be captured. For the vigorous troops will arrive first and the feeble straggle along behind, so that if this method is used only one-tenth of the army will arrive.

---

[96] It should be noted that the Chinese text of verse VII.9 points to the army's being lost if *any* of the logistics elements listed there is lacking (a disjunctive concept), *not* (as Griffith's translation implies) that the army will be lost if *all* of those elements are lacking (a conjunctive concept).

VII.8. In a forced march of fifty li the commander of the van will fall, and using this method but half the army will arrive. In a forced march of thirty li, but two-thirds will arrive.

VII.9. It follows that an army which lacks heavy equipment, fodder, food and stores will be lost.

Passage #1.16. Planning Factors Involving Lead Times and Their Concatenation

III.8. [In attacking a walled city] [t]o prepare the shielded wagons and ready the necessary arms and equipment requires at least three months; to pile up earthen ramps against the walls an additional three months will be needed.

Passage #1.17. Further Quantitative Planning Factors, Some Suggestive of Information Aggregation on a Country-wide Level

(Focus here is on Sun Tzu material having quantitative logistics or economic content. Verse II.12 suggests presence of a price system with money – itself, of course, a quantitative phenomenon whose existence may also encourage use of further quantitative planning factors based on amounts of money.[97])

II.1–II.2. (Translation from Lau's 1965 article, p. 324, which starts with a caption supplied by Sun Tzu: *"The method of employing troops."*[98]) "When one thousand fast four-horse chariots, one thousand four-horse wagons covered in leather, and one hundred thousand armour-clad troops are used, if provisions have to be transported over a distance of a thousand li, then what with expenditure at home and in the field, on the guest advisers, on materials such as glue and lacquer, and on the supply of chariots and armour, it will cost one thousand pieces of gold every day before one hundred thousand troops can be raised."

. . .

II.9. Those adept in waging war do not require a second levy of troops nor more than one provisioning.

. . .

II.12. Where the army is, prices are high; when prices rise the wealth of the people is exhausted. When wealth is exhausted the peasantry will be afflicted with urgent exactions.

II.13. With strength thus depleted and wealth consumed the households in the central plains will be utterly impoverished and seven-tenths of their wealth dissipated.

---

[97] Lest Sun Tzu's allusion to a price system be given an unduly modern twist, note should be taken of a cautionary observation of Lewis regarding Warring States economic life: "it would be easy to exaggerate the role of cash in what was still primarily a barter economy. . . . cash in the early periods was largely used within the sphere of government." (Lewis, Background footnote 7, p. 607.)

[98] Rationale for using Lau's translation rather than Griffith's is given in Appendix 1 footnote 18. The italics are Lau's.

II.14. As to government expenditures, those due to broken-down chariots, worn-out horses, armour and helmets, arrows and crossbows, lances, hand and body shields, draft animals and supply wagons will amount to sixty per cent of the total.

II.15. Hence the wise general sees to it that his troops feed on the enemy, for one bushel of the enemy's provisions is equivalent to twenty of his; one hundredweight of enemy fodder to twenty hundredweight of his.

Passage #1.18. Pitfalls of Plausible but Overly Narrow Planning Factors

(This verse harmonizes with Passage #1.1, which envisions troop strength as but one factor in predicting military success. A general principle suggested by verse IX.45 is a warning against reliance on other simple but naive planning factors – say, ones relying on some technologcal advantage that is potent yet at the same time fragile, like the famous swords of Wu in early Chinese warfare.[99])

IX.45. In war, numbers alone confer no advantage. Do not advance relying on sheer military power.

Passage #1.19. Limits of Calculation (1): Appropriate Empirical Data As Crucial

(Although foreknowledge commonly requires more than raw data or observation – analysis is needed too – these verses underscore the crucial need for accurate, timely empirical data relevant to the situation at hand.[100])

XIII.3. Now the reason the enlightened prince and the wise general conquer the enemy whenever they move and their achievements surpass those of ordinary men is foreknowledge.

XIII.4. What is called "foreknowledge" cannot be elicited from spirits, nor from gods, nor by analogy with past events, nor from calculations.[101] *It must be obtained from men who know the enemy situation.* [Emphasis supplied.]

---

[99] See a passage from another Warring States text, the Zhanguo ce (translated under the title Chan-Kuo T'se by J. I. Crump, Jr.; Oxford, UK: Clarendon, Press, 1970, p. 330): "The sword Wugan when used on flesh could cleave an ox or a horse asunder and against metal it could split bronze vessels. But if you struck it flat-sided against a pillar it would break into three pieces; if you chopped with it against a stone it would shatter into a hundred pieces." For background on the Zhanguo ce see Early Chinese Texts (cited on p. xxii above), pp. 1–11; Wilkinson §58.6.3.2, p. 777.

[100] Verse XIII.4 is often treated as Sun Tzu's categorical rejection of all reliance on mantic practices as guides to political–military decision making. However, the situation is not quite so clearcut. For further discussion see p. A-48f. (where Passage #1.19 also appears).

[101] Ames, p. 169 translates Griffith's "calculations" as "astrological calculations" and Mair, p. 129 as "astrological verification." Those qualifiers should put modern audiences on notice that the concept of calculation appearing here should not be unthinkingly imbued with modern trains of thought.

Passage #1.20. Limits of Calculation (2): Warfare's Indeterminate Aspect

("These" refers to a Sun Tzu kitbag of tactical or strategic moves, most displaying some type of cunning, gathered in verses I.17–I.26. See Passage #5.1. In putting any of those stratagems into practice, minutiae matter, necessitating that every engagement be approached on its own terms as a unique, highly dynamic situation not navigable by general precepts alone, or by information restricted to knowledge possessed before events start to unfold.[102])

I.27.   These are the strategist's keys to victory. It is not possible to discuss them beforehand.

Passage #1.21. Limits of Calculation (3): Combinatorics Overwhelming Capacity to Calculate, by the Same Token Affording a Bottomless Quiver of Opportunities

(While not the only possible reading, a combinatorial complexity interpretation is natural here, envisioning the possibilities of which Sun Tzu speaks in combinatorial ways. One context that vividly illustrates a combinatorial profusion of strategic possibilities outstripping human information processing is the game of *weiqi*, some version of which was probably already extant in Sun Tzu's time.[103])

V.6.   Now the resources of those skilled in the use of extraordinary forces [qi] are as infinite as the heavens and earth; as inexhaustible as the flow of the great rivers.[104]

V.7.   For they end and recommence; cyclical, as are the movements of the sun and moon. They die away and are reborn; recurrent, as are the passing seasons.

V.8.   The musical notes are only five in number but their melodies are so numerous that one cannot hear them all.

V.9.   The primary colours are only five in number but their combinations are so infinite that one cannot visualize them all.

---

[102] See Griffith, p. 70 (related observation of traditional commentator Mei Yaochen). A modern perspective is Ames, p. 90, noting that "[c]omplex systems such as battle conditions are rich in information." For further relevant background see Lau & Ames, pp. 53ff. Compare John Lewis Gaddis, *The Landscape of History: how historians map the past* (Oxford, UK/New York: Oxford University Press, 2002), focusing on phenomena of chaos in war (chaos here being understood in its modern mathematical sense). A kindred practitioner thought comes from Rommel: "Because of the great variety of tactical possibilities that motorisation offers it will in the future be impossible to make more than a rough forecast of the course of a battle" (*The Rommel Papers*, Preliminaries footnote 6, p. 517). For an alternative (operations security) reading of verse I.27 see Passage #8.25A.

[103] For further discussion of a combinatorial interpretation of Passage #1.21, along with alternatives to it, see pp. A-50f. and Appendix 1 anchor footnote 29 there.

[104] Qi and zheng appearing in verses V.6, V.11, and V.12 are Chinese military concepts for which there are no truly adequate equivalents in English. See Theme #14 analysis. For now, Griffith's translation of them in verses V.6 and V.11 is left unchanged.

V.10. The flavours are only five in number but their blends are so various that one cannot taste them all.

V.11. In battle there are only the normal [zheng] and extraordinary [qi] forces, but their combinations are limitless; none can comprehend them all.

V.12. (Translation following Lau, 1965 article, p. 331.[105]) "[The qi and the zheng] produce each other endlessly like a ring and who is there that can exhaust the possibilities?"

Passage #1.22. General's Natural Sphere As Differing from That of Ruler

III.18. Now the general is the protector of the state. If this protection is all-embracing, the state will surely be strong; if defective, the state will certainly be weak.

III.19. Now there are three ways in which a ruler can bring misfortune upon his army.

III.20. When ignorant that the army should not advance, to order an advance or ignorant that it should not retire, to order a retirement. This is described as "hobbling the army".

III.21. When ignorant of military affairs, to participate in their administration. This causes the officers to be perplexed.[106]

III.22. When ignorant of command problems [quan] to share in the exercise of responsibilities. This engenders doubts in the minds of the officers.

III.23. If the army is confused and suspicious, neighbouring rulers will cause trouble. This is what is meant by the saying: "A confused army leads to another's victory."

Passage #1.23. Division of Labor between General and Ruler

XII.16. And therefore it is said that enlightened rulers deliberate upon the plans, and good generals execute them.

Passage #1.24. Widespread Underestimation of Successful Strategists

IV.8. To foresee a victory which the ordinary man can foresee is not the acme of skill;

IV.9. To triumph in battle and be universally acclaimed "Expert" is not the acme of skill, for to lift an autumn down requires no great strength; to distinguish between the sun and moon is no test of vision; to hear the thunderclap is no indication of acute hearing.

IV.10. Anciently those called skilled in war conquered an enemy easily conquered.

---

[105] Rationale for substituting Lau's translation for Griffith's is given in Appendix 14 footnote 3.

[106] Verse III.21 points to military management issues, where calculations are commonly done in non-combat environments very different from those of battlefield command decision. See Eccles, Background footnote 62, pp. 132–33 on the basic distinction between command and management.

IV.11. And therefore the victories won by a master of war gain him neither reputation for wisdom nor merit for valour.

IV.12. For he wins his victories without erring. "Without erring" means that whatever he does insures his victory; he conquers an enemy already defeated.

THEME #2 TERRAIN CLASSIFICATION AND RULES

Sun Tzu's nine terrains:
Passage #2.1 (verses XI.1–XI.23) (nine types of terrain and associated prescriptive rules)

espionage networks (background for extending Sun Tzu's terrain rules in social network directions):
Passage #2.2 (verses XIII.3–XIII.6 + XIII.15–XIII.16) (espionage networks and their targets, with urban terrain aspect)

★★★★★★

The Sun Tzu text brims with terrain classifications, by Griffith verse count a little under 20 percent of total verses. In engaging with Sun Tzu that level of attention should not be overlooked; it is a key that unlocks many of Sun Tzu's ideas on other topics, deception among them (Theme #7). A basic military reason for Sun Tzu's terrain emphases is not hard to discern. Sun Tzu's text is primarily about land warfare, and its terrain foci reflect a land warfare soldier's natural preoccupations (far more than those of a sailor or, in modern times, an airman, albeit that they too must confront versions of "terrain"). In addition, the warfare of Sun Tzu's era was evolving away from chariot warfare towards mass-infantry armies. As E. Bruce Brooks has put it, the "new issue for a non-chariot army will have been fighting on non-level ground."[1]

Concern with types of terrain shows Sun Tzu grappling with the concept and feasibility of doctrine built from an inventory of prescriptive terrain rules.[2] Despite some limitations, analyzed below, this strand of the Sun Tzu text adds to its intellectual heft, helping carve out a niche for Sun Tzu that is more concrete than the generalities of a philosophical tract on efficacy or gaining advantage, and more enduring than nuggets of advice tailored to the technical military conditions of Sun Tzu's time. Much of Sun Tzu's terrain classification content is tactical or operational rather than truly strategic in focus. Yet some of his terrain prescriptions plainly

---

[1] Brooks, p. 60. Brooks & Brooks, Introduction footnote 36, p. 37 place early manifestations of this military revolution, "not at the beginning of the Warring States … but earlier [around 510 BC]" (i.e., around the time of the putative historical Sun Tzu).

[2] In translating di, "ground" (Griffith, Minford) = "terrain" (Ames, Mair, Sawyer) = "territory" (Needham & Yates). "Terrain" works well for Sun Tzu (3) explorations and is also used here in referring to Sun Tzu's terrain rules in general. Following Griffith, specific terrain types are called "ground."

aspire to offer strategic advice. Moreover, a well-crafted system of tactical doctrine, not unlike a law code, may itself constitute a strategic resource.

Several overlapping terrain classifications appear in the Sun Tzu text, a pattern consistent with an evolving body of doctrine. In order to avoid becoming enmired in the over-many details of attempting to coordinate and reconcile all of them, the most developed of those classifications is chosen here as a flagship example (Passage #2.1).[3] A summary of that classification, along with associated action advice, appears in Table 1. A basic goal of Table 1 is to encourage modern audiences to grapple with Sun Tzu's nine terrains as a package, spotting both sources of coherence and possible loose ends (some of them logistical). A further goal is to establish a context for situating alternative interpretations of particular rules (say, following one commentator rather than another) within the context of the full set. Sun Tzu's other terrain classifications are gathered and analyzed in Appendix 2.

In approaching the terrain rules there is a fork in the interpretive road. One path would treat those rules simply as Sun Tzu's acknowledgment of a role for context – physical, psychological, political, etc. – in which military or strategic action unfolds.

That interpretation downplays the system-building aspects of Sun Tzu's terrain ideas. An alternative interpretation – not inconsistent with the first, but more ambitious – would point to a nascent system-building instinct.[4] More specifically, it would envision Sun Tzu's terrain thinking as an ancient foreshadowing of the modern rule-based expert system (RBES) concept, a computing idea rooted in "if . . . then . . ." reasoning with complex rule systems that has elicited keen interest in some modern military quarters. A Sun Tzu/RBES comparison is worth exploring, for both the similarities and the differences it reveals.

Starting on the RBES side, the motivation for developing a modern RBES application typically starts with recognition that a particular domain of human activity or knowledge is governed by many rules – commonly too many for any non-expert human to keep in mind and work with fluently.[5] Oftentimes, a further stumbling block to their use is that those rules overlap and interlock in complicated ways. Conjoined with external inputs (e.g., observer data fed to the system), inferences from one or more rules commonly activate further rules, spawning possibly lengthy logic chains. Mobilizing all the rules, as needed – working through their implications rapidly yet surefootedly – is an approximate description of how human experts operate, utilizing their knowledge base to produce action recommendations.

A talented human expert can put on a virtuoso performance here. Yet modern computing, which can readily navigate a truly vast number of rules, beckons as a marvelous alternative path to this same goal, automating the process and even coping with domains complex enough so that no true human experts exist. Furthermore, once such a computer-based RBES capability is extant it can be distributed far and wide, surmounting widespread (and often profoundly frustrating) limitations on number, quality, availability, and trustworthiness of human experts. Also importantly, that capability can always be upgraded by adding new rules or refining old ones (or by

---

[3] Brooks, p. 60 characterizes Sun Tzu XI's terrain classification as Sun Tzu's most "advanced."

[4] A system-building instinct is certainly not unknown in other early Chinese military texts. For an example see McNeal, Background footnote 58, pp. 120–21.

[5] For RBES as a modern computing area see Nils J. Nilsson, *The Quest for Artificial Intelligence: a history of ideas and achievements* (Cambridge, UK/New York: Cambridge University Press, 2010).

Table 1 *Sun Tzu's rules for nine terrain types*

Apart from present placement of "serious ground" – which highlights (1)–(3) as a natural series, tracing an invader's path – order of types (1)–(9) conforms to Sun Tzu's own verse XI.1 list (Passage #2.1). The grouping into three sets of three types each is a modern analyst's move, not found in Sun Tzu.

Positing that these terrain rules grew out of evolution of early Chinese warfare toward operations on complex terrain unsuitable for chariots, "communicating ground" has a special status as a type of non-complex terrain (in effect, a contrast case). Unobstructed level ground would be a prototype. Cao Cao (Minford, p. 269) indicates that a road network may also give rise to this type of ground.

For all nine types the second set of Sun Tzu XI prescriptions (verses XI.15–XI.23) is listed in second place here – i.e., as (1.2), (2.2), etc. With minor changes, Griffith's translation is used.

Type of terrain	Associated prescriptive rules
Types of terrain defined by depth of the army's incursion into hostile territory:	
(1) dispersive ground (no incursion, i.e., operations in one's own territory)	(1.1) do not fight here (1.2) unify the determination of the army
(2) frontier ground (shallow incursion)	(2.1) do not stop in the frontier borderlands (2.2) keep one's forces closely linked (note a)
(3) serious ground (deep incursion);	(3.1) plunder (3.2) ensure a continuous flow of provisions
Types of terrain having contingent aspects of both danger and opportunity:	
(4) key ground	(4.1) do not attack an enemy who occupies key ground (4.2) hasten up one's army's rear elements (note b)
(5) communicating ground (Sawyer: "traversable"; Giles: "open")	(5.1) do not allow your formations to become separated (alternatively: "do not try to block the enemy's way") (5.2) pay strict attention to one's defenses
(6) focal ground	(6.1) ally with neighboring states (6.2) strengthen one's alliances
Types of adverse terrain (Sun Tzu's serious ground category [3] could also fit here):	
(7) difficult ground	(7.1) press on (7.2) press on over the roads
(8) encircled ground	(8.1) devise stratagems (8.2) block the points of access and egress
(9) death ground	(9.1) fight! (9.2) activate the courage of desperation in one's troops

Note a. Mair, p. 122 translates: "I shall cause my troops to maintain a command structure," which could point to physical concentration of troops or to connectedness maintained through signals. It is also possible that Sun Tzu is here calling for steps to control impetuous or disloyal subordinates.

Note b. Alternatively, see Mair, p. 122: "I shall rush to the rear." Following traditional commentator Chen Hao (see Giles, p. 136), the idea may be cutting the enemy's supply line or exploiting his rear area vulnerabilities.

deleting problematic rules), more smoothly (and frequently with less resistance!) than the human variety of expert can update or extend that expert's knowledge base. Therein lies an attractive feature of RBES since much expert knowledge is inherently dynamic (e.g., as laws and administrative rules change, scientific and engineering frontiers advance, and previously unimportant types of knowhow become crucial).

Shifting focus to Sun Tzu and taking Passage #2.1 as a reference point (see again Table 1), there is a certain family resemblance between its assemblage of terrain rules and the rules comprising the knowledge base of an RBES. On the most basic level, the fundamental goal of systematically mobilizing knowledge to inform action is palpably similar. So too is a more specific agenda of framing that problem as one of working with a specific domain – in Sun Tzu's case defined by the concept of di (terrain or ground) – and an associated universe of prescriptive rules. Likewise shared is a willingness to probe beyond the conventional boundaries of that domain. Here Sun Tzu's willingness to enlarge his rule set beyond topographic categories – e.g., introducing a concept of "focal ground" rooted in political geography – also finds parallels in imaginative, often heuristic, rules found in certain RBES applications that, conventionally viewed, get "over into what might appear to some to be a different field."[6] From a modern standpoint, Sun Tzu's intellectual flexibility in this regard is one of the most appealing features of his approach to terrain issues.[7]

A notable point of Sun Tzu/RBES similarity (and a way in which both differ from many formulations of military doctrine) is an aspiration to present the rules in a standardized way, yet one that also smoothly accommodates interlocking rules.[8]

To that end, Sun Tzu's Passage #2.1 advice comes packaged not as one-liners but as two-liners, comprising, first, a description of a particular type of terrain; second, a prescription (in fact two prescriptions) for dealing with it. Sun Tzu's rules are actually *more* uniform in format than those found in most modern RBES, since each Passage #2.1 rule may be expressed (coded) as involving a single antecedent condition (specifying a particular kind of terrain the army finds itself in) plus an action

---

[6] Words quoted are from 17 October 1949 letter to Henry Eccles from Rear Admiral Worrall R. Carter, USN (Ret.), commenting on the scope of the first (1949) edition of Eccles's *Operational Naval Logistics* (Washington, DC: NAVPERS 10869, April 1950). See Box 27 folder 1 in the Eccles Papers in the Naval War College Naval Historical Collection, US Naval War College, Newport, RI.

[7] Lionel Giles (1875–1958) – one of Sun Tzu's earliest (1910) English-language translators, writing in the era of the treaty ports – showed a remarkably tin ear for this facet of Sun Tzu. Taking Sun Tzu to task with language redolent of a bygone time, Giles, p. 100 remarks of the terrain classification in Sun Tzu X (Passage #2.4A): "It is hardly necessary to point out the faultiness of this classification. A strange lack of logical perception is shown in the Chinaman's unquestioning acceptance of glaring cross-divisions such as the above." In point of fact, some of the most compelling applications of modern RBES draw strength from a knowledge base that incorporates rules of mongrel provenance, whose coordinated application to the problem at hand is precisely what delivers much value-added!

[8] While in Sun Tzu's case the rule "interlocks" are simpler than those commonly encountered in modern RBES, they are still a basic feature of his framework. Passage #2.1's terrains come equipped with two action recommendations, so that two different Sun Tzu terrain rules interlock through a common antecedent. In the case of verse XI.8, several distinct types of "difficult ground" (mountains, forests, ravines, marshlands, etc.) are itemized as giving rise to that type of ground. One possible coding would be as several rules, interlocking through sharing a common consequent.

recommendation triggered when that condition is met.[9] But that regularity should not be regarded as undermining a Sun Tzu/RBES parallelism. Pioneering formalisms in many mathematics or computing areas often impose stringent regularity conditions that more developed formalisms later relax or do away with.

At the same time, there are also basic Sun Tzu/RBES differences. A core motivation for RBES in computing is the sheer number of rules that must be navigated, swamping human cognition. That is precisely the rationale for RBES reliance on inference engines, algorithms that thread back and forth among the many rules, generating conclusions from logic chains that lumbering humans cannot keep pace with.

Yet Passage #2.1 has a mere nine terrain types, and one modern accounting points to only twenty or a few more distinct terrain configurations mentioned in Sun Tzu.[10] That small tally suggests that the motivation for Sun Tzu's terrain rules is rather different than that for a modern RBES. On this key question of "why the terrain rules?" Sun Tzu is silent. One response to it would envision those rules as prompts to generals of Sun Tzu's era to avoid certain errors of judgment to which many of them were prone (possibly in part for emotional reasons of the sort Sun Tzu plays up in Passages #1.8–#1.10). Frontier ground or encircled ground are examples – the first because generals might be tempted for many reasons to dally near friendly territory; the second because it is a "bad scenario" that only the most intrepid would be likely to approach in a creative frame of mind. A focus on avoiding simple, if costly, judgment errors helps explain Sun Tzu's lack of interest in developing the terrain rules in directions calling for more complex logic chains and, along with them, the inference engines that are the workhorses of modern RBES.[11]

True to his predilection for exaggerated certitude (see pp. 64–65 above), Sun Tzu notably makes no attempt to allow for effects of uncertainty in the statement or application of his terrain rules. From a modern RBES standpoint this is an analytical weakness, albeit some modern pioneers of RBES likewise tended to pass over the need for RBES to surmount thorny difficulties of judgment under uncertainty (and thus in a backhanded way stood in Sun Tzu's corner!). However, such a lacuna in Sun Tzu

---

[9] To avoid potential confusion it should be noted that some of Passage #2.1's verses bundle more than one prescriptive rule in a single verse (a flaw, but not a serious one, in Griffith's versification).

[10] Sawyer, p. 141. Considerably closer to the familiar complexity-of-rules rationale for a modern RBES is the type of list-making exercise found in Mohist work on city defense. See Yates, Introduction footnote 28, pp. 244–45:

> As the number of details is positively daunting, the specialist reader who is to lead the defence and his underlings are no longer expected to remember all the possibilities by sheer feat of memory. Rather, using the lists in the texts, they are required to choose which of the items presented seem to fit the geographical location of the town or city, the size of the population and the level of its technical expertise, and the amount of material already available and ready for the defence.

[11] Logic chains more complex than those found in Sun Tzu's terrain rules were certainly not foreign to early Chinese thought. A case in point comes from commentary on verse XI.52 (Passage #4.17A) by traditional commentator Mei Yaochen, putting forward a four-step logic chain. See Giles, p. 141.

For a further example from the *Zhuangzi* text, see translation by Watson, #1 footnote 59, pp. 311–12 (thought experiment involving ruler of Han in a conflict with Wei, two of the principal warring states).

should not be surprising if the raison d'etre of the terrain rules, as just suggested, is to galvanize a general to face up to unpalatable or unnoticed realities. Doing so effectively might indeed call for advice to be expressed with unwavering certitude. At the same time, the one-step nature of the reasoning provides a type of transparency, shielding the user of Sun Tzu's advice from consequences of failure to engage with uncertainty that might arise were the logic chains longer.

There is a further facet of Sun Tzu's terrain rules that an error avoidance perspective illuminates. If error avoidance is indeed a basic raison d'etre for those rules, it would be by no means unnatural for lists of such rules to grow, side by side with accumulating stories of generals who had made terrain blunders (as mass-infantry warfare became the new norm). Synergy of rules might then take hold: a reluctant general might be more amenable to paying heed to a given rule in Sun Tzu's rule set because *other* rules there had proven their worth in the past. In short, different rules may come to reinforce one another's authority in the mind of command.

By modern yardsticks, certainly computing ones, Sun Tzu's terrain classifications are very simple stuff. Where they are less simple is the nature of entities – really profiles of overall military situations – being analyzed. Whatever its motivation, Sun Tzu's approach to terrain faces many of the same basic difficulties arising in any body of prescriptive rules intended to guide high-level decision making. The foci of Sun Tzu's terrain advice, reflecting not merely topographic but also psychology-of-an-army and even diplomatic considerations, suggests advice pitched to a high command level, starting with battlefield command and in some cases ranging up to the ruler's level (e.g., Passage #2.1's verses XI.13 and XI.19, pertaining to alliances). Those are much loftier levels than those on which most modern military applications of RBES have taken root (generally involving relatively low-level, albeit essential, logistics, engineering, administrative, etc. tasks). Relatedly, informed opinion in modern military thinking allows a significant role for prescriptive rules in tactics, logistics, and communications; a narrower role at the operational level of war; and little if any role in strategy and certainly in grand strategy.[12] The complexity of strategic situations usually renders them unique in an almost clinical sense, with each situation requiring its own customized analysis that resists being shoehorned into a coarse-grained classificatory scheme like "rules for terrain types." Strategic analysis certainly calls for calculation (Passage #1.1 is an example), but the pertinent type of calculation, some quantitative, some qualitative, does not reduce at all well to application of "if . . . then . . ." rules like those in Passage #2.1 and Table 1.

Viewed from that standpoint, Sun Tzu's terrain rules exhibit imagination, hubris, insight, and naivety in roughly equal measure.[13] Perhaps more importantly than any specific prescriptive content, the terrain rules contribute to advancing military

---

[12] See Eccles, Background footnote 62, p. 71. A related perspective, pointing to a niche for doctrine sandwiched between technology and politics, is Hew Strachan, *The Direction of War: contemporary strategy in historical perspective* (Cambridge, UK: Cambridge University Press, 2013), pp. 221–22.

[13] Judea Pearl, *Heuristics: intelligent search strategies for computer problem solving* (Reading, MA: Addison-Wesley, 1984), p. 115 notes that "Heuristics are discovered by consulting simplified models of the problem domain" (e.g., models that remove constraints present in the actual situation). Interpreting "model" broadly, here may be a useful perspective on the genesis of Sun Tzu's terrain rules.

knowledge through their encouragement of searching reflection on the nature and classification of military situations. To put the point another way, the process of developing those rules may at times be more valuable than the rules themselves.[14]

Although Passage #2.2 is from a part of Sun Tzu (the espionage chapter) not usually associated with Sun Tzu's terrain thinking, it represents the Sun Tzu text's closest approach to exploring social network "terrain" and is included for that reason.

Roads Not Taken by Sun Tzu in Developing Theme #2

Sun Tzu's substantial interest in terrain might plausibly have led him to discuss military operations involving modification of terrain itself – as in diverting a river or (on a more modest scale) felling timbers to trap an enemy army.[15] Less visible expedients might include steps to contaminate water supplies.[16]

One might think that such terrain-modifying operations would have interested Sun Tzu, perhaps even prominently so. The twenty-first-century Chinese artificial-island-building initiatives in the South China Sea give modern incentive to try to spot relevant strands of Sun Tzu's thought. Yet apart from a lone, rather ambiguous verse XII.14 (Passage #9.7) on using water to isolate or possibly fragment an enemy force Sun Tzu makes no allusion to what could be conceived as terrain modification.

Also undeveloped by Sun Tzu, yet clearly relevant to at least some of his terrain types, is the basic issue of how a commander should cope with the challenge of correctly identifying what type of terrain his army (or the enemy army) is in, or is about to enter. In some cases this would be obvious – but in others not (and for a variety of reasons).[17] Native guides could be a resource here, but might lack necessary specialist expertise. They might also be untrustworthy.[18]

---

[14] Cf. Paul Bracken's observation that, by contrast to widespread tendencies in military analysis to "model complex and think simple" (in a manner of speaking), net assessment thinking at its best does just the opposite: "to model simple and think complex. The spirit is one of using relatively simple models, numbers, and trends, and to think long and hard about what they mean." See Bracken's article cited in Part A footnote 2 (p. 54 above), p. 100 (emphasis in original).

[15] A version of the latter possibility, sourced by the Zuozhuan (see p. xxiii above) to a 518 BC time close to that of the putative historical Sun Tzu, is discussed by Mark Elvin, The Retreat of the Elephants: an environmental history of China (New Haven, CT/London: Yale University Press, 2004), p. 45. For examples of manmade flooding for military purposes, from assorted periods of Chinese history, see Ralph D. Sawyer (with the collaboration of Mei-chün Lee Sawyer), Fire and Water: the art of incendiary and aquatic warfare in China (Boulder, CO: Westview Press, 2004), chapter 9, "Aquatic attack," pp. 275–315. See also Wilkinson, §13.6.1, p. 209 (1128 AD opening of dikes to block Jurchen cavalry, leading to change of course of the Yellow River, one of the world's major river systems).

[16] See Giles, pp. 82–83, noting pertinent Sun Tzu commentator opinion on verse IX.6 (Passage #9.6). See also Needham & Yates, pp. 269–70 (examples from Chinese military history of water poisoning or deliberate water pollution).

[17] A fine example from the Malayan campaign early in World War II involved ill-considered British focus on defending roads, and roads alone, treating nearby mangrove swamps and water-covered rice fields as impassable to their Japanese adversary (in effect, as a case Sun Tzu's "difficult ground" that would compel the Japanese to stick to the roads). That British assumption proved very faulty. See Cecil Brown, Suez to Singapore (New York: Random House, 1942), pp. 421, 494.

[18] Oddly, Sun Tzu – who twice refers to use of native guides (see Passage #2.8A's verse XI.51; also verse VII.11 in Passage #2.9A) – never raises the issue of their trustworthiness. For an

Likewise important, but not provided by Sun Tzu, would be terrain guidance tailored to hybrid situations, involving an army that finds itself in *two* of Sun Tzu's kinds of ground at once. For example, parts of Sun Tzu XI depict scenarios that could involve both death ground and serious ground (the latter referring to an army deep in enemy territory with hostile strong points in its rear; see verse XI.7 in Passage #2.1).[19] Sun Tzu's death ground prescriptions have a psychological thrust. By contrast, Sun Tzu's serious ground prescriptions center on logistics issues (for a summary of both sets of prescriptions see Table 1 above). If coordinated, those different emphases might suggest interesting insights into areas where psychology meets logistics. But Sun Tzu does not explore this promising direction.

Furthermore, and significantly, nothing in the Sun Tzu text suggests that Sun Tzu himself contemplated any far-reaching analogical extension of the idea of terrain (say, a "human terrain" concept). Such extrapolation must await Sun Tzu (3) work.

Sun Tzu (2) ("Sun Tzu Extended") and Sun Tzu (3) ("Sun Tzu Analogical") Frontiers of Theme #2

On the stage of world history, armies come and go but the physical realities of terrain remain much the same, as (at least until the contemporary digital age, and even then to a sizable extent) do broader constraints of space, place, and distance. Many of Sun Tzu's terrain conundrums would have been thoroughly familiar to Robert E. Lee and the Army of Northern Virginia – among them, verse XI.11 (Passage #2.1) advice, if interpreted as pointing to endemic headaches of desertion or of soldiers going AWOL; logistics woes; and using roads to advantage in crossing difficult terrain (advice that by the 1860s would extend to railroads).[20]

At the same time, Sun Tzu's terrain rules, as standardly interpreted, plainly fit some land warfare contexts better than others. An example is Sun Tzu's "dispersive ground" advice against fighting in home territory. Perspicacious that advice may have been for early Chinese conscript armies whose members were mostly farmers in civilian life, continually feeling the tug of the need to attend to their land and who

---

example of some of the perils of relying on guides, from Rome's struggle with Carthage, see Passage #12.33A.

[19] See also p. 322 below, raising this possibility in connection with analysis of Passage #9.14.

[20] Thomas B. Buell, *The Warrior Generals: combat leadership in the Civil War* (New York: Three Rivers Press, 1997), p. 255 touches on both the need for continuous flow of provisions (see Passage #2.1's verse XI.20, whose focus is deep incursions into enemy territory) and speedy movement when traversing difficult ground (verse XI.21):

> When we consider these [supply] circumstances, we begin to understand why railroads dominated military strategy in the [American] west. Armies follows rail lines as surely as water follows a riverbed. Both the Federal and Confederate armies depended upon them, the Federal armies perhaps to a somewhat greater extent. A corollary to railroads was the need to accumulate supplies at depots on the rail lines, primarily at cities and towns. Every prudent army commander predicated his operations on having one or more depots in his rear and a rail line connecting his army to them. Thus we may state a military theorem of the war in the west: wherever a major battle was fought, there was either a railroad, or a river that served the same purpose, or both.

often had scant if any stake in the rulers' wars of the day.[21] It seems less applicable to certain other kinds of armies – among them, armies whose ranks are filled with professional soldiers bred to follow the flag; or troops fired by some great cause (religion, ideology, nationalism, etc.). Crafting generalized versions of Sun Tzu's terrain rules therefore requires judgment. Table 2 presents a bird's-eye view of suggested enduring military content of these rules, emphasizing land warfare.

From a land warfare standpoint, some of Sun Tzu's most interesting terrain thinking pertains to ravines (xian), a concept representing one type of Sun Tzu's difficult ground that appears in numerous places in the text, in both concrete topographic and more abstract senses.[22] A hallmark of ravines is their dual aspect. On the one hand, over much of the history of land warfare, ravines are a tactical resource. The narrowness of some ravines can provide a defense against an attacker who is trying to attack up a ravine.[23] In modern warfare ravines can also serve as corridors for radio waves to avoid radio intercepts.[24] They are also a natural hiding place for troops to be used in an ambush (especially when, as in Sun Tzu's time, reconnaissance capabilities are limited). On the other hand, if a military force is inside a ravine and the enemy is atop it on one or both sides, then that force is at a natural disadvantage that could place it in an encircled or even a death ground situation. Yet Sun Tzu's thinking encourages not terminating the logic chain prematurely. In particular (and importantly depending on the gradient of the sides) trapping an enemy army in a ravine may trigger a ferocious death ground response, turning the tables on the apparent victor. Relatedly (drawing an idea from a different part of Sun Tzu's terrain thinking), the top of a ravine may also be a case of Sun Tzu's "entrapping" ground in the sense of verse X.3 (Passage #2.4A). Scrambling down the sides of that ravine to engage a cornered foe at close quarters (and perhaps also aiming to capture valuable supplies or equipment) may be easy – but scrambling back up, should that attack fail, may be difficult, if feasible at all, when fighting at close quarters is underway. Defeat may then be snatched from the jaws of victory!

---

[21] Wilkinson, §24.11, p. 354 profiles Warring States armies as "huge well-drilled armies based on peasant conscription." Brooks & Brooks, Introduction footnote 36, p. 93 describe the "new soldiers" of fifth-century BC China as commoners who "did not have the elite warriors' lifelong training and ingrained dedication. They had to be forced to serve, and they did not like it much." They then quote an ancient poem, possibly based on an early marching song, whose plaintive tone still carries force even after all the ages that have passed (and the unavoidable further distance imposed by any translation). That poem reads in part:

> Worn down, alack; worn down, alack,
> Why do we not go back?
> Were it not to serve our sire,
> What would we be doing in the mire?

[22] For a roundup of occurrences of xian in Sun Tzu see Appendix 2 anchor footnote 31 (also noting that an abstract counterpart to ravines arises in, and is a source of difficulty for, machine learning optimization problems). The modern Sun Tzu literature commonly gives ravines scant attention (an exception, albeit brief, is Sawyer, p. 324 note 152).

[23] Cf. Sun Tzu's verse X.5 (Passage #2.4A).

[24] A possibility not lost on Chechen separatists fighting the Russian army. See p. 743 of Timothy L. Thomas, "Russian tactical lessons learned from fighting Chechen separatists," *Journal of Slavic Military Studies*, 2005, 18(4), 731–66.

Table 2 *General military principles building on Sun Tzu's rules for nine terrains*

Sun Tzu's nine terrains rules are presented here in generalized form. Numbering (1)–(9) conforms to order of these terrains in Table 1 (p. 89 above). Sun Tzu terrain names are shown in parentheses.

To recognize and avoid or surmount:	To promote and exploit:
(1) social forces eroding army's cohesion ("dispersive ground") (note a)	forces building army's sense of unity and fighting spirit
(2) passivity or indecisiveness when encroaching on enemy territory ("frontier ground") (note b)	taut command control of one's own forces
(3)	sustained logistics support and logistics flexibility during deep incursions into enemy territory ("serious ground")
(4) pig-headed attacks on natural strong points ("key ground")	"get there first with the most men" (note c); attack on enemy's rear (e.g., supply line)
(5) dispositions risking defeat in detail ("communicating ground") (note d)	connectivity and mutual support of own forces
(6)	alliance-building steps ("focal ground") (note e)
(7)	speedy movement ("difficult ground") (note f)
(8) command and control breakdowns ("encircled ground") (note g)	resourcefulness in adverse situations (note h)
(9)	supranormal performance elicited when soldiers face desperate circumstances ("death ground") (note i)

Note a. Such social forces often grow out of ethnic, regional, linguistic, or other social cleavages. Some could be of types foreign to Sun Tzu. See Karl W. Deutsch, *Nationalism and Social Communication: an inquiry into the foundations of nationalism* (2nd ed.; Cambridge, MA: MIT Press, 1966).

Note b. Sun Tzu's frontier ground warning calls to mind the French Army hunkered down in its Maginot Line fortifications during the so-called phoney war that preceded the fall of France in 1940. That period's ambiance of faux equilibrium was epitomized by French reluctance to use their advanced fire control systems to shoot (save in their imaginations!) at German troops and positions plainly visible a few miles away. See Harold Nicolson, *Diaries and Letters* (Nigel Nicolson ed.; New York: Atheneum), Vol. II: *The War Years 1939–1945* (1967), pp. 40–43.

Note c. Observation attributed to Nathan Bedford Forrest. See Colonel Robert Debs Heinl, Jr., US Marine Corps (Retired), *Dictionary of Military and Naval Quotations* (Annapolis, MD: United States Naval Institute, 1966), p. 63 (correcting a common misquote). Compare Cao Cao's "Do not attack, but concentrate on being the first to occupy the advantage" (Minford, p. 275). An alternative interpretation of Sun Tzu's key ground advice (see Table 1 note b) would be as counseling attack on the enemy's

Table 2 (cont.)

rear. A Sun-Tzu-esque expression of the underlying idea might be exploiting enemy logistics weakness to counter enemy tactical strength. Cf. p. 113 (xu/shi principle).

Note d. Sun Tzu pays considerable attention to the idea of cutting off or being cut off. See #2 footnote 69. A basic focus was slow-moving infantry forces, where cut-off troops could prove easy prey. Generalizing the point needs to allow for varying conditions. Potent long-range modern weapons make signals connectivity widely preferred to physical concentration, which may be frankly unwise.

Note e. Sun Tzu's focal ground as a place of strategically vital intersections (Ames, p. 153) invites creative extensions to modern warfare. One candidate for such intersections might be the mid-section of the "logistics bridge" connecting the economy of a nation with combat operations of its armed forces (Eccles, Background footnote 52, pp. 53–57). At that midpoint, where civilian and military jurisdictions overlap and mindsets meet, there is a natural place for organizational alliances to harmonize often conflicting civilian and military criteria of judgment and associated organizational cultures. For further logistics bridge analysis see Passage #12.14 (pp. 417–19) and Figure 6 (p. 435).

Further afield from warfighting, but crucial for military innovation, a further candidate for focal ground would be R&D activities situated at an intersect of multiple strategic technologies, creating need for alliances among diverse technology pioneers, entrepreneurs, venture capitalists, and others.

Note f. Sun Tzu's speedy movement focus would nowadays extend to fast algorithms (Theme #3, in particular pp. 123–24).

Note g. Cf. verse XI.22's "block the points of access and egress" – so that enemy infiltrators or spies cannot enter nor can one's own troops drift away or desert. For related modern analysis see "Operations of Encircled Forces: German Experiences in Russia," Department of the Army Pamphlet No. 20-234, 14 January 1952 (published January 25, 2011), prepared by a committee of former German officers under the supervision of Historical Division, EUCOM (U.S European Command):

> Experience has shown that only seasoned troops, in the best fighting condition and under the firm control of their commanders, are able to withstand the mental strain of combat in encirclement and are likely to retain the high degree of physical fitness needed … But even with troops that satisfy these requirements it is necessary to apply stern measures in order to prevent any slackening of control, which would inevitably result in lowering their morale. It is surprising how fast the bonds of discipline will disintegrate in an encirclement. Mobs of unarmed soldiers trying to proceed on their own, captured horses loaded down with superfluous equipment, and other similarly depressing sights were not uncommon in some of the larger German pockets in Russia. They had a contaminating effect and called for swift and drastic countermeasures. … More than ever the place of the commander, under such circumstances, is in the midst of his troops. Their minds will register his every action with the sensitivity of a seismograph.

Note h. On a level of small unit tactics a comment on the US Marines at Chosin Reservoir during the Korean War captures some of the spirit of resourcefulness for which verse XI.14 advocates:

> Both the enlisted men and their officers had stored up hundreds of small combat ruses and ploys in World War II: When the enemy makes a noisy probing attack, he is probably trying to locate your machine guns, so respond only with grenades and rifle fire if possible [etc.].

See Thomas E. Ricks, The Generals: American military command from World War II to today (New York: Penguin Press, 2012), p. 159.

Table 2 (cont.)

In the digital age, the depth of training and doctrine development made possible by digital technologies creates possibilities for giving far fuller play to Sun Tzu's verse XI.14 counsel than ever before – e.g., by developing catalogs of different kinds of specific "encircled ground" situations and crafting customized doctrine for each, prepared beforehand to be accessed digitally as needed.

Note i. As David Graff has noted, there is "nothing uniquely Chinese about this idea." See p. 58 of David A. Graff, "Brain over brawn: shared beliefs and presumptions in Chinese and Western strategemata," Extrême-Orient Extrême-Occident, 2014, No. 38, 47–64. See also p. 342 (quoting Napoleon).

Yet as the annals of world military thought also corroborate, there are deep splits of opinion regarding the soundness of death ground thinking. For example, Graff, p. 59 notes "Byzantine aversion to 'deadly ground' stratagems," attributing it in part to their limited reserves of trained manpower. A case of successful exploitation of a death ground situation was a battle of Gaius Flaminius, a Roman commander (died 217 BC) who placed his army in death ground and went on to win a victory. But his judgment was roundly criticized by Polybius after the fact. See M. J. V. Bell, "Tactical reform in the Roman Republican Army," Historia: Zeitschrift für Alte Geschichte, 1965, 14(4), 404–22.

Sun Tzu's death ground concept also finds echoes in grand strategy, including the Bolshevik concept of a foe becoming "more secretive, devious and vicious the closer it came to defeat" (a theory for which Stalin advocated). See p. 98 footnote 20 of Iain Lauchlan, "Young Felix Dzerzhinsky and the origins of Stalinism," in Discussing Stalinism: problems and approaches (Markku Kangaspuro and Vesa Oittinen, eds.; Helsinki: Aleksanteri Institute, University of Helsinki, 2015), pp. 93–113.

One modern place to look for some of the characteristics of Sun Tzu's ravines is urban combat taking place in modern cities with their often intricate street patterns and many multi-story built structures, frequently lofty. This is a type of warfare of substantial twenty-first-century importance. Because capabilities for constructing such structures were limited in Sun Tzu's time, here is an example of a Sun Tzu (2) frontier that has no exact Sun Tzu (1) counterpart.[25]

Now turn from Sun Tzu (2) to Sun Tzu (3). Many ages of war have found fascination in some new or emerging class of developments (sometimes labeled "revolution in military affairs") seen as capable of drastically changing how wars are fought and won or lost. In Sun Tzu's time a leading case in point would have been the crossbow in the hands of ordinary infantry, making possible far more extensive operations on non-level terrain inaccessible to chariots. A place to look for a twenty-first-century counterpart would be technology supporting social networks and the developments made possible by it. Social networks are sometimes conceptualized as defining a "social space" generalizing physical space and making it more irregular and complex by inserting cross-cutting, space-transcending network ties.

---

[25] Describing the Warring States period as an "age when techniques for true multilayered structures had not yet been fully developed," see p. 672 of Wu Hung, "The art and architecture of the Warring States period," in Loewe & Shaughnessy (eds.), Background footnote 7, pp. 651–744.

Contemporary communications capabilities have opened the floodgates to arbitrary levels of network complexity. Such social network intricacies replicate on a new level the complex terrain issues that have intrigued and bedeviled land warfare commanders from time immemorial and that were a likely formative influence on the genesis of the Sun Tzu text.

For these reasons, social network "terrain" offers a natural venue for finding a twenty-first-century living legacy for Sun Tzu's terrain rules. Such a legacy has grand strategy aspects, for all the reasons that grand strategy decisions, including decisions to use military force, often grow out of social network dynamics unfolding in national capitals far removed from any battlefield. The following discussion briefly analyzes selected social network situations identified as sharing structural characteristics with (a) encircled ground, (b) difficult ground, (c) serious ground, and (d) death ground.[26] That exercise requires bridging two very different bodies of thought – Sun Tzu's military thinking and social network analysis. Parts of Sun Tzu's espionage chapter (Passage #2.2) help with this task, though only to a limited extent. Disciplined analogical thinking is therefore crucial here. The present goal is not systematic treatment but opening up an area for inquiry.[27]

Sun Tzu's encircled ground concept (verse XI.9) has affinities with small-world networks in the sense pioneered by Stanley Milgram and reinvigorated as a research area by Duncan Watts in the late 1990s.[28] Both concepts have a topological flavor. For Sun Tzu, "encircled ground" is ambush country.[29] A hallmark of small-world networks is that they harbor network ties that bring into close proximity seemingly remote network neighborhoods, thus setting a stage for sociological counterparts to ambush: leakages of secrets, flows of juicy gossip (true or false!), and backchannels of social influence and pressure of many kinds.[30] Everyday life in national capitals abounds in examples.

---

[26] For suggested social network counterparts to the remaining five of Sun Tzu's nine terrains see pp. A-8off. in Appendix 2.

[27] It is worth noting that the network situations proposed here as counterparts to Sun Tzu's terrain types all involve, to a greater or lesser extent, not only social networks, but also institutional and possibly other types of social structure. This points to a research frontier that might be labeled "networks plus," coordinating social network analysis with analyses of structures and dynamics of other kinds that may not reduce to applied graph theory, the standard métier of modern network analysts.

[28] For review of what has become a major literature, with pioneering contributions by Stanley Milgram (a social psychologist) and, separately, Ithiel de Sola Pool (a communications researcher) and Manfred Kochen (a mathematician), see Sebastian Schnettler, "A structured overview of 50 years of small-world research," *Social Networks*, 2009, 31(3), 165–78.

[29] As traditional commentator Du Mu observes bluntly (Griffith, p. 131), "Here it is easy to lay ambushes and one can be utterly defeated."

[30] Something of that ambush potential is captured by an observation of Duncan J. Watts, "Networks, dynamics, and the small-world phenomenon," *American Journal of Sociology*, 1999, 105(2), 493–527:
>What is new is the idea that the structural changes required can be very subtle. Unlike the difference between a chain and a star graph or a ring lattice and a random graph, the difference between a big- and a small-world graph can be a matter of only a few randomly rewired edges – *a change that is effectively undetectable at the level of individual vertices.*
>
>(p. 522, emphasis supplied)
>
>In Sun Tzu's categories, this point also calls to mind *wu xing* (formlessness) thinking (Theme #8).

Sun Tzu's thinking challenges existing small world analysis to delve more deeply into strategic and tactical possibilities created by sometimes unexpected "shortcuts" in social networks.[31] Some such shortcuts start out innocuously and indeed may serve a constructive purpose as informal liaison ties, facilitating coordination between complex organizations that cuts through layers of bureaucracy and its lengthy and cumbersome "official channels."[32] At the same time, shortcuts of that type are commonly few in number, and for that reason are akin to narrow conduits (a network counterpart to ravines!) evoking Sun Tzu's definition of encircled ground as "ground to which access is constricted, where the way out is tortuous" (verse XI.9). Reliance on such shortcuts could lead to real harm when even one shortcut tie "goes bad," as when a useful liaison channel mutates into a conduit for espionage.[33] For that reason, shortcut ties are natural targets for manipulation and need to be monitored with utmost care.

In addressing the perils of encircled ground, verse XI.22 (Passage #2.1), calling for blocking points of access and egress, translates in a network context to a call for vigilance backed by concrete measures geared to preventing the importing or exporting of undesirable effects of the kind familiar in some applications of small world analysis (e.g., information leakage to unfriendly quarters, or deceptive or underhanded overtures emanating from such quarters). In order to do so, however, network paths – commonly short if sometimes tortuous – creating opportunities for such undesirable effects must be spotted. Here Sun Tzu's verse XI.22 thinking calls to mind what Jon Kleinberg has dubbed the *algorithmic* small world problem, whose focus is on actually finding short network paths (as contrasted with merely knowing that they exist).[34] Read strategically, verse XI.22 points to a race, often having data collection as well as algorithmic aspects, to carry out that task before an enemy does.

Complementing such blockage steps, which are defensive in nature, a small world counterpart to verse XI.14 also points to a role for resourcefulness in devising

---

[31] For overview (though not addressing strategic aspects or implications) see Gilbert Strang, "Random shortcuts make it a small world indeed" (review of Duncan Watts, *Small Worlds: the dynamics of networks between order and randomness* [Princeton, NJ: Princeton University Press, 1999]), published in SIAM News, December 1999, 32(10).

[32] See Rear Admiral H. E. Eccles, USN, Ret'd, "Logistics, systems analysis and military management" (working paper prepared under sponsorship of The George Washington University Logistics Research Project, Contract Nonr761(05) Project NR 047 001, 20 October 1967), p. 32. See Box 47 Folder 11 in the Eccles Papers in Naval Historical Collection, US Naval War College, Newport, RI.

[33] Cf. Stephen Kotkin, *Stalin*, Vol. 1: *Paradoxes of Power, 1878–1928* (New York: Penguin Press, 2014), p. 703, noting what "might look like shop talk but did cross the line over to espionage."

[34] For pioneering analysis of algorithmic small world issues see Jon Kleinberg, "Navigation in a small world," *Nature*, 24 August 2000, 406, p. 845 (noting that "it is easier to find short chains between points in some networks than others"). Illustrating later work, see Massimo Franceschetti & Ronald Meester, "Navigation in small-world networks: a scale-free continuum model," *Journal of Applied Probability*, 2006, 43(4), 1173–80. It should be cautioned that this type of search problem depends on the information conditions assumed. This in turn raises issues that most network formal modeling literature does not engage with, starting with what should be regarded as a "network tie" (which may vary with the nature of the search). Network ties may also wax or wane over time (e.g., reflecting stories their audiences tell about them), and in response to being used or probed. Questions like these enter an area where social network modeling meets ethnography.

aggressive counter-measures. In a spirit Sun Tzu might have applauded, some of those counter-measures might turn small world predicaments to advantage. They might, for example, involve setting up a situation where a would-be intruder (spy, saboteur, etc.), possibly rather pleased with himself for exploiting a low-visibility network path, winds up gaining access to a pot of highly privileged information – which, however, is actually false information planted there by the defender for just that purpose. Or the game might be more fully turned around, now taking the initiative and mobilizing known small world channels as conduits for aggressively putting out false or misleading information, thereby taking cues from Sun Tzu's verse XIII.19, but now substituting an expendable network tie for Sun Tzu's expendable spy!

> XIII.19.
> And it is by this means that the expendable agent, armed with false information, can be sent to convey it to the enemy.
>
> (verse from the espionage chapter, see Passage #11.6)

Shifting focus to difficult ground in Sun Tzu's sense (verse XI.8), a quite different application of Sun Tzu's terrain thinking points to a class of issues at an intersect of social networks and career or life-course patterns.

A connection with Sun Tzu starts by taking note of a twentieth-century US politician's wise observation about politics ("All politics is local"), restating it as the proposition that "all network situations are local." As grand strategy careers unfold, but before they reach the highest national level, they often involve spells of participation – temporary but for extended time periods (possibly measured in years) – in "local" social networks rooted in remote or specialized parts of a far-flung organization, its allied organizations, and associated geographic communities. From the vantage of a transient participant, social networks found there commonly constitute "difficult ground" because of fights or lingering residues of past fights involving permanent denizens of those local networks. A transplanted outsider pursuing ambitious larger career goals while trying to operate in this type of social network terrain is often clueless as to the actual reasons for those fights or even their existence (e.g., because their toxic residue is unspoken, hence invisible).[35] Unseen vulnerabilities may then all too easily ensue, all the trickier because of a veneer of civility. Under those conditions, whose dynamics no outsider is likely to be able to decipher and master in the time available, the thrust of Sun Tzu's advice "in difficult ground, press on" commends itself as a way of minimizing accumulation of social and professional baggage that might risk derailing a promising career.[36] A practical

---

[35] The Chinese phrase Griffith translates as "difficult ground" is *pi di*, "defeated" or "subverted" ground (Sawyer, pp. 320–21 note 125). Some Sun Tzu commentators would narrow the concept to focus on ground that is muddy or swampy or easily inundated, always an infantryman's nightmare (see C. E. Wood, *Mud: a military history*, Washington, DC: Potomac Books, 2006). Mud imagery certainly captures something about the class of situations proposed here as social network analogs to *pi di*. For further discussion of *pi di* in a land warfare context see Appendix 2 footnote 48.

[36] A larger framing of this risk exposure issue comes from Karl Ulrich Mayer, "New directions in life course research," *Annual Review of Sociology*, 2009, 35, 413–33, noting (p. 424): "Exposure to risk, measured by its incidence and duration, can be a powerful concept in mapping and measuring life courses. It also has the virtue of tying life course research to public policy."

recommendation would be for the newcomer to keep relevant projects and initiatives moving ahead briskly, preferably in well-established, overt channels (a counterpart to verse XI.21's "over the roads"), holding to barest minimum any need to engage in depth with idiosyncrasies of provincial networks and their often elephantine local memories. Indeed there is strategic value in being clearly perceived by long-term denizens of a local network as being merely in a transient status there – in effect, as just passing through their environment rather than becoming a long-term stakeholder in it.

In order to arrive at their ultimate high-level decision-making positions, every rising generation of grand strategy practitioners must at some point in their budding careers navigate a tricky local network situation of the type just described (an experience commonly repeated more than once in a career). Because grand strategy careers are routinely intertwined through friendships or interpersonal alliances, marriage, patronage, etc., creating yet other layers of social network entanglements, success or failure on the part of one young practitioner in crossing a patch of difficult ground may have spillover effects on the career of a different practitioner. Grand strategy in practice is often inseparable from the personal trajectories of a small population of elite actors from whose ranks the actual players at any one moment are drawn. As a result, the network traps of the sort just described are of relevance to more than individuals alone. Their cumulative upshot may have an effect on the substance and quality of national-level grand strategy itself.

Shifting focus to Sun Tzu's serious ground (verse XI.7), the concept of fragile position in a social network structure commends itself as an analog to the transparently precarious "serious ground" situation of Sun Tzu's army embarked on a deep incursion into hostile territory, whose vulnerability there is plain for all to see.[37] An example of a fragile social position is suggested by the awkward situation of a successor to a now-deceased charismatic leader, often conjoining a deficit of perceived legitimacy or credibility with an inherited baggage of many social network ties, some friend, some foe, both kinds contributing to loss of flexibility.[38] Read from that vantage, Sun Tzu's "look to your logistics" advice for surmounting the challenges of operating in serious ground (verses XI.13, XI.20) merits scrutiny as guidance for how a fragile social network position might be bolstered. As Sun Tzu recognized for his military case of serious ground, a key to doing so lies in stabilizing the logistics. In a like spirit, a charismatic leader's ill-omened successor would be well advised to use his status to nail down control of some type of valuable logistics or economic resource – by means akin to Sun Tzu's plundering, if more genteel options are not available – before that successor's fragile social network

---

[37] The concept of fragile network position, foreshadowed in Max Weber's as well as Erving Goffman's writings, is analyzed in Matthew S. Bothner, Edward B. Smith, and Harrison C. White, "A model of robust positions in social networks," *American Journal of Sociology*, 2010, 116(3), 943–92. A key assumption of their model (pp. 982–83) is that the network is transparent enough to make all actors "aware of the means by which positions in [a prevailing status] ordering are determined – whether these positions are durably built through diversification or exist vulnerably because of concentration."

[38] See Max Weber, *Economy and Society: an outline of interpretive sociology* (Berkeley: University of California Press, 1978), pp. 248, 1135–41 (work cited in Bothner et al., previous footnote, p. 946).

position implodes and his authority is "gone with the wind."[39] In a formulation Sun Tzu might have appreciated, logistics strength then compensates for social network weakness.[40]

Sun Tzu's death ground presents a special challenge for social network thinking. While many civilian situations involve bitter, long-running conflicts, they often lack a counterpart to violent death and certainly none to large-scale military combat. If death ground is to be given a generalized meaning, a shift of analytical register is needed. A starting point for doing so comes from the fact that targeted attacks on network hubs – roughly, network nodes having outsized numbers of connections – can often break up a network in short order, making it fall apart into disconnected fragments.[41] Such an outcome offers a candidate for network "death." "Occupying death ground" would then describe a network situation where one or more major hubs are at risk of being irreversibly compromised, disabled, or destroyed.[42]

The central dynamic animating Sun Tzu's death ground thinking is the connection he discerns between truly desperate military situations and psychological responses they elicit, galvanizing soldiers placed there to put forth extraordinary efforts, even leading to triumph against all odds. Positing the interpretation of "network death" just offered, Sun Tzu's death ground principle suggests exploring mechanisms – psychological, social, or other – through which highly exposed, *apparently* vulnerable strategic actors in network hub positions may be able to rise to the occasion, snatching an implausible victory from the jaws of defeat (or oblivion). An analytical lead comes from work of Padgett & Ansell in their classic social network case study of Cosimo de' Medici, the great Florentine banker and politician (1389–1464). The mechanism they identify and analyze is termed "robust action and multivocality," and its crux is that "single actions [by the central network node or hub, in their case Cosimo de' Medici himself] can be interpreted coherently from multiple perspectives simultaneously, *the fact that single actions can [therefore] be moves in many games at once*" (emphasis supplied).[43] In their analysis, that style of control was key to how their

---

[39] A relevant structural perspective on the concept of "precarious position" in a social order is David Howarth, "Removing precariousness: a conceptual note," in Céline Lageot and Nathalie Martin-Papineau (eds.), *Approches franco-britanniques de la précarité* (Paris: Librairie générale de droit et de jurisprudence, 2016), pp. 165–73.

[40] *Xu/shi* (emptiness/fullness), one of Sun Tzu's basic concept pairs, fits well here. For more on *xu/shi*, which in the title of Sun Tzu VI Griffith, p. 96 renders as "weaknesses and strengths," see p. 113.

[41] A concept much discussed in network science, network hubs may be multiple. Their presence renders certain types of networks – commonly called scale-free – "highly sensitive to sabotage of a small fraction of the sites" (i.e., network nodes). See Reuven Cohen, Keren Erez, Daniel ben-Avraham, & Shlomo Havlin, "Breakdown of the Internet under intentional attack," *Physical Review Letters*, 2001, 86(16), 3682–85. This type of question has deep roots in the analysis of military command control systems. See classic work of Paul Baran, "On a distributed command and control system configuration" (US Air Force Project RAND Research Memorandum RM-2632, December 31, 1960).

[42] A different direction for generalization of Sun Tzu's death ground thinking, further afield from social network analysis as usually understood, is proposed and analyzed in Theme #10 "frontiers" analysis. See pp. 344–45 below.

[43] See p. 1263 of John F. Padgett & Christopher K. Ansell, "Robust action and the rise of the Medici, 1400–1434," *American Journal of Sociology*, 1993, 98(6), 1259–1319. A hub-and-spokes model of the structural position of the Medici is described by Padgett & Ansell, pp. 1278–79, 1285–86.

focal figure was able to "[ride] herd on vast macropolitical and macroeconomic forces far beyond his control," founding a dynasty that would dominate Florence for three centuries. To use military terminology, multivocality operates as a kind of force multiplier in a communications sphere. It is a network counterpart to the physical multiplier arising from extraordinary efforts put forth by troops in death ground. Even as the psychology involved in Sun Tzu's death ground may make feasible sophisticated military steps ranging well beyond simply "going berserk" or comparably primal responses,[44] the psychology of multivocality gives scope for applying consummate communications skills to achieve strategic ends.[45]

There may also be an affinity between robustness (non-fragility) of an actor's network position and the ability of that actor to engage successfully in robust action and multivocality.[46] Putting the point in Sun Tzu's own categories, success in surmounting perils of network "serious ground" may provide precisely the extra edge needed to rise to the unforgiving challenges of network "death ground." Here, however, a limitation on the analogy should be noted. Various of Sun Tzu's traditional commentators advocate for the wanton destruction of logistics resources ("Put our stores to the torch," etc.) as a means to "inflame the officers and incite the rank and file so they have no thought of living" and thereby catalyze the type of behavior needed to triumph in death ground.[47] A network analog would be deliberately wrecking one's own robust social network position the better to achieve robust action – a stratagem at odds with both network analysis and common sense.

An insightful observation of D. C. Lau finds application here. An analogy, he noted, may at times shed light through analyzing ways in which it breaks down.[48]

In that spirit, the issue just noted illuminates the difference between a robust action and multivocality "game" played over and over and a "death ground" game played but once (or only rarely). A second difference is felicitously expressed by Pogo: "We have met the enemy and he is us." In fact, in many – perhaps most – cases, a core audience

---

Regarding robust action and multivocality see also John F. Padgett & Walter W. Powell, with contributions by others, *The Emergence of Organizations and Markets* (Princeton, NJ/Oxford, UK: Princeton University Press, 2012), pp. 24–26. For further context see review of this book by Scott A. Boorman, *Acta Sociologica*, 2014, 57(4), 363–67.

[44] For a concrete example of a complex death ground response, involving an episode from the military career of Cao Cao (one of Sun Tzu's eleven traditional commentators) see #1 footnote 92 above. For further analysis see Passage #9.14.

[45] For a related analytical direction see Barry O'Neill, *Honor, Symbols, and War* (Ann Arbor: University of Michigan Press, 1999), pp. 43–44: "*a symbolic message can be more ambiguous than a conventional communication. This is often an advantage.*" (Emphasis in original, which also provides examples.) As literary critic William Empson might have put it: "When in social network death ground, fight ambiguously!" More broadly, the concept of multivocality finds a natural ally in Empson's seminal work *Seven Types of Ambiguity* (3rd ed., rvd.; Norfolk, CT: New Directions Books, 1953). Regarding Empson's China connection see Li Cao, "Cambridge critics and China: an introduction," *The Cambridge Quarterly*, 2012, 41(1), 4–25.

[46] See Bothner, et al., #2 footnote 37, pp. 984–85, in particular p. 985: "We expect robust action to be more likely among robustly positioned actors because of the capacity of their social positions to absorb, much like an insurance policy, the costs of accidents as they behave strategically."

[47] Ames, p. 221, translating Sun Tzu–related material in the *Tongdian* (compiled by Sun Tzu traditional commentator Du You). Minford, pp. 274–75 translates remarks to similar effect by He Yanxi, another traditional commentator.

[48] See D. C. Lau, *Mencius* (Harmondsworth, UK/Baltimore, MD: Penguin Books, 1970), p. 262.

for a multivocal action or communication is the strategic actor's *own* social network, and to wreck that network, or a major part of it, would be antithetical to the raison d'etre of seeking multivocality in the first place.[49]

✶✶✶✶✶✶

Set of principal passages used to illustrate Sun Tzu Theme #2 (terrain)

Passage #2.1. Nine Types of Terrain and Associated Prescriptive Rules

(Nine terrain types – some psychological or political as much as or more than topographic – are distinguished here. This material is the textual basis for the Table 1 overview of Sun Tzu's nine terrains rules on p. 89 above.[50]

As also indicated there, Passage #2.1's advice is double-barreled, stating *two* prescriptive rules for each terrain type, the first found in verses XI.11–XI.14, the second in verses XI.15–XI.23.[51])

XI.1. In respect to the employment of troops, ground may be classified as dispersive, frontier, key, communicating, focal, serious, difficult, encircled, and death.

XI.2. When a feudal lord fights in his own territory, he is in dispersive ground.[52]

XI.3. When he makes but a shallow penetration into enemy territory he is in frontier ground.[53]

XI.4. Ground equally advantageous for the enemy or me to occupy is key ground.[54]

---

[49] A third difference between Sun Tzu's death ground scenario and Cosimo de' Medici's situation is that the former, as Sun Tzu describes it, centers on supranormal combat performance of soldiers en masse; the latter, on exceptional strategic/communications performance of *one* key actor.

[50] **Note on Passage #2.1 analysis:** Passage #2.1's main text footnotes 52–60, nine in number (one for each terrain type), provide basic orientation to each of Sun Tzu's nine terrains. Footnotes 61–66 further elaborate various aspects. Further textual or interpretive details are assigned to Passage #2.1 footnotes in Appendix 2 (where, following this book's practice, Passage #2.1 again appears).

[51] For Griffith's editing of Sun Tzu XI see Appendix 2 discussion of Passage #2.1 (p. A-64).

[52] A common traditional explanation for the "dispersive ground" label is recited by Giles, p. 114:

> So called because the soldiers, being near to their homes and anxious to see their wives and children, are likely to seize the opportunity given by a battle and scatter in every direction. "In their advance," observes [traditional commentator] Du Mu, "'they will lack the valour of desperation, and when they retreat, they will find harbours of refuge."

A more colorful snapshot comes from Needham & Yates, p. 46, referring to this type of terrain as a "place where the army and the soldiers' minds easily disintegrate." Other interpretations can also be proposed. See Appendix 2 anchor footnote 4.

[53] *Qing di,* lit. "light ground." Griffith's "frontier ground" represents an interpretation, one that is broadly supported by various of the traditional commentators. See Minford, pp. 267–68.

[54] Griffith, p. 130 footnote 4: "as Du Mu says, 'strategically important.'" The traditional commentators play up strategic mountain passes as a source of key ground examples. See Minford, pp. 268–69. See also Wilkinson §24.4, "Strategic topography," pp. 343–44, listing a selection of the most contested passes. Surveying the full sweep of Chinese history, Wilkinson, §24.6, p. 347 further observes that "warfare before the advent of railways tended to be concentrated at strategic natural features, such as mountain passes or river crossings and also at strategically placed walled cities."

XI.5. Ground equally accessible to both the enemy and me is communicating.[55]

XI.6. When a state is enclosed by three other states its territory is focal. He who first gets control of it will gain the support of All-under-Heaven.[56]

XI.7. When the army has penetrated deep into hostile territory, leaving far behind many [walled] enemy cities and towns, it is in serious ground.[57]

XI.8. When the army traverses mountains, forests, precipitous country, or marches through defiles, marshlands, or swamps, or any place where the going is hard, it is in difficult ground.[58]

XI.9. Ground to which access is constricted, where the way out is tortuous, and where a small enemy force can strike my larger one is called "encircled".[59]

XI.10. Ground in which the army survives only if it fights with the courage of desperation is called "death".[60]

XI.11. And therefore, do not fight in dispersive ground; do not stop in the frontier borderlands.

XI.12. Do not attack an enemy who occupies key ground; in communicating ground do not allow your formations to become separated.

XI.13. In focal ground, ally with neighbouring states; in deep ground,[61] plunder.

XI.14. In difficult ground, press on; in encircled ground, devise stratagems;[62] in death ground, fight.

XI.15. In dispersive ground I would unify the determination of the army.

---

[55] Communicating ground – another term might be "open ground" – is ground presenting few topographic obstacles to communication and maneuver (a twenty-first-century update might be the Internet, if little regulated). See Sawyer, p. 326 note 178: "Army movement is unhampered." See also Table 1 (p. 89 above). A related concept in verse X.2 (Passage 2.4A) Griffith, p. 124 renders as "accessible ground."

[56] Needham & Yates, p. 47 identify focal ground as a "place of intersecting highways, important for international relations," a characterization that points squarely toward realms of grand strategy.

[57] Logistics vulnerabilities do much to define serious ground, since long supply lines are awkward under the best of circumstances and vulnerable to depredations by any unfriendly forces along the way (including enemy holdouts in fortified strong points, which many Warring States cities were).

To underscore the point, the word "walled" has been inserted in verse XI.7 (following Mair, p. 118).

For further discussion of logistics aspects of serious ground see #2 footnote 64 below.

[58] As noted earlier (see #2 footnote 35), some traditional commentators envision "difficult ground" (pi di) as ground that is muddy, swampy, easily inundated, etc. See Griffith, p. 131 footnote 1; Giles, p. 117. However, Sun Tzu's definition in verse XI.8 plainly aspires to greater generality.

[59] The character for "encircled" (wei) appearing here also appears in the name of the game of weiqi, suggesting a point of connectivity between these two strands of Chinese strategic thought.

[60] Lau, 1965 article, p. 328 notes another death ground definition: "Terrain from which there is no way out" (Sun Tzu XI material omitted by Griffith, which appears on p. A-67 in Ames's translation).

[61] For translation consistency with verses XI.1, XI.7, and XI.20 (where the same term zhong di appears) Griffith's "deep ground" in verse X.13 should read "serious ground."

[62] The basic term here is mou. Griffith's translation of it as "stratagems" resonates with Sun Tzu's deception emphases and brings the encircled ground concept into dialogue with Theme #7 (deception).

XI.16. In frontier ground I would keep my forces closely linked.[63]
XI.17. In key ground I would hasten up my rear elements.
XI.18. In communicating ground I would pay strict attention to my defences.
XI.19. In focal ground I would strengthen my alliances.
XI.20. In serious ground I would ensure a continuous flow of provisions.[64]
XI.21. In difficult ground I would press on over the roads.[65]
XI.22. In encircled ground I would block the points of access and egress.[66]
XI.23. In death ground I could make it evident that there is no chance of survival. For it is in the nature of soldiers to resist when surrounded; to fight to the death when there is no alternative, and when desperate to follow commands implicitly.

Passage #2.2. Espionage Networks and Their Targets, with Urban Terrain Aspect

(This material is Sun Tzu's closest approach to explicit cognizance of social networks and the type of "terrain" they represent. Particularly relevant is verse XIII.16 describing targets of espionage. Although the tasks described in verse XIII.16, which might best be read as examples rather than an exhaustive

---

[63] As noted in Table 1 (p. 89 above), Mair, p. 122 translates ". . . I shall cause my troops to maintain a command structure," which aligns well on Sun Tzu's top-down command control focus. Cf. verse X.13 (Passage #13.33), read as a warning about impetuous subordinate commanders.

[64] The call here is for an ongoing flow of grain or other foodstuffs. The Chinese term Griffith renders as "continuous" is *ji* and the military idea thereby expressed aligns well on the all-important word "sustained" in the comprehensive concept of logistics stated on p. 49 above (see also p. 34). A core attribute of good logistics, one that many a general throughout history has overlooked to his great cost, is precisely such sustainability or continuity of logistics support – e.g., keeping the army adequately fed for the full duration of a campaign, not just for a few days at its outset, or sporadically.

Sun Tzu does not specify how such continuity is to be achieved. One option – plainly congenial to Sun Tzu (e.g., verse XI.13, verse II.15 [Passage #3.5], and other verses) and favored by most of the traditional commentators (see Giles, p. 137) – is to live off the enemy (or the enemy states' population) to the greatest feasible extent. But this is likely to be only a short-term solution unless the army continually changes location in quest of fresh supplies, a "solution" quite possibly incompatible with tactical goals or needs. An alternative possibility is to create and defend a supply line, possibly leading all the way back to friendly territory. Verse X.2 (Passage #2.4A) makes a clear if glancing allusion to some sort of supply line. Yet neither verse XI.20 nor other parts of Sun Tzu provide clarifying specifics. In short, verse XI.20 points to, but does not resolve, a core challenge of logistics generalship.

[65] A visceral sense of this difficult ground advice is beautifully conveyed by British travel writer Eric Newby, *A Short Walk in the Hindu Kush* (London: Secker & Warburg, 1958), p. 129, quoting a comment of a native guide in Afghanistan: "In Panjshir and in Parian there is only one way and that is the way of the road . . . The roads were made long ago; if there were another way we should use it."

[66] Mair, p. 122: "I shall seal up the openings." Rationales offered by traditional commentators include: veiling break-out plans (Meng shi); blocking opportunities for soldiers to desert (Wang Xi); triggering courage of despair in troops who are trapped (Du Mu, Mei Yaochen). See Minford, pp. 293–94 (all four commentators); Giles, pp. 137–38 (all four commentators); Griffith, pp. 132–33 (Du Mu).

list, need not be limited to an urban environment, social networks rooted in urban settings – where many officials are based – are often both dense and messy, presenting both need and opportunity for imaginative information-gathering.

The social roles Sun Tzu names in verse XIII.16 refer, of course, to the social structures of his day, hence are now antiquarian. Some aspects, however, are timeless. After the defending general himself, the first category of human targets for information-gathering mentioned in the Chinese text of verse XIII.16 is *zuo you*, lit. "left right," sometimes translated as "trusted associates."[67] Giles, p. 171 describes *zuo you* as a "comprehensive term for those who wait on others, servants and retainers generally" – i.e., the underbelly of a hierarchical social structure through which information has always tended to leak profusely. A further category is *ye zhe*, which Mair, p. 130 translates as "appointment secretary," likewise a natural espionage target as one who controls access to his principal.[68] The *ye zhe* reference is a candidate for being Sun Tzu's closest approach to explicit recognition of a topological aspect of social networks.[69]

Traditional commentator Du Mu's verse XIII.16 commentary envisions collecting information about abilities and personalities of individuals of interest – "Are they wise or stupid, clever or clumsy," etc.[70] Collecting human intelligence of that type would naturally extend to analyzing persons of interest through their social networks.[71]

---

[67] Mair, p. 130. This category is Griffith's "staff officers," a terminology that may be too narrowly military and also too exclusively official. Another possible translation of *zuo you* would be "entourage."

[68] Giving *ye zhe* a slightly different twist, Du You refers to it as involving "those whose duty it is to keep the general supplied with information." See Giles, p. 171. The *ye zhe* role also surfaces in the military part of the *Mozi* text, where again it has a connotation of personnel who control access, e.g., as at the gates of a headquarters. See Johnston, Background footnote 12, pp. 866–67, 894–97; Li Ling, Background footnote 84, pp. 130–33.

Griffith's translation of *ye zhe* as "ushers" risks extraneous modern connotations. As alternative translation possibilities Griffith, p. 10 also suggests "chamberlain" or "receptionist."

[69] Although not involving social networks, there are strands of Sun Tzu's thinking that have a type of topological cast. Sun Tzu shows much awareness of the military perils of being "cut off" and, mutatis mutandis, the advantages of placing enemy forces in a cut-off position. That is an instinct that military commanders and *weiqi* players share. For examples from Sun Tzu's terrain chapters, involving the Chinese term *jue* in a sense of "cut off," see Appendix 2 footnotes 8 and 50. Although the term *jue* is lacking, concerns about being cut off are also recognizable in Sun Tzu's concepts of "serious ground" (e.g., verse XI.20's call for maintaining a continuous flow of provisions for an army deep in enemy territory) and "encircled ground" (verse XI.9). An example of cutting off, involving water as a weapon of war, is verse XII.14 (Passage #6.13). More examples – now communications-focused – appear in verse XI.58, advocating for closing passes and breaking tallies to cut communications when a war is starting (steps discussed from various perspectives in Passages #4.20A, #8.26A, and #11.20A).

[70] See Griffith, p. 148 (translating a longer Du Mu commentary segment).

[71] A different Warring States text, the *Lüshi chunqiu*, makes explicit a focus on gathering information about social networks, proposing to evaluate a person of interest through scrutinizing their friends and associates (a task that might nowadays be described as ego network analysis). See *Lüshi chunqiu* translation entitled *The Annals of Lü Buwei* (complete translation and study by John Knoblock & Jeffrey Riegel; Stanford, CA: Stanford University Press, 2000), p. 621: "Thus your

Also noteworthy is verse XIII.15, pointing to a type of network failure. There is a double-edged sword here: the more personable and well networked the spy, a desirable attribute for many espionage purposes, the worse the potential downside and the larger the cleanup job should that spy ever go bad.)

XIII.3. Now the reason the enlightened prince and the wise general conquer the enemy whenever they move and their achievements surpass those of ordinary men is foreknowledge.

XIII.4. What is called "foreknowledge" cannot be elicited from spirits, nor from gods, nor by analogy with past events, nor from calculations. It must be obtained from men who know the enemy situation.

XIII.5. Now there are five sorts of secret agents to be employed. These are native, inside, doubled, expendable, and living.

XIII.6. When these five types of agents are all working simultaneously and none knows their method of operation, they are called "The Divine Skein" and are the treasure of a sovereign.[72]

. . .

XIII.15. If plans relating to secret operations are prematurely divulged the agent and all those to whom he spoke of them shall be put to death.

XIII.16. Generally in the case of armies you wish to strike, cities you wish to attack, and people you wish to assassinate, you must know the names of the garrison commander, the staff officers [zuo you], the ushers [ye zhe], gate keepers, and the bodyguards. You must instruct your agents to inquire into these matters in minute detail.

★★★★★★

servant says that he is unable to judge a man's fate by his appearance, but only by examining his friends." For Lüshi chunqiu background see Early Chinese Texts (cited on p. xxii above), pp. 324–30; Wilkinson, §58.6.3, p. 778.

[72] Making a bow to a similar concept appearing in the Laozi Daoist classic, Ames, p. 169 and p. 296 note 224 renders "Divine Skein" as "imperceptible web" (a fine image for a clandestine network!).

Strategist Should Be Cheap

This topic comprises Themes #3 and #4, whose central thrusts are, respectively, cheap military successes and paths to the same larger political end using civilian approaches – i.e., winning without major fighting (at least in a classic military sense). Although it does not capture the sum total of Sun Tzu's Theme #3 thinking, a core part of that thinking focuses on extremes of both benefits and costs – reaping the former and avoiding the latter.

Theme #4 thinking finds classic expression in verse III.3 (Passage #3.3):

III.3.

For to win one hundred victories in one hundred battles is not the acme of skill. To subdue the enemy without fighting is the acme of skill.

This idea, which is vintage Sun Tzu, stands as a stimulus for soldiers everwhere (and some civilians too) to "lift their sights" to glimpse vistas of more creative paths to success than bloodshed and mayhem. Such a perspective is especially important in pushing back on a persistent fallacy, found in all ages of war but often associated with appearance of new destructive technologies, that equates strategy with destruction, replacing it by a more nuanced concept of strategy as the art of control (pp. 187–94 below). Because a strategy-as-control standpoint is common to both, Themes #3 and #4 are natural complements in Sun Tzu's way of war.

Yet it is also essential to be clear that Sun Tzu's Themes #3 and #4 ideas remain open to use of deadly force. Apocryphal it may be, but the story of Sun Tzu's drilling of the ruler's concubines, with outcomes fatal to two of them (see p. 13 above), is itself ancient. That story must be, as Roger Ames says, "one of the best-known anecdotes in Chinese military lore."[1] It is part of Sun Tzu's persona as it has appeared over two millennia and more of Chinese history, where "winning without fighting" has often involved deadly force of types other than those associated with armies.[2] Cheap victories may also arise from uses of conventional military power in ways short of major combat. Logistics chokeholds have a part to play here.[3]

---

[1] Ames, p. 32.

[2] In the concubines story, note that "winning" means Sun Tzu's success in getting hired by the ruler!

[3] See Shiji 92 (Watson translation cited on p. xxiii above), Han Dynasty I, p. 174, quoting an unnamed adviser: "The Han soldiers have marched a thousand miles into a strange land. If the cities of Qi all turn against them, they will find themselves with no way to get food for their forces, and they can be overcome without a fight." (The advice was not followed, and things didn't pan out that way.)

Above all, for a strategist in Sun Tzu's mold the political goal is utterly unencumbered by any felt need to accumulate battlefield victories as inherently meritorious feats. To the contrary, great kudos is accorded to finding paths to that political goal which circumvent hard fighting and in that sense come cheap.[4]

THEME #3 SEEKING LARGE, CHEAP GAINS AND AVOIDING PROTRACTED WAR

fundamentals of cheap success:

Passage #3.1 (verse V.16) (crossbow metaphor)

Passage #3.2 (verse IV.4) (reality check: cheap paths to success are not always immediately available)

Passage #3.3 (verses III.1–III.11) (winning with greatest profit)

Passage #3.4 (verses II.1–II.8) (sine qua non for winning with profit: avoiding protracted war)

Passage #3.5 (verses II.9–II.21) (fighting cheaply by living off the enemy)

Passage #3.6 (verses XI.7 + XI.13 + XI.31–XI.32) (more about living off the enemy; one's troops as resource to be husbanded)

sources of outsized advantage:

Passage #3.7 (verse XI.29) (sources of outsized advantage (1): speed)

Passage #3.8 (verses VII.12–VII.13) (imageries for speed and the advantages it provides)

Passage #3.9 (verse VII.5) (ponderousness of a large conventional army, foreclosing advantages of speed)

Passage #3.10 (verses VI.13–VI.17) (sources of outsized advantage (2): information superiority)

Passage #3.11 (verses IV.16–IV.19) (sources of outsized advantage (3): winning by calculation)

Passage #3.12 (verses V.3–V.5) (sources of outsized advantage (4): combined use of qi ("extraordinary forces") and zheng ("normal forces"))

Passage #3.13 (verse XI.9) (sources of outsized advantage (5): encircled ground)

avoiding high price tag options:

Passage #3.14 (verses VII.22 + VII.25–VII.33) (avoiding military options with a high price tag)

Passage #3.15 (verses VIII.7–VIII.8) (more high price tag steps to avoid)

Passage #3.16 (verses III.18–III.23) ("faux cheap" expedients imposed by a ruler)

Passage #3.17 (verses XII.15–XII.19) (avoiding wasteful delay in exploiting success; avoiding catastrophically costly moves)

---

[4] Traditional Sun Tzu commentator Wang Xi sums up the idea, in many ways a cultural stance: "However great or small the entity involved, be it state, army, regiment, detachment, or company, taking it intact enhances prestige and power; destroying it loses prestige and power." (Minford, p. 132)

apparent exception that is no exception:

Passage #3.18 (verses XIII.12 + XIII.17 + XIII.21) (exception to rule of cheap: taking good care of secret agents)

Passage #3.19 (verses XIII.1–XIII.2) (why this exception is no exception: espionage as fantastically cheap by comparison with conventional warfare)

******

Theme #3 unites under one heading a set of observations and emphases distributed widely throughout the Sun Tzu text. Taking cues from Sun Tzu, who never attempts to systematize this material (by contrast, for example, to his terrain lists), the present roundup of it follows an open-textured approach that does not seek to cram this strand of Sun Tzu into a propositional inventory. Theme #3 is about identifying and exploiting courses of strategic and tactical action that advance a strategic actor's goals in a major way yet are also, to put it vividly if non-rigorously,

cheap! cheap! outrageously cheap (relative to the effects produced)!

This "rule of cheap" is best viewed as a mindset.[1] On the most basic level it embodies Sun Tzu's keen awareness that the nature of war itself, and the massive instabilities it creates, harbors both opportunities for fantastic gain and possibilities for egregious costs on many scopes and scales. To keep emphasis where it belongs – and lest the entire Sun Tzu text be imported here! – the passages assembled to illustrate Theme #3 are those where the cheapness motif is most explicit.

The text contains enough specifics to enable the meaning of "cheap Sun Tzu" to be explored in some detail. Sun Tzu pays much attention to extremes of strategic and tactical advantage and disadvantage, a focus on outliers that differentiates Theme #3 from Theme #1 (calculation more generally). That attention takes both qualitative and quantitative forms, the latter including alertness to order-of-magnitude disparities (in resources, costs, combat potential, effectiveness, etc.) and the outsized advantages for one side or the other they often yield. Because modern audiences may find congenial the quantitative tonality of an orders-of-magnitude focus, it is a natural (though not the only) point of entry into Sun Tzu's Theme #3 thinking.[2]

A sampler of Sun Tzu's relevant thinking is found in the following verses:[3]

---

[1] Ames, p. 87 expresses Sun Tzu's preoccupation with cheapness in optimization language: "The fundamental question the Sun-tzu seeks to respond to is how does the enlightened ruler achieve victory at the minimum cost?" While a modern operations researcher might resonate with this formulation, it does not capture Sun Tzu's more specific preoccupation with cost–benefit extremes.

[2] Wilkinson §38.9, p. 535 notes that in pre-modern Chinese texts the "numbers 10, 100, 1,000, and 10,000 are frequently used to indicate orders of magnitude, not exact numbers." The Chinese text of verse VI.13 (Passage #3.10), referring to *shi* ("ten" or "ten-fold"), is illustrative. See pp. 133–34 below.

[3] In engaging with these examples it is important to recognize that order-of-magnitude statements are inexact by definition, hence straddle qualitative and quantitative thinking, partaking of aspects of both. For this reason, developing Theme #3 calls for a flexible frame of mind, one that accommodates three conceptual boxes (literal/order-of-magnitude/figurative), not just two (literal/figurative).

II.15.

Hence the wise general sees to it that his troops feed on the enemy, for one bushel of the enemy's provisions is equivalent to twenty of his; one hundredweight of enemy fodder to twenty hundredweight of his. (from Passage #3.5)

XIII.1.

Now when an army of one hundred thousand is raised and dispatched on a distant campaign the expenses borne by the people together with the disbursements of the treasury will amount to a thousand pieces of gold daily. ...

XIII.2.

One who confronts his enemy for many years in order to struggle for victory in a decisive battle yet who, because he begrudges rank, honours and a few hundred pieces of gold [to pay spies] remains ignorant of his enemy's situation, is ... no general ... (from Passage #3.19)

(Taking the length of a campaign as, say, 100 days, the "military option" here is deemed to cost 100,000 pieces of gold, while an espionage/covert operations alternative would cost only a few hundred pieces of gold – a two to three orders of magnitude disparity in the spies' favor.[4])

IV.19.

Thus a victorious army is as a hundredweight balanced against a grain; a defeated army as a grain balanced against a hundredweight. (from Passage #3.11)

(Unpacking the Chinese expression for "hundredweight," Ames estimates this disparity in combat potential as amounting to a 600 to 1 differential – again two to three orders of magnitude.[5])

A strategist of Sun Tzu's persuasion is ever on the lookout for just such situations, and ways of engineering them, where these kinds of outsized disparity – with the attendant cost-reducing results – operate in his favor.

Clearly those instances where Sun Tzu explicitly quantifies an orders-of-magnitude disparity, or (as in verse IV.19) uses a quantifiable word picture, are easy to spot. But there are further examples in Sun Tzu where similar thinking appears in qualitative garb. An image redolent of cheap success is the age-old Chinese military/strategic principle, well represented in Sun Tzu, of "striking with 'fullness' (shi) against 'emptiness' (xu)."[6] Sometimes who or what is "full" versus "empty" calls for clarifying comments. For example, Sun Tzu's death ground situation (Passage #1.12) envisions a force in death ground having its strength multiplied many times over

---

[4] This campaign duration may be on the short side for the Warring States period (a point that suggests a yet greater cost disparity in favor of espionage). See Lewis, Background footnote 7, p. 628, observing that "Warring States campaigns often lasted for more than a year."

Of course, what Sun Tzu says in verses XIII.1–XIII.2 falls short of making a spy option a full-fledged alternative to conventional war. However, in context of verses III.10 (Passage #3.3), XIII.22 (Passage #5.19), and everything else we know about Sun Tzu, such an interpretation seems a reasonable bet.

[5] Ames, p. 286 note 139.

[6] Needham & Yates, p. 35. Sawyer, p. 306 note 17 observes that the xu/shi concept pair "is closely identified with Sun Tzu's thought, though the terms may have predated him." Sun Tzu refers to both xu and shi in verses V.4 (Passage #3.12) and VI.27 (Passages #3.20A and #4.14). The same concept pair also gives Sun Tzu VI its title. For roundup of further occurences of xu and shi in the Sun Tzu text see Appendix 3 footnote 10.

To be clear, shi 實 (fullness) is a different character and concept from shi 勢 (strategic advantage).

by the psychological responses unleashed by the desperate situation.[7] A different source of outsized advantage – or way of avoiding outsized costs – comes from Sun Tzu's terrain material and is medical in thrust:

> IX.13.
> An army prefers high ground to low; esteems sunlight and dislikes shade. Thus, while nourishing its health, the army occupies a firm position [shi].[8] An army that does not suffer from countless diseases is said to be certain of victory.[9]

What Griffith here translates as "countless diseases" the Chinese text describes as, literally, "the one hundred diseases" (bai ji).[10] While this is plainly a figure of speech (it also appears in early Chinese medical writings),[11] to anyone with practical public health experience – which would include experienced commanders of large infantry forces, ancient or modern – the phrase also conjures the ravages that disease can all too easily wreak on an army, swiftly eviscerating that army's combat potential by an order of magnitude or more.[12]

On a strategic plane, perhaps the loftiest expression of the ambitions of Sun Tzu's Theme #3 thinking comes from the first verses of Sun Tzu III, whose focus is on capturing intact an entire enemy army or even an enemy state (by implication, at bargain-basement cost to one's own side):[13] Passage #3.3 begins:

> III.1.
> Generally in war the best policy is to take a state intact; to ruin it is inferior to this.
> III.2.
> To capture the enemy's army is better than to destroy it [etc.].

Conversely, Sun Tzu's deepest strategic worry is becoming bogged down in a protracted war, a concern conveyed in Sun Tzu's well-known injunction against it in Passage #3.4. That concern about protracted war has something in common with wariness about effects of the one hundred diseases, since both have potential to foreclose even the possibility of "outrageously cheap" advantageous outcomes.

On a tactical plane, a compelling illustration of Theme #3 thinking is Sun Tzu's interest in fire attacks, the only type of special weapon Sun Tzu discusses in any detail (see verses XII.1–XII.13 in Passage #9.7). Fire attacks, to whose user Sun Tzu gives kudos in verse XII.13, amount to a cheap tool of warfare, cheaper still when

---

[7] The Wei Liaozi, one of the Seven Military Classics (also of Warring States vintage), points to a multiplier of two or more orders of magnitude: "Ten thousand men who fight are not as good as one hundred men who are truly aroused." See Sawyer, Seven Military Classics, p. 276.

[8] This shi is the "fullness" (or "solidity") half of the xu/shi concept pair just noted.

[9] Depending on where the Sun Tzu text took shape, possible context for verse IX.13 is the southerly geographic location of the state of Wu (which the putative historical Sun Tzu is said to have served) and that of Yue, Wu's eventual conqueror. See de Crespigny, Background footnote 95, p. 278, noting the Yangzi basin's warm climate and wetlands as a breeding ground for disease.

[10] Griffith, p. 117 footnote 6.

[11] Paul U. Unschuld, Huang Di nei jing su wen: nature, knowledge, imagery in an ancient Chinese medical text (Berkeley, CA: University of California Press, 2003), p. 183.

[12] Compare Jeffrey S. Sartin, "Infectious diseases during the Civil War: the triumph of the 'Third Army'," Clinical Infectious Diseases, 1993, 16(4), 580–84: "Altogether, two-thirds of the approximately 660,000 deaths of soldiers were caused by uncontrolled infectious diseases, and epidemics played a major role in halting several major campaigns" (p. 580; erratum in same journal, 2002, 34(2), p. 292). See also Hans Zinsser, Rats, Lice and History (Boston, MA: Little, Brown, 1963; first published in 1935).

[13] A radically different understanding of verses III.1–III.2 has been proposed by Lau, 1965 article, pp. 333–35. It is analyzed and rejected in Appendix 3, pp. A-97ff.

weather or other conditions favor rapid spread of the blaze (a possibility of which Sun Tzu's fire attack verses take due note). Reinforcing its cheapness, fire can be admirably self-sustaining, albeit sometimes prone to get out of control.[14]

On both strategic and tactical levels, sheer speed – speed of decision, speed of execution – is a basic tool for realizing Sun Tzu's lofty Theme #3 aspirations. Verse XI.29 (Passage #3.7) is a basic statement, but the text continually returns to, indeed hammers on, the same idea (Passages #3.4, #3.8, #3.21A–#3.28A).

Expanding on an earlier point about Sun Tzu's mindset (p. 112 above), a unifying thread running throughout these and other Theme #3 examples is a Sun Tzu belief that military situations are shot through and through with potential for extremes of advantage or disadvantage: take just the right cheap step, reap a truly outsized gain; make the wrong move, incur an exorbitant cost. It is worth observing that crossbows and their trigger mechanisms (in Sun Tzu's time, a state-of-the-art advanced military technology) are mentioned several times in the Sun Tzu text.[15] The fundamental principle of a well-aimed crossbow – tiny force on the trigger, swift, deadly effect at a distance – might be taken as a Warring States Chinese symbol of Sun Tzu Theme #3 thinking (Passage #3.1). Moving from metaphor to operational practice, a menu of military sources of outsized advantage is collected in Passages #3.7–#3.13.[16] From a contemporary perspective, particularly noteworthy is Passage #3.10, pointing to outsized advantage growing out of information superiority.

Viewed from a different angle, Sun Tzu's Theme #3 thinking raises a challenge to settling for merely "good enough," the mantra of satisficing made famous by Herbert Simon and behavioral decision theory. In modern times (and throughout much of Chinese history too) a tolerance for at most fair-to-middling successes is associated with the bureaucratic phenomenon, Chinese no less than Western.[17] Theme #3 pushes back against that bureaucratic mindset and the committee-attenuated versions of strategic thinking for which it often stands. At the same time, of course, Sun Tzu the pragmatist would scarcely reject opportunities for modest gains of the sort associated with economy or being economical (whether in a military or a civilian management sense). That is a useful lens for examining some of what Sun Tzu has to say about logistics (e.g., Passage #3.6's advice to attend to nourishment and other well-being of one's troops, a scarce resource worth careful husbanding).[18]

In the environment of Warring States China, Sun Tzu's emphasis on well-planned (see Passage #1.1) offensive wars of quick decision meshes excellently with Theme

---

[14] For further analysis of Sun Tzu's treatment of fire attacks see Appendix 15.

[15] See verses V.16 and II.14 (Passages #3.1 and #3.5) and, via an allusion to a crossbow trigger mechanism, also verse XI.48 (Passage #10.13). For origins and development of the crossbow in China see Background footnote 17 and Needham & Yates, pp. 135–46 (also referring to Sun Tzu).

[16] In part because it is often logistically cheap, psychological warfare is also commonly cheap warfare. From a Theme #3 standpoint it is no accident that Sun Tzu favors psychological paths to advantage, which underscores connectivity between Themes #3 and #6 (attacking the adversary's strategy). See, for example, Passages #6.7–#6.10.

[17] A case for existence of well-developed early Chinese bureaucracy from a time long before Sun Tzu is developed in Li Feng's book cited in Introduction footnote 50.

[18] This facet of Sun Tzu's Theme #3 bears a first-cousin resemblance to modern attention to "economy of force" as one of the Principles of War (whose provenance as a military concept is discussed in Liddell Hart's letter to Griffith of 9 April 1959, see LH 1/333/43 in the Liddell Hart Papers in Liddell Hart Centre for Military Archives, King's College London). But, at least as usually

#3. Sun Tzu pays noticeably little attention to defensive wars. By contrast to Clausewitz's balanced attention to defense and attack, there is no Sun Tzu chapter on defensive warfare providing a counterpart to Sun Tzu III on offensive warfare. Even if successful (from the defender's standpoint), wars of a defensive sort are likely to be costly, foreclosing possibilities for reaping extreme advantage (if for no other reason than that defensive operations are likely to unfold on one's state's own territory rather than the enemy's, with economic and human costs that run high). A de-emphasis on defensive warfare sets Sun Tzu apart from harmonics present in Chinese strategy of later times, where a defensive orientation, in particular vis-à-vis Inner Asian peoples, was a longstanding and prominent pattern.[19] Therein lies a major challenge to monolithic understandings of "the Chinese strategic tradition."

Sun Tzu's relentless focus on cheapness comes with an important flip side, involving exceptional wariness about becoming entrapped in high-price-tag military options – including ones whose high cost may not be initially apparent. Protracted war is a prominent strategic case, but there are other examples too. Sun Tzu's well-known concern about getting bogged down in costly attacks on cities fits here. While Sun Tzu does not make the distinction explicit, there are at least two different reasons why attacks on cities may be a high-price-tag option. One pertains to the cost, in lives as well as in other opportunities forgone, of capturing a city; the second, to possible destruction of valuable resources contained in that city and whose existence may be the reason for trying to capture it in the first place.[20]

Deep wariness about attacking cities, much of whose rationale did not outlast the military conditions of Sun Tzu's time, surfaces multiple times in the Sun Tzu text spread over three chapters (verses III.7–III.9 in Passage #3.3; verse II.3 in Passage #3.4; verse VIII.7 in Passage #3.15).[21] Again with an eye on their avoidance, further high-price-tag options not specifically involving cities appear in Passages #3.14–#3.16. Most notably, Passages #3.15 and #3.16 suggest that Sun Tzu's view that a ruler's orders may at times be disobeyed by the general – surely among Sun Tzu's most provocative and controversial positions (see Theme #13 for analysis) – is, at least in part, a manifestation of a Theme #3 mindset. To a ruler remote from the realities of a military situation, decisions rendered by that ruler may be all too easily be "faux cheap" – i.e., anything but cheap.

formulated, economy of force tends to be both broader and blander in focus than Sun Tzu's keen alertness to outliers.

[19] See Needham & Yates, pp. 44–45.

[20] For extensive further coverage of issues pertaining to attack and defense of cities in early Chinese warfare – the latter a topic of particular interest to the Mohist school of Warring States thought – see Robin Yates's richly detailed "early poliorcetics" discussion in Needham & Yates, pp. 241–485.

[21] Lau & Ames, p. 31 describe an evolving Warring States posture pertaining to attacks on cities, noting military trends favoring successful assault. See also Wilkinson, §24.11, p. 354. The archaeologically recovered Sun Bin military text, a text later than Sun Tzu, actually recommends attacks on certain types of fortifications. See Ames, pp. 28–29; Lau & Ames, pp. 167–68 (translating from the Sun Bin work). Meanwhile, long-term economic trends favored "development of walled cities as centers of wealth and commerce that made siege more profitable" (Ames, p. 29).

In short, there are grounds to believe that both ongoing military and larger societal evolution soon rendered the thrust of Sun Tzu's verse III.7 advice outdated, certainly as a general principle.

That vein of thinking leads back to grand strategy, putting in play verse XII.18 and other Passage #3.17 verses. A basic concern here is flawed decisions that launch exorbitantly costly (now or later) wars or other projects. A foe's machinations could, of course, egg on the ruler or general to fall into just such a trap (see Passages #6.8–#6.10).[22] However, such pitfalls may also emanate from a decision maker's own psychological dynamics, in whose inner workings he becomes entrapped.

Shifting focus to possible limits of "cheap Sun Tzu," there are two instances in the Sun Tzu text, both involving payments to people, where Sun Tzu carves out what might appear to be flagrant exceptions to his "rule of cheap." Yet those apparent violations of the cheapness principle are, on scrutiny, calculated efforts aimed at realizing far greater gain. Framed another way, the exceptions are faux expensive, not faux cheap!

A first exception arises in connection with espionage, where Sun Tzu plays up the immense importance of "liberality" in rewards to secret agents (Passage #3.18). As Sun Tzu takes pains to make clear (Passage #3.19, whose verses have already been quoted on p. 113), the absolute costs incurred here are tiny, even negligible, as compared with the costs of conducting a major conventional military operation.[23]

A second apparent exception is only glancingly mentioned in the Sun Tzu text and hence is easy to overlook. Yet it is enlightening as a corroboration of Sun Tzu's utter pragmatism and willingness to incur limited costs in quest of far greater benefits. That second exception is found in the allusion at the start of Passage #3.4 to largesse extended to a favored circle of onlookers or hangers-on whose presence, for one reason or another, is deemed valuable.[24] Context there suggests that such payments could be substantial, yet Sun Tzu is plainly willing to accept them as a fact of life, presumably because they may cut down on far larger costs of warfighting.

Roads Not Taken by Sun Tzu in Developing Theme #3

Given Sun Tzu's focus on living off the enemy as one basic variety of cheap logistics, it is worth noticing Sun Tzu's lack of mention of water transport, which in a pre-digital era would have been a prominent type of cheap logistics. That lack is all the more striking because the historical Sun Tzu is said to have served

---

[22] A non-military example of such a stratagem (noted by traditional Sun Tzu commentator Du You) involved the state of Han's "friendly" advice to Qin to squander its resources on digging a canal. In that case, however, the stratagem backfired badly. For the story see #4 footnote 98.

[23] Needham & Yates, p. 54 sum up Sun Tzu's thinking: "For spies there should be no economy of money, gifts and emoluments, because it is both cheaper and easier to achieve victory with their help than by use of the full force of an army."

[24] Understandings of these roles vary. Griffith, p. 72 refers to "stipends for the entertainment of advisers and visitors." Minford, p. 120 refers to "dealings with envoys and advisers," while Ames, p. 107 refers to these worthies as "foreign envoys and advisors." For purposes of Theme #3 analysis these differences are not all that important. The key point is that Sun Tzu is alluding to persons whose presence is tolerated (or invited!) for reasons other than purely military essentiality (possibly because they are expected to spread their impressions, accurate or otherwise, to friends, enemies, and neutrals in ways seen as likely to be advantageous). An important clarifying comment comes from Brooks & Brooks, Introduction footnote 36, p. 121. Rendering the relevant Sun Tzu text as "necessities for visitors and guests" they observe that "[d]iplomacy, both between states and between armies, was part of war" (see their p. 121 footnote 1).

the state of Wu, a major part of whose surface area was wet (marshes, rivers, lakes, and ponds).[25]

In many ages of war, for one side or the other asymmetric warfare has been associated with outsized gains at a low cost, certainly by some yardsticks.[26] The long-running warfare between Chinese polities and Inner Asian peoples holds many examples. Great technological superiority over a foe also holds promise of outsized gains (albeit that military history suggests that this type of edge is often unreliable). Sun Tzu does not touch on any of these situations. While Sun Tzu's ideas may have relevance here too – there are certainly asymmetric warfare cases (notably in cyber warfare areas) where they work superbly – it is important to be open to the possibility that Sun Tzu's ideas may need interpretation and tailoring to fit particular asymmetric conflicts. Here is a case where the difference between Sun Tzu (1), on the one hand, and Sun Tzu (2) or (3) on the other, needs to be kept clearly in view.

These omissions are in keeping with, and may in part reflect, Sun Tzu's larger lack of interest in different types of military forces and their performance characteristics (a striking deficit in a military manual on land warfare). Relatedly, there is also nothing in Sun Tzu about the merits of versatile, multifunctional forces, cross-training, etc. (which in many military contexts can yield multiplier effects that advance Theme #3's core agenda).

Sun Tzu's emphasis on speed also shows some signs of limited analytical horizons. Its focus is heavily on speed of execution supported by speed of command decision. One is hard put to spot any attention to speed of learning, a harder-to-measure variety of speed that is also crucial in warfare's complex environment, perhaps particularly so at the start of a war.[27]

A further set of Sun Tzu's limitations has a methodological cast. While the Sun Tzu text shows a basic awareness of order-of-magnitude differences, it never takes the next step, actually deriving or working with different orders of magnitude rather than

---

[25] Sawyer, p. 89 (citing figure of 15%). Waterborne logistics definitely had a place in early Chinese warfare. Yates, Background footnote 53, p. 304 notes Qin use of boats to transport military supplies. See also de Crespigny, Background footnote 95, p. 278, noting that water "offers excellent means of transport, by ship or barge along rivers or canals." Joseph Needham (with the collaboration of Wang Ling and Lu Gwei-Djen), *Science and Civilisation in China*, Vol. 4: *Physics and Physical Technology*, Part III: *Civil Engineering and Nautics* (Cambridge: Cambridge University Press, 1971) sums up pithily (p. 441), "And of course, as ever, inland water transport."

[26] An image capturing something of the challenge of asymmetric warfare is attributed to Warring States military thinker Sun Bin by his biography in Shiji 65 (see p. xxii above): "One set on unraveling a jumbled tangle of silk threads does not strike at it with his closed fist." (See Lau & Ames, p. 9.)

[27] See p. 4 of Michard Howard, "Military science in an age of peace," *The RUSI Journal*, 1974, 119(1), 3–11: "when everybody starts wrong, the advantage goes to the side which can most quickly adjust itself to the new and unfamiliar environment and learn from its mistakes." Of course, given that Warring States China was a polar opposite to an age of peace, it is conceivable that Sun Tzu saw no need to develop speed of learning as a separate category in the warfare he knew.

Deficit of attention to learning by the Sun Tzu text should be distinguished from how the development of the text may itself be evidence of a type of long-term learning process, galvanized by harsh lessons of Warring States warfare. Brooks, p. 61 may be read as a sketch of such a process, in the form of a running commentary on a proposed sequence of chapter accretions.

simply pointing to order-of-magnitude disparities.[28] Doing so would have called for significantly deeper engagement with quantitative thinking. However, in fairness to Sun Tzu, such engagement is not necessary to recognize that order-of-magnitude differences exist and that some of them have basic military or strategic implications.

Scrutinizing Sun Tzu's Theme #3 thinking from the standpoint nowadays labeled cost–benefit analysis, many – though not all – of Sun Tzu's relevant concepts gravitate toward what Mark Elvin has felicitously framed as "war and the logic of short-term advantage."[29] At times it seems that Sun Tzu's predilection for short-term advantage may lead to choices that, in a longer run, are "faux cheap" – i.e., not cheap at all. A telling case in point is Sun Tzu's advice to support an invading army by plunder (verse XI.13 in Passage #3.6; see also verse II.10 in Passage #3.5). Extensive plundering of the sort needed to sustain a mass-infantry army will create civilian victims, complicating governance of conquered territory and possibly triggering abandonment of formerly productive tilled land by victims of plunder.[30]

Meanwhile it is also unclear whether Sun Tzu's thinking recognizes how small gains, cumulated over a long period, may sometimes translate into a major gain. Capturing an enemy state intact by subversion would be a basic example. Although it is certainly possible to identify quite a few strands of Sun Tzu that align well on subversion motifs (see pp. 145, 147–49; Table 3 on pp. 151–52; and Passages #4.12– #4.16), a drawn-out, cumulative process of winning-by-subversion is nowhere explicitly spelled out in the Sun Tzu text.

Shifting focus from benefits to costs, Sun Tzu's thirteen chapters show little trace of any counterpart to a tax administrator's perspective, eternally concerned about how myriads of tiny siphons may allow major resources to dribble away (possibly belying first impressions of a low and very tolerable cost to the state).[31] One has to

---

[28] A forerunner in Western tradition would be *The Sand Reckoner* of Archimedes (third century BC).

In modern times working with different orders of magnitude is illustrated by so-called Fermi problems. The latter are so named after physicist Enrico Fermi (1901–1954), who famously estimated the energy release at the 1945 TRINITY nuclear test by the distinctly low-tech step of throwing torn-up scraps of paper into the air just before the explosion, then noting how far the blast wave carried them before they fell to earth. Regarding Fermi's style of estimation, defined as the "process of coming up with estimates of the correct order of magnitude for various real-world quantities," see Timothy Gowers, "How should mathematics be taught to non-mathematicians?" (June 8, 2012), available at Gower's weblog https://gowers.wordpress.com (visited October 6, 2021).

[29] "War and the logic of short-term advantage" is the title of chapter 5 in Elvin, #2 footnote 15. Sun Tzu's short-termism tendency is just that – a tendency, not a definitive doctrinal position. There are some parts of Sun Tzu, notably the initial segment of Sun Tzu IV (Passages #4.4, #5.4), that point to versions of a "long game," privileging patience. Theme #5 analysis returns to this point. See pp. 178. 183, 196–97 below; also pp. 190–91 (analysis of Sun Tzu's relevant thinking from Eccles–Rosinski standpoint).

[30] Pitfalls of a focus on plundering are revisited in Passage #13.6 analysis. See pp. 447–48 below.

[31] The "myriads of tiny siphons" image is from Joseph F. Fletcher, Jr., "Turco-Mongolian monarchic tradition in the Ottoman empire," *Harvard Ukrainian Studies*, 1979–80, 3/4 (Part 1), 236–51. See p. 243 there. The closest approach to Sun-Tzu-qua-tax-expert – not from Sun Tzu's thirteen-chapter core but from archaeologically recovered Sun Tzu–related material that Ames, pp. 36–37 labels "commentarial" – is briefly noted in #1 footnote 42. As a contribution to grand strategic thinking it has puzzling, in fact downright problematic aspects. See critique by Petersen, Introduction footnote 32, p. 13.

read between Sun Tzu's lines to turn up similar concerns.[32] Yet tax administrators, and their ways of thinking, had to exist to sustain Sun Tzu's armies on anything like the 100,000-man scale contemplated in verses II.1–II.2 (Passage #3.4) and XIII.1 (Passage #3.19). Here is evidence that the basic perspective of the Sun Tzu text aligns on concerns of field commanders and is divorced from those of civilian administrators, even ones whose roles are critical to mounting a military effort.

Perhaps most strikingly, there is nothing in Sun Tzu endorsing (or even noting) the possible desirability of incurring major "up front" costs in anticipation of reaping yet greater deferred benefits.[33] An example that would have been pertinent in Sun Tzu's time no less than in many other ages of war would be steps to create or improve a road network as preparation for future strategic contingencies. Other early Chinese thinkers and doers took cognizance of projects necessitating long-term planning, many having relevance to war potential.[34] But not Sun Tzu.

These several omissions underscore that Sun Tzu's relentless focus on costs and benefits, in some ways so reminiscent of thought patterns of modern economics,[35] stops well short of a modern cost–benefit analysis perspective.

Sun Tzu (2) ("Sun Tzu Extended") and Sun Tzu (3) ("Sun Tzu Analogical") Frontiers of Theme #3

In an alternative framing – not inconsistent with a focus on amazingly cheap successes but with a distinct anchoring – Sun Tzu Theme #3 unifies around a deep and abiding interest in what a statistician might call outliers and their strategic implications.[36] From the history of contrarian investing in finance, that facet of the spirit of Theme #3 is illustrated by an often quoted (if apocryphal)

---

[32] Underlying Sun Tzu's dispersive ground advice in verses XI.11 and XI.15 (Passage #2.1) one may infer possible concerns about how small leakages of personnel – through troops deserting or going AWOL – could cumulate to eviscerate an army. Of course, this dispersive ground advice might have different roots (e.g., concern about mass desertions precipitated by a battle). See #2 footnote 52.

[33] This deficit is part of Sun Tzu's lack of attention to long-range planning. See pp. 66–67 above.

[34] Warring States examples would include urban fortifications, along with elaborate observation towers, gates, etc.; water control projects; and "great walls" erected for defensive purposes against hostile neighbors (regarding the latter see Background footnote 74). On city planning and urban architecture of the Warring States period, along with the rise of the "double city" with its two distinct walled enclosures (one for rulers and administrators, the other for merchants, artisans, and peasants), see Mark E. Lewis, The Construction of Space in Early China (Albany: State University of New York Press, 2006), pp. 8, 150ff. (from his chapter entitled "Cities and Capitals"). On water control projects – "one of the great themes of Chinese history" – see Wilkinson, §13.6, pp. 208–11, especially §13.6.4, "Major ancient water control works."

[35] See Oliver E. Williamson, Markets and Hierarchies: analysis and antitrust implications (New York: Free Press, 1975), p. 37: "The power of economics, in relation to the other social sciences, is to be traced in no small part to its unremitting emphasis on net benefit analysis."

[36] A Sun Tzu emphasis on outliers finds support in verse XIII.12 (Passage #3.18), which may be read as characterizing spies and relations with spies as outliers on three dimensions (closeness, rewards, confidentiality); and in verse XIII.23 (Passage #4.7). In statisticians' language comes a Sun Tzu–compatible comment: "It is still informative, however, and may be important, to examine samples and residuals for the presence of 'outliers' or 'exotic values' because their unexpected behavior may indicate failure of a model or point to an unexpected phenomenon." See

saying attributed to Baron Rothschild, "the time to buy is when there is blood in the streets."[37]

An exceptionally compelling example of Sun Tzu's Theme #3 comes from a context where Sun Tzu's thinking is rarely cited, at least by the terrorists who most directly stand to profit from it. In an analysis of selected major terrorist events (Kenya and Tanzania bombings; USS *Cole* bombing; September 11 attacks; Bali nightclub bombings; Madrid subway bombing; London Underground bombing) Martin Shubik and Aaron Zelinsky have calculated ratios expressing

$$\frac{\text{cost inflicted on society}}{\text{cost to terrorist organization}}$$

of perpetrating particular terrorist attacks.[38] The ratios they calculate range from 3220:1 to a truly remarkable 1,270,000:1.[39] Spreads like these involving so many orders of magnitude – all very much favoring terrorist attacker over non-terrorist defender – are outstanding examples of Sun Tzu Theme #3 (even though, as noted earlier, Sun Tzu himself does not explore asymmetric warfare topics). They should sound a dramatic warning concerning vulnerabilities of twenty-first-century societies worldwide to certain types of asymmetric warfare.

As memories of the 9/11 attacks start to fade, here is one first-rate starting point for launching further analysis inspired by asking: *"What did Master Sun know that we still don't (or have yet to absorb adequately)?"* Echoing Shubik and Zelinsky, there is a need to find ways of defining an acceptable level of damage terrorists can inflict – and must be expected to continue to inflict – on target societies and states, and then to work to confine that damage to within that acceptable level. A more open-endedly ambitious counter-terrorism agenda risks furthering a terrorist goal of inducing the target societies, possibly for emotional reasons (Passage #3.17), to take on unacceptable and/or unsustainable costs, leading to distortion of their core social values and priorities. From a perspective of terrorists and their allies and state sponsors, that upshot would be felicitous indeed, an apex of Sun Tzu (3) achievement. Such an upshot might also invite opportunistic third parties to try to profit from the situation (Passage #3.16's verse III.23; Passage #3.4's verse II.5).

---

p. 991 of David C. Hoaglin, Boris Iglewicz, & John W. Tukey, "Performance of some resistant rules for outlier labeling," *Journal of the American Statistical Association*, 1986, 81(396), 991–99.

Returning to Theme #3's root idea of cheap approaches, it should, however, be noted that not all strategic traditions share Sun Tzu's enthusiasm for that focus. Alfred Thayer Mahan famously warned against solutions "presented in the fascinating garb of cheapness." See Mahan's *The Influence of Seapower upon History, 1660–1783* (12th ed.; Boston, MA: Little, Brown & Company, 1918), p. 539.

[37] Sometimes quoted with additional clause "... even if the blood is your own."

[38] See Martin Shubik & Aaron Zelinsky, "Terrorism damage exchange rates: quantifying defender disadvantage," *Defense & Security Analysis*, 2009, 25(1), 7–20.

[39] Shubik & Zelinsky, previous footnote, p. 14, table 5, "Damage exchange rate estimates." Their article notes (p. 10) that "virtually all the figures are utilized only to provide order of magnitude estimates."

This article also discusses "structural funding requirements," comparing annual spending by terrorist groups with annual war on terror costs (again finding a vast disparity in favor of terrorists). The analysis does not address some types of non-monetary costs to terrorists who perpetrate major attacks (e.g., hard-to-replace losses by death or capture of key terrorist personnel resulting from counter-terrorist initiatives, possibly galvanized by a high-profile terrorist event).

The cyber domain, which should be conceived broadly, holds a further cornucopia of twenty-first-century examples of truly extreme cost–benefit disparities that are grist for Theme #3, in many cases involving order-of-magnitude spreads greater than the one to three orders-of-magnitude ones noted or implicit in the Sun Tzu text (and even greater than the six orders-of-magnitude one found by Shubik and Zelinsky). Relevant examples are diverse and arise on multiple levels of digital technology and its applications. Only a sampler will be noted here. Many of the disparities center on the unprecedented unlocking of the mobility of information in digital form, whereby a simple software command may unleash a software virus – or a juicy piece of gossip – or a fake news item – that, sometimes abetted by small-world social network structures and processes (see pp. 99–101 above), spreads far and wide, often unpredictably and uncontrollably. Or a firm or government agency may be gutted by a single employee exiting the building with crucial non-public data or software on a portable hard drive or other inconspicuous storage medium. Discovery of even a single injudicious email, meticulously preserved in electronic memory, may be enough to bring down a major business firm. Social media usage spawns yet further possibilities of like ilk.

A separate affinity between Sun Tzu's Theme #3 thinking and software environments also merits major attention.[40] It stems from the fact that a carefully chosen minuscule change in software, possibly even changing a single alphanumeric character – say, an exponent in a mathematical formula – can degrade or modify that software's performance in ways that have major effects.[41] In some cases those effects are also inconspicuous, in the sense that even the existence of an untoward performance problem is discovered only much later (if it ever is). In a pre-digital computing context, some of the possibilities are presaged by (non-sabotage) naval warfare experience during the battle of the Falkland Islands in World War I, where salvos fired at extreme range by British warships continually fell 100 yards to the left of their German targets. The targets were eventually destroyed, but it took a long time. It was later discovered that the problem came from effects of the Earth's rotation. A correction to reflect that rotation had in fact been programmed into the design of

---

[40] Compare an observation of traditional commentator Zhang Yu on verse V.25 (Passage #9.3): "The force applied is minute but the results are enormous" (Griffith, p. 95).

[41] Further examples arise in situations where computing meets natural language. In the 1980s a newspaper ran an article erroneously describing a criminal defendant as a "vicious informer." The reporter had intended to label that individual a "vicious enforcer," a concept having utterly different implications in the pertinent criminal subculture. According to the reporter who wrote the story this was simply a typo, sourceable to his lack of familiarity with the word processing system he was using.

As shown by the following schema (inspired by one shown on p. 377 of Katherine Stovel, Michael Savage, and Peter Bearman, "Ascription into achievement: models of career systems at Lloyds Bank, 1890–1970," *American Journal of Sociology*, 1996, 102(2), 358–99), transforming "enforcer" into "informer" requires only two letter substitutions:

enforcer
↓ ↓
informer

The *Wall Street Journal* (not the newspaper where the original error occurred) ran a followup story headlined: "On Trips, He Takes a Toothbrush, Extra Clothes and His Correction" (July 13, 1987, p. 25).

British gunsights, but with a tacit assumption encoded there that naval battles take place around latitude 50 degrees north rather than 50 degrees south. A correction factor was present – but it was the wrong correction for the location.[42]

Translated into a context of sabotage, this type of phenomenon points to a class of computing threats distinct from well-known and widely cited software virus or worm concepts. *By contrast to widespread understandings of "cyber warfare," the Internet may not be involved here.* In application to twenty-first-century weapon and other military systems that all crucially depend on software, the principle thus expressed has basic implications extending throughout the life-cycle of a system. Especially since the price tags of advanced systems run so high, scrutiny of those implications with a weather eye on Sun Tzu needs to start with the acquisition decision, including rigorous thinking ab initio about possibilities for sabotage by those who wish to thwart the system (or even better, to harness it to further their own agendas).[43]

Shifting to a different (though overlapping) focus, a major, oft-repeated Sun Tzu emphasis is on speed as a source of cheap advantage – strategic, tactical, or logistical (Passages #3.7–#3.8, #3.21A–#3.28A).[44] Such emphasis dovetails with many twenty-first-century military problems where speed is of utmost relevance, not only in kinetic realms of warfare but also in decision support and its associated computational needs. Sun Tzu's treatment of speed accommodates both strands. In particular, it is not a large leap to see parts of Sun Tzu's "soft counter-strategist" thinking (Theme #6) as advice about slowing down enemy decision-making processes.[45] That is not a new challenge: a sixth-century BC historical incident recounted by the Zuozhuan employs a *weiqi* image to depict an irresolute ruler as "holding a stone [i.e., *weiqi* piece] and not being able to decide [where to place it]."[46] Fast forward to the twenty-first century, that same *weiqi* imagery evokes computational issues associated with algorithm run-time – in particular, algorithms that run too slowly to serve

---

[42] See pp. 67–68 of *Littlewood's Miscellany* (Béla Bollobás, ed.; Cambridge, UK: Cambridge University Press, 1986) (writings of twentieth-century Cambridge mathematician John Littlewood).

[43] For analysis of software warfare – a cousin of what is now called cyber warfare, but placing emphasis on software and algorithm vulnerabilities to sabotage by subverted insiders rather than the Internet – see Scott A. Boorman & Paul R. Levitt, "Deadly Bugs," *Chicago Tribune* magazine Sunday, May 3, 1987, pp. 19ff., and Scott A. Boorman & Paul R. Levitt, "Software warfare and algorithm sabotage," *Signal*, May 1988, 42(9), pp. 75ff. (both cited in Boorman, Introduction footnote 4, p. 114 note 59).

[44] Interestingly, "speed" – whose military value Sun Tzu so emphatically plays up – fails to appear, at least by name, on many lists of the so-called Principles of War. A sizable menu of such lists is gathered by Hughes, Introduction footnote 11, pp. 290–95. It shows "speed" (or "rapidity" or "time"; see p. 297 on Nimitz's emphasis on avoiding loss of time) mentioned explicitly on just six of forty-seven lists of those principles. Holding aside one list that is sourced, sans specifics, to "USSR 1953," the five lists that include speed are ones attributed to Bedford Forrest (1864), Nimitz (1923), Giáp (1960), Montgomery (1968), and Sergey Gorshkov (1976). It is worth noting that neither of two distinct lists sourced to Sun Tzu's thinking (Hughes, pp. 290, 292) mentions speed as such.

[45] That is a useful lens through which to analyze the substance of Passages #6.7–#6.10.

[46] See p. 643 footnote 2 of Zu-Yan Chen, "The art of black and white: *wei-ch'i* in Chinese poetry," *Journal of the American Oriental Society*, 1997, 117(4), 643–53. Chen's analysis also makes reference to Sun Tzu verse VII.13 (Passage #3.8). For translation of the pertinent *Zuozhuan* passage see Durrant, Li, & Schaberg, *Zuo Tradition (Zuozhuan)* (cited on p. xxiii above), Vol. 2, pp. 1156–57.

a particular human decision-making purpose. Here Sun Tzu's contribution centers on integrating emphasis on speed into a larger way of war, a task having basic relevance to a wide range of twenty-first-century advanced military technologies.[47]

A step away from pure high tech, though still intimately related to it, are the Sun Tzu Theme #3 vistas opened by modern industrial espionage or its more genteel sibling "competitor intelligence." Faced with the many frustrating challenges presented by the continual need to innovate, one option for a modern firm is to invest heavily in R&D capabilities – always expensive, never guaranteed to produce usable results (certainly not on any fixed timeframe). Another possibility is to let rivals (or sometimes governments) shoulder that costly burden, while investing oneself in a cutting-edge espionage capability to capture the fruits of their labors. Such a stratagem amounts to a modern application of Sun Tzu Theme #3, echoing the spirit of verses XIII.1–XIII.2 (Passage #3.19). The essence of Sun Tzu lives on here. Pay hundreds of millions for pharmaceutical drug development or steal the winning formula from a rival, using the same cheap human intelligence means conjured by Sun Tzu millennia ago? Holding aside modern niceties like the rule of law, a dismissive perspective on which seems inherent in Sun Tzu Theme #10 (no scruples and controlled viciousness), the choice seems elementary!

Over and above any technological capers, Sun Tzu's Theme #3 also has important implications for thinking about patterns of social organization and ways of avoiding (or endrunning) organizational failures. Bureaucracy certainly existed in Sun Tzu's time; then as now much bureaucratic structure grows out of need to rise to the challenges of war. Yet bureaucracies the world over have always been costly to run and are famously prone to be slow. Informal social networks can often get good results both fast and cheaply, certainly by comparison with bureaucratic alternatives. By that same token, astute positioning vis-à-vis informal network structures and processes can be a major asset to a strategic actor. At the same time, however, bureaucracies, certainly in their modern Weberian incarnations, have always tended to look askance at informal networks, viewing them as more threat than resource. That type of thinking is deeply entrenched in a bureaucratic mindset, often impervious to the best efforts of those who advocate for the positive contributions social networks can offer (e.g., as ways of bypassing the excessive complexity and frequent inflexibility of massive organizations). In a world where complex formal organizations are often in charge, yet coexist with digitally empowered social network capabilities, there is scope here for Sun Tzu (3) thinking, bringing together Sun Tzu's fascination with outliers and comparative analysis of performance characteristics of informal social networks and bureaucracies.[48]

★★★★★★

---

[47] Plucking examples from a longer list, candidates include autonomy, biotechnology, directed-energy weapons, hypersonic vehicles, and quantum science (order of listing is alphabetical).

[48] For research leads see Mie Augier, James G. March, & Andrew W. Marshall, "The flaring of intellectual outliers: an organizational interpretation of the generation of novelty at the RAND Corporation," *Organization Science*, 2015, 26(4), 1140–61 (see p. 1143 for a list of key individuals – not exhaustive – who played a significant part at RAND in the early post–World War II period).

In Sun Tzu (2) and Sun Tzu (3) settings, as in Sun Tzu (1) contexts, the flip side of seeking outsized gains is avoiding outsized costs (which are, of course, often the outsized gains of some adversary!). For Sun Tzu, such outsized costs are epitomized by becoming bogged down in costly attacks on walled cities, something of a bête noire for Sun Tzu about which he warns repeatedly. But here a broader perspective is in order, one that recognizes divergence between a Sun Tzu (1) focus and natural foci of Sun Tzu (2) and (3) analysis. In warfare and other strategic competitions there may indeed be (as Passage #3.3's verse III.7 warning against attacking cites suggests) a "worst policy," which might have seductive appeal (cf. verse III.9's impatient general!). However, *the specific nature and content of that worst policy will vary greatly, depending on prevailing technology as well as military, political, economic, and social conditions.* It may have nothing to do with attacking walled cities or with urban warfare more broadly. In some twentieth-century historical contexts, for example, a candidate for a "worst policy" might be to sink national talent and treasure into pushing the state of the art to create and deploy an ultra-ambitious missile defense shield that, when deployed, is vulnerable to being circumvented at low cost.

Building on the doctrinal spirit though not the letter of Passages #3.14–#3.16, a durable piece of Sun Tzu advice is to be alert to courses of action that look cheap but aren't and hence have potential to generate egregious violations of Sun Tzu's cheapness principle. At least three kinds of factors may contribute to that result:

(a) Flawed decision-making processes reflecting psychological or political forces;

(b) Analytical blind spots (e.g., adverse dynamics present in a situation that are not clearly understood by a decision maker choosing a course of action);

(c) Adversary stratagems (see Theme #7, deception).

Illustrating (a), the annals of military history are rife with examples of the enormous costs often attendant on "no retreat" orders, many of them issued by dictators far from the battlefield with limited experience in operational military command.[49]

Illustrating (b), one of the most important, and in some ways still underappreciated, doctrinal lessons of World War II is the way in which a major logistics effort is constantly on the brink of escaping rational control. A key insight here is that the actors involved – of whom there are many – may each be taking steps that, viewed separately, are well intentioned, sensible, and indeed *locally* productive – but whose cumulative upshot is to produce an unbelievable snarl and degradation of system performance.[50] This is cheapness failure born of coordination failure, arising in a system that inevitably has major non-market aspects because money prices have a limited role and frequently do not align well on actual scarcities.[51] At the root of this violation of Theme #3 commonly stands a failure to understand a logistics system comprehensively, with an eye to achieving effective control of it.

---

[49] E.g., Hitler's December 1941 "Halt Order" to German forces facing Soviet counter-attack around Moscow, leading to a confrontation between Hitler and the experienced panzer leader Guderian who favored flexibility and selective retreat (in fact was already conducting it). There were also counterpart inflexibilities on the Soviet side, emanating from Stalin.

[50] Eccles, Background footnote 52, chapter 7, "The Logistic Snowball," pp. 102–14.

[51] For analytical perspective, informed by civilian experience in the late Soviet era, see János Kornai, "Bureaucratic and market coordination," *Osteuropa Wirtschaft*, 1984, 29(4), 306–19.

Also illustrating (b), the US decision to dismantle the Baathist army and state in the aftermath of the 2003 Iraq War is a candidate for being a horrendous violation of Sun Tzu Theme #3 and in particular of Passage #3.3's first verse:

III.1.

Generally in war the best policy is to take a state intact; to ruin it is inferior to this.

That dismantling was in some ways a plausible, even obvious, step (as attacking walled cities might have seemed to some generals in Sun Tzu's time). But as a knowledgeable participant observer of the 2003 war's aftermath later wrote:[52]

Bremer did not believe there were credible Iraqi leaders who could assume power, and he decided that the CPA [Coalition Provisional Authority] had to directly administer the country for an undefined period. America was going to rebuild Iraq, as it had rebuilt Germany and Japan after World War II. In order to overhaul the basis on which the state was run under Saddam, Bremer decreed that the CPA was dissolving the Baath party and disbanding the Iraqi security forces. *But in so doing, the US was removing the sinews of the state that had held the country together. There was now a power vacuum and a free-for-all.*

[Emphasis supplied.]

In the slightly longer run these consequences, born of lack of insight into pertinent social and political forces, proved immensely costly in US and allied blood and treasure. To these must be added costs in a currency of lost US credibility, always crucial in a part of the world where perception often creates its own reality.

Illustrating (c) is the operational concept – at whose heart was a controlled retreat presenting a faux opportunity for cheap gain by the other side (see Passage #7.14) – that underlay the great Qin victory at Changping and helped clear the way for China's first imperial unification, bringing the Warring States era to a close.

Reflecting on a course of action that became exorbitantly costly on a longer timescale, one where factors (a), (b), and (c) all played a part, Colonel Harry G. Summers, Jr. wrote of the Vietnam War:[53]

Although not normally considered as part of the principle of Simplicity, living conditions, and particularly the contrast in living conditions between the headquarters (particularly in Saigon but also at the elaborate base camps throughout the country) and soldiers in the field, were at least a partial cause of American public antipathy toward the war. [Emphasis in original.]

...Quartered mostly in Saigon, [the reporters] saw first-hand the contrast in life-styles, and this experience made many of them cynical. Michael Herr commented on the contradiction between the "Dial Soapers in Saigon" and the "grungy men in the jungle." This contradiction most certainly influenced their reporting. In American living rooms the television nightly news could and did portray the fleshpots of Saigon in juxtaposition with the tattered infantryman in the jungle. The contrast was deadly.

In a schematic telling (and with the caveat that this is but one strand of a complex war that went on for a long time and had many strands), the relevant story starts with well-intentioned attempts to try to boost morale of troops (and civilians serving with US government agencies) far from home by supplying them with creature comforts where feasible, as it was in Saigon and some other locations. In

---

[52] See Emma Sky, *The Unraveling: high hopes and missed opportunities in Iraq* (New York: PublicAffairs, 2015), p. 11. One vignette, from her time as CPA Governorate Co-ordinator of Kirkuk, Iraq, vividly conveys, in characteristically understated language, some of the practical effects on the ground (p. 36): "The only person who understood how the budget worked was a Christian woman who had been de-Baathified. We managed to get an exemption to bring her back to work."

[53] See Summers's *On Strategy: a critical analysis of the Vietnam War* (Foreword by Jack N. Merritt, Major General, US Army; New York: Dell, 1984), pp. 217, 219.

the abstract, that was a laudable, sensible ambition and may well have initially appeared bureaucratically as a "cheap" option. But supplying amenities to a large and growing number of US personnel called into being a need for additional layers of personnel and support reminiscent of the supply line problems in the background of verse II.15 (Passage #3.5; see p. 63 above). The process got out of control and started to hobble the war effort, both on a level of operations and, yet more importantly, on a level of mindset. As Colonel Summers continues:[54]

> A final impact of this violation of the principle of Simplicity may have even more serious long-term effects. The "business as usual" approach to the Vietnam war at the Washington level spilled over into our combat operations in Vietnam. Base camps were constructed that were even more elaborate than the training camps constructed in the United States during World War II and attempts were made to provide all the amenities of home. The effect was an inordinate number of soldiers tied down in base camp operations and a reduction in our ability to rapidly deploy our forces.

[Emphasis in original.]

Unintended consequences for the civilian Vietnamese population then also entered the brew. As a military staff study of the period put it bluntly:[55]

> Many of the problems were economic in nature. The introduction into the South Vietnamese economy, already in a perilous state after long years of war, of many free-spending U.S. troops armed with what the Vietnamese considered vast quantities of money was a near-disaster. Since the U.S. troops were the primary source of demand for many items, local merchants pushed prices steadily upward until they were far above the buying power of the average Vietnamese. Often this caused resentment towards the Americans in the local community and some doubt as to the sincerity of their intent to help the local people. ... Another unpleasant sociological manifestation was that the normal earning patterns changed drastically. Thus, for example, a wife could earn more as a laundress, mess hall worker, or PX employee than her husband in his everyday job.

Dynamics like these reincarnate, in a twentieth-century war, those that Sun Tzu depicts in verses II.12 and II.13 (Passage #3.5):

> II.12.
> Where the army is, prices are high; when prices rise the wealth of the people is exhausted. When wealth is exhausted the peasantry will be afflicted with urgent exactions.
> II.13.
> With strength thus depleted and wealth consumed the households in the central plains will be utterly impoverished and seven-tenths of their wealth dissipated.

Meanwhile the Communist adversary was doing a far better job at keeping their own side's logistics austere, streamlined, and thoroughly sustainable – and well out of the glare of publicity.

In part transmitted by TV war reporting, in part by stories brought back home by returning service personnel, ripple effects of all these in-country Vietnam dynamics then came back home to the US, fueling domestic antipathy to the war which deepened into active domestic opposition (which, of course, the Communist foe actively sought to exploit). So too did the monetary costs of the war. Effects of a "business as usual" mentality, initially propagated from Washington to Vietnam, now ramified in a reverse direction as well, helping set off and perpetuate a vicious cycle.

---

[54] Summers, previous footnote, p. 219.
[55] See Staff Study, "The impact of military build-up on the advisory function" (Department of the Army Headquarters, 300th Civil Affairs Group, Riverdale, MD, transmittal to Commanding Officer, United States Army Combat Development Command dated 5 April 1970), pp. B-4 to B-5.

These dynamics then combined with others – born of failures of US grand strategy, command control, intelligence, and public relations, many redolent of trying to fight the Vietnam War on a faux cheap basis (e.g., no declaration of national emergency; misrepresentation of war-related budget deficits) – to produce what has been aptly labeled the "Vietnam Hurricane."[56] Pursuit of a business-as-usual approach to fighting a land war in Asia on the cheap culminated in contributing to one of the most expensive debacles in modern US foreign relations history – precisely the sort of costly, protracted war against which Sun Tzu took so firm a stand.

★★★★★★

Set of principal passages used to illustrate Sun Tzu Theme #3 (cheap approaches)

Passage #3.1. Crossbow Metaphor

(This passage and the next serve as prologue to the main Theme #3 development starting with Passage #3.3. Verse V.16's crossbow image stands here for the principle that outsized benefits in a military sense – tiny cost, huge gain – may be possible.[57] While such benefits are a laudable ambition, verse IV.4 cautions that only sometimes is it possible to realize them. Continuing the image, the crossbow may be ready to fire, but it still needs a suitable target and a supply of bolts.)

V.16. His potential [shi] is that of a fully drawn crossbow; his timing, the release of the trigger.

Passage #3.2. Reality Check: Cheap Paths to Success Are Not Always Immediately Available

IV.4. Therefore it is said that one may know how to win, but cannot necessarily do so.[58]

Passage #3.3. Winning with Greatest Profit

(This passage, from the beginning of Sun Tzu's chapter on offensive warfare, expresses Sun Tzu's "cheapness" motif on multiple scopes and scales, from grand strategy all the way down to minor tactics.)

---

[56] Henry E. Eccles, "The Vietnam Hurricane," *Shipmate* (United States Naval Academy Alumni Association), July–August 1973, 36, 23–26. This analysis of the Vietnam War is also reprinted in Eccles, Background footnote 62, pp. viii–xv. See especially Eccles's analytical diagram on p. xi there.

[57] A crossbow metaphor for shi (strategic advantage) is further developed in the archaeologically recovered Sun Bin military text, a later text than Sun Tzu. See Lau & Ames, pp. 30, 38–39, 62–63, 118–20.

[58] Mair, p. 88 translates pithily: "Therefore, it is said, 'Victory can be foretold, but cannot be forced.'"

III.1. Generally in war the best policy is to take a state intact; to ruin it is inferior to this.

III.2. To capture the enemy's army is better than to destroy it; to take intact a battalion, a company or a five-man squad is better than to destroy them.

III.3. For to win one hundred victories in one hundred battles is not the acme of skill. To subdue the enemy without fighting is the acme of skill.

III.4. Thus, what is of supreme importance in war is to attack the enemy's strategy;

III.5. Next best is to disrupt his alliances;

III.6. The next best is to attack his army.

III.7. The worst policy is to attack [walled] cities.[59] Attack cities only when there is no alternative.

III.8. To prepare the shielded wagons and ready the necessary arms and equipment requires at least three months; to pile up earthen ramps against the walls an additional three months will be needed.

III.9. If the general is unable to control his impatience and orders his troops to swarm up the wall like ants, one-third of them will be killed without taking the city. Such is the calamity of these attacks.

III.10. Thus, those skilled in war subdue the enemy's army without battle. They capture his cities without assaulting them and overthrow his state without protracted operations.

III.11. Your aim must be to take All-under-Heaven intact.[60] Thus your troops will not be worn out and your gains will be complete. This is the art of offensive strategy.[61]

---

[59] Insertion of "walled" follows Mair, p. 85 and supplies an important piece of tactical context.

[60] In China's pre-imperial period the contours of the concept of "All-under-Heaven" (tianxia) – here invoked by Sun Tzu – were often loosely defined, having variable mixes of cultural, political, and other content (though the idea tended to evolve in a political direction). For analysis see Yuri Pines, "Changing views of tianxia in pre-imperial discourse," Oriens Extremus, 2002, 43, 101–16. In a separate study cited in Appendix 5 footnote 22 Pines notes (p. 82 there) that "peace through unifying All-under-Heaven ... was the common desideratum of the vast majority of preimperial thinkers."

[61] Sun Tzu's spotlight in verse III.11 is on a broad grand strategy canvas, pertaining to a given state's posture vis-à-vis ongoing strategic competitions with all the numerous polities and actors struggling for advantage in the twilight of the Zhou dynasty. This should be noted as a distinct, much more far-reaching, strategic agenda than how best to triumph over any particular rival state.

Depending on what facet of this complex process is emphasized, verse III.11 can be infused with various shades of meaning. In a context of an agenda as ambitious as "taking All-under-Heaven," there is certainly something to be said for an interpretation that places emphasis on self-preservation, enabling a strategic actor to stay in the game indefinitely. A basic way to accomplish this end is to keep one's own military forces intact. See Ames, pp. 111–12 (building on Lau's 1965 article, p. 335). At the same time, the goal of preservation of conquered states to which Griffith's translation points also makes geostrategic sense – if feasible. Perhaps Mair, p. 86 gets it best by somewhat freely translating the start of verse III.11 "Instead, with a comprehensive strategy, he contends before all under heaven" and concluding it with "This is the method of attack by stratagem" (a translation of the Chinese word mou, which Griffith translates as "strategy," that plays up its overtones of cunning or deception).

Passage #3.4. Sine qua Non for Winning with Profit: Avoiding Protracted War

(Here and in Passage #3.5 there is much antiquarian material tied to conditions of war in Sun Tzu's age. Some of that material has provided bones of contention for commentators and translators. The larger ideas are both easier to translate and timeless. The largest is avoiding protracted war with its devastating costs.)

II.1–II.2. (Translation from Lau's 1965 article, p. 324, which starts with a caption supplied by Sun Tzu: "*The method of employing troops.*"[62]) "When one thousand fast four-horse chariots, one thousand four-horse wagons covered in leather, and one hundred thousand armour-clad troops are used, if provisions have to be transported over a distance of a thousand li, then what with expenditure at home and in the field, on the guest advisers, on materials such as glue and lacquer, and on the supply of chariots and armour, it will cost one thousand pieces of gold every day before one hundred thousand troops can be raised."

II.3. Victory is the main object in war. If this is long delayed, weapons are blunted and morale depressed. When troops attack cities, their strength will be exhausted.

II.4. When the army engages in protracted campaigns the resources of the state will not suffice.

II.5. When your weapons are dulled and ardour damped, your strength exhausted and treasure spent, neighbouring rulers will take advantage of your distress to act. And even though you have wise counsellors, none will be able to lay good plans for the future.

II.6. Thus, while we have heard of blundering swiftness in war, we have not yet seen a clever operation that was prolonged.

II.7. For there has never been a protracted war from which a country has benefited.

II.8. Thus those unable to understand the dangers inherent in employing troops are equally unable to understand the advantageous ways of doing so.[63]

Passage #3.5. Fighting Cheaply by Living off the Enemy

(A gearshift may be detected here, with the broad-brush grand strategy considerations of Passage #3.4 transitioning into concrete logistics issues. The durable point with which Passage #3.5 leads off is simple but compelling: Sun Tzu is cautioning against overstretching a logistics system and the economic base

---

[62] Rationale for using Lau's translation rather than Griffith's is given in Appendix 1 footnote 18.

[63] There is a germ of general insight here, one that an organizational failures analyst should appreciate: To understand how to wield any instrument of power, you must know ways in which it can fail.

on which it rests, through need for repeated infusions of fresh logistics support, manpower included. In short, both logistics and economic limitations must be recognized.

Sun Tzu's attention next turns to ways of successfully navigating the difficulties. Procurement, manpower & personnel, supply, transport, maintenance & repair, and other concerns familiar to logisticians appear here. However, rather than seeking to formulate general logistics management principles – a modern agenda – Sun Tzu stays focused on extracting dramatic benefits or avoiding extreme costs. Given that standpoint, it is noteworthy – and perhaps somewhat surprising – that Sun Tzu makes no allusion to possibilities of water transportation as a way of easing hardships imposed by verse II.11's "distant transportation.")

II.9. Those adept in waging war do not require a second levy of conscripts nor more than one provisioning.

II.10. They carry equipment from the homeland; they rely for provisions on the enemy. Thus the army is plentifully provided with food.

II.11. When a country is impoverished by military operations it is due to distant transportation; carriage of supplies for great distances renders the people destitute.

II.12. Where the army is, prices are high; when prices rise the wealth of the people is exhausted. When wealth is exhausted the peasantry will be afflicted with urgent exactions.[64]

II.13. With strength thus depleted and wealth consumed the households in the central plains will be utterly impoverished [xu] and seven-tenths of their wealth dissipated.[65]

II.14. As to government expenditures, those due to broken-down chariots, worn-out horses, armour and helmets, arrows and crossbows, lances, hand and body shields, draft animals and supply wagons will amount to sixty per cent of the total.

II.15. Hence the wise general sees to it that his troops feed on the enemy, for one bushel of the enemy's provisions is equivalent to twenty of his; one hundredweight of enemy fodder to twenty hundredweight of his.

---

[64] There is context here Sun Tzu does not clarify, namely, whose peasants are they – those of one's own state or those of another state being invaded (and hopefully conquered). Depending on where the army is deployed, "Where the army is, prices are high" might point to either. From that vantage, verse II.12 might be read as support for moving the army, once mobilized, out of its home state as soon as possible, lest inflation take hold – i.e., for launching an offensive campaign without delay.

By extension, it is also an argument for bringing that campaign to an early successful conclusion.

[65] The xu half of the xu/shi concept pair (see p. 113 and #3 footnote 6 above) appears here. Mair's translation brings out the idea with clarity (Mair, p. 82): "The exhaustion of the troops' strength and the depletion of resources lead to emptiness [xu] within the households of the Central Plains."

II.16. The reason troops slay the enemy is because they are enraged.[66]

II.17. They take booty from the enemy because they desire wealth.

II.18. Therefore, when in chariot fighting more than ten chariots are captured, reward those who take the first. Replace the enemy's flags and banners with your own, mix the captured chariots with yours, and mount them.

II.19. Treat the captives well, and care for them.

II.20. This is called "winning a battle and becoming stronger".

II.21. Hence what is essential in war is victory, not prolonged operations. And therefore the general who understands war is the Minister of the people's fate and arbiter of the nation's destiny.

Passage #3.6. More about Living off the Enemy; One's Troops As Resource to Be Husbanded

(Focus here is on an invader's supply problems. In the background loom the difficulties of feeding an army on campaign that led one early Chinese military counselor to observe: "I have heard it said that, when provisions must be transported a thousand miles, the soldiers have a hungry look."[67])

XI.7. When the army has penetrated deep into hostile territory, leaving far behind many enemy cities and towns, it is in serious ground.

. . .

XI.13. . . . in deep [i.e., serious] ground, plunder.[68]

. . .

XI.31. Plunder fertile country to supply the army with plentiful provisions.

XI.32. Pay heed to nourishing the troops; do not unnecessarily fatigue them. Unite them in spirit; conserve their strength. Make unfathomable plans for the movements of the army.

Passage #3.7. Sources of Outsized Advantage (1): Speed

(Speed in conduct of military operations is intimately related to cost reduction, reducing or even eliminating costs of many kinds. Not coincidentally, the Sun Tzu text contains a bumper harvest of passages calling for speed.[69] Traditional commentator Mei Yaochen points to logistics advantages that speed affords: "Swiftness saves expenses and materials, and spares the people's labor."[70])

---

[66] Griffith, p. 75 footnote 1 suggests that verse II.16 seems out of place. An alternative understanding, rooted in Sun Tzu's Theme #3 emphases, may make it seem less so. Exploiting troops' hostile feelings may be a cheap way of motivating them in combat, even as harnessing their greed (verse II.17) may also be, with prospects of booty providing natural incentives. See Passage #9.11 analysis.

[67] See Shiji 92 (Watson translation cited on p. xxiii above), Han Dynasty I, p. 168. Traditional commentator Zhang Yu echoes the observation. See Minford, p. 125.

[68] For translation consistency, "deep ground" here should read "serious ground." See #2 footnote 61.

[69] Beyond Passages #3.4, #3.5, #3.7, and #3.8 see also Passages #3.21A–#3.28A.

[70] Mair, p. 142 note 8 (Mei's commentary pertains to verse II.21 in Passage #3.5). Although not the only reason for the importance of speed, logistics considerations can be a major factor here. The Zuozhuan records the case of a battle precipitated by one side's concerns that their chariots' leather

XI.29. Speed is the essence of war.[71] Take advantage of the enemy's unpreparedness; travel by unexpected routes and strike him where he has taken no precautions.

Passage #3.8. Imageries for Speed and the Advantages It Provides

(From the world of natural phenomena verse VII.13 offers several imageries for speed in executing military operations. Deception, highlighted in verse VII.12, can be a facilitator here, enabling unopposed, hence speedy, military moves.[72])

VII.12. Now war is based on deception. Move when it is advantageous and create changes in the situation by dispersal and concentration of forces.

VII.13. *When campaigning, be as swift as the wind; in leisurely march, majestic as the forest; in raiding and plundering, like fire; in standing, firm as the mountains. As unfathomable as the clouds, move like a thunderbolt.* [Emphases supplied.]

Passage #3.9. Ponderousness of a Large Conventional Army, Foreclosing Advantages of Speed

(A large army – especially a mass-infantry one – is cumbersome and slow-moving under the best of circumstances.[73] Cao Cao puts it simply: "You may arrive late."[74] By implication, verse VII.5 points to roles for smaller, highly agile forces.)

VII.5. One who sets the entire army in motion to chase an advantage will not attain it.

Passage #3.10. Sources of Outsized Advantage (2): Information Superiority

(Order-of-magnitude thinking, with its characteristic mix of definiteness and imprecision, is evident in the Chinese text of verse VI.13. Sawyer's translation of

---

was at risk of rapid deterioration in wetland operations against the state of Wu (a state the putative historical Sun Tzu is said to have served), whose own chariots used wood (a less vulnerable material). See Sawyer, p. 307 note 23; Durrant, Li, & Schaberg, *Zuo Tradition (Zuozhuan)* (cited on p. xxiii above), Vol. 3, p. 1755 and explanatory footnote 68 on p. 1754.

[71] A Chinese proverb, whose relevance to Sun Tzu is anchored by Cao Cao's role as a Sun Tzu commentator, points to an epitome of speed that merges thought and action: *Shuo Cao Cao, Cao Cao jiu dao* ("Speak of Cao Cao, and Cao Cao is here"). That saying – which blends exceptionally well with levels of speed possible in the digital age – is noted by Giles, p. xxxv and by de Crespigny, Background footnote 95, p. 201 and his accompanying footnote 124. The sense approximates that of the English-language idiom "Speak of the Devil."

[72] Compare verse VI.6 in Passage #6.18 (passage titled "Sun Tzu's First Symphony").

[73] Griffith's translation of verse VII.5 is mildly free, though the military idea is plain enough. The annals of world military history are replete with cases where a military operation failed because the commander could not see his way clear to avoid toting along a ponderous supply train.
A classic case in point involves Burgoyne's Saratoga campaign. See James A. Huston, *The Sinews of War: Army logistics, 1775–1953* (Washington, DC: Office of the Chief of Military History, United States Army, 1966), chapter IV, "Logistics of the Saratoga campaign," pp. 44–57, observing (p. 53) that Burgoyne "was anxious to move forward, but not anxious enough to lighten his load."

[74] Minford, p. 198.

the relevant segment is more literal: "If we are concentrated into a single force while [the enemy] is fragmented into ten, then we attack him with ten times his strength."[75])

VI.13. If I am able to determine the enemy's dispositions while at the same time I conceal my own then I can concentrate and he must divide. And if I concentrate while he divides, I can use my entire strength to attack a fraction of his. There, I will be numerically superior. Then, if I am able to use many to strike few at the selected point, those I deal with will be in dire straits.

VI.14. The enemy must not know where I intend to give battle. For if he does not know where I intend to give battle he must prepare in a great many places. And when he prepares in a great many places, those I have to fight in any one place will be few.

VI.15. For if he prepares to the front his rear will be weak, and if to the rear, his front will be fragile. If he prepares to the left, his right will be vulnerable and if to the right, there will be few on his left. And when he prepares everywhere he will be weak everywhere.

VI.16. (Translation from D. C. Lau, 1965 article, p. 330.[76]) "It is the one who has to prepare against his enemy who is few and the one who makes his enemy prepare against him who is many."

VI.17. If one knows where and when a battle will be fought his troops can march a thousand li and meet on the field. But if one knows neither the battleground nor the day of battle, the left will be unable to aid the right, or the right, the left; the van to support the rear, or the rear, the van. How much more is this so when separated by several tens of li, or, indeed, by even a few!

Passage #3.11. Sources of Outsized Advantage (3): Winning by Calculation

(Verse IV.19 offers a vivid image of outsized advantage. If read as a package, these verses open a window on Sun Tzu's thinking about "winning by calculation."[77] Although the details are much different, the idea of winning by calculation is epitomized by the game of weiqi's formidable combinatorial complexity.)

---

[75] Order-of-magnitude issues might also enter in another way here. Passage #3.10's high uncertainty situation facing the enemy could compel attention to a combinatorial explosion of scenarios, each plausible and each begging for its own tailored contingency plan. The number of such plans might easily get out of control, swamping a foe's planning and decision-making capabilities.

[76] Rationale for using Lau's translation rather than Griffith's is given in Appendix 3 footnote 9.

[77] Appendix 1 anchor footnote 10 and footnote 11 provide several alternative interpretations of verses IV.16–IV.18, which Sun Tzu's commentators and translators have understood in widely differing ways. An interpretation proposed by Needham & Yates, p. 47, anchored in terrain and troop dispositions, particularly lends itself to integrating the thrust of verses IV.16–IV.18 with that of verse IV.19.

IV.16. Now the elements of the art of war are first, measurement of space; second, estimation of quantities; third, calculations; fourth, comparisons; and fifth, chances of victory.

IV.17. Measurements of space are derived from the ground.

IV.18. Quantities derive from measurement, figures from quantities, comparisons from figures, and victory from comparisons.

IV.19. Thus a victorious army is as a hundredweight balanced against a grain; a defeated army as a grain balanced against a hundredweight.

Passage #3.12. Sources of Outsized Advantage (4): Combined Use of Qi ("Extraordinary Forces") and Zheng ("Normal Forces")

(Through its positioning here, verse V.4 – whose thrust is reminiscent of that of verse IV.19, though invoking a different set of images – signals a belief, affirmed and reaffirmed at numerous points in the Sun Tzu text, that the right choice of strategy and tactics can yield, not merely advantage, but *outsized* advantage.)

V.3. That the army is certain to sustain the enemy's attack without suffering defeat is due to operations of the extraordinary [qi] and the normal [zheng] forces.

V.4. Troops thrown against the enemy as a grindstone against eggs is an example of a solid [shi] acting upon a void [xu].[78]

V.5. Generally, in battle, use the normal force [zheng] to engage; use the extraordinary [qi] to win.

Passage #3.13. Sources of Outsized Advantage (5): Encircled Ground

(Here it is the enemy, not one's own side, that stands to harvest cheap gains.[79])

XI.9. Ground to which access is constricted, where the way out is tortuous, and where a small enemy force can strike my larger one is called "encircled".

Passage #3.14. Avoiding Military Options with a High Price Tag

(These verses depict assorted military moves to be avoided because of their probable exorbitant price tag, in some cases because they might trigger a formidable "death ground" enemy response of the sort profiled in Passage #1.12.)

---

[78] While more difficult to quantify than the imagery in verse IV.19 (Passage #3.11) – grindstone weights in particular can be very variable – Sun Tzu's grindstone-and-eggs imagery also suggests a multiple order of magnitude disparity in weight (with obvious implications for fate of the eggs!).

[79] The Wuzi military text invokes a two orders of magnitude idea to describe a comparable terrain advantage: "Where roads are constricted and dangerous, where there are famous mountains and great bottlenecks and *where if ten men defend, a thousand cannot pass*, this is potential in respect to terrain." See Griffith, p. 162 (emphasis supplied). On the Wuzi text see Background footnote 27.

VII.22. And therefore those skilled in war avoid the enemy when his spirit [qi 氣][80] is keen and attack him when it is sluggish and his soldiers homesick. This is control of the moral factor [qi 氣]. [Emphasis supplied.]

. . .

VII.25. They do not engage an enemy advancing with well-ordered banners nor one whose formations are in impressive array. This is control of the factor of changing circumstances.

VII.26. Therefore, the art of employing troops is that when the enemy occupies high ground, do not confront him; with his back resting on hills, do not oppose him.

VII.27. When he pretends to flee, do not pursue.

VII.28. Do not attack his élite troops [rui zu, lit. "sharp troops"]

VII.29. Do not gobble proffered baits [er bing, lit. "bait troops"]

VII.30. Do not thwart an enemy returning homewards.

VII.31. To a surrounded enemy you must leave a way of escape.

VII.32. Do not press an enemy at bay.

VII.33. This is the method of employing troops.

Passage #3.15. More High-Price-Tag Steps to Avoid

VIII.7. There are some roads not to follow; some troops not to strike; some cities not to assault; and some ground which should not be contested.

VIII.8. There are occasions when the commands of the sovereign need not be obeyed.

Passage #3.16. "Faux Cheap" Expedients Imposed by a Ruler

(These steps may appear to be cheap or efficient to a ruler and his entourage, yet prove to be anything but cheap, indeed may prove catastrophically costly.)

III.18. Now the general is the protector of the state. If this protection is all-embracing, the state will surely be strong; if defective, the state will certainly be weak.

---

[80] *Qi* is a traditional Chinese military concept closely related to fighting spirit or morale (Ames, p. 131), but also having other aspects intertwined with traditional Chinese belief systems where "vapors" [qi 氣] are visible manifestations of invisible cosmic potential, not to be confused with naturally occurring phenomena like clouds. See Appendix 11 footnote 32 (citing the *Shiji* and modern sources). Sawyer leaves qi 氣 untranslated, observing (pp. 318–19 note 116) that a "separate monograph on the psychology of qi [氣] in battlefield contexts would be required to fully address the subject." On a point of clarification, qi 氣 is a distinct concept from qi 奇 ("crafty," "unconventional," "surprise," etc.) which appears in Passage #3.12 and takes center stage in Theme #14 analysis.

Engaging with qi 氣 on a philosophical plane, it is hard to improve on an imaginative comment on it by Zhang Dainian, *Key Concepts in Chinese Philosophy* (translated and edited by Edmund Ryden; New Haven, CT/London/Beijing: Yale University Press/Foreign Languages Press, 2002), p. 45:

Perhaps the best translation of the Chinese word qi [氣] is provided by Einstein's equation, $e = mc^2$. According to this equation matter and energy are convertible. In places the material element may be to the fore, in others, what we term energy. Qi embraces both.

III.19. Now there are three ways in which a ruler can bring misfortune upon his army.

III.20. When ignorant that the army should not advance, to order an advance or ignorant that it should not retire, to order a retirement. This is described as "hobbling the army".

III.21. When ignorant of military affairs, to participate in their administration. This causes the officers to be perplexed.

III.22. When ignorant of command problems to share in the exercise of responsibilities. This engenders doubts in the minds of the officers.

III.23. If the army is confused and suspicious, neighbouring rulers will cause trouble. This is what is meant by the saying: "A confused army leads to another's victory."

Passage #3.17. Avoiding Wasteful Delay in Exploiting Success; Avoiding Catastrophically Costly Moves

(Particularly noteworthy from a Theme #3 vantage is verse XII.18, expressing in lightly veiled terms a concern that a small perceived slight to a ruler might trigger a catastrophic war leading to extinction of that touchy ruler's state.[81] A larger principle is that decisions taken for emotional reasons are often costly.)

XII.15. Now to win battles and take your objectives, but to fail to exploit these achievements is ominous and may be described as "wasteful delay".

XII.16. And therefore it is said that enlightened rulers deliberate upon the plans, and good generals execute them.

XII.17. If not in the interests of the state, do not act. If you cannot succeed, do not use troops. If you are not in danger, do not fight.

XII.18. A sovereign cannot raise an army because he is enraged, nor can a general fight because he is resentful. For while an angered man may again be happy, and a resentful man again be pleased, a state that has perished cannot be restored, nor can the dead be brought back to life.

XII.19. Therefore, the enlightened ruler is prudent and the good general is warned against rash action. Thus the state is kept secure and the army preserved.

---

[81] This is not a contrived contingency. Stories from Spring and Autumn China tell of a world where

> casual social intercourse of the nobles provided a fertile ground for quarrels, any perceived slight was answered with force, the segmentation of authority turned appeals to force into wars, and the probability and severity of reprisals placed great pressure on anyone involved in a quarrel to strike first with full military force in order to avoid falling victim himself to an adversary's assault.

(Lewis, Background footnote 11, p. 40)

Compare Schelling, #1 footnote 71, p. 23 ("strike first with full force" incentives operating on gunfighters in the American Old West).

Passage #3.18. Exception to Rule of Cheap: Taking Good Care of Secret Agents

(Sun Tzu's advocacy for generously rewarding secret agents – in Sun Tzu's world, a cast of characters who seem to be mostly motivated by money or material rewards – provides a kind of codicil to the Theme #3 doctrine of cheapness.)

XIII.12. Of all those in the army close to the commander none is more intimate than the secret agent; *of all rewards none more liberal than those given to secret agents*; of all matters none is more confidential than those relating to secret operations.

. . .

XIII.17. It is essential to seek out enemy agents who have come to conduct espionage against you and to *bribe them to serve you*. Give them instructions and *care for them*. Thus doubled agents are recruited and used.

. . .

XIII.21. The sovereign must have full knowledge of the activities of the five sorts of agents. This knowledge must come from the doubled agents, and therefore *it is mandatory that they be treated with the utmost liberality*. [Emphases supplied.]

Passage #3.19. Why This Exception Is No Exception: Espionage As Fantastically Cheap by Comparison with Conventional Warfare

(Previously quoted in an abbreviated form on p. 113 above, this pair of verses is one of the most compelling illustrations of Theme #3 thinking in all of Sun Tzu.)

XIII.1. Now when an army of one hundred thousand is raised and dispatched on a distant campaign the expenses borne by the people together with the disbursements of the treasury will amount to a thousand pieces of gold daily. There will be continuous commotion both at home and abroad, people will be exhausted by the requirements of transport, and the affairs of seven hundred thousand households will be disrupted.

XIII.2. One who confronts his enemy for many years in order to struggle for victory in a decisive battle yet who, because he begrudges rank, honours and a few hundred pieces of gold, remains ignorant of his enemy's situation, is completely devoid of humanity. Such a man is no general; no support to his sovereign; no master of victory.

★★★★★★

THEME #4 WINNING WITHOUT FIGHTING, BY NON-MILITARY WAYS AND MEANS

affirmations of principle of winning without fighting:

Passage #4.1 (verses III.1– III.3 + III.10–III.11) (winning without fighting)

Passage #4.2 (verses III.4–III.7) (hierarchy of kinds of strategy, privileging non-military means)

Passage #4.3 (verses XII.16–XII.19) (thinking through decisions for war with utmost care)

Passage #4.4 (verses IV.1–IV.15) (winning against an enemy already defeated)

non-military tools and ways of wielding them:

Passage #4.5 (verses XI.6 + XI.13 + XI.19) (alliance-building as a tool)

Passage #4.6 (verses V.19 + V.21–V.22) (people picking as a tool)

Passage #4.7 (verses XIII.3–XIII.4 + XIII.13–XIII.14 + XIII.16 + XIII.19–XIII.20 + XIII.22–XIII.23) (spies and covert action as tools)

Passage #4.8 (verses IX.25–IX.28) (words as weapons)

Passage #4.9 (verses VIII.17–VIII.24) (ad hominem stratagems: exploiting personality traits)

Passage #4.10 (verses I.15–I.27) (menu of stratagems, here viewed through a non-military lens)

Passage #4.11 (verses VIII.14–VIII.15) (further stratagems, including burdening adversary or rival with a task)

ideas relevant to subversion campaigns:

Passage #4.12 (verses III.18–III.23) (situation conducive to mounting an effective subversion campaign)

Passage #4.13 (verse II.5) (situation conducive to an effective subversion campaign, cont.)

Passage #4.14 (verses VI.26–VI.29) (water image for opportunistic dynamics of subversion)

Passage #4.15 (verses IX.37 + IX.41) (social dynamics providing traction for a subversion campaign)

Passage #4.16 (verse XI.44) (defender's standpoint: countering subversive dynamics)

★★★★★★

Winning without fighting dovetails with the fundamental idea of Theme #3, seeking sources of cheap gains – the more outsized the better. Certainly many of the basic ideas (spying is cheap; ideal of capturing the enemy intact rather than destroying that enemy; etc.) that animate Theme #3 also animate Theme #4. However, Sun Tzu Theme #4 has its own distinct identity as a cluster of ideas. In its purest form, it is about displacing conflict onto an altogether different plane from war in its military sense, one where the setting, the immediate goals, the tools, the criteria of judgment, and often the actors are fundamentally different from those present in the warfare that traditional soldiers fight.[1] Moving away from ideal type, military means – possibly through their sheer existence – may still have a role in some Theme #4 contexts, though the emphasis lies elsewhere. Because non-military success commonly draws on a mix of ingredients, the Sun Tzu of Theme #4 is a grand strategist in Liddell Hart's mold, for whom[2]

> fighting power is but one of the instruments of grand strategy – which should take account of and apply the power of financial pressure, of diplomatic pressure, of commercial pressure, and not least of ethical pressure, to weaken the opponent's will.

Theme #4 paths to success are profiled with a traditional Chinese tilt by Needham & Yates, observing that "victory without fighting"[3]

> may be achieved by skilful diplomacy, by sending agents and envoys to foreign countries, by making the enemy withdraw his plans, or merely by clever calculation, 'defeat by plans' ... In the last cases some fighting might be allowed, though victory was achieved in reality by the competition of minds rather than by force. ... Such ideas were propagated in Sun Tzu ...

There is a strong and highly developed strand of Chinese strategic thinking that prizes precisely this type of path to success – not only holding a door open to the possibility of winning by non-combat means but actively esteeming such a pathway, looking on it with great favor for reasons that are in part instrumental (fighting is expensive!) but also have broader cultural roots. This strand is a very old one, as old as Sun Tzu (indeed older), and is rooted in many facets of Chinese tradition – philosophical, ethical, bureaucratic, historiographical, and more, as well as in more narrowly political–military realms.

Much of Sun Tzu's thinking is certainly compatible with a modus operandi that extends smoothly from military to civilian realms of conflict.[4] What is *not* found in Sun Tzu, however, is explicit development of the specifics of *how* to win by using

---

[1] At the outset of Theme #4 development it is helpful to distinguish among several (possibly overlapping) categories of users of non-military tools (all of whom are potential consumers of Sun Tzu):

    (a) state actors;

    (b) non-state actors aspiring to play in the major leagues alongside or against state actors, for example, some multinational enterprises, revolutionaries, some terrorist groups;

    (c) political or bureaucratic factions or interest groups based within, or working alongside of, state actors and their instrumentalities;

    (d) private individuals, social networks, or organizations.

Many strategic actors in the last two categories lack access to military tools, or have only marginal access to them, and for that reason how they draw on Sun Tzu may differ in important ways from how state actors do. In developing Theme #4 state actors will remain the default focus, until attention shifts to selected non–state actor cases starting in Theme #4 "frontiers." See pp. 152ff. below.

[2] B. H. Liddell Hart, *Strategy* (2nd rvd. ed.; New York: Praeger, 1967), p. 336, also p. 31 (further summary statement). Ethical pressure is often seen as altogether absent from Sun Tzu's repertoire. However, Giles, p. 162 makes the intriguing suggestion that Sun Tzu XIII's famous case for spying (Passage #3.19) amounts to accusing a ruler or general who ignores it of a crime against humanity!

[3] Needham & Yates, p. 38.

[4] Here it is relevant that in Warring States China the "distinction between grand strategy and rhetorical persuasion, on the one hand, and combat, on the other, was not crucial." See Lewis, Background footnote 11, p. 101. Yet as Lewis also emphasizes, the Warring States period saw the military commander's role emerge with its own identity and sphere of action separate from that of the ruler – a distinction echoed in Sun Tzu's explicit delineation of situations where the general

non-military approaches. Combing the Sun Tzu text for passages elaborating paths to success by means that don't center on a general and an army yields at best a slender harvest. To borrow a phrase from Ames's translation, Sun Tzu's mindset remains that of an "expert in use of the military," and with that mindset comes a cluster of emphases centering around armies and combat power. One way to sum up Sun Tzu's focus is that, of all the tools and weapons of conflict and, relatedly, of alternative forms of power, Sun Tzu systematically discusses only two: first, conventional armies; second, spies (and then only in the close-to-stand-alone last chapter, quite possibly a latecomer to the text and which provides limited insight into how spies are actually utilized). That lacuna may be surprising to some Western audiences unwittingly indoctrinated by a modern popular culture that has grown up around Sun Tzu to envision Sun Tzu as some sort of social or political philosopher (or at least a great theorist of palace intrigue!). Concrete approaches to winning by non-military means are indeed set forth in detail in various other early Chinese texts – but not in Sun Tzu.

The key to successful engagement with Theme #4 therefore starts by recognizing that Sun Tzu's own coverage of it is like an unfinished painting – there is just enough suggestive material there to envision, at least in bare outline, what the finished artwork might look like. That process of envisioning calls for active use of the viewer's imagination, and is itself part of Sun Tzu's contribution.[5] There could be many ways of conjuring the completed artwork depending on specific interests and assumptions and how lively the conjuror's imagination is. A lawyer's skill-set versed in spotting, critiquing, and at times flat-out rejecting analogies has a part to play here. Emphases on analogical thinking, already well developed in early Chinese culture, support a notion that some traditional Chinese readers of Sun Tzu would have welcomed, indeed reveled in, that type of intellectual exercise.[6]

---

may flout the orders of the ruler (see Theme #13 analysis). Such role differentiation need not, of course, preclude military/civilian concept sharing. Indeed it may tacitly promote it, for all the reasons that divided elites the world over often keep a wary eye on one another!

[5] Here Sun Tzu's legacy and tradition has resonance with strands of art history. As *New York Times* art critic Roberta Smith has written, "Unfinished paintings are enticing cracks in the façade of art history, lures along the path to a deeper understanding of artistic processes and impulses." See her "The fascination of the unfinished" (Critic's Notebook), *New York Times*, January 10, 2014, p. C27. See also Peter Brooks, *Enigmas of Identity* (Princeton, NJ: Princeton University Press, 2011), p. 175 (commenting on work of Cézanne often labeled "unfinished"): "The eye is being retrained, called upon to learn anew to picture landscape as a production of the mind." That concept of "retraining the eye" captures the challenge of transposing Sun Tzu's ideas to non-military applications.

Thanks are owing to Paul Solman, Scott Boorman's colleague in Yale's Grand Strategy program, for this reference to Peter Brooks's work (personal communication, April 19, 2020).

[6] There is evidence that in Warring States times the Sun Tzu text already had an appreciative audience among literate civilians. Han Feizi (d. 233 BC) – a major Warring States thinker of Legalist persuasion (which the Sun Tzu text is not) – averred that, in his time, "in every household there is one who keeps concealed the books of Sun and Wu" (translation from Petersen's article cited below). Bai Gui, a late fourth-century BC businessman is reported to have said that "I manage my business affairs in the same way that . . . the military experts Sun Zi and Wu Zi deployed their troops." See *Shiji* 129 (Watson translation cited on p. xxiii above), *Han Dynasty II*, p. 439. On the *Wuzi* text see Background footnote 27.

It bears mentioning that in both these cases the Chinese text refers just to "Sun and Wu," thus opening a door to a possibility that "Sun" is actually Sun Bin, a later Warring States military thinker (on whom see Lau & Ames, pp. 4ff.). That possibility is further explored by Jens Petersen, "On the

Against this backdrop, it is helpful to discuss the nature and role of what might be labeled "dual use" Sun Tzu passages. These are passages that lend themselves, with few interpretive demands upon the reader, to being read on one level that is military and on a second level that is civilian (and, by virtue of the latter reading, may provide useful advice to an "off-battlefield general").[7] In effect, such passages help build a "doppelgänger" theory of civilian conflict that coexists with Sun Tzu's military theory. There are quite a few such passages. Of course, such a theory is an analytical construct distilled from a modern reading of Sun Tzu. We do not know whether Sun Tzu's readers in early China would have engaged with the text in a similar way.

Forays into a dual use reading of Sun Tzu may be usefully organized along a conceptual continuum. That continuum starts with Sun Tzu material of fully dual use type – passages where even a literal-minded reader can effortlessly toggle back and forth between military and civilian applications. Much of Sun Tzu's thinking about spies would be an example of fully dual use material. A step beyond that would be other Sun Tzu passages that support a dual use reading with only minor effort. A case in point would be Passage #4.9's advice to exploit character flaws in an enemy general, which requires little retooling to find use in spotting comparable pressure points on civilian leaders, decision makers, or other influentials (where indeed Passage #4.9 may find more fertile opportunities for application than in dealing with a foe's grizzled, suspicious battlefield generals). A fractionally more ambitious step would bring into the same fold Sun Tzu passages like Passage #4.4 whose allusions to "battle" are the chief impediment to a civilian reading. Many civilian contexts harbor some sort of counterpart to "battle." What that counterpart is depends, of course, on context. A place to seek examples would be some form of crystallizing event involving a major irreversible commitment in a civilian setting of intense conflict. An example might be an adviser's audience with a ruler involving a passionate plea against a ruler's misguided decision – the sort of "remonstrance" activity, well known in traditional Chinese political culture, whereby an underling takes the ever-risky step of speaking truth to power in a context of deep disagreement.[8]

Ranging further out along the dual use continuum involves progressively more ambitious analogical exercises aiming to squeeze civilian content from military ideas. Examples have already been suggested in connection with reinterpreting Sun Tzu's terrain thinking in social network settings (Theme #2 "frontiers," pp. 98–105 above). Precisely because so much of the Sun Tzu text is oriented around issues of

---

expressions commonly held to refer to Sun Wu, the putative author of the *Sunzi bingfa*," *Acta Orientalia*, 1992, 53, 106–21. However, rigorous distinction between Sun Tzu of the thirteen chapters and Sun Bin seems largely beside the point for purposes of gauging Sun Tzu's civilian influence. As Sawyer, p. 161 points out, over the centuries "people conflated the works of the two Suns, rather than viewing them as distinct individuals and authors."

[7] It is worth noting an affinity between the present concept of dual use Sun Tzu passages and the classical game theory tradition, which – as befits a formal area – is fundamentally indifferent as to whether a particular abstract game is given a civilian or a military interpretation (or possibly both).

[8] "Remonstrance" as a mode of action is analyzed in Charles O. Hucker, *The Censorial System of Ming China* (Stanford, CA: Stanford University Press, 1966), noting (pp. 6–7) that its roots go back to pre-imperial times. Regarding the institutionalized censorial system in imperial China see Wilkinson, §17.1.3.3, p. 274.

psychology and information, albeit geared toward military settings, much – even most – of what Sun Tzu says might find a place somewhere along the dual use continuum. Yet the further out one ventures on it, the further removed is the interpretation from the letter of the text and the more issues of validity in interpretation (to borrow a useful concept of literary scholar E. D. Hirsch) intrude.[9]

There is certainly Sun Tzu content whose civilian transposability is limited. For example, as is only to be expected of a treatise on land warfare, much of Sun Tzu presupposes a Euclidean spatial context, two- or three-dimensional.[10] While spatial factors (e.g., ones involving architecture or urban planning) can be important in some major civilian conflict settings too, the issues and emphases there are usually remote from those arising in land warfare. Meanwhile Sun Tzu's references to biophysical problems like hunger, thirst, or physical fatigue impinging on soldiers – while also relevant in some civilian settings – seem largely extraneous in many others (e.g., struggles between well-placed elites far from any margin of subsistence).

A greater barrier to dual use is illustrated by Passage #1.32A, whose military focus is force size ratios. While comparative counts of people, units of wealth, etc. may shape civilian conflict outcomes too, they cannot be expected to play a role in civilian situations in a way that aligns at all closely on how relative army sizes affect battle outcomes.[11] Still less promising as candidates for dual use are Sun Tzu passages tied to characteristics of a Warring States Chinese army. Many of Sun Tzu's organizational assertions found there generalize guardedly at best.[12]

A more subtle source of limitation on dual use is the fact that the timescales of civilian conflicts, and steps taken as those conflicts unfold, are not uncommonly more protracted than those associated with active conventional warfare, a point with ramifications for the types of problems a strategist must face and navigate. Contextual factors not contemplated by Sun Tzu's thirteen chapters – such as demography, leadership succession, and sociopolitical identity – start to become important.

---

[9] See E. D. Hirsch, *Validity in Interpretation* (New Haven, CT: Yale University Press, 1967). Analysis of criteria of judgment to guide and discipline Sun Tzu dual use exploration has received scant attention in the modern Sun Tzu literature. Much of it seems to treat the task of transposing Sun Tzu's content into civilian realms as unproblematic (or as a problem already solved by someone else unspecified).

[10] A three-dimensional example would be verse V.14's raptor and its prey (Passage #1.4). Further 3-D examples come from Sun Tzu's interest in ravines as a terrain type. See p. 95 above.

[11] E.g., gambler's ruin problems in finance, where outcomes depend, inter alia, on initial endowments.

[12] Attempts to do so run up against the need to gain insights resilient enough to survive a *double* major change of context: (a) military → civilian plus (b) Warring States China → another society having a different social structure and culture. Unsurprisingly, efforts at such double transfer often turn out, on scrutiny, to involve a bridge too far. Some of the obstacles come into focus by reflecting on the fact that Warring States armies were deeply authoritarian, centralized organizations (yet ones that existed prior to availability of writing on paper), whose rank and file was mostly comprised of illiterate peasant conscripts. Some of Sun Tzu's thinking can indeed be harnessed to convey complex organizations insights, but developing them calls for more than simply pointing to dual use verses.

Keeping these caveats in mind, spotting promising candidates for dual use passages is a kind of textual detective exercise. The present roundup of passages anchoring Theme #4 adopts a relatively conservative approach to dual use. It also excludes much "dual use" material that fits readily under other Sun Tzu themes.[13]

A first group of passages, anchored in Passages #4.1 and #4.2, centers on general affirmations of Theme #4 thinking. To the warfare-minded, those passages have a horizon-broadening aspect: they place Sun Tzu's credibility behind the existence of a kind of "parallel universe" (another description would be "alternative sphere of reality") where you can still get what you really want without incurring anything like the costs of war! Encouraging that line of thought, Passage #4.3 warns those contemplating combat operations that warfare is fraught with peril and should be weighed with utmost care before committing to it. Passage #4.4's verse IV.14 is a further affirmation of core Theme #4 ideas. Passage #4.4 as a whole develops a line of thought, centered around the idea of lying in wait for vulnerabilities of foes or rivals to show themselves, applicable with little or no editing in many civilian conflict settings.

A second group of passages (Passages #4.5–#4.11) provides concrete grist for Theme #4 by pointing to means or tools that are non-military (or at least not necessarily military). Candidly, those passages run short on detail. A few more details might have come in handy in fleshing out what sorts of non-military conflicts (if any) Sun Tzu himself contemplated; what his key assumptions about those conflicts were; and what his specific ideas for achieving success there may have been. Such details are more needed in many civilian settings than in a context of conventional land warfare, which has a well-defined and widely known structure.

Illustrating Sun Tzu's repertoire, this second passage group includes Passage #4.5 (building alliances); verse I.25 in Passage #4.10 (attacking alliances; see also verse III.5 in Passage #4.2);[14] #4.6 (picking the right people); #4.7 (spies and covert action); #4.8 (words as weapons, especially suited for deception purposes); #4.9 (psychological warfare targeting a person of interest); #4.10 (assorted measures, often psychological, all having natural civilian interpretations); #4.11 (burdening an adversary or rival with a task, a particularly productive line of thinking in civilian social structures already rife with social or administrative obligations).[15]

---

[13] For example, various ideas in Theme #8's "ladder of formlessness" (p. 270 below) have natural civilian uses, some centering on non-state actors. See Nüjie analysis, pp. 153–57 below. Separately, Theme #12 development puts a battery of Sun Tzu ideas to work in a dual use way. See in particular Table 6 and pp. 404ff. discussion below.

[14] A further part of Sun Tzu's alliance thinking is Passage #4.17A, though there are substantial ambiguities in its interpretation. Beyond that, there are reasons to regard that material as problematic. It may be an interpolation (see Appendix 8 anchor footnote 19 as well as Table 13A on p. A-37).

[15] Interestingly, Sun Tzu never takes true cognizance of economic warfare as a tool of conflict (as contrasted with noting war's economic costs for one's own state, see Passage #3.4). Passages #3.5 and #3.6 point to advantages of living off enemy resources, a distinct albeit possibly overlapping idea.

A third group of Theme #4 passages roughs out a nascent Sun Tzu theory of winning by subversion.[16] Anchored by verse IV.10 (Passage #4.4),[17]

IV.10.
Anciently those called skilled in war conquered an enemy easily conquered.

and drawing material from five more of Sun Tzu's thirteen chapters, this third group comprises Passages #4.12–#4.16 (also relevant is earlier Theme #4 material about spies or creating estrangements).[18] With its evocative word picture of an army and polity divided against itself, Passage #4.12 serves in a kind of flagship role here.[19]

Roads Not Taken by Sun Tzu in Developing Theme #4

Adherence to the discipline of the text makes clear certain major holes and limitations in Sun Tzu's Theme #4 outreach. The single most basic hole is Sun Tzu's lack of explicit attention to how non-state actors might use, and greatly profit by, Sun Tzu's ideas.

A further basic limitation is the pervasive presence in Sun Tzu of what, in a modern framing, might be termed a "zero-sum" mentality.[20] Zero-sum thinking transposes awkwardly, if at all, to nuanced civilian environments, where "win–win" situations often exist and a blend of cooperation and conflict is normal. Recognizing and building on that basic fact is commonly a sine qua non for an astute civilian strategist to advance her interests with greatest efficacy. Yet Sun Tzu makes no effort at developing his ideas in this type of direction.[21] Especially in a context of a densely networked world (a ruler's court would be a good example), where friend and foe are both entangled in complex and often contradiction-riddled larger networks of ties, here is a truly major omission in Sun Tzu's strategic thought.[22] It also has practical consequences. For example, a Sun Tzu–inspired strategist might

---

[16] A pertinent Chinese strategic image comes from sericulture, and involves "conquering or subduing the enemy step by step, i.e. 'eating a whole [leaf] like the silkworms do, swallowing like the whales' (canshi jingtun)." See Needham & Yates, p. 39.

[17] Glossing verse IV.10, Griffith, p. 87 footnote 1 observes: "The enemy was conquered easily because the expert previously had created appropriate conditions" – a summary that excellently captures how a successful subversion campaign should work.

[18] Verses VI.9 and VI.24 (from Passage #8.1,"Formlessness") also resonate well with subversion ideas.

[19] One of Sun Tzu's contributions to a theory of subversion may actually reside in something Sun Tzu does *not* say, but could be read as implying: namely, a warning to keep subversive acts low-profile so as to avoid triggering some type of desperate "death ground" response by an endangered foe (Passage #1.12). Cf. relevant strands of Dzerzhinsky's and other Soviet Communist thought (p. 98 above).

[20] On Sun Tzu's zero-sum tendencies see p. 541 of Robert W. Dimand & Mary Ann Dimand, "von Neumann and Morgenstern in historical perspective," *Revue d'économie politique*, 1995, 105(4), 539–57.

[21] Even the "truce" in verse IX.28 (Passage #4.8) is a sham, a cover for a nefarious scheme.

[22] For algebraic (semigroup) approaches to analyzing this type of situation, probing questions like "are my enemy's friends also my enemies?," see Scott A. Boorman & Harrison C. White, "Social structure from multiple networks. II. Role structures," *American Journal of Sociology*, 1976, 81(6), 1384–1446.

be prone to overlook the possibility that two adversaries could work out a win–win accommodation between themselves, thereby emerging as a more formidable foe.[23]

Relatedly, Sun Tzu shows no interest in engaging with the very real challenge of designing and building enduring structures, including intangible ones (such as crafting a lasting alliance) widely associated with successful statecraft and statesmanship. In particular, Sun Tzu shows no interest in the ever-thorny problem of how to craft stable incentives for a mixed bag of allies (along with assorted proxies, protégés, etc.). Were Sun Tzu to be reincarnated as an engineer he would be a combat engineer, not a civil engineer or urban planner.[24]

More broadly, there is nothing in Sun Tzu that explicitly sets forth features of any specific civilian conflict environment where Sun Tzu's strategic ideas might find use. On the most basic level, one finds in Sun Tzu no menu of types of civilian tools of conflict on which a Sun Tzu–inspired strategist might draw; the assemblage found in Passages #4.5–#4.11 is a Sun Tzu analyst's construct, not Sun Tzu's own. Yet a categorizing instinct is hardly foreign to whoever wrote or compiled the Sun Tzu text. Like so many other early Chinese texts it is replete with short lists on many topics, some of them numbered.[25] Absence of a list of types of civilian tools seems a telltale sign that the civilian applicability of Sun Tzu's ideas is to some significant extent an audience invention, ranging beyond any self-identified remit of the text.

There is also a large difference between simply mentioning a particular tool of conflict or form of power and insights into how to wield it adroitly. Here too Sun Tzu's civilian coverage shows limitations. Even in the case of his favorite civilian tool – spies – much of the treatment is given over to a typology of spies and description of their activities. How spies might be actually employed to achieve strategic ends, possibly purely civilian ones, is only sparsely covered.

"How am I doing?" is a basic evaluative question that all practicing strategists the world over must continually pose to themselves. Answers to that question – at least adequate ones – tend to be considerably more complex and elusive in non-military contexts than in conventional warfare ones. Yet Sun Tzu shows no signs of recalibrating his military thinking for applications to civilian environments, where constraints and criteria of success might, even in Warring States China, be quite different from battlefield ones.

Again underscoring the relatively undeveloped character of Sun Tzu's Theme #4 thinking, it is also worth noticing that Sun Tzu draws on a rich array of non-military metaphors to illuminate aspects of successful warfare (in Sun Tzu V alone, images of a grindstone smashing eggs; the flow of great rivers; torrential water; a raptor and its prey; logs and stones rolling down a mountain). By contrast, the text is devoid of uses of metaphor deployed for an inverse purpose, drawing on military imagery to illuminate facets of non-military life or activities.[26] Doing so might be one natural

---

[23] A candidate for an exception to this Sun Tzu blind spot, though introduced only glancingly by Sun Tzu, is verse XI.39 (Passage #9.8). See #1 footnote 40 above.

[24] This facet of Sun Tzu again illustrates a deficit of long-term planning foci. See pp. 66–67 above. One of Sun Tzu's few mentions of built structures of a physical type – verse VI.11 (Passage #6.18), alluding to "high walls and deep moats" – positively scoffs at their efficacy.

[25] For relevant background see Wilkinson, §38.11, pp. 537–38, "Mnemonic number[ed] sets."

[26] An example comes from a renowned calligrapher, Wang Xizhi (321–79 AD): "paper represents the troops arrayed for battle; the writing brush, sword and shield; ink represents the soldier's armor;

starting point for a more systematic foray into analyzing the continuations of war by civilian means with which Sun Tzu is so famously associated. Yet Sun Tzu takes no step in that direction.

Sun Tzu (2) ("Sun Tzu Extended") and Sun Tzu (3) ("Sun Tzu Analogical") Frontiers of Theme #4

Focusing first on conflicts between state actors, possibly the single most important insight to be gleaned from Sun Tzu Theme #4 centers on the cheap successes that a well-orchestrated campaign of subversion and political warfare may be able to realize against even a militarily top-notch adversary – undermining that adversary's psychological and political coherence, even to a point of ultimately toppling that actor without a fight.[27] The traditional reading of verse III.1 (Passage #4.1) is certainly consistent with such a concept. Yet it is also important to recognize that the Sun Tzu text itself plays up only the end result (and without any affirmation that it is always feasible), giving only barest hints regarding steps along the path leading to it. The closest the text comes to giving a word picture of those intermediate steps comes from the army-as-water images found in verses VI.27–VI.28 (Passage #4.14).

Parsed as strategic theory, that water imagery evokes a seminal 1952 article of then-Captain (later Rear Admiral) J. C. Wylie, Jr., developing an analytical direction he credits to Herbert Rosinski.[28] Wylie distinguishes between two basic kinds of strategy (which, importantly, can operate in concert): *sequential strategy*, involving a sequence of steps each building on previous ones, as might happen during an invasion of enemy territory; and *cumulative strategy*, comprising many lesser actions that are not sequentially connected but whose cumulative impact may spell success. As an example of the latter Wylie cites the "tonnage war" waged by US submarines against Japanese merchant shipping (and thereby also the Japanese economy) during World War II. He also mentions psychological warfare as a possible further vein of examples, though without developing specifics.

In view of Sun Tzu's psychological emphases, it is in the psychological varieties of strategy – often cumulative, though in some cases sequential – where Sun Tzu, Wylie, and subversion campaigns share much common ground. Wylie's article suggests that the naval form of cumulative strategy that he identifies as one key ingredient of US Pacific War success was marginalized in American strategic thinking, both during World War II and after it. Rather, the focus was on sequential strategy, as famously illustrated by the Pacific War's "island hopping" and "leapfrogging" operations.[29] In contemporary times, given the vagaries of collective

---

the ink-slab, a city's wall; while the writer's ability is the chief commander." See Tsien, #1 footnote 1, p. 175. In an era of Internet-enabled communications technology, Wang's observation should be a reminder that the human writer's ability, like command talent, remains a basic limiting factor.

[27] Griffith, p. xi offers a succinct word picture.

[28] Captain J. C. Wylie, Jr., USN, "Reflections on the war in the Pacific," United States Naval Institute *Proceedings*, April 1952, 78(4), 351–61. Rosinski's role is acknowledged in Wylie's p. 360 footnote 4.

[29] For brief description of each of these two strategies see p. 7 of Samuel E. Morison, "Thoughts on naval strategy, World War II," *Naval War College Review*, March 1968, pp. 3–10. The latter strategy's

memory, clarity of US thinking about what was once termed "termite warfare" (a phrase of Cold War vintage used to point to a cumulative Communist strategy with political warfare emphases[30]) may have dwindled rather than sharpened since the early post–World War II era in which Wylie's ideas took shape.[31]

Subversion ideas harmonize outstandingly with Sun Tzu's signature emphasis on cheapness (Theme #3). Subversion may in fact be cheap for two different kinds of reasons, each supported by verses XIII.1–XIII.2 (Passage #3.19). It is cheap because the carnage of major battles is averted, encounters in which success – if it occurs at all – is won "not by personnel and materiel in prime condition, but by the debris of an organization worn by the strain of campaign and shaken by the shock of battle."[32] Subversion may also be cheap because it reduces the need for an attacker to maintain (and, of course, pay for) a large, costly military establishment in the first place.

A subversion-compatible strand of Sun Tzu helps buffer Sun Tzu's overall thinking against what might otherwise be a major criticism, targeting his predilection for ways of warfighting that aspire to achieve major successes but without hard-fought battles. Taken by itself, such an aspiration could open Sun Tzu to much the same vein of criticism that Clausewitz levied, with great clarity and force, at tendencies in eighteenth-century European generalship aspiring to replace sanguinary encounters by clever, largely bloodless maneuvers – an aim whose flaws the advent of Napoleonic warfare exposed and swept aside "with the effect of pent-up waters which, suddenly released, plunge into a bottomless abyss" (Sun Tzu verse IV.20, see Passage #9.4).

For another reason too, subversion as a facet (albeit an implicit one) of Sun Tzu's thinking and its applications should hold particular interest for his twenty-first-century US audiences. For many of those US audiences the surprise attack legacy of Pearl Harbor, which the 9/11 attacks did much to reinforce for later generations, still looms large. For them, strategic surprise attacks have a visceral reality that a campaign of subversion, with its long-term orientation, low-visibility, and general

---

signature "hitting 'em where they ain't" concept is evocative of Sun Tzu's *xu/shi* ideas (Theme #3, p. 113).

[30] See William R. Kintner, *The Front is Everywhere* (Norman: University of Oklahoma Press, 1950).

[31] A key paragraph in Wylie's article (p. 361) describes the effect of cumulative strategy in terms of a "minute accumulation of little items piling one on top of the other until at some unknown point the mass of accumulated actions may be large enough to be critical." That analytical image is a prescient anticipation of physicist Per Bak's more recent formal model of "sandpile" phenomena, where grains of sand are added one by one to a sandpile – without any visible effect until, abruptly, an avalanche is triggered. See Per Bak, *How Nature Works: the science of self-organized criticality* (New York: Copernicus, 1996). Compare observation of traditional commentator Zhang Yu quoted in #3 footnote 40.

A different vein of formal modeling relevant to cumulative strategy and Sun Tzu draws on percolation theory. See Serge Galam, "Terrorism and passive supporters: an approach from physics," in Katalin Martinás, Dario Matika, and Armano Srbljinović (eds.), *Complex Societal Dynamics: security challenges and opportunities* (NATO Advanced Research Workshop, Zagreb, Croatia, 2009; Washington, DC: IOS Press, 2010), pp. 175–89. Focus here is on passive supporters of clandestine activities – terrorism, for one, but also other activities such as tax evasion, corruption, black markets, etc. – who "constitute an invisible and unnoticeable social group, which produces a naturally friendly social space potentially open to terrorist [or other clandestine actor] moves." "Invisible and unnoticeable" has considerable resonance with Sun Tzu's formlessness thinking. See Passage #8.1.

[32] *Sound Military Decision* (Newport, RI: US Naval War College, 1942), p. 198.

amorphousness – amounting to a type of formlessness (Theme #8) – lacks. Raising awareness of just how potent a subversion campaign can be is one area where Sun Tzu's ideas stand to make a major contribution. Here is a leading contemporary contender for "*What did Master Sun know that we still don't (or have yet to absorb adequately)?*"

No discussion of Sun Tzu's subversion-related thinking should overlook verse XIII.16 (Passage #4.7), underscoring Sun Tzu's alertness to the value of (and possibility of acquiring through espionage) detailed personal or administrative information regarding an adversary's key people, lowly as well as higher-ranking. Harmonizing with Sun Tzu's emphases on achieving mastery "upstream" of any battle,

> IV.12.
> . . . he conquers an enemy already defeated.
>
> (from Passage #4.4)

attack and defense focused on acquiring access to personal data defines a basic twenty-first-century conflict arena. Seemingly remote from conventional warfare, it is an arena where knowledge obtained there may hold the key to ultimate victory or defeat.

One subversion-related idea encountered again and again in Sun Tzu – to which Passages #4.2, #4.7–#4.10, and #4.12 all make contributions – is dividing, indeed splintering into many pieces, an adversary's camp. Estrangement as a strategic move is an oft-recurring harmonic in Chinese history.[33] In analyzing use of estrangement techniques it is helpful to distinguish conceptually between two tasks. The first centers on spotting existing fault lines in human relationships on the enemy side, as a prelude to exploiting them (an example would be conflicts between a foe's military and civilian high command, see Passage #4.12). The second centers on that exploitation itself and, where feasible, creation of still further fissures.

The rise of modern computing has seen significant advances in regard to the first task. These center on computer-assisted capabilities – for which open sources are increasingly providing all the data necessary – to reveal fault lines in complex social structures. Illustrating this type of pattern-recognition capability is the blockmodeling area of social network analysis.[34] Blockmodeling can be succinctly described as a form of social network cartography, a framing that brings with it a natural tie-in to military applications (since mapmaking has long been a basic activity of general staffs).[35] Importantly, blockmodeling not only analyzes how a social network might be partitioned into "blocs" (typically more than two) based on patterns of "structural equivalence" of network actors, but also provides an interpretable roadmap of the

---

[33] A Chinese term for "estrangement" (see Sawyer, #1 footnote 3, p. 115) is *lijian*. The character for *li* (whose meaning is also "estrange," "sow discord," etc.) appears in Passage #4.10's verse I.25 ("When he is united, divide him"). For related Theme #6 analysis see p. 209 and also p. 211 below.

[34] See Ronald L. Breiger, Scott A. Boorman, and Phipps Arabie, "An algorithm for clustering relational data with applications to social network analysis and comparison with multidimensional scaling," *Journal of Mathematical Psychology*, 1975, 12(3), 328–83.

[35] See Jürgen Espenhorst, "A good map is half the battle! The military cartography of the Central Powers in World War I," in Elri Liebenberg, Imre Josef Demhardt, and Soetkin Vervust (eds.), *History of Military Cartography* (Cham: Springer International Publishing, 2016), pp. 83–130. For further perspective, focusing on eighteenth-century Qing mapmaking initiatives as Qing dynasty power expanded westward into Central Asia, see Perdue, Introduction footnote 52, "Maps and power," pp. 442–57.

relationships between those blocs.[36] Such interbloc relationships, which commonly involve more than one type of network tie, along with their changes over time, can give significant insight into patterns in the social structure, some of them grist for strategic manipulation. A candidate for a Sun Tzu–inspired blockmodeling application might, for example, map cleavages in social networks of a ruler and entourage along with high civil and military officials in that ruler's regime (cf. Passage #4.12).

The analytical direction represented by blockmodeling is about recognizing fault lines in a social network, giving insight into where social or interpersonal conflicts, often latent, exist. Creating or intensifying conflicts, or devising stratagems for reaping gains from existing ones, is a separate challenge. A flavor of possibilities, with Sun Tzu anchoring, may be gleaned from Table 3 whose focus is cyber warfare as a potentiator of political warfare.[37] All thirteen Sun Tzu chapters are represented here.

★★★★★★

Immeasurably facilitating extensions of Sun Tzu to non-military conflicts is the fact that there is very little in Sun Tzu's way of war that necessitates a mass of specific supporting machinery, either institutional (like a general staff) or technological. Many core Sun Tzu emphases – among them, calculation (Theme #1), botanizing situations (Theme #2), a predilection for being cheap (Theme #3), and (never to be forgotten!) "know the enemy and know yourself" (Themes #11 and #12) – encourage transposing Sun Tzu's brand of high-end military thinking deep into the heart of civilian life. Even some seeming impediments to such transposition dwindle on scrutiny (e.g., references to "battle" can often be replaced by watershed events of other kinds).[38] Those flexibilities suggest that there is little inherent reason why an applier of Sun Tzu must be a state actor or even an aspiring peer of such an actor (say, major multinational enterprise, revolutionary, terrorist, etc.).[39]

---

[36] For pioneering statement of the structural equivalence concept foundational to this type of social network analysis see François Lorrain and Harrison C. White, "Structural equivalence of individuals in social networks," *Journal of Mathematical Sociology*, 1971, 1(1), 49–80.

[37] In keeping with Sun Tzu's orientation toward offensive warfare Table 3 is oriented to offensive aspects of cyber-political warfare, by contrast to defensive ones (e.g., information security measures).

[38] Modern candidates for civilian analogs to decision for battle are diverse. Pertinent attributes include: (a) committing to a major, high-stakes choice that (b) initiates a qualitatively new, if temporary, heightened intensity of conflict and (c) once initiated is hard to reverse, if reversible at all. On going to trial as a counterpart to decision for battle see David C. Nelson, "On military strategy and litigation," *Vermont Law Review*, 2007, 31(3), 557–92 (focus is applicability of Sun Tzu's *Art of War* to legal practice; p. 558 there observes that relevant analysis "must proceed by analogy and metaphor because the [Sun Tzu] material was not written about the practice of law").
A counterpart to battle of another kind might be an inflammatory social media post. Note, however, that unlike a trial and many battles in military history, an ensuing social media "battle" often lacks a well-defined terminus – unless, of course, the post is taken down (and even then controversy it triggered may still rage, then perhaps sputter on and off). A pre-modern Chinese civilian counterpart to "battle" is noted on p. 142 above.

[39] To be clear, Sun Tzu himself is unwavering in his own focus on state actors. Even when Sun Tzu allows a general to defy orders from his ruler, what legitimates such disobedience is always the best interests of the state/ruler/people (albeit that this inherently complex concept is left unanalyzed). See p. 434 below and Passages #13.15–#13.18. See especially verse X.19 in Passage #13.15.

Table 3 *Cyber warfare as a potentiator of political warfare, correlated with ideas from Sun Tzu*

All these stratagems may be harnessed in service of bureaucratic conflict no less than partisan or other political warfare. There are also foreshadowings in labor strife of a modern but pre-digital era.

1. Clandestinely operating from the sidelines – exploiting the low visibility of many forms of digital attack, conjoined with the difficulty of reliably attributing such attacks to their actual sources – to stir up new, or exacerbate pre-existing, conflicts between an adversary and third parties. (note a)
2. Activating elephantine digital memories to revive (for emotional or other impact on adversaries, rivals, or third parties) old jealousies, unwise statements, missteps of all kinds. (note b)
3. Helping adverse (or, to help an ally, favorable) information "go viral" in digital context. (note c)
4. Illuminating vulnerabilities of actors and fragile network positions, for strategic gain. (note d)
5. Disrupting an adversary's poise on the eve of a carefully timed major political step. (note e)
6. Creating situations that compel an adversary to operate outside that adversary's "comfort zone" (relevant measures might be as simple as steps to burden an adversary with an unfamiliar or uncongenial task, thereby disrupting that adversary's foci of high-quality attention). (note f)
7. Placing adverse information in precisely the right hands for aggressive exploitation. (note g)

Note a. The concept of a troublemaker operating from the sidelines is at the heart of Passage #4.12's verse III.23. See also Passage #4.13's verse II.5. Troublemaking mechanisms are epitomized by verse XIII.19 (Passage #4.7), "And it is by this means [a double agent positioned to know what information to plant, when, how] that the expendable agent, armed with false information, can be sent to convey it to the enemy." Nowadays the "expendable agent" may not be a human being but a machine.

Note b. "Know the enemy" (Passages #11.1 and #11.2) sets a stage here. Verses III.5 (Passage #4.2) and I.25 (Passage #4.10) point to creating estrangements. Verses VIII.17–VIII.24 (Passage #4.9), I.22–I.24 (Passage #4.10), and XII.18 (Passage #4.3) home in on assorted human frailties, all good grist for manipulation.

Note c. Verse XI.29 (Passage #3.7) states, "Speed is the essence of war." Verse IX.41 (Passage #4.15) in Griffith's reading is about disgruntled individuals whispering amongst each other, suggesting that spread of information can be both viral and sotto voce. Digital information also shares some characteristics with fire, including possibility of swift, unpredictable spread (Passage #3.8's verse VII.13). Verses XII.4 and XII.5 (Passage #9.7) are about timing fire attacks to exploit favorable environmental conditions. Digital counterparts exist (e.g., some kinds of phishing attacks).

Note d. Quest for enemy vulnerabilities animates Passage #4.4. Its verses IV.2–IV.4 could apply to individuals no less than to armies. Passage #4.7's verse XIII.16 call for detailed information-gathering about human targets also fits well here. Meanwhile the army-as-water image of verse VI.27 (Passage #4.14) evokes a spirit of probing opportunistically for weaknesses in the adversary's defenses, nowadays often digital, worming through crevices found there to obtain sensitive, actionable data.

Table 3 (cont.)

Note e. In the background of verses I.22–I.24 (Passage #4.10) and VIII.17–VIII.24 (Passage #4.9) stands the concept of disrupting an enemy leader's state of mind in ways that might disrupt poise. The importance Sun Tzu attaches to exquisite timing – the foe's as well as one's own – is conveyed by Passage #3.1's verse V.16 crossbow image and by Passage #1.4's verse V.14 image of raptor and prey.

Note f. Ideas from verses I.23 and I.24 (Passage #4.10) and VIII.15 (Passage #4.11) find application here.

Note g. Verse XIII.19 (Passage #4.7) suggests relevant thinking.

Once this connection to the state is severed, *the cast of actors, great and small, who might profitably draw on Sun Tzu becomes vast indeed*.[40] Here, of course, is a basic reason for Sun Tzu's broad modern appeal and base of fans. The analytical vistas thus opened could be explored almost without limit, spanning innumerable civilian contexts of widely diverse types. To repeat a point made in the Introduction (p. 3), Sun Tzu is highly conducive to thinking analogically, in fact on many levels. Yet all is not smooth sailing. A nagging obstacle to free-form transposition of Sun Tzu's ideas to arbitrary civilian contexts is Sun Tzu's proclivity noted earlier (p. 145) to approach all strategic problems in starkly adversarial ("zero-sum") terms.[41] Certainly altruism or prosocial behavior and Sun Tzu's ideas mesh poorly, if at all. If Sun Tzu's ideas are too faithfully or literally applied, many opportunities for mutual benefit in conflict resolution, especially over a long run, seem destined to be missed.

This consideration acts as a constraint, seriously narrowing the non-state-actor civilian contexts where Sun Tzu's thinking stands to offer real value. The remainder of the present Theme #4 discussion sketches two very different examples where the difficulties just noted are at least partially in abeyance, so that prospects for applying Sun Tzu's thinking gain traction.

---

[40] Worth noting, because it harmonizes with other parts of Sun Tzu (e.g., Theme #8, "formlessness") is a case where it it unclear where a state actor's strategy leaves off and private initiative begins.

[41] A foray into a non-traditional application of Sun Tzu's ideas, one that merits a thoughtful reading for the light it sheds on both possibilities for and obstacles to discovering a "cooperative Sun Tzu," is Khoo Kheng-hor, *Applying Sun Tzu's Art of War in Managing Your Marriage* (Subang Jaya, Selangor, Malaysia: Pelanduk Publications, 2002), cited in Lawrence Freedman, *Strategy: a history* (Oxford, UK: Oxford University Press, 2013), pp. 509 and 698 note 14. On pp. 7–8 this essay commences with a clear illustration of the dual use idea (p. 142 above):

To use [Sun Tzu] positively, a person needs to have an open mind and be clever enough to substitute some of the seemingly negative words used in the original context. An example is the word "enemy." Should you come across this word, don't think of an "enemy" as someone to watch out for lest he brings harm to you. Instead, substitute the word with "the person you deal with." In this way, he could be your boss, your colleagues, employees, customers, spouse, children, etc.

It goes on to offer imaginative textual anchoring, with a dash of poetic license, for marriage-relevant advice from Sun Tzu (of *c.* 400 Griffith verses, around 10 percent are cited). However, illustrating the difficulty of distilling advice from Sun Tzu about building a cooperative relationship, Sun Tzu IV and V – both candidates for standing at the heart of Sun Tzu (cf. Passage #4.4 and Theme #5) – are not represented in the verses selected. Sun Tzu's espionage chapter is represented only by part of verse XIII.1 (which says nothing about spying).

The first example centers on a Han dynasty advice manual for elite young women. That work is the *Nüjie* – the title is variously translated as "Precepts for Girls" or "Lessons [or "Admonitions"] for Women" – written by Ban Zhao (c. 48–c. 120 AD).[42] One of the most prominent women in Chinese history, Ban Zhao was a person of distinguished scholarly accomplishment and credentials who spent her life around the Han court and was the teacher of a future empress there.[43] In all likelihood, she would have been well aware of the Sun Tzu text and its ideas, if only through her two eminent brothers, respectively a general and a historian (in whose history of the Han dynasty Sun Tzu is catalogued).[44] However, her *Nüjie* – which, along with one earlier work also of Han vintage, has been described as having "set the tone of girls' education for the next two millennia"– has been traditionally viewed through a Confucian lens as guidance to elite young women on how to accept a subservient place in a patriarchal Confucian social order.[45]

Challenging that interpretation, a modern analysis of the *Nüjie* by Yu-shih Chen has made a case that its substance aligns better on ideas found in two distinctly non-Confucian texts.[46] One is Sun Tzu; the other is the *Laozi*, the Daoist classic (whose thinking has considerable common ground with Sun Tzu).[47] According to this revisionist interpretation, the *Nüjie* is an advice manual geared to teaching young married

---

[42] For brief *Nüjie* background – authorship, intended audience, and political and cultural context – see pp. 175–79 of Olivia Milburn, "Instructions to women: admonitions texts for a female readership in early China," *Nan Nü: Men, Women, and Gender in China*, 2018, 20(2), 169–97. Milburn's important article also translates the *Jiaonü*, an instructions-to-elite-women text believed to be substantially earlier than the *Nüjie*, possibly of late Warring States vintage. As Milburn notes (p. 177), advice found in that text contains similarities with the *Nüjie*. From the present standpoint, particularly noteworthy are the *Jiaonü*'s points of resonance with Sun Tzu, e.g.: "Today a wise daughter-in-law trains herself to think carefully" (Milburn, p. 182), which aligns well on Theme #1 ideas.

For *Nüjie* translation see Wilt Idema & Beata Grant, *The Red Brush: writing women of imperial China* (Cambridge, MA/London: Harvard University Press Asia Center, 2004), pp. 36–42.

[43] For background on Ban Zhao and the *Nüjie*, see Wilkinson, §10.3.2, p. 185; Idema & Grant, previous footnote, pp. 17–42. A pioneering study is Nancy Lee Swann, *Pan Chao: foremost woman scholar of China* (New York/London: Century Co., 1932), which contains a *Nüjie* translation on pp. 82–99.

[44] Ban Zhao was the sister of historian Ban Gu and general Ban Chao (who asserted Chinese control over a major part of what is now Xinjiang province). She contributed to bringing Ban Gu's history of the Western Han dynasty, the *Hanshu*, into its ultimate form following the latter's 92 AD death in prison (he had been involved with the wrong court faction). See Wilkinson §59.2.1, pp. 793–94 (Ban Gu's life); §59.2, pp. 793–96 (background on the *Hanshu*, a landmark of Chinese historiography).

Regarding mention of Sun Tzu in the *Hanshu*, see Griffith, p. 13 and Gawlikowski & Loewe, Background footnote 79, pp. 447–48.

[45] The "next two millennia" quotation is from Anne Behnke Kinney, *Representations of Childhood and Youth in Early China* (Stanford, CA: Stanford University Press, 2004), p. 149.

[46] Yu-shih Chen, "The historical template of Pan Chao's 'Nü Chieh'," *T'oung Pao*, 1996, 82(4/5), 229–57. To be clear, identifying affinities in substance or standpoint between the *Nüjie* and Sun Tzu is a product of modern analysis. The *Nüjie* text itself does not quote or directly allude to Sun Tzu.

[47] See, e.g., Brooks, p. 66 regarding the "inference that both these texts accumulated side-by-side in [the state of Lu] during the same half-century," with associated cross-influences (albeit that their primary agendas were different). For background on the *Laozi* text see Introduction footnote 37.

women the strategic moves needed "to play the survival game well" in context of the consummately tricky and violence-prone politics of the Han imperial court. Its poisonous atmosphere of intrigue is perhaps best summed up by the stark observation that "Those were treacherous surroundings indeed. Few empresses in the Han dynasty lived to a ripe old age. Most of them died between the ages of 17 and 27."[48]

Approaching the Nüjie as a Sun-Tzu-esque training manual for elite young women, it represents a case where the most relevant Sun Tzu insights and emphases are in key ways very different from those that Sun Tzu offers in classic military settings, where routine use of deadly force is taken for granted by all participants. The cultured elegance of an imperial court seems worlds apart from any battlefield (especially when viewed through the eyes of an inexperienced but enormously privileged young person coming of age in those surroundings).[49] More than any specific strategic action advice, the value-added of teaching Sun Tzu's ideas in such a setting resides in inculcating a mentalité that thoroughly grasps and accepts, on an intuitive level, just how deeply deceptive the court environment is, when set alongside the often deadly reality that lurks just beneath its cultivated veneer.[50] To borrow a concept from sociologist W. I. Thomas, inculcating Sun Tzu's ideas helps inject a much-needed note of strategic realism into an inexperienced newcomer-participant's "definition of the situation" – a situation that, to invoke Bernard Fall's evocative title theme, could all too easily become "hell in a very small place."[51]

---

[48] Chen, #4 footnote 46, pp. 233 footnote 10, 257. Kinney, #4 footnote 45, p. 148 sums up Chen's argument: "[The Nüjie] can also be read as a survival manual for women under the close and at times hostile scrutiny of their husbands' families." See also Kinney, p. 150 ("the Nüjie provided important survival skills for girls"). The radical difference between this line of interpretation and one informed by a Confucian value system is noted by Paul R. Goldin, The Culture of Sex in Ancient China (Honolulu: University of Hawai'i Press, 2002), p. 177 note 117. Other interpretations exist too. See, e.g., Wilkinson, §10.3.2, p. 185, noting work of Lisa Raphals which makes a case that "Ban [Zhao] chose a conservative tone to deflect criticism from her radical proposal that girls should be educated along with boys."

[49] Chen, #4 footnote 46, p. 257 profiles the Nüjie's intended audience as follows: "The girls so advised were most likely members of Ban Zhao's own family, destined to marry into other upper-class families or even to be chosen to enter the imperial court as the emperor's concubines."

[50] A window into the toxic environment in which Ban Zhao's young women were thrust as very junior participants comes from another source, Zhang Heng (c. 73–139 AD), a polymath notable for both his physical science accomplishments (he is credited with inventing the first seismoscope) and an ethnographer's instincts (an unusual mix in any era). Work of Zhang Heng points to an arrogant mindset among the aristocracy of the age, ever open to hearsay and "able to prove anything to his [the aristocrat's] own satisfaction" – a profile of a personality type that Ban Zhao's young women would do well not to cross, inadvertently or otherwise. See p. 172 of E. R. Hughes, Two Chinese Poets; Vignettes of Han Life and Thought (Princeton, NJ: Princeton University Press, 1960), chapter X, "Critique of Chang Heng's fu on the Eastern Capital," pp. 172–220.

A further flavor of what a newcomer to court life in Ban Zhao's time might have been up against comes from de Crespigny's psychological portrait of Cao Cao, who lived in a slightly later period of Chinese history. Of Cao Cao's mentality, he writes, "a man in Cao Cao's position could not afford to ignore any show of disrespect or abuse of favour. There was a definite, albeit undrawn line, and those who crossed it were in dangerous territory" (de Crespigny, Background footnote 95, p. 371).

[51] See Bernard Fall's Hell in a Very Small Place: the siege of Dienbienphu (Philadelphia: Lippincott, 1967). Also setting relevant tonality is Fall's p. 51 ("deadly guessing game that is grand strategy").

Such a goal points to prioritizing a different set of Sun Tzu verses, and drawing rather different lessons from them, than those most relevant to advising generals. One basic lesson is the need for Ban Zhao's young women to rein in youthful spontaneity and high spirits, in favor of careful, deliberate, often delicate calculation in traversing an environment brimming with rules and constraints, some more visible than others, never straying from Sun Tzu calculating mode (Theme #1).[52] For Ban Zhao's advisees – far more than in the case of competent military professionals – such a calculating mindset may be far from congenial. But it is essential, indeed vital. Also crucial for these women is presenting a bland, innocuous front for which the motivation is overridingly defensive: "Do not invite harm to yourself by appearing as a threat to others; stay alive."[53] There are strong echoes here of Sun Tzu Theme #8 (formlessness).[54] Formlessness (wu xing) might indeed have been more attainable for Ban Zhao's young women than for many a land warfare general who, bedeviled by the tyranny of logistics, finds attaining formlessness elusive on terrestrial battlefields, even a conceit of armchair theorists.[55] An important factor here is that Ban Zhao's young women can operate as individuals, looking out for themselves, by contrast to a general in charge of an army.[56]

Occupying social roles and positions that many different audiences will scrutinize unmercifully, Ban Zhao's young women find themselves in a kind of domestic sociological counterpart to Sun Tzu's focal ground, a "place of intersecting

---

[52] See Chen, #4 footnote 46, p. 232: "Families like the extended royal family were an extremely dangerous environment for their members. Correct positioning of oneself in that labyrinthine complex of human relationships, and masterful manipulation of power alliances within the extended familial hierarchy, just to survive," in fact "required as much calculation, intelligence, and political and military skills as [those required for] a minister at court or a general in the field."

Chen's pp. 255–56 quote from various calculation-oriented parts of Sun Tzu, among them verse I.28 (Passage #1.2) and verses IV.16–IV.18 (Passage #1.7). Verse IV.15 (Passage #4.4) is taken as underscoring a need to take cognizance of constraints, latent as well as obvious, and adapt to the *dao* of the situation. See Chen, p. 256: "Not knowing or not observing the *dao* and its rules and standards of measurements (*fa-du*), as in the case of Ban Gu's household [which ran afoul of the emperor, see #4 footnote 44 above], results in destruction and death."

[53] Chen, #4 footnote 46, p. 256. There is also resonance between Ban Zhao's first precept to girls, "[Appear] inferior and powerless" (Chen, p. 256), and Sun Tzu verse I.23, "Pretend inferiority and encourage his arrogance" (quoted by Chen, p. 255).

[54] Milburn, #4 footnote 42, p. 177 observes that the "Nüjie suggests that making oneself unobtrusive was a full-time occupation for Han dynasty elite wives." Although verse XI.61 (Passage #7.11) is not cited by Chen, its first part ("Therefore at first be as shy as a maiden") depicts just the sort of formless blandness that Ban Zhao's young women need as cover. That demeanor is exceptionally well captured by a somewhat discursive French translation of verse XI.61 by Father Amiot, an eighteenth-century Jesuit and pioneering Sun Tzu translator. See Passage #12.28A, p. A-219 (quoting Amiot).

To that end, Ban Zhao's young women are well advised to avoid violating rules of etiquette and similar norms present in their environment, thereby calling undue attention to themselves – a point that again underscores basic differences between this type of civilian milieu and a military one. Cf. Cao Cao's comment on verse III.21 (Passage #4.12), translated by Griffith, p. 81: "An army cannot be run according to rules of etiquette."

[55] See Sawyer, Introduction footnote 55, p. 60.

[56] See pp. 273–77 below on challenges facing a general striving for individual attributes of inscrutability or "formlessness" under the ever-watchful eyes of an entourage.

highways, important for international relations."[57] In one of his rare explicit departures from a zero-sum mindset, Sun Tzu's focal ground advice (Passage #4.5) is to build alliances. Ban Zhao's advice to her protagonist is likewise to build alliances, ones strong enough to veil intramural tensions or mistakes as effectively as any mask of command donned by a Sun Tzu general. "[I]f members of the same household live in harmony," Ban Zhao writes, "slander can be covered up, but if relatives by birth and by marriage are not united, then evil will spread far and wide."[58]

The Nüjie counsel just sketched deviates markedly from Sun Tzu's emphasis on offensive warfare. More aggressive stratagems by a young woman along lines pointed by Passage #4.10's menu[59]

> All warfare is based on the *dao* of deception.
> When able, feign not able;
> When active, feign not active [etc.]

might, of course, be feasible, particularly when backed by eloquence of speech and arts of rhetorical persuasion. But more aggressive stratagems are also risky, especially when put into effect by an inexperienced young individual dealing with far more experienced elders.[60] Sun Tzu's abhorrence of unnecessary risk comes into its own here. As verse VII.4 (Passage #1.15) observes,

> VII.4.
> Now both advantage and disadvantage are inherent in manoeuvre.

To this Sun Tzu thought Cao Cao adds a telling warning: "One skilled will profit by it; if he is not, it is dangerous."[61] Substituting "communication" for "maneuver," that warning transposes well to the Nüjie's young women. Therein lies a compelling reason as to why the Nüjie contains no counterpart to Sun Tzu's exhortations to a general to display virtuoso tactical creativity (e.g., Passage #4.14's verse VI.26).[62]

Weighing a direct equivalence between the role of a young woman (who, of course, may herself ultimately come into a strategically important role, e.g., as empress, empress dowager, etc., albeit most commonly in an offstage advisory capacity) and that of a Sun Tzu general, further parts of Sun Tzu come into play. An illuminating Nüjie passage pertains to commands received from on high:[63]

> If your mother-in-law says, "Do that" and she is wrong, still it is appropriate that you go along with (*shun*) her command. Do not act contrary to what your parents-in-law say, be it right or wrong, nor dispute with them what is crooked and what is straight: this is what is meant by "bend and follow."

---

[57] Needham & Yates, p. 47 (previously quoted in Theme #2 context, see #2 footnote 56).

[58] Nüjie translation by Idema & Grant, #4 footnote 42, p. 41.

[59] Quoted from Chen, #4 footnote 46, p. 255. He uses Griffith's translation with minor modifications.

[60] Cf. Chen, #4 footnote 46, p. 257: "It is clear that Ban Zhao is not denying the girls their own sense of right and wrong or their ability to judge. But she sees it as dangerous for them to exercise such judgment when they are in a situation in which they are completely overpowered." Although that point is geared toward moral judgments, not strategic steps, its force readily extends to the latter.

[61] Griffith, p. 103.

[62] As Anne Behnke Kinney puts it: "'Speaking truth to power' is always a delicate business, and even more difficult for those crossing gender boundaries." See p. xxxi of her *Exemplary Women of Early China: the Lienü zhuan of Liu Xiang* (Anne Behnke Kinney, trans. and ed.; New York: Columbia University Press, 2014). (Regarding the Lienü zhuan text see Wilkinson §10.3.2, p. 185.)

[63] Chen, #4 footnote 46, p. 257 (also translated by Idema & Grant, #4 footnote 42, p. 40).

Here the Nüjie shares with Sun Tzu a strategist's keen awareness of the costs that may attend irreversible steps.[64] There is, to be sure, an apparent clash between the Nüjie's mother-in-law counsel and Sun Tzu's advice to the decision maker of primary interest to Sun Tzu, namely, the commanding general, to the effect that

VIII.8.

There are occasions when the commands of the sovereign need not be obeyed.

(from Passage #3.15)

But Ban Zhao's advice is to an individual and, what's more, one who is in a fragile social structural position as a young entrant in a sophisticated, quite possibly highly hostile, environment.[65] By contrast, the perspective of verse VIII.8 is that of the state at war – "the road to survival or ruin" (as verse I.1 in Passage #1.1 puts the case). Once allowance is made for this basic difference, any tension with Sun Tzu's counsel largely evaporates. Indeed the Nüjie advice just quoted lends support to a Sun Tzu–inspired strategic interpretation of Ban Zhao's work. A traditional Confucian interpretation of the Nüjie might indeed be faulted for overreaching, distorting what are in fact nuanced Confucian ethical positions by unqualified pronouncements like the mother-in-law one.[66] But once one takes into account the vulnerabilities of the audience to whom the Nüjie is addressed, a Sun Tzu–inspired reading of the content passes a test of strategic realism with flying colors.

A second and very different example (more exactly, class of examples), again illustrating both the value of Sun Tzu's strategic thinking and the need to tailor it for non-military applications, involves economic activity. Exploration of this direction appropriately starts with finance. Lending practices, of course, have deep, indeed probably universal, roots in reliance on social trust and the social stability it represents.[67] Yet especially in modern times, financial activity has taken on a set of very different structural characteristics that create a superb hunting ground for aggressive applications of Sun Tzu's thinking.[68] Among those characteristics are:

- tendencies toward extremes of leverage (in a standard business or economics sense), with the potential it creates for cheap, outsized gains (along, of course, with major, even systemic, risks should things go wrong);

---

[64] See Nüjie translation by Idema & Grant, #4 footnote 42, p. 40 (warning of irretrievable consequences "if you say what you should not be saying, watch what you should not be watching") and pp. 38–39 (tracing downward spiral in a marital relationship). A sampler of Sun Tzu material illustrating Sun Tzu's alertness to irreversibility issues is assembled in #13 footnote 4.

[65] For fragility of social position as a network concept see Bothner et al. cited in #2 footnote 37.

[66] Cf. Goldin, #4 footnote 48, pp. 99–103, critiquing Ban Zhao's Nüjie along similar lines.

[67] See Lien-sheng Yang, Money and Credit in China: a short history (Cambridge, MA: Harvard University Press, 1952), pp. 51–52; D. C. Twitchett, Financial Administration under the T'ang Dynasty (2nd ed.; Cambridge, UK: Cambridge University Press, 1970), pp. 72–73 (both discussing feiqian, lit. "flying money," a Chinese name for a social-trust-based value transfer system commonly known in the West by its Arabic name ḥawāla). For a thumbnail sketch of ḥawāla see Sood, #1 footnote 60, pp. 242ff.

[68] However, it needs to be noted that Sun Tzu's ideas, while certainly relevant, do not get us very far in spotting specific profitable ways of juggling streams of costs and benefits spread over time, a basic analytical move at the heart of many types of financial strategizing. See pp. 119–20 above.

- frequently frenetic pace of activity, nowadays conjoined with reliance on complex, often opaque calculations (both conducive to manipulations of many types);
- high level of abstraction (which, by adding yet a further layer of opacity, sets a stage for nuanced and often manipulative uses of information);
- impersonal ambiance, rooted in big organizations and markets, nowadays commonly global in scope, and abetted by the abstract nature of daily activity there, all being features that tend to erode any residual inhibitions of human players.

Especially when taken as a package these characteristics can be harbingers of great systemic risk and societal instability (and by the same token are good grist for attention of Sun Tzu–inspired grand strategists!).

Affinity of Sun Tzu for finance need not, of course, stop short with aggressive, yet still legal, practices. A case can be made for harmony between Sun Tzu and insider trading practices, viewed as an economic version of winning without fighting:[69]

IV.8.
To foresee a victory which the ordinary man can foresee is not the acme of skill.

(from Passage #4.4)

XIII.3.
Now the reason the enlightened prince and the wise general conquer the enemy whenever they move and their achievements surpass those of ordinary men is foreknowledge.

(from Passage #4.7)

For all these reasons, economic eras when finance is "in command" particularly lend themselves to Sun Tzu–inspired ways of thought and action, more so than do other eras where non-finance business skill-sets (e.g., production, general management, marketing, etc.) are ascendant. Aggressive practices and corruption will exist in those other eras too, but seem less likely to epitomize "Sun Tzu unbound." Opportunities for "thinking like Sun Tzu" may be especially pronounced when financial institutions and practices are themselves rapidly evolving and in a state of great flux, much as was Chinese warfare in the historical context in which the Sun Tzu text took shape. Both contexts present great opportunities for a decision maker whose style is animated by Sun Tzu to run circles around ones having a more sedate, traditional, rule-abiding mentality. A comment on the time of Sulla in the late Roman Republic applies well here: "The most ruthless and clear sighted of the younger generation had been presented with an unparalleled opportunity to leapfrog their elders."[70]

Shifting attention away from high finance, especially in its modern digital incarnations, to other varieties of economic activity, the picture changes. Efforts to profit

---

[69] Cf. ideas of Henry Manne, a pioneer of law-and-economics who took a maverick position challenging on basic economic grounds the rationale for rules against insider trading. For perspective on Manne's intellectual legacy see Richard W. Painter, "Insider trading and the stock market thirty years later," *Case Western Reserve Law Review*, 1999, 50(2), 305–11. Of course, a basic difference between Manne and Sun Tzu is that the former is espousing a certain policy standpoint, whereas the latter remains wedded to the perspective of a strategist who, sharp as he may be, is looking out only for his parochial individual interests (perhaps including those of his boss).

[70] Tom Holland, *Rubicon: the triumph and tragedy of the Roman Republic* (London: Little, Brown, 2004), p. 90.

from a Sun Tzu mindset in general commercial settings find fewer opportunities. Maneuvers there tend to be less nimble; opportunities for manipulation, less intricate. Of course, such opportunities will exist there too, as will some individuals with the knowledge and ruthlessness to exploit them.[71] There is certainly room for businesses to take cues from Passage #4.7 (Sun Tzu XIII) and invest in competitor intelligence, which need not always stray into corporate espionage. Quite separately, there are echoes of Sun Tzu's canniness about terrain in astute uses of localization in business deals and enterprises – say, retaining subsidiaries' foreign leadership.

Yet efforts at enlightened commercially oriented readings of Sun Tzu may well end up bearing a tenuous relationship to the ideas that give the text its famously sharp edge. It seems hard to escape the conclusion that a major part of what Sun Tzu has to offer in general commercial life hovers around what in business usage are sometimes called sharp practices (technically legal, dubiously ethical).[72] Depending on the pertinent legal regime, those practices may cross a line into downright illegality. Sharp practices may be efficacious on a one-time basis, or in some settings where a business actor's exploitative behavior is unconstrained (though deep strategic insight seems scarcely necessary there!). Yet regular recourse to highly aggressive business practices is scarcely a good guide to how to succeed in most ordinary business contexts where reputation and social trust are a sine qua non for sustainable success. Indeed engaging in such practices risks a clash with Sun Tzu Theme #3, for all the reasons that social trust lowers costs of doing business and violating trust raises those costs.[73]

Sun Tzu's mentality is worlds apart from such thinking. A "zero trust" outlook, at least where communications from "the other side" are involved, is palpable in Passage #4.8's four verses, from a part of Sun Tzu that Brooks's analysis (Table 12A on p. A-35) suggests is the oldest layer of the text.

There remains a chasm between high finance–oriented areas (where Sun Tzu's ideas, especially in their digital incarnations, indeed pack a punch, albeit often accompanied by systemic risks) and workaday commerce (where those ideas deliver less value-added and may backfire). How to analyze that divide and the forces that shape it, possibly using mathematical or computational models to do so, is a worthy Sun Tzu (3) research frontier.

★★★★★★

---

[71] Compare the case of the rapacious Bai Gui, a late fourth-century BC businessman (and self-proclaimed Sun Tzu fan) noted in #4 footnote 6.

[72] A label in some ways even more redolent of Sun Tzu because of its distinctively manipulative flavor is "creative compliance" – i.e., "using the law to escape legal control without actually violating legal rules." See p. 848 of Doreen McBarnet & Christopher Whelan, "The elusive spirit of the law: formalism and the struggle for legal control," *Modern Law Review*, 1991, 54(6) (Law and Accountancy), pp. 848–73; David Howarth, "Lawyers and the rule of law," in Christopher May & Adam Winchester (eds.), *Handbook on the Rule of Law* (Cheltenham, UK: Edward Elgar, 2018), pp. 271–88.

[73] See Kenneth J. Arrow, *The Limits of Organization* (New York/London: Norton, 1974), p. 23, "Trust is an important lubricant of a social system."

Set of principal passages used to illustrate Sun Tzu Theme #4 (winning without fighting, by non-military ways and means)

Passage #4.1. Winning without Fighting

(Although civilian means are not specifically named in it, Passage #4.1 is a pillar of Theme #4. An important, sometimes overlooked, corollary of verse III.3 is worth noting. By reason of its reference to victory *in battle*, Sun Tzu's oft-quoted dictum – "Know the enemy and know yourself; in a hundred battles you will never be in peril" (verse III.31 in Passage #11.1] – actually depicts only a second-best level of achievement. For Sun Tzu, verse III.3 makes clear that the actual apex of achievement is gaining victory without fighting any battles at all.[74]

On a point of clarification, Griffith's verse III.10 reference to "protracted operations" should be understood as a reference to protracted *combat* operations.[75] Protracted operations of other types – diplomatic, intelligence, etc. – are entirely consistent with Sun Tzu's basic aim of prevailing without battle and at a low cost. Indeed they may be essential to doing so.)

III.1. Generally in war the best policy is to take a state intact; to ruin it is inferior to this.

III.2. To capture the enemy's army is better than to destroy it; to take intact a battalion, a company or a five-man squad is better than to destroy them.

III.3. For to win one hundred victories in one hundred battles is not the acme of skill. *To subdue the enemy without fighting is the acme of skill.* [Emphasis supplied.]

    . . .

III.10. Thus, those skilled in war subdue the enemy's army without battle. They capture his cities without assaulting them and overthrow his state without protracted operations.

III.11. Your aim must be to take All-under-Heaven intact. Thus your troops will not be worn out and your gains will be complete. This is the art of offensive strategy.[76]

---

[74] What constitutes a "no battles" victory may, of course, depend on how broad a canvas, temporal as well as spatial, is assumed. As Maoist-era military theorist Guo Huaruo observed,"Victory without fighting a battle is sometimes the result of a long preceding period of preparatory fighting . . . For example, we [the People's Liberation Army] would not have been able to take Beijing without a fight during the War of Liberation, unless we had already won the Battle of Tianjin [January 1949] that came before it." See Guo Huaruo 郭化若, Sunzi yi zhu 孫子譯註 (Sun Tzu, Translated and Annotated) (Shanghai: Guji, 1984), p. 100 (translated by Minford, p. 133). For a brief account of the Communist capture of Tianjin, then of Beijing, see Odd Arne Westad, Decisive Encounters: the Chinese civil war, 1946–1950 (Stanford, CA: Stanford University Press, 2003), pp. 225–27.

[75] Sawyer, p. 177 helps dispel any ambiguity on this point by translating the ending of verse III.10 as ". . . without prolonged fighting." An example would be a successful subversion campaign.

[76] Interpretation issues pertaining to verse III.11 are analyzed in #3 footnotes 60 and 61.

Passage #4.2. Hierarchy of Kinds of Strategy, Privileging Non-military Paths

(Although fighting is not ruled out, Sun Tzu's two most favored choices here are each compatible with non-military paths to achieving strategic objectives.)

III.4. Thus, what is of supreme importance in war is to attack the enemy's strategy;

III.5. Next best is to disrupt his alliances;

III.6. The next best is to attack his army.

III.7. The worst policy is to attack cities. Attack cities only when there is no alternative.

Passage #4.3. Thinking through Decisions for War with Utmost Care

(These verses encourage advancing a state's goals by non-military means if at all possible. There is also advice here – perhaps even more relevant in many civilian settings than in military ones, since the cast of characters there is often larger, with diverse interests – to be careful not to pick unnecessary fights.)

XII.16. And therefore it is said that enlightened rulers deliberate upon the plans, and good generals execute them.

XII.17. If not in the interests of the state, do not act. If you cannot succeed, do not use troops. If you are not in danger, do not fight.

XII.18. A sovereign cannot raise an army because he is enraged, nor can a general fight because he is resentful. For while an angered man may again be happy, and a resentful man again be pleased, a state that has perished cannot be restored, nor can the dead be brought back to life.

XII.19. Therefore, the enlightened ruler is prudent and the good general is warned against rash action. Thus the state is kept secure and the army preserved.

Passage #4.4. Winning against an Enemy Already Defeated

(This is a basic "dual use" passage. Its emphasis on strategic patience fits many civilian contexts as well as, or better than, most wars. Winnowing out the military content mostly requires coming up with a serviceable substitute for Griffith's "battle" in verses IV.9 and IV.14, which is not hard to do in many civilian contexts.

Verse IV.11's notion, supported by verse IV.8, of a successful strategist as having a subtlety that is opaque to ordinary onlookers may also fit some civilian contexts better than conventional warfare.)

IV.1. Anciently the skilful warriors first made themselves invincible and awaited the enemy's moment of vulnerability.

IV.2. Invincibility depends on one's self; the enemy's vulnerability on him.

IV.3. It follows that those skilled in war can make themselves invincible but cannot cause an enemy to be certainly vulnerable.

IV.4. Therefore it is said that one may know how to win, but cannot necessarily do so.[77]

IV.5. Invincibility lies in the defence; the possibility of victory in the attack.

IV.6. One defends when his strength is inadequate; he attacks when it is abundant.

IV.7. The experts in defence conceal themselves as under the ninefold earth; those skilled in attack move as from above the ninefold heavens. Thus they are capable both of protecting themselves and of gaining a complete victory.

IV.8. To foresee a victory which the ordinary man can foresee is not the acme of skill;

IV.9. To triumph in battle and be universally acclaimed "Expert" is not the acme of skill, for to lift an autumn down requires no great strength; to distinguish between the sun and moon is no test of vision; to hear the thunderclap is no indication of acute hearing.

IV.10. Anciently those called skilled in war conquered an enemy easily conquered.

IV.11. And therefore the victories won by a master of war gain him neither reputation for wisdom nor merit for valour.

IV.12. For he wins his victories without erring. "Without erring" means that whatever he does insures his victory; he conquers an enemy already defeated.

IV.13. Therefore the skilful commander takes up a position in which he cannot be defeated and misses no opportunity to master his enemy.

IV.14. Thus a victorious army wins its victories before seeking battle; an army destined to defeat fights in the hope of winning.

IV.15. Those skilled in war cultivate the *Dao* and preserve the laws [*fa*] and are therefore able to formulate victorious policies.[78]

Passage #4.5. Alliance-Building As a Tool

(These verses from Sun Tzu's terrain thinking stand apart from much of Sun Tzu in that their focus is constructive, not destructive: on building alliances, not undermining them. A basic takeaway is that alliances are a resource for which a particular need arises when operating in highly exposed focal ground. That

---

[77] Traditional commentator Du Mu (Minford, p. 150): "All I can know is whether my own strength is sufficient to defeat the enemy. I cannot cause the enemy to slacken and provide me with the opening necessary for victory." This comment (even if not intended that way) seems especially applicable to often sloppy civilian settings where strategic actors are rarely vigilant all (or even most of) the time.

[78] Regarding Daoistic overtones of verse IV.15 see Appendix 13 footnote 13. Verse IV.15's reference to *fa* need not be to "law" in a civilian sense of that term, at least one that is familiar to modern audiences. *Fa* also appears in verse I.8 (Passage #1.1) in a sense of "method" (Griffith's translation is "doctrine"). For further analysis of *fa* see Appendix 1 anchor footnote 5.

thinking could be generalized in social network directions.[79] However, Sun Tzu provides no further specifics fleshing out verse XI.13.)

XI.6.   When a state is enclosed by three other states its territory is focal. He who first gets control of it will gain the support of All-under-Heaven.[80]

...

XI.13.   In focal ground, ally with neighbouring states ...

...

XI.19.   In focal ground I would strengthen my alliances.

Passage #4.6. People Picking As a Tool

(Although Sun Tzu here may be mainly thinking about choice of subordinate commanders or perhaps staff officers – a Warring States counterpart to the title theme of Douglas Southall Freeman's *Lee's Lieutenants* – these verses articulate a concept that transposes well to civilian settings where choice of allies, associates, assistants, etc. must also be made.[81] A strategist must be able to size up people accurately and make use of each according to his or her talents, a task that has attracted much attention in many eras of Chinese thought.[82]

Verses V.21 and V.22 are best envisioned as a package: "motivation" or "talent" are not absolute concepts but are *relative* to a situation and its strategic possibilities.[83] The middle segment of verse V.19 points to a kindred thought and invites a dual use civilian reading. For example, a sense that an organization, which may be civilian, is in a shi – strategically advantageous – situation percolates through its ranks, creating soaring morale and, with it, often enhanced effectiveness.)

V.19.   Order or disorder depends on organization [shu]; courage or cowardice on circumstances [shi]; strength or weakness on dispositions [xing].

...

[79] For an example suggested by scholarship on public sector unions see Appendix 2, p. A-85.

[80] For the "All-under-Heaven" concept see #3 footnote 60. The concept as used in verse XI.6 seems to point to the influence that may arise from control of a central node in a geographic network. Vulnerabilities of such a position are seemingly not considered. Cf. Appendix 2 footnote 9 (example).

[81] Also relevant here is verse I.15 (Passage #4.10), if read broadly as advocating for picking key associates and allies with an eye on compatibility of strategic styles.

[82] For an example from a time long after Sun Tzu see Michael C. McGrath, "Frustrated empires: the Song-Tangut Xia war of 1038–44," in Don J. Wyatt (ed.), *Battlefronts Real and Imagined: war, border, and identity in the Chinese middle period* (New York: Palgrave Macmillan, 2008), pp. 151–90, in particular p. 159: "One common response to statecraft problems was to try to find the right men."

[83] For further development of this line of thought, together with analysis of textual and interpretive issues arising in verses V.21 and V.22, see Appendix 4 anchor footnote 4.

V.21. Therefore a skilled commander seeks victory from the situation [shi] and does not demand it of his subordinates [ren].[84]

V.22. He selects his men [ren] and they exploit the situation [shi].

Passage #4.7. Spies and Covert Action As Tools

(It would be easy to include Sun Tzu's entire espionage chapter here. Because that chapter appears in full as Passage #11.6 below, present focus is on the more versatile parts of Sun Tzu XIII, in particular ones that could find ready application in many purely civilian conflicts remote from conventional warfare, possibly centering on non-state actors.[85] In that spirit, verse XIII.16 is included for its overtones suggestive of detailed personal or administrative information-gathering deep in the heart of a civilian social structure. Verse XIII.19's deception/ disinformation stratagem could certainly see applications there. Verse XIII.20 points to the role of time and timing factors in successful espionage. In highly schedule-bound civilian settings, no less than in military ones, the value of information often crucially depends on when it becomes available. Verse XIII.22, which underscores how information obtained from well-placed defectors may suffice to produce regime change, might find application in cases of regime change not involving warfare, including ones occurring in the private sector. Verse XIII.23's call for using the "most intelligent people as agents" harmonizes with the cognitive and stylistic demands of spying in elite civilian circles.[86]

Sun Tzu's double agent [fanjian] material is downplayed in the present verse selection. However, a question worth posing is whether Sun Tzu's fanjian category – which in its original Sun Tzu context has problematic aspects, discussed on pp. 368–71 below – might find renewed vitality in non-state-actor civilian settings where, from a Simmelian "conflict and the web of group-affiliations" perspective, everyone has potential to become a double agent of a sort.[87])

---

[84] Lit., "men" or (degendering the concept) "people." Translating in that way (Sawyer, p. 188), which avoids presupposing a formal hierarchy, provides flexibility for civilian applications of Sun Tzu that Griffith's translation of ren as "subordinates" (with its overtones of formal organization) does not. For example, members of a strategic actor's existing ego network may be "selected" (activated) by that actor to do tasks, favors, etc., yet are not that actor's subordinates in any sense. A similar observation applies to adding ties to, or dropping them from, that network.

[85] Rationale for omitting particular Sun Tzu XIII verses in Passage #4.7's selection is given in Appendix 4, pp. A-104f. (where Passage #4.7 also appears).

[86] A Chinese phrase found in verse XIII.23 is shang zhi (Minford, p. 324's "highest intelligence"), a concept that illustrates Sun Tzu's attention to outliers (p. 120 above). Its appearance may shed light on Sun Tzu's focus there. Such intelligence seems a highly desirable trait for a spy who is observing everyday life of elites where all is smooth on the surface but valuable clues may lurk in subtle changes in behavior or mood – say, an awkward pause or a switch of conversation topic or register, or an unobtrusive change in the regularity with which a social network tie is maintained. Yet such high intelligence may be irrelevant in other contexts, e.g., situations where an information windfall depends on simply being in the right place at the right time (say, to overhear a conversation about some basic fact like troop strength or quantity of logistics reserves).

[87] Georg Simmel, Conflict and The Web of Group-affiliations (translated, respectively, by Kurt Wolff & Reinhard Bendix; foreword by Everett C. Hughes; New York: Free Press, 1955).

XIII.3. Now the reason the enlightened prince and the wise general conquer the enemy whenever they move and their achievements surpass those of ordinary men is foreknowledge.

XIII.4. What is called "foreknowledge" cannot be elicited from spirits, nor from gods, nor by analogy with past events, nor from calculations. It must be obtained from men who know the enemy situation.[88]

. . .

XIII.13. He who is not sage and wise, humane and just, cannot use secret agents. And he who is not delicate and subtle cannot get the truth out of them.

XIII.14. Delicate indeed! Truly delicate! There is no place where espionage is not used.

. . .

XIII.16. Generally in the case of armies you wish to strike, cities you wish to attack, and people you wish to assassinate, you must know the names of the garrison commander, the staff officers, the ushers, gate keepers, and the bodyguards. You must instruct your agents to inquire into these matters in minute detail.[89]

. . .

XIII.19. And it is by this means [the double agent] that the expendable agent, armed with false information, can be sent to convey it to the enemy.

XIII.20. It is by this means also that living agents can be used at appropriate times.

. . .

XIII.22. Of old, the rise of Yin [Shang dynasty] was due to Yi Zhi, who formerly served the Xia [dynasty]; the Zhou [dynasty] came to power through Lü Ya, a servant of the Yin.

XIII.23. And therefore only the enlightened sovereign and the worthy general who are able to use the most intelligent people as agents are certain to achieve great things. Secret operations are essential in war; upon them the army relies to make its every move.

Passage #4.8. Words As Weapons

(The stratagems gathered here are scarcely subtle. They align better on what Rupert Smith has labeled "low cunning" than they do on the higher realms of diplomacy.[90] Yet crude as they are, these stratagems represent a point of contact

---

[88] For further analysis of verse XIII.4 in a dual use spirit see Table 5 on pp. 405–406 below.

[89] Griffith's translation "in minute detail" is somewhat free, though clearly that kind of idea is in play. Mair's translation of the latter part of verse XIII.16 (Mair, p. 130) conveys an open-ended agenda of personal data gathering that readily generalizes to many civilian settings: "I must first know the names and surnames of the defending general, his trusted associates, the appointment secretary, the gatekeepers, his retainers, then *order my spies to find out as much as they can about them.*" (Emphasis supplied.) Social network information has a natural place here. See pp. 107–109 above.

[90] General Rupert Smith, *The Utility of Force* (New York: Knopf, 2007), p. 16. For discussion of verses IX.25–IX.28 from a deception vantage see Passage #7.16 (where these verses also appear).

between Sun Tzu and arts of rhetoric and persuasion that were extensively developed and practiced in Warring States China.[91] Given limitations on literacy and written mediums in Sun Tzu's time, the moves depicted here should be envisioned as involving oral, not written, communications.

It should be noted that these four verses appear in non-contiguous locations in the traditionally transmitted Chinese text of Sun Tzu IX, which Griffith has rearranged. Athough this rearrangement does not affect the point-by-point substance, it could wrongly suggest that Sun Tzu envisioned verbal stratagems as a distinct tactical category, when in fact he may have entertained no such concept.)

IX.25. When the enemy's envoys speak in humble terms, but he continues his preparations, he will advance.

IX.26. When their language is deceptive but the enemy pretentiously advances, he will retreat.

IX.27. When the envoys speak in apologetic terms, he wishes a respite.

IX.28. When without a previous understanding the enemy asks for a truce, he is plotting.

Passage #4.9. Ad Hominem Stratagems: Exploiting Personality Traits

(As suggested previously [p. 142 above], civilian leaders – who are often far less carefully vetted and trained, and in some cases may be outright amateurs through-and-through – may be more vulnerable to manipulation by the kinds of approaches Sun Tzu contemplates here than are military professionals.[92] In contemplating civilian applications of this material, particularly noteworthy is verse VIII.22, if read as advice from Sun Tzu to exploit altruistic or prosocial impulses to provide leverage for personal attacks.)

VIII.17. There are five qualities which are dangerous in the character of a general.

VIII.18. If reckless, he can be killed;

VIII.19. If cowardly, captured;

---

[91] From around the time of the putative historical Sun Tzu, a sense of the heights achieved by some skilled tongues is illustrated by a comment about a certain lawyer of the state of Zheng, one Deng Xi (c. 546–501 BC), of whom an early Chinese source writes: "He could make the false to be true, and the true to be false. The standard of true and false ceased to exist, so that the permissible and the non-permissible fluctuated from day to day." See Kung-chuan Hsiao, *History of Chinese Political Thought*, Vol. 1: *From the Beginnings to the Sixth Century, A.D.* (F. W. Mote, trans.; Princeton, NJ: Princeton University Press, 1979), p. 369 (quoting from *Lüshi chunqiu*, on which see #2 footnote 71).

[92] Mao certainly placed in the amateur category Duke Xiang of Song (died 637 BC), the subject of a famous story also noted by traditional commentator He Yanxi (Griffith, p. 116; Minford, p. 228). According to that story, Duke Xiang's gentlemanly instincts led him to defer battle until a stronger enemy had crossed a river and was ready to fight. He went down to defeat, was badly wounded, and died the next year. Duke Xiang's generalship elicited scathing comments from Mao: "We are not Duke Xiang of Song and have no use for his asinine ethics" (*Selected Military Writings of Mao Tse-tung*, Peking: Foreign Languages Press, 1963, p. 240). For Duke Xiang's tale see Durrant, Li, & Schaberg, *Zuo Tradition (Zuozhuan)* (cited on p. xxiii above), Vol. 1, p. 357. The story also appears in *Shiji* 38, translated by Nienhauser, Vol. V.1 (cited on p. xxiii), p. 283.

VIII.20. If quick-tempered you can make a fool of him;

VIII.21. If he has too delicate a sense of honor you can calumniate him;

VIII.22. If he is of a compassionate nature you can harass him.

VIII.23. Now these five traits of character are serious faults in a general and in military operations are calamitous.

VIII.24 The ruin of the army and the death of the general are inevitable results of these shortcomings. They must be deeply pondered.

Passage #4.10. Menu of Stratagems, Viewed Here through a Non-military Lens

(There is little intrinsically military here. The strategist to whom these recommendations are addressed does not have to be a general. Likewise, the opposing general in verse I.22 could be replaced by a civilian adversary or rival.

Some kinds of civilian contexts may actually be more conducive than conventional military ones to giving practical effect to Sun Tzu's Passage #4.10. For example, Sawyer, p. 305 note 15 expresses skepticism about the feasibility of verse I.19's advice, as commonly understood, in conventional warfare settings. Big infantry forces, situated close at hand, are hard to miss! By contrast, small world social networks may foster radical misperceptions as to what network locations are "near" versus "far," thereby opening a door for surprise attacks.[93])

I.15. If a general who heeds my strategy is employed he is certain to win. Retain him! When one who refuses to listen to my strategy is employed, he is certain to be defeated. Dismiss him![94]

I.16. Having paid heed to the advantages of my plans, the general must create situations which will contribute to their accomplishment. By "situations" I mean that he should act expediently in accordance with what is advantageous and so control the balance.

I.17. All warfare is based on deception.

I.18. Therefore, when capable, feign incapacity; when active, inactivity.

I.19. When near, make it appear that you are far away; when far away, that you are near.

I.20. Offer the enemy a bait to lure him; feign disorder and strike him.[95]

I.21. When he concentrates, prepare against him; where he is strong, avoid him.

I.22. Anger his general and confuse him.

I.23. Pretend inferiority and encourage his arrogance.

I.24. Keep him under a strain and wear him down.

---

[93] See pp. 99–101 above, citing small-world analyses of Duncan Watts, Jon Kleinberg, and others.

[94] A civilian analog to "Dismiss him!" might involve a decision to drop a social network tie because the other party was deemed not good enough at "thinking like Sun Tzu," hence not a worthy ally.

[95] Griffith's translation is one way of reading this second clause (for an alternative way see #7 footnote 59 and Appendix 7 footnote 5). In civilian contexts, feigning "disorder" may draw on diverse capabilities. For an individual, acting talent may be needed; for an organization, organizational resources suitable for engineering visible *apparent* confusion (e.g., "sham fights") that can rapidly disappear once no longer relevant.

I.25. When he is united, divide him.[96]

I.26. Attack where he is unprepared; sally out when he does not expect you.

I.27. These are the strategist's keys to victory. It is not possible to discuss them beforehand.

Passage #4.11. Further Stratagems, Including Burdening Adversary or Rival with a Task

(Verse VIII.14 points to a version of diplomacy, albeit one that is neither nice nor subtle. As Thomas Schelling once observed, "The power to hurt ... is a kind of bargaining power."[97]

Verse VIII.15 contains the germ of an interesting concept, which might be framed as "burdening an adversary with a task."[98] That idea has particularly rich potential in non-military settings involving complex social structures – already highly developed in Warring States China – where everyday life even for high-ranking elites requires breasting an endless stream of tasks and obligations, ceremonial and practical alike. On the part of those so burdened, those tasks divert precious high-quality attention and other resources away from creative strategic thinking and tactical follow-through, thereby laying groundwork for exerting control over those who must continually carry out such tasks.)

VIII.14. He who intimidates his neighbours does so by inflicting injury upon them.[99]

VIII.15. He wearies them by keeping them constantly occupied, and makes them rush about by offering them ostensible advantages.

---

[96] Traditional commentator Zhang Yu gives verse I.25 a civilian twist: "Sometimes drive a wedge between a sovereign and his ministers; on other occasions separate his allies from him. Make them mutually suspicious so that they drift apart. Then you can plot against them" (Griffith, p. 69). For Theme #6 perspective on Zhang Yu's comment see p. 211 below.

[97] Schelling, #1 footnote 71, p. v.

[98] Needham & Yates, p. 33. This is Principle (5) on a list of ten principles of the "classical Chinese theory of war" (see their pp. 31–37, from a part of the book identified on their p. xxvii as mostly drafted by Krzysztof Gawlikowski). A fine example of Principle (5) comes from traditional commentator Du You (Minford, p. 222): "He prevents them from having any rest, as the state of Han did to the state of Qin, by advising them to dig a canal." For the story of that project – which also involved a switch of loyalties by a key project engineer that ultimately greatly benefited Qin, the intended victim of the stratagem – see Shiji 29 (Watson translation cited on p. xxiii above), Han Dynasty II, pp. 54–55.

[99] Traditional commentator Jia Lin develops the verse VIII.14 concept further, emphasizing civilian means possibly involving injuries that do not advertise themselves as such. Some of his ideas – such as presenting the enemy with "licentious musicians [yin yue] and dancers to change his customs" – are ingenious and their potential effectiveness might have elicited a weary nod of agreement from mid-twentieth-century Western establishmentarians wary of a burgeoning counterculture ("We're gonna rock around the clock tonight," etc.). Relevant Jia Lin commentary is translated by Griffith, pp. 113–14 and by Minford, p. 221.

Passage #4.12. Situation Conducive to Mounting an Effective Subversion Campaign

(Passages #4.12–#4.16 scout the Sun Tzu text for a nascent theory of subversion. Its roots are already found in verses like III.1 and III.10 [Passage #4.1] and IV.10 and IV.12 [Passage #4.4]. In a kindred spirit, traditional commentator Du Mu points to a strategist who "sees the potential dynamic of a situation, a weakness in the enemy camp, and seizes the opportunity. It is like pushing over a rotten tree."[100]

Verse III.23 suggests the possibility that deep intramural antagonisms – here ones set off by a ruler's inept interference with his own military establishment – may give traction to a subversion campaign by a foreign adversary.)

III.18. Now the general is the protector of the state. If this protection is all-embracing, the state will surely be strong; if defective, the state will certainly be weak.

III.19. Now there are three ways in which a ruler can bring misfortune upon his army.

III.20. When ignorant that the army should not advance, to order an advance or ignorant that it should not retire, to order a retirement. This is described as "hobbling the army".

III.21. When ignorant of military affairs, to participate in their administration. This causes the officers to be perplexed.

III.22. When ignorant of command problems to share in the exercise of responsibilities. This engenders doubts in the minds of the officers.

III.23. If the army is confused and suspicious, neighbouring rulers will cause trouble. This is what is meant by the saying: "A confused army leads to another's victory."

Passage #4.13. Situation Conducive to an Effective Subversion Campaign, cont.

II.5. When your weapons are dulled and ardour damped, your strength exhausted and treasure spent, neighbouring rulers will take advantage of your distress to act. And even though you have wise counsellors, none will be able to lay good plans for the future.

Passage #4.14. Water Imagery for Opportunistic Dynamics of Subversion

(It takes no great leap of imagination to discern in these verses a word picture of a successful subversion campaign.[101] Sun Tzu's water image here offers no obvious counterpart to battle, but in a context of subversion none is needed.)

---

[100] Minford, p. 139 (translating Du Mu commentary on verse III.10).

[101] Cf. *Laozi* LXXVIII: "In the world there is nothing more submissive and weak than water. Yet for attacking that which is hard and strong nothing can surpass it. This is because there is nothing that can take its place." See Lau, #1 footnote 34, p. 140. As Sawyer, #2 footnote 15, p. 243 puts it, "The crucial insight [is] that water, the softest and most flexible of things, can paradoxically subvert and even destroy the hardest." It is worth noting that at least two of Sun Tzu's five

VI.26. Therefore, when I have won a victory I do not repeat my tactics but respond to circumstances in an infinite variety of ways.

VI.27. Now an army may be likened to water, for just as flowing water avoids the heights and hastens to the lowlands, so an army avoids strength [shi 實] and strikes weakness [xu 虛].[102]

VI.28. And as water shapes its flow in accordance with the ground, so an army manages its victory in accordance with the situation of the enemy.

VI.29. And as water has no constant form [xing 形], there are in war no constant conditions [shi 勢].[103]

Passage #4.15. Social Dynamics Providing Traction for a Subversion Campaign

(These verses, while pointing to ongoing social processes rather than a specific outcome, show Sun Tzu taking cognizance of decentralized social dynamics on which a subversion campaign might build. Although the focus of these two verses is on a general and his troops, it is a short step to transposing the concepts to contemplate subversion undermining a civilian authority structure, exploiting gossip wafting through the halls of the palace or the bureaucracy.)

IX.37. When his troops are disorderly, the general has no prestige.

. . .

IX.41. When the troops continually gather in small groups and whisper together the general has lost the confidence of the army.[104]

Passage #4.16. Defender's Standpoint: Countering Subversive Dynamics

(Verse XI.44 generalizes as a call for counter-measures against rumors or fake news spread by a subversion campaign, whose target might be a civilian population. In Warring States China, of course, verse XI.44 implementation would most likely have centered on punishment, not on education or persuasion.)

XI.44. He [the general] prohibits superstitious practices and so rids the army of doubts. Then until the moment of death there can be no troubles.

★★★★★★

---

categories of spy ("inside agents" and "double agents") may be viewed as subverted insiders. See p. 378 below.

[102] Sun Tzu's xu/shi (emptiness/fullness) concept pair introduced on p. 113 again appears here.

[103] The concepts of xing and shi appearing in verse VI.29, both central to Sun Tzu, are introduced on pp. 175–76 below. Chinese characters are included in verses VI.27 and VI.29 for disambiguation reasons, in particular to clarify that shi 實 in verse VI.27 is a different character and concept from shi 勢 in verse VI.29.

[104] A different understanding of this verse is possible (though the upshot of the situation depicted seems likely to be much the same). For analysis see Appendix 4 anchor footnote 16.

Strategist Should
Find Advantage

A core contention woven into the fabric of Sun Tzu's thinking is that all situations faced by a strategic actor, even those that appear on their face to be losing ones, hold seeds of opportunity that, if grasped correctly, can be parlayed into strategic advantage.[1] An illustrative statement starts off Passage #5.1 below.

When stated without qualification or nuance this is strong stuff, albeit in keeping with Warring States cultural ideas regarding the surpassing importance of the talent of the commanding general.[2] It is also an optimistic stance and conceivably a confidence-building one (though just the opposite might be true if Sun Tzu's advice fails to deliver on promise). In any case, such a position is bound to elicit controversy – indeed skepticism – in many quarters, even those inclined to be sympathetically disposed toward Sun Tzu. E. Bruce Brooks captures such skepticism well when he notes Sun Tzu VI as making "nearly magical claims."[3] It is therefore important

---

[1] Ames, p. 78: "All determinate situations can be turned to advantage" (statement directly follows quotation of verses V.18–V.19 [Passage #5.9] in Ames's translation). While there is a temptation to treat Sun Tzu's statements of the core contention as rhetorical flourishes, not to be taken literally, the basic idea is too broadly and thoroughly grounded in Sun Tzu to be passed off so easily. It is closely related to a positive stance vis-à-vis war's indeterminate or chaotic factor. Lau & Ames, p. 55 quote a Zhuangzi story that makes vivid both the positive contribution and the distinctiveness of "chaos":

> The ruler of the North Sea was "Swift," the ruler of the South Sea was "Sudden," and the ruler of the Central Sea was Lord Hundun – "Chaos." Swift and Sudden had on several occasions encountered each other in the territory of Chaos, and Chaos had treated them with great hospitality. Swift and Sudden, devising a way to repay Chaos' generosity, said: "Human beings all have seven orifices through which they see, hear, eat and breathe. Chaos alone is without them." They then attempted to bore holes in Chaos, each day boring one hole. On the seventh day, Chaos died.

The hundun concept appears in verse V.17 in Passage #5.9, a linchpin of Theme #5 development.

[2] Of course, Passage #1.1 lists a general's ability as but one of several factors determining the outcome of a war – but the context there is warfare as waged by conventional armies. Much of the comparative advantage of a supremely talented general (or strategist) resides elsewhere, in realms of grand strategy and spies. See pp. 178ff. below (four paths to shi). For a supporting stance taken by traditional commentator Li Quan see Yates, Introduction footnote 51, p. 74.

[3] Brooks, p. 61, also suggesting that Sun Tzu's more extreme claims were toned down in chapters added later in his proposed sequence of chapter accretions (summarized in Table 12A on p. A-35).

that less intellectually aggressive variants of the core contention can also be found in the Sun Tzu text. A leading example is Passage #5.4, which is noteworthy for its recognition of asymmetries between attack and defense and also for its emphasis on the need to wait for enemy vulnerability to materialize. Exploration of the idea of finding advantage is the focus of Theme #5. The companion Theme #6 concentrates on what, by Sun Tzu's own affirmation (verse III.4 in Passage #6.1), is a preferred path to achieving that end, at least in a context of offensive warfare: "Thus, what is of supreme importance in war is to attack the enemy's strategy."

The present approach to analyzing Sun Tzu's core contention draws on the strategy/tactics distinction, which suggests a kind of scope condition for it (albeit one that is not found in the Sun Tzu text).[4] Simply put, the core contention is more compelling on a strategic plane than on a tactical one. Indeed in many tactical military situations it seems demonstrably false, even bizarre, as much military operations research and gaming attests (as indeed do some of Sun Tzu's own affirmations).[5] That is not to gainsay that various of Sun Tzu's battlefield-level observations shed light on his thinking about creating advantage. They will be harnessed to that end below.

One way of connecting the core contention just advanced to the experience and presuppositions of modern audiences is to anchor it in an appreciation of the nature and role of strategic creativity.[6] An excellent summary of what strategic creativity means comes from a description of a character in *Water Margin* (like the *Romance of the Three Kingdoms*, one of China's great pre-modern vernacular novels): "Never dismayed by difficult situations, he always has a plan – simple, elegant, efficient, and, if need be, treacherous."[7] The core contention points to a fundamentally greater scope for such creativity than is acknowledged in many non-Chinese strategic traditions.[8] In Sun Tzu's way of war basic scope for creativity grows out of the primacy of war's human dimensions, which give rise to an almost infinitely rich set of pressure points and opportunities.

---

[4] As noted on pp. 47–48 above, the Sun Tzu text lacks cognizance of a strategy/tactics distinction.

[5] See verses III.17 (Passage #5.27A) and VI.13 (Passage #5.28A), both of which may be read as implying that tactical superiority can be so great that an inferior foe is doomed. However, nothing Sun Tzu says there excludes a possibility that distinguished generalship might yet turn defeat into victory. Cf. verse VI.19 (Passage #5.2). Passage #6.18 contains this and much other relevant Sun Tzu material.

[6] A more anodyne phrase would be "strategic resourcefulness." See Griffith, p. 111. Note an affinity between a focus on strategic creativity and a focus on outliers, here referring to outliers in strategic ability of a general or strategist. Cf. Augier, March, & Marshall, #3 footnote 48.

[7] See p. 162 of Robert Ruhlmann, "Traditional heroes in Chinese popular fiction," in Arthur F. Wright (ed.), *The Confucian Persuasion* (Stanford, CA: Stanford University Press, 1960), pp. 141–76. For background on the *Shuihu zhuan* (*Water Margin*), conventionally attributed to Shi Nai'an (late Yuan/early Ming), see Wilkinson, §31.1.6, p. 451.

[8] Needham & Yates, p. 67 note a strand of Soviet-era Russian opinion that went so far as to assert that "on account of their different education, Chinese military men had an entirely different understanding of strategy." See V. M. Primakov, *Zapiski volontera: grazhdanskaya voyna v Kitaye* (Moscow: Nauka, 1967), p. 67. At the same time, limits on Sun Tzu's brand of "strategic creativity" need to be recognized, particularly its deficit of attention to military innovation (see Theme #6 analysis on pp. 215–16 below).

A privileged place for strategic creativity (as distinguished from technological ingenuity and tactics growing out of it) is actually quite rare in most times and places worldwide, both historical and contemporary.[9] Sometimes an appearance of such creativity is cultivated in some quarters, but that often turns out to be only a veneer, with scant impact on crafting strategy used for operational purposes. In modern times, one of the most important factors contributing to sidelining strategic creativity is the bureaucratic and other organizational climate in which much "strategic" thinking takes shape. Such an environment commonly fosters a skeptical, even corrosive, attitude toward "strategic creativity," at times coming close to questioning the need for, or possibility of, "strategy" itself in any serious sense.[10]

By contrast to this naysaying, a central place for the creative strategist, incarnated as commanding general, had already put down roots in Chinese thinking by the time of the Warring States, whose military writings (Sun Tzu's among them)

> presented combat as an intellectual discipline in which the powers of mind and textual mastery of the commander, along with the unthinking obedience and uniform actions of the troops, guaranteed victory. It was no longer combat itself but the control of men and the manipulation of combat for higher ends that now constituted the essence of war.[11]

Chinese elite and popular culture – more than most Western culture – has long known a role that might be dubbed the "iconic strategist."[12] Such strategists and their ruses and manipulations, great and small, from which their reputations are inseparable, pop up all over Chinese history, from Tai Gong in the misty past surrounding the fall of Shang and birth of the Zhou dynasty (well before the time of Homer in the West, and centuries before Sun Tzu) all the way down to Mao Zedong and Deng Xiaoping in modern times.[13]

Part and parcel of this way of thinking is that the role of "strategist" becomes more than a title or job description. It is a calling, one whose performance has both

---

[9] Cf. an evaluation (vintage 1916) of a talented subordinate by his French Army superior: "He should however be wary, as the head of a service, of any tend[ency] to originality." See Claude Brezinski & Dominique Tournès, *André-Louis Cholesky: mathematician, topographer and army officer* (Cham: Springer International Publishing, 2014), p. 215.

[10] Perhaps paradoxically in light of its aura of mathematical creativity, game theory is not much of a counterweight here. Indeed previously unimagined strategic options are more likely to turn up at a pre-formal stage of analysis, when the characterization of the relevant "game" is still a work in progress, than after the fact, once that game has crystallized and its formal analysis commences.

[11] Lewis, Background footnote 11, p. 11. See also his p. 98: "In the writings of the military specialists [Sun Tzu among them] the key to warfare lay in the man who perceived, calculated, and decided, while those who actually carried arms and spilled blood were reduced to secondary roles."

[12] Cf. Needham & Yates, pp. 70, 80–88 ("The Great Popularity of Military Thought among the People"). The notion of the iconic strategist has a practical side: that very status may itself be harnessed as a strategic or tactical asset. Such an idea underpins the efficacy of Zhuge Liang's "empty city stratagem," one of the most famous ruses in the history of Chinese strategy of any era. See pp. 249–50 below.

[13] Candidates for pre-imperial examples would include historical figures like Wu Zixu or Wu Qi (to whom the *Wuzi* military treatise is attributed) as well as Sun Bin. Biographies of the last two figures rub shoulders with Sun Tzu's biography in *Shiji* 65; see Nienhauser, Vol. VII (1994), pp. 37–45; Nienhauser, Vol. VII (2021), pp. 69–90 (both cited on pp. xxiii–xxiv). Further examples can be culled from traditional commentaries on Sun Tzu as well as from the Three Kingdoms period (notably Zhuge Liang and Cao Cao, his archenemy and Sun Tzu commentator). It is frankly immaterial for Zhuge Liang's iconic status that many, perhaps almost all, of his supposed accomplishments lack a factual basis.

intellectual depth and aspects of a major art form.[14] Sun Tzu himself explicitly gives kudos to a strategist who attains his aim in an "artful and ingenious manner."[15] Of course, placing the strategist role in center stage leads straight on to contemplation of ways of neutralizing its occupant, which is where Theme #6 analysis starts.

THEME #5 CONCEPT OF SHI (STRATEGIC ADVANTAGE) AND FOUR PATHS TO SHI

statements of Sun Tzu's core contention:

Passage #5.1 (verses I.15–I.27) (basic statement of core contention; exhortation to strategist to create advantageous situations and some tips for doing so)

Passage #5.2 (verses VI.18–VI.19) (creating advantage: further statement of core contention)

Passage #5.3 (verse X.17) (command of information as opening pathway to victory)

Passage #5.4 (verses IV.1–IV.15) (tempered version of core contention: need to wait for an opening)

Passage #5.5 (verses II.4–II.7) (conceding limits to the core contention?)

Sun Tzu statements regarding the fundamental relations of strategy, logistics, and tactics:

Passage #5.6 (verses XII.16–XII.19) (subordination of military strategy to grand strategy)

Passage #5.7 (verses II.1–II.2 + II.9–II.11 + II.14) (dependence of military strategy and tactics on logistics)

Passage #5.8 (verses VII.1–VII.9) (dependence of military strategy and tactics on logistics, cont.)

four paths to advantage anchored in Sun Tzu's ideas:

Passage #5.9 (verses V.17–V.20) (starting point for conjuring paths to advantage: "surface impression ≠ actual situation")

Passage #5.10 (verses III.3–III.7) (first path to advantage: expansive definition of the situation)

Passage #5.11 (verses VI.24–VI.30) (second path to advantage: inspiration from nuanced ways of water)

Passage #5.12 (verses XIII.1–XIII.2 + XIII.13–XIII.14 + XIII.16 + XIII.23) (first and second paths to advantage: espionage exemplifies both)

Passage #5.13 (verses V.3–V.12) (third path to advantage: superior complexity management)

Passage #5.14 (verses VII.12–VII.13) (third path to advantage, cont.: mastery of switches of posture or style)

Passage #5.15 (verse VII.21) (fourth path to advantage: awaiting a propitious moment)

---

[14] Note Passage #5.13, which may be read as situating strategy on a plane with other artistic pursuits.

[15] Verse XI.57 (Passage #5.26A). This is Mair, p. 124's "using cleverness to achieve great things."

Passage #5.16 (verse VI.31) (fourth path to advantage, cont.: inevitability of large-scale change)

Passage #5.17 (verses V.13–V.16) (fourth path to advantage: mastery of time and timing factors)

Passage #5.18 (verse XII.15) (avoiding tardy exploitation of success)

history as a reserve army of strategic ideas:

Passage #5.19 (verse XIII.22) (history as quiver of ideas for informing strategic doctrine or style)

★★★★★★

In many respects Theme #5 is the "master theme" in the Sun Tzu text – the most central, the most strategic in the fullest sense of that term, an intellectual centerpiece that helps place the other themes in clear perspective and relationship, helping to fuse them into a coherent whole.

Notwithstanding Theme #5's centrality in Sun Tzu's thinking, its organized analytical development in the text – certainly by comparison with Sun Tzu's detailed and concrete development of each of Themes #1, #2, and #3 – remains generic, sparse, even elusive, almost like a work of post-modern art. There are good reasons for this, since Sun Tzu had little more than metaphors to work with in grappling with Theme #5 concepts that challenge our best modern analytical capabilities, including mathematical and computational ones. Predictably, this situation spawns many questions regarding what, in modern parlance, might be called the scope conditions of Sun Tzu's thinking. For these reasons, shedding light on Sun Tzu's Theme #5 must be prepared to engage with significant challenges and difficulties on a level of ideas, not of text alone.

At the heart of Theme #5 development stand two distinct but intimately related early Chinese military concepts, each having applications in both strategy and tactics. The first, which gives Sun Tzu IV its title, is xing. Griffith describes xing as "meaning 'shape', 'form', or 'appearance' or in a more restricted sense 'disposition' or 'formation'," while also noting that the term "connotes more than mere physical dispositions."[1] An integral part of xing is pattern: how assets or other elements are positioned vis-à-vis one another – in the first instance spatially (the game of weiqi suggests illustrations) but possibly in other ways too (e.g., via location in some network). For this reason, xing takes a basic step beyond the bean counter's mentality found in cruder versions of a net assessment agenda.

A second and yet more fundamental concept – which is the central focus of the present discussion – is shi. Shi gives Sun Tzu V as well as the present Theme #5 analysis its title and is often translated as "strategic advantage." Ames, p. 71 describes shi as the "key and defining idea" in Sun Tzu and offers a four-pronged focus for efforts to understand it:[2]

---

[1] Griffith, p. 85 footnote 1. Li Ling, #1 footnote 84, p. 167 observes that the character 形 in Sun Tzu IV's title and 刑 (both pronounced xing) are related, and that 刑 was "originally written as xing 型, which means 'mold'" (for creating bronze objects). From this meaning derives a new meaning, "model," in a sense of something we should imitate.

[2] See Ames, p. 73. For analysis of evolution of the shi concept, tracing it back to shi's earliest known paleographic form and tracking meanings of shi lost and added in subsequent development,

We must struggle to understand how *shi* can combine in one idea the following cluster of meanings:

1. "aspect," "situation," "circumstances," "conditions"
2. "disposition," "configuration," "outward shape"
3. "force," "influence," "momentum," "authority"
4. "strategic advantage," "purchase".

Ames's four-pronged conceptualization underscores a basic tension at the heart of the translator's art between attempts to arrive at a one-word (or even a single-phrase) translation of a multidimensional concept like *shi* and the deeper goals of translation. Ames puts his finger on the ensuing dilemma:

The irony is that we serve clarity in highlighting what makes sense in our own conceptual vocabulary only to bury the unfamiliar implications that, in themselves, are the most important justification for the translation.

(Ames, p. 71)

In fact, it is this effort to reconstitute the several meanings as an integrated whole and to fathom how the character in question can carry what for us might well be a curious, often unexpected, and sometimes even incongruous combination of meanings that leads us most directly to a recognition of difference.

(p. 72)

Overlaid on these challenges are reasons to believe that the concept of *shi* in Sun Tzu's time was not fully stable, and that, in particular, the meanings of *xing* and *shi* as military/strategic concepts were undergoing a long-term process of divergence along with efforts at clarification.[3]

A starting point for conveying the relationship between *xing* and *shi* comes from two different descriptions offered by modern scholars. One comes from Ames, p. 82: "Where *xing* is limited to the tangible and determinate shape of physical strength, *shi* includes intangibles such as morale, opportunity, timing, psychology, and logistics." A different perspective – importantly dialectical in ways that Ames's is not – comes from Li Ling. Building on parts of the Sun Tzu text that may be read as offering definitions of *shi*, *xing*, or both,[4] Li Ling summarizes his analysis as follows:[5]

Xing is the visible thing with shape, whereas *shi* is the invisible thing; things which can be seen are *xing*; *xing* is the static thing, whereas *shi* is the dynamic one; *xing* is what you can achieve through preparation, while *shi* is what you need to do to deal with this or that specific enemy. There is *xing* in *shi*, *shi* in *xing*. When these two words are used together as a compound [which is not the case in the Sun Tzu text], it means a situation created by man.

---

see Boqun Zhou, #1 footnote 11, pp. 88–97 (a study that importantly builds on Qiu Xigui's 2012 pathbreaking etymological analysis of *shi*, not available when Ames did his earlier work). For the Warring States period Zhou's analysis (Zhou, p. 96) classifies *shi*'s meanings into four groups, recognizably related to Ames's categories 1–4 though differing somewhat in specifics.

[3] See Yates, Introduction footnote 28, p. 220. From within the Sun Tzu text, an example of *shi* not being fully clarified comes from its usages in verses X.7 and X.10 where, as Giles, p. 105 points out, "[i]t is obvious that *shi* cannot have the same force" in both locations. See Appendix 2 footnote 43.

[4] In particular, verses I.16 (Passage #5.1) and V.19 (Passage #5.9).

[5] Li Ling, #1 footnote 84, p. 168. For further details of Li Ling's analysis – which also critiques assorted English translations of *xing* and *shi* by Griffith, Ames, Sawyer, and Minford, noting that none of them captures the dialectical *xing*/*shi* relationship – see Li Ling, #1 footnote 84, pp. 164–72. His own proposed English translations of *xing* are "visible dispositions" or "potential energy"; of *shi*, "invisible dispositions" or "released energy." The fact that he offers not one but two translations of each concept brings us back to Ames's observations on tensions at the heart of translation.

Against this backdrop, attention naturally turns to identifying *usable* (or, in modern military jargon, actionable) strategic insights that may be gleaned from Sun Tzu's Theme #5 thinking. It is plainly unsatisfactory merely to recite attributes of *shi* or *xing* or their relationship, calling upon strategists to turn those ideal types into wondrous accomplishments by mechanisms left unspecified. Theme #2 offers a useful control case here, since whether one agrees with them or not, Sun Tzu's terrain rules are chock full of actionable advice.

In a quest for actionable Theme #5 leads, game theory provides a useful foil. Game theory routinely starts with a well-defined formalized "game" whose analysis is the game theorist's task. So basic is that formulation that it is usually taken for granted without comment. Insight into how Sun Tzu might deviate from this familiar, comfortable, formalist paradigm (which, of course, would have been terra incognita in his time) is suggested by the famous opening lines of the *Laozi* Daoist classic:

> The way [*dao*] that can be told
> Is not the constant way.[6]

Adapting the spirit of the *Laozi*, a parallel strategic principle might read:

> The game that can be specified
> Is not the actual game.

In other words, *no attempt to enumerate options and attendant payoffs is actually a reliable bedrock description of the latent possibilities of the situation. The better the strategist the less good the model!* To put the point affirmatively, *there is always at least one more strategic option, a kind of trap door in the situation, that an adroit strategist can spot and turn to advantage.*[7] To return to a game theory framing, the main value of specifying "the game" may be to encourage figuring out a way of circumventing it: in short, replacing it by a different, more advantageous game that better describes the actual situation at hand![8] Especially against a backdrop of the tidal wave of success enjoyed by modern game theory, here is a candidate for "*What did Master Sun (and Laozi too) know that we still don't (or have yet to absorb adequately)?*"

---

[6] Translation by D. C. Lau, #1 footnote 34, p. 57.

[7] A hint that Sun Tzu (who had no inkling of formal game theory) might have intuitively grasped possibilities of mismatch between a game specification and the actual conflict it is supposed to represent – mismatch that an adroit antagonist might be able to exploit – comes from a classical game theory analysis of Sun Tzu. See Emerson M. S. Niou & Peter C. Ordeshook, "A game-theoretic interpretation of Sun Tzu's *The Art of War*," *Journal of Peace Research*, 1994, 31(2), 161–74:

> a strategy is regarded by your opponent as *qi* [奇] *if your opponent has no idea whether it is possible or feasible.* If we accept this interpretation, then we cannot so easily (if at all) set this part of Sun Tzu into a game theory mold, since game theory assumes that all players are aware of all strategies available to an opponent.
>
> [p. 172, emphasis supplied]

A 1990 Caltech working paper version of this article is cited, with brief summary, on p. 1029 of Barry O'Neill, "Game theory models of peace and war," in Robert J. Aumann & Sergiu Hart (eds.), *Handbook of Game Theory with Economic Applications*, Vol. II (Amsterdam: Elsevier, 1994), pp. 995–1053.

To be clear, Niou & Ordeshook's interpretation of *qi* is one possibility. For others see Theme #14 analysis.

[8] Cf. Schelling's dictum: "the inevitable lot of a definitive survey is to serve as a definitive target." See Thomas C. Schelling, *The Strategy of Conflict* (Cambridge, MA: Harvard University Press, 1960), p. vi.

To get further analytical purchase, there is a need for ways of connecting the abstract strategic concept of *shi* with practical paths to its achievement. Not surprisingly, early Chinese military writing from a time after Sun Tzu contains efforts to do just that.[9] However, a close reading of Sun Tzu can perhaps do better, by developing a suite of ideas about achieving *shi* that are both anchored in the text and also point beyond it, capturing something of the larger spirit of strategic creativity with which Sun Tzu has long been identified. Four such paths to *shi* are proposed here:[10]

First path: exploiting an expansive understanding of "the situation";

Second path: exploiting nuanced or fine-grained understandings of the situation, ones that disaggregate received conventional accounts of it;

Third path: exploiting situational complexity and mastery of switchings;

Fourth path: strategic patience (waiting for the situation to redefine itself, conjoined with superior skills at spotting nascent trends before others do).

All four paths may be said to squeeze pathways to advantage (*shi*) out of "reinventing" the prevailing situation, discerning more degrees of freedom in it than initially appear.[11] While none of these paths is easy to put into practice, at the heart of each is a practical concept that has no need for invoking near-magical qualities of efficacy which modern Western accounts, possibly reflecting traces of a lingering orientalism, still sometimes impute to ancient Chinese strategy or stratagems it inspires.

The active ingredient of the first path – which draws inspiration from the expansive spirit of *shi* – pivots on replacing a narrower view of a conflict situation by broader vistas, then exploiting new networks of possibility thereby opened.[12]

Importantly, doing so need not – indeed generally does not – involve escalation in a classic military sense, thereby risking a clash with Theme #3 (indeed some types of de-escalation may occur, as juxtaposing verses XIII.1 and XIII.2 in Passage #5.12 suggests). Pursuing this first path to *shi* on what nowadays would be called a strategic plane is facilitated by the fact that strategic situations are commonly much more open-ended than battles or other tactical ones. The definition of a strategic

---

[9] For example, the *Huainanzi*, an early Han work, offers a three-way typology of sources of *shi*. Ames – calling this typology "the most lucid explanation of *shi* as a military term available to us in any of the early texts" – translates it as (1) "morale advantage" (*qi shi*), (2) "terrain advantage" (*di shi*), and (3) "opportunity advantage" (*yin shi*) (e.g., exploiting failures of enemy operational readiness like fatigue, hunger and thirst, ill-preparedness and disorder, etc.). See Roger T. Ames, *The Art of Rulership: a study in ancient Chinese political thought* (Honolulu: University of Hawai'i Press, 1983), pp. 70–71. On the *Huainanzi* see *Early Chinese Texts* (cited on p. xxii above), pp. 189–95; Wilkinson, §59.5.7, p. 799.

[10] To be clear, these four paths, which should not be regarded as mutually exclusive or exhaustive, are a modern analyst's construct based on analysis of Sun Tzu's ideas and emphases. By contrast to the *Huainanzi*'s treatment of *shi* (see previous footnote), whose categories have a tactical flavor, the four paths as presently delineated are more strategic in focus.

[11] This type of thinking accords with an intellectual posture that Goldin has termed the "primacy of the situation" (essentially an acute awareness of context). Describing it as "pervasive in the received literature of the Warring States period," he points specifically to Sun Tzu as providing its clearest expression. See Goldin, Introduction footnote 7, pp. 4, 15.

[12] There are continuities here with broadly based strands of traditional Chinese thought. See, e.g., Ames, p. 56, observing that an "important factor in classical Chinese 'knowing' is comprehensiveness."

"situation" is often in the eye of the beholder, with any boundaries being as much (or more) set by perception or convention as by any "objective" factors.[13]

Relevant paths to shi commonly start by recognizing that intangibles – realms of cognition, emotion, and information – are as much a part of "the situation" as are any physical realities. Opening those paths frequently involves lifting constraints that the strategist himself has imposed on the definition of the situation, unwittingly and possibly for cultural reasons. "No scruples Sun Tzu" (Theme #10) has a basic place here (cf. the sad story of Duke Xiang of Song!).[14]

To grasp the centrality of this first path to advantage in Sun Tzu's thinking one need look no further than his exhortation to win without fighting (verse III.3 in Passage #5.10), which suggests operating in conflict arenas and with a métier remote from the wars conventional soldiers fight. The spirit of the first path is also captured by Cao Cao's commentary on verse I.16 (Passage #5.1), pointing to the magnificently extensible category of methods "outside the regular methods" for creating advantage.[15] Epitomizing the first path to shi is the espionage chapter, for which Passage #5.12 serves as placeholder. Expanding the definition of "the situation" in a different direction, a second example – more exactly, an entire sheaf of possibilities – involves achieving military success, not through defeat of the enemy army in the field but through exploiting vulnerabilities in the logistics support on which that army crucially depends (cf. Passages #5.7 and #5.8).[16] Such vulnerabilities may arise naturally, or

---

[13] A sympathetic perspective from a game theory quarter comes from Richard K. Guy, "What is a game?," pp. 43–60 in Richard J. Nowakowski (ed.), *Games of No Chance* (Mathematical Science Research Institute Publications No. 29; Cambridge, UK/New York: Cambridge University Press, 1996). See in particular his p. 45:

> It is hard to draw the line between mathematics and psychology. There are even cases where you should prefer a bad move to a good one! Your opponent might be learning to play a game with which you're already familiar. In this case you'll probably be able to win a few times despite the bad moves you deliberately make so as not to give away your strategy. Or one move, theoretically the best, might gain you only a dollar, while another, which loses a dollar, might win you a few hundred if your opponent fails to find the subtle winning reply. Or you may be a baby-sitter, whose job is much more peaceful if your opponent wins. Or a card-sharp who's losing while the stakes are low, in anticipation of winning later when the stakes are higher.

[14] For the tale of Duke Xiang, and what happened to him, see #4 footnote 92.

[15] Mair, p. 139 note 19. An area of game theory relevant to Cao Cao's thinking is hypergames, whose key idea is one of expanding focus to a "larger game that is really being played whether or not both players are aware of it." See Nicholas S. Kovach, Alan S. Gibson, & Gary B. Lamont, "Hypergame theory: a model for conflict, misperception, and deception," *Game Theory*, Vol. 2015, Article ID 570639, available at www.hindawi.com/journals/gt/2015/570639 (visited August 22, 2021). For an illustration of hypergame thinking see P. G. Bennett & M. R. Dando, "Complex strategic analysis: a hypergame study of the fall of France," *Journal of the Operational Research Society*, 1979, 30(1), 23–32. That article offers a memorable dictum: "if you start off with a fundamentally inadequate view of the situation, then the best technical assistance that does not challenge this will succeed only in turning blunders into better-organized blunders" (Bennett & Dando, p. 31).

[16] Broadening the definition of the situation in logistics directions aligns closely on Admiral Eccles's comprehensive concept of logistics, which envisions logistics as encompassing the full range of the means of war writ broadly, not merely supply issues or any other limited set of logistics functions (Boorman, Introduction footnote 4, pp. 99–102). In any age of war this Eccles concept – and the great range of logistics vulnerabilities to which it points – offers one of the most militarily

they may be created (a special case of the latter would be using fire attack against enemy supplies; see Passage #6.13).

Whereas the first path looks outward, seeking qualitatively new realms of action where advantage may be had, the second path to *shi* looks inward, taking a characteristically "fine-grained" view of the situation.[17] Its hallmark is seeking nuanced options and opportunities there, often of a sort overlooked or devalued by high-level planners and "big picture" intellectuals alike.[18] Sun Tzu's Passage #5.11 water imagery fits well here, evoking water's uncanny way of worming its way through the slightest crevice, frustrating all attempts to dam it. Less metaphorically, so too does Sun Tzu's emphasis on psychological factors in warfare, centered in the ambition of asserting control over an entire military situation by control of the mind of just one person – namely, that of the opposing general. A fine-grained perspective is also captured by Sun Tzu's remarks on delicacy of espionage dealings (verse XIII.14 in Passage #5.12). A kindred spirit of attention to minute detail shows in verse XIII.16 (also Passage #5.12) pertaining to human intelligence targets.[19]

Extending the concept beyond generals and spies, human beings (perhaps especially in their relationships and interactions with one another) have more degrees of freedom – to borrow the statistician's phrase – than any inanimate technology (certainly one of any pre-digital variety!).[20] That complexity renders human situations more "porous" – to use an image inspired by ways of water, Sun Tzu's favorite

---

compelling supports for the importance of Sun Tzu's first path. Sun Tzu's thinking carries Sun Tzu part-way there – Passage #1.1 on net assessment is relevant, as is verse V.19 (Passage #5.9) if read as offering Sun Tzu's definition of logistics – but emergence of the full concept would have to await modern times.

[17] A fine-grained perspective is admirably conveyed by traditional commentator Du Mu's verse IV.13 observation (Passage #5.4): "Letting slip no chance means keeping a close watch on every slightest crack in the enemy's guard, not missing the tiniest shred of a chance" (Minford, p. 156). Glossing verse IV.9 Giles shows his Victorian era roots by quoting poet Robert Browning to similar effect:

> Sun Tzŭ [Giles notes] reserves his approbation for things that
>
> > "the world's coarse thumb
> > And finger fail to plumb."
> > (as quoted by Giles, p. 29)

[18] Continuing to use game theory as a foil for developing Sun Tzu's concepts, it is worth noting that game-theoretic formulations of conflict tend to exclude, ab initio, precisely those seemingly small, even minuscule details where roots of *shi* are often found. Adequate incorporation of such details into the description of the game, even positing that one knew which details to add, might so complicate it as to waterlog efforts at formal analysis, even rendering such analysis downright impractical.

[19] Note the possibility, illustrated by the story of Zhuan Zhu (noted by Sun Tzu in verse XI.37, see Passage #6.5), that the belly of a braised fish might hide a knife used in an assassination – in Zhuan Zhu's case, a killing that brought to the throne of Wu the patron of the putative historical Sun Tzu. See also Passage #10.8 (short summary) and Appendix 6 footnote 2 (further details).

[20] Putting aside the modern phraseology, Sun Tzu would almost certainly have concurred with this point – but then might have limited its application to those in high places (ruler, general, spymaster, etc.). Warring States military thought (Sun Tzu's included) tended to attribute little if any independent volition or agency to rank-and-file soldiers. For discussion and analysis see #9 footnotes 7 and 8.

source of imagery – than first appearances often indicate.[21] For that reason such situations lend themselves to nuanced manipulations, often having an emotional twist (see parts of Passage #5.1; also verse XII.18 in Passage #5.6). That line of thinking suggests that Sun Tzu's more ambitious aspirations about achieving *shi* may find greater scope for practical realization amid the manifold complexities and ambiguities of civilian life than in any kinetic form of warfare.

A third path to advantage, for which Passage #5.13 provides basic anchoring, obtains its results from what might be termed superior complexity management. Some of that management has a combinatorial thrust, emphasizing being better than one's foe in sure-footed maneuvering through a forest of combinations (Passage #5.13's verse V.8 allusion to musical notes provides relevant imagery). By envisioning the quiver of *qi* stratagems available to a master strategist as being inexhaustible, verse V.6 points to a slightly different type of complexity management, picking and choosing from a large "reserve army of stratagems," possibly in recombinant ways ("some of one ruse, some of another"), to match the situation at hand: VI.26.

> Therefore, when I have won a victory I do not repeat my tactics but respond to circumstances in an infinite variety of ways.

(Passage #5.11)

One source of that "reserve army" might be Passage #5.1's *ruses de guerre*. Taken individually, they are familiar from many parts of military history. What matters is the skill with which they are chosen and deployed (with tailoring as needed, one place complexity may enter) in specific situations. Meanwhile, double agentry (see Passage #7.10, there labeled Sun Tzu's "Symphony of Spies") generates its own versions of combinatorial and other complexity that can easily run off the scale.[22] Yet, as these several strands of Sun Tzu's thinking also suggest, something other than "crisp" combinatorics in a conventional mathematical (or *weiqi* game) sense is at work here. Much of what Sun Tzu has in mind might be better envisioned as a type of "soft combinatorics" wherein both the elements being combined (e.g., specific strategic or tactical steps) and ways of combining or sequencing them are less precisely defined than, say, moves in *weiqi* or combinatorial game theory analyses.

A core part of superior complexity management is mastery of *switching* strategic or tactical posture (in some cases better described as "style"). The importance that Sun Tzu's thinking attaches to such mastery is suggested by the attention Passage #5.13 accords to the foundational concept of *bian*, mentioning that concept (whose basic

---

[21] An anecdote from the *Zhanguo ce* (a Warring States text previously mentioned in #1 footnote 99), as recounted by Goldin, Introduction footnote 7, p. 23 describes a relevant civilian scenario:

> a servant girl is told by her mistress, who is having an affair and wishes to remove the inconvenience of a husband, to carry a goblet of poisoned wine to her master. The servant-girl knows that the wine is deadly, and so she is faced with a classic Dilemma: she cannot disobey her mistress, but she can hardly connive at the murder of her master. Her solution is to drop the goblet deliberately.

(Also translated by Crump, #1 footnote 99, p. 531.)

[22] For context, however, it should be noted that Sun Tzu's thinking about double agents (*fanjian*) is not airtight. See pp. 368–71 below.

meaning is "changes") four times in verses V.8–V.11.[23] Verse VII.13 (Passage #5.14), from a different part of Sun Tzu, provides further grist for this line of thinking. Using vivid imagery, that verse invites thinking about the general's task – the quest for shi – in what might be labeled stylistic terms.[24] In military strategy and tactics, as in fine arts like painting and music (as well as in mathematics and in martial arts), styles run in families. Observe, for example, that verse VII.13 is framed, not in terms of a single "style", but of a family of styles – in this case six in total.[25] That multiplicity underscores that a basic challenge facing a general is one of switching styles to meet exigencies of a changing situation. Six styles give rise to $6 \times 5 = 30$ possible switches; n styles yield $n(n-1)$ switches. Such switches are where mistakes often get made.[26] The moment at which a switch takes place is also a point where the mask of command may slip a little, enabling an astute adversary to glean valuable shards of information.[27] Traversing this complexity, often under great stress, without stumbling (and also capitalizing on enemy stumbles in style switching; see Passage #5.1's verse I.24) stands at the heart of the third path to shi.[28]

---

[23] Griffith's translation of Passage #5.13 masks the salience of a notion of change (bian) in what Sun Tzu says there. The role of bian is clarified in Appendix 5, where Passage #5.13 again appears and is annotated to show each occurrence of bian. For further analysis, weighing combinatorial versus non-combinatorial ways of reading verses V.6–V.12, see Appendix 1 anchor footnote 29.
Illustrating a switching idea in a concrete military situation – which Passage #5.13's high level of abstraction does not provide – Sun Tzu's treatment of fire attack (Passage #9.7's verses XII.7 and XII.8) emphasizes promptly following up on fire attack, once fully launched, by use of troops. A warning not to procrastinate or shy away from the need for such a switch comes through clearly in verses XII.7–XII.9.

[24] For an approach to the analysis of style, noting work of Nathan Leites cited in Introduction footnote 45, see Scott A. Boorman, "A memo on style: reflections on 'style' as a sociological concept," Yale Journal of Sociology, Fall 2011, 8, 181–94.

[25] Ames's translation shows more plainly than Griffith's that verse VII.13 encompasses six "styles." To clarify the point, markers [a]–[f] are inserted in Ames's translation of verse VII.13:
> Thus, [a] advancing at a pace, such an army is like the wind; [b] slow and majestic, it is like a forest; [c] invading and plundering, it is like fire; [d] sedentary, it is like a mountain; [e] unpredictable, it is like a shadow; [f] moving, it is like lightning and thunder."
> (Ames, p. 130)
It is worth noticing that [e], while capable of being envisioned as a "style," differs from the other five alternatives listed because it is harder to relate to a specific military deployment or course of action.

[26] A maxim attributed to Napoleon makes a bow to this point, characterizing the transition from defense to offense as "one of the most delicate operations in war" (Maxim #19, see Operational Level of War – Its Art, Preliminaries footnote 8, p. 1–9). Although that transition has other aspects too (e.g., logistics ones), a defense-to-offense transition involves a style switch, as Sun Tzu's vivid imagery of the "ninefold earth" (defense) and "ninefold heavens" (offense) suggests (see verse IV.7 in Passage #5.4).

[27] See, e.g., verse IX.43 (Passage #9.16). It points to an enemy style switch ("cruel" to "fearful," to use Mair's p. 111 language) in the enemy's dealings vis-à-vis his own troops that has potential to give insight into what the enemy general is really afraid of.

[28] Illustrating the point, Sun Tzu's exhortation to a commander in encircled ground to devise stratagems (verse XI.14 in Passage #2.1) can be read as a call to switch mindset seamlessly – from straightforward to imaginative and cunning – if plunged without warning into a dire ambush situation. Sun Tzu's larger emphasis on the importance of speed in war (see Theme #3) further underscores a need for fast yet poised switchings.

A fourth path to advantage exploits the time dimension. Verse VII.21 (Passage #5.15) offers a starting point, pointing to Theme #9's concept of building on natural forces and momentum. Verse VI.31 (Passage #5.16) provides further images of change; its language ties to other major strands of traditional Chinese thought.

Passage #5.4 is Sun Tzu's most developed analytical statement of the fourth path. Timescales in strategy or grand strategy tend to be longer (even by orders of magnitude) than those in tactics. Read from that perspective, Passage #5.4 underscores a need to wait, conceivably for a very long time, for the situation as it presently exists to change in a way that brings opportunities for advantage, as holes in the enemy's armor reveal themselves.[29] Doing so calls for a strategist to possess the skill-sets of a wily survivor (see Passage #5.4's verse IV.7; also Passage #3.3's verse III.11). Although Sun Tzu does not say so – possibly reflecting a mindset that treats identity as fixed and stable – exercising such patience may place enormous demands on the patient strategist's sense of personal and political identity.

Seeds of advantage often start tiny.[30] By the same token they are easy to miss (and also to overestimate, a major pitfall on which Sun Tzu touches only glancingly).[31] A description by Mark Lewis puts a finger on the observational challenge:[32]

> The central act that translated [expedient] assessment into successful maneuver was the identification of the "pivots" (ji 機). This word concretely signified the firing mechanism of the crossbow, but it had the more abstract meaning of a moment of change, or more precisely the moment just before a new development or a shift in direction became visible. These "pivots," which are also translated as "seeds" when applied to processes where an organic metaphor seems more suitable, constituted the nodal points of a situation in flux, and it was a characteristic of the "sage" or "superior man" to recognize them before they became manifest.

To that end, a Sun Tzu–inspired strategist needs a mindset supple enough to grasp seeds of fresh opportunity that even seemingly unlucky events may produce (see verse VII.2 in Passage #5.8; verses VIII.12–VIII.13 in Passage #5.21A). Equally necessary, once events start to break favorably, is the drive and follow-through to exploit the situation to the hilt to achieve fullest realization of *shi* (Passage #5.18).[33]

Using these four paths – profiting by an expansive concept of the situation, drilling down into fine-grained but possibly game-changing detail, managing complexity and mastering the ceaseless switches of style and posture it entails, and building on time

---

[29] Such a focus harmonizes with strands of Daoist thought emphasizing inevitability of eventual decline (here, that of a foe). See D. C. Lau, "The treatment of opposites in *Lao Tzŭ* 老子," *Bulletin of the School of Oriental and African Studies, University of London*, 1958, 21(2), 344–60, in particular p. 353.

[30] See Lau & Ames, p. 57: "The capacity of the small, incipient, and seemingly incidental to control the large by virtue of its pivotal position is an enduring theme in classical Chinese philosophy, and underlies [the] notion of getting the most from a situation while minimizing loss." See also Ames, p. 90.

[31] Verses IX.45–IX.46 (Passage #11.14) convey a relevant warning, albeit rather generically phrased.

[32] Lewis, Background footnote 11, p. 119. See also analysis of *ji* in Boqun Zhou, #1 footnote 11, pp. 74–88, part of a larger discussion of what he terms the "lever schema" in early Chinese thought (see his chapter 3, pp. 59–98). Looking beyond the firing mechanism of the crossbow, and the physical technology it represents, in some ways a deeper idea comes from Cao Cao's commentary on verse IV.8 (Passage #5.4): "the thing is to see the plant before it has germinated" – a concept that places the great general in company of modern developmental biologists. See Giles, p. 28.

[33] A warning about perils of overreaching is implicit in Cao Cao's commentary on Passage #5.21A. See pp. A-127f. and A-54.

and timing factors – a sufficiently resourceful strategic actor, supported by excellent intelligence, will in Sun Tzu's view indeed be able to find or create veins of advantage (*shi*), even ones that turn defeat into victory. Should those four paths to *shi* fail to do the job by themselves, deception (Theme #7) combines well with all of them.[34]

******

Choice of passages to illustrate and develop Theme #5 calls for selectivity to avoid undue duplication of material assigned to other themes (all of which, in one way or another, are also about promoting *shi*). Most of Sun Tzu V (which has *shi* in its title) appears in the present Theme #5 passage selection.[35]

Theme #5 passage selection leads off with passages containing expressions – and in some cases qualifications – of Sun Tzu's core contention (p. 171 above), which asserts that any determinate strategic situation can be turned to advantage (Passages #5.1–#5.4, with Passage #5.5 adding a limitation). Because that core contention is at the very least controversial, possibly tinged with strategic mysticism, focus next turns to examining how Sun Tzu engaged with a thoroughly non-mystical, structural level of war, by exploring Sun Tzu's nearest approaches to recognizing what Admiral Eccles termed the "fundamental relations" among strategy, logistics, and tactics (Passages #5.6–#5.9).[36] This Sun Tzu material, which is incompletely developed by modern standards, is an important lens on the state of military theory in Sun Tzu's time, illuminating both what it grasped and what it failed to grasp. Among the best bits are those that show cognizance of logistics factors and limitations (albeit that Sun Tzu's working out of logistics concepts and principles remains fragmentary). Also noteworthy here is Sun Tzu's subordination of military strategy to grand strategy that is a natural reading of verse XII.16 (Passage #5.6).

Next, with the exceptionally multifaceted Passage #5.9 in a gearshift role, Passages #5.10–#5.18 supply textual anchoring for the four paths to *shi* profiled earlier.

In a shift of register, Passage #5.19 rounds out Theme #5 passages on a Chinese cultural note, suggesting a role for history as a major source of strategic inspiration.

Roads Not Taken by Sun Tzu in Developing Theme #5

Reiterating a basic point made earlier (pp. 47–48), Sun Tzu himself lacked a clear sense of a strategy/tactics distinction. Even Sun Tzu V, which – for reasons rooted in its title theme of *shi* – might seem a prime candidate for finding an unequivocal focus on strategy, is interspersed with material oriented, at least in the first instance, toward battlefield problems and events (e.g., verse V.17 in Passage #5.9). Because of this blurring of levels, and because Sun Tzu's core contention that all determinate

---

[34] See also pp. 507–508 below, revisiting this role of deception from a broader information warfare vantage.

[35] For roundup of Sun Tzu V material not included in this selection see Appendix 5 footnote 3.

[36] For overview of Eccles's concept of the fundamental relations among strategy, tactics, and logistics see Eccles, Background footnote 62, pp. 61–66. Those fundamental relations include:
- subordination of military strategy to grand strategy (a topic also relevant to Sun Tzu Theme #4);
- subordination of both tactics and logistics to strategy;
- dependence of both strategy and tactics (at all levels of strategy and command) on logistics.

situations can be turned to advantage may fall flat in tactical realms, Sun Tzu may not always be his own best advocate – a point that needs to be recognized more clearly than it usually is.

A key conundrum in developing Sun Tzu Theme #5 – especially for an environment where Sun Tzu's ideas have widely taken root (as some sources suggest was already the case in Warring States times[37]) – involves clarifying the dynamics of an encounter between two strategic actors, each of whom has thoroughly absorbed Sun Tzu's teachings. Nothing Sun Tzu directly says clarifies what might then transpire.[38]

A very important omission in Sun Tzu's Theme #5 thinking is a deficit of advice regarding how a victorious state should go about reaping the full conqueror's prize of a recent conquest – in effect, a "day after the war" question. What little Sun Tzu has to say on this topic will be gathered under Theme #13, where Passages #13.2–#13.6 round up relevant material. Especially as regards the sensitive topic of plunder, quite possibly a problematic course of action that receives repeated Sun Tzu endorsements, there are loose ends in Sun Tzu's thinking here.[39] Those loose ends are especially noteworthy precisely because a major, and intellectually powerful, strand of Sun Tzu points to reaping strategic advantage from an expansive "definition of the situation" (i.e., the first path to advantage discussed on pp. 178–80). Surely one type of expansive definition would take note of the situation in a conquered country after a war's successful conclusion, including the welfare (and, of course, attitudes toward the victor) of its civilian populace. Yet Sun Tzu has nothing to say that would forbid or at least rein in plunder once the war is won.[40]

Illustrating that even classics are not always internally consistent there is also unresolved tension between Sun Tzu on shi and Sun Tzu on terrain (Theme #2). The former comes close to rejecting all formulaic approaches – the "globalizing generalizations" of which Lau and Ames write (see their p. 56). Intimations of such rejection appear in verses I.27 (Passage #5.1) as well as in VI.26 and VI.29 (Passage #5.11). By contrast, Sun Tzu's terrain material brims with definite prescriptive rules. Here is at least a fault line in Sun Tzu's thinking, if not an outright contradiction.[41] There are,

---

[37] See well-known observation of Han Feizi (died 233 BC) quoted in #4 footnote 6 above.

[38] For some ideas regarding a possible "Sun Tzu vs. Sun Tzu" scenario see p. 273 below.

[39] Unqualified advice to plunder was already identified as a soft spot in Sun Tzu's thinking by traditional commentator Li Quan, writing in the Tang dynasty. In response, he went so far as to propose that Sun Tzu's advice for an invader to plunder be stood on its head, as advice not to plunder, a radical reversal of message that elicited a blast from Giles, p. 119 ("Alas, I fear that in this instance the worthy commentator's feelings outran his judgment").

The present study also scrutinizes Sun Tzu's stance on plunder from the perspectives of multiple themes, including Themes #3, #6, #8, #10, and #13. For overview, including a list of Sun Tzu's endorsements of plunder, see Passage #13.6 discussion in Appendix 13 (pp. A-223f.).

[40] Plunder following victory in war may have ranged well beyond foodstuffs needed to sustain an army. An archaeologically recovered source describes a Qin government boat loaned to an ordinary citizen and used to transport tiles, possibly from "one or more Chu cities that the Qin had previously destroyed, revealing one way in which the victors despoiled their former enemies." See Yates, Background footnote 53, p. 304.

[41] In fairness to Sun Tzu, it should be noted that related tensions persist in our own time. Professionals of many stripes constantly struggle to strike the best balance in their areas of expertise between the very real advantages of prescriptive rules, starting with the fact that such rules can be conveniently recorded, studied, taught, and updated with an eye to useful applications, and the sometimes much greater achievements of the holistic, often intuitive,

of course, ways of trying to glue these parts of Sun Tzu together. One approach would resurrect an interpretive project of discovering in Sun Tzu a strategic level vs. a tactical level, with prescriptive rules being treated as pertinent to the latter but not the former context. But such an effort runs afoul of the fact shi can also find expression on a tactical level, while some of the terrain rules proffer advice that is fundamentally strategic. A more promising alternative (taking cues from E. Bruce Brooks's accretionist theory of the text) would be to treat Sun Tzu's thirteen chapters as itself an evolving corpus, groping toward a more refined sense of doctrine and its limits. For example, the verses just noted (I.27, VI.26, VI.29) are from chapters (Sun Tzu I and VI) to which Brooks assigns a relatively late date, whereas the terrain rules bunch in chapters assigned an early date.

It is perhaps best to concede existence of rough edges here, understanding that Sun Tzu's analytical unity is in fact incomplete, even partly flawed (quite possibly reflecting the fact that the text we have is a compilation, not work of one person).

A further deficit in Sun Tzu's Theme #5 material involves absence of pointers for educating those who fill, or who are destined to fill, high command or other strategist roles.[42] Mastery of Theme #5, certainly at the level of proficiency Sun Tzu demands, is plainly a high order of human accomplishment. Passage #5.13's allusions to assorted artistic pursuits suggest as much. However, as a coup d'oeil of the Theme #5 passages makes plain, there are also inherent difficulties in capturing and conveying in *words* the actual insights or reflexes such mastery requires, suggesting that more is needed here than doctrinal pronouncements alone.[43] Study of theory must be blended with exposure to some form of practice. A question then arises: how and where to fit direct or vicarious exposure to practical applications of Sun Tzu's ideas about shi into the life course of the strategist (possibly starting in early adolescence or even earlier).[44] Sun Tzu does an admirable job of setting a stage for recognizing these educational challenges, but leaves us in the dark about how to rise to them. Sun Tzu takes no interest in the education of strategists.

judgment of "grandmaster" professionals. The latter, like strategists in Sun Tzu's mold, are few in number, for good or ill (depending on who they are and what cause they serve!), and hard to replace when they die or retire, if replaceable at all.

[42] For example, Sun Tzu says nothing about one important way of fostering strategic or tactical creativity, which is immersion in an idea-rich environment (even if seemingly chaotic or unfocused). In an unplanned way, the flourishing of insightful military thought in Warring States China, Sun Tzu's included, may well have profited immensely from the larger creative intellectual ferment of that age.

[43] Mair, p. 139 note 19 makes the intriguing observation that Sun Tzu "must have been aware" of conceptual difficulties associated with elucidating shi, "[o]ne of the most important concepts in the Sun Zi [text] ... [yet] also an extremely elusive idea." A comment attributed to Emperor Tang Taizong (r. 626–49) observes, "'Probably military strategy can be transmitted as ideas but cannot be handed down as words.'" See Sawyer, *Seven Military Classics*, p. 341. A millennium and more on, Alfred North Whitehead posed a related challenge: "I do not think in words," he said, "I begin with concepts, then try to put them into words, which is often very difficult." See *Dialogues of Alfred North Whitehead, as recorded by Lucien Price* (Boston, MA: Little, Brown, 1954), p. 150.

[44] A contemporary Japanese adaptation of Sun Tzu for children is analyzed on pp. 203–14 of Paul van Els & Frank Witkam, "Schoolyard soldiers: the art of adapting the *Art of War*," *Journal of Chinese Military History*, 2019, 8(2), 191–224.

Sun Tzu (2) ("Sun Tzu Extended") and Sun Tzu (3) ("Sun Tzu Analogical") Frontiers of Theme #5

Despite the chasm separating two very different civilizations, glimpsed at time periods separated by over two millennia, there is resonance between Sun Tzu on shi (backed by Sun Tzu's advocacy for taking the enemy intact in preference to destroying the enemy, see Passage #3.3's verses III.1 and III.2) and the strategy-as-control motif at the heart of a classic 1955 strategy essay by Herbert Rosinski (1903–1962). A historian by education, Rosinski was a refugee from Nazi Germany who wrote a seminal book on the German army and had a productive association with the US Naval War College in Newport, Rhode Island during the 1950s, where his ideas were an important intellectual influence on both Admirals Eccles and Wylie.[45]

Entitled "New Thoughts on Strategy," Rosinski's essay's defining concept is that strategy is the art of control. Written with a universalistic thrust, Rosinski's essay shows its twentieth-century Western provenance by its references to both air power and modern science (each viewed through a lens of the strategy-as-control idea, as is sea power). A clue that there may be good grist here for Sun Tzu (2) and (3) developments is that serendipitous parallels to Rosinski appear in several observations in Roger Ames's analytical introduction to his Sun Tzu translation, alluding to a talented commander as being one who is able to exert situation control.[46]

Rosinski's strategy-as-control analysis can be elaborated along lines mapped by Admiral Eccles, distinguishing seven dimensions of any strategic control project:[47]

    (i) What to control (i.e., the object or objects of control);

    (ii) What is the purpose of this control;

    (iii) What is the nature of the control;

    (iv) What degree of control is necessary;

    (v) When the control is to be initiated;

    (vi) How long the control is to be maintained;

    (vii) What general method or scheme of control is to be used.

For clarity, it should be emphasized that this seven-part list is a *structural framework*, whereas Sun Tzu is advancing a *theory* with the concept of shi as its centerpiece. Bearing this caveat in mind, there are insights to be had from exploring how the two analyses may relate to one another. Harking back to the second of two questions posed on p. 1 of the present study – *"What are Sun Tzu's limitations or blind spots?"* – among the most useful of those insights are instances where application of the framework turns up limitations in Sun Tzu, thereby clarifying implicit scope conditions on the applicability of Sun Tzu's ideas that range beyond Warring States Chinese context.

---

[45] Rosinski's terse (c. 600 word) essay is reproduced in Eccles's *Military Concepts and Philosophy* (1965, see Preliminaries footnote 5), pp. 46–47. Wylie's important distinction between sequential and cumulative kinds of strategy was also in part inspired by Rosinski. See #4 footnote 28 above.

[46] See Ames, p. 78: "By developing a full understanding of those factors that define one's relationship with the enemy, and by *actively controlling and shaping the situation* so that the weaknesses of the enemy are exposed to one's acquired strength, one is able to ride the force of circumstances to victory." For further allusions to situation control in a similar vein see Ames, pp. 81, 86. Theme #6 passage selection further anchors situation control ideas in the Sun Tzu text. See pp. 208, 214–15 below.

[47] See Boorman, Introduction footnote 4, p. 103. The seven-part classification (i)–(vii) is from Eccles's 1965 book (cited in Preliminaries footnote 5), p. 48.

Start with dimension (i) ("object of control"). Given emphases already prominent in Sun Tzu's first chapter (Passage #5.1), it is clear that, for Sun Tzu, the most basic focus of control is the mind of the enemy general or ruler (or, generalizing the idea to broader settings, the mind of the principal actor of interest). That concept is a signature Sun Tzu contribution to the art of control, even though Sun Tzu himself never quite said it in so many words.[48] It plainly fits many contexts where the fundamental adversary is a single individual or is otherwise embodied in a well-defined center (e.g., ruler and entourage; general and staff). It has less relevance when an adversary's "will" is decentralized and diffuse, as in the case of a people in arms or a politically engaged citizenry (or certain terrorist social movements); or is vested in a decentralized computing system or network. There will be sources of *shi* there too – analysts of modern public opinion or social media could attest to that – but Sun Tzu's direct relevance somewhat recedes. In those latter contexts, one could even conceive a Sun Tzu–inspired actor being led astray, investing in futile efforts to locate an adversary's "central mind" or master source of direction that would be a worthy object for a Sun-Tzu-esque control project. *Such a target may not exist.*

Turning to dimension (ii) ("purpose of control"), Sun Tzu's core goal is plainly conquest of other states. As already noted (p. 185 above) Sun Tzu says nothing about specific steps to be taken the day *after* the conquest or how to realize *shi* then. That deficit could easily give rise to a raft of strategic errors, including a tendency of a Sun Tzu–inspired actor to proliferate enemies needlessly in the aftermath of conquest (or en route to it). Sun Tzu's oft-repeated advice to plunder could in fact be a stimulus to such errors, goading members of a destitute conquered population to push back with all means available to them with the same sort of desperate fury Sun Tzu attributes to soldiers in death ground.[49]

Sun Tzu's dimension (iii) ("nature of control") preference is for achieving *shi* via pathways that are, first and foremost, psychological and informational (albeit that a role for kinetic means is never ruled out). This has the makings of a deep and generalizable insight, giving Sun Tzu legitimate claim to a title of "pioneering information warfare theorist" (see pp. 504ff.). But it is not a universal one. Especially in modern times there are some strategic competitions where a logistics or technological edge (e.g., access to strategic airlift or sealift, or to oil or other raw materials), of a sort not closely centered on psychological or information advantages, may make all the difference to the outcome.[50] Absent compensating stimuli from other quarters, a Sun Tzu–inspired conflict actor may be prone to underestimate physical logistics and military hardware (tanks, ships, planes, etc.) as sources of *shi*.[51]

---

[48] See Theme #6 for roundup of relevant Sun Tzu statements, which include Passages #6.7–#6.10.

[49] In the world's political–military history, there are more scenarios here than peasant rebellions alone. Critiquing the Treaty of Versailles as a "Carthaginian peace" imposed on Germany after its World War I defeat, John Maynard Keynes warned of how "[a]n inefficient, unemployed, disorganized Europe faces us, torn by internal strife and international hate, fighting, starving, pillaging, and lying." See Keynes's *The Economic Consequences of the Peace* (London: Macmillan, 1919), p. 141.

[50] For the case of oil, see Anand Toprani, *Oil and the Great Powers: Britain and Germany, 1914–1945* (Oxford, UK: Oxford University Press, 2019).

[51] Cf. Needham & Yates, p. 66: "overestimation of the psychological factors, manifested in many different forms, was a constant characteristic of Chinese military thought."

Sun Tzu's most important posture vis-à-vis dimension (iv) ("degree of control") is crisp and (mostly) unequivocal: bet only on a sure thing.[52] Einstein is famously quoted as saying: "God does not play dice." Neither does a Sun Tzu–inspired strategist pursuing shi.[53] Overlaid on this precept is a Sun Tzu predilection for exceptionally precise control, epitomized by the image of the raptor and its prey (verse V.14 in Passage #5.17). Yet as any combat-experienced commander (or political operative, for that matter) knows full well, there are many circumstances where gambles do need to be taken because adequate information is lacking. Here Sun Tzu shows basic analytical limitations (see Theme #1 analysis, pp. 64–66).

Jumping ahead to dimension (vii) ("scheme of control"), much of Sun Tzu's most important thinking can be anchored in one or more of the four paths to shi sketched on pp. 178–84, supplemented by Passage #5.1's menu of stratagems. That thinking is further fleshed out by ideas assigned to other themes. There is a remarkably high density of smart ideas here. Yet one must still return – at best – the Scots law verdict of "not proven" to Sun Tzu's "core contention" (p. 171 above) that strategic advantage (shi) can be found in all determinate situations. Furthermore, absent a more definitive model of the pertinent situation (which in many ways is exactly what Sun Tzu's theory rejects) there is also ambiguity surrounding how far that "advantage" goes – for example, is it local or global (as a mathematician might understand those terms)?[54] Despite these various doubts and ambiguities, the present position is that it is more productive to take the core contention seriously, and to explore what its contribution might be to catalyzing strategic creativity, than to reject it out of hand.[55]

Sun Tzu's emphases pertaining to the Eccles–Rosinski framework's time and timing dimensions (v) and (vi) – when to initiate control and how long to maintain it – merit particular scrutiny, both for what they capture and for what they omit.

---

[52] A few possible concessions by Sun Tzu can be spotted. See #1 footnote 36 for candidates.

[53] A flavor of the performance standard Sun Tzu expects comes from an observation quoted in the Shiji, to the effect that "among the schemes of the wisest man one in a thousand will end in error, while among those of the greatest fool one in a thousand will succeed." See Shiji 92 (Watson translation cited on p. xxii above), Han Dynasty I, p. 171. This comment (not from the Sun Tzu text) illustrates the chance-denying mindset also found in Sun Tzu (see p. 64 above), along with the level of calculational prowess that sets Sun Tzu's ideal strategist apart from ordinary people.

A modern context for such numbers comes from an observation of Andrew Ng, an artificial intelligence researcher, noting that "Most people underestimate the difference between 95% and 99% accuracy – 99% is a game-changer" (The Economist, September 9, 2017, pp. 59–60 article on face recognition technology).

[54] A comment of mathematician S. S. Chern transposes well here: "The future lies in the study of relations between local and global phenomena." (Chern, American Scientist, 1978, 66(4), 506.)

[55] Here it is important that Sun Tzu's core contention can be useful without being literally accurate. Cf. "quixotic defense of a theory" heuristic discussed in Willim J. McGuire, "Creative hypothesis generating in psychology: some useful heuristics," Annual Review of Psychology, 1997, 48, 1–30, pointing to examples from history of physics as well as psychology (and also suggesting [p. 21] that this "odd-sounding heuristic ... probably receives more use as a creative technique than is recognized").

Shifting focus to computational arenas, there is a hint of a Sun Tzu perspective in the observation that "there is no universal method or package applicable to every fluids problem. In fact I would go further to say that even for a particular problem there is no best method, one should always be able to dream up something better." See E. J. Hinch, Think before You Compute: a prelude to computational fluid dynamics (Cambridge, UK: Cambridge University Press, 2020), p. xiii (emphasis supplied).

Regarding dimension (v) ("when the control is to be initiated"), a signature and notably insightful Sun Tzu principle is that the time to find (and often plant) seeds of *shi* is *early on* in a flow of events – certainly long before any pitched battle.[56] That is a basic message of Passage #5.4. Passage #5.12 on espionage suggests ways to do this. Applications of this line of thinking are diverse. Especially in modern times, when logistics systems have grown vastly complex and wars are technology-driven, this Sun Tzu idea puts a spotlight on "upstream" logistics roots of battlefield outcomes – on qualities of a logistics system rooted in the strengths or weaknesses of a nation's economy. The capabilities and limitations of such a system, which should be broadly conceived, take shape long before a war starts, nowadays fueled by strategic programming decisions and R&D project management for military innovation (both involving lengthy lead times, sometimes spanning decades). Sun Tzu has nothing directly to say about such activities, an omission that reflects not only his ancient time but also his lack of interest in building lasting infrastructure (see pp. 67, 146 above). However, an easy extension of Sun Tzu's "upstream" emphasis is that a major logistics system, including its R&D component, *itself* presents an inviting target of attack long before outbreak of any war.[57] Other potentially war-winning upstream interventions pivot on psychology, public opinion, and domestic politics. Particularly consonant with Sun Tzu's emphases are steps to ensure that an enemy leadership or elite, and possibly a broader civilian population too, enters a war riven by bitter divisions, possibly born of fights over extraneous matters.

On a grand strategic timescale, an intriguing if unstated corollary of Sun Tzu's positioning with respect to dimension (v) – one that does not seem to have attracted as much attention in the Sun Tzu literature as it merits – points to steps to control an enemy leader or key decision maker by pressure points or other influences from an earlier stage in the leader's career or life course, well before that leader rose to current prominence (even as far back as, say, that person's young adulthood).

Dimension (vi) turns to the "how long to exert control" question. Sun Tzu's posture on this issue is complex and several layers need to be peeled back. First of all, Sun Tzu exhibits deep, unequivocal distaste for protracted war. Focus here is on avoiding prolonged conduct of conventional military operations, a costly activity by almost any yardstick (Passages #5.12, #3.4, etc.). At the same time, verses IV.2–IV.4 (Passage #5.4) underscore a need to wait for a moment when an exploitable vulnerability of the enemy opens up.[58] The counsel here is patience – possibly lots of it. This part of Sun Tzu's thinking lays intellectual foundations for a depth of strategic planning horizon that is a signature feature of much subsequent Chinese strategic thought down to the present day.[59] Such ideas are compatible with

---

[56] As traditional commentator Du Mu observed in connection with verse III.4 (Passage #5.10): "He who excels at resolving difficulties does so before they arise. He who excels in conquering his enemies triumphs before threats arise." (Griffith, p. 77)

[57] See Theme #1 "frontiers," pp. 72–74 (attack on R&D activities) and Theme #6 "frontiers," pp. 219ff. (attack on a logistics system).

[58] This Sun Tzu emphasis puts a spotlight on the nature, sources, and limitations of sanctuary.

[59] In different ways, both Deng Xiaoping's "hide and bide" and Mao's "protracted war" concepts provide illustrations. Deng Xiaoping's famous dictum, commonly translated as "hide and bide" (*tao guang yang hui*), is actually an old proverb with quite complicated meanings. The protracted war concept is developed in one of Mao Zedong's most famous political–military essays, "On protracted war" (1938). See Mao, #4 footnote 92, pp. 187–266.

acceptance of protracted conflict, even a taste for it so long as costly fighting can be avoided. Time itself becomes the ultimate source of shi.[60]

Yet there are also ways in which Sun Tzu wobbles a little here, certainly if the focus is Sun Tzu (2) or (3). One manifestation of that wobble is Sun Tzu's lack of a long-term planner's instincts (see pp. 66–67 above). A more specific deficit, already noted in connection with dimension (ii), is that Sun Tzu's thinking about shi shows little interest in how a control project born of war might need to be extended beyond the moment of the "complete victory" (quan sheng) of verse IV.7 (Passage #5.4). As already noted (p. 185 above) there are loose ends in Sun Tzu's relevant thinking. When confronting a Sun Tzu–inspired foe, those loose ends suggest canvasing for omissions or "holidays" in that foe's planning pertaining to what happens after a successful war (as contrasted with achieving military victory). Especially in light of Sun Tzu's popularity in many US military circles, the 2003 Iraq War, viewed through the lens of a US adversary, offers an instructive case in point. In that conflict, magnificent control by the US and allies of the military situation during the war's active phase swiftly led to quan sheng in a military sense – only to be followed by underplanned, disorganized, often ineffectual, and immensely costly control projects in the war's aftermath.[61]

A different facet of dimension (vi) finds expression in verse XIII.22 (Passage #5.19), amounting to a Sun Tzu affirmation of the long strategic shadow of the remote past (a version of "control in deep time" that involves looking backward in time to seek applicable precedents). Verse XIII.22 is an early (and somewhat stylized) expression of longstanding and broadly based Chinese cultural awareness of China's past, its history and historiography. China's long and often richly documented history provides a truly massive inventory of strategic precedents and examples. At times those precedents may be more cognitively "real" to Chinese strategists than are episodes from far more recent Western political–military history to their Western counterparts.[62] The weight thus accorded to Chinese history encourages treating such historical cases as a type of scenario "reserve" to be drawn on to suggest or inform ways of tackling current problems – in effect, a culturally grounded version of

---

[60] This is a way of stating the root principle of the fourth path to advantage (pp. 178, 183 above).

[61] For overview and analysis see Emma Sky's book The Unraveling (2015), cited in #3 footnote 52.

[62] "In no other country," Henry Kissinger wrote in his On China (New York: Penguin Press, 2011) (an observation Jonathan Spence notes in his New York Review of Books review of Kissinger), "is it conceivable that a modern leader would initiate a major national undertaking by invoking strategic principles from a millennium-old event," as Mao was known to do in discussing policy matters. For his part, daring to disagree with Mao in the aftermath of the disastrous Great Leap Forward of the 1950s, prominent Chinese Communist military leader Peng Dehuai famously cast himself as Zhang Fei, a Romance of the Three Kingdoms character well-known for his adversarial relation with Cao Cao, to whom Mao was widely compared. For his pains, Peng was purged and ultimately died, a victim of the Cultural Revolution, only to receive posthumous rehabilitation. For a brief account see Jürgen Domes, Peng Te-huai: the man and the image (Stanford, CA: Stanford University Press, 1985), p. 91.

Regarding the general phenomenon see also John K. Fairbank, "The grip of history on China's leadership," in Fairbank's China Watch (Cambridge, MA/London: Harvard University Press, 1987), chapter 11, pp. 86–94 (essay cited by Ames, pp. 66–67).

case-based decision.[63] Uses of that type of tool can be relatively subtle, as when historians commissioned by a Chinese emperor may have sought to use historical examples to convey, sotto voce, an implicit warning to their patron.[64] Although the Sun Tzu text is not a work of history, the larger Sun Tzu tradition – comprising not only the text itself but also its reception over the centuries – also harbors a related form of looking back. At least from Tang dynasty times forward, through work of Sun Tzu's traditional commentators and others, many of Sun Tzu's strategic and tactical precepts have been brought to life for later generations by cases culled from many eras of Chinese military history – from the Zuozhuan, the Shiji, the Chinese dynastic histories, or (shifting attention to historical fiction) the *Romance of the Three Kingdoms* (*Sanguo*), among other sources.[65]

Of course, according weight to history in this way can very easily shade over into approaching the historical record as a source of opportunities for shaping the past to control the future, a move found in Chinese cultural and strategic repertoires from very early times.[66] Verse XIII.22 (Passage #5.19) may in fact be read in that way, creating a respectable precedent for spying activities out of presumed exploits of illustrious individuals who, if both existed in historical fact, would be more accurately characterized as defectors.[67] Parts of Chinese historiographical tradition contain many further examples of creating "great men" and sculpting their lives and achievements for current purposes, at times on scant evidence.[68] The Maoist era showed the Sun Tzu work itself not immune from kindred manipulations of its message.

★★★★★★

Sun Tzu Theme #5 thinking taps a singularly rich vein of strategic ideas. Only a few further examples will be mentioned here, emphasizing Sun Tzu (2) and (3) directions. On a grand strategic canvas the spirit of Theme #5 in action may be spotted in

---

[63] For relevant formal model-building, replacing a standard rational choice (expected utility theory) framework by a case-based one, see Itzhak Gilboa & David Schmeidler, A Theory of Case-Based Decisions (Cambridge, UK: Cambridge University Press, 2001).

[64] For this Tang dynasty case see Twitchett, #1 footnote 44, p. 5 and his accompanying footnote 2.

[65] For orientation to the first two works see pp. xxiii–xxiv; for the Sanguo see p. xxiv and pp. 45–46. Regarding the dynastic histories, see Wilkinson §49, pp. 689–713. His table 112 (p. 694) provides an overview.

[66] See, e.g., Brooks & Brooks, Introduction footnote 36, pp. 47–48 (case of "invent[ing] an already bureaucratic antiquity," involving c. 320 BC insertion of an anecdote into the Zuozhuan to give fourth-century BC managers of a then-new land policy the imprimatur of "the minister of a great state, two centuries earlier"). A further layer of the phenomenon is illustrated by accusations directed by one early Chinese school of thought against another: "Your antiquity isn't old enough" (Brooks, p. 74).

[67] Sawyer, #1 footnote 3, pp. 7, 12, 134–35.

[68] For the Zhuge Liang legend – counterfactually creating a place for him in Chinese history as a surpassingly great strategist – see Eric Henry, "Chu-ko Liang in the eyes of his contemporaries," Harvard Journal of Asiatic Studies, 1992, 52(2), 589–612. For balanced perspective note should be taken of major streams of Chinese historical writing whose allegiance was to "impartial reconstruction of reliable and accurate history," not to any didactic or manipulative agenda. See Hsu Kwan-san, "The Chinese critical tradition," The Historical Journal, 1983, 26(2), 431–46 (quoting from p. 435).

the World War II Allies' "culture of encouragement" described in Paul Kennedy's *Engineers of Victory* (2013).[69] That culture rested, not on superior ways of allocating existing Allied material resources, nor yet on simply producing more of the same, but rather on fostering an organizational climate conducive to *creating* new science and new military technology that had never previously existed anywhere. Much of that science and technology was the handiwork of young, very junior, previously unrecognized scientists and engineers (contributions of physicist R. V. Jones, immortalized by Winston Churchill's post-war account, would be one case in point).[70] The Axis powers were fundamentally less good at mobilizing that type of technical creativity, thereby giving the Allied powers a major, and arguably war-winning, technological edge. Although, to be clear, Sun Tzu's own focus is not on promoting military innovation, certainly not in any technological sense, the "culture of encouragement" that Kennedy portrays harmonizes excellently with the "expansive understanding of the situation" path to *shi* (see pp. 178–80 above).

Complementing Kennedy's title theme, a separate Sun Tzu (3) incarnation of Theme #5 is suggested by David Howarth's *Law as Engineering*.[71] Its fundamental concept is that transactional and legislative lawyers (who comprise a major, if often unsung, part of the legal profession) are in the business of designing and building social structures and devices in much the same sense that engineers design and build physical ones. With the proviso that Sun Tzu (1) is *not* at all a "builder" in a relevant sense, the creative institutional design direction to which Howarth's book points – whose tools are those of legal draftsman, but whose problem-solving style is an engineering one – gives scope for ambitious and nuanced control projects worthy of a modern Sun Tzu (3). In Sun Tzu's terms, at the heart of Howarth's concept is achieving *shi* (strategic advantage) through design of the legal structures involved in legislative law as well as in private transactions (e.g., setting up business organizations, doing M&A deals, etc.). Crucially, such structures are not only constraints, whittling down available courses of action, but also, in the best spirit of verse VIII.13 ("By taking into account the favourable factors, he makes his plan feasible ...") *enablers*, greatly enlarging the realm of the possible (compare again Sun Tzu's first path to *shi*, pp. 178–80).[72]

Two basic features of Sun Tzu's control thinking mesh particularly well with the law-as-engineering concept. The first is Sun Tzu's preference for planting seeds of control early on, long before any "battle" (in legal context, litigation or perhaps alternative dispute resolution approach) is even contemplated. A forte of legal drafting tools is planning for some contingency that materializes, if at all, only many years later (compare Sun Tzu traditional commentator Du Mu: "He who excels at resolving difficulties does so before they arise").[73] A second feature is recognition of just how much leverage for control can flow from even a seemingly

---

[69] See Paul Kennedy, *Engineers of Victory: the problem solvers who turned the tide in the Second World War* (New York: Random House, 2013).

[70] Winston S. Churchill, *The Second World War*, Vol. II: *Their Finest Hour* (5th ed.; London: Cassell, 1955), chapter XIX, "The wizard war," pp. 337–52.

[71] This Howarth book is cited in Introduction footnote 13.

[72] Verse VIII.13 appears in Passage #5.21A and is discussed, with further context, in Passage #1.28A.

[73] This Du Mu observation was previously quoted (with further specifics) in #5 footnote 56.

minute step – say, single clause or sentence added or deleted.[74] Exquisite titration of wording choices in legal drafting is a superb exemplar of the spirit of precise and nuanced yet surefooted control for which Sun Tzu's thinking about *shi* stands (e.g., raptor and prey image in Passage #5.17's verse V.14).[75]

Notwithstanding the formidable strengths of legal draftsmanship as a tool of control, many executives, among them many military professionals, still tend to view legal structures purely as sources of constraint – mostly as nuisances, blocking them from doing things they would otherwise do. For that reason, the idea of law as an enabler – in Sun Tzu's terms, a source of *shi* that can reshape a situation in ways that further the strategic aims of the designer – is a major twenty-first-century contender for "*What did Master Sun know that we still don't (or have yet to absorb adequately)?*"[76] It is a worthy modern update to Du Mu's commentary observation just quoted.

Returning full circle to the starting premise – essentially a cultural/philosophical stance – articulated on p. 171 above, the essence of that premise is that (as Roger Ames has put it) "All determinate situations can be turned to advantage." Hyperbole aside, in the biophysical world as we know it that proposition strains credulity, indeed may be simply false. But enter a world of ubiquitous and complex software, algorithms, and digital networks and this observation gains traction. Indeed for such a world, *and all that it in turn controls*, the proposition as stated by Ames may be substantially accurate, and stands to become yet more so as the twenty-first century unfolds.

★★★★★★

Set of principal passages used to illustrate Sun Tzu Theme #5 (paths to *shi*)

Passage #5.1. Basic Statement of Core Contention; Exhortation to Strategist to Create Advantageous Situations (and Some Tips for Doing So)

(Verse I.15 affirms p. 171's "core contention," to the effect that a sufficiently able commander can find seeds of success in all determinate situations. Passage #5.1's stratagems are a menu of ways to do precisely that. Many of them exploit one or more of the four paths to *shi* set out on pp. 178–84. Take verse I.23, for example. Encouraging a foe to preen himself on his superiority in tactical forces may foster a misplaced arrogance that overlooks logistics vulnerabilities. Taking a larger view of the situation that gives logistics its due – a case of the first path to advantage [pp. 179–80] – his more astute opponent may be able to exploit those vulnerabilities and prevail. Or take verse I.18. Actively exploiting sources of

---

[74] A relevant observation of traditional commentator Zhang Yu is quoted in #3 footnote 40.

[75] Shifting register away from mathematical or related tools of control to natural language ones, cf. the case noted in #3 footnote 41 (involving substitution of two letters of the alphabet for other letters). In legislative drafting, a counterpart might be changing a couple of words or perhaps a clause.

[76] A modern source of examples with Sun-Tzu-esque aspects, partaking of both legislative craftsmanship and strategic-political manipulations in roughly equal measure, would be the case of Robert Moses, profiled by Robert A. Caro, *The Power Broker: Robert Moses and the fall of New York* (New York: Knopf, 1974). See, e.g., Caro's chapter 10, "The best bill drafter in Albany," pp. 172–77.

advantage revealed by a fine-grained reading of the situation – i.e., the second path to advantage [pp. 180–81] – may appear like blatant inactivity to an unperceptive foe, lulling him into a complacency that proves his undoing.)

I.15. *If a general who heeds my strategy is employed he is certain to win. Retain him! When one who refuses to listen to my strategy is employed, he is certain to be defeated. Dismiss him!* [Emphasis supplied.]
I.16. Having paid heed to the advantages of my plans, the general must create situations [*shi*] which will contribute to their accomplishment. By "situations" [*shi*] I mean that he should act expediently in accordance with what is advantageous and so control the balance.[77]
I.17. All warfare is based on deception.
I.18. Therefore, when capable, feign incapacity; when active, inactivity.
I.19. When near, make it appear that you are far away; when far away, that you are near.
I.20. Offer the enemy a bait to lure him; feign disorder and strike him.
I.21. When he concentrates, prepare against him; where he is strong, avoid him.
I.22. Anger his general and confuse him.
I.23. Pretend inferiority and encourage his arrogance.
I.24. Keep him under a strain and wear him down.
I.25. When he is united, divide him.
I.26. Attack where he is unprepared; sally out when he does not expect you.[78]
I.27. These are the strategist's keys to victory. It is not possible to discuss them beforehand.

Passage #5.2. Creating Advantage: Further Statement of Core Contention

(In some ways verse VI.19 is a yet stronger affirmation of Sun Tzu's p. 171 core contention, since it lacks verse I.15's possible overtones of hucksterism.[79] A very important general principle to which verse VI.19's second sentence points is that power and force are not always the same as *usable* power and force.[80])

VI.18. Although I estimate the troops of Yue as many, of what benefit is this superiority in respect to the outcome?
VI.19. *Thus I say that victory can be created.* For even if the enemy is numerous, I can prevent him from engaging. [Emphasis supplied.]

---

[77] It is possible to view the latter part of verse I.16 as a definition of *shi*. See Mair, p. 139 note 19.
[78] Lau, 1965 article, p. 321 footnote 9 translates: "Attack where the enemy is not prepared; go by way of places *where it never occurred to him you would go*." (Emphasis supplied.) Lau's translation injects a touch of hypergame thinking. See #5 footnote 15 above (citing sources on hypergames).
[79] One of the traditional commentators, Zhang Yu, bluntly observes of verse I.15, "Master Sun is trying to persuade the king of Wu to employ his services." See Minford, p. 111.
[80] For modern statement of this usable power principle see Eccles, Background footnote 62, p. 56.

Passage #5.3. Command of Information As Opening Pathway to Victory

(Drawn from one of the oldest layers of the Sun Tzu text,[81] this verse helps to bolster the plausibility of the core contention through its suggestion that knowledge, an informational and cognitive attribute that might be within the grasp of even a seemingly weak strategic actor, can be made to lead on to victory, a physical world event. Long before the age of software, access to the right information and its exploitation is here being identified as a decisive determinant of success in war. Passage #11.1 – starting with verse III.31, "Know the enemy and know yourself; in a hundred battles you will never be in peril" – and Passage #11.2, which is also from Sun Tzu X, reinforce the point.)

X.17. Conformation of the ground [*xing* 形] is of the greatest assistance in battle. Therefore, to estimate the enemy situation and to calculate distances and the degree of difficulty of the terrain so as to control victory[82] are virtues of the superior general. *He who fights with full knowledge of these factors is certain to win; he who does not will surely be defeated.* [Emphasis supplied.]

Passage #5.4. Tempered Version of Core Contention: Need to Wait for an Opening

(Several verses here point to the virtue of strategic patience. By acknowledging limitations on a strategist's ability to prevail by unilateral action or on a predictable time horizon, these verses modulate the core contention found in Passages #5.1–#5.3.[83] But for Passage #5.4, Sun Tzu's stance might indeed verge on what Brooks, p. 61 has termed "nearly magical claims," to detriment of credibility.)

IV.1. Anciently the skilful warriors first made themselves invincible and awaited the enemy's moment of vulnerability.
IV.2. Invincibility depends on one's self; the enemy's vulnerability on him.
IV.3. It follows that those skilled in war can make themselves invincible but cannot cause an enemy to be certainly vulnerable.
IV.4. Therefore it is said that one may know how to win, but cannot necessarily do so.[84]
IV.5. Invincibility lies in the defence; the possibility of victory in the attack.

---

[81] According to Brooks's proposed order of chapter accretions, Sun Tzu X is next-to-earliest in the sequence, preceeded only by Sun Tzu IX. For further specifics see Table 12A on p. A-35.

[82] "Control victory" is evocative, but "creating the conditions for victory" (Mair, p. 115) is clearer.

[83] It is worth noting a family resemblance between verse IV.3 and Clausewitz's famous principle (*On War*, Preliminaries footnote 1, pp. 357–59) that defense is the stronger form of warfare (a structural point that, as Clausewitz also notes en passant, finds civilian parallels in litigation).

[84] Mair, p. 88 sums up verse IV.4's key idea in eight words: "Victory can be foretold, but cannot be forced."

IV.6. One defends when his strength is inadequate; he attacks when it is abundant.[85]

IV.7. The experts in defence conceal themselves as under the ninefold earth; those skilled in attack move as from above the ninefold heavens. Thus they are capable both of protecting themselves and of gaining a complete victory.

IV.8. To foresee a victory which the ordinary man can foresee is not the acme of skill;

IV.9. To triumph in battle and be universally acclaimed "Expert" is not the acme of skill, for to lift an autumn down requires no great strength; to distinguish between the sun and moon is no test of vision; to hear the thunderclap is no indication of acute hearing.

IV.10. Anciently those called skilled in war conquered an enemy easily conquered.

IV.11. And therefore the victories won by a master of war gain him neither reputation for wisdom nor merit for valour.

IV.12. For he wins his victories without erring. "Without erring" means that whatever he does insures his victory; he conquers an enemy already defeated.

IV.13. Therefore the skilful commander takes up a position in which he cannot be defeated and misses no opportunity to master his enemy.

IV.14. Thus a victorious army wins its victories before seeking battle; an army destined to defeat fights in the hope of winning.

IV.15. Those skilled in war cultivate the Dao and preserve the laws [fa] and are therefore able to formulate victorious policies.

Passage #5.5. Conceding Limits to the Core Contention?

(Verse II.5 opens a door to acknowledging limits on Sun Tzu's core contention.)

II.4. When the army engages in protracted campaigns the resources of the state will not suffice.

II.5. When your weapons are dulled and ardour damped, your strength exhausted and treasure spent, neighbouring rulers will take advantage of your distress to act. *And even though you have wise counsellors, none will be able to lay good plans for the future.* [Emphasis supplied.]

II.6. Thus, while we have heard of blundering swiftness in war, we have not yet seen a clever operation that was prolonged.

II.7. For there has never been a protracted war from which a country has benefited.

---

[85] Verse IV.6's idea is turned on its head in the Han strips (not known to Griffith, because discovered only in 1972), a literal translation of which reads: "One defends, then he will be abundant; he attacks, then he will be inadequate." This Han strips version of verse IV.6 may actually express a deeper idea, aligning on Clausewitz's concept of defense as the stronger form of warfare (#5 footnote 83 above).

Passage #5.6. Subordination of Military Strategy to Grand Strategy

XII.16. And therefore it is said that enlightened rulers deliberate upon the plans, and good generals execute them.

XII.17. If not in the interests of the state, do not act. If you cannot succeed, do not use troops. If you are not in danger, do not fight.

XII.18. A sovereign cannot raise an army because he is enraged, nor can a general fight because he is resentful. For while an angered man may again be happy, and a resentful man again be pleased, a state that has perished cannot be restored, nor can the dead be brought back to life.

XII.19. Therefore, the enlightened ruler is prudent and the good general is warned against rash action. Thus the state is kept secure and the army preserved.

Passage #5.7. Dependence of Military Strategy and Tactics on Logistics

(Sun Tzu's awareness of the dependence of military strategy and tactics on logistics shows here and in Passage #5.8, even though his treatment falls well short of a systematic theory of logistics.)

II.1–II.2. (Translation from Lau's 1965 article, p. 324, which starts with a caption supplied by Sun Tzu: "*The method of employing troops.*"[86]) "When one thousand fast four-horse chariots, one thousand four-horse wagons covered in leather, and one hundred thousand armour-clad troops are used, if provisions have to be transported over a distance of a thousand li, then what with expenditure at home and in the field, on the guest advisers, on materials such as glue and lacquer, and on the supply of chariots and armour, it will cost one thousand pieces of gold every day before one hundred thousand troops can be raised."

. . .

II. 9. Those adept in waging war do not require a second levy of conscripts nor more than one provisioning.

II.10. They carry equipment from the homeland; they rely for provisions on the enemy. Thus the army is plentifully provided with food.

II.11. When a country is impoverished by military operations it is due to distant transportation; carriage of supplies for great distances renders the people destitute.

. . .

II.14. As to government expenditures, those due to broken-down chariots, worn-out horses, armour and helmets, arrows and crossbows, lances, hand and body shields, draft animals and supply wagons will amount to sixty per cent of the total.

---

[86] Rationale for using Lau's translation rather than Griffith's is given in Appendix 1 footnote 18.

Passage #5.8. Dependence of Military Strategy and Tactics on Logistics, cont.

(Logistics issues dominate these verses, which constitute a large slice of Sun Tzu's most extensive discussion of mobile warfare.[87] It is worth noticing that verses VII.7 and VII.8 have significant medical logistics aspects.[88])

VII.1. Normally, when the army is employed, the general first receives his commands from the sovereign. He assembles the troops and mobilizes the people. He blends the army into a harmonious entity and encamps it.[89]

VII.2. Nothing is more difficult than the art of manoeuvre. What is difficult about manoeuvre is to make the devious route the most direct and to turn misfortune to advantage.

VII.3. Thus, march by an indirect route and divert the enemy by enticing him with a bait. So doing, you may set out after he does and arrive before him. One able to do this understands the strategy of the direct and the indirect.[90]

VII.4. Now both advantage and danger are inherent in manoeuvre.

VII.5. One who sets the entire army in motion to chase an advantage will not attain it.[91]

VII.6. If he abandons the camp to contend for advantage the stores will be lost.

VII.7. It follows that when one rolls up the armour and sets out speedily, stopping neither day nor night and marching at double time for a hundred li, the three commanders will be captured. For the vigorous troops will arrive first and the feeble straggle along behind, so that if this method is used only one-tenth of the army will arrive.

VII.8. In a forced march of fifty li the commander of the van will fall, and using this method but half the army will arrive. In a forced march of thirty li, but two-thirds will arrive.

VII.9. It follows that an army which lacks heavy equipment, fodder, food and stores will be lost.

---

[87] See verses VII.2–VII.16 (Passage #6.29A). Most of these verses also appear in Passage #7.9.

[88] For a vivid eyewitness account of the pathetic condition of many common foot soldiers in Kuomintang-era provincial Chinese armies on the march, see Graham Peck, *Two Kinds of Time* (Boston, MA: Houghton Mifflin, 1950), pp. 216–17.

[89] For a split of opinion (not particularly important in the present context) regarding a verse VII.1 interpretation issue see Appendix 13 footnote 8.

[90] In a glaring case of a Sun Tzu scenario standing in need of clarification, verse VII.3 has been subject to diametrically opposite interpretations, depending whether it is one's *own* army or the *enemy's* army that is to take the circuitous route, in the former case reaping advantage, in the latter inducing enemy disadvantage thereby. For discussion see Appendix 7 anchor footnote 16.

[91] Bracketing a few translation or interpretation issues (see Appendix 5 footnotes 8–10), there is little mystery as to the mobile warfare dilemma to which Sun Tzu is pointing here: If you lug all your impedimenta with you, you'll be too weighed down to maneuver effectively; if you leave them behind, you'll lose them and you'll be sorry! Cf. #3 footnote 73 (noting Burgoyne's problem in his Saratoga campaign). See also Passage #1.15, discussing verses VII.2–VII.9 from a calculation perspective.

Passage #5.9. Starting Point for Conjuring Paths to Advantage: "Surface Impression ≠ Actual Situation"

(Passage #5.9's verses fuse several strands of thinking. One is verse V.17's concept that "surface impression ≠ actual situation," which is a jumping off point for all four paths to advantage sketched on pp. 178–84. Sun Tzu's immediate focus in verse V.17 is on chaotic battle scenes. Yet the idea expressed there also invites application to strategic or grand strategic situations which can be chaotic in their own way, often accompanied by elaborate storylines and rhetorics that have little to do with what is actually going on and indeed commonly work to obfuscate it.

A second strand is deception, a handmaiden to all four paths to advantage.

A third strand involves a suite of basic concepts (*shu/shi/xing*), each of which is centrally relevant to military or strategic success. Shi and *xing* have already been discussed on pp. 175ff. Shu appears in Passage #1.13, where a meaning of "enumeration" is noted. Verse V.19's occurrence of *shu*, which Ames, p. 120 translates as "logistics," is noteworthy as an implicit definition of logistics – in fact a rather good one! – as an activity mediating between order and chaos.[92])

V.17. In the tumult and uproar the battle seems chaotic, but there is no disorder; the troops appear to be milling about in circles but cannot be defeated.[93]

V.18. Apparent confusion is a product of good order; apparent cowardice, of courage; apparent weakness, of strength.[94]

V.19. Order or disorder depends on *shu* 數 [Griffith's "organization"]; courage or cowardice on *shi* 勢 [Griffith's "circumstances"]; strength or weakness on *xing* 形 [Griffith's "dispositions"].[95]

---

[92] "Organized confusion" – a phrase that crystallized out of US logistics experience in the Pacific in World War II – captures a relevant nuance here. A visceral sense of the meaning of organized confusion is excellently conveyed by a photograph on the journal page facing Henry E. Eccles, "Naval logistics," US Naval Institute *Proceedings*, 1948, 74(4), 483. Its caption (p. 482) reads: "This picture shows just one little strip of beach during the Saipan invasion. Multiply it by thousands, and add the planes, the fleets, and the transports and you begin to get an idea of what Naval Logistics means."

Closer to the kind of warfare Sun Tzu knew, a depiction of the Japanese army transport corps in the 1904–1905 Russo-Japanese War also conveys an imagery of organized confusion, "bringing order out of a chaos of mules, donkeys, cows, and carts, and a babel of moos, squeals, brays, and coolies' shouts." See Frederick Palmer, *With Kuroki in Manchuria* (New York: Charles Scribner's Son, 1904), p. 309.

[93] Verse V.17 refers to *hundun* ("chaos"), a concept already encountered in Part C footnote 1 (p. 171). For more on *hundun*, emphasizing civilian situations, see Passage #12.5 below.

[94] Opinions differ as to whether or not to read a focus on deception in verse V.18. Weighing the interpretive issues in the case of verse V.18, and finding a deception interpretation more compelling, see Appendix 7 anchor footnote 14. For a discussion of deception-focused readings of verses V.17 and V.20 see Passage #7.7 in main text and Appendix 7 (where verses V.17–V.20 also appear).

[95] Ames's translation of verse V.19 (Ames, p. 120) conveys Sun Tzu's idea in a flexible way: "The line between disorder and order lies in logistics [*shu*]; between cowardice and courage, in strategic advantage [*shi*]; and between weakness and strength, in strategic positioning [*xing*]."

V.20. Thus, those skilled at making the enemy move do so by creating a situation to which he must conform; they entice him with something he is certain to take, and with lures of ostensible profit they await him in strength.

Passage #5.10. First Path to Advantage: Expansive Definition of the Situation

(For a grizzled soldier schooled in hard combat, verses III.3–III.5 are a nudge to raise his sights and contemplate perhaps previously unimagined paths to advantage, some qualitatively different from fighting in its military sense. Sun Tzu is pointing here, not to a single path to expanding the definition of the situation and thereby obtaining advantage, but to an entire sheaf of such paths, introduced in an open-textured way that invites play of strategic imagination.)

III.3. For to win one hundred victories in one hundred battles is not the acme of skill. To subdue the enemy without fighting is the acme of skill.

III.4. Thus, what is of supreme importance in war is to attack the enemy's strategy;

III.5. Next best is to disrupt his alliances;

III.6. The next best is to attack his army.

III.7. The worst policy is to attack cities. Attack cities only when there is no alternative.

Passage #5.11. Second Path to Advantage: Inspiration from Nuanced Ways of Water

(The second path to advantage may be illustrated by water's ability to worm around obstacles and seep its its way through even the tiniest crevices.[96])

VI.24. The ultimate in disposing one's troops is to be without ascertainable shape. Then the most penetrating spies cannot pry in nor can the wise lay plans against you.

VI.25. It is according to the shapes that I lay the plans for victory, but the multitude does not comprehend this. Although everyone can see the outward aspects, none understands the way in which I have created victory.

VI.26. Therefore, when I have won a victory I do not repeat my tactics but respond to circumstances in an infinite variety of ways.

VI.27. Now an army may be likened to water, for just as flowing water avoids the heights and hastens to the lowlands, so an army avoids strength and strikes weakness.

VI.28. And as water shapes its flow in accordance with the ground, so an army manages its victory in accordance with the situation of the enemy.

---

[96] Passage #5.11 verses contain numerous occurrences of *xing* (concept introduced on p. 175 above). They are rounded up and analyzed in Appendix 5, pp. A-122f., where Passage #5.11 also appears.

VI.29. And as water has no constant form, there are in war no constant conditions.

VI.30. Thus, one able to gain the victory by modifying his tactics in accordance with the enemy situation may be said to be divine.[97]

Passage #5.12. First and Second Paths to Advantage: Espionage Exemplifies Both

(The first two paths to *shi* coalesce here. Espionage by its nature enlarges the definition of the situation, the first path to *shi*, beyond what conventional military operations offer. At the same time, the tools of espionage are capable of being wielded with great delicacy, especially by spies in high places, to identify and profit from nuanced sources of advantage – the second path to *shi*.[98] A basic contrast between "coarse-grained" tools like conventional armies and "fine-grained" tools like spies, or at least some varieties of spies, is on display here.)

XIII.1. Now when an army of one hundred thousand is raised and dispatched on a distant campaign the expenses borne by the people together with the disbursements of the treasury will amount to a thousand pieces of gold daily. There will be continuous commotion both at home and abroad, people will be exhausted by the requirements of transport, and the affairs of seven hundred thousand households will be disrupted.

XIII.2. One who confronts his enemy for many years in order to struggle for victory in a decisive battle yet who, because he begrudges rank, honours and a few hundred pieces of gold, remains ignorant of his enemy's situation, is completely devoid of humanity. Such a man is no general; no support to his sovereign; no master of victory.

. . .

XIII.13. He who is not sage and wise, humane and just, cannot use secret agents. And he who is not delicate and subtle cannot get the truth out of them.

XIII.14. Delicate indeed! Truly delicate! There is no place where espionage is not used.

. . .

XIII.16. Generally in the case of armies you wish to strike, cities you wish to attack, and people you wish to assassinate, you must know the names of the garrison commander, the staff officers, the ushers, gate keepers, and the bodyguards. You must instruct your agents to inquire into these matters in minute detail.

. . .

XIII.23. And therefore only the enlightened sovereign and the worthy general who are able to use the most intelligent people [*shang zhi*] as agents are certain to achieve great things. Secret operations are essential in war; upon them the army relies to make its every move.

---

[97] Ames, p. 127 (with supporting evidence as provided in his p. 289 note 159) renders *shen* (Griffith's "divine") as "inscrutable." Ames's word choice avoids possible religious overtones – extraneous here – while also harmonizing with formlessness ideas in Passage #8.1 (where verse VI.30 also appears).

[98] For verse XIII.23 as harmonizing with high-end spying on elites see #4 footnote 86.

Passage #5.13: Third Path to Advantage: Superior Complexity Management

(This passage encourages strategic creativity that thrives on mastery of complexity. An example from a different strand of Chinese strategic tradition would be strategy of the game of *weiqi*. Like *weiqi* strategy Passage #5.13 has natural combinatorial aspects.[99] For further analysis of *qi* and *zheng* see Theme #14.)

V.3. That the army is certain to sustain the enemy's attack without suffering defeat is due to operations of the extraordinary [*qi*] and the normal [*zheng*] forces.

V.4. Troops thrown against the enemy as a grindstone against eggs is an example of a solid acting upon a void.

V.5. Generally, in battle, use the normal force [*zheng*] to engage; use the extraordinary [*qi*] to win.

V.6. Now the resources of those skilled in the use of extraordinary forces [*qi*] are as infinite as the heavens and earth; as inexhaustible as the flow of the great rivers.

V.7. For they end and recommence; cyclical, as are the movements of the sun and moon. They die away and are reborn; recurrent, as are the passing seasons.

V.8. The musical notes are only five in number but their melodies are so numerous that one cannot hear them all.

V.9. The primary colours are only five in number but their combinations are so infinite that one cannot visualize them all.

V.10. The flavours are only five in number but their blends are so various that one cannot taste them all.

V.11. In battle there are only the normal [*zheng*] and extraordinary [*qi*] forces, but their combinations are limitless; none can comprehend them all.

V.12. (Translation following Lau, 1965 article, p. 331.[100]) "The *qi* and the *zheng* produce each other endlessly like a ring and who is there that can exhaust the possibilities?"

Passage #5.14. Third Path to Advantage, cont.: Mastery of Switches of Posture or Style

VII.12. Now war is based on deception. Move when it is advantageous and create changes in the situation by dispersal and concentration of forces.[101]

VII.13. When campaigning, be as swift as the wind; in leisurely march, majestic as the forest; in raiding and plundering, like fire; in standing, firm as the mountains. As unfathomable as the clouds, move like a thunderbolt.

[99] See Appendix 1 anchor footnote 29 as well as footnote 30 (combinatorial strands in early Chinese and early Indian thought).

[100] Rationale for using Lau's translation rather than Griffith's is given in Appendix 14 footnote 3.

[101] Modern weapons require rethinking of verse VII.12's last part. See Table 2 note d (p. 97 above).

Passage #5.15. Fourth Path to Advantage: Awaiting a Propitious Moment

(Verse VII.21 need not be confined to changes in mood occurring over the course of a single day, thus freeing Sun Tzu's thinking to be more broadly applied to changes occurring on a longer timescale.[102] There is resonance here with Passage #5.4 ideas about awaiting the enemy's moment of vulnerability.)

VII.21. During the early morning spirits are keen, during the day they flag, and in the evening thoughts turn toward home.

Passage #5.16. Fourth Path to Advantage, cont.: Inevitability of Large-Scale Change

(By challenging any assumption of a static or immutable "game," verse VI.31 speaks to the fourth path to advantage.[103])

VI.31. Of the five elements, none is always predominant; of the four seasons, none lasts forever; of the days, some are long and some short, and the moon waxes and wanes.

Passage #5.17: Fourth Path to Advantage: Mastery of Time and Timing Factors

(One idea present here focuses on a small, precise action, epitomized by a tiny force on a crossbow trigger, exerted at exactly the right moment. A second idea points to the favorable momentum such an action unleashes. For elaboration of Sun Tzu's thinking about momentum see Theme #9 and especially Passage #9.3.)

V.13. When torrential water tosses boulders, it is because of its momentum [shi];
V.14. When the strike of a hawk breaks the body of its prey, it is because of timing.
V.15. Thus the momentum [shi] of one skilled in war is overwhelming, and his attack precisely regulated.
V.16. His potential [shi] is that of a fully drawn crossbow; his timing, the release of the trigger.

Passage #5.18. Avoiding Tardy Exploitation of Success

(This verse again involves the fourth path to shi, now glimpsed from a time point further along it. As Lau & Ames, p. 63 note: "Shi is not a given; it must be created and carefully cultivated." Failure of exploitation versus overreaching are

---

[102] See Preliminaries footnote 4, noting interpretation of traditional commentator Mei Yaochen.
[103] Brooks & Brooks, Introduction footnote 36, pp. 84–85 give a thumbnail sketch of the Chinese five elements or five phases theory (water, fire, metal, wood, earth; order of listing varies), comparing it with ancient Indian and Greek theories. See also Mair, p. 148 note 7; Wilkinson, §37.1, table 75, pp. 515–16 ("Five-phase theory correspondences"); Zhang Dainian, #3 footnote 80, pp. 95–103.

Verse VI.31 may be read as counseling patience of a sort familiar to many underdogs the world over. Context is ever-changing. Biding one's time may see emergence of substantially more favorable conditions.

Scylla and Charybdis here.[104] A natural reading of verse XII.15 finds Sun Tzu favoring aggressive exploitation, an emphasis in keeping with shi's dynamic nature.[105] One might like to see Sun Tzu go further, and say something about converting a local, temporary advantage into a far-reaching qualitative restructuring of the situation. Such strategic exploitation is one of the most difficult challenges in warfare, requiring command judgment of a high order often backed by logistics prowess. Here, however, it is important not to impute to Sun Tzu a level of conceptual sophistication he did not possess.[106] Verse XII.15 is not stated or developed by Sun Tzu in a way that truly captures a strategic exploitation idea.)

XII.15. Now to win battles and take your objectives, but to fail to exploit these achievements is ominous and may be described as "wasteful delay".

Passage #5.19. History As Quiver of Ideas for Informing Strategic Doctrine or Style

(Sun Tzu is here drawing on events of a far earlier time to support a doctrinal point underscoring the value of spies, especially ones in high places. The underlying history cited is, at the very least, skimpy, controversial, and quite probably in part mythological. Details need not be developed here. The important points are, first, that Sun Tzu is reaching out to allegedly historical cases as a source of strategic precedent, thereby foreshadowing a pattern that would flourish throughout later Chinese history; second, that the dramatis personae of those cases, Yi Zhi and Lü Ya [also known as Tai Gong], have been cast in drastically contradictory lights by admirers and detractors down through the centuries – i.e., the historical record itself has become a field for strategic action, with rival interpreters striving for a historiographical version of shi.[107])

---

[104] A lens on some of the issues is provided by the long-running historical debate over whether or not the Union commander at Gettysburg, George Meade, passed up an opportunity to extinguish Lee's army – and therefore prolonged the Civil War by an extra two years – as a result of Meade's cautious and limited post-battle pursuit. See A. Wilson Greene, "From Gettysburg to Falling Waters: Meade's pursuit of Lee," pp. 161–201 in Gary W. Gallagher (ed.), *The Third Day at Gettysburg and Beyond* (Chapel Hill, NC/London: University of North Carolina Press, 1994),

    For corroboration from outside the Sun Tzu text that comparable issues were recognized and analyzed in early Chinese military and strategic thinking see Appendix 5 footnote 22.

[105] Highlighting the dynamic nature of shi see Li Ling's analysis quoted on p. 176 above. For further analysis of verse XII.15 textual and interpretation issues see Appendix 5 anchor footnote 21.

[106] Relevant here is lack of a strategy/tactics distinction in Sun Tzu. See pp. 47–48 above.

[107] Illustrating the point in the case of Yi Zhi (better known as Yi Yin) – a semi-legendary figure of a time long before Sun Tzu who "eventually became recognized as China's first covert agent because Sun Tzu thus dubbed him so" – Sawyer, #1 footnote 3, pp. 7–12 conveys a sense of historical interpretation battleground. Depending whose word you put stock in, Yi Yin was a "virtuous, self-sacrificing minister" (as the Confucians sought to portray him) or a skilled spy versed in the black arts. See Sawyer, #1 footnote 3, p. 10.

XIII.22. Of old, the rise of Yin [Shang dynasty] was due to Yi Zhi, who formerly served the Xia [dynasty]; the Zhou [dynasty] came to power through Lü Ya, a servant of the Yin.

★★★★★★

THEME #6 ATTACKING THE ADVERSARY'S STRATEGY

basic statement and context:
Passage #6.1 (verses III.3–III.7) (attacking the enemy's strategy – basic statement of principle)

counter-strategist mode:
Passage #6.2 (verse III.18) (setting a stage for counter-strategist mode: competent general as natural target)
Passage #6.3 (verse X.19) (setting a stage, cont.: rarity of generals with unimpeachable motives)
Passage #6.4 (verse XIII.16) (hard counter-strategist mode: targeted killing as tool of conflict)
Passage #6.5 (verse XI.37) (hard counter-strategist mode: precedents from Chinese history)
Passage #6.6 (verse I.15) (hard counter-strategist mode: sidelining enemy talent)
Passage #6.7 (verse VII.20) (soft counter-strategist mode: enemy commander's mind as object of attack)
Passage #6.8 (verses I.22–I.25) (soft counter-strategist mode: degrading quality of enemy commander's thinking)
Passage #6.9 (verse XIII.17) (soft counter-strategist mode: venality to control information sources)
Passage #6.10 (verses VIII.17–VIII.24) (counter-strategist mode: exploiting character flaws)

counter-plan mode:
Passage #6.11 (verse XI.51) (counter-plan mode on a plane of grand strategy)
Passage #6.12 (verses VI.20–VI.23) (counter-plan mode, emphasizing active reconnaissance steps)
Passage #6.13 (verses XII.1 + XII.13–XII.14) (counter-plan mode: targeting the enemy's logistics support)
Passage #6.14 (verse VIII.16) (counter-plan mode: focus on enemy capabilities, not intentions?)

(footnote 107 cont.)
For a larger context see Wilkinson §18.6, p. 290, "Bad Last Rulers," noting that in Chinese history the last ruler of a dynasty is often painted in a drastically adverse light by later generations (tyrannical, self-indulgent, licentious, etc.). Exceptionally negative press was accorded to the final rulers of Xia and Shang, the two dynasties to whose fall verse XIII.22 alludes.

situation control mode:
Passage #6.15 (verse V.20) (situation control mode: introduction to concept)
Passage #6.16 (verses XI.25–XI.29) (asserting situation control: elaboration of concept)
Passage #6.17 (verses VIII.14–VIII.15) (asserting situation control: grand strategy illustration)
Passage #6.18 (verses VI.1–VI.19) (asserting situation control: Sun Tzu's "First Symphony")
Passage #6.19 (verses XIII.22–XIII.23) (asserting situation control: intelligence-oriented version)

★★★★★★

By Sun Tzu's explicit affirmation (see verse III.4 in Passage #6.1), "attacking an adversary's strategy" is Sun Tzu's summum bonum, standing at the core of Sun Tzu's strategic theory.[1] Yet this is a concept – at once thought-provoking and ambiguous – whose meaning students of Sun Tzu are largely left to puzzle out for themselves.

Illustrating some of the difficulties, if an enemy's strategy involves alliances (scarcely an uncommon pattern, as the history of Warring States China and some of Sun Tzu's own advice corroborates), measures to disrupt and break down those alliances might appear to qualify as an attack on the adversary's strategy. Yet Passage #6.1 explicitly differentiates an attack on strategy from an attack on alliances, rating the former more highly at least in a setting of offensive warfare (the title topic of Sun Tzu III).[2] This suggests that, for Sun Tzu, attacking the enemy's strategy had a

---

[1] The Chinese term for what is to be attacked is *mou*, which (depending on context) may be variously translated as "strategy," "plans," or "stratagem" (the latter term playing up a psychologically oriented playbook, a focus that the Sun Tzu text privileges in many places).

The Chinese term in verse III.4 that Griffith, joined by Ames, Minford, and Sawyer, translates as "attack" is *fa*, which could also be rendered "baulk" (Giles) or "stymie" (Mair). The present choice is to opt for "attack," accepting Giles's observation (Giles, p. 18) that "Perhaps the word 'baulk' falls short of expressing the full force of *fa*, which implies not an attitude of defence, whereby one might be content to foil the enemy's stratagems one after another, but an active policy of counter-attack."

[2] A scope condition conceivably setting limits on verse III.4 is that – at least according to its chapter title – Sun Tzu III is about *offensive* warfare (its title is *mou*, lit. "strategy/plan/stratagem" + *gong*, lit. "attack"). However, Lau's 1965 article, p. 327 cautions against "tak[ing] chapter titles too seriously" in Sun Tzu, noting that (as is common in other early Chinese works) such titles "give only a rough indication of the contents of chapters." Lau's comment is certainly true of Sun Tzu III, which includes verses III.16 and III.17 about making a good getaway (see Passage #1.32A). It seems quite possible that this and other Sun Tzu chapter titles may have been supplied only after intergenerational transmission of the text was well underway. Relatedly, as Wilkinson §70.6.1 ("Book and Chapter Titles") notes, "Until the Han, the titles of most Chinese works were not chosen by their authors."

Supporting an idea that any offensive-warfare-based limitation on verse III.4 may be more apparent than real, an observation of Sawyer bears repeating here: "In general, unlike much Western military doctrine, the principles and concepts of Chinese military science are not rigidly constrained nor limited to one level or sphere of application . . ." (previously quoted in Preliminaries footnote 3).

more specific meaning, or cluster of meanings, which it is a challenge for modern analysts to try to reconstruct.

Attempting to piece the concept together in a textually anchored way, the present analysis identifies in Sun Tzu three "modes" of attacking the enemy's strategy. These modes are not mutually incompatible and often blend. Indeed there is little reason to believe that Sun Tzu himself crisply distinguished them, so that this three-way distinction is best viewed as an analytical crutch to help clarify Sun Tzu's Theme #6 thinking, rather than as a direct statement of it. By way of context, it should again be noted that the Sun Tzu text took shape in an era prior to the advent of the modern incarnation of a general staff whose work generates mounds of detailed information set down on paper (a communications medium that did not exist in Sun Tzu's time). In Sun Tzu's time it therefore seems likely that most "strategies" or "plans" were never committed to written form. This sets practical limits on the complexity and specificity of the "strategies" possible in Sun Tzu's time, and – by the same token – also the options available for attacking an opponent's strategy.

The three modes of attack presently proposed may be summed up as follows:

- attacking the enemy's strategy by attacking the enemy strategist ("counter-strategist mode");
- attacking the enemy's strategy by attacking (or nullifying) that strategist's work product, "strategy," "plan," etc. ("counter-plan mode");
- attacking the enemy's strategy by asserting control over the situation ("situation control mode").

Counter-strategist mode is well summarized by Sun Tzu traditional commentator Li Quan, "Attack strategy at its very source."[3] It is one manifestation of a "personalized" focus that has been a longstanding harmonic in Chinese strategy and warfare, one that never loses sight of that fact that a winning *strategy* is authored by a winning *strategist*.[4]

A basic (if notably unsubtle) way of implementing this first mode of attack – one redolent of "decapitation strike" thinking that achieved prominence early in the nuclear age and has more recently resurfaced in drone warfare – pivots on killing or capturing the strategist-qua-person. Early Chinese history is richly endowed with assassination stories, to some of which verse XI.37 (Passage #6.5) alludes.[5] Verse XIII.16 (Passage #6.4) points in the same direction, as does verse VIII.18

---

[3] Minford, p. 134, translating Li Quan commentary on verse III.4 (Passage #6.1).

[4] See also p. 64 above. Such emphasis harmonizes with a tendency, reflected in Chinese biographical traditions (especially in earlier times), to view "the truly important events in history [as] ... the result of actions or inactions of people, not of divine laws, let alone institutions" (Wilkinson, §9.3, p. 158).

On a side note, it is worth observing that classical game theory de-emphasizes, indeed widely frowns on, a focus on strategist-qua-person, especially a fallible one.

[5] Note should also be taken of verse IX.18 (Passage #7.5) where (in a departure from other leading translators) Mair, p. 110 translates jian 姦 (Griffith's "spies") as "snipers." In Sun Tzu's way of war, with its relentless focus on the importance of the talented general, Mair's version makes considerable military sense, since the military value of successful killing of an enemy general – a possibly priceless asset to the foe – by a hidden sniper seems likely to exceed, possibly by far, that of any information about a mass-infantry army gleaned by a hidden observer (spy) lurking in the undergrowth. Bow and crossbow weapons available in Sun Tzu's time certainly lent themselves to such strikes. Commenting on city defense (from a later point in Chinese history), Needham &

(Passage #6.10).[6] The concept of a decapitation strike was certainly not lost on Sun Tzu's traditional commentators. One of them, Li Quan, recites a telling anecdote about how a top planning officer was dispatched as envoy to a parley and wound up being beheaded there by the enemy. Looking back on his military success that ensued, the comment of the beheader was that this envoy "was [the enemy ruler's] intimate counselor. If I had spared [his] life, he would have accomplished his schemes, but when I killed him, [the ruler] lost his guts."[7] Here is an early Chinese case where performing diplomatic duties proved no shield against a counter-strategist strike![8]

The salience of the capture possibility in early Chinese military thinking should not be underestimated (verse VIII.19 in Passage #6.10; verse VII.7 in Passage #7.9, "Sun Tzu's Second Symphony"). A case in point, from a time later than Sun Tzu, is the denouement of the Han Xin battle story recounted earlier where an enemy adviser is captured and in fact given favorable treatment (see p. 32 and note verse II.19 in Passage #3.5 on treating captives well).

If the enemy is fortunate enough to possess an outstanding commander (or perhaps adviser), an alternative path to decapitation, potentially no less effective, would be getting that commander or adviser sidelined or dismissed outright – say, on the basis of a false rumor or what might nowadays be called fake news. Reading slightly between the lines, verse I.15 (Passage #6.6) suggests that Sun Tzu would have entertained that possibility. The (often) bloodless, insidious, yet effective quality of such an alternative path to decapitation is indeed vintage Sun Tzu.[9]

Efficacy of a counter-strategist mode of attack draws on two assumptions.

One is the principle, ingrained in a Warring States mindset and succinctly expressed in Passages #6.2 and #6.3, that the quality (including loyalty) of the commanding

Yates, p. 384 observe: "the main function of the [crossbow] marksmen was to shoot down generals leading the attack."

On a point of clarification, *jian* 姦 is a different character from *jian* 間, which is the term used to refer to spies throughout Sun Tzu's espionage chapter (where *jian* 姦 does not appear).

[6] Note that Passage #6.10 distinguishes between killing the enemy general (verse VIII.18) and capturing him (verse VIII.19), corroborating Sun Tzu's alertness to different types of decapitation.

[7] Griffith, pp. 77–78. The relevant Li Quan commentary passage is also translated by Minford, p. 134.

[8] As this Li Quan example also suggests, a strategist being targeted need not always be in a commanding general role. Having a ruler's or commanding general's ear may suffice. The *Han Feizi* text, a Legalist classic, notes that "The lord of Wei has a minister named Zhao Jia whom he consults in matters of policy, and the lord of Han has a similar minister named Duan Gui. Both *these men have the power to talk their lords into changing their plans.*" See *Han Feizi: Basic Writings* (Burton Watson, trans.; New York: Columbia University Press, 2003), p. 61 (emphasis supplied).

[9] Recourse to "estrangement methods ... employed on the strategic level to eliminate competent and therefore dangerous generals" (Sawyer, Introduction footnote 55, p. 353) is a stratagem that has been used time and again throughout Chinese history.

A pertinent fifth-century BC example, involving a contemporary colleague of the putative historical Sun Tzu, concerns the fate of Wu Zixu. Wu Zixu was a distinguished general of Wu who fell into disfavor and was induced to commit suicide by the Wu ruler, who heeded a false rumor of Wu Zixu's disloyalty slanderously spread by one of his high officials (who is said to have been subverted by the state of Yue, the Wu state's archenemy of which verse VI.18 [Passage #6.18] takes note). For this tale, one of the more famous of its age, see Wu Zixu's *Shiji* 66 biography in Nienhauser, Vol. VII (1994), pp. 49–59; Nienhauser, Vol. VII (2021), pp. 91–106, especially pp. 103–104 (both cited on pp. xxiii–xxiv).

general is a sine qua non of a ruler's political–military success.[10] In Sun Tzu's net assessment exercise (Passage #1.1), the commanding general's ability is listed second only to the *dao* of the ruler on a list of seven factors predicting victory in war.

A second assumption, implicit in Passage #6.3's description of the general as a "jewel," is that truly top-notch (and loyal!) strategic talent is rare, and by that token far from easy to replace. Annals of Chinese and other military history certainly support the thrust of this second assumption: great captains are indeed rare. Its specific fit to Warring States China is, of course, an empirical question, one that substantially outruns our data. Warring States China did produce many generals of high caliber.[11] Yet that is not the same yardstick as asserting that, at a given time, any particular state possessed a "deep bench" of high command talent sufficient to buffer against assassination or other attrition contingencies. Most probably did not, thus leaving them vulnerable to counter-strategist attacks.[12] Also lurking in the background is the loyalty dimension. It is noteworthy that Passage #6.3 applies the term "jewel" to a general who steadfastly serves the best interests of the sovereign; that general's military competence is treated elsewhere.

The assorted counter-strategist expedients just described might be labeled "hard counter-strategist measures": one day the strategist is available as a resource, the next day he is not. There are also "soft counter-strategist" measures that center on degrading the quality of the enemy general's strategic and tactical thinking. "Soft" should by no means be confused with being ineffective or "strategy of second choice."[13] There is much Sun Tzu textual support for a belief that defective decision making can be induced and lead on to catastrophic consequences for the decision maker.

Geared toward the mind of the enemy commander, a "soft counter-strategist" attack aims to induce one or more of:

- (a) a wrong action
- (b) an untimely action
- (c) loss of poise and coordination (undermining that commander's efficacy).

---

[10] For context see Lewis, Background footnote 11, chapter 3, "The Art of Command," pp. 97–135.

[11] See Kierman, Background footnote 31, p. 321 note 54: "It is an impressive list" (citing examples).

[12] As traditional commentator He Yanxi observes, "the scarcity of fine generals has always been a source of calamity" (Minford, p. 130). A structural consideration to note is that the increasing scale of warfare as the Warring States period progressed made strategy (and, of course, the logistics underpinning it) more demanding, placing ever greater burdens on commanders. An upshot was that strategy became "so complex that the replacement of a general could, and frequently did, result in an army's defeat and the endangerment of an entire nation." See Sawyer, *Seven Military Classics*, p. 11.

[13] One early text, the *Yi Zhou shu*, describes a strategy session contemplating "a series of aggressive moves called 'soft strikes' (*rou wu*), which take aim against ineptitude and indolence within the Zhou court." See p. 116 of Michael Nylan, "The many Dukes of Zhou in early sources," in Benjamin Elman & Martin Kern (eds.), *Statecraft and Classical Learning: The Rituals of Zhou in East Asian history* (Leiden/Boston, MA: Brill, 2009), pp. 94–128. For Yi Zhou shu background see Appendix 1 anchor footnote 8.

To be clear, the present hard/soft counter-strategist distinction is a modern Sun Tzu analyst's construct. It need not align on the phrase *gang ruo* in verse XI.41's Chinese text, frequently translated "hard and soft" (e.g., Ames, p. 159; Mair, p. 121; Sawyer, p. 222; see Passage #14.28A discussion). The hard/soft counter-strategist distinction, as developed here in Theme #6 context, should also not be equated with the well-known modern distinction between "hard" and "soft" power.

Read from a soft counter-strategist vantage, verse I.25 (Passage #6.8) suggests attacks, not against an enemy strategist-qua-individual but against that strategist's entourage or social network, disrupting human relations and social trust within that "inner ring." As traditional commentator Zhang Yu puts the case (Griffith, p. 69), "Sometimes drive a wedge between a sovereign and his ministers." Belief in the efficacy of such measures runs early and deep in Chinese strategy. Unless the enemy possesses a commander of pre-eminent ability and authority, that type of soft counter-strategist strike, and the ongoing frictions it produces, may actually do more to degrade the quality of enemy thought and action than killing the individual who happens to be formally in command – a very Sun-Tzu-esque line of thought! Note that the target in this genre of attack is not the same as an alliance between states, hence suggesting a possible conceptual division of labor between verse III.4 (attacking the enemy's strategy) and verse III.5 (attacking the enemy's alliances).

Animating parts of Passages #6.8 and #6.10 is a Sun Tzu–inspired insight that might be labeled the "relevance of the irrelevant": the idea that the active ingredient of an effective attack on a foe's decision making may pivot on that foe's preoccupation or entanglement with thoroughly irrelevant matters. In casting around for irrelevancies that could do this type of useful work, the creative and expansive definition of the situation animating Theme #5 comes into full play.

A second mode of attacking the enemy's strategy may be labeled "counter-plan." Much of counter-plan mode – particularly illustrated by verse VI.20 in Passage #6.12, though other Sun Tzu VI verses also suggest the same idea – boils down to learning or ascertaining the enemy's plan and then devising responses to stymie it (or better yet, to turn it to advantage).[14] Of course, not all enemy plans are equally worth peeking at. The greatest gain here seems likely to arise when there is a major secret dimension to the enemy plan, crucial to its effectiveness, that offers a target to be ferreted out, either by reconnaissance (Passage #6.12) or by espionage.[15] Verse XIII.23 (Passage #6.19)

---

[14] Because steps that stymie an adversary's pet plan, project, etc. often lead to great frustration and loss of poise, counter-plan mode may overlap with soft counter-strategist mode. Indeed here may be one reason Sun Tzu favored attacks on strategy over attacks on alliances as a general rule (Passage #6.1). In many successful counter-plan attacks, a sizable part of the payoff may reside in robbing the enemy general of his self-confidence and the courage that comes with it (see Passage #6.7's verse VII.20). When alliances – especially opportunistic alliances that are unstable under the best of circumstances – come unglued, damage to the ego of a key enemy actor may be less severe. Cf. modern analyst's observation quoted in #7 footnote 42 on the investment of ego in crafting a deception plan.

[15] A contrast case would be the so-called Longzhong Plan of 207 AD, put forward by Zhuge Liang as a strategy for defeating Cao Cao and restoring the Han dynasty. For the plan as described in the *Sanguo zhi* (a historical work, to be distinguished from the fictionalized *Romance of the Three Kingdoms*, see Background footnote 102) see Ralph D. Sawyer & Mei-chün Lee Sawyer, *Zhuge Liang: strategy, achievements, and writings* (North Charleston, SC: CreateSpace Independent Publishing Platform, 2014), pp. 28–30.

Its illustrious authorship notwithstanding (for more on Zhuge Liang see Henry, #5 footnote 68), the Longzhong plan has been faulted by later critics for reasons that include its lack of an ingredient of stratagem/deception and related failure to incorporate both halves of Sun Tzu's qi/zheng concept pair where qi has overtones of crafty, unorthodox, etc. steps (see Theme #14). See John Killigrew, "Zhuge Liang and the Northern Campaign of 228–234," *Early Medieval China*, 1999, 5(1), 55–91, in particular pp. 59–60 (plan's highlights) and pp. 80–82 (critique of it by late Ming/early Qing soldier-scholar Wang Fuzhi [1619–1692]). Killigrew's p. 81 brings in Sun Tzu.

suggests that Sun Tzu may have thought similarly. An obvious example of a secret plan would be an attempted deception by the enemy, though there might be other motives for secrecy too (e.g., how the enemy is planning to break a tie among equally attractive strategic or tactical options). If the enemy's strategy or plan (mou) is complex, so much the better for counter-plan mode, for all the reasons that complex machinery is a natural sabotage target.[16]

A notably versatile subspecies of counter-plan mode involves attacking a plan by attacking its logistics support, exploiting the universal dependence of military plans on logistics. A case in point is Passage #6.13, noting fire attacks directed against logistics resources. A related idea involves disrupting the enemy's plan at a vulnerable point in its execution. Passage #6.25A on river crossings is an example.[17]

A different facet of Sun Tzu's counter-plan mode involves verse VIII.16 (Passage #6.14):

> VIII.16.
>
> It is a doctrine of war not to assume the enemy will not come, but rather to rely on one's readiness to meet him; not to presume that he will not attack, but rather to make one's self invincible.

Read expansively, verse VIII.16 takes a stand in favor of basing military decisions on enemy capabilities only, thus minimizing leverage the enemy can derive from tricky plans. Such a stance could apply to situations other than defensive ones and to multiple levels of war ranging from grand strategy down to operations or tactics.[18] It would treat Sun Tzu as weighing in – indeed in no uncertain terms! – on a long-running debate in military circles between those who favor basing military decisions on enemy capabilities alone, versus others who would allow a role for an estimate of enemy intentions to enter command decisions (in modern times, often a minority position).[19] Favoring an exclusive focus on adversary capabilities is risk-averse military conservatism: in a nutshell, a desire to nip wishful thinking in the bud and thereby avoid falling prey to deception. Favoring allowance for adversary intentions is the lure of achieving far, far greater gain – or escaping near-certain defeat when the foe possesses overwhelming material advantages.[20]

Of course, by that same token of lacking an ingredient of deception the plan would have been relatively leak-proof!

[16] Mou often has further overtones of "complex plots." See Sawyer, Introduction footnote 55, p. 18.

[17] A further offshoot of counter-plan mode, again timing-focused, comes from Cao Cao's commentary on verse III.4: "While the enemy is still formulating his strategy, it is easy to attack" (Minford, p. 134). That line of thinking is, of course, another rationale for Sun Tzu's predilection for speed (Theme #3).

[18] This appears to be Griffith's understanding of verse VIII.16, as evidenced by his invoking it in his analysis of the disastrous battle of Savo Island in the Guadalcanal campaign, in which the Imperial Japanese Navy defeated a US/Australian naval force with heavy loss of Allied ships and personnel. See Griffith, Introduction footnote 20, p. 60.

[19] See analysis in Eccles, Preliminaries footnote 5, chapter IX, "Command and decision," pp. 118–49. As also noted there, the verdict of military history is mixed as regards the extent to which great captains have eschewed reliance on enemy intentions. See Eccles, p. 126, quoting William A. Reitzel, *Background to Decision Making* (Newport, RI: US Naval War College, 1958).

[20] An example of intentions-based military decision making comes from early in the Korean War: The fame attained by General Michaelis in the first year of the Korean struggle rested to a large extent on his ability to estimate the pattern of thought of opposing North Korean generals. Time and again he left his front manned by a skeleton force because he estimated that the Reds would attack from the flank or rear; and he was right. His actions were not

This is not a minor debate and Sun Tzu's position on it warrants further probing. Extrapolating from Sun Tzu's fascination with degrading the quality of an adversary's thinking or otherwise pushing an enemy's hot buttons (e.g., Passages #6.8– #6.10) it is not difficult to impute to Sun Tzu a willingness, even eagerness, to take courses of action based on estimates of enemy intentions.[21] Reinforcing that line of thinking is the pivotal role Sun Tzu accords to espionage, a classic window into enemy intentions.[22] If read as categorically rejecting any reliance on enemy intentions, verse VIII.16 could seem quite odd, a source of tension if not outright contradiction in Sun Tzu. It could also be startling to Western audiences accustomed to endless affirmations of the exceptionalism of ancient Chinese military thought.[23]

Alternatively, verse VIII.16 could be read as merely a prudent call for military preparedness – in simplest terms, for the state to maintain a strong military establishment.[24] This too is a type of counter-plan idea, but now more narrowly focused on a grand strategy level. A call for preparedness on that level need not preclude some degree of reliance on estimates of enemy intentions (e.g., in guiding diplomatic or intelligence operations in time of peace).[25] It is also agnostic on whether or not to rely on intentions at lower levels of decision making (e.g., military strategy, operations, or tactics).

Which, if either, understanding of verse VIII.16 is correct? Or, as is sometimes useful in dealing with ambiguities in Sun Tzu, should we treat each as a valid part

---

based on unconsidered rashness. Rather he recognized that his troops were so outnumbered that he had to deploy strength only in areas where he expected attack. He could not afford the luxury of a conservative decision.

See p. 377 of O. G. Haywood, Jr., "Military decision and game theory," *Journal of the Operations Research Society of America*, 1954, 2(4), 365–85.

[21] This interpretation of Sun Tzu is bolstered by some readings of verse XI.56 (Passage #6.27A), which (for example) Sawyer translates "The prosecution of military affairs lies in according with and [learning] in detail the enemy's intentions." (Sawyer, p. 224; bracketed word "learning" is Sawyer's.) For further discussion of this somewhat problematic verse see Appendix 7 anchor footnote 24.

[22] Possibly also relevant here are strands of early Chinese thinking that play up the transparency of human intentions to a shrewd observer. For further discussion (citing sources) see #8 footnote 26.

[23] For much the same reason, however, verse VIII.16 has at times provided a useful peg for smuggling a dash of Sun Tzu into the often doctrinally conservative halls of conventional Western warfare. For example, verse VIII.16 in the Griffith translation appears as a stand-alone quote – without accompanying comment – in *Naval War College Review*, May–June 1973, p. 72 (juxtaposed there with p. 73 of article by William Reitzel, "Mahan on the use of the sea," pp. 73–82 in that issue).

[24] Sawyer, pp. 321–22 note 133 discerns a grand strategy/preparedness focus in verse VIII.16. Apparently so too did Lord Roberts, a prominent British soldier of the nineteenth century, who pointed to verse VIII.16 in the Giles translation as "one that the people of this country [Britain] would do well to take to heart." See Giles, p. xlii footnote 6. (Regarding Lord Roberts's military career see p. 6 above.)

[25] See Reitzel work cited in #6 footnote 19 (as quoted by Eccles, Preliminaries footnote 5, p. 127): There appears to be little doubt, for example, that at levels where national security decisions are made, judgments about enemy intentions are not only heavily weighted, but are practically essential to making choices. ... Doctrine with respect to capabilities and intentions cannot apply in a rigid form over the full range of uncertainties.

of Sun Tzu's thinking? Some support for a narrower (preparedness) interpretation of verse VIII.16 stems from noting its location immediately following verses VIII.14–VIII.15 (Passage #6.17), which may be read as grand strategy advice focusing on a Warring States state actor muscling around lesser actors (zhu hou, commonly rendered "feudal lords"). For what it is worth, the traditional commentators (e.g., He Yanxi, quoted on p. 230 below) also largely align on a preparedness interpretation.

However, this is one Sun Tzu puzzle that seems likely to elude crisp solution.

A third mode of attacking the enemy's strategy might be labeled "situation control."[26] This third mode has a distinctly dynamic flavor, making it the most difficult of the three modes to pin down doctrinally or textually in a tidy way. A basic idea, one well-known to weiqi players, is compelling the adversary constantly to respond to one's actions, forcing him to play "catch up" (thereby derailing the foe's attempt to follow his own preferred strategy and maneuvering him into an unfavorable situation, often incrementally). An active ingredient here, well conveyed by Griffith's "Seize something he cherishes and he will conform to your desires" (from Passage #6.16's verse XI.28), is one of exerting a type of behavioral control over the enemy. As with successful counter-plan attacks, damage to the morale of the enemy army and particularly its general may be a major part of the payoff (see verse VII.20 in Passage #6.7).

Although deception combines well with situation control, and often facilitates it, situation control need not always involve deception. For example, the segment of verse XI.28 just quoted could be read as advocating for hostage-taking.[27]

Situation control in its fullest sense is commonly not achieved by a single discrete act or event but rather involves flows of behavior in time, both one's own and the enemy's. Modern behavioral psychology, and its first cousins in machine learning, involves kindred concepts and emphases founded on behavior control by both positive and negative reinforcement means, thereby evoking Passage #6.18's verse VI.3:

> VI.3.
> One able to make the enemy come of his own accord does so by offering him some advantage. And one able to prevent him coming does so by hurting him.

However, Sun Tzu lacked the analytical wherewithal to develop this dynamic behavioral level further.[28]

The Sun Tzu passage that comes closest to summing up the third mode of attack on the enemy's strategy is Passage #6.18, whose rich texture merits the appellation

---

[26] Mair's translation of Passage #6.18's verse VI.2 conveys the key idea: "he who is skilled in battle controls the movements of others but does not allow his own movements to be controlled by others" (Mair, p. 95). Sawyer, p. 315 note 85 labels this "one of Sun-tzu's fundamental principles." As is often the case with Sun Tzu's abstract lines of thinking, it is not hard to spot terrain roots of the concept. See, e.g., Cao Cao on verse X.6 (Passage #7.6): "The particular advantage of securing heights and defiles is that *your actions cannot then be dictated by the enemy*" (Giles, p. 103, emphasis supplied).

[27] For general background see Lien-sheng Yang, "Hostages in Chinese history," *Harvard Journal of Asiatic Studies*, 1952, 15(3/4), 507–21. This article notes (p. 507) a hostage exchange in 720 BC, a time long before Sun Tzu.

[28] For background on operant conditioning and schedules of reinforcement see Howard Rachlin, *Introduction to Modern Behaviorism* (2nd ed.; San Francisco: W.H. Freeman, 1976).

"Sun Tzu's First Symphony."[29] As this symphonic image suggests, Passage #6.18 is best approached in its entirety, rather than fractionating it into component ideas (albeit that several different topics may be identified). Reading Passage #6.18's evocative word pictures of successful situation control ("subtle and insubstantial, the expert leaves no trace," etc.) might indeed earn it the further label of "near-magical" thinking affixed by Brooks to Sun Tzu VI, of which Passage #6.18 forms a large part. However, Passage #6.18 can also be read a little less literally, as an impressionistic profile of the tempo and rhythms of a successful effort to implement "situation control." So understood, Passage #6.18 may perhaps be exonerated from charges that it claims too much. Indeed it does a rather effective job of conveying to Sun Tzu's audiences a vivid intuitive sense of what situation control means. What might help clinch that sense, but is lacking in Passage #6.18 (as it is throughout Sun Tzu), would be anchoring via examples from Chinese military history.[30]

A separate part of Sun Tzu, also displaying "situation control" instincts, is Passage #6.19 whose focus is on use of secret agents (or perhaps defectors) to exercise control over events at the highest level. More could be said here, in particular about the special role Sun Tzu ascribes to double agents (fanjian) in achieving such felicitous outcomes, but that discussion will be deferred to Theme #11 (where Sun Tzu's concept of the fanjian role is analyzed and critiqued; see pp. 368–71 below).[31]

Roads Not Taken by Sun Tzu in Developing Theme #6

Because Sun Tzu gives little explicit guidance as to the meaning of an attack on an adversary's strategy it is hard to pronounce authoritatively about limits of his Theme #6 thinking. One set of clues regarding limits centers on Sun Tzu's lack of attention to major military innovation and its disruptive effects on established patterns of warfare. Military innovation, of course, takes many forms, generally a blend of tangibles and intangibles. The former commonly center on new technology (weapons, logistics, communications, and more); the latter, on new concepts of organization, doctrine, or command and control. Whatever its active ingredients, major innovation has potential to upend an enemy's strategy entirely, scrambling his plans, unraveling his poise, possibly leading on to his downfall. Such effects of innovation should be classic grist for Theme #6. Yet there are no traces of this level in Sun Tzu, an omission all the more striking because, as two scholars of early China put it, "Classical China was a time when innovation was urgently relevant: when the states most successful in innovating were the ones most likely to survive."[32]

In a limited but real sense – one that centered on copying initially unfamiliar methods of their adversaries – the Mongols of the time of Chinggis Khan (r. 1206–27) and his early successors epitomized openness to military innovation. Commonly

---

[29] This image of a symphony builds in part on Ames, pp. 69–70, with a key observation being that "[t]here is a sense in which the value and meaning of each note can only be understood within the context of the entire symphony." (Ames, p. 69.)

[30] An important though still limited step in that direction is taken by traditional commentators Li Quan and Du Mu who provide assorted historical examples to anchor and illustrate various parts of Passage #6.18. See Minford, pp. 184, 188; Griffith, pp. 97–98, 99 footnote 1.

[31] See also Passage #7.10 (passage designated Sun Tzu's "Symphony of Spies").

[32] Brooks & Brooks, Introduction footnote 36, p. 38.

relying on expertise of former enemies now working for them, the Mongols were very good at learning new things. They illustrated a formidable capacity, not once but many times over (against numerous different antagonists across Asia), to master ways of war that were initially outside their military experience.[33] One strand of that capacity reached deep into the roots of logistics. For example, as one authority has observed, "[w]herever the Mongols appeared they sought out artisans whose skills had direct or indirect military applications. . . . The Mongols regularly moved artisans of all types from one cultural zone of their vast empire to another."[34]

That type of far-reaching adaptive ability is nowhere in evidence in the Sun Tzu text. What is found in Sun Tzu is encouragement to build on a deep quiver of existing tricks, made inexhaustible by endless adjustments and variants to match current conditions.[35] But Sun Tzu does not take the next step of nudging his general to master (or to be prepared to defend against) any radically new ways of war. One hint that such broader horizons may have been outside Sun Tzu's ken is lack of asymmetric warfare allusions in the text. Sun Tzu also takes no visible steps to advocate for (or peddle!) any particularly novel technology or type of military capability. Fire and water as weapons of war (topics of Sun Tzu XII) would have been scarcely new in Sun Tzu's time. Cavalry is notably unmentioned (see pp. 29–30 above). It is conceivable that, by the standards of his time, Sun Tzu stood for some degree of innovation in command, control, and communications and, separately, in use of incendiary warfare methods (see Appendix 15 for further analysis of Sun Tzu's thinking on both topics). But advocacy for either class of methods on Sun Tzu's part seems unlikely to represent a major break with military practices already extant.

All in all, Sun Tzu's general seems best envisioned as a highly creative applier of existing military knowhow, not as an originator or pioneer of new knowhow.[36]

---

[33] See May, Introduction footnote 52, chapter 7, "The opponents of the Mongols," pp. 100–114. See also Thomas T. Allsen, "The circulation of military technology in the Mongolian empire," in Di Cosmo, Introduction footnote 52, pp. 265–93.

[34] Allsen (cited in previous footnote), p. 266; May, Introduction footnote 52, p. 64. For further background see also Allsen's *Commodity and Exchange in the Mongol Empire: a cultural history of Islamic textiles* (Cambridge, UK/New York: Cambridge University Press, 1997), pp. 30–45. In a kindred vein, echoing Allsen's description of Mongol practices, one would certainly like to know Sun Tzu's ideas, if any, relating to "systematic identification, mobilization, and redistribution of military technicians" of a formerly enemy state, whose militarily valuable skill-sets would be a basic part of the booty of a successful war. (Quoted words are from Allsen, cited in previous footnote, p. 267.)

But reflecting his next-to-nil coverage of the "day after victory" Sun Tzu offers nothing on point.

[35] See Passage #5.11's verse VI.26 and Passage #5.13 (especially verse V.6 there).

[36] Minford, p. 219 translates an observation of traditional commentator Jia Lin: "A general should prize adapting to change" (the context is Passage #8.28A's verse VIII.10). However, that idea stops short of asserting that the general should take a hand in *initiating* innovation.

Yet military innovation is not all about new hardware or even new skill-sets. It is *also* about innovations in a realm of concepts. Subject to the limits of our knowledge, which are substantial, it is tantalizing to cast Sun Tzu himself (an entity to be distinguished from the commanding general on which the text focuses) as a major conceptual innovator in his time. For example, the Sun Tzu text is the earliest traditionally transmitted text to invoke a distinction between *qi* and *zheng* methods in a military context (Passage #5.13; see Theme #14 for further analysis).

Shifting attention to counter-plan mode, it is worth pointing out that (strikingly for a major advocate of deception!) Sun Tzu never squarely tackles the all-important question of how to tell the real enemy plan from a fake plan – a timeless intelligence conundrum, as striking examples from modern experience attest.[37]

Also never truly addressed by Sun Tzu is the reality that *effective* efforts at attacking the adversary's strategy can be exceedingly hard to pull off (note again Brooks's criticism of Sun Tzu VI [Brooks, p. 61] as making "nearly magical claims"). In the warfare of his own time that point was thoroughly appreciated by Mao Zedong.[38]

Sun Tzu (2) ("Sun Tzu Extended") and Sun Tzu (3) ("Sun Tzu Analogical") Frontiers of Theme #6

Sun Tzu's Theme #6 counter-strategist mode, both hard and soft, is timeless and illustrations of its many possibilities could be culled from all chapters of world military history. Yet this concept also needs updating to reflect the ever more institutionalized character of modern high command. Trends in warfare worldwide, from Napoleonic times to the present day, mean that quality of high command has increasingly become an attribute of a system, rather than of any individual person, whether commanding general or other. That fact both generates new counter-strategist opportunities and places limitations on efficacy of counter-strategist mode.

First, new opportunities: Abetted by the capabilities of contemporary computing and databases, one of the most basic of those opportunities is almost embarrassingly simple. It builds on the kind of routinized information-processing tasks, coupled with systematic case-by-case followup, that modern bureaucracies are often very good at. First, develop a list – often notably short – of human actors in the enemy's camp with a flair for strategy (and perhaps also the leadership talents and organizational roles to give practical effect to it).[39] Then whittle down that list, one person at a time, by taking steps to delete that person from it, by either hard or soft counter-strategist measures.

An opportunity of a different kind arises when a modern bureaucracy faces a powerful individual. Not being constrained by the human life course and its limits, well-established bureaucracies – whose capabilities for ceaseless self-renewal call to mind the "flow of the great rivers" of which verse V.6 speaks (Passage #5.13) – can

---

[37] In Sun Tzu's coverage of counter-deception, Passage #7.16 comes closest to addressing the fake plan issue, but its focus on interpreting communications from the enemy is specialized and the scope seems limited to easy cases. Barton Whaley's modern study (#7 footnote 4, pp. 121–22) found that: of our 114 [modern] cases involving surprise or deception, 10 are known to have involved receipt by the victim of detailed *documents* about the attacker's plans. Of these, five were deliberately planted misinformation and all were gullibly swallowed. The other five cases involved inadvertent loss of these documents. It is remarkable that of those five true "plans," four were *discredited* by the victim and only one was accepted as the genuine warning it was. [Emphases in original; page references are from 2007 Artech reissue, which corrects a page order garble in 1969 MIT edition.]

[38] For cautionary observations from Mao see Griffith, pp. 51–52. Still more pointed are comments in Mao's essay "Problems of strategy in China's revolutionary war" (1936), #4 footnote 92, pp. 129–30.

[39] The idea of this type of list-making, though not stated in so many words, is implicit in verse XIII.16 (Passage #6.4).

be exceedingly good at outwaiting individual human beings. Although Sun Tzu does not say anything about it, such patience may amount to a counter-strategist mode of attack.[40] The dynamics of the human life course (including natural processes of health and disease and aging, cf. ideas of French philosopher-physician Georges Canguilhem) will work their inexorable toll on even the most astute and resilient human strategist. Among those effects is the possibility that certain kinds of aggressive moves or highly energized courses of action (Passage #6.18) that would have been appealing to an adversary strategist at one stage of his life course come to have lesser appeal at a later life-stage.[41]

Second, a new kind of limitation (perhaps more exactly, a modification of concept): A separate implication of modern bureaucracy is the way it may challenge, or at least profoundly redefine, the *meaning* of a counter-strategist mode of attack. In many quarters, worldwide, the institutionalization of modern high command has now reached a point where the strategic concepts and imagination of the "person at the top" are but one of a raft of inputs to a vast, impersonal, bureaucratized apparatus whose cumulative output generates the actual decisions. The person at the top may not even furnish the most important input. Under such conditions, Sun Tzu's counter-strategist mode needs to shift focus, away from degrading the decision-making quality of the titular leader (see Passages #6.7–#6.10) and in a direction of *degrading the quality (including timely performance) of bureaucratic algorithms that have to a significant extent supplanted thought processes of the individual at the top.* That too is a Theme #6 agenda – but the object of the attack is now fundamentally different, for all the reasons that (only half in jest) modern bureaucracies have been likened to non-human life forms.[42] In the twenty-first century many of those bureaucratic algorithms may themselves be increasingly machine-based, at least in part, as machine learning and artificial intelligence extend their sway.

In a modern world where bureaucracies come not as stand-alone entities but in battalions (so to speak), a further new kind of leverage – a modern update to Sun Tzu verse I.25 (Passage #6.8) – involves getting the adversary's bureaucracies fighting one another, even on a level of existential intensity. Internecine bureaucratic conflict, which will always exist to some extent, has the makings of a plum target for outside troublemakers (note verses II.5 [Passage #5.5] and III.23 [Passage #7.8]). If existing fights in and between bureaucracies can be made to escalate, the resulting warfare can last a very long time (bureaucracies are long-lived and can muster a lot of resources!) and can be very hard to de-escalate (bureaucracies have elephantine memories!). US precedents include experience with the long-lasting, even inter-generational, fallout of poisonous interservice and civilian/military controversies

---

[40] An exception proving the rule – where the man outwaited the bureaucracy (in this case that of the British Raj, whose existence was cut short by 1947 partition and birth of modern India and Pakistan) – see Milan Hauner, "One man against the empire: the Faqir of Ipi and the British in Central Asia on the eve of and during the Second War World," *Journal of Contemporary History*, 1981, 16(1), 183–212.

[41] In play here is the fourth path to advantage (p. 183 above), with the bureaucracy "waiting out" an existing situation until it is replaced by a different, more favorable one.

[42] James G. March & Herbert Simon, *Organizations* (New York: Wiley, 1958), p. 4 (a "biological analogy is apt here, if we do not take it too literally or too seriously") – words written long before the twenty-first-century rise of artificial intelligence.

arising in the aftermath of World War II and the 1947 birth of the modern US defense establishment.[43]

Shifting focus to counter-plan steps, in many twentieth- or twenty-first-century contexts a major attack on the enemy's complex and frequently fragile logistics rises to a level of an attack on that enemy's strategic plans.[44] Supporting that concept, it is sometimes said that at sufficiently high levels of command strategy and logistics tend to coalesce.[45] More than a few echoes of the spirit of Sun Tzu's verse I.24 (Passage #6.8), but now geared more toward counter-plan than counter-strategist mode, may be found in T. E. Lawrence's World War I initiative to bottle up the Turkish Army contingent occupying Medina – and then to allow the rail line supplying Medina to keep functioning, but only barely![46] A twenty-first-century information logistics counterpart to such a stratagem would be bandwidth throttling of the Internet. That is a tool of control that can at times be more effective than shutting down Internet access altogether, since would-be Internet users then become preoccupied with navigating technological hurdles which, like the supply problems of the Turkish Army in Medina, are surmountable but only barely (with implications for user behavior both on and off the web). When such technological constraints are lifted – which in a backhanded way is exactly what happens when Internet access is shut down entirely – people have time on their hands and motivation to be creative in finding alternative communication channels, including face-to-face contact. The upshot may be protest activity on a dispersed basis hard for an authoritarian regime to damp, if indeed doing so proves possible at all.[47]

As these examples also suggest, attacks on an enemy's logistics smack not only of counter-plan mode but also (and more broadly) of situation control mode. Logistics, comprehensively conceived, is a horizon-broadener, evoking Hamlet's observation that "There are more things in heaven and earth, Horatio, than are dreamt of in your philosophy." It is also the ultimate "details" subject, one that goes

---

[43] See Anand Toprani, "'Our efforts have degenerated into a competition for dollars'. The 'revolt of the admirals', NSC-68, and the political economy of the Cold War," *Diplomacy & Statecraft*, 2019, 30(4), 681–706, and the same author's "Budgets and strategy: the enduring legacy of the revolt of the admirals," *Political Science Quarterly*, 2019, 134(1), 117–46. See also James Forrestal, *The Forrestal Diaries* (edited by Walter Millis with the collaboration of E. S. Duffield; New York: Viking Press, 1951).

[44] As an old saying goes, "He who controls the spare parts controls the operation."

[45] See Boorman, Introduction footnote 4, p. 102 and p. 113 note 50 (citing writings of Admiral Eccles). Note that major attacks on logistics not involving attacks on allies or alliances would naturally fall under a rubric of verse III.4 but not verse III.5 (thereby possibly shedding light on a puzzle posed at the start of Theme #6, querying the division of labor between Passage #6.1's verses III.4 and III.5). Of course, complexities wrought by twenty-first-century supply chains may well blur anew the distinction between these two Sun Tzu verses.

[46] See p. 110 of *Evolution of a Revolt: early postwar writings of* T. E. *Lawrence* (edited and with an introduction by Stanley & Rodelle Weintraub; University Park/London: Pennsylvania State University Press, 1968), pp. 100–19 (from article by Lawrence first published in 1920).

[47] For further analysis, including both formal modeling and empirical anchoring in Arab Spring events, see Navid Hassanpour, *Leading from the Periphery and Network Collective Action* (Cambridge, UK/New York: Cambridge University Press, 2016). A key insight from his analysis (see his p. 28) is that *reducing* connectivity by cutting electronic communications may actually operate as a catalyst for effective revolutionary political mobilization, rather than serving to retard and dampen it.

hand in hand with spotting sources of advantage or vulnerability, at times major, in seeming minutiae.[48] Both characteristics commonly favor the attacker who has done his homework (see first and second paths to shi described on pp. 178–81). Some of the most effective attacks on logistics systems are ones where genuine strengths of the adversary's logistics are turned, jujitsu-like, against themselves. Logistics is not a "thing" or static entity but a marvelously intricate ensemble of flows of many kinds – among them, of materiel, information, personnel, and more.[49] One major disruption in those flows can easily beget others. Versions of this type of dynamic (though reflecting deficits in US planning, not deliberate hostile action) are well documented in US World War II experience (experience also echoed during parts of the Vietnam War).[50] From there, it is a short step to conjuring opportunities for creative disruption or sabotage in a future war.[51] It is unfortunate that Sun Tzu himself was not enough of a systems thinker, at least in ways that focused on logistics systems (which in his time would, of course, have been relatively rudimentary), to explore this direction of thinking, which has both Sun Tzu (2) and Sun Tzu (3) implications.

A technological frontier of twenty-first-century warfare – one that seems suited par excellence to unlocking the potential of Sun Tzu's concept of attacking a foe's strategy – is algorithm sabotage. Algorithm sabotage may be broadly characterized as an attack on the complex mathematical-logical algorithms that stand at the heart of the control of a vast range of modern military systems (among them, weapon, command and control, logistics, navigation, communications, and, of course, intelligence).[52] As is the case with Sun Tzu's Theme #6 counter-strategist mode, which may involve (a) degrading the performance of an enemy strategist currently in place or (b) getting that strategist replaced by a worse strategist, two levels of algorithm sabotage may be distinguished:

- editing a foe's existing algorithm to modify it in a way advantageous to one's own interests or deleterious to the enemy's (what that algorithm does, how long it takes to do it, etc.);

---

[48] A relevant observation of Admiral Eccles – informed by deep wartime background in surmounting challenges of advanced base development on isolated Pacific islands – warns against the pitfalls of "overlook[ing] the apparently insignificant details that almost invariably are the difference between success and failure." See p. 12 of Eccles's report "The Establishment of Advanced Naval Bases in the Central Pacific Area, as Seen by the Advanced Base Section, Service Force, U.S. Pacific Fleet" (10 December 1945), filed in Box 85, Folders 2–4 in Eccles Papers held in the Naval Historical Collection, US Naval War College, Newport, RI (cited in Boorman, Introduction footnote 4, p. 110 note 31). See also Report, p. 18 ("lack of a small tool or piece of equipment might create conditions all out of proportion to the normal value of this equipment"); Report, Annex A, p. 9 (machine tool case).

[49] Illustrating some of these issues is the all-important difference between nominal control over petroleum resources in the ground and actual ability to exploit those petroleum resources for strategic ends. See Anand Toprani, "The first war for oil: the Caucasus, German strategy, and the turning point of the war on the Eastern Front," Journal of Military History, 2016, 80(3), 815–54.

[50] Boorman, Introduction footnote 4, p. 113 note 48 (citing sources for both wars).

[51] See Eccles, #2 footnote 6, p. 151 (statement written around 1950): "Sabotage, in the past never more than a nuisance, may well be serious."

[52] For algorithm sabotage as a military concept see two papers by Boorman & Levitt, #3 footnote 43.

- influencing for one's own advantage the adversary's choice of a particular type of algorithmic approach (selected from among multiple candidates to address a particular problem of interest).

Sun Tzu Theme #6 has relevance to both.[53]

A technically competent subverted insider in a sensitive job may be well positioned to carry out algorithm sabotage, thereby suggesting a twenty-first-century update to Sun Tzu's emphasis on uses of double agents (*fanjian*) in planting false information. A devastating algorithm modification can often be of low visibility so as not to attract premature in-house attention to the change. Specific possibilities are legion and there is a spectrum of degrees of technical sophistication ranging from very elementary (say, supplying a slightly wrong value for an exponent in a simulation program) all the way up to forms of sabotage building on higher mathematics.[54] Opportunities for algorithm sabotage may grow when the workings of an algorithm are thoroughly opaque to the victim of sabotage (which may, of course, be regarded as a case of failure to "know yourself" in Sun Tzu's illustrious "know the enemy and know yourself" dictum).[55] Such a situation is extremely common, at the current state of the art, in machine learning applications (though that same opacity may also be a double-edged sword, making it more difficult for a would-be saboteur to achieve a specific desired effect).[56] One type of algorithm sabotage that may elude detection, because its active ingredient involves only an act of omission, is suggested by a

---

[53] A Sun Tzu verse that captures something of the spirit of algorithm sabotage is verse XI.56, which Griffith, p. 139 translates (following traditional commentator Du Mu): "Now the crux of military operations lies in the pretence of accommodating oneself to the designs of the enemy" (Passage #6.27A).

　　Accepting Griffith's reading, the active ingredient here is one of *accommodating, indeed promoting, a foe's tactical or strategic design on one level, even while frustrating it on a more basic level that opens a door to his downfall.* In the twenty-first century, that "design" could be computer-algorithm-based.

　　For textual and interpretive issues arising in verse XI.56 see Appendix 7 anchor footnote 24.

[54] A flavor of some of the possibilities, involving a non-sabotage case, is a computational instability issue – not known at the time – lurking in the background of a pioneering numerical weather forecasting approach (*c.* 1913) undertaken by the English Quaker scientist Lewis F. Richardson (1881–1953, also known for his mathematical analysis of arms races). For the history see George W. Platzman, "A retrospective view of Richardson's book on weather prediction," *Bulletin of the American Meteorological Society*, 1967, 48(8), 514–50, in particular pp. 514, 527–28; Peter Lynch, "The origins of computer weather prediction and climate modeling," *Journal of Computational Physics*, 2008, 227(7), 3431–44. (Fortuitously, Richardson's work sidestepped the problem because it took only one time step, i.e., actual implementation was very limited. See p. 514 of Platzman's article.)

[55] See verse III.31 (Passages #11.1 and #12.1).

[56] An instructive example antedating the current machine learning era is the case of "Algorithm K" as described by Donald Knuth, *The Art of Computer Programming*. Vol. 2: *Seminumerical Algorithms* (3rd ed.; Reading, MA: Addison-Wesley, 1998), pp. 4–5. On its face, that algorithm should be an enormously effective means of generating pseudo-random numbers – its excruciating opacity would almost seem to guarantee that! – but it actually turns out under some circumstances to be highly *ineffective* for that purpose, generating a ludicrously non-random result (as an example provided by Knuth shows). The thrust of Sun Tzu Passage #1.7 (see pp. 59–60 discussion) is relevant here since pseudo-random-number algorithms are usually buried several tiers down in a computational hierarchy of algorithms, remote from what most end users see, evaluate, or even understand. Accordingly, flaws located there can be insidious indeed.

non-sabotage case that came to light in the course of British naval operations during the 1982 Falklands War.[57] It involved a basic (but previously overlooked) "hole" in an algorithm on which British naval air defense software crucially depended. Specifically, that software failed to provide for a contingency involving two aircraft attacking on parallel courses, thereby presenting equally attractive targets. The upshot was that the software on which a British warship was relying to take air defense steps against two attacking Argentine planes sized up the situation – and, after the fashion of Buridan's famous ass that starved to death between two equally attractive bundles of hay, proceeded to take no action![58] A British naval disaster was fortuitously averted by a dud Argentine bomb.

Algorithm sabotage is a rich growing point for Sun Tzu (3) frontiers of Theme #6, combining in varying mixes aspects of counter-strategist, counter-plan, and situation control modes. A counter-plan standpoint is particularly well suited to conveying why algorithm sabotage can be so potent. In a traditional biophysical world, military plans may be carried out, usually imperfectly, and in any case are commonly soon discarded for reasons summed up by the old military adage that "no plan survives first contact with the enemy." By contrast, plans coded into software will be carried out faithfully and often for the indefinite future by machines specifically programmed to do just that. As a result, a software saboteur's editing of an algorithm can often work its nefarious but hidden effects uninterruptedly until the problem is detected and rectified, which can take a long time (if it ever happens).

Prospects for effective general defensive measures against algorithm sabotage seem limited, certainly when trying to stymie subverted insiders with authorized access to algorithms. So great is the dependence of twenty-first-century individuals and organizations, worldwide, on digital capabilities of many types that algorithm

---

Generalizing this class of issues, see George V. Neville-Neil, "Know your algorithms," *Communications of the ACM* (Association for Computing Machinery), 2019, 62(4), 22–23, noting (p. 23):

> Algorithms are at the heart of what we as software engineers do, even though this fact is now more often hidden from us by [software] libraries and well-traveled APIs [application programming interfaces]. The theory, it seems, is that hiding algorithmic complexity from programmers can make them more productive. If I can stack boxes on top of boxes – like little Lego bricks – to get my job done, then I do not need to understand what is inside the boxes, only how to hook them together. The box-stacking model breaks down when one or more of the boxes turns out to be your bottleneck. Then you will have to open the box and understand what is inside, which, hopefully, does not look like poisonous black goo.

[57] This example is described in more detail in the first of two Boorman & Levitt papers cited in #3 footnote 43. Were this type of omission to be the work of a saboteur, it would be a near-ideal candidate for meeting three attributes of an effective attack as set forth by computer security expert Bruce Schneier: "low chance of discovery, high deniability if discovered, and minimal conspiracy to implement." See February 27, 2014 Bruce Schneier post in "Schneier on Security" blog.

[58] Cf. Nicholas Rescher, "Choice without preference: a study of the history and logic of the problem of Buridan's Ass," *Kant-Studien*, 1959, 51, 142–75. Versions of the dilemma arise in Islamic tradition (associated with al-Ghazālī, born in Tūs, Iran c. 1058, died 1111); in parts of pre-modern Chinese literary tradition; and in computer design (regarding the latter see Leslie Lamport, "Buridan's principle," *Foundations of Physics*, 2012, 42(8), 1056–66).

sabotage can very easily shade over into, indeed be indistinguishable from, ways of reshaping human or organizational behavior that range far beyond the traditional connotations of "military attacks" or even "acts of sabotage."[59] From one standpoint, this is "warfare according to *Principia Mathematica*." From another, it is a part of verse III.4's living legacy to twenty-first-century conflicts and a candidate for a case of "*What did Master Sun know that we still don't (or have yet to absorb adequately)?*"[60]

In the puzzle palaces of contemporary cyber warfare, Sun Tzu's "combinatorial" passage (Passage #5.13) – conjoined with ways (noted on pp. 122–23 above) in which a tiny change in software may produce a vastly outsized effect – invites a type of combinatorial approach to situation control. Following in Sun Tzu's situation control footsteps, a cyber strategist may be able to devise *coordinated attacks on distinct, even seemingly unrelated, computing systems whose joint impact on the biophysical world far exceeds the sum of their effects taken singly.* Drawing inspiration from D. C. Lau's translation of verse I.26 (Passage #5.1), "Attack where the enemy is not prepared; go by way of places *where it never occurred to him you would go*" (emphasis supplied), the extent of coordination in such attacks might even go undetected where the systems being attacked were not only unconnected but geared to purposes that would not usually be considered in the same breath. One place to seek examples would be urban environments rich in diverse computing applications whose outages or foibles could exert superadditive effects on the humans who rely on them. Such "combinatorial attack" is a candidate for a cyber warfare counterpart to operational art in conventional warfare.[61] In many cyber warfare analyses that combinatorial level of thinking remains underdeveloped, if it is considered at all, with the lion's share of attention flowing either to the countless technological details of cyber attack, addressing one system or

---

[59] Cf. Martzloff, Preliminaries footnote 19, pp. 58–60 describing the role of what he terms "resolutory rules" in traditional Chinese mathematics. Those rules are described as

> dogmatic "stratagems of action," which if followed mechanically should lead automatically to the expected result, but which in practice may permit a certain degree of freedom in action, since not all the stages of the calculations are rigorously specified, as they should be in the case of true algorithms.

Such quasi-algorithms, which have counterparts in many bureaucratic and other administrative settings worldwide, are often natural targets for algorithm sabotage in a generalized sense, combining an aura of authoritative precision and expertise with low-visibility latitude to "edit" what the algorithm actually does in specific cases.

[60] In Chinese tradition, a case of data integrity compromise – which the centrality of the emperor's role in imperial China imbues with a larger flavor of algorithm sabotage – appears in a vignette from the sixteenth-century Chinese novel *Xiyou ji* (*Journey to the West*). In that story, "one of the Tang emperors visits the underworld, where the officials, looking into their dossiers, kindly add 20 years to his life by marking two extra strokes in the figure for the length of his reign as previously allotted. This was a change from 13 to 33, i.e. —||| to ≡|||." See Needham, Preliminaries footnote 14, p. 17.

Translated in part by Arthur Waley under the title *Monkey*, *Xiyou ji* is sometimes bracketed with the *Sanguo yanyi* as one of China's "six greatest vernacular novels" (see Wilkinson, §31.1.6, pp. 451–52).

Regarding the very real problem – dating from early times – of easy-to-falsify Chinese notations for writing numerals, and efforts to combat such abuse, see also Wilkinson, §38.3, p. 533.

[61] For military background see *Operational Level of War – Its Art* cited in Preliminaries footnote 8.

application at a time; or else to cyber warfare's "big picture" grand strategic plane. What is widely lacking is attention to an intermediate level interpolating between these two. For reasons related to the impossibility of planning for, or even envisioning, all of the combinatorial possibilities that might arise there (see again Passage #5.13), that intermediate level is where talents of a modern Sun Tzu seeking digital *shi* might flourish.

★★★★★★

At its deepest level – towards whose inner recesses Sun Tzu's Theme #6 thinking points, though Sun Tzu only glancingly engages with it – an attack on strategy rises to the level of an attack on the roots of the adversary's self-concept and sociopolitical identity from which all human strategy ultimately flows. Developing that line of thinking – which bespeaks one of the most potent forms of situation control but also commonly involves counter-strategist and counter-plan modes of attack – touches deep into the heart of civilian life. In Sun Tzu's playbook, a natural starting point for attack on sociopolitical identity builds on the estrangement ideas that are common ground of Themes #6 and #4.[62] The art form here is turning against themselves the strong social and psychological forces out of which identity and political will are forged.[63]

A Vietnam War anecdote told by Colonel Harry G. Summers, Jr. is pertinent:[64]

> "You know you never defeated us on the battlefield," said the American colonel.
>
> The North Vietnamese colonel pondered this remark a moment. "That may be so," he replied, "but it is also irrelevant."

(Conversation in Hanoi, April 1975)

★★★★★★

---

[62] Part of verse XI.25 (Passage #6.16) could be imbued with the notion of engineering conflicts between social classes or status groups, and in that sense might be read as pointing to destabilizing an enemy's sociopolitical identity. Mair's translation (Mair, pp. 118–19) reads: "cause ... the nobles and the commoners to be unable to aid each other; the superiors and inferiors to be unable to cohere."

[63] Yet resilience of some sociopolitical identities should not be underestimated, especially if anchored in a story or set of stories that capture the imagination of a broad audience, making destabilization no easy feat. One case having relevance to Sun Tzu centers on King Goujian of Yue (r. 496–465 BC), who (as Milburn puts it) "became the cornerstone of Yue identity." Goujian's life story is intertwined with the struggle between Yue and Wu in (or shortly after) the era of the putative historical Sun Tzu, who served Wu (Giles, pp. xxvii–xviii; Lewis, Background footnote 7, p. 601). Even in Han dynasty times, long after the Warring States Yue polity became extinct, Yue identity embodied in the figure of King Goujian was so strong as to lead Han officials to advocate for travel restrictions barring Bai Yue rulers from visiting a location associated with Goujian, "lest they be seen as laying claim to his legacy." See Milburn, #1 footnote 41, p. 29 (both quotations). (Bai yue, lit. "hundred Yue," was a label applied to diverse ethnicities in southern China.)

For influence of the Goujian story in modern times see Paul A. Cohen, *Speaking to History: the story of King Goujian in twentieth-century China* (Berkeley, CA: University of California Press, 2009).

[64] Summers, #3 footnote 53, p. 21.

Set of principal passages used to illustrate Sun Tzu Theme #6 (attacking the adversary's strategy)

Passage #6.1. Attacking the Enemy's Strategy – Basic Statement of Principle

III.3. For to win one hundred victories in one hundred battles is not the acme of skill. To subdue the enemy without fighting is the acme of skill.

III.4. Thus, what is of supreme importance in war is to attack the enemy's strategy [mou];

III.5. Next best is to disrupt his alliances;

III.6. The next best is to attack his army.

III.7. The worst policy is to attack cities. Attack cities only when there is no alternative.

Passage #6.2. Setting a Stage for Counter-strategist Mode: Competent General As Natural Target

III.18. Now the general is the protector of the state. If this protection is all-embracing, the state will surely be strong; if defective, the state will certainly be weak.

Passage #6.3. Setting a Stage, cont.: Rarity of Generals with Unimpeachable Motives

(As traditional commentator Du Mu observes [Minford, p. 260], "Jewels of this kind are rare." For exactly that reason, they are also natural targets.)

X.19. And therefore the general who in advancing does not seek personal fame, and in withdrawing is not concerned with avoiding punishment, but whose only purpose is to protect the people and promote the best interests of his sovereign, is the precious jewel [bao] of the state.

Passage #6.4. Hard Counter-strategist Mode: Targeted Killing As Tool of Conflict

(A targeted killing focus is unequivocal here. Note that while a key enemy strategist may be one target, verse XIII.16 leaves a door open to further possibilities, say, eliminating personnel on whom that strategist crucially relies for one reason or another, professional or personal. Picking an optimal set of targets within feasibility constraints may itself present a combinatorial problem that, in a backhanded way, suggests a further application of verse V.22 [Passage #4.6].)

XIII.16. Generally in the case of armies you wish to strike, cities you wish to attack, and people you wish to assassinate, you must know the names of the garrison commander, the staff officers, the ushers, gate keepers, and the bodyguards. You must instruct your agents to inquire into these matters in minute detail.

Passage #6.5. Hard Counter-strategist Mode: Precedents from Chinese History

(These are well-known examples from early Chinese history, both involving a targeted killing or threat of it.[65] It should be noted that these cases are invoked by Sun Tzu to make a point quite different from advocacy for targeted killing, focusing instead on responses of troops who find themselves in extreme peril ["death ground"]. However, a relevant *mentalité* is definitely conveyed.)

XI.37. But throw them into a situation where there is no escape and they will display the immortal courage of Zhuan Zhu and Cao Gui.

Passage #6.6. Hard Counter-strategist Mode: Sidelining Enemy Talent

(Reading verse I.15 in mirror image, it suggests the possibility of manipulating an enemy ruler, possibly with falsified information, into firing or sidelining his own top-notch talent that could win his war, to be replaced by mediocrity – in a sociological framing, a vacancy chain gone bad.[66] The upshot may be much the same as a targeted killing of the enemy's talented strategist.[67])

I.15. If a general who heeds my strategy is employed he is certain to win. Retain him! When one who refuses to listen to my strategy is employed, he is certain to be defeated. Dismiss him!

Passsage #6.7. Soft Counter-strategist Mode: Enemy Commander's Mind As Object of Attack

(Verse VII.20 illustrates the concept of "attacking the mind of the enemy commander" for which Sun Tzu is justly famous.[68] The Chinese concept of *xin*

---

[65] For sources on verse XI.37's two cases see Appendix 6 footnote 2. For a thumbnail sketch of specifics of those two escapades see also Passage #10.8 discussion on p. 349 below.

[66] An alternative reading of verse I.15, discussed in Appendix 6 anchor footnote 3, casts Sun Tzu himself as the one who would stay or depart. Counter-strategist attack might then focus on inducing Sun Tzu to perceive his counsel as being violated by the ruler, creating a rupture in their relationship.

[67] A Warring States historical example, involving the decisive Qin victory over the rival state of Zhao at Changping – sometimes called the Chinese Cannae (see Passage #7.14 below) – is recounted by the traditional commentator Li Quan, drawing on the Shiji (Minford, p. 116). According to the Shiji, Qin secret agents convinced the suggestible Zhao ruler to replace a competent commander by a substitute general (Zhao Kuo, himself the son of a famous general). Zhao Kuo was such a poor choice that even his own mother tried to warn the Zhao ruler that Zhao Kuo should never command an army (!).

This story is famous and has given rise to a modern Chinese proverb *zhishang tanbing* (lit. "talking about stratagems on paper" – i.e., armchair strategist). Zhao Kuo had learned the art of war from books and discussions of theory, but did not know how to apply that knowledge in actual warfare.

For the Shiji 81 account see Nienhauser, Vol. VII (1994), pp. 269–70; Nienhauser, Vol. VII (2021), pp. 489–91 (both cited on pp. xxiii–xxiv). See also Giles, p. 166.

[68] One of the traditional commentators, Zhang Yu, makes what amounts to a structural observation regarding verse VII.20 (Minford, p. 207): "The mind is the general's chief asset."

has deep roots in Chinese thought, both Confucian and Daoist. It comes endowed with both emotional and rational-cognitive content. Some translators accordingly suggest rendering xin as "heart-mind." An important perspective on xin comes from Zhang Dainian, noting that xin "is not just one organ among others but something that coordinates the work of all organs of thinking."[69] In short, an attack on xin amounts to an attack on the enemy commander's ability to integrate patterns of thought and, by extension, patterns of action too.)

VII.20. Now an army may be robbed of its spirit and its commander deprived of his courage [xin].

Passage #6.8. Soft Counter-strategist Mode: Degrading Quality of Enemy Commander's Thinking

(This passage may be read as treating capacity to calculate as itself an object of attack, in ways geared to undermining an enemy general's poise and judgment; or, as traditional commentator Zhang Yu discussing Passage VII.20 puts it, the attacker "robs him of his presence of mind, his ability to think strategically."[70] The steps Sun Tzu notes here could have social or political as well as psychological levels. For example, verse I.25 suggests a class of stratagems whose implementation might involve either soft or hard counter-strategist moves. Factional struggles often siphon off scarce high-quality attention and effort. They can also lead to sidelining of talent that chose the wrong side in an intramural fight.[71])

I.22. Anger his general and confuse him.[72]

I.23. Pretend inferiority and encourage his arrogance.[73]

---

[69] See Zhang Dainian, #3 footnote 80, p. 391 (pp. 391–409 further discusses xin). Griffith, p. 108 footnote 1, suggests "deprived of his wits" as an alternative translation of verse VII.20's latter half.

[70] Minford, p. 208. The quoted observation again comes from Zhang Yu's verse VII.20 commentary.

[71] To be clear, lack of scope conditions in verse I.25 allows for other types of division – for example, division achieved by driving a wedge between allied states (Griffith, p. 69, quoting one of Zhang Yu's ideas) or by inducing parts of the enemy army, formerly united, to become separated (Giles, p. 7).

[72] Although Griffith's verse I.22 translation is slightly free, it succinctly conveys how anger may easily induce confusion and, with it, exploitable judgment errors. Drawing on his US Navy warship command experience during World War II, an observation of George C. Homans, "The small warship," *American Sociological Review*, 1946, 11(3), 294–300, is pertinent:

> He [the commanding officer] is subject to strains unlike those met by an executive in civil life. He is on call at all times; his judgment may be warped by lack of sleep. He is still more dangerous when he is angry. Much of the work of a warship is of an emergency nature. If a mistake is made at a crisis, it is only human for him to be furious; his fury will be in proportion to his desire to do a good technical job, and it will take the form of an overwhelming urge to bawl somebody out. It is then that he must watch himself.
>
> (p. 300)

[73] A story from the *Romance of the Three Kingdoms* provides a (fictional) example:

> Lord Guan had an ailing arm [from a poisoned arrow wound, later cured by good medicine], and an ailing attitude as well – namely, his overestimation of himself and his arrogance toward all others. Lu Xun . . . had a method for aggravating Lord Guan's [latter] ailment – namely, rich

I.24. Keep him under a strain and wear him down.[74]
I.25. When he is united, divide him.

Passage #6.9. Soft Counter-strategist Mode: Venality to Control Information Sources

(Given the extent of dependence of Sun Tzu's general – and ruler – on their spies, a dependence that Sun Tzu XIII takes pains to spell out [see Passage #6.19], the type of bribery depicted in verse XIII.17 easily rises to a level of counter-strategist attack, attacking the strategist by attacking his sources of information.[75])

XIII.17. It is essential to seek out enemy agents who have come to conduct espionage against you and to bribe them to serve you. Give them instructions and care for them. Thus doubled agents are recruited and used.

Passage #6.10. Counter-strategist Mode: Exploiting Character Flaws

(This is a further menu of counter-strategist moves, some hard, others soft.)

VIII.17. There are five qualities which are dangerous in the character of a general.
VIII.18. If reckless, he can be killed;
VIII.19. If cowardly, captured;
VIII.20. If quick-tempered you can make a fool of him;
VIII.21. If he has too delicate a sense of honor you can calumniate him;[76]
VIII.22. If he is of a compassionate nature you can harass him.
VIII.23. Now these five traits of character are serious faults in a general and in military operations are calamitous.

---

gifts and honeyed words. Lü Meng resigned his office, and Lord Guan assumed he was free of a problem, a greater relief than the cure of his arm. And so he pulled out his southern defenses ... But by so doing, Lord Guan was more severely poisoned than by the arrow!
    (See Moss Roberts translation cited on p. xxiv above, pp. 1068–69, chapter 75 note 8.)

Work of a Song poet suggests the game of *weiqi* as a testbed for exploring arrogance and the inescapable helplessness, culminating in a merciless denouement, to which it so often leads. See "The arrogant fail," p. 219 of Zu-yan Chen, "Shao Yong's (1011–77) 'Great Chant on Observing Weiqi': an archetype of neo-Confucian poetry," *Journal of the American Oriental Society*, 2006, 126(2), 199–221.

[74] Useful background here is the medical sociology concept of stressors, defined as "*conditions of threat, demands, or structural constraints that, by the very fact of their occurrence or existence, call into question the operating integrity of the organism.*" See Blair Wheaton,"The domains and boundaries of stress concepts," in Howard B. Kaplan (ed.), *Psychosocial Stress: perspectives on structure, theory, life-course, and methods* (San Diego, CA: Academic Press, 1996), p. 32 (emphasis in original).

[75] Cf. Stefano Musco, "Intelligence gathering and the relationship between rulers and spies: some lessons from eminent and lesser-known classics," *Intelligence and National Security*, 2016, 31(7), 1025–39, in particular pp. 1032–33, "King's eyes and ears: spies as 'ramifications' of the ruler."

[76] For a perspective see Lewis, Background footnote 11, "Warfare and Honor," pp. 36–43 (analyzing patterns of sanctioned violence in the Spring and Autumn period that preceded the Warring States era). See also his pp. 94–96 (snapshot of subsequent Warring States developments).

VIII.24. The ruin of the army and the death of the general are inevitable results of these shortcomings. They must be deeply pondered.

Passage #6.11. Counter-plan Mode on a Plane of Grand Strategy

(Presence of a counter-plan idea here is strengthened if the plans of the neighboring states have a hostile duplicitous aspect, which in Warring States China was often the case.)

XI.51. One *ignorant of the plans of neighbouring states cannot prepare alliances in good time;*[77] if ignorant of the conditions of mountains, forests, dangerous defiles, swamps and marshes he cannot conduct the march of an army; if he fails to make use of native guides he cannot gain the advantages of the ground. A general ignorant of even one of these three matters is unfit to command the armies of a Hegemonic King. [Emphasis supplied.]

Passage #6.12. Counter-plan Mode, Emphasizing Active Reconnaissance Steps

(While other means of intelligence-gathering may play a part, a major focus here is on collecting information by active reconnaissance measures.[78] That task would have presented significant challenges if, as seems likely, Sun Tzu's armies lacked cavalry, a basic tool of reconnaissance over much of the history of land warfare. Such a deficit would only to tend to reinforce Sun Tzu's emphasis on spies.)

VI.20. Therefore, determine the enemy's plans and you will know which strategy will be successful and which will not;[79]

VI.21. Agitate him and ascertain the pattern of his movement.

VI.22. Determine his dispositions [xing] and so ascertain the field of battle.[80]

VI.23. Probe him and learn where his strength is abundant and where deficient.

---

[77] Mair, p. 123 gives this segment a more specific twist (though a counter-plan idea persists): "he who does not know the plans of the feudal lords should not forge diplomatic ties with them."

[78] The *Wuzi* – another of the *Seven Military Classics* (see Background footnote 27) – offers a relevant scenario, considerably more specific than anything in verses VI.20–VI.23. See #11 footnote 79.

[79] See, e.g., *Romance of the Three Kingdoms* (Moss Roberts translation cited on p. xxiv above), chapter 18, p. 141: "Jia Xu saw through Cao Cao's plan and prepared countermeasures."

[80] A kindred spirit appears in an observation of Marshal of the Soviet Union G. K. Zhukov:

Past experience shows that the outcome of a battle depends, in the final analysis, on how well – to the point, strictly and attentively – the CO [commanding officer] and his staff organize the attack. *Of paramount importance in this intricate effort is intelligence. When one is aware of enemy dispositions, strength and resources, and of the specific features of enemy-held terrain, one can forecast the enemy reaction unerringly.*

(See G. K. Zhukov, *Reminiscences and Reflections* [Vic Schneierson, trans.; Moscow: Progress Publishers, 1985], Vol. 1, p. 150, emphasis supplied.)

Passage #6.13. Counter-plan Mode: Targeting the Enemy's Logistics Support

(Attack on logistics is a versatile, even near-universal, counter-plan measure. Note that Sun Tzu contemplates not only attacks that involve destruction of supplies and equipment [verse XII.1] but also attacks whose effect is to sever transportation arteries [verse XII.14; also, some understandings of verse XII.1's fifth type of target].)

XII.1.   There are five methods of attacking with fire. The first is to burn personnel; the second, to burn stores; the third, to burn equipment; the fourth, to burn arsenals; and the fifth, to use incendiary missiles.[81]

. . .

XII.13. Those who use fire to assist their attacks are intelligent; those who use inundations are powerful.

XII.14. Water can isolate an enemy but cannot destroy his supplies or equipment.

Passage #6.14. Counter-plan Mode: Focus on Enemy Capabilities, Not Intentions?

(As previously discussed on pp. 212–14 above, this verse may be read as staking out a doctrinal position with regard to enemy capabilities vs. intentions as a basis for planning, coming down in favor of the former on multiple scopes and scales of military action. It could also be given a narrower reading as simply a call for pre-war preparedness – or, as traditional commentator He Yanxi puts it more elegantly, "When the world is at peace, a gentleman keeps his sword by his side."[82])

VIII.16. It is a doctrine of war not to assume the enemy will not come, but rather to rely on one's readiness to meet him; not to presume that he will not attack, but rather to make one's self invincible.

Passage #6.15. Situation Control Mode: Introduction to Concept

(Lau, 1965 article, p. 332 translates the first part as "Hence when one who is good at setting the enemy in motion shows himself [xing, lit. "form"], the enemy is sure to follow after him," which conveys a behavioral control idea with clarity.[83])

---

[81] Understandings of verse XII.1 vary. Mair, p. 125 gives a crisp reading that aligns well on familiar logistics categories (personnel, supply, transportation, facilities): "In all there are five kinds of incendiary attack. They are to use fire against: "1. men, 2. grain, 3. carts, 4. storehouses, 5. supply lines." For further discussion of the fifth category see Appendix 6 footnote 8.

[82] Griffith, p. 114. The other traditional commentators support this understanding.

[83] A case can be made that Du Mu, the civilian, captures the spirit of verse V.20 better than Cao Cao, the archetypical military man. See Minford, p. 173:

Cao Cao: "Provides the outward signs of weakness" [i.e., weakness should be feigned].

Du Mu: "Not just weakness. What this means is, if we are strong and the enemy is weak, then we should exhibit weakness, and this will cause the enemy to come; if, on the other hand, if we are weak and the enemy is strong, then we should exhibit strength, and this will cause the enemy to depart. *The enemy's movements all follow ours.*" [Emphasis supplied.]

V.20. Thus, those skilled at making the enemy move do so *by creating a situation to which he must conform*; they entice him with something he is certain to take, and with lures of ostensible profit they await him in strength.[84] [Emphasis supplied.]

Passage #6.16. Asserting Situation Control: Elaboration of Concept

(One path to situation control is mapped by verses XI.25 and XI.26, whose focus is on disrupting the enemy's coordination, in fact on multiple levels. A further path appears in verse XI.28, which resembles the Passage #6.15 concept but is more specific. By pointing so emphatically to speed, verse XI.29 suggests speed as a basic tool of situation control, which connects with Theme #3 ideas.)

XI.25. Anciently, those described as skilled in war made it impossible for the enemy to unite his van and his rear; for his elements both large and small to mutually co-operate; for the good troops to succour the poor and for superiors and subordinates to support each other.

XI.26. When the enemy's forces were dispersed they prevented him from assembling them; when concentrated, they threw him into confusion.

XI.27. They concentrated and moved when it was advantageous to do so; when not advantageous, they halted.

XI.28. Should one ask: "How do I cope with a well-ordered enemy host about to attack me?" I reply: *"Seize something he cherishes and he will conform to your desires."* [Emphasis supplied.]

XI.29. Speed is the essence of war. Take advantage of the enemy's unpreparedness; travel by unexpected routes and strike him where he has taken no precautions.

Passage #6.17. Asserting Situation Control: Grand Strategy Illustration

(Explicit focus here is on controlling the behavior of leaders of other states or political entities – Sun Tzu's *zhu hou* terminology is commonly translated as the "feudal lords" – though the idea could be applied to other targets. Griffith's translation of verse VIII.14, with its focus on doing injury and thereby intimidating an audience, is consistent with the temper of Sun Tzu's times. Verse VIII.15 is actually more interesting, with its overtones of keeping a foe or rival constantly burdened with extraneous tasks or rushing around in quest of perceived advantages, ever off-balance, ceaselessly and futilely playing catch-up.[85])

---

[84] Mair, p. 93 translates verse V.20's last part as ". . . manipulates the enemy with empty advantages, while his main force lies in wait."

[85] Griffith's "makes them rush about [qu]" can also be translated as "makes them come [to the place where you want them to come]." That concept underscores a situation control idea yet more plainly.

VIII.14. He who intimidates his neighbours does so by inflicting injury upon them.

VIII.15. He wearies them by keeping them constantly occupied, and makes them rush about by offering them ostensible advantages.[86]

Passage #6.18. Asserting Situation Control: Sun Tzu's "First Symphony"

(If the situation control motifs in the preceding passages are likened to single notes in music, Passage #6.18 is the symphony. It is one of the most coherent extended expressions of the rhythms of Sun Tzu's way of war found anywhere in the text, albeit geared to a level of military operations rather than being fully strategic. Verses VI.1–VI.18 are deliberately reproduced here without elisions in order to underscore their fundamental unity as expressing a certain operational style.[87] Major harmonics of that style include achieving information superiority; disrupting the enemy's poise; and perhaps most fundamentally, exerting control over a foe by controlling the wellsprings of that foe's conduct.[88])

VI.1. Generally, he who occupies the field of battle first and awaits his enemy is at ease; he who comes later to the scene and rushes into the fight is weary.[89]

---

[86] Griffith's "ostensible" has no counterpart in the Chinese text. Some commentators and translators (e.g., Griffith, Giles) interpret the end of verse VIII.15 in a way that suggests deception. Others avoid introducing a deception motif, instead playing up use of a profit incentive as a means of control (e.g., Mair, p. 106: "that which causes the feudal lords to give allegiance is advantage"). That ambivalence may be the insight: no brightline distinguishes control by deception (Theme #7) from situation control by other means. Dangling actual advantages (but very small ones!) may actually work better as a means of control than advantages that are entirely illusory. A similar issue arises in interpreting verse V.20 (Passage #6.15). Cf. #6 footnote 84.

Such blurring of boundaries finds a modern market-oriented expression in work of Arthur Leff, *Swindling and Selling* (New York: Free Press, 1976), whose analysis points to a sometimes gossamer line between those respective activities, at least when selling is one-on-one (i.e., not mass selling).

[87] Additional verses of Sun Tzu VI appear in Passage #8.1.

[88] This strand of Sun Tzu merits comparison with ideas in Timothy L. Thomas, "Russia's reflexive control theory and the military," *Journal of Slavic Military Studies*, 2004, 17(2), 237–56, describing reflexive control (p. 237) as "conveying to a partner or opponent specially prepared information to incline him voluntarily to make the predetermined decision desired by the initiator of the action."

[89] Conditions of war in Sun Tzu's age may have made verse VI.1 an obvious choice to lead off Sun Tzu VI. In Western military tradition, work of a French soldier (Jean V de Beuil, 1405–1478), based on his experiences in the Hundred Years' War, states with Sun-Tzu-esque certitude: "*Everywhere and on all occasions that foot-soldiers march against their enemy face to face, those who march lose and those who remain standing still and holding firm win.*" (Quoted in Philippe Contamine, *War in the Middle Ages* [Michael Jones, trans.; New York: B. Blackwell, 1984], p. 231, emphasis supplied.) A nineteenth-century analysis concurs: at least in ancient battles between troops of similar morale, "the least fatigued always won." See Ardant du Picq, *Battle Studies; Ancient and Modern Battle* (trans. from 8th French edition by Colonel John N. Greely & Major Robert C. Cotton; New York: Macmillan, 1921), p. 52.

For further background on Jean de Beuil's life and work see Matthieu E. Chan Tsin, "Jean de Beuil: Reactionary Knight," Ph.D. dissertation, Purdue University, 2005.

VI.2. And therefore those skilled in war bring the enemy to the field of battle and are not brought there by him.

VI.3. One able to make the enemy come of his own accord does so by offering him some advantage. And one able to prevent him coming does so by hurting him. [Emphases supplied.]

VI.4. When the enemy is at ease, be able to weary him; when well fed, to starve him; when at rest, to make him move.[90]

VI.5. Appear at places to which he must hasten;[91] move swiftly where he does not expect you.

VI.6. That you may march a thousand li without wearying yourself is because you travel where there is no enemy.

VI.7. To be certain to take what you attack is to attack a place the enemy does not protect. To be certain to hold what you defend is to defend a place the enemy does not attack.[92]

VI.8. Therefore, against those skilled in attack, an enemy does not know where to defend; against the experts in defence, the enemy does not know where to attack.

VI.9. Subtle and insubstantial, the expert leaves no trace [wu xing]; divinely mysterious, he is inaudible. Thus he is master of his enemy's fate.

VI.10. He whose advance is irresistible plunges into his enemy's weak positions; he who in withdrawal cannot be pursued moves so swiftly that he cannot be overtaken.

VI.11. When I wish to give battle, my enemy, even though protected by high walls and deep moats, cannot help but engage me, for I attack a position he must succour.[93]

VI.12. When I wish to avoid battle I may defend myself simply by drawing a line on the ground; the enemy will be unable to attack me because I divert him from going where he wishes.[94] [Emphases supplied.]

VI.13. If I am able to determine the enemy's dispositions [xing] while at the same time I conceal my own [wu xing] then I can concentrate and he must divide. And if I concentrate while he divides, I can use my entire strength to attack a fraction of his. There, I will be numerically superior. Then, if

---

[90] Ames, p. 123 links verses VI.4 and VI.5 in a causal way that reinforces a situation control idea: "Thus being able to wear down a well-rested enemy, to starve one that is well-provisioned, and to move one that is settled, lies in going by way of places where the enemy must hasten in defense."

[91] Here is an unusual place in Sun Tzu where a textual variant leads to a reversal of the basic military advice. For analysis, which supports Griffith's understanding, see Appendix 6 anchor footnote 12.

[92] A similar textual issue also arises in the second sentence of verse VI.7. See Appendix 6 footnote 14.

[93] Situation control thinking is particularly vivid in this verse and the following one.

[94] Giles, p. 46 renders the last clause: "All we need to do is to throw something odd and unaccountable in his way." That understanding builds on traditional commentator Li Quan's "we puzzle him by strange and unusual dispositions" and also on Du Mu's roundup of illustrative anecdotes from Chinese military history (one being Zhuge Liang's famous empty city ruse, see pp. 249–50 below).

Sun Tzu's psychological emphases are on full display here.

I am able to use many to strike few at the selected point, those I deal with will be in dire straits.[95]

VI.14. The enemy must not know where I intend to give battle. For if he does not know where I intend to give battle he must prepare in a great many places. And when he prepares in a great many places, those I have to fight in any one place will be few.

VI.15. For if he prepares to the front his rear will be weak, and if to the rear, his front will be fragile. If he prepares to the left, his right will be vulnerable and if to the right, there will be few on his left. And when he prepares everywhere he will be weak everywhere.

VI.16. (Translation from Lau, 1965 article, p. 330.[96]) "It is the one who has to prepare against his enemy who is few and the one who makes his enemy prepare against him who is many."

VI.17. If one knows where and when a battle will be fought his troops can march a thousand li and meet on the field. But if one knows neither the battleground nor the day of battle, the left will be unable to aid the right, or the right, the left; the van to support the rear, or the rear, the van. How much more is this so when separated by several tens of li, or, indeed, by even a few!

VI.18. Although I estimate the troops of Yue as many, of what benefit is this superiority in respect to the outcome?

VI.19. Thus I say that victory can be created. For even if the enemy is numerous, I can prevent him from engaging.[97]

Passage #6.19. Asserting Situation Control: Intelligence-Oriented Version

(By achieving access to the highest enemy circles – cf. verse XIII.22 – verse XIII.23's highly intelligent secret agent may be a three-in-one asset, able to achieve results in counter-strategist, counter-plan, and situation control modes alike.)

XIII.22. Of old, the rise of Yin [Shang dynasty] was due to Yi Zhi, who formerly served the Xia [dynasty]; the Zhou [dynasty] came to power through Lü Ya, a servant of the Yin.

XIII.23. And therefore only the enlightened sovereign and the worthy general who are able to use the most intelligent people as agents are certain to achieve great things. Secret operations are essential in war; upon them the army relies to make its every move.

★★★★★★

---

[95] In this and the following verses the core military idea is one of "defeating the enemy in detail."

[96] Rationale for using Lau's translation rather than Griffith's is given in Appendix 3 footnote 9.

[97] Cf. traditional commentator Jia Lin: "The enemy may be many in number, but if he does not know our military circumstances, *we can always keep him busy with his own preparations*, so that he has no time to plan for an engagement." (Minford, p. 190, emphasis supplied) That conveys a situation control idea, for cognitive, administrative, or logistics reasons (or all three together!).

Strategist Should Enact Stratagems and Formlessness

There are two Sun Tzu verses which, by Sun Tzu's own affirmations, may be seen as summations of the active ingredient of his way of war. One is Theme #6's centerpiece verse III.4 (Passage #6.1):

III.4.
Thus, what is of supreme importance in war is to attack the enemy's strategy.

The second, expressing one of the most famous ideas in all world military thought, is:[1]

I.17.
All warfare is based on deception.

Both concepts engage with the all-important question of how to create and shape the advantageous situations for which Theme #5 calls. Their substance is related, but not identical, a point that repays further exploration.

It is not difficult to envision attacks on an enemy's strategy that do not automatically involve deception. Among them might be some counter-strategist attacks; attacks on logistics (see Passage #6.13); or (ranging beyond Sun Tzu's explicit foci) economic warfare or, as a modern offshoot, economic sanctions. Some forms of algorithm sabotage (see pp. 220–23 above) open opportunities for reshaping an adversary or rival's (or ally's!) behavior having little of the ambiance of deception, at least as commonly understood.

Probing the converse possibility – i.e., deception that does not amount to an attack on the enemy's strategy – is more challenging, partly because (as noted at the outset of Theme #6 analysis) Sun Tzu's concept of "attacking the enemy's strategy" is itself not easy to pin down in precise terms. It would, of course, be possible to define such an attack broadly, as a game theorist might be inclined to do, so as always to subsume deception, hence making the issue moot. But invoking definitional fiat forecloses lines of analysis that could shed useful light on Sun Tzu's thinking.

---

[1] This idea is actually endorsed twice in the Sun Tzu text – once in verse I.17 (Passage #7.1), once in verse VII.12 (Passage #7.9). The Chinese character used to signify deception is *gui* in verse I.17, *zha* in verse VII.12, but no material difference in substantive thrust should be inferred. Some readings of verse XI.56 (Passage #7.18A), including Griffith's, make it a candidate for yet a third endorsement.

One way of tackling the question starts by noting that an attack on the enemy's strategy presumes the *existence* of that strategy as an object of attack. Game theory's ecumenical concept of "strategy" comes close to asserting this condition universally, elevating any course of action to the dignity of constituting a "strategy." Yet (partly because game theory tends to overlook bureaucracy at war, partly because it commonly ignores the high-quality focused attention needed to devise coherent strategy) it may at times be more accurate to describe an actor's conduct as indicative of an *absence* of strategy. A hint of such possibility appears in an observation of traditional commentator Du You (Minford, p. 135): "The highest form of warfare is to overcome the enemy while he is still making his plans [mou]." In some cases too, a so-called "strategy" amounts to no more than a haphazard series of improvisations, diluting the concept so drastically that the purported "strategy" is one in name only.[2] The annals of world political–military history are rife with examples, some born of seesaw factional infighting that never arrives at usable consensus (a possibility to which Passage #7.1's verse I.25 could be read as pointing).

Yet even if an adversary lacks a coherent strategy (possibly because a previous attack on strategy – or on its strategist author – has been successful!), major opportunities for deception often exist. Indeed a rudderless foe may be a prime candidate for falling prey to deceptions of many sorts, because prone to grasp compulsively at shards of perceived gain or for sources of coherence where none exist.[3]

Meanwhile the scopes of application, and also the areas of overlap, of verses III.4 and I.17 wax and wane with changes in warfare itself. To draw on Clausewitz's illustrious "what kind of war" question,[4] in some kinds of wars the enemy's strategy may unravel straight away once a key strategist is removed. But in other kinds of wars a vacancy chain dynamic may operate to replace, with little or no disruption, a strategist thus eliminated by one of similar views and capabilities.[5] Then a decapitation move may have psychological or morale effects, but the enemy's strategy is likely to survive unscathed. To take a different example, some kinds of war depend crucially on complex logistics that are superb backdoor targets for attacking a foe's strategy; other kinds, whose logistics is simpler, may not be nearly so vulnerable.

On the deception side, some environments afford scope for ambitious, even deep, forms of deception, others far less so. Even in Warring States military texts a little later than Sun Tzu there is reason to believe that the role assigned to deception in warfare may have somewhat diminished.[6] In a context of late Napoleonic land warfare, Clausewitz famously (and controversially) dismissed, with limited exceptions, a major

---

[2] Cf. Hugh M. Arnold, "Official justifications for America's role in Indochina, 1949–67," *Asian Affairs: An American Review*, 1975, 3(1), 31–48 (itemizing no fewer than twenty-two separate rationales!).

[3] There is a hint of such a possibility in Sun Tzu verse VIII.15 (Passage #6.17): "He wearies them by keeping them [adversaries or rivals] constantly occupied, and *makes them rush about by offering them ostensible advantages*." (Emphasis supplied.) See also verse V.20 (Passage #7.7).

[4] Clausewitz, *On War*, Preliminaries footnote 1, pp. 88–89, calling for attention by the statesman and commander to "the kind of war on which they are embarking."

[5] For the vacancy chain concept see Harrison C. White, *Chains of Opportunity: system models of mobility in organizations* (Cambridge, MA: Harvard University Press, 1970).

[6] Commenting on emphases in the military text ascribed to Sun Bin (Warring States text later than Sun Tzu), Lau & Ames, p. 39 observe that

Although the [archaeologically recovered] Sun Bin text is fragmentary, when compared with the *Sunzi*, there seems to be a decrease in the emphasis upon deceit as a major factor in warfare.

role for "cunning" (deception, stratagem, etc.) in war.[7] In the twenty-first century, the ever-deepening intertwining of warfare and the cyber arena unleashes profound possibilities for deception unrivaled in any earlier historical era. War by algorithm stands to expand, simultaneously, the role of attacks on enemy algorithms (and by extension, attacks on the enemy strategy those algorithms support and may increasingly also dictate); and the role of deception in concealing the targets, nature, and often even the existence of such attacks. As the twenty-first century unfolds, verses III.4 and I.17 may accordingly be expected to stand in an ever closer symbiotic relationship.

The preceding discussion has concentrated on one flank of Sun Tzu's Theme #7 (deception) thinking, involving its positioning vis-à-vis Theme #6 ideas. A second flank of Theme #7 points ahead to Theme #8 (formlessness). Ralph Sawyer has suggested a succinct way of placing Themes #7 and #8 in clear perspective and relationship: "While unstated, there is an essential continuum that ranges from concrete deceptive methods that create false appearances to being completely formless."[8] Creating a false appearance, leading a foe to take a specific step based on a crisp but false image of reality, exemplifies deception ("create an uproar in the east and strike in the west").[9] A situation where a plethora of "realities" appear equally likely to that foe, perhaps prompting the sorts of unfocused, futile "catchup" efforts that Sun Tzu profiles so acutely (see Passage #8.1), exemplifies formlessness. The difference is between getting the enemy to bet wrong, decisively, versus presenting that enemy with a myriad of choices, just one of which happens to be correct (and that one may be continually mutating).

Despite verse I.17's fame, formlessness – provided that circumstances make it feasible – may be the deeper conceptual direction. A Daoistic perspective on Sun Tzu lends support to that position as do many parts of the Sun Tzu text (see Theme #8 passage selection beginning with Passage #8.1). Yet Theme #8 analysis suggests that a close approximation to formlessness may be realistically attainable only in certain forms of warfare, almost always remote from that of mass-infantry armies. Here are the makings of a paradox. The most perfect realization of Sun Tzu's way of war may actually exist only in contexts remote from the conventional land warfare of mass-infantry armies where Sun Tzu's core military identity and intellectual focus resided.

Therein lies one of Sun Tzu's great ironies.

★★★★★★

There is also less reliance on extraordinary [qi] operations. The formalization of warfare, both in Europe and China, made battle more open, having the effect of reducing in some degree dependence upon covert operations and surprise tactics.

However, they are also careful to note that this point can be overstated.

[7] See, e.g., Clausewitz, On War, Preliminaries footnote 1, pp. 202–3. For a modern perspective on Clausewitz's stance see Handel, Introduction footnote 17, pp. 224–25.

[8] Sawyer, Introduction footnote 55, pp. 59–60; also his p. 239. Sun Tzu himself makes no attempt to clarify a deception/formlessness distinction, by contrast to cases where the text takes affirmative steps to try to clarify other basic distinctions (e.g., verses I.16 [Passage #7.1], V.5 [Passage #7.4], V.19 [Passage #7.7], etc.). Such omission supports envisioning the deception/formlessness distinction in Sun Tzu as being one of degree, not of kind.

[9] Griffith, pp. 79–80, translating traditional commentator Zhang Yu on verse III.13 (Passage #1.32A).

THEME #7 "ALL WARFARE IS BASED ON DECEPTION"

centrality of deception in Sun Tzu's way of war:
Passage #7.1 (verses I.16–I.27) (Sun Tzu's way of deception)
Passage #7.2 (verses XIII.2 + XIII.14 + XIII.23) (ubiquity of deception in Sun Tzu's way of war: reliance on spies)
Passage #7.3 (verses XI.42–XI.43 + XI.45–XI.46 + XI.55) (license to deceive one's own people)

ecological vantage on deception:
Passage #7.4 (verses V.3–V.12) (deception combined with straightforward measures)
Passage #7.5 (verses IX.16-IX.22 + IX.35) (affinity between deception and complex terrain)
Passage #7.6 (verses X.4 + X.6) (affinity between deception and complex terrain, cont.)
Passage #7.7 (verse V.17–V.20) (more general forms of complexity and deception potential)
Passage #7.8 (verses III.18–III.23) (complex situations born of a ruler's ill-advised meddling, creating opportunities for deceptions fomented by third parties)

Sun Tzu's "symphonies" (Sun Tzu's "First Symphony" is verses VI.1–VI.19, see Passage #6.18):
Passage #7.9 (verses VII.2–VII.9 + VII.12–VII.16) (Sun Tzu's "Second Symphony": deception, calculation, and operational style)
Passage #7.10 (verses XIII.17–XIII.21) (Sun Tzu's "Third Symphony" or "Symphony of Spies": deception and calculation, espionage version)
Passage #7.11 (verse XI.61) (surprise switch of style, relevant to all Sun Tzu's symphonies)

counter-deception:
Passage #7.12 (verse VIII.16) (basic counter-deception mindset)
Passage #7.13 (verses VII.27–VII.29) (avoiding feigned retreats and other baits)
Passage #7.14 (verse IX.31) (avoiding Cannae scenario, Warring States China version)
Passage #7.15 (verse IX.44) (basic counter-deception reasoning tool: incongruity testing)
Passage #7.16 (verses IX.25–IX.28) (further counter-deception tips: puncturing deceptive language)
Passage #7.17 (verse XI.44) (countering disinformation on a population level)

******

Deception, in its many incarnations by both self and adversary, is the one of the fourteen themes with which modern audiences, certainly Western ones, most closely associate Sun Tzu's ideas.[1] That sentiment is, of course, summed up by Sun Tzu's "All warfare is based on deception" (verse I.17), a stark affirmation that is not only pithy but also happens to translate exceptionally well into English. This in turn sets a high bar all around – for practitioners, ancient and modern, who are expected to be ingenious enough to live up to Sun Tzu's perceived standard of craftiness; for commentators and analysts, who are expected to contribute their own worthy thoughts about deception; and, not least, for Sun Tzu himself, whose treatise is expected to harbor incisive ideas expanding on verse I.17's promising beginning.[2]

Against this backdrop of great expectations, challenges arise. Theme #7 is in many ways the most difficult of the fourteen themes to develop in a balanced way as well as to structure effectively. One reason for this is that Sun Tzu's core deception insight is embarrassingly simple. In four words: *Try deception. It works.* As Li Ling has pointed out, Sun Tzu's endorsement of deception in effect established the deception principle as a cardinal doctrinal principle of Chinese warfare.[3] To get a sense of just how important is that Sun Tzu insight, it is helpful to get outside Chinese tradition and to recognize precisely how unstable and often tenous has been the status of the deception principle in much modern warfare – in a nutshell, how difficult it has often been to get modern generals to think like Sun Tzu.[4]

One thing is certain: Without exception, deceptions explicitly spelled out in the Sun Tzu text are simple, even starkly so. Here Sun Tzu's proclivity for extremely compressed, sparse statements – however understandable in light of characteristics of the classical Chinese language and the physical constraints imposed by bulky bamboo strip books – creates real barriers to intellectual and textual understanding. More than once in the text one would like to know more about exactly what scenario or scenarios Sun Tzu had in mind and why deception would be expected to work so well there. What's more, Sun Tzu's comments unequivocally focused on deception are scattered, seldom allocated more than a verse or two in any single location – a marked contrast with the sustained development found in, say, the material on terrain or on net assessment or, for that matter, the dynamic level of situation

---

[1] See Scott A. Boorman, "Deception in Chinese strategy," pp. 313–37 in William W. Whitson (ed.), *The Military and Political Power in China in the 1970s* (New York: Praeger, 1972), reprinted in part under the title "Stratagem: the Chinese view in the Sun Tzu," pp. 203–206 in Hy Rothstein & Barton Whaley (eds.), *The Art and Science of Military Deception* (Boston, MA/London: Artech House, 2013). For further general background on deception topics, see Handel, Introduction footnote 17, pp. 423–24.

[2] Worth noting here is a tendency on the part of Sun Tzu's commentators and audiences, from early times down to the present day, to read Sun Tzu with deception in the forefront of their minds, seeing deception motifs at every turn (even when they are not explicit in what Sun Tzu says).

[3] That principle marked a cultural sea change between military ideals of the Spring and Autumn era, which frowned on ungentlemanly measures (as the famous case of Duke Xiang of Song, see #4 footnote 92, illustrates so well!), albeit with imperfect practical effect; and the Warring States, which moved toward wholehearted embrace of deception. See Li Ling, #1 footnote 84, pp. 66–68, a book whose title derives from Sun Tzu's affirmation of a deception principle in verse VII.12 (Passage #7.9).

[4] See Barton Whaley's massively documented *Stratagem: deception and surprise in war* (Cambridge, MA: Center for International Studies, Massachusetts Institute of Technology, 1969; reprinted by Artech House, Norwood, MA, 2007, with new introduction by Whaley on pp. xiii–xxi). *Page references cited in the present study are to the 2007 edition, which also corrects some typos in the 1969 work.*

control profiled in Passage #6.18. The espionage chapter is, of course, an exception – but Sun Tzu's treatment of that topic really stands alone, with little visible coupling to the military part of the text (in particular, how the fruits of successful espionage might be integrated with military deception efforts is left untouched).

The following effort to reconstruct a Sun Tzu theory of deception tackles the difficulties just described, but cannot eliminate them.[5] Perhaps more than with any other of the fourteen Sun Tzu themes, judgment needs to be exercised to avoid grafting onto Sun Tzu's sparsely presented deception thinking any of a wide range of ideas borrowed from other bodies of thought foreign to Sun Tzu.[6] Avoiding this pitfall calls for being alert to the probable Sun Tzu (1) context of Sun Tzu's observations relating to deception. That should not preclude venturing generalizations to other Sun Tzu (2) or (3) contexts, but basic facts like absence of paper as a medium of communications in Sun Tzu's time should always be kept in mind.[7] A complicating factor here is that deception is a military and intelligence topic where, in part because of the inherent nature of the subject matter, in part because devotees of deception tend to be practitioners, not theorists, terminology is often fluid and its application yet more so, rarely being fully consistent across users or settings.[8]

A basic first step in the present analysis is to situate Sun Tzu's deception thinking within Sun Tzu's larger analysis of war. Passage #7.1 containing verse I.17 is a major source of Sun Tzu ideas about deception, about which more will be said below. Unstated there, but also fundamental, is a deep-seated harmony between Themes #7

---

[5] Close reading of the Sun Tzu text does not turn up nearly as much material with an explicit deception focus as might be expected. Particularly noteworthy is absence of a Sun Tzu passage on military deception topics that might qualify as a natural "flagship" passage of the sort that exists for most of the other fourteen Sun Tzu themes. The closest contender is Passage #7.1, but only a minority of its verses point to deception directly. That deficit could be taken to mean that deception is less important to Sun Tzu than commonly averred; or (more likely) it could be taken as saying that deception is so deeply woven into the fabric of Sun Tzu's military thinking that no flagship passage is required.

[6] An example would be attempting to distinguish between "strategic" and "tactical" deception content in Sun Tzu. In some ways Sun Tzu's lack of a distinction between strategy and tactics (see pp. 47–48 above) is an intellectual advantage for Theme #7, encouraging spotting applications for his versatile deception dicta all the way from grand strategy to minor tactics. Erecting barriers to such exploration seems artificial. For supporting thoughts (not specific to Sun Tzu) see Whaley, #7 footnote 9, p. 149.

[7] It is worth noting that Sun Tzu makes no reference to secret communications methods of any sort.

[8] Even fundamental definitional questions, starting with what constitutes "deception," can go unresolved. A modern definition of deception comes from Whaley, #7 footnote 4, p. 82, defining it as an "act intended by its perpetrator to dupe or mislead a victim" (emphasis in original). "Surprise" refers to "instances where a sudden military action by one antagonist has not been predicted, much less anticipated, by its intended victim" (again p. 82). Deception thus pivots on its perpetrator's intent, surprise on its victim's perceptions. Note that these definitions of deception and surprise say nothing about features of the environment that may facilitate (or retard) deception or surprise efforts.

It should be cautioned that not all Sun Tzu material assigned to Theme #7 fits neatly under Whaley's deception rubric. Indeed there may be insight from identifying and reflecting on Sun Tzu ideas that don't quite fit there. See #6 footnote 86 relating to verse VIII.15 (and citing work of Arthur Leff). Roger Ames's observations on the translator's task (p. 176 above) are also relevant here.

(deception) and #3 (cheap measures).[9] A further fundamental is that, in Sun Tzu's way of war, deception is ubiquitous and its targets may include not only foes but also one's own people (Passages #7.2 and #7.3). Theme #10 (no scruples) enters here.

Even as belief in the efficacy of a counter-strategist mode of attack rests on certain assumptions (see pp. 209–10 above), so too Sun Tzu's faith in deception's efficacy likewise rests on a set of assumptions.

A first assumption is that unfettered political–military authority exists to conceive and implement deception plans without external editing or constraint. That first assumption, which meshes seamlessly with Sun Tzu's view of the general as the ultimate arbiter of all military choices, even at times overruling the ruler (see Theme #13), accords with the warfare of Sun Tzu's time.

A second assumption, implicit in the starring role Sun Tzu accords to deception, is that deception will work not just once or occasionally, but time and again, *even against a deception-minded antagonist*. Modern support for such a position comes from an insight that – in honor of its formulator, a pioneering student of deception across many realms – might be called Barton Whaley's (empirical) Theorem:[10]

> Indeed, this is a general finding of my [Whaley's] study – that is, the deceiver is almost always successful *regardless* of the sophistication of his victim in the same art. On the face of it, this seems an intolerable conclusion, one offending common sense. Yet it is the irrefutable conclusion of the historical evidence.

> [Emphasis in original.]

In Chinese warfare and strategy more broadly, an affinity for deception, while not universal, has certainly been a longstanding persistent harmonic.[11] That affinity is already in evidence from many episodes in the *Zuozhuan*, sometimes described as China's oldest narrative history, that chronicles events of the Spring and Autumn period (a time largely antedating the Sun Tzu text even when that text is assigned an early date).[12] It has spilled over into Chinese popular culture through the *Romance of*

---

[9] "By any purely economic accounting, deception can be relatively cheap – indeed ridiculously cheap." See p. 131 of Barton Whaley's "The one percent solution: costs & benefits of military deception," in John Arquilla & Douglas A. Borer (eds.), *Information Strategy and Warfare: a guide to theory and practice* (New York: Routledge, 2007), pp. 127–59.

[10] Whaley, #7 footnote 4, p. 76. One limitation of Whaley's quoted statement is that it supplies no explicit scope conditions. The empirical cases his study marshals focus on conventional warfare.

[11] Evaluation of deception's place in Chinese strategic thinking has been greatly complicated by effects of Confucian tradition, which (somewhat to oversimplify) found deception abhorrent. See Lisa Raphals, "Sunzi versus Xunzi: two views of deception and indirection," *Early China*, 2016, 39, 185–229, analyzing sharply contrasting views in Sun Tzu and the Xunzi text, the latter of which, as she puts it, "marks the beginning of a long history of Confucian criticism of the Sunzi" (Raphals, p. 185).

Focusing on a different major Chinese belief system, Lewis, Background footnote 11, pp. 124–25 notes that the Laozi Daoist classic could be read as asserting "that the true Way was in some sense a Way of deception, since it always appeared to be the opposite of the actual case." That observation has relevance to Sun Tzu, a text that has been characterized as Daoistic (see Mair, p. 47).

[12] See Zhang Wenru 張文儒, "Zuozhuan li de bingxue wenhua linian" 左傳裏的兵學文化理念 (The Military Expert School's cultural concepts in the Zuozhuan), in Binzhou xueyuan xuebao 滨州学院学报 (*Journal of Binzhou University*), 2009(1), pp. 9–15. See also Kierman, Background footnote 31, a study heavily based on the Zuozhuan, observing (p. 40) that "we find a good deal of covert action and deception" there. The salience of deception in the Zuozhuan should sound a note of caution about

the Three Kingdoms depictions of military practitioners who never tired of using deception on one another, often with great success, even though their victims were surely adepts in the same art.

A third basic assumption regarding deception is that the military (or other conflict) situations in which a Sun Tzu–inspired strategist operates will provide a suitable flow of raw materials for mounting successful deceptions. Here it is significant that Sun Tzu's treatments of deception typically come equipped with a larger context involving broader forces and factors that constitute the "environment" of deception and that in some cases (e.g., because of that environment's complexity) promote its success. *For that reason Sun Tzu's deception thinking is, in a broad sense, ecological.* Therein lies a key source of purchase for coming to analytical grips with Sun Tzu's relevant ideas and contributions.[13]

Adapting an imaginative concept of the Yale Law School's Arthur Leff, who wrote a seminal law essay titled simply "Law and,"[14] one might therefore envision Sun Tzu's theory of deception as pivoting, not on deception per se, but rather on

deception and

One way of completing this avowedly incomplete thought (and the need to do so is, of course, precisely the point) points to mixing and matching deceptions with strategic or tactical moves and measures involving active ingredients of other kinds:

deception and other operational measures

Reading Passage #7.1 from a standpoint that does not become too mesmerized by verse I.17, one finds Sun Tzu's deception thinking there interwoven with a variety of further strategic or tactical expedients that burst the bounds of even a broad understanding of deception – enough such ideas so that it is not even entirely clear that "deception" should be treated as Passage #7.1's dominant harmonic.[15] A basic reason why deception stands to be so effective is precisely that Sun Tzu intends for it to be directed against an adversary who, to varying degrees, is piqued and rattled

---

too simple a vision of Spring and Autumn period warfare as populated mainly by the likes of Duke Xiang of Song (#4 footnote 92) and similarly above-board honorable gentlemen.

[13] In the background of Sun Tzu's faith in efficacy of deception stands a further factor, possibly quite important and again of broadly ecological type (now understanding "ecology" in an information ecology sense). In a paperless world, the information environments in which Sun Tzu's generals and rulers operated would have had few available channels feeding information to decision makers, and for that reason might have been conducive to successful deceptions built on hijacking those few channels. Note verse XIII.16's attention to the *ye zhe* role (see p. 108 above). A modern analyst notes:

> The bias favoring a small amount of consistent information over a large body of less consistent data supports the common maxim in deception operations that the deceiver should control as many information channels as possible in order to reduce the amount of discrepant information available to the target. *Deception can be effective with even a small amount of information as long as the target does not receive credible contradictory data.*

(pp. 308–309 of Richards J. Heuer, Jr., "Strategic deception and counterdeception: a cognitive process approach," *International Studies Quarterly*, 1981, 25(2), 294–327; emphasis supplied)

[14] Arthur Allen Leff, "Law and," *Yale Law Journal*, 1978, 87(5), 989–1011, also available on the web at: http://digitalcommons.law.yale.edu/fss_papers/2817 (visited September 1, 2021).

[15] Handel, Introduction footnote 17, p. 217: "Sun Tzu's definition of deception is very broad indeed."

(verse I.22), self-absorbed and out-of-touch (verse I.23), stressed-to-the-limit (verse I.24), and internally conflicted (verse I.25). There is resonance here with Sun Tzu's alertness to the impact of emotions on military calculations (see Theme #1 analysis).

Further developing this perspective is Passage #7.4 which, read from a deception angle, highlights the importance of combining deception with other measures of a straightforward type (see Theme #14 analysis).[16] A predilection for this kind of "mixological" vantage on deception – mixing deception in with diverse other measures, from which it draws efficacy – finds many echoes in later eras of Chinese history. A pertinent spirit has rarely been better captured than by John Fairbank, describing a sixteenth-century Ming dynasty episode involving a Chinese general's use of

> the imperial prestige, offers of pardon, patronizing friendship, subornation of colleagues, poisoned wine, moral principles, false intelligence, procrastination, beautiful women, solemn and fair promises, bribery, banquets, threats, intimidation, lies, cajolery, assassination, and deployment of troops to undo his opponents.[17]

To get a visceral sense of the Sun-Tzu-esque spirit on display here, Fairbank's list can be put in perspective by juxtaposing it with two other lists, each of modern provenance and each having a somewhat similar thrust, though lacking the intricately mixological spirit – blending the crafty with the straightforward at every turn – found in Fairbank's description of his Ming dynasty Chinese case:

> (Lenin's version[18])
> We must be able to stand up to all this, agree to make any sacrifice, and even – if need be – to resort to various stratagems, artifices and illegal methods, to evasions and subterfuges ...
> ... shows very clearly ... the importance of *combining* legal and illegal struggle.
> (a modern US version[19])
> Espionage involves peeking at the other fellow's hand, marking the cards, cooking the books, poisoning the well, breaking the rules, hitting below the belt, cheating, lying, deceiving, defaming, snooping, eavesdropping, prying, stealing, bribing, suborning, burglarizing, forging, misleading, conducting dirty tricks, dirty pool, skulduggery, blackmail, seduction, everything not sporting, not kosher, not cricket.

But Sun Tzu's mixological propensity does not exhaust the full measure of how Sun Tzu's thinking embeds deception in a larger context. A further Sun Tzu focus, growing out of what some scholars identify as the earliest layers of the text, is:[20]

deception and complex terrain

A placeholder for this focus is Sun Tzu's notable attention to the concept of *xian*, which may refer to a type of topography – often translated as "ravines" – but may also convey more abstract ideas conveying a mix of danger and strategic value.[21]

---

[16] Although Passage #7.4's qi (translated in Lau's 1965 article as "crafty") does not always involve deception, in military context it frequently does. For relevant discussion of qi see p. 463 below.

[17] See p. 23 of John K. Fairbank, "Introduction. Varieties of the Chinese military experience," in Kierman & Fairbank, Background footnote 31, pp. 1–26, profiling steps by which the Ming dynasty general Hu Zongxian (1512–1565) sought to suppress seacoast marauders.

[18] "'Left-wing Communism: an infantile disorder," in V. I. Lenin, *Collected Works*, Vol. 31: April–December 1920 (Moscow: Progress Publishers, 1966), pp. 55, 61 (emphasis in original). See also pp. 45, 46 there.

[19] Joseph E. Persico, *Roosevelt's Secret War: FDR and World War II espionage* (New York: Random House, 2001), pp. 17–18.

[20] See Table 12A on p. A-35 (Brooks's proposed order of Sun Tzu chapter accretions).

[21] See earlier discussion of ravines in Theme #2 context, starting on p. 95 above. For a roundup of occurrences of *xian* in Sun Tzu see Appendix 2 anchor footnote 31.

A plausible case can be made that the evolution of Chinese warfare, moving emphasis away from the level ground needed for chariot fighting and in a direction of complex terrain like ravines, contributed in no small measure to inspiring Sun Tzu's deception focus. Troops can be hidden in ravines, facilitating ambush. They can also be trapped there by surprise appearance of a foe atop the ravine or by enemy steps to block its exits. While Sun Tzu should definitely not be considered an "environmentalist" in a modern sense, he might be described as "environmentally aware." Sun Tzu's many terrain verses (Theme #2) flesh out the meaning of that perspective.

On an elemental level, complex terrain commonly favors hiders over seekers (verse IX.18; in a different way possibly verse IX.20 too [Passage #7.5]), thereby favoring certain kinds of deception (ambush, snipers, clandestine observation).[22] Passage #7.5's verse IX.21 contemplates a situation where a layer of artificial complexity has been added to an already somewhat complex terrain situation, thereby aiming to throw the adversary off balance, not because of insufficient attention to deception possibilities but through a surfeit of it. Verse IX.21 cautions against falling into such a trap. Verse X.4 (Passage #7.6) points to another, in some ways more devious, complex terrain scenario whose concept may be diagrammed as follows:

> complex terrain → lure → enemy undertakes complex project → striking when project is half-done, at moment of greatest vulnerability

The more complex and fragile the project is, the better suited it is to be a trap. In the warfare of Sun Tzu's time, river crossings would be a further major source of examples (see verse IX.5 in Passage #14.7).

Importantly, each of these scenarios embodies a generalizable concept not confined to topography. Yet further ways in which complexity may be a handmaiden of deception also appear in verses V.17 and V.18:

> V.17.
> In the tumult and uproar the battle seems chaotic, but there is no disorder; the troops appear to be milling about in circles but cannot be defeated.
> V.18.
> Apparent confusion is a product of good order; apparent cowardice, of courage, apparent weakness, of strength.

(see Passage #7.7)

Shifting to a fully abstract plane, a further strand of deception thinking implicit in Sun Tzu takes cues from Passage #7.4's verse V.11:

> V.11.
> In battle there are only the normal [zheng] and extraordinary [qi] forces but their combinations are limitless; none can comprehend them all.

and proceeds to envision a combinatorial explosion of possible sequences of qi and zheng measures as itself a kind of complex "terrain," sharing with complex

---

[22] Some of the deception opportunities that even mildly complex terrain can afford are made vivid by a training photograph distributed by the British War Office c. 1916, purporting to show a sniper lurking in tall, weedy grass. So effective is the camouflage that a natural question arises: Is there a sniper at all? Or just an empty field (or, as it were, a "pretense of a pretense")? See reproduction of that photo as figure P.1, "Sniper in grass," facing p. 9 of Hanna Rose Shell, *Hide and Seek: camouflage, photography, and the media of reconnaissance* (New York: Zone Books, 2012) and pp. 9–11 there.

physical terrain vistas on deception possibilities unimaginable in simpler, sparser environments. Consider, for example, the challenge facing a strategic actor engaged in a struggle with a Sun Tzu–inspired foe that has already led to a lengthy series of enemy moves, some straightforward (zheng), some involving deception (qi).[23] Is a fresh move by that adversary a straightforward one (as it may initially appear to be) or is it deception? Or may it seem like deception but actually be straightforward? The more intricate the sequence of prior moves, the more taxing and psychologically disorienting to the human evaluator the task of making a correct judgment call here may be. There is simply too much "history," never helpful enough.

An affinity between complex terrain and deception also extends to "terrain" in senses closer to human social structures and processes:

deception and complex sociopolitical situations

The basic idea here is that complex individual psyches and, perhaps even more so, complex sociopolitical situations harbor rich deception potential.[24] Grist for that line of thinking comes from Passage #7.8's word picture of a meddling ruler, his conflicted military establishment, and his troublemaking neighbors:

III.23.

If the army is confused and suspicious, neighbouring rulers will cause trouble. ...

In an environment where everyone may already be lying, just a little, an ounce of false information fed into the situation (possibly building on a double agent, see Passage #7.10's verse XIII.19) may deliver a pound of ensuing distortion – the multiplier arising from deception victims' vigorous pursuit of their own designs:

XI.56.

Now the crux of military operations lies in the pretence of accommodating oneself to the designs of the enemy.

(Passage #7.18A)

Those victims have then become enthusiastic (if unwitting) accomplices in their own deception![25]

Although recognizing it calls for some sleuthing, since what Sun Tzu has to say is not couched in modern categories, a further Sun Tzu deception focus is:

deception and calculation

For Sun Tzu, calculation is a basic part of the "environment" of deception. Calculation facilitates deception.[26] It is the basis of counter-deception. It may also become itself an object of deception, as when one or the other side attacks

---

[23] As noted earlier (#7 footnote 16), qi need not involve deception (but often does). See p. 463 below.

[24] A sense of what might be achievable in a strategic environment populated by complex psyches is conveyed by Sawyer, Introduction footnote 55, p. 358 (discussing a Song dynasty military work):

The more competitive and complex the relationships among the parties – be they commanders, politicians, officials, or armies – the greater the possibility of exploiting the often deliberately induced confusion and then maneuvering in the chaos, contrary to the enemy's expectations.

[25] See also #6 footnote 53 (relation to algorithm sabotage ideas) and #6 footnote 86 (relation to concepts in verses VIII.15 and V.20). There is also common ground here with Theme #9.

[26] A supporting idea appears in verse XI.32 (Passage #8.3). See Appendix 8 footnote 12.

the adversary's calculations (a move that harks back to Theme #6 and counter-strategist mode).[27]

Illustrating calculation in service of deception, a famous Warring States ruse began by taking note of some of Passage #7.9's force attrition statistics (likely well known to generals of the time), then enacted dwindling numbers of cooking fires to mimic heavy attrition on successive days of advance into enemy territory. Seeing those dwindling fires, the enemy general fell into an inference trap not unknown to modern "quants": if your preferred model or story (here, that the enemy army is disintegrating) fits the observations, that story must be correct! Eager to exploit the opportunity, the enemy general ditched his slow-moving infantry and raced on ahead with lightly armed shock troops to crush the invaders. He was ambushed, his own army crushed. He himself committed suicide.[28]

Another type of calculation serving deception underpins (though is not explicit in) Passage #7.10, a part of Sun Tzu XIII that merits the title "Symphony of Spies."

Insight into Sun Tzu's deception-and-calculation thinking comes from a commentary observation of Cao Cao on verse I.17: "Warfare is without fixed form [xing]; it takes cunning and deceit [guizha] as its Way [dao]."[29] Amplifying Cao Cao's thought, water's lack of form – also a Sun Tzu focus (see Theme #9) – offers a view of deception that molds itself on characteristics of the deceived, and the highly customized deception opportunities that may then arise.

Such a water imagery for deception points in a quite different direction from widespread modern analyses of deception that seek to pin down deception in a fixed structural framework.[30] Pushing back against such aspirations, Cao Cao's line of thinking offers a "disruptive" conceptualization of deception – challenging many conventional ways of thinking about it – and is worth attention for that reason.

As a water imagery also underscores, however, calculations in service of deception are often error-prone: flow of water can be hard to predict! Certainly that was true in Sun Tzu's time, given the extremely limited supporting resources available (starting with lack of paper as a medium of communication and memory). This analytical direction points to a further harvest of deception opportunities, now targeting the calculations on which the foe's thinking rests (see Passage #7.1's verses I.22–I.25, though other parts of Passage #7.1 also stand to contribute ideas).

Sun Tzu's coverage of counter-deception is, candidly, something of a miscellany (Passages #7.12–#7.17, as well as parts of Passage #7.5). It is plainly geared around wartime struggles between opposing generals in the field, not complex long-running civilian competitions. It is concentrated in the earlier layers of the text as identified by

---

[27] Lest modern expectations intrude, it is worth reiterating the p. 57 caveat that calculation in Sun Tzu is best envisioned in modest terms, aligning more closely on commonsense notions of "thinking the situation through carefully" than on any elaborate supporting machinery, human or mechanical.

[28] This was the battle of Maling (c. 341 BC), whose description appears in the Shiji 65 biography of Sun Bin (said to be a descendant of Sun Tzu) that directly follows Sun Tzu's biography. For further specifics on Sun Bin and Maling see Lau & Ames, pp. 4ff.

[29] Passage #7.19A (verses VI.27–VI.29) provides further background for Cao Cao's observation. The translation quoted here is from M. A. Butler's work cited in Appendix 7 footnote 25.

[30] One such framework (clearly geared to be an example only) comes from Lau's 1965 article, pp. 330–31 (quoted on pp. A-236f.). Its gist is indicated on p. 463 below. Game theory provides many more structural frameworks for analyzing deception, some far more elaborate than Lau's.

Brooks (see Table 12A on p. A-35) and perhaps for that reason tends to be concrete in focus. With the possible exception of verse VIII.16 for other reasons (discussed under Theme #6, see pp. 212–14 above), the most interesting of Sun Tzu's counter-deception comments is verse IX.44 (Passage #7.15), pointing to the need for timely analysis to avoid falling prey to deception:

IX.44.

When the enemy troops are in high spirits, and, although facing you, do not join battle for a long time, nor leave, you must thoroughly investigate the situation.

In modern categories verse IX.44 is suggestive of incongruity testing (albeit focused on one specific situation), evoking Whaley's observation that "a sure sign of a competent deception analyst is one who systematically applies incongruity analysis."[31]

Passage #7.16, where a central harmonic is "words not matching actions" (again, a type of incongruity), involves elementary, in some ways naive, yet possibly useful communications analysis. Like much of Sun Tzu's terrain thinking it is expressed in a set of "if ... then ..." rules. The simplicity of those rules could open a door for missteps born of enemy manipulation (hence falling into a deception trap after all, just not the one Sun Tzu's stated rule aims to avoid!), though the basic idea of harnessing communications analysis for counter-deception purposes is sound and durable.

A fifth broader context for Sun Tzu's deception thinking harks back to a concept of style (see Theme #5) and is textually anchored by Passage #7.9's verse VII.13 setting forth a menu of six operational "styles":

deception and operational style switches

As previously noted in discussing verse VII.13 (see p. 182 above), multiplicity of styles entails a need to switch styles, perhaps often, as changing strategic or tactical situations warrant (read from that angle, Passage #7.11's verse XI.61 is a case in point). Any major style switch, certainly one involving a large army, is never entirely seamless and the general who undertakes it must be alert to errors of both under-correction (e.g., spillovers, physical or psychological, from a previous style's "momentum") and over-correction (often born of impulsive neglect of a new style's new logistics requirements).[32] Errors of both kinds figure in Passage #7.9. Verse VII.5 warns of lugging too much paraphernalia along (for example, in verse VII.13 categories, in transitioning from "majestic march" to what is supposed to be a lightning raid "like fire"). In counterpoint, verses VII.6 and VII.9 warn of the ill consequences of rash decisions to abandon stores and equipment, as (for example) when an army switches too precipitously from standing "firm as a mountain" to moving "like a thunderbolt" (see again verse VII.13).

---

[31] See p. 399 of Barton Whaley, "Deception planning in 145 different disciplines: lessons from behind other hills," pp. 397–412 in Rothstein & Whaley, #7 footnote 1, p. 399.

Presence of verse IX.44's concept – read as a fragment of systematic counter-deception doctrine – in a text as early as Sun Tzu may be regarded as a significant achievement on Sun Tzu's part.

[32] Switches of style commonly bring with them new logistics constraints, inadequate planning for which is one way major errors can occur. See Boorman, #5 footnote 24, p. 188.

Deception cannot eliminate all pitfalls of style switches. Yet it can often mask and thereby mitigate them. It can also help induce style switch errors by the enemy (the losing general's error that culminated in the Maling battle defeat [p. 246 above] might be interpreted in this way). Therein lies a possible military rationale for why verse VII.12 (deception) rubs shoulders with verse VII.13 (styles) in Passage #7.9.

Roads Not Taken by Sun Tzu in Developing Theme #7

With the possible (but undeveloped) exception of the "false information" of verse XIII.19 (Passage #7.10), concrete examples of deception provided by Sun Tzu are all elementary as well as sparsely described – the ambush, the sniper or other operative lurking in rough terrain, the feigned retreat, deception as to location, and a few more. There is not even a passing mention of more imaginative ruses like creating doppelgängers of the commanding general.

Perhaps relatedly, one striking omission in Sun Tzu's treatment of deception is that Sun Tzu – who undertakes other classificatory exercises (e.g., types of terrain or types of spy) – makes no effort at all, even a rudimentary one, to catalog military ruses or stratagems (which is what Sun Tzu's audiences might expect to find in a world classic "deception manual"). Yet absence of a typology of stratagems makes sense if Sun Tzu's interest aligns more closely on spotting situations conducive to deception than on specific scenario development for the deceptions they enable (a task to which Sun Tzu's ever-sparse expository style is not well suited).

A further, arguably even more striking, omission is lack of any attention on Sun Tzu's part to multiple subterfuges, either simultaneous or sequential.[33] It might, of course, be argued that someone as deeply invested in deception as is Sun Tzu would take use of multiple, possibly interlocking, deceptions entirely for granted, hence might not see any particular reason to mention them. Yet there are certainly issues here that merit discussion – for example, how to ensure that overly many, or overly intricate, deceptions don't undermine each other (as suggested by the famous modern saying "Oh what a tangled web we weave, when first we practice to deceive").[34]

Continuing with an Arthur Leff–inspired framing (p. 242 above), a menu of other Theme #7 roads-not-taken is developed here using a "deception and" format.

---

[33] Some of the traditional commentators gesture in this direction (e.g., Meng shi on verse XI.26 [Passage #6.16], see Mair, p. 155 note 8; Wang Xi on verse I.24 [Passage #7.1], see Minford, p. 115), but what they offer falls far short of indicating how a full-fledged military deception campaign involving multiple coordinated deceptions might be put together. Verses V.20 (Passage #7.7) and VII.3 (Passage #7.9) might be read as pointing to multiple deceptions, but neither develops this direction.

    Sun Tzu's closest approach to sketching a "campaign" level of deception is Passage #7.10 from the espionage chapter.

[34] A possible hint of awareness of this issue appears in a statement in the Sun Bin military text, an archaeologically recovered text that is later than Sun Tzu (Mair, p. 12 assigns its "coagulated" form to early Han times). As Lau & Ames, p. 175 translate it: "an excess of surprise operations, however, will overshoot the mark of victory." But see their pp. 234–35 note 377 (noting interpretive ambiguities).

deception and higher-order beliefs

Perhaps contrary to some modern reader expectations, Sun Tzu shows no apparent interest in stratagems aiming to capitalize on the opportunities and pitfalls presented by "he thinks I think . . ." and related logic chains in strategic reasoning.[35] Indeed one reading of verse VIII.16 (Passage #7.12) suggests that Sun Tzu would have rejected them out of hand (see pp. 212–14 above). As a shrewd soldier, Sun Tzu perhaps instinctively did not trust the reliability of strategic or tactical moves based on the ramifications of such reasoning, seductive as they may be. Yet there are also places where Sun Tzu might do well to take cognizance of vulnerabilities of his own advice. For example, Passage #7.16's "cookbook" rules (whose focus is on interpreting verbal communications from the enemy) risk falling into elementary traps if an alert foe has read Sun Tzu's playbook and exploits naive reliance on it to pull off deceptions.[36] Comparable pitfalls lurk in various other verses in Sun Tzu IX (see Table 6, pp. 407ff.), certainly in their originally intended military intelligence applications (see Passage #11.26A). Even a sympathetic reading of Sun Tzu raises warning flags as to the soundness of parts of his advice here. Because sophisticated deception is only to be expected in double agent situations, even more pitfalls of "he thinks I think . . ." type loom in the background of Sun Tzu's double agents emphasis (Passage #7.10). Of course, here as in many other places in the text, Sun Tzu merits kudos as a pioneer – but even famous pioneers have blind spots.

A further deficit in Sun Tzu's deception thinking involves lack of attention to the crucial "how" of successful deception or, put in a slightly different way,

deception and its implementation and logistics

Implementation of deception shares much common ground with challenges of putting on a successful performance – deception's Stanislavski level.[37] That level is epitomized in Chinese military lore and legend by the "empty city" ruse of Zhuge Liang (*fl.* early third century AD), in which the great strategist (with pitifully few supporting troops) was able to back down an enemy host while visibly sitting with but a few retainers nearby an open gate of an apparently undefended city, burning

---

[35] D. C. Lau's interpretation of the *qi/zheng* distinction (quoted on pp. A-236f.) takes some steps in that direction, but this is an interpretation of Sun Tzu, not anything Sun Tzu says directly. Importantly, there is nothing in either Sun Tzu or Lau's interpretation that engages with a question posed by Morris F. Friedell, "On the structure of shared awareness," *Behavioral Science*, 1969, 14(1), 28–39:

> we have seen that it can be quite profitable for an actor to induce another to believe a polynomial proposition of high degree [reference here is to higher-order beliefs and related propositions], but this may require much adroitness. . . . The general question arises: Is there a structure of props necessary or sufficient to induce a reasonable man to believe a given such proposition, and, if so, what is it? Findings on this topic might perhaps be directly useful only to confidence men or espionage agencies, but valuable sidelights might be shed on the geometry and furniture of the social world.

(p. 39)

[36] Compare ideas of Song military thinker Xu Dong (*fl. c.* 1000 AD) on "Contrary Employment of Ancient Methods" (from Xu Dong's military treatise noted in Introduction footnote 55).

[37] See Constantin Stanislavski, *An Actor Prepares* (Elizabeth Reynolds Hapgood, trans.; introduction by John Gielgud; New York: Theatre Arts Books, 1948).

incense while playing a zither.[38] As this case suggests, putting on a successful deception performance is facilitated by props, tangible or intangible.[39] Yet such logistics of deception gets scant attention from Sun Tzu, save for a glancing mention in Passage #7.5's verse IX.21 alluding to "blinds in the bushes" (translation by Ames, p. 142) and possibly one other case.[40] There is nothing at all in Sun Tzu about distinctively clever props, such as the childcare items deliberately left behind at a campsite by a reconnaissance detachment of the great Mongol general Subutai's forces, as a ruse to make the enemy think that the Mongols were accompanied by wives and children, hence were fugitives – thereby inducing fatal overconfidence on the part of the foe leading to a smashing defeat of the enemy.[41]

In view of the crucial importance Sun Tzu unequivocally accords to deception, also notable – and again involving deception implementation issues – is Sun Tzu's apparently unanalyzed faith that a deceptive message or step will be interpreted by the enemy in the way the deceiver intends.[42] Beyond this, many an attempted deception operation in history has failed to deceive because of its telltale spoor.[43]

---

[38] *Romance of the Three Kingdoms* (Moss Roberts translation cited on p. xxiv above), chapter 95, pp. 734–35. The story is noted by traditional Sun Tzu commentator Du Mu. See Griffith, pp. 97–98.

[39] See Kierman, Background footnote 31, p. 30 (provision of special implements for particular stratagems). The "empty city ruse" has been given a game-theoretic interpretation by Christopher Cotton and Chang Liu, "100 horsemen and the empty city: a game theoretic examination of deception in Chinese military legend," *Journal of Peace Research*, 2011, 48(2), 217–23. A key to their model is λ, defined as the ex ante probability that Zhuge Liang (who might be either weak or strong in his available military assets) is in fact strong. A basic reason for treating that parameter as large (in which case their game theory analysis points to Zhuge Liang's always staying in the city, as a bluff if he is weak, and the enemy's retreating) is his "reputation as a masterful strategist" (p. 222). Their analysis does not go further to probe where λ comes from (a Goffmanesque analysis task where received game theory is of limited help). λ amounts to an intangible prop, potentiated by tangible ones (incense, zither, etc.).

[40] See verse IX.20 (Passage #7.5), "When the trees are seen to move the enemy is advancing," which might suggest camouflage measures. However, traditional commentators Cao Cao and Zhang Yu interpret otherwise, e.g., focusing on a path being cut for the enemy army (Minford, p. 237).

[41] For the story see p. 53 of Stephen Pow and Jingjing Liao, "Subutai: sorting fact from fiction surrounding the Mongol empire's greatest general (with translations of Subutai's two biographies in the *Yuan shi*)," *Journal of Chinese Military History*, 2018, 7(1), 37–76.

[42] See Heuer, #7 footnote 13, p. 300: "Planning and implementing a deception typically involves a major investment of time, energy, and ego. When people make such an investment in preparing a message, they tend to overestimate how clear this message will be to the receiver."

[43] A fine example of unraveling comes from an unstable period of Tang dynasty history. The basic concept of the "Sweet Dew plot of 835" engineered by one court faction was to "lure the eunuchs [a different faction] away from the protection of the Shen ce Army [*shen ce jun*, an elite unit responsible for protecting the emperor] and then to annihilate them quickly, a strategy that remained a well-kept secret until the last minute." But then things rapidly went wrong:

> Just as they [Qiu Shiliang and a party of eunuchs] reached the courtyard where the trap was to be sprung, however, a gust of wind blew aside a flap of the tent in which Li Xun's armed men were hiding. The clank of their weapons alerted the eunuchs, most of whom were able to rush panic-stricken back into the inner palace before the gates were closed against them. Inside they forced Wenzong [the Tang emperor] to return to the harem, beyond the reach of Li Xun and the other government officials. Qiu Shiliang and the other eunuchs immediately summoned the aid of the dreaded Shen ce Army, whose detachments were sent to massacre suspected courtiers in the official precincts.

Sun Tzu shows some cognizance of deception failure possibilities (e.g., verse IX.22 in Passage #7.5, "Birds rising in flight is a sign the enemy is lying in ambush"). Yet the closest Sun Tzu comes to giving advice about how to recover from a failed deception effort is the crude (and possibly counter-productive) prescription of verse XIII.15:[44]

> XIII.15.
> If plans relating to secret operations are prematurely divulged the agent and all those to whom he spoke of them shall be put to death.

(Passage #8.8, also Passage #11.6)

In sum, Sun Tzu is not at all a technician of deception. He is a broad strategic thinker who approaches deception from an essentially philosophical vantage, as a tributary to larger military or strategic art.

A further Sun Tzu omission – consistent with Sun Tzu's lack of attention to alliance politics noted under Theme #1 (p. 66 above) – is absence of exploration of:

deception and ambiguous identity ("what side am I on?" question)

Illustrating this issue are situations where interdependent actors – including ones nominally on the same side – betray, or at least seriously contemplate betraying, one another, in a setting of multiple or ambiguous or fluid loyalties.[45] This might be labeled the William Empson level of deception, the sort of murky situation evocatively described by Martin Shubik's image of "games within the game."[46] Here is precisely the kind of arena where one would expect insights from Sun Tzu about deception to cluster densely. On the contrary, Sun Tzu delves not at all into the complex machinations that situations of this sort may spawn. Even Sun Tzu's double agent material (pp. 368–71 below) has limited use here, because Sun Tzu never engages with conflicted motivations of double agents and implications for credibility.

See p. 657 of Michael Dalby, "Court politics in late T'ang times," in Denis Twitchett (ed.), *The Cambridge History of China*, Vol. 3: *Sui and T'ang China, 589–906, Part I* (Cambridge, UK: Cambridge University Press, 1979), pp. 561–681.

[44] Making a case that verse XIII.15's advice could easily prove counter-productive see #11 footnote 73.

[45] An excellent example of the sort of content *not* found in Sun Tzu are the machinations of the mobile persuader/diplomats of whom Lewis, Background footnote 7, p. 633 writes:

> Awarded the office of minister by one ruler, they proceeded to other states to secure alliances that would be sealed by the granting of another ministerial seal. Thus, several independent states came to share a single minister, whose primary loyalty to any state remained questionable. The fluidity of loyalties was intensified as persuaders dismissed from one state sought office in another, only to return to the state that had expelled them with offers of new alliances that they would secure if restored to their former position.

These muddy practices are epitomized by the career of Su Qin (died 284 BC) – "said to have simultaneously held the ministerial seals of six states" – who on his deathbed "by making a true confession of his treason disguised as a clever lie, manipulated his mortal enemy into avenging his death" (!). See Lewis, Background footnote 7, pp. 633–34. Su Qin's name appears in the Han strips version of verse XIII.22 (Passage #5.19), though not in the traditionally transmitted text of Sun Tzu. It seems likely to be a later interpolation there. See Background footnote 90.

[46] See Martin Shubik, *A Game-Theoretic Approach to Political Economy* (Volume 2 of Martin Shubik, *Game Theory in the Social Sciences*; Cambridge, MA: MIT Press, 1984), pp. 643–53.

For citation to Empson's *Seven Types of Ambiguity* see #2 footnote 45.

Here is an important case where the Sun Tzu of legend needs to be differentiated from Sun Tzu (1). Once again, Sun Tzu is found to be writing as a land warfare expert, not as a diplomat or political analyst and certainly not as a courtier.

Sun Tzu (2) ("Sun Tzu Extended") and Sun Tzu (3) ("Sun Tzu Analogical") Frontiers of Theme #7

It is easy to see a basis for Sun Tzu's towering military reputation in verse I.17's unadorned message, "All warfare is based on deception." At the heart of the case stands one of the most compelling logic chains in all military theory. Wars happen. They frequently get out of control, and wind up being exorbitantly costly, often for the "victor" as well. Properly conceived and executed – a challenge that is first and foremost one for a deceiver's imagination – deception can open a path to winning cheaply (see Theme #3, especially Passage #3.19, Sun Tzu's classic case for espionage as a cheap alternative to conventional warfare). Even if a deceiver is far weaker by conventional strength metrics, the military historical record suggests that deception, in some cases broadly conceived, has an impressive track record as an (often) very cost-effective ticket to military success.[47] Yet (and herein lies a paradox!) that vast military potential of deception has *also* commonly been underrated, marginalized, or ignored outright. Reasons for such a blind spot are diverse. They may include some kinds of warrior ethos; ethical barriers; bureaucratic inertia and infighting (often involving status competitions between regular military forces and their less orthodox rivals in the same military establishment, who are more likely to be open to uses of deception); simple failures of imagination; and more.

Offsetting those factors, Sun Tzu's advocacy for deception – its credibility rooted in Sun Tzu's unique reputation and influence, now enduring for some twenty-five centuries – can itself be a resource, galvanizing a strategic actor to embrace deception as a path to attaining large ends while sidestepping torrential outflows of blood and treasure. Yet there remains much reason to believe that, to this day, many otherwise highly informed and intendedly rational actors continue to underestimate deception's potential, as a tool both for advancing their own cause and for that of their foes or rivals. Herein is a powerful response to the probing question: *"What did Master Sun know that we still don't (or have yet to absorb adequately)?"*

In the twentieth century Liddell Hart, who as a young man had seen firsthand the tragically costly warfare on the Western Front in World War I, was a distinguished advocate of the value of deception in war, as (with somewhat different interests and emphases) was Barton Whaley. As such, both men contributed greatly to advancing Sun Tzu (2) and (especially in Whaley's case) Sun Tzu (3) intellectual agendas.

Of course, nothing in the applied human sciences is quite so simple and there is also need for attention to lack of scope conditions on verse I.17 (Passage #7.1). Sun Tzu's verse I.17 counsel is indeed sharp – but it can be a double-edged sword. *Once a*

---

[47] For a major assemblage of military-historical examples (not primarily East Asia–focused) whose analysis supports efficacy of military deception, see Whaley's *Stratagem* monograph (cited in #7 footnote 4). Whaley, #7 footnote 31, p. 412 provides a summary.

An important scope limitation on Whaley's case studies is his exclusion (*Stratagem*, p. 84) of "all cases at the extreme of the small-scale, local actions of the commando type, although many of these achieved surprise and were intended as diversionary raids or served political-strategic ends."

deception-oriented mindset takes full root the capabilities it creates have potential to be unleashed far and wide, indiscriminately directed against ill-advised or simply inappropriate targets (including legitimate rivals in one's polity, people on one's own side, innocent bystanders, etc.).[48] As Sun Tzu himself says (Passage #7.2):

XIII.14.
Delicate indeed! Truly delicate! There is no place where espionage is not used.

Theme #10 examines related "no scruples" pitfalls from a Sun Tzu standpoint.

★★★★★★

As "roads not taken" analysis makes clear, what Sun Tzu *does not* do is to offer hot tips that might teach an aspiring deceiver how to deceive more effectively. What Sun Tzu *does* do, first and foremost, is to inculcate heightened awareness of the *kinds of environments* whose structural characteristics afford major deception opportunities. The dwindling of chariot warfare and the rise of mass-infantry armies – the military revolution of Sun Tzu's age – gave newfound opportunities for operations involving complex terrain (e.g., ravines) and with them new scope for deception. The place to seek a living legacy of Sun Tzu's deception thinking is "ecologies" – some unimagined in Sun Tzu's time – that are also conducive to deception.[49]

The ecologies relevant to military deception may be strategic, logistical, tactical, or a blend thereof.[50] In conventional land warfare in many eras of military history, deception opportunities growing out of terrain in its narrow topographic sense have often been *relatively* simple, primarily involving opportunities for ambush or related forms of surprise attack. Logistics is a different matter. The complexity of even a primitive logistics system, on both strategic and tactical levels, can hide much, including sophisticated clandestine steps to degrade or misrepresent its performance.[51] In modern times, military logistics has evolved into a

---

[48] As Max Weber, #2 footnote 38, p. 987 famously observed: "Once fully established, bureaucracy is among those social structures which are the hardest to destroy." That comment might apply, possibly with particular force, to bureaucracies (or components thereof) whose raison d'etre is deception operations. See also Peter Fleming & Stelios C. Zyglidopoulos, "The escalation of deception in organizations," *Journal of Business Ethics*, 2008, 81(4), 837–50 (analyzing organizational dynamics whereby "an initial lie can begin a process whereby the ease, severity, and pervasiveness of deception increases over time so that it eventually becomes an organization level phenomenon").

[49] With verse IX.18 (Passage #7.5) pointing the way, biological ecology and evolutionary biology can usefully inform this Sun Tzu–inspired line of deception thinking. See Mikael Mokkonen and Carita Lindstedt, "The evolutionary ecology of deception," *Biological Reviews*, 2016, 91(4), 1020–35, concluding with a list of eleven propositions (pp. 1031–32). One of those propositions (p. 1032) is especially worth noting here: "Ecological factors affecting the costs and benefits of deceptive acts, or of responding to deceptive signals, may also be more important than previously thought . . ."

[50] A classic source of connection between military deception and ecology in its biological sense comes from the comparative study of camouflage in animals and human warfare. See Hugh B. Cott, *Adaptive Coloration in Animals* (introduction by Julian S. Huxley; London: Methuen, 1966), p. xii.

[51] An example of logistics deception in Han times is recounted by Sawyer, Introduction footnote 55, pp. 131–32, describing a hungry army lured by a bonanza of food abandoned by an (apparently) fleeing foe who then proceeded to counter-attack, inflicting a resounding defeat – a defeat rendered "even more ignominious . . . [by] the stratagem underlying it because the conspicuous piles of beans were mere chimera, a coating applied atop mounds of sand."

supremely complex ecosystem where many things can be hidden and deception opportunities abound.

Generalizing to Sun Tzu (3) realms, Sun Tzu should teach us to scout out where are the truly complex environments conducive to providing cover for deception – and where new ones are emerging.[52] In the fast-paced twenty-first century there is much grist here for cases of "*What did Master Sun know that we still don't (or have yet to absorb adequately)?*"

To this point in the twenty-first century the leading contender for ecologies conducive to unleashing the power of Sun Tzu's verse I.17 is digital structures and devices – algorithms, software, hardware, and computer networks, along with all the many physical and human systems they support, integrate, and control.[53] As the twenty-first century advances, those digital structures are starting to take on an almost biological complexity. For that reason, they are superb incubators for Sun-Tzu-esque deceptions.

On a different plane of structure, urban environments present a type of high complexity represented by sprawling "megacities" with their rickety infrastructures (starting with available water supplies). World cities are a distinct, if overlapping, category of complex environment with somewhat different vulnerabilities, often rooted in the requirements of the sophisticated international firms (banks, law firms, accounting firms, advertising firms, etc.) that congregate there.[54] Both types of environment provide excellent cover for deceptions. Trends toward "smart cities," now worldwide, stand to multiply those deception opportunities.

A challenge growing out of Sun Tzu's deception thinking points to a kind of meta-deception skill: cultivating an ability to spot, more quickly than adversaries or rivals, when a particular environment is *about to enter* a qualitatively new phase of complexity, thereby opening a door to possibilities for new types and levels of deception (in some cases realizing previously impractical deception ideas).[55] There can be much payoff from being a first mover in exploiting such creative opportunities before others spot them or devise effective strategies to operate there. That is a type of skill-set found among some financial innovators, accounting innovators, tax evaders, criminal innovators, and a few other sorts of like-minded strategic actors.

---

[52] For a relevant perspective, noting the "crucial importance of environmental complexity as the basis of many ecological phenomena," see Scott A. Boorman, "Mathematical ecology and its place among the sciences. II. Analogues in the social sciences," *Science*, October 27, 1972, 176(4059), 391–94 (second part of two-part review by Thomas W. Schoener and Scott A. Boorman of Robert H. MacArthur, *Geographical Ecology* [New York: Harper and Row, 1972]).

[53] Sun Tzu's cognizance (Passage #1.7 and pp. 59–60 above) of how military calculations may form a pyramid, with each calculation depending on results one layer down, gives a hint of the vistas opened by the possibility of "layering" deceptions-within-deceptions. That layering concept suggests designing one set of bugs you are *supposed* to find (compare verse IX.21 in Passage #7.5) along with a second set of more crafty bugs you are *not* supposed to find (or only much later, often by a fluke). It comports even better with creating bugs that, once spotted and fixed, trigger other more insidious bugs.

[54] For a relevant Braudel-inspired conceptual framework see Peter J. Taylor, "World cities and territorial states under conditions of contemporary globalization," *Political Geography*, 2000, 19(1), 5–32.

[55] Twenty-first-century biology and its applications is likely to serve up many advances of that type.

Barton Whaley proposed what he termed the Principle of Naturalness – by which he meant feigned naturalness, the art that magicians invoke to make "each sleight or other conjuring move simulate some normal gesture, action, or posture" and thereby seem natural to the audience.[56] As Whaley wrote, "Success in this is, perhaps, the highest test of their art," observing that "[e]xamples of systematic and comprehensive application of the Principle of Naturalness are rare in military practice and almost entirely overlooked by military theory and military doctrine."

Through its complex terrain roots Sun Tzu's deception thinking has resonance with Whaley's Principle of Naturalness. By veiling much, and confusing the viewer's attention, a complex environment is a major resource for simulating naturalness. Carrying the point a step further, precisely because they contain so much structure and process that is so very artificial (with all the quirks and snafus that, unsurprisingly, come along with it), complex software and algorithm environments afford especially great scope for a deceiver to simulate "naturalness" while at the same time working the many levers of control also present there to advance that deceiver's true goals. Few actions taken in, or with respect to, those digital worlds qualify as truly "natural." That extreme artificiality, and the expectations and mindsets it engenders in all participants, gives enormous leeway for credibly passing off as natural even highly atypical effects or actions taken by a deceiver.[57]

★★★★★★

[56] Whaley, "The principle of naturalness," pp. 41–43 in Rothstein & Whaley, #7 footnote 1. Whaley's Principle of Naturalness harmonizes with some of the major emphases of Daoism, a tradition with which Sun Tzu has affinities (Mair, pp. 47–49; Brooks, p. 66). Although deception is not a focus there, a spirit of "naturalness" pervades one of the most famous passages in the Zhuangzi, a core Daoist text. It describes a Cook Ding who was so in tune with the finer details of an ox's anatomy that he was able to cut up thousands of oxen without his knife becoming dulled, whereas less adept cooks were forced to change knives continually because their less natural ways of cutting blunted them. See Watson, #1 footnote 59, pp. 50–51 (the same footnote 59 gives further sources on the Zhuangzi text). The case of Cook Ding is noted by Minford, pp. 182–83 in context of verse VI.10 (Passages #6.18 & #8.1).

[57] Everyday life in stochastic networks routinely sees ambiguous situations having strategic aspects (e.g., is a typo in an email part of somebody's deception operation, say, geared to flying under the radar of screening or surveillance software – or the sender's subtle message – or is it just a typo, even as Freud's famous cigar is sometimes just a cigar?).

Unquestionably there is a basic tension between Sun Tzu's denial of chance (at least where outcomes of interest to Sun Tzu are involved!) on which Lau & Ames comment (see p. 64 above) and stochastic process thinking as a language for describing complex environments conducive to the deceptions Sun Tzu so prized. Depending on standpoint, that tension is an interpretive loose end – or a research frontier. There is room here for fresh thinking about conceptual foundations of deception. For a relevant mathematician's perspective see David Mumford, "The dawning of the age of stochasticity," in Vladimir I. Arnold et al. (eds.), Mathematics: frontiers and perspectives (Providence, RI: American Mathematical Society, 2000), pp. 197–218.

Set of principal passages used to illustrate Sun Tzu Theme #7 (deception)

Passage #7.1. Sun Tzu's way of deception

(With verses I.16–I.17 affirming the basic deception principle, the verses that follow provide a kind of intellectual matrix for Sun Tzu's deception thinking. There is substantial variability here as to whether, and in what sense, deception is involved. For example, verses I.22–I.24 do not directly involve deception but – as any con man knows full well – each profiles a psychological state that predisposes to falling prey to deception. In a class by itself, inasmuch as the central focus of manipulation is now the interacting motivations of multiple parties, is the dynamics envisioned by verse I.25. By many usual understandings of deception, verse I.25 overflows the boundaries of the deception category. But that indeed may be the insight: for Sun Tzu there is no crisp line separating "deception" per se and steps to undermine and fractionate – and thereby to spoil, to apply Erving Goffman's compelling word – the fundamental identity of an enemy or rival.)

> I.16. Having paid heed to the advantages of my plans [ji], the general must create situations [shi] which will contribute to their accomplishment. By "situations" [shi] I mean that he should act expediently in accordance with what is advantageous [yin li] and so control the balance [quan].[58]
> I.17. All warfare is based on deception. [Lit.: "War is the Way [dao] of deception [gui]."]
> I.18. Therefore, when capable, feign incapacity; when active, inactivity.
> I.19. When near, make it appear that you are far away; when far away, that you are near.
> I.20. Offer the enemy a bait to lure him; feign disorder and strike him.[59]
> I.21. When he concentrates, prepare against him; where he is strong, avoid him.
> I.22. Anger his general and confuse him.
> I.23. Pretend inferiority and encourage his arrogance.
> I.24. Keep him under a strain and wear him down.
> I.25. When he is united, divide him.
> I.26. Attack where he is unprepared; sally out when he does not expect you.
> I.27. These are the strategist's keys to victory. It is not possible to discuss them beforehand.

---

[58] Ideas found in this verse are analyzed from a Theme #1 (calculation) standpoint on p. 59 above.
[59] Reflecting an ambiguity in the Chinese text, Mair, p. 79 translates differently:

> When one's opponents are greedy for advantage, tempt them. *When one's opponents are in chaos, seize them.*
>
> [Emphasis supplied.]

Taken by itself, Mair's understanding of the second part appears to move away from a deception focus. Notice, however, that an integrated reading of verse I.20 along lines pointed by Mair has the enemy rising to a bait, becoming disorganized (a natural enough upshot, especially in the armies Sun Tzu knew), then being toppled – a serviceable description of a feigned retreat stratagem common in the warfare of Sun Tzu's time, indeed one of the oldest and best known of all military ruses. See #7 footnote 90 below (citing sources); Appendix 7 footnote 5 (further analysis). For civilian interpretations of verse I.20 see #4 footnote 95.

Passage #7.2. Ubiquity of Deception in Sun Tzu's Way of War: Reliance on Spies

XIII.2. One who confronts his enemy for many years in order to struggle for victory in a decisive battle yet who, because he begrudges rank, honours and a few hundred pieces of gold, remains ignorant of his enemy's situation [i.e., who fails to make effective use of spies], is completely devoid of humanity. Such a man is no general; no support to his sovereign; no master of victory.

. . .

XIII.14. Delicate indeed! Truly delicate! *There is no place where espionage is not used.* [Emphasis supplied.]

. . .

XIII.23. And therefore only the enlightened sovereign and the worthy general who are able to use the most intelligent people as agents are certain to achieve great things. *Secret operations are essential in war; upon them the army relies to make its every move.* [Emphasis supplied.]

Passage #7.3. License to Deceive One's Own People

(Armies are small, gossipy worlds. For that and logistics planning reasons movements of a mass-infantry army, and the plans they portend, may be hard to veil. Giving practical effect to Sun Tzu's counsel in verse XI.43 may hence leave few alternatives but for a general commanding such an army to put in place affirmative deceptions of his own officers and men – deceptions possibly less needed in more agile forms of warfare centered on smaller, elite units.[60] On a different plane, ideas in verses XI.45 and XI.46 mesh well with modern aspirations to harness unpredictability as a tactical tool. Although "mixed strategy" is nowadays associated with calculus of probabilities and game theory – both unknown to Sun Tzu – it should be noted that implementing simple versions of a mixed strategy idea require neither formal game theory nor high-tech support.[61])

XI.42. It is the business of a general to be serene and inscrutable, impartial and self-controlled.
XI.43. He should be capable of keeping his officers and men in ignorance of his plans.[62]
. . .

---

[60] For further analysis in context of Theme #10 (no scruples) see pp. 335–38 below.
[61] For an example, involving a commander (described as never having heard of game theory) who would ask one of the other members of his team "to decide which hand held behind his back contained a blade of grass" – with the upshot determining the day's choice of formation (Sun Tzu's *xing!*) – see p. 175 of R. S. Beresford & M. H. Peston, "A mixed strategy in action," *OR, Operational Research Quarterly*, 1955, 6(4), 173–75. (This example is also noted by O'Neill, #5 footnote 7, p. 1020.)
[62] The Chinese text has no discontinuity here. The apparent discontinuity reflects Griffith's editing.

XI.45. He changes his methods and alters his plans so that people have no knowledge of what he is doing.

XI.46. He alters his camp-sites and marches by devious routes, and thus makes it impossible for others to anticipate his purpose.

...

XI.55. Set the troops to their tasks without imparting your designs; use them to gain advantage without revealing the dangers involved. Throw them into a perilous situation and they survive; put them in death ground and they will live. For when the army is placed in such a situation it can snatch victory from defeat.

Passage #7.4. Deception Combined with Straightforward Measures

(Although qi, which Lau's 1965 article translates as "crafty," is not the same concept as verse I.17's gui or verse VII.12's zha, both signifying deception, they are plainly related ideas.[63] Strategists adept at qi are also likely to be skilled deceivers. For detailed analysis of the qi/zheng concept pair see Theme #14.)

V.3. That the army is certain to sustain the enemy's attack without suffering defeat is due to operations of the extraordinary and the normal forces.

V.4. Troops thrown against the enemy as a grindstone against eggs is an example of a solid acting upon a void.

V.5. Generally, in battle, use the normal force to engage; use the extraordinary to win.

V.6. Now the resources of those skilled in the use of extraordinary forces [qi] are as infinite as the heavens and earth; as inexhaustible as the flow of the great rivers.

V.7. For they end and recommence; cyclical, as are the movements of the sun and moon. They die away and are reborn; recurrent, as are the passing seasons.

V.8. The musical notes are only five in number but their melodies are so numerous that one cannot hear them all.

V.9. The primary colours are only five in number but their combinations are so infinite that one cannot visualize them all.

V.10. The flavours are only five in number but their blends are so various that one cannot taste them all.

V.11. In battle there are only the normal [zheng] and extraordinary [qi] forces, but their combinations are limitless; none can comprehend them all.[64]

V.12. (Translation following Lau, 1965 article, p. 331.[65]) "The qi and the zheng produce each other endlessly like a ring and who is there that can exhaust the possibilities?"

---

[63] Illustrating how a qi strategy or tactic may not involve deception see p. 463 below.

[64] Griffith's translation problematically inverts the Chinese text's word order, which mentions qi first, zheng second (on Sun Tzu's part, a possibly deliberate choice of order of mention).

[65] Rationale for substituting Lau's translation for Griffith's is given in Appendix 14 footnote 3.

Passage #7.5. Affinity between Deception and Complex Terrain

(What Sun Tzu says here can, of course, be given a narrow reading limited to challenging or treacherous physical terrain. It may also be given a generalized reading pointing to a broader "ecology of deception" and how complex environments – both naturally occurring [verses IX.16, IX.18] and engineered [verse IX.21, possibly IX.20 too] are natural handmaidens of the deceiver's art.)

IX.16. Where there are precipitous torrents, "Heavenly Wells", "Heavenly Prisons", "Heavenly Nets", "Heavenly Traps", and "Heavenly Cracks", you must march speedily away from them. Do not approach them.[66]

IX.17. I keep a distance from these and draw the enemy toward them. I face them and cause him to put his back to them.

IX.18. When on the flanks of the army there are dangerous defiles or ponds covered with aquatic grasses where reeds and rushes grow, or forested mountains with dense tangled undergrowth you must carefully search them out, for these are places where ambushes are laid and spies are hidden.

IX.19. When the enemy is near by but lying low he is depending on a favourable position.[67] When he challenges to battle from afar he wishes to lure you to advance, for when he is in easy ground he is in an advantageous position.

IX.20. When the trees are seen to move the enemy is advancing.

IX.21. When many obstacles have been placed in the undergrowth, it is for the purpose of deception.[68]

IX.22. Birds rising in flight is a sign that the enemy is lying in ambush; when the wild animals are startled and flee he is trying to take you unaware.

. . .

IX.35. When birds gather above his [i.e., the enemy's] camp sites, they are empty.[69]

---

[66] Clarification of these six terrain features comes from Ames, p. 141, who translates verse IX.16: "Encountering steep river gorges, natural wells [ground vulnerable to inundations, as from rainwater runoff], box canyons, dense ground cover, quagmires, or natural defiles, quit such places with haste." See also Appendix 2 footnote 30 (citing further explanatory material).

[67] The Chinese word Griffith translates as "favourable position" is *xian* (a topographic meaning of which is "ravine," "narrow pass," etc.), whose presence points to this sentence as involving complex terrain. For further analysis see Appendix 7 footnote 8 and Appendix 11 footnote 33.

[68] Sawyer, p. 324 note 153 clarifies the stratagem: "Easily visible obstacles have been made deliberately detectable in order to create the suspicion of ambush or the emplacement of entangling devices and thereby beguile the ordinary commander to divert his forces to the enemy's advantage."

[69] Traditional commentator Chen Hao reads the birds as a sign of a secret enemy change of position, in effect harnessing animal behavior in service of counter-deception. See Giles, p. 93; Griffith, p. 120.

Passage #7.6. Affinity between Deception and Complex Terrain, cont.

(Albeit conveyed in a plain vanilla way, verse X.4 harbors a significant further idea. Complex environments not only afford cover for operatives to hide in nooks and crannies, awaiting an opportune moment to spy or strike [see Passage #7.5], but also can elicit complicated projects which, while being carried out, involve major, if temporary, flexibility loss. If the enemy can be lured into making that type of commitment there will come a point when he is highly vulnerable.[70])

X.4. Ground equally disadvantageous for both the enemy and ourselves to enter is indecisive. The nature of this ground is such that although the enemy holds out a bait I do not go forth but entice him by marching off. When I have drawn out half his force I can strike him advantageously.[71]

   . . .

X.6. In precipitous ground I must take position on the sunny heights and await the enemy. If he first occupies such ground I lure him by marching off; I do not follow him.[72]

Passage #7.7. More General Forms of Complexity and Deception Potential

(This overly terse but rich and evocative passage moves focus away from terrain issues per se while maintaining focus on a complexity/deception nexus. Read through a deception lens, verse V.18 sets forth three types of deceptive façade.[73] All might involve contrived complexity, but it is the first façade there – using chaos to veil order – that most directly evokes contrived complexity as a

---

[70] There are basic connections between this line of thinking and Theme #14. See p. 466 below.

[71] Commentators construe this maneuver as a feigned withdrawal (a concept explicitly used in Mair's translation, p. 114). For pertinent commentator excerpts see Griffith, p. 124; Minford, pp. 251–52.
 An easily visualized example of the same concept is a feigned withdrawal that lures the enemy to commit to a river crossing, then picks a moment to strike when the enemy army is split between the two banks. See Passage #14.7's verse IX.5 (though use of deception is not explicit there).

[72] By contrast to verse X.4 (where a deception idea is unequivocal), presence of a deception idea in verse X.6 (as suggested by Griffith's "lure") is possible but less clear. A sizable body of scholarly opinion, pre-modern and modern, weighs against it. For further analysis see Appendix 7 footnote 12.

[73] Du Mu's commentary (Minford, p. 171) spells out a view of the specifics:
 In other words, if you wish to feign disorder and entice your enemy into the fray, you must first achieve a high degree of order; if you wish to feign fear and lure the enemy into a trap, you must first achieve a high degree of courage; if you wish to feign weakness and lead the enemy into hubris, you must first achieve a high degree of strength.
 Note overlaps with various of Passage #7.1's verses. Sun Tzu's thinking is cited as inspiration for a game theory argument in Branislav L. Slantchev, "Feigning weakness," International Organization, 2010, 64(3), 357–88 (his p. 376 quotes Passage #7.1's verses I.22 and I.23 in Giles's translation).
 Non-deception interpretations of verse V.18 are also possible. They are weighed in Appendix 7 anchor footnote 14, concluding that a deception interpretation is more compelling.

deception tool.[74] Separately, as regards verse V.17, while the battle situation it depicts is a prototype, the kinds of deception opportunities it suggests might find counterparts in other complex situations, say, the tangled social network landscape of a ruler's court. Verse V.20 likewise has overtones of deception. Verses V.17–V.20 are sandwiched between other Sun Tzu V material having no obvious deception aspect.[75] But that seemingly inartful positioning could be informative: a clue that Sun Tzu did *not* see deception as an analytically discrete category or topic, by contrast to many modern treatments of deception issues.[76])

V.17. In the tumult and uproar the battle seems chaotic, but there is no disorder; the troops appear to be milling about in circles but cannot be defeated.

V.18. Apparent confusion is a product of good order; apparent cowardice, of courage; apparent weakness, of strength.

V.19. Order or disorder depends on organization [shu]; courage or cowardice on circumstances [shi]; strength or weakness on dispositions [xing].

V.20. Thus, those skilled at making the enemy move do so by creating a situation to which he must conform; they entice him with something he is certain to take, and with lures of ostensible profit they await him in strength.[77]

Passage #7.8. Complex Situations Born of a Ruler's Ill-Advised Meddling, Creating Opportunities for Deceptions Fomented by Third Parties

(The sorry state of civil–military relations Sun Tzu here depicts is rife with opportunities for unfriendly outsiders to reach into the ensuing complex situation and use a variety of deceptions to manipulate it for their own profit.)

III.18. Now the general is the protector of the state. If this protection is all-embracing, the state will surely be strong; if defective, the state will certainly be weak.

III.19. Now there are three ways in which a ruler can bring misfortune upon his army.

III.20. When ignorant that the army should not advance, to order an advance or ignorant that it should not retire, to order a retirement. This is described as "hobbling the army".

---

[74] For a concrete example from early Chinese military history, involving the famous battle of Chengpu (632 BC) where one side made adroit use of a contrived dust cloud to mask a crucial stratagem, see Appendix 11 footnote 30 (quoting from analysis of Kierman, Background footnote 31, p. 54).

[75] For example, verses V.17–V.20 are preceded and followed by Theme #1 (calculation) material: Passage #1.4 (verses V.14–V.16) + Passage #1.40A (verses V.21–V.22). They are also straddled by Passage #9.3's verses V.13–V.16 + V.21–V.25 from Theme #9 (building on natural forces and momentum).

[76] See also #7 footnote 5 noting lack of a natural candidate for a "flagship" Sun Tzu passage squarely addressing deception.

[77] Griffith's "ostensible" lacks a counterpart in the Chinese text, though that is a possible interpretation.

III.21. When ignorant of military affairs, to participate in their administration. This causes the officers to be perplexed.[78]

III.22. When ignorant of command problems to share in the exercise of responsibilities. This engenders doubts in the minds of the officers.

III.23. *If the army is confused and suspicious, neighbouring rulers will cause trouble. This is what is meant by the saying: "A confused army leads to another's victory."* [Emphasis supplied.]

Passage #7.9. Sun Tzu's "Second Symphony": Deception, Calculation, and Operational Style

(If Passage #6.18 is likened to Sun Tzu's First Symphony, the present set of verses, read as a package, might be called Sun Tzu's "Second Symphony." The extent of Passage #7.9's attention to logistics issues, not always well represented in treatments of operational art or mobile warfare, is noteworthy. Read through a verse VII.12 deception lens, Passage #7.9 is a call for deception steps to be thoroughly integrated with maneuver in mobile warfare.[79] The military active ingredient is well captured by an observation of Wavell: "The general aim of all strategic maneuver is, in fact, to upset the equilibrium of the enemy commander."[80])

VII.2. Nothing is more difficult than the art of manoeuvre. What is difficult about manoeuvre is to make the devious route the most direct and to turn misfortune to advantage.

VII.3. Thus, march by an indirect route and divert the enemy by enticing him with a bait. So doing, you may set out after he does and arrive before him.[81] One able to do this understands the strategy of the direct and the indirect.[82]

---

[78] For atmospherics pertinent to this and the next two verses see also verse IX.41 which Griffith, pp. 121–22 renders "When the troops continually gather in small groups and whisper together the general has lost the confidence of the army." Verse IX.41 also appears in Passage #4.15 (where it is analyzed under a rubric of social dynamics providing traction for a subversion campaign).

[79] Liddell Hart, #4 footnote 2, p. 337 conveys a relevant idea:
Movement generates surprise, and surprise gives impetus to movement. For a movement which is accelerated or changes its direction inevitably carries with it a degree of surprise, even though it be unconcealed; while surprise smooths the path of movement by hindering the enemy's counter-measures and counter-movements.
Whaley, #7 footnote 4, p. 72 quotes Liddell Hart's remarks, adding that "This type of case where one surprise is piled upon another is one of the great opportunities provided by highly mobile warfare."

[80] Wavell, Introduction footnote 44, p. 158.

[81] As earlier noted (#5 footnote 90) verse VII.3 can be read in radically different ways. Griffith's translation represents one interpretation. For discussion see Appendix 7 anchor footnote 16.

[82] Griffith's translation of the yu/zhi concept pair as "strategy of the direct [zhi] and indirect [yu]" in verses VII.3 and VII.16 may reflect influence of Liddell Hart's "indirect approach" concept. Verse VII.3 (in Giles's translation) appears on p. 11 of Liddell Hart's *Strategy* (cited in #4 footnote 2). However, alignment of Liddell Hart's ideas on Sun Tzu's should not be overstated; see #14 footnote 2. Mair's translation of yu/zhi (Mair, p. 100) as "circuitous" (yu) and "straight" (zhi) is

VII.4. Now both advantage and danger are inherent in manoeuvre.

VII.5. One who sets the entire army in motion to chase an advantage will not attain it.

VII.6. If he abandons the camp to contend for advantage the stores will be lost.

VII.7. It follows that when one rolls up the armour and sets out speedily, stopping neither day nor night and marching at double time for a hundred li, the three commanders will be captured. For the vigorous troops will arrive first and the feeble straggle along behind, so that if this method is used only one-tenth of the army will arrive.

VII.8. In a forced march of fifty li the commander of the van will fall, and using this method but half the army will arrive. In a forced march of thirty li, but two-thirds will arrive.

VII.9. It follows that an army which lacks heavy equipment, fodder, food and stores will be lost.
        ...[83]
        ...

VII.12. *Now war is based on deception.* Move when it is advantageous and create changes in the situation by dispersal and concentration of forces.[84] [Emphasis supplied.]

VII.13. When campaigning, be as swift as the wind; in leisurely march, majestic as the forest; in raiding and plundering, like fire; in standing, firm as the mountains. As unfathomable as the clouds, move like a thunderbolt.

VII.14. When you plunder the countryside, divide your forces. When you conquer territory, divide the profits.

VII.15. Weigh the situation, then move.

VII.16. He who knows the art of the direct and the indirect approach will be victorious. Such is the art of manoeuvring.

---

preferred. See also *yu/zhi* discussion in connection with Passage #14.27A, where Passage #7.9's verses also appear.

[83] Material elided here may be an interpolation. See Table 13A note a on p. A-37.

[84] Because it rubs shoulders with the first half that so greatly plays up deception, verse VII.12's second half is worth more attention than it generally gets, as a lens on the practical content of Sun Tzu's military deception thinking. Keeping in mind that the context is land warfare, it seems possible that the idea covers detaching one or more parts of an army to serve as "surprise troops" (*qi bing*) for giving effect to some stratagem (see battle description on p. 31 above as well as #14 footnote 12).

From annals of Western military history, verse VII.12's second half also calls to mind Napoleon's use of the *bataillon carré* ("battalion square") in his grand tactics, using "controlled dispersion" to achieve maximum flexibility in a way disorienting to the enemy. See William J. Wood, "Forgotten sword," *Military Affairs*, October 1970, 34(3), 77–82, describing the system (p. 77) as "the indispensable link between [Napoleon's] strategy and his battlefield tactics." See also Azar Gat, "The hidden sources of Liddell Hart's strategic ideas," *War in History*, 1996, 3(3), 293–308, noting (p. 299) Napoleon's "mastery of deception, feints, and diversions to create surprise, disorientation, and miscalculation on the enemy's part." However, any verse VII.12 parallelism with the *bataillon carré* concept must be handled with caution because (as Wood, p. 80 notes) the latter depended crucially on field intelligence provided by reconnaissance cavalry, a capability most likely lacking in Sun Tzu's armies.

Passage #7.10. Sun Tzu's "Third Symphony" or "Symphony of Spies": Deception and Calculation, Espionage Version

(These verses cast the spymaster in a kind of conductor's role for a symphony of spies. Passage #7.10 is Sun Tzu's closest approach to describing a full-fledged espionage campaign.[85] It assigns a starring role to double agents [fanjian].[86])

XIII.17. It is essential to seek out enemy agents who have come to conduct espionage against you and to bribe them to serve you. Give them instructions and care for them. Thus doubled agents [fanjian] are recruited and used.

XIII.18. It is by means of the doubled agent that native and inside agents can be recruited and employed.

XIII.19. And it is by this means that the expendable agent, armed with false information, can be sent to convey it to the enemy.

XIII.20. It is by this means also that living agents can be used at appropriate times.

XIII.21. The sovereign must have full knowledge of the activities of the five sorts of agents. This knowledge must come from the doubled agents [fanjian], and therefore it is mandatory that they be treated with the utmost liberality.

Passage #7.11. Surprise Switch of Style (Relevant to All Sun Tzu's Symphonies)

(Verse XI.61's action sequence, centering on a surprise switch of styles, could find application in both conventional warfare and espionage settings.[87])

XI.61. Therefore at first be shy as a maiden. When the enemy gives you an opening be swift as a hare and he will be unable to withstand you.

Passage #7.12. Basic Counter-deception Mindset

(Taken at face value, verse VIII.16 is a blunt statement of a fundamental "no trust" principle, one that takes on added force if the "enemy" is construed broadly

---

[85] A variant interpretation (cf. Mair, pp. 130–31) gives verses XIII.17–XIII.20 a cumulative thrust, with espionage steps profiled in verse XIII.18 building on steps discussed in XIII.17, those in verse XIII.19 building on XIII.18's, etc. While the role of fanjian (commonly translated as "double agents") remains basic, that variant interpretation somewhat narrows their role and helps avoid expecting too much of fanjian (see next footnote). However, lack of detail limits our ability to clarify Sun Tzu's ideas here.

[86] However, Sun Tzu's fanjian category has problematic aspects. See pp. 368–71 below.

[87] A candidate for a civilian example of manipulative use of style switches is Paul Frillman's recollection of Chiang Kai-shek and Madam Chiang's World War II receptions, alluding to "their curious mixture of Chinese opulence, Methodist austerity, and YMCA heartiness." See Frillman & Peck, #1 footnote 48, pp. 162–63.

to include potential enemies and rivals.[88] Verse VIII.16's counter-deception relevance is put in sharp focus by a comment of William Harris, a modern analyst:[89]

> Part of the skill in deception work is to coax one's enemy into an assessment of intentions (perhaps by dangling information that confirms preconceptions) rather than an assessment of capabilities, in a situation where the adversary should know better than to expect much luck with an intentions estimate.)

VIII.16. It is a doctrine of war not to assume the enemy will not come, but rather to rely on one's readiness to meet him; not to presume that he will not attack, but rather to make one's self invincible.

Passage #7.13. Avoiding Feigned Retreats and Other Baits

(Feigned retreats have been described as "apparently one of the most common ruses employed in antiquity."[90] Warring States generals probably often operated as their own intelligence analysts, certainly in battlefield situations. Because these verses do not come equipped with any advice on how to distinguish a ruse from an opportunity, Sun Tzu's message appears to be a very simple one, akin to a "take a moment to think before you act" caution to a commander. Although verse VII.28's advice might seem extraneous to counter-deception thinking, a foe's elite troops seem a natural candidate for being chosen to put on a "baited" performance. The Chinese term that Griffith renders as "elite" is *rui*, lit. "sharp," which in the armies of Sun Tzu's time evokes a contrast between sullen peasant levies who made up the bulk of the force and the professional soldiers at its core.)

VII.27. When he pretends to flee, do not pursue.
VII.28. Do not attack his *élite* troops.
VII.29. Do not gobble proffered baits.

Passage #7.14. Avoiding Cannae Scenario, Warring States China Version

(Brief as it is, this verse, framed as a warning to a potential victim of deception, foreshadows in barest outline the basic concept of the battle of Changping between Qin and Zhao *c.* 260 BC.[91] This was one of the decisive battles – more exactly a campaign – of the Warring States period, indeed in all of Chinese history,

---

[88] Sun Tzu's intended scope of application of verse VIII.16 is not entirely clear. It could be read broadly, as advice against relying on an estimate of enemy intentions anywhere on a scale spanning from grand strategy to minor tactics; or more narrowly, as a call to prudent preparedness on level of a country's overall military readiness. For further analysis see pp. 212–14 above.

[89] See p. 563 of William R. Harris, "Counter-deception planning," in Rothstein & Whaley, #7 footnote 1, pp. 551–75.

[90] Sawyer, p. 319 note 119 (feigned retreats in various *Seven Military Classics* works). Wavell, Introduction footnote 44, p. 157 describes such retreat as "probably the oldest of all stratagems."

[91] For a sense of scale see Lewis, Background footnote 7, p. 629: "The great battle of Changping, though more geographically concentrated than the preceding campaigns [profiled by Lewis, pp. 628–29], involved two armies deadlocked across a front that stretched for hundreds of li."

helping pave the way to China's imperial unification under Qin in 221 BC.[92] Sometimes described as the Chinese Cannae, Changping involved a controlled retreat by a part of the Qin army, drawing the enemy Zhao army after them, with the rest of the Qin army held back to spring a trap.[93] The Zhao army was split in two and its supply lines cut.[94] The troops thus cut off held out for some six weeks, making desperate, futile attempts at break-out. To keep the encirclement intact, the Qin ruler famously delved deep into his manpower pool calling up "all those over fifteen years of age," a departure from regular practice.[95])

IX.31. When half his force advances and half withdraws he is attempting to decoy you.

Passage #7.15. Basic Counter-deception Reasoning Tool: Incongruity Testing

(This verse also appears in Passage #11.17, where it is discussed in more detail.)

IX.44. When the enemy troops are in high spirits,[96] and, although facing you, do not join battle for a long time, nor leave, you must thoroughly investigate the situation.

Passage #7.16. Further Counter-deception Tips: Puncturing Deceptive Language

(As Clausewitz observed: "words, being cheap, are the most common means of creating false impressions."[97])

---

[92] See Wilkinson §24.1.2, p. 340. On a point of chronology, even Brooks's relatively late dating of the Sun Tzu text assigns all thirteen chapters to dates prior to Changping. See Table 12A on p. A-35. If one accepts Brooks's analysis (Brooks, p. 70) that assigns Sun Tzu IX a mid-fourth-century BC date, verse IX.31 would antedate Changping by close to a century.

[93] On Changping see Shiji, chapters 73 and 81, translated in Nienhauser, Vol. VII (1994), pp. 170–71, 269–70; Nienhauser, Vol. VII (2021), pp. 311–12, 489–91 (both cited on pp. xxiii–xxiv above). Resemblance to Cannae is well brought out in summary by Lewis, Background footnote 7, p. 640: "Bo Qi [the Qin commander] allowed Zhao's forces to advance in the center, encircled them on the flanks, cut their supply lines, and seized the fortifications they had left behind." (For a thumbnail sketch of Hannibal's crushing Cannae victory over the Roman army see Liddell Hart, #4 footnote 2, pp. 48–49.)
  A major foundation for Qin's Changping battlefield victory had already been put in place by a prior off-battlefield stratagem, involving a ruse perpetrated by Qin deception agents that catalyzed a crucial (and exceedingly ill-advised!) change of Zhao's commanding general. See #6 footnote 67.

[94] See Wicky W. K. Tse, "Cutting the enemy's line of supply: the rise of the tactic and its use in early Chinese warfare," Journal of Chinese Military History, 2017, 6(2), pp. 131–56.

[95] See pp. 206, 217 of Robin D. S. Yates, "Social status in the Ch'in: evidence from the Yün-meng legal documents. Part One: Commoners," Harvard Journal of Asiatic Studies, 1987, 47(1), 197–237.

[96] What Griffith renders as "high spirits" (an English-language phrase that might suggest "boisterousness," "animal spirits," or the like) is nu, more standardly rendered "angry" (see, e.g., Mair, p. 111).

[97] Clausewitz, On War, Preliminaries footnote 1, p. 202.

IX.25. When the enemy's envoys speak in humble terms, but he continues his preparations, he will advance.

IX.26. When their language is deceptive but the enemy pretentiously advances, he will retreat.[98]

IX.27. When the envoys speak in apologetic terms, he wishes a respite.[99]

IX.28. When without a previous understanding the enemy asks for a truce, he is plotting.[100]

Passage #7.17. Countering Disinformation on a Population Level

(Given a mildly generalized reading, verse XI.44 suggests that Sun Tzu was well aware of the ravages that enemy disinformation can wreak once it takes hold in a population of soldiers, and the need for decisive measures to stamp it out.[101] Eradicating disinformation, however, is easier said than done, starting with all the reasons that armies the world over have always been rumor mills.[102])

XI.44. He prohibits superstitious practices and so rids the army of doubts. Then until the moment of death there can be no troubles.

★★★★★

THEME #8 ACHIEVING FORMLESSNESS AND OPACITY

basic statement regarding formlessness:
Passage #8.1 (verses VI.8–VI.19 + VI.24–VI.31) (formlessness [wu xing])

wu xing 1 (for descriptions of wu xing 1–6 see Figure 3 formlessness ladder on p. 270):
Passage #8.2 (verses II.10 + II.15 + II.18–II.20) (logistics flexibility)
Passage #8.3 (verses XI.7 + XI.13 + XI.20 + XI.31–XI.32) (more about logistics flexibility)

---

[98] Ames, p. 142, translates Griffith's "deceptive" as "belligerent." That word choice is closer to the Chinese text, which literally refers to "strong" language. Mair, p. 110 translates: "when the enemy's language is aggressive and he makes a show of rushing forward, he is getting ready to retreat."

[99] Traditional commentator Du Mu envisions the enemy as coming and "presenting gifts" (possibly hostages?), which goes beyond what verse IX.27 literally says but is a reasonable illustration.

[100] Traditional commentator Li Quan (Minford, p. 240) provides useful clarification: "If there is talk of peace, but no proper treaty, there is a plot going on."

[101] See Kierman, Background footnote 31, p. 41, noting "numerous occasions [in early Chinese warfare] when, impelled by fatalism or orders or omens, men back into battle convinced that they cannot win and will in all likelihood die for no good cause." He further observes (citing a Zuozhuan case), "This is hardly the high and positive morale that an army should hope to instill in its officers."

[102] Cf. observation of Henry, #5 footnote 68, p. 609: "Popular cults generally have a stubborn life of their own. Chinese history in particular affords many examples of governments at all levels attempting rather futilely to stamp out local cultic practices that were deemed undesirable."

wu xing 2 (builds on Passage #8.1's verse VI.8 on presenting the enemy no good focus for attack or defense):
Passage #8.4 (verses XI.38–XI.39) (presenting no salient tactical focus for the enemy)

wu xing 3:
Passage #8.5 (verses V.17–V.20) (formlessness out of apparent confusion, masking form)
Passage #8.6 (verses XIII.6 + XIII.12) (epitome of a formless strategic capability: a well-functioning spy network; relates to *wu xing* 3 but has relevance to the full *wu xing* ladder)
Passage #8.7 (verse IX.18) (avoiding unfriendly eyes that could negate formlessness)
Passage #8.8 (verse XIII.15) (information denial measures; relates to *wu xing* 4 as well as *wu xing* 3)

wu xing 4:
Passage #8.9 (verses XI.42–XI.43 + XI.45–XI.46 + XI.49) (inscrutability of commander)
Passage #8.10 (verses IV.8–IV.12) (depth of strategic calculation, together with secrecy, as source of opacity for the strategist)
Passage #8.11 (verse XI.54) (opacity in administrative posture)
Passage #8.12 (verses XIII.21 + XIII.23) (defending against a kind of espionage that might pose a particular threat to *wu xing* 4 formlessness)

wu xing 5 (builds on Passage #8.1's verse VI.26 on never repeating the same tactic):
Passage #8.13 (verses V.5–V.7) (strategic or tactical imagination as source of formlessness)
Passage #8.14 (verses XI.9 + XI.14) (compensating in a conceptual realm for loss of flexibility in a physical one)
Passage #8.15 (verse XI.29) (speed as facilitator of formlessness; may also compensate for past *wu xing* failures by making them irrelevant)

wu xing 6 (topmost rung of the Figure 3 formlessness ladder):
Passage #8.16 (verses XII.17–19) (rejecting extraneous or inessential reasons for action)
Passage #8.17 (verses XI.27–XI.28) (implicit advice not to cherish much)
Passage #8.18 (verses X.19 + X.21) (divesting of extraneous motivations)
Passage #8.19 (verses VIII.17–VIII.24) (character traits inimical to formlessness)
Passage #8.20 (verses I.22–I.25) (states of mind undercutting formlessness)

extension of formlessness thinking:
Passage #8.21 (verse XI.55) (concept for reaping advantage by utilizing formlessness on one level to engineer, then profit by, utter inflexibility on a different level)

******

Succinctly put, Theme #8 prizes a strategic or tactical posture that exhibits "form-lessness," a quality of being without "form" or to use the Chinese-language phrase, *wu xing*, lit. "no form."[1] A modern scholar conveys *wu xing*'s meaning in evocative terms: "*Wu xing* is therefore the ultimate counterintelligence."[2]

Many writers on strategy play up some version of the importance of flexibility – limiting the enemy's, enhancing one's own (though possibly with certain qualifica-tions, in modern times often associated with ideas of Thomas Schelling; in Sun Tzu's time, with death ground thinking). The point could be put more strongly: one basic way in which people who "think strategically" differ from most other people is a greatly heightened awareness of steps, by self or others, that, once taken, would have irreversible effects greatly limiting future flexibility.

Sun Tzu's *wu xing* goes further. *Wu xing* stands in somewhat the same relation to affirmations of a need for flexibility widely found in the doctrines of modern militaries and managers that Sun Tzu's Theme #3 emphasis on outrageously cheap moves bears to standard cost–benefit thinking – namely, no basic inconsistency, but a different order of ambition entirely. Verse VI.24 (Passage #8.1) is a manifesto for formlessness:

VI.24.

The ultimate in disposing one's troops is to be without ascertainable shape [*wu xing*]. Then the most penetrating spies cannot pry in nor can the wise lay plans against you.

What such "formlessness" might mean in practical terms, and how to go about attaining it (including in non-military situations), is a question left largely open by Sun Tzu – a situation that, predictably, has frustrated many Sun Tzu disciples and has spawned numerous conceptual exercises, some of them unduly recondite. Under the best of circumstances, there is a whiff of paradox in seeking to place systematic structure on a concept that is all about rejection of any structure whatsoever. Unstated by Sun Tzu, a basic conundrum intrudes: how to reconcile lofty *wu xing* aspirations with mundane realities of mass-infantry warfare.[3]

Building in part on ideas in Lau's 1965 article, clues to clarification may be pieced together from the Sun Tzu text. Some of these involve explicit references to *wu xing* or *xing* ("form"); others do not. The resulting analysis envisions Sun Tzu's thinking about formlessness as a six-rung heuristic "ladder" (Figure 3 on p. 270).[4] Its bottom two

---

[1] *Wu xing* 無形 (formlessness) should not be confused, by virtue of its identical pinyin romanization, with *wu xing* 五行 (the so-called five elements or five phases). The latter is a different early Chinese concept that appears widely in military, medical, and other early texts, including in verse VI.31 (Passage #8.1) at the end of Sun Tzu VI. For further background on *wu xing* 五行 see #5 footnote 103.

[2] Rand, Background footnote 95, p. 115 (from article on Sun Tzu traditional commentator Li Quan).

[3] A historical puzzle is how trends in the warfare of Sun Tzu's time, which are widely acknowledged to involve rise of mass-infantry armies (e.g., Brooks, p. 59; Lewis, Background footnote 11, pp. 60–61), can be reconciled with Sun Tzu's fascination with formlessness. One catalyst might have been sheer frustration, growing out of the tension between the very limited flexibilities of any such army and the soaring military imaginations of the talented professional commanders then taking center stage on the Chinese military scene (see Lewis, Background footnote 11, pp. 97–98 and his chapter 3 passim).

[4] The Figure 3 ladder has affinities with analysis in JeeLoo Liu, "Was there something in nothingness? The debate on the primordial state between Daoism and Neo-Confucianism," in JeeLoo Liu & Douglas L. Berger, *Nothingness in Asian Philosophy* (New York: Routledge, 2014), pp. 181–96:

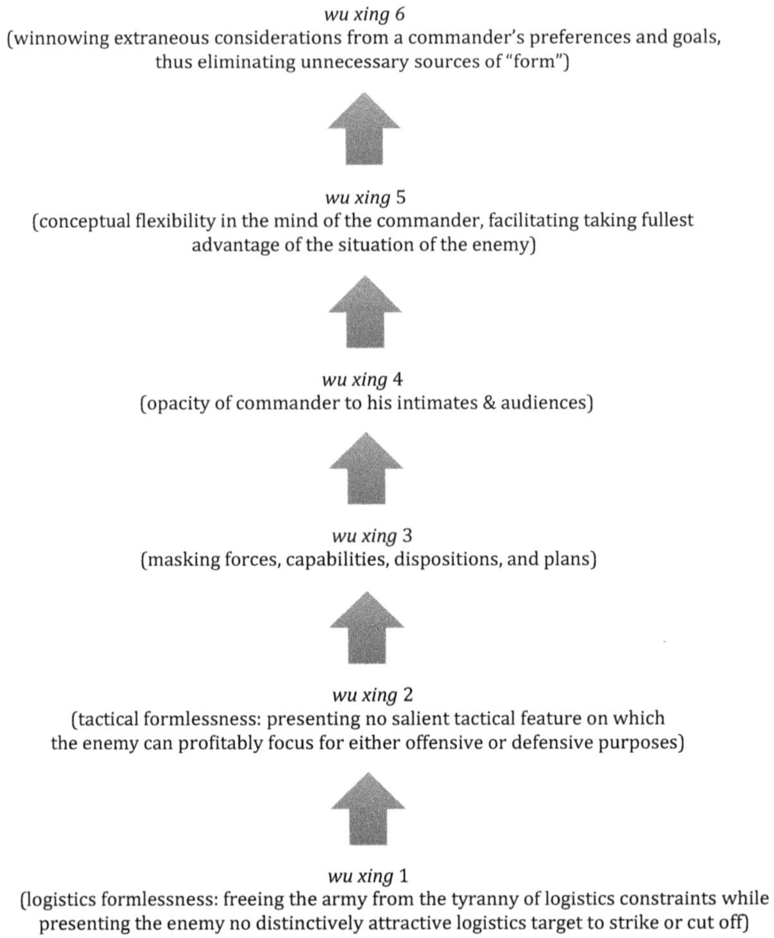

wu xing 6
(winnowing extraneous considerations from a commander's preferences and goals,
thus eliminating unnecessary sources of "form")

wu xing 5
(conceptual flexibility in the mind of the commander, facilitating taking fullest
advantage of the situation of the enemy)

wu xing 4
(opacity of commander to his intimates & audiences)

wu xing 3
(masking forces, capabilities, dispositions, and plans)

wu xing 2
(tactical formlessness: presenting no salient tactical feature on which
the enemy can profitably focus for either offensive or defensive purposes)

wu xing 1
(logistics formlessness: freeing the army from the tyranny of logistics constraints while
presenting the enemy no distinctively attractive logistics target to strike or cut off)

Figure 3. Heuristic ladder of formlessness (*wu xing*), inspired by ideas from Sun Tzu.
    The goal of this figure is to put basic structure on the multifaceted and possibly elusive concept of *wu xing*. Six types of formlessness are assembled here, labeled *wu xing* 1, *wu xing* 2, etc.* Moving up the ladder is broadly correlated with an increasing focus on intangibles. No inference should be drawn that $x > y$ means that attaining *wu xing x* necessarily implies or requires attaining *wu xing y*. For example, a commander may be excellent at keeping his intimates in the dark as to his thinking, yet his mass-infantry army, as currently deployed, may present a "form" highly visible to the enemy.

*Labeling inspired in part by Vankeerberghen's (separate) conceptual distinction among three types of *quan* (concept analyzed in Sun Tzu Theme #1 context, see pp. 58–59 above), designated *quan* A, *quan* B, and *quan* C. See her *Early China* article cited in #1 footnote 11, pp. 65, 67, and 73, respectively.

rungs center on manifestations of "formlessness" that (at least for conventional armies) are still relatively tangible and concrete, anchored in space and place.[5] Climbing higher, more intangibles start to enter the picture as focus shifts to versions of invisibility (or perhaps more accurately, low visibility or stealth).[6] Upon reaching the top two rungs of the ladder, "formlessness" comes close to becoming a fully abstract quality, centering on a strategist's conceptual flexibility and imagination (rung 5) and a type of inner control exercised by the strategist over himself (rung 6).[7] To be clear, this six-rung conceptual "ladder" is a modern analyst's construct, not a concept found as such in the Sun Tzu text.

Treating logistics as the physical foundation of military flexibility, the bottom rung of the ladder (wu xing 1) pertains to formlessness in a logistics sense. This type of formlessness is epitomized by Sun Tzu's aspiration (e.g., Passages #8.2 and #8.3) for the army to live off the enemy, freeing it from dependence on a supply line that imposes a type of logistics form – quite possibly a draconian one given the massive food and other supply needs of the 100,000-man armies Sun Tzu talks about.[8]

The Chinese term for formless, 'wuxing,' means being invisible, imperceptible, and literally, without a concrete form. Compatible with Chinese usage, the predicate "is formless" can literally mean several things:
1. x is formless if it has no definite form; or
2. x is formless if it is invisible and intangible; or
3. x is formless if it does not have material existence.

Something can be formless in the first sense without being formless in the second sense. Something can be formless in the second sense without being formless in the third sense. We can say that a cloud or fog is formless in the first sense, while air is formless in the second sense. Spirits or ghosts would be formless in the third sense.

(Liu, pp. 191–92)

[5] In engaging with a complex, elusive concept it is often productive to juggle its several meanings, much as Ames encourages us to do for shi (pp. 175–76 above). In that spirit, it is a useful exercise to relate Liu's three-pronged conceptualization (previous footnote) to Sun Tzu's ideas. The bottom two rungs of the formlessness ladder have affinities with Liu's meaning #1 (formlessness as absence of definite form). Sun Tzu's likening of an army to water (Passage #8.1) is also a pertinent image.

Of course, in any such proposed correspondence limitations need to be recognized. Figure 3 is geared to clarifying wu xing as a military or strategic concept, whereas Liu's focus is philosophical.

[6] Continuing with footnote 5's exercise, the third and fourth rungs of the formlessness ladder have affinities with Liu's meaning #2 in its formlessness-as-invisibility aspect (intangibility, not so much; weapons and soldiers are tangible!). Unwritten information secretly percolating via spy networks (Passage #8.6) also illustrates Liu's meaning #2, in many ways better than any hidden army (though, of course, the human actors involved are both visible and tangible, if only one knew who they were!).

[7] The top two rungs have affinities with Liu's meaning #3 (formlessness as absence of material existence). Traits of flexibility in the mind of a ruler or commander – which in Sun Tzu's pre-neuroscience age might be treated as having no material existence whatever – would illustrate Liu's meaning #3, as would a general's introspective steps geared to pruning his own extraneous preferences.

Modern English-language usage equating spies with spooks also gestures to meaning #3, though for Sun Tzu's human spies Liu's meaning #2 seems the relevant one.

[8] Illustrating the idealized (even faintly parochial?) quality of Sun Tzu's advice to an army to live off the land, supply limitations long presented barriers to Chinese armies attempting to operate on Inner Asian steppe where living off the land was impractical, certainly for any extended period of time. See Perdue, Introduction footnote 52, pp. 522–53.

Logistics formlessness would also have intangible aspects, e.g., associated with logistics organization and planning (topics Sun Tzu does not address as such). But, centering as it would have done in Sun Tzu's time (as indeed in more modern times) on matters like an army's supply of food and ammunition, *wu xing* 1 formlessness is never far removed from highly tangible concerns.[9]

One rung up the conceptual ladder is formlessness in a tactical sense (*wu xing* 2). A fine statement of the requisite tactical profile – presenting no salient tactical focus for the enemy – comes from Passage #8.1's first verse:

> VI.8.
> Therefore, against those skilled in attack, an enemy does not know where to defend; against the experts in defence, the enemy does not know where to attack.

Achieving such a posture is facilitated by a high level of tactical flexibility and mutual support. Through its felicitous image of the "simultaneously responding snake" Passage #8.4 is a word picture of this state of affairs. Details of how to achieve relevant force dispositions in the warfare of his time are not spelled out by Sun Tzu; they would be antiquarian today and need not be explored here.

Wu xing 3 is a prominent part of Sun Tzu's formlessness thinking. Passage #8.1 is an anchor here. Read on the most elementary level it plays up need for steps that, in modern terminology, promote operations security (in a military sense), denying the enemy information that could give insight into one's own army's capabilities, dispositions, and plans. Passage #8.5, especially verse V.18 there, suggests that a contrived appearance of confusion may be a significant resource here – possibly at times even more of a resource than measures to promote secrecy.

World historical experience with fog of war suggests that Sun Tzu's *wu xing* 3 aspirations, lofty as they are, were at least within a realm of possibility in his era even for large conventional armies, given the primitive means of surveillance and reconnaissance then available and especially the absence of cavalry.[10] Passage #8.7 gives a flavor of some of those means and, by implication, also their limitations. Passage #11.26A profiles the sorts of information that efforts to pierce *wu xing* 3 might yield. (Passage #12.26 mines the same vein of ideas from a broader perspective.)

---

[9] A situation of unchallenged logistics "plenty" might be regarded as a kind of logistics formlessness, since the commander's usual headache of logistics limitations (i.e., tyranny of "logistics form"!) is thereby alleviated, if only temporarily. A famous (if probably apocryphal or exaggerated) *Romance of the Three Kingdoms* anecdote involves a ruse by which Zhuge Liang tricked Cao Cao's men into bombarding Zhuge's river flotilla concealed in a fog with thousands upon thousands of arrows and crossbow bolts – which harmlessly accumulated in bundles of straw placed there for just that purpose on Zhuge's light craft. Hastening downriver, Zhuge Liang's men mockingly called out "Thanks to the prime minister [Cao Cao] for the arrows!" See Moss Roberts translation (cited on p. xxiv above), chapter 46, pp. 353–55. Logistics formlessness has here become deception's progeny!

[10] Noting that even small groups of horsemen still have basic limitations as means of reconnaissance, Martin van Creveld, *Command in War* (Cambridge, MA/London: Harvard University Press, 1985), p. 23 observes: "Given these limitations, there appear to be few cases in history before 1800 in which armies were subject to direct observation by the enemy when more than twenty or so miles away." Further relevant context comes from observations of Waterfield quoted in #11 footnote 87 below.

But see Sawyer, p. 305 note 15 (skeptic's position, playing up difficulty of blocking enemy eyes). For further specifics see p. 167 above.

Yet innumerable constraints operate to make achieving wu xing 3 elusive in large-scale conventional land warfare. Espionage – conceived as an alternative strategic capability to mass armies – knows far fewer such constraints and, if done adroitly, can provide an exemplary case of wu xing 3. Passage #8.6 indicates Sun Tzu's awareness of this potential. Indeed parts of Passage #8.1, however military in motivation, might almost have been written with espionage in mind (e.g., verse VI.9, "Subtle and insubstantial, the expert leaves no trace [wu xing]," etc.).

A basic aspect of wu xing 3 – essential if a war is not to drag on indefinitely, violating other basic Sun Tzu principles (Passage #3.4) – consists in knowing when to abandon this type of formlessness, to clinch an actual victory. A doctrinal answer to that question resides in Passage #8.1, which countenances abandonment of formlessness (wu xing) after, but only after, the foe has revealed his own form (xing). This in turn suggests one of the few textually grounded insights we have regarding a probable course of events when one competent Sun Tzu–inspired actor squares off militarily against another. A plausible scenario would start with a possibly drawn-out sequence of stratagems and counter-stratagems – perhaps Byzantine but (largely) bloodless – as each side awaits flexibility-losing commitments (showing xing) by the other, while itself remaining formless (wu xing).[11] As Whaley has pointed out, parity in skill in deception seems unlikely – a point with which Sun Tzu would probably have concurred – meaning that advantage in this preliminary contest would be likely to flow to the more skilled deceiver.[12] That contest would then be followed by an effort at a decisive stroke by the side that felt it had achieved the upper hand. As a conjecture (albeit one that greatly outruns available Warring States battle information), the more effective the deception preceding that stroke, the less likely it would be highly sanguinary, at least for the victor and possibly for both sides.[13]

Whereas wu xing 3 centers on the army as a whole, its dispositions, movements, etc., wu xing 4 centers on happenings in the immediate vicinity of the commanding general, a focus that again underscores Sun Tzu's strongly personalized vantage on the higher direction of war (a standpoint already encountered in Theme #6's discussion of counter-strategist attacks).[14] Wu xing 4 centers on the need for

---

[11] Cf. Goldin, Introduction footnote 7, p. 16: "'Formlessness' . . . is a code-word for avoiding any type of committed formation until the enemy has already disclosed his [form]. It is the enemy who determines how one will destroy him . . ." (Emphasis supplied.) Sawyer, Introduction footnote 55, p. 239 makes the clarifying point that "the formless can also be concealed within false appearances, so that being formless does not necessarily mean being amorphous."

Lau, 1965 article, p. 332 translates the start of verse VI.13 (Passage #8.1) in a way that points to an active duel between xing and wu xing: "Hence by making the enemy show himself while remaining invisible, I shall be concentrated while the enemy will be divided." Similarly, see Sawyer, p. 316 note 92 (noting possible causal reading of verse VI.13, involving steps to induce a foe to betray his "form").

[12] Whaley, #7 footnote 4, p. 76.

[13] Cf. verse II.19 (Passage #8.2) recommending good treatment of captives.

[14] While denying the enemy knowledge of military secrets is a shared goal of rungs 3 and 4, the enemy's métier to be guarded against may be very different – e.g., in the case of rung 3 involving spies lurking in dense tangled undergrowth who observe one's army's movements (Passage #8.7's verse IX.18); in that of rung 4 involving the "most intelligent people" who as secret agents worm their way into a general's inner circle of confidants (Passage #8.12's verse XIII.23).

the general himself to be opaque about his intentions and plans even in dealings with his own people, high-level and menial alike (from whose impressions and gossip spies in his own camp can all too easily relay that general's innermost thoughts and worries back to unfriendly quarters). Passage #8.9's verse XI.43 makes the basic point:[15]

XI.43.
He should be capable of keeping his officers and men in ignorance of his plans.

Read from a *wu xing* 4 standpoint, verse XIII.5's draconian advice (Passage #8.8) points to an effort to reassert *wu xing* 4 after it has been compromised (a most unusual case of Sun Tzu playing catch-up). An arguably more efficacious, and certainly more interesting, line of defense against leakers may reside in the natural opacity of the calculations of a deep strategic thinker (see verses VI.25 [Passage #8.1] and IV.8 [Passage #8.10]).

For Sun Tzu, "inscrutability" is a *personal* attribute of the general (verse XI.42 in Passage #8.9). While such a quality, available whenever needed, is likely to be a general's asset in any age of war, in times more recent than Sun Tzu's a spy's sneaking a peek at written orders or files could be a viable route to puncturing information security at the top, one whose effectiveness has little to do with the general's inscrutability. In Sun Tzu's time – when paper was not yet a medium of written communication, and its alternatives were awkward and limited at best – written orders would not exist in most situations; the "files" (certainly in any sense approximating a modern Weberian one), not at all.[16] Here is a case where there is need to be alert to how conditions of a bygone time might shape Sun Tzu's counsel.

With the flourishing of the *bingfa* (military methods) genre of texts in Warring States China, the instinct that animated quest for *wu xing* 4 would in all likelihood also have fueled a desire to keep such doctrine statements closely held (though, to be clear, there is nothing directly about this in Sun Tzu). Such a desire would have been reinforced by a tradition of respect for the potency of doctrine statements, at times invested with a magical aura of esoteric wisdom, that had taken hold by Warring States times (interwoven with trends toward military specialization).[17] Efforts to deny

---

[15] The possibility of making an exception for certain high-level subordinate commanders or staff is raised in #8 footnote 71, though there is no textual support in Sun Tzu to suggest making such an exception. Much closer to our own times, some successful generals have taken a practice of information denial to seeming extremes. Stonewall Jackson is a case in point: "The infinite pains with which Jackson sought to conceal, even from his most trusted staff officers, his movements, his intentions, and his thoughts, a commander less thorough would have pronounced useless." See G. F. R. Henderson, *Stonewall Jackson and the American Civil War* (London/New York: Longmans, Green, and Co., 1909), Vol. I, p. 421 (quoted by Giles, p. 131).

Taking verse XI.43 at face value, Sawyer, Introduction footnote 55, pp. 58–59 comments that Sun Tzu is here out of step with "the modern emphasis upon conveying the commander's intent to the subordinate commanders." In the annals of Western military history a classic case of the latter would be the "Nelson doctrine," a term applied to Lord Nelson's practice of candidly sharing his thinking and plans with senior officers prior to naval action, so as to ensure unified effort when the moment for action arrived. See p. 76 of A. T. Mahan, "The Naval War College," *North American Review*, July 1912, 196(680), 72–84.

[16] For background on mediums of written communication in early China see #1 footnotes 1 and 3.

[17] See Lewis, Background footnote 11, pp. 98–103 on "The commander and texts" (noting "stories of revealed, magically potent military treatises").

access to such privileged knowledge to potential adversaries, who might use it to upend one's own side's strategy, appear in later Chinese history.[18] A case in point is a Tang dynasty minister who objected to a 730 AD Tibetan request for access to Chinese texts regarded as having military or strategic doctrine content.[19]

Sui generis in Sun Tzu, but again pointing to a general's formlessness of a sort, is verse XI.54 (Passage #8.11), urging disregard for the tyranny of precedent. There is a potential tip-of-the-iceberg aspect here, since verse XI.54 might appear to endorse a radical precedent-defying stance applicable to a great range of customary or institutional practices, civilian as well as military. Yet verse XI.54 lacks support from elsewhere in the text, a clue that a more limited interpretation is in order.[20] One such interpretation is that Sun Tzu's focus here is on military administration, which in Warring States thinking was emerging as a "separate sphere" divorced from civilian life (including in its legal and administrative aspects).[21] Applied to verse XI.54, such separation suggests, not a call by Sun Tzu for precedent-defying thinking in general, but a fairly narrow warning that too predictably adhering to established administrative norms might undercut a general's *wu xing*, possibly opening a door to a foe's machinations.[22] An observation of Admiral Eccles applies: a commander, he wrote, "can be either the master or the prisoner of the institutions."[23]

Unstated by Sun Tzu, but compatible with his holistic leanings, is that some information needing protection may center, not on specific military secrets (important as those surely are), but rather on the sorts of insights a good ethnographer in the general's entourage might pick up: the texture of that general's daily life; his relations with those around him; how he spends his time; what's on his mind; what is his emotional state; recent changes in any of these; etc. Such "soft" or

---

[18] Needham & Yates, p. 88 points out an interesting cultural tension between Chinese tendencies to cloak military thought in secrecy, on the one hand, and its widespread popularization, on the other. Hints of the latter already appear by the time of Han Feizi (died 233 BC). See #4 footnote 6 above.

[19] The official's memorandum warned against sharing those works with the Tibetans who "though ... of a wild and warlike nature ... are firm in their plans, intelligent, industrious and resolved to learn with rapt attention." The texts were shared anyway. See Needham & Yates, p. 73; see also p. 12 of Thomas A. Marks, "Nanchao and Tibet in South-western China and Central Asia," *The Tibet Journal*, Winter 1978, 3(4), 3–26.

[20] Sensing something awry, Griffith, p. 139 footnote 3 suggests that verse XI.54 is "obviously out of place." There is certainly a harmonic in Sun Tzu emphasizing administrative consistency that clashes with the liberating counsel of verse XI.54 if too literally or broadly understood. See, e.g., verse X.14 (Passage #13.33) pointing to need for "consistent rules" (Mair, p. 115's "constant regulations").

[21] See Lewis, Background footnote 11, p. 127 (observation is quoted in #13 footnote 26 below).

[22] Compare *Laozi* XXXVI, observing that the "instruments of power in a state must not be revealed to anyone." Glossing that observation, D. C. Lau adds: "in the wrong hands, *even the knowledge of how they are dispensed can be turned into a source of power*." (Emphasis supplied.) See Lau, #1 footnote 34, p. 95 and his accompanying explanatory footnote.

[23] Eccles, Preliminaries footnote 5, p. 201. A case can also be made for a still narrower reading of verse XI.54, based on positing that verses XI.51–XI.53 are an interpolation. For that reading, which would focus verse XI.54's counsel on a death ground situation, see Appendix 8 anchor footnote 19.

"atmospheric" information can be of real military and strategic value (even at times of greater value than intelligence on supposedly "harder" matters).[24] There could be opportunities here for the kinds of exceptionally intelligent spies (now meaning enemy spies) of which verse XIII.23 (Passage #8.12) speaks. The best defense against such penetrations is for a general to enforce a kind of information blackout around his person. Because such a blackout requires countless day-to-day judgment calls, great and small, by the general, it also assumes aspects of a leadership style – one that pivots on that general's poise and self-control, never providing telltale crumbs of information revealing his actual thinking and emotions.[25] Realizing such a style in practice, amid all the demands and stresses of high command, is no mean accomplishment in any age of war. But given early Chinese theories of personality and character – ones that assigned credence to, and placed emphasis on, an ability to read the inner person behind the lacquered mask – what Sun Tzu Passage #8.9 is demanding of the general is a truly extraordinary performance by the standards of his own Warring States time.[26]

The spirit of Sun Tzu's *wu xing* 4 thinking has never been better summed up than by the inspired metaphor of an enemy who stands "as if on a lighted stage," visible to all the theatergoers who fill the darkened theater – to wit (in Mao's revolutionary warfare context), the common people who observe and provide information on the

---

[24] E.g., enabling the enemy to get an inside reading on whether a general has lost his courage (see verse VII.20 in Passage #6.7).

[25] A comparison with Daoist ideas is instructive here. On first impression, there is kinship between Sun Tzu's *wu xing* 4 and the tonality of a list from the *Zhuangzi*, a core Daoist text: "In his movement [the sage] is like water; in his stillness he is like a mirror; in his response he is like an echo. Indistinct, he seems shadowy; silent, he seems limpid." (Lau, #1 footnote 34, p. 48.)

However, by contrast to a Daoist sage whose command control responsibilities are nil, and who therefore has enormous flexibility as to what posture or lack of posture to assume at a given moment, a Sun Tzu general leading an army in the field – or, for that matter, a grand strategist – must aim for *wu xing*, including *wu xing* 4, under often difficult circumstances not of his own choosing.

[26] Relevant traditions and methodologies, having broad roots in early Chinese thought, are analyzed by Sawyer, #1 footnote 3, chapter 11, "Knowing men," pp. 309–59. See also Eric Henry, "The motif of recognition in early China," *Harvard Journal of Asiatic Studies*, 1987, 47(1), 5–30. The key term zhi ("to know/knowing") has definite "disguise-piercing" aspects (Henry article, pp. 8, 15). Zhi also appears in Sun Tzu's call for "men who know the enemy situation" (Passage #11.6's verse XIII.4).

An underlying premise, which Sawyer terms the "problematic theory of transparency," is that "the major and minor aspects of human behavior invariably express an individual's true character, motives, and intent. It does not necessarily deny the ability to assume a façade, but it does assert that such fabrications will eventually be pierced by the knowledgeable." See Sawyer, #1 footnote 3, p. 316. As Sawyer's label "problematic" suggests, this early Chinese cultural-philosophical position seems over-optimistic (albeit that rapidly evolving twenty-first-century digital technology – face recognition, big data, etc. – may be making it more credible than ever before!). It is, however, not necessary to go that far to appreciate the challenge faced by a Sun Tzu general living and operating under the ever-watchful (or more exactly, prying) scrutiny of an entourage. Compare Sawyer, #1 footnote 3, p. 318 (noting examples of "concealed motives being deduced from observing someone so self-absorbed in his plans for mounting an attack that he failed to perceive his immediate surroundings").

enemy's situation.[27] The essence of Passage #8.9's counsel is that Sun Tzu's general must take care never to sit on such a lighted stage himself.[28]

Wu xing 5 and 6 now bring in the enemy more directly – a concrete foe having his own xing. Wu xing 5 aspirations are conveyed by one of Sun Tzu's water images:

VI.28.

> And as water shapes its flow in accordance with [yin] the ground, so an army manages its victory in accordance with [yin] the situation of the enemy.

(Passage #8.1)

Here water's "formlessness" is linked to the Chinese philosophical concept yin whose basic meaning is responsiveness to context. In a military setting yin has connotations of both readiness and flexibility in exploiting openings presented by the enemy.[29] As such it has both tangible and intangible aspects. It is the latter that are emphasized in Figure 3's wu xing 5 rung. One basic tool for realizing yin is a commander's having a "bottomless quiver" of strategic and tactical ideas and expedients (see Passage #8.13's verse V.6). A second is deep unpredictability and originality in their use (Passage's #8.1's verse VI.26), thereby denying the foe opportunity to narrow down the options he needs to plan for by studying stratagems past.[30]

Achieving wu xing 5 has everything to do with imagination (Passage #8.13 makes the point with vivid imagery; Passage #8.14 is an illustrative scenario). Doing so also draws on sheer speed in operational thinking and execution (see Passage #8.15 as well as the "swift as a hare" imagery of verse XI.61 [Passage #8.26A]).

The sixth, topmost, rung of the Figure 3 formlessness ladder refocuses Sun Tzu's formlessness project from formless means to ends that likewise exhibit as little form as possible (albeit that having coherent strategic goals unavoidably sets limits, imposing a type of form). In a nutshell, in Sun Tzu's way of war it pays to be indifferent across many outcomes – indeed all but the truly essential ones, few in number.[31] An ideal general should cultivate the kind of formlessness that "cherishes" (or conversely, "abhors") as little as possible. Any other posture gives unnecessary traction to the enemy, who must be expected to manipulate it ruthlessly.

---

[27] Samuel B. Griffith, Mao Tse-tung on Guerrilla Warfare (New York: Praeger, 1961), p. 23. Wu xing 4 military thinking also finds civilian echoes in early Chinese rulers' penchants for observation towers or similar lofty structures that not only "provided visible signs of the ruler's power [but] were also tools of invisibility. They masked the ruler's person even while making his presence known, demonstrating his ability to scrutinize his people and his foes without himself being observed." See Lewis, #3 footnote 34, p. 155.

[28] Of course, tradeoffs and judgment calls arise. A case of a general risking loss of wu xing 4, at least to some degree, is that of Wang Jian, a late Warring States general whose military prowess played a key part in unification of China under Qin. At a critical point in the final campaign against Chu, Wang Jian is said to have shared meals with his troops in order to obtain an unfiltered sense of their morale. See Shiji 73 (relevant passage is translated in Nienhauser, Vol. VII material cited in #12 footnote 44).

[29] For conceptual discussion of yin see Ames, pp. 83–84; Lau & Ames, p. 72.

[30] A study of Sun Tzu that emphasizes these strands of his thinking is Derek M. C. Yuen, Deciphering Sun Tzu: how to read the "Art of War" (New York: Oxford University Press, 2014), building on Yuen's article of the same title in Comparative Strategy, 2008, 27(2), 183–200.

[31] This vein of thought also harmonizes with strands of Daoist thinking. See Laozi XLVI: "There is no crime greater than having too many desires." (D. C. Lau translation, #1 footnote 34, p. 107.)

Wu xing 6 (which, on a psychological level, entails a type of control exercised by the general over himself) is never spelled out in so many words by Sun Tzu, but has broad anchoring in the text through special cases. Passage #8.16 highlights a case of failed formlessness of a plane of grand strategy, centering on failure to escape the tug of extraneous emotional forces (possibly precipitating an unnecessary, costly, even catastrophic war!). Passage #8.17 sets forth a more general formulation. Passage #8.18 extols a general whose conduct is thoroughly divested of assorted extraneous motivations Sun Tzu describes there. Sun Tzu's support for a general's defying ill-conceived orders from the ruler (more sources of extraneous *xing*!) also illustrates a *wu xing* 6 mindset.[32]

Wu xing 6 formlessness applies to military assets, in the sense that no such asset should be "cherished" for reasons that range beyond its prosaic military or strategic value (cf. Passage #8.17's verse XI.28). This line of thought also warns of a general's (or ruler's) personal entanglements with the trappings of prestige and prominence of social position – all being badges of "form," not formlessness, and flagged as potential targets by verse I.23, "Pretend inferiority and encourage his arrogance" (Passage #8.20). Passage #8.19 extends the point to include other character traits that import extraneous considerations. More broadly, *wu xing* 6 points to casting a cold, appraising eye over the many entities (people, places, assets, . . .) on which human actors (even some otherwise great strategists!) so commonly place an out-sized value, possibly for reasons bound up with personal identity – *any one of which can and will become a pressure point for that reason.* This minimalistic mindset, devoid of sentimentality, augurs willingness to make sacrifice plays whenever strategically or tactically advantageous. Sun Tzu's death ground thinking points in that direction even though his script predicts victory (e.g., Passage #8.21's verse XI.55). Adopting the perspective of the ruler (who is observing the proceedings), the spirit of *wu xing* 6 also pervades the story of Sun Tzu and the concubines. A moral of that story (one that the ruler has some difficulty grasping!) is need for the ruler to take in stride the killing of a concubine or two as an incidental cost of a training demonstration.

Pushing this line of thinking yet one step further, a corollary is that *a strategist in Sun Tzu's mold should not even cherish deception for its own sake,* over and above what instrumental uses it may have (which, of course, are often great). Again just to be clear, there is nothing directly in the Sun Tzu text about taking such a position, but that is precisely where Sun Tzu's *wu xing* thinking, pursued to its logical conclusion, points. Indirect support for such a position may be drawn from the Janus-faced nature of the *qi/zheng* concept, which in present context may be treated as a warning not to become so enamored of deception and other "crafty" measures (*qi*) as to lose sight of the never-ending need for further measures that are *zheng* – i.e., "straight-forward."[33] In this aspect, "pure" Sun Tzu stands apart from the world portrayed in the *Romance of the Three Kingdoms*, where deception often seems prized for its own sake by the dramatis personae (quite possibly, of course, because in a work of fiction such a proclivity is an audience-builder!).

******

---

[32] See Theme #13 analysis, approaching this topic from Sun Tzu (1), (2), and (3) standpoints.
[33] These translations of *qi* and *zheng* are from Lau's 1965 article. For further analysis see Theme #14.

Given his investment in the concept, it is only natural that Sun Tzu should look for clever ways to develop formlessness thinking a step or two further. That is one way to look at verse XI.55's thrust (Passage #8.21). From a formlessness standpoint, the essence of verse XI.55 is that, in special circumstances that Sun Tzu spells out there, transforming a type of formlessness into a certain polar opposite of formlessness may (paradoxically!) be a source of great advantage. Starting with deliberate opacity of the general vis-à-vis his own people (i.e., exploiting *wu xing* 4 formlessness),

> XI.55.
>
> Set the troops to their tasks without imparting your designs; use them to gain advantage without revealing the dangers involved. ...

[Emphasis supplied.]

the idea is to engineer a zero-flexibility situation ("death ground") for one's *own* troops, which in turn elicits "nonlinear behavior" from them providing the leverage that snatches victory from the jaws of defeat. A little contorted as it is (certainly by some ethical standards; see Theme #10 analysis), this has the makings of an insightful strategic logic chain. Sun Tzu here sets a foot in the intellectual realm explored in modern times by work of Thomas Schelling, one of whose signature concepts is precisely the engineering of strategic situations that lock in a commitment, seemingly to the disadvantage but (when the logic is pushed a step or two further) possibly greatly to the advantage of its engineer.[34] In some respects verse XI.55's scenario is actually richer than many of those sketched by Schelling, since Sun Tzu's troops being committed to grave peril are seen as endowed with their own psychological dynamics – perhaps not those of a rational actor but nonetheless real and essential to the viability of the scenario (as Sun Tzu is well aware). The death ground scenario Sun Tzu sketches is high risk indeed, and might not work out as intended (a feature shared with various Schelling scenarios). However, verse XI.55 sets forth an imaginative shard of doctrine of genuine value, if only to alert a general to the possibility that the enemy might attempt a similar approach – and pull it off!

Roads Not Taken by Sun Tzu in Developing Theme #8

Not explored by Sun Tzu (his zero-sum orientation is felt here!) is whether there might be yet a further type of "formlessness," assignable to a new top rung ("*wu xing* 7") on the "formlessness" ladder because it implicates issues of not only strategy, logistics, and tactics but also of sociopolitical identity. In purest expression, *wu xing* 7 would treat the overarching strategic goal as itself flexible – for example, admitting grand strategic compromises that lay foundations for a lasting peace, rather than the conquest of other states (and survival of one's own) that seems to be Sun Tzu's sole preoccupation; or for that matter, conceivably even a strategic surrender by one's own state![35]

---

[34] There are many variants, some pushing Sun-Tzu-esque thinking further than Sun Tzu ever did. See, e.g., Schelling, #5 footnote 8, pp. 195–96: "to compel by threat an enemy's retreat I have to be committed to move forward, and this requires setting fire to the grass behind me with the wind blowing toward the enemy."

[35] Of course, *wu xing* 7 strategic surrender by a major state actor will unleash a host of lesser *xing* ("forms") previously muted or marginalized: "cultures of defeat," often multiple; surge in black markets; new social movements; and more. For analysis of the manifold ramifications and subtleties of a major strategic surrender case, touching the lives of ordinary people, not just ruling

Relatedly, Sun Tzu touches not at all on the nature and desirability of "formlessness" on a level of statecraft – e.g., what formlessness might mean in treaty negotiations or other diplomatic contexts (perhaps Sun Tzu's *wu xing* thinking favors secret treaties?).[36] Sun Tzu the military thinker is once again in plain view here.

Returning to the Figure 3 ladder, a noteworthy instance where Sun Tzu does not push the outer bounds of *wu xing* 6 as far as possible (or as a "pure" doctrinal interpretation of Sun Tzu might suggest) pertains to self-defilement or similar behavior. One of Passage #8.19's verses, suggesting that a sense of honor could be a liability,

VIII.21.
If he has too delicate a sense of honor you can calumniate him.

might be extended to encourage a Sun Tzu–inspired strategic actor to carry out acts of self-defilement – arguably one of the purest expressions of formlessness by an individual, and in some situations a potent strategic and tactical tool. Certainly self-defilement (*ziwu*) is a known expedient in Chinese history, used to establish certain kinds of credentials (say, as a device by which a too successful general deliberately paints a corrupt image of himself, thereby giving credibility to the claim that he poses no threat to a suspicious ruler).[37] Under some circumstances such a tactic could serve as a counter-counter-strategist measure, reducing the odds that a strategist is sidelined or dismissed (see p. 209 above). Yet Sun Tzu is silent here.

Meanwhile Sun Tzu says nothing to encourage a commander to admit errors and missteps, as a prelude to rectifying them, which may be regarded as a desirable type of ego-denying formlessness (a facet of *wu xing* 6). About the closest Sun Tzu comes is noting that arrogance in an *adversary* may be an exploitable trait (verse I.23 in Passage #8.20), though this observation is plainly a step away from readiness candidly to admit one's own errors. Given the ambitiousness of Sun Tzu's way of war, where even the best commander must expect to err from time to time – Sun Tzu's seeming denial of chance (p. 64 above) to the contrary – there is a hole here.

On a specifically military level, there is also no basis in the Sun Tzu text for believing that Sun Tzu – here meaning Sun Tzu (1) – endorsed or even contemplated a version of "formlessness" arising from intentionally giving junior officers latitude to undertake independent initiatives within limits set by their superior's overall intent, along the lines of modern "mission command" thinking. For similar reasons, Sun Tzu is not a pioneer of guerrilla warfare, with its almost inevitable tendencies toward decentralization. Sun Tzu by instinct is a centralizer, not a decentralizer.[38]

---

elites, see John W. Dower, *Embracing Defeat: Japan in the wake of World War II* (New York: W.W. Norton, 1999).

[36] For a modern perspective on the secret treaty phenomenon see Megan Donaldson, "The survival of the secret treaty," *American Journal of International Law*, 2017, 111(3), 575–627.

[37] For an example from Western Han times not long after the end of the Warring States – where an overly popular general deliberately undertook to cultivate a reputation for financial corruption, so as to seem less of a threat to the Han emperor – see Shiji 53 (Watson translation cited on p. xxiii above), *Han Dynasty I*, pp. 95–96. The episode is also recounted in Chen, #4 footnote 46, pp. 235–36.

[38] For further analysis of Sun Tzu's stance see pp. 437–38, 501, and pp. A-26ff. in Appendix 15.

Sun Tzu (2) ("Sun Tzu Extended") and Sun Tzu (3) ("Sun Tzu Analogical") Frontiers of Theme #8

Whether, and in what sense, attainment of wu xing falls within the realm of possibility is a highly contextual question. Modern high tech (and long before it, biological evolution!) sets off many "arms races" whose outcomes, depending on the balance of factors at a given moment, may either facilitate or frustrate Sun Tzu's wu xing aspirations. Yet many structural tendencies of modern life seem clearly on the side of wu xing. In particular, there are far more options for "fighting stealthily," often using clandestine rather than merely covert means, than Sun Tzu ever dreamt of. Without aiming for an exhaustive list, the following types of warfare – one both ancient and modern, one modern, one hypermodern, and all in their own way possessing track records of high efficacy – stand out as conducive to approximating Sun-Tzu-esque formlessness under realistic operational conditions:

    (a) guerrilla warfare;

    (b) naval undersea warfare;

    (c) cyber warfare.

There are insights to be had from analyzing the wu xing scorecard of each of these types.[39] This will be done here, first, for guerrilla warfare (one of Sam Griffith's other military analysis specialties!), then for the other two cases. Of particular interest are interference patterns where achieving one type of formlessness on the wu xing ladder may frustrate or undercut formlessness of another type.[40]

"War in the shadows," as guerrilla warfare has been evocatively labeled,[41] represents a type of warfare that draws immense strength from its qualities of formlessness. Guerrillas commonly make do with only the most rudimentary logistics, living off the enemy just as Sun Tzu encourages (wu xing 1) or, alternatively, off the largesse of a (sometimes) sympathetic civilian population. Guerrilla forces are tactically

---

[39] Also meriting mention here is "swarming" as a military concept. Although the Sun Tzu text does not point to swarming as such, several verses in Passage #8.1 are suggestive of the concept, among them Passage #8.1's verse VI.24 ("without ascertainable shape") and verse VI.28 (water image).

    While swarming is not so much a type of warfare as an operational style shared by a number of different types, swarming has much in common with Sun Tzu's wu xing 2. Exemplified by tactics of Chinggis Khan's Mongol army, swarming is

> seemingly amorphous, but it is a deliberately structured, coordinated, strategic way to strike from all directions, by means of a sustainable pulsing of force and/or fire, close-in as well as from stand-off positions. It will work best – perhaps it will only work – if it is designed mainly around the deployment of myriad, small, dispersed, networked maneuver units . . .

See John Arquilla and David Ronfeldt, *Swarming & the Future of Conflict* (Santa Monica, CA: RAND, 2000, p. vii). Swarming phenomena are also familiar from some non-human biological contexts. Cf. Mark W. Moffett, "Ants & the art of war," *Scientific American*, 2011, 305(6), 84–89 (mentioning Sun Tzu in context of swarm intelligence, though the verse cited is XI.29 [Passage #8.15] on speed as the essence of war, which captures only one facet). A core military benefit of swarming is the extreme disorientation it commonly produces in its victims (e.g., making them feel as if the enemy's forces are everywhere yet nowhere). That kind of psychological impact is vintage Sun Tzu.

[40] For some candidates (not an exhaustive list) see #8 footnotes 43, 47, and 51 below.

[41] Robert B. Asprey, *War in the Shadows: the guerrilla in history* (2 vols.; Garden City, NY: Doubleday, 1975).

amorphous. Verse VI.16 (Passage #8.1) captures with uncanny accuracy – all the more striking since Sun Tzu nowhere alludes to guerrilla warfare! – the frustrations of regular force commanders trying to cope with enemy irregulars:

> VI.16.
> (Translation from Lau, 1965 article, p. 330.) "It is the one who has to prepare against his enemy who is few and the one who makes his enemy prepare against him who is many."

The amorphous quality of guerrillas facilitates *wu xing* 2 in both offense and defense (the latter evoking Sun Tzu's simultaneously responding snake [Passage #8.4]). Guerrillas often hide in complex terrain (jungles, mountains, etc.) (*wu xing* 3), frequently the more complex the better, also melting away when hard-pressed into the arms of a friendly populace – complex terrain of another sort! (again *wu xing* 3). Formlessness is abetted when few if any outsiders have contact with guerrilla leadership based in the hinterland or a foreign sanctuary, so that reliable knowledge of that leadership and its limitations or vulnerabilities is scant (*wu xing* 4). Drawing on excellent intelligence and bound by few organizational or planning rigidities of the sort deeply ingrained in conventional militaries, guerrilla commanders are well positioned to think and act imaginatively (*wu xing* 5).

The circumstances that so often launch and sustain a guerrilla war – e.g., growing out of an ideology-driven revolutionary movement, or a populace battling a foreign invader – are often ideally conducive to *wu xing* 6 formlessness, with its characteristically ruthless disregard for all incidentals (among them, often human rights).

Yet all these harmonies with Sun Tzu should not obscure the fact that many of guerrilla warfare's *wu xing* qualities are largely possible because of the decentralization that so often goes hand in hand with guerrilla operations – a pattern of command control that Sun Tzu does not explicitly discuss in any context and which he might well have rejected out of hand had he thought about it, at least for the armies he knew. Put simply, decentralization motifs find scant support in Sun Tzu's thinking, where the commanding general's will is all-important and his army is but the instrument for its expression, the more unified the better (as the latter part of Passage #8.11's verse XI.54 illustrates and verse XI.41a [Passage #13.24] reinforces).[42] The Sun Tzu text also contains no trace of the all-important political active ingredient of Maoist "people's war" rooted in mobilization of the civilian masses. Those omissions suggest that Sun Tzu has significant limitations as a guide to guerrilla warfare, certainly in its Maoist incarnation. Following Sun Tzu too faithfully might actually become a source of guerrilla weakness, e.g., promoting military overcentralization as well as a disregard of mass political mobilization steps.

The six rungs of the formlessness ladder also provide a template for analyzing ways in which guerrilla formlessness may fail, undercutting the guerrilla's efficacy and identifying possible pressure points for an adversary. For example, *wu xing* 1 failure arises when a guerrilla effort becomes dependent on resupply from an external sanctuary and the associated supply line, which imposes a version of

---

[42] Even across a chasm of centuries, there is a certain comradeship between Sun Tzu's predilection for centralization and Stalin's posture at an early stage of his war with Germany. With his regime's survival hanging in the balance, Stalin nonetheless "was scarcely likely to encourage 'guerrilla' initiative, with its inevitably decentralized emphasis, during a period of high political tension." See John Erickson, *Stalin's War with Germany*, Vol. 1: *The Road to Stalingrad* (New Haven, CT/London: Yale University Press, 1999), p. 240.

form.[43] Some types of terrain may also give rise to wu xing 3 failure, because conducive to surveillance from the skies and use of airpower against guerrillas. High-tech intelligence, surveillance, and reconnaissance tools – a case in point would be drones – make wu xing 3 formlessness much more difficult for guerrillas (though their effects may be blunted in congested urban environments, where supports for wu xing 3 formlessness are as much social as physical). Extraneous entanglements by key guerrilla leaders in human social networks – family, other intimates, etc. – may also give rise to wu xing 4 failure. Wu xing 5 failures (and often wu xing 6 failures too) are illustrated by cases where a social movement with a guerrilla arm allows itself to be fettered by unrealistic theories or agendas (a situation in which the Chinese Communists found themselves in 1927, when a then-ascendant leadership coterie insisted on attempting urban uprisings, with disastrous results for the Communist cause, rather than launching guerrilla warfare in China's vast rural hinterland).

Worlds apart from the ragged guerrilla, another kind of warfare that also comes close to epitomizing formlessness is naval undersea warfare. The ocean is a huge, amorphous place, well suited to concealment. For that reason over much of human history blue water naval warfare has often had opportunities to achieve much more perfect realizations of wu xing than has conventional land warfare. But a naval version of "the world according to Sun Tzu" truly comes into its own when technology allows naval warfare, as Bernard Brodie put it well, to enter a new dimension underseas.[44] From a Sun Tzu formlessness perspective, a basic breakthrough in giving operational expression to undersea formlessness came about through Admiral Rickover's 1950s nuclear Navy, with its submarines' then-newfound capabilities of operating submerged for fabulously long periods and across vast distances without refueling or resupply (wu xing 1, wu xing 3).[45] Fostering wu xing 4 opacity are "silent service" traditions – sometimes described as cult-like! – of extreme reticence on the part of submariners to discuss any aspect of their service. Wu xing 4 has roots in personnel choices (epitomized by Rickover's famed highly personalized approach to officer selection[46]); in the natural physical isolation of submarines; and in the intensity and comprehensive nature of submarine force communications security.[47]

---

[43] As guerrilla or other irregular forces evolve toward ever greater reliance on high-tech weapons and other advanced capabilities, wu xing 1 may become a casualty of achieving wu xing of other types.

[44] A modern echo of the spirit of Sun Tzu's wu xing comes from Melville: "Consider the subtleness of the sea; how its most dreaded creatures glide under water, unapparent for the most part, and treacherously hidden beneath the loveliest tints of azure." See Herman Melville, Moby-Dick, or The Whale (edited with an introduction by Alfred Kazin; Cambridge, MA: The Riverside Press, 1956), p. 222.

[45] In 1960 the USS Triton traveled 35,979 nautical miles submerged. See Thomas B. Allen & Norman Polmar, Rickover: father of the nuclear navy (Washington, DC: Potomac Books, 2007), pp. 50–51.

[46] For a vignette, see Allen & Polmar (previous footnote), pp. 71–72.

[47] A suggestion, from an article written early in the nuclear submarine era, is that this same "silent service" characteristic may at times have actually retarded development of broadly based understanding of submarines as a strategic weapon system and the insights regarding best utilization of such a capability that might emerge from informed broader debate. See pp. 8–9 of Eugene M. Emme, "Technical change and Western military thought – 1914–1945," Military Affairs, Spring 1960, 24(1), 6–19. Here is a candidate for a subtle but important tension between wu xing 3

Much more could be said about particular *wu xing* structural characteristics of undersea warfare, including the long-running arms race between submarines' *wu xing* 3 and technological attempts to negate it.[48] For present Sun Tzu analytical purposes, however, perhaps the most important point to emphasize is that, over its history, there have been persistent tendencies, in many military and naval establishments and civilian leaderships, to fail to grasp the true potential of undersea warfare, and the magnitude of the threat it poses to naval surface warfare assets along with civilian merchant shipping and ways of war based on them. In present terms this is a failure of *wu xing* 5 and perhaps *wu xing* 6 as well. Whether that failure of imagination holds seeds of opportunity – or of disaster – depends, of course, on whether it is occurring in an adversary's defense establishment or one's own. Illustrating the point, naval mine warfare (along with counter-measures to it) has often tended to be an "orphan" area, neglected or marginalized by leading naval thinkers. From a Sun Tzu *wu xing* vantage, it deserves better. Likewise attention-worthy are the many twenty-first-century possibilities for unmanned autonomous or partially autonomous undersea craft with a wide range of capabilities. A failure to grasp the *wu xing* potential of undersea warfare advances is a strong candidate for "*What did Master Sun know that we still don't (or have yet to absorb adequately)?*"

In the twenty-first century, in some ways the closest approximation to Sun Tzu's formlessness is warfare conducted in the cyber realm. Illustrating *wu xing* 1, most logistics support requirements familiar from warfare over most of human history (even including guerrilla warfare) – so heavily anchored in the needs of flesh-and-blood human beings, often in large numbers – are irrelevant there. The primary asset required to launch a devastating attack is simply the knowledge that resides in human cognoscenti (e.g., in the brains of knowledgeable hackers). That is a population notoriously hard to spot, hard to track. Evoking Sun Tzu's "simultaneously responding snake" of verse XI.38 (Passage #8.4) an attack on one hacker may elicit hostile responses from many others (*wu xing* 2). Meanwhile the sheer complexity of software – and of many algorithms too – makes it a magnificent hiding place for camouflaging all manner of tools and weapons of software warfare (*wu xing* 3). Taking cues from verse VI.27, "Now an army may be likened to water ..." (Passage #8.1), the mobility over computer networks and "fluidity" (modifiability) of software in cyber environments give rise to yet further possibilities for concealment.[49]

---

and *wu xing* 5 goals – one that might find parallels in other areas of twenty-first-century strategic importance and high sensitivity.

[48] The dramatic early successes of German World War II submarine warfare in the Atlantic against Allied merchant shipping came close to epitomizing a triumph of formlessness over form – but only for a time, until (as Paul Kennedy puts it) "newer elements," many of them technological, entered the fight, laying bare the "form" of attacking submarines. For the story see his *Engineers of Victory* (cited in #5 footnote 69), chapter 1, "How to get convoys safely across the Atlantic," pp. 5–73.

[49] Building on a familar distinction of Donald Rumsfeld, a computer security analysis differentiates cyber security threats capitalizing on "known unknowns" (things we know we do not know) and others that exploit "unknown unknowns" (things we don't know we don't know). See Derek E. Bambauer, "Ghost in the network," *University of Pennsylvania Law Review*, 2014, 162(5), 1011–91, in particular pp. 1089–90. The "unknown unknowns" category evokes verse VI.24 (Passage #8.1).

Greatly simplifying the attacker's task of achieving wu xing 3 (wu xing 4 too!) is the well-known attribution problem: the difficulty, at times insurmountable with available resources, of accurately tracing a known cyber warfare attack back to the person or organization who initiated it (or sometimes even more relevantly, to whoever paid for it to happen, a further challenge in a world of advanced money laundering methods and technology).[50]

Meanwhile the same internet infrastructure that facilitates software action-at-a-distance also supports great flexibility in choice of targets for attack as well as specific means of attack, giving vast scope for an attacker's imagination (wu xing 5). Attacks can be tailored to the situation of the adversary even as Sun Tzu verse VI.28 (Passage #8.1) prescribes, with an eye to opportunistic exploitation of vulnerabilities. Diversity of possible attacks can also realize a cyberspace counterpart to verse VI.26's "Therefore, when I have won a victory I do not repeat my tactics ..."

Yet there remain basic differences between software environments and oceans. All the opacity and the remarkable suppleness of modern software coexists with layer upon layer of enormously precise structure – an attribute not shared with the sea, albeit that it too involves structure of physical and biological varieties. From a standpoint of software attacks, this further characteristic of the conflict environment is a double-edged sword. On the one hand, it affords a technically competent attacker all the advantages of a cornucopia of vulnerable targets as well as hiding places. On the other hand, that same precise structure means that cyber attacks can sometimes be traced back to their source, placing the attacker on the "lighted stage" of which Griffith wrote (see pp. 276–77 above) and losing the kind of invisibility and anonymity that is a hallmark of wu xing 3 and 4 cyber formlessness. Cyber environments also come richly endowed with memory capabilities – formidably detailed, dispersed, often next-to-impossible to erase, thereby tending to make memories of all sorts of human communications and actions taken there hard to eradicate fully. Here is a fact of software-dependent life – and basic feature of twenty-first-century individual and collective memory – that cuts markedly against achieving anything like a full measure of Sun Tzu's "formlessness." Much of cyber warfare's unique texture comes from the way it threads a path between manifestations of formlessness and form. As with a hider/seeker dynamic in naval undersea warfare, arms races are born.[51]

As is often the case with Sun Tzu, Sun Tzu's most basic contribution here may be on a psychological plane. In the case of oceans there is little need for humans who have sailed on (or simply viewed!) the Seven Seas to be intuitively convinced of their vast size, great depth, and attendant potential for concealment. By contrast, non-specialist human actors are frequently lulled into a false sense of security by the familiar structures – so apparently well bounded and so solid! – visible to them in the restricted quarters of cyberspace with which they have familiarity. For those

---

[50] Anonymous attacks, and the type of formlessness they represent, are not mentioned by Sun Tzu. Redolent of wu xing as such attacks are, Sun Tzu's focus on two-sided conflicts almost by definition sharply de-emphasizes, even if not entirely excluding, an anonymous attacks contingency.

[51] A cyber warfare capability's harmonious alignment on wu xing is also not guaranteed. For example, software's inherent flexibility may encourage creating ever more elaborate software to serve wu xing purposes. Yet that seductive ease of elaboration may itself come to undercut wu xing goals, ending up with monstrously complex, inflexible systems, almost an antithesis of wu xing.

non-specialists Sun Tzu's *wu xing* thinking may serve an educational purpose, helping overcome such optical illusions and judgment errors to which they often lead.

******

With the expansion of algorithm-controlled warfare into novel spheres, twenty-first-century conflicts will also see other kinds of new developments on the frontiers of *wu xing*. There are many possibilities, and only one example will be noted here.

Some twenty-first-century forms of human conflict involve the important hider–seeker problem facing insurgents or terrorists who seek to preserve the security of a safe house by placing calls, not from the house itself, but from nearby locations which they attempt to keep spread out, with not too many calls being placed from the same location – thereby hoping for a random-appearing profile that avoids alerting their watchful adversaries.[52] Unsurprisingly, the upshot is likely to be anything but random, for much the same reason that human efforts at producing "random" numbers commonly yield far from random results. But then too, with a dash of paradox, even a more perfectly random pattern, perhaps emerging from reliance on a good-quality random number generator, might *itself* look suspicious to surveillance software! Human beings tend to fall into routine, and absence of any apparent routine whatsoever is weird. Randomness then becomes its own badge of form, inviting a fresh look at verse V.18 (Passage #8.5): "Apparent confusion is a product of good order ..."

Carrying this line of thinking a step further suggests that the ultimate *wu xing* (or at least *wu xing* 3) means that the adversary's suspicions are not even aroused, so that the foe never poses to himself the question "Is there anything here to find out?" This is not a question of veiling *xing*. Rather, it is one of whether there is even any *xing* to be veiled. Putting the challenge in Sun Tzu's frame of reference,

> VI.24.
> (Translation by Mair, p. 99.[53]) "Therefore, the extreme skill in showing one's positions may reach to the degree of there seeming to be no position. ..."

While such an aspiration may be unattainable in the conventional warfare of Sun Tzu's (or any other) time, it need not be out of reach in some civilian environments, where even the presence of a hostile foe, or at least one relevant to them, may not have crossed the minds of many participants. From the vantage of such a foe, the challenge then becomes one of ensuring that no flicker of suspicion (or of curiosity) is ever prematurely aroused. Part of that challenge could be framed as an imperative: never to set off attention-grabbing coincidences, which a modern statistical study describes as "a surprising concurrence of events, perceived as meaningfully related, with no apparent causal connection." As that study further observes:[54]

---

[52] See Rupert Smith, #4 footnote 90, p. 284 (guerrillas "using public phone boxes at random"). See also "Shrinking the haystack. Software is helping the search for guerrillas' and terrorists' safe houses and weapons caches," *The Economist*, January 16, 2016, pp. 86–87. As relevant technology continues to advance, new versions of this hider/seeker scenario seem likely to pop up again.

[53] For Griffith's verse VI.24 translation see p. 269 above. Verse VI.24 also appears in Passage #8.1.

[54] See p. 853 of Persi Diaconis and Frederick Mosteller, "Methods for studying coincidences," *Journal of the American Statistical Association*, 1989, 84(408), 853–61. For further psychological analysis of coincidences see Thomas L. Griffiths & Joshua B. Tenenbaum, "Randomness and coincidences: reconciling intuition and probability theory," *Proceedings of the 23rd Annual Conference of the Cognitive Science Society* (2001), available via http://cocosci.berkeley.edu/tom/index.php (visited July 15, 2021).

Coincidences abound in everyday life. They delight, confound, and amaze us. They are disturbing and annoying. Coincidences can point to new discoveries. They can alter the course of our lives; where we work and at what, whom we live with, and other basic features of daily existence often seem to rest on coincidence.

What does or does not get perceived as a "serious" coincidence – a hint of important unseen *xing* in a formless ocean of detail, arousing the observer's curiosity or suspicion – is crucial in deciding whether further digging is called for to try to shed light on it. Psychological factors influencing that perception may be as important as (or more important than) any more objective ones, a point on which Carl Jung (who was actively interested in coincidences, from a standpoint influenced by Daoism) and Sun Tzu might find common ground. The role played by such factors in turn creates an arena for distinctively psychological Sun-Tzu-esque manipulations.[55]

Wu xing once more lives up to its billing as the ultimate counterintelligence.

★★★★★★

Set of principal passages used to illustrate Sun Tzu Theme #8 (formlessness)

Passage #8.1. Formlessness (*Wu Xing*)

(All six rungs of the Figure 3 ladder may be found or inferred here, with *wu xing* 3 or *wu xing* 4 ideas – the latter clearly apparent in verse VI.9 – populating the first part of Passage #8.1, *wu xing* 5 surfacing at its end with verse VI.26 providing a sharp example.[56] The core idea of tactical formlessness, *wu xing* 2, appears in verse VI.8. Echoes of *wu xing* 6 are present in verse VI.25, pointing to strategic accomplishments beyond the ken of the masses and thereby implying that their strategist author must be prepared to accept an inconspicuous elite status, not expecting mass adulation and the "form" that inevitably accompanies it.[57] Verse VI.9 goes further, if read as suggesting that a strategist might at times prefer to position himself in an advisory role outside the formal chain of command, still further away from the limelight – as famously did Sun Tzu's clansman and putative descendant Sun Bin in a well-known Warring States campaign.[58]

---

[55] There is overlap here between Theme #8 and Theme #7 (see discussion of Whaley's Principle of Naturalness, p. 255 above). However, differences in focus should be noted. For example, the Principle of Naturalness does not provide any obvious counterpart to the "fun" coincidence (Diaconis & Mosteller, previous footnote, p. 860), which might at times provide a buffer against arousing types of suspicion that lead on to deeper digging. Conversely, the *Zhuangzi* text's image of Cook Ding (#7 footnote 56) epitomizes naturalness, yet his skills (or the lack thereof in lesser chefs) do not particularly evoke coincidence ideas.

[56] References to *wu xing* in Passage #8.1 are identified in brackets. Other mentions of *xing* (or, in one instance, *shi* used as a near-synonym) are flagged in Appendix 8 (where Passage #8.1 also appears).

[57] Verses IV.8–IV.12 (Passage #8.10 = Passage #1.24) amplify this line of thinking. See also pp. 64, 71 discussions above.

[58] The ruler of Qi had offered Sun Bin (said by his *Shiji* biography to be a descendant of Sun Tzu) army command, but Sun Bin respectfully declined, saying "A man crippled by punishment is not qualified" (he had previously lost his feet as a judicial punishment). So the ruler made Sun Bin

The closest to an omission is logistics formlessness, *wu xing* 1, though its presence offstage might be surmised as support for other types of *wu xing*.[59])

VI.8. Therefore, against those skilled in attack, an enemy does not know where to defend; against the experts in defence, the enemy does not know where to attack.

VI.9. Subtle and insubstantial, the expert leaves no trace [*wu xing*]; divinely mysterious, he is inaudible. Thus he is master of his enemy's fate.

VI.10. He whose advance is irresistible plunges into his enemy's weak positions; he who in withdrawal cannot be pursued moves so swiftly that he cannot be overtaken.

VI.11. When I wish to give battle, my enemy, even though protected by high walls and deep moats, cannot help but engage me, for I attack a position he must succour.

VI.12. When I wish to avoid battle I may defend myself simply by drawing a line on the ground; the enemy will be unable to attack me because I divert him from going where he wishes.

VI.13. If I am able to determine the enemy's dispositions while at the same time I conceal my own [*wu xing*] then I can concentrate and he must divide. And if I concentrate while he divides, I can use my entire strength to attack a fraction of his. There, I will be numerically superior. Then, if I am able to use many to strike few at the selected point, those I deal with will be in dire straits.

VI.14. The enemy must not know where I intend to give battle. For if he does not know where I intend to give battle he must prepare in a great many places. And when he prepares in a great many places, those I have to fight in any one place will be few.

VI.15. For if he prepares to the front his rear will be weak, and if to the rear, his front will be fragile. If he prepares to the left, his right will be vulnerable and if to the right, there will be few on his left. And when he prepares everywhere he will be weak everywhere.

VI.16. (Translation from Lau, 1965 article, p. 330.[60]) "It is the one who has to prepare against his enemy who is few and the one who makes his enemy prepare against him who is many."

VI.17. If one knows where and when a battle will be fought his troops can march a thousand li and meet on the field. But if one knows neither the battleground nor the day of battle, the left will be unable to aid the right,

chief adviser to a different commanding general, and acting on Sun Bin's advice a great victory was won (battle of Maling, noted earlier in a different connection on p. 246). The story is told by Lau & Ames, pp. 9–10, translating from the same chapter of the *Shiji* where Sun Tzu's biography appears. See also Nienhauser, Vol. VII (1994), pp. 40–41; Nienhauser, Vol. VII (2021), pp. 72–76 (both cited on pp. xxiii–xxiv).

[59] Less charitably to Sun Tzu (and also raising a caution flag regarding Griffith's kudos to Sun Tzu's logistics acumen, see Introduction footnote 23), even Sun Tzu may not be immune to tendencies found in much strategic writing to underplay the importance and difficulty of logistics!

[60] Rationale for using Lau's translation rather than Griffith's is given in Appendix 3 footnote 9.

or the right, the left; the van to support the rear, or the rear, the van. How much more is this so when separated by several tens of li, or, indeed, by even a few!

VI.18. Although I estimate the troops of Yue as many, of what benefit is this superiority in respect to the outcome?

VI.19. Thus I say that victory can be created. For even if the enemy is numerous, I can prevent him from engaging.

. . . (in a shift of perspective, in the omitted verses [see Passage #6.12] Sun Tzu's focus swings away from achieving formlessness and its advantages to ways of reducing the enemy's formlessness to form.)

VI.24. The ultimate in disposing one's troops is to be without ascertainable shape [wu xing]. Then [wu xing] the most penetrating spies cannot pry in nor can the wise lay plans against you. [Emphasis supplied.]

VI.25. It is according to the shapes that I lay the plans for victory, but the multitude does not comprehend this. Although everyone can see the outward aspects, none understands the way in which I have created victory.[61]

VI.26. Therefore, when I have won a victory I do not repeat my tactics but respond to circumstances in an infinite variety of ways.

VI.27. Now an army may be likened to water, for just as flowing water avoids the heights and hastens to the lowlands, so an army avoids strength and strikes weakness.

VI.28. And as water shapes its flow in accordance with the ground, so an army manages its victory in accordance with the situation of the enemy.

VI.29. And as water has no constant form, there are in war no constant conditions.

VI.30. Thus, one able to gain the victory by modifying his tactics in accordance with the enemy situation may be said to be divine [shen].[62]

VI.31. Of the five elements, none is always predominant; of the four seasons, none lasts forever; of the days, some are long and some short, and the moon waxes and wanes.[63]

---

[61] Mair, p. 99 translates: "Though victory based on my positions [xing] may be displayed before the masses, the masses cannot know them. The people may all know the positions [xing] whereby I achieve victory, but no one knows how I utilize those positions [xing] to produce victory."

[62] Ames, p. 127 renders *shen* as "inscrutable," which conveys a Theme #8 thrust more clearly. Ames, p. 289 note 160 also notes that in the Han strips the same character *shen* appears at the end of Sun Tzu VI, which "concludes with a round dot and an additional two characters *shen yao* – 'the essentials of inscrutability.'" He speculates that this might be an "alternative chapter title, or possibly a reader's summary of the contents of the chapter." Even though *wu xing* as such does not enter explicitly here, either possibility is support from an early Chinese quarter for aligning Passage #8.1 on *wu xing* ideas.

[63] For background on the five-elements theory to which verse VI.31 alludes see #5 footnote 103.

Passage #8.2. Logistics Flexibility (Relates to *Wu Xing* 1)

(Sun Tzu never ventures a general statement about logistics flexibility, and in that respect must be regarded as not-quite-modern in treating this core military topic. While imaginative and aggressive, the Passage #8.2 ideas fall well short of being comprehensive since living off the enemy cannot be a universal prescription.)

II.10. They carry equipment from the homeland; they rely for provisions on the enemy. Thus the army is plentifully provided with food.

. . .

II.15. Hence the wise general sees to it that his troops feed on the enemy, for one bushel of the enemy's provisions is equivalent to twenty of his; one hundredweight of enemy fodder to twenty hundredweight of his.

. . .

II.18. Therefore, when in chariot fighting more than ten chariots are captured, reward those who take the first. Replace the enemy's flags and banners with your own, mix the captured chariots with yours, and mount them.

II.19. Treat the captives well, and care for them.

II.20. This is called "winning a battle and becoming stronger".

Passage #8.3. More about Logistics Flexibility (Relates to *Wu Xing* 1)

XI.7. When the army has penetrated deep into hostile territory, leaving far behind many enemy cities and towns, it is in serious ground.

. . .

XI.13. In focal ground, ally with neighbouring states; in deep [i.e., serious] ground, plunder.[64]

. . .

XI.20. In serious ground I would ensure a continuous flow of provisions.[65]

. . .

XI.31. Plunder fertile country to supply the army with plentiful provisions.

XI.32. Pay heed to nourishing the troops; do not unnecessarily fatigue them. Unite them in spirit; conserve their strength. Make unfathomable plans for the movements of the army.

Passage #8.4. Presenting No Salient Tactical Focus for the Enemy (Relates to *Wu Xing* 2)

(Concepts of tactical flexibility and mutual support are on display here.[66] Verse VI.8, which leads off Passage #8.1, gives a general statement of the key idea.)

---

[64] For translation consistency, "deep ground" here should read "serious ground." See #2 footnote 61.

[65] Although verse XI.20 could be a reference to a supply line, dependence on which might limit flexibility, most of the traditional commentators treat it as a further reference to living off the enemy. See #2 footnote 64.

[66] Ensuring that different parts of one's army can offer mutual support and never find themselves cut off is also an emphasis in Sun Tzu's Passage #2.1 terrain rules (for overview see Table 1 on p. 89). For other examples of Sun Tzu's focus on "cut off" situations see #2 footnote 69.

XI.38. Now the troops of those adept in war are used like the "Simultaneously Responding" snake of Mount Ch'ang. When struck on the head its tail attacks; when struck on the tail, its head attacks, when struck in the centre both head and tail attack.

XI.39. Should one ask: "Can troops be made capable of such instantaneous coordination?" I reply: "They can." For, although the men of Wu and Yue mutually hate one another, if together in a boat tossed by the wind they would co-operate as the right hand does with the left.

Passage #8.5. Formlessness out of Apparent Confusion, Masking Form (Relates to Wu Xing 3)

(While Passage #8.1 is Sun Tzu's main repository of *wu xing* 3 thinking, Passage #8.5's verses supplement those thoughts, clarifying that there can be more to *wu xing* 3 than simple invisibility. In one modern terminology, the pertinent concept is using noise to mask signal – or even existence of any coherent signal at all.)

V.17. In the tumult and uproar the battle seems chaotic, but there is no disorder; the troops appear to be milling about in circles but cannot be defeated.

V.18. *Apparent confusion is a product of good order;* apparent cowardice, of courage; apparent weakness, of strength.[67] [Emphasis supplied.]

V.19. Order or disorder depends on organization; courage or cowardice on circumstances; strength or weakness on dispositions.

V.20. Thus, those skilled at making the enemy move do so by creating a situation to which he must conform; they entice him with something he is certain to take and with lures of ostensible profit they await him in strength.

Passage #8.6. Epitome of a Formless Strategic Capability: A Well-Functioning Spy Network (Relates to Wu Xing 3 but Has Relevance to the Full Wu Xing Ladder)

XIII.6. When these five types of agents are all working simultaneously and *none knows their method of operation,* they are called "The Divine Skein" and are the treasure of a sovereign.[68] [Emphasis supplied.]

. . .

XIII.12. Of all those in the army close to the commander none is more intimate than the secret agent; of all rewards none more liberal than those given to secret agents; of all matters none is more confidential than those relating to secret operations.

---

[67] Appendix 7 anchor footnote 14 weighs alternative understandings of verse V.18, coming down in favor of a deception interpretation.

[68] Sun Tzu's five categories of spies are profiled in verses XIII.7–XIII.11 (Passage #11.6).

Passage #8.7. Avoiding Unfriendly Eyes That Could Negate Formlessness (Relates to Wu Xing 3)

IX.18. When on the flanks of the army there are dangerous defiles or ponds covered with aquatic grasses where reeds and rushes grow, or forested mountains with dense tangled undergrowth you must carefully search them out, for these are places where ambushes are laid and spies are hidden.

Passage #8.8. Information Denial Measures (Relates to Wu Xing 4 and Wu Xing 3)

XIII.15. If plans relating to secret operations are prematurely divulged the agent and all those to whom he spoke of them shall be put to death.

Passage #8.9. Inscrutability of Commander (Relates to Wu Xing 4)

(These verses shift focus from wu xing as an army-wide project to wu xing as a general's personal project. To that end, verses XI.42–XI.43 profile what military historian John Keegan has called the mask of command – but not just any mask, a mask capable of veiling, and perhaps actively diverting attention from, the ceaseless mental activity of the sort Sun Tzu plainly expects from his general.[69])

XI.42. It is the business of a general to be serene and inscrutable, impartial and self-controlled.[70]

XI.43. He should be capable of keeping his officers and men in ignorance of his plans.[71]

---

[69] See John Keegan, The Mask of Command (New York: Viking, 1987), p. 11: "The leader of men in warfare can show himself to his followers only through a mask, a mask he must make for himself, but a mask made in such form as will mark him to men of his time and place as the leader they want and need." In Sun Tzu's worldview the first two clauses would be accepted, probably with little comment; the third perhaps as well – but tempered by a Warring States Chinese view tending to equate common soldiers with mindless objects, as verse XI.49 implicitly suggests. See #9 footnotes 7 and 8.

[70] The Chinese term you that Griffith here translates (joined by Minford) as "inscrutable" is not verse VI.30's shen (Passage #8.1), which also can be translated in that way. Other translators variously render you as "reserve[d]" (Mair), "remote" (Ames), "obscure" (Sawyer). The basic idea here may be understood as a general's maintaining a professional and social distance between himself and others in his army, including those who might be regarded as (or see themselves as) his "intimates."

[71] The Chinese term corresponding to "officers and men" is shizu, which could be read as referring just to lower-ranking officers and men, hence possibly allowing for an exception for high-level subordinate commanders and staff (who might be briefed on the commanding general's thinking). However, nothing Sun Tzu says explicitly points to any such exception. In commenting on verse XI.43, the traditional commentators likewise do not point to one. The thrust of Passage #8.9, with implicit support from verse XIII.23 (Passage #8.12), favors an expansive understanding of the set of people on one's own side to be kept in the dark. A profile of Cao Cao captures the spirit: "He has a sardonic streak and keeps his own counsel. His characteristic gesture is a laugh or a smile, but without warmth. He seems isolated from even his closest advisers and kinsmen."

...72

XI.45. He changes his methods and alters his plans so that people have no knowledge of what he is doing.

XI.46. He alters his camp-sites and marches by devious routes, and thus makes it impossible for others to anticipate his purpose.73

...

XI.49. He burns his boats and smashes his cooking pots; he urges the army on as if driving a flock of sheep, now in one direction, now in another, and none knows where he is going.

Passage #8.10. Depth of Strategic Calculation, Together with Secrecy, As Source of Opacity for the Strategist (Relates to Wu Xing 4)

(These verses focus on high strategic achievement and reasons why its strategist author is destined to remain an inconspicuous elite. The role of secret information is one obvious factor contributing to that strategist's opacity. But another, perhaps distinctively leak-proof, basis for that opacity could be the inherent depth of a master strategist's calculations, capitalizing on non-secret factors and forces – possibly of low visibility – whose subtle effects most people would overlook.74 A novice *weiqi* player up against a master might well appreciate this point.

A modern analyst has noted that this part of Sun Tzu also points to a type of deception opportunity: "A huge advantage of making victories *seem* easy is that potential enemies will gain the false impression that victory is gained *because* it was easy, and they will not examine it with care."75 [Emphasis in original.])

IV.8. To *foresee* a victory which the ordinary man can foresee is not the acme of skill; [Emphasis supplied.]

IV.9. To triumph in battle and be universally acclaimed "Expert" is not the acme of skill, for to lift an autumn down requires no great strength; to distinguish between the sun and moon is no test of vision; to hear the thunderclap is no indication of acute hearing.

IV.10. Anciently those called skilled in war conquered an enemy easily conquered.

See Roberts's profile of the Cao Cao character in *Romance of the Three Kingdoms* (translation cited on p. xxiv above), p. 940.

One way of interpreting verse XI.43's advice would treat it as setting forth a *presumption* that any given individual, whether high-ranking or not, will be excluded from the general's confidences.

72 The Chinese text has no discontinuity here. The apparent discontinuity reflects Griffith's editing.

73 In the background here stands the "calculating Sun Tzu" of Theme #1. Mair's translation of the latter part of verse XI.32 (Passage #8.3) is on point: "... When you put your forces in motion, do so with calculated planning, so that the enemy will not be able to discern your aims" (Mair, p. 119).

74 Commenting on verse IV.10, traditional commentator Mei Yaochen refers to one skilled in war as one "who looks below the surface of things." See Giles, p. 29. Cf. also verse VI.25 (Passage #8.1).

75 See Yuen,"Deciphering Sun Tzu" article in *Comparative Strategy* cited in #8 footnote 30, p. 193.

IV.11. And therefore the victories won by a master of war gain him neither reputation for wisdom nor merit for valour.[76]

IV.12. For he wins his victories without erring. "Without erring" means that whatever he does insures his victory; he conquers an enemy already defeated.

Passage #8.11. Opacity in Administrative Posture (Relates to Wu Xing 4)

(The general's unpredictability that Sun Tzu counsels in this sui generis verse helps build wu xing 4. Indeed to a bureaucrat – or a precedent-minded lawyer – the substance of verse XI.54, if read literally without limiting conditions, could represent an epitome of formlessness. In reality, the advice seems narrower.[77])

XI.54. Bestow rewards without respect to customary practice; publish orders without respect to precedent. Thus you may employ the entire army as you would one man.[78]

Passage #8.12. Defending against a Kind of Espionage That Might Pose a Particular Threat to Wu Xing 4 Formlessness

(In Sun Tzu's theory of war the overridingly important intelligence prize is finding out what is going on in the mind and heart of the enemy general. From that vantage the emphasis in verse XIII.23 – the final verse in the Sun Tzu text – comes into focus. It may indeed require the highest level of human talent to penetrate a general's inner circle and, once in position there, to succeed in stripping away the general's mask veiling his true thoughts and feelings even from his intimates.[79] Verse XIII.21 suggests that double agents have uses in avoiding this kind of perilous breach in wu xing 4 security.)

XIII.21. The sovereign must have full knowledge of the activities of the five sorts of agents. This knowledge must come from the doubled agents, and therefore it is mandatory that they be treated with the utmost liberality.

. . .

XIII.23. And therefore only the enlightened sovereign and the worthy general who are able to use the most intelligent people as agents are certain to achieve great things. Secret operations are essential in war; upon them the army relies to make its every move. [Emphasis supplied.]

---

[76] Traditional commentator Du Mu amplifies: "Inasmuch as his victories are gained over circumstances that have not come to light, the world at large knows nothing of them, and he wins no reputation for wisdom; inasmuch as the hostile state submits before there has been any bloodshed, he receives no credit for courage." See Giles, p. 29, also volunteering that "Du Mu explains this very well."

[77] For one such narrower reading see Appendix 8 anchor footnote 19.

[78] Mair, p. 123 conveys Sun Tzu's verse XI.54 advice in categories familiar to modern audiences: "Offer rewards that exceed the legal limits; deliver orders that go beyond administrative norms – you will be able to drive the hosts of the three armies as though you were directing a single man."

[79] For further analysis of verse XIII.23's substance and implications see #4 footnote 86.

Passage #8.13. Strategic or Tactical Imagination As Source of Formlessness (Relates to Wu Xing 5)

(In Passage #8.1 a basic statement of wu xing 5 is verse VI.26. Verse V.6 offers supporting images, most notably one involving a renewable water resource.)

V.5. Generally, in battle, use the normal force [zheng] to engage; use the extraordinary [qi] to win.
V.6. *Now the resources of those skilled in the use of extraordinary forces [qi] are as infinite as the heavens and earth; as inexhaustible as the flow of the great rivers.* [Emphasis supplied.]
V.7. For they end and recommence; cyclical, as are the movements of the sun and moon. They die away and are reborn; recurrent, as are the passing seasons.

Passage #8.14. Compensating in a Conceptual Realm for Loss of Flexibility in a Physical One (Relates to Wu Xing 5)

XI.9. Ground to which access is constricted, where the way out is tortuous, and where a small enemy force can strike my larger one is called "encircled".
. . .
XI.14. In difficult ground, press on; *in encircled ground, devise stratagems*; in death ground, fight. [Emphasis supplied.[80]]

Passage #8.15. Speed As Facilitator of Formlessness (Relates to Wu Xing 5 and May Also Compensate for Past Wu Xing Failures by Making Them Irrelevant)

(Although it serves other purposes too, speed in military operations is a basic facilitator of formlessness, enabling the card deck, to use a modern metaphor that might have appealed to Deng Xiaoping, sometime honorary president of China's National Bridge Association, to be reshuffled in the twinkling of an eye.)

XI.29. Speed is the essence of war. Take advantage of the enemy's unpreparedness; travel by unexpected routes and strike him where he has taken no precautions.

Passage #8.16. Rejecting Extraneous or Inessential Reasons for Action (Relates to Wu Xing 6, the Topmost Rung of the Formlessness Ladder)

(A common core of Passages #8.16–#8.20 is advice to a strategic decision maker to prune his own overdeveloped preferences. Of course, some form [xing] will unavoidably remain – but it should be no more than the situation requires.)

---

[80] Ames, p. 155 replaces Griffith's "stratagems" by "contingency plans" as a translation of the Chinese text's mou. From a Theme #8 vantage Ames's choice has some advantages because it conjures an image of a "reserve army" of pre-existing plans or tactical moves constituting a key resource for achieving wu xing 5 (cf. Table 2 note h on p. 97 above for a modern military history example). The commonsense point is that occupying "encircled ground' is not a good time to start thinking from scratch.

XII.17. If not in the interests of the state, do not act. If you cannot succeed, do not use troops. If you are not in danger, do not fight.

XII.18. A sovereign cannot raise an army because he is enraged, nor can a general fight because he is resentful. For while an angered man may again be happy, and a resentful man again be pleased, a state that has perished cannot be restored, nor can the dead be brought back to life.

XII.19. Therefore, the enlightened ruler is prudent and the good general is warned against rash action. Thus the state is kept secure and the army preserved.

Passage #8.17. Implicit Advice Not to Cherish Much (Relates to Wu Xing 6)

(Reading verse XI.28 in mirror image, it contains a call for pruning one's own preferences as far as possible, lest the enemy exploit them.[81])

XI.27. [Anciently, those described as skilled in war] concentrated and moved when it was advantageous to do so; when not advantageous, they halted.

XI.28. Should one ask: "How do I cope with a well-ordered enemy host about to attack me?" I reply: "*Seize something he cherishes and he will conform to your desires.*" [Emphasis supplied.]

Passage #8.18. Divesting of Extraneous Motivations (Relates to Wu Xing 6)

X.19. And therefore the general who in advancing does not seek personal fame, and in withdrawing is not concerned with avoiding punishment, but whose only purpose is to protect the people and promote the best interests of his sovereign, is the precious jewel of the state.[82]

. . .

X.21. If a general indulges his troops but is unable to employ them; if he loves them but cannot enforce his commands; if the troops are disorderly and he is unable to control them, they may be compared to spoiled children, and are useless.

Passage #8.19. Character Traits Inimical to Formlessness (Relates to Wu Xing 6)

VIII.17. There are five qualities which are dangerous in the character of a general.

VIII.18. If reckless, he can be killed;

VIII.19. If cowardly, captured;

VIII.20. If quick-tempered you can make a fool of him;

---

[81] In a fine example of the traditional commentators crossing swords, Cao Cao interprets verse XI.28 as being about "seiz[ing] some strategic advantage he [the enemy] depends on." A later commentator, Chen Hao, rejoins: "This refers not just to strategic advantages, but to any person or thing of importance to the enemy." See Minford, pp. 278–79. Chen Hao better captures the spirit of *wu xing* 6, since persons or things may be subjectively "important" even though they are strategically irrelevant!

[82] Mair, p. 115 understands this verse in an imperative sense ("Therefore, when advancing do not seek fame," etc.), which brings out a *wu xing* 6 harmonic still more explicitly.

VIII.21. If he has too delicate a sense of honor you can calumniate him;

VIII.22. If he is of a compassionate nature you can harass him.

VIII.23. Now these five traits of character are serious faults in a general and in military operations are calamitous.

VIII.24. The ruin of the army and the death of the general are inevitable results of these shortcomings. They must be deeply pondered.

Passage #8.20. States of Mind Undercutting Formlessness (Relates to Wu Xing 6)

(Especially relevant to wu xing 6 are verses I.23 and I.25. Arrogant people tend to have superfluous preferences reflecting an appetite for deference. Factional politics often crystallizes attitudes and positions on all sorts of purely intramural issues, thereby proliferating preferences and undercutting formlessness.)

I.22. Anger his general and confuse him.

I.23. Pretend inferiority and encourage his arrogance.

I.24. Keep him under a strain and wear him down.

I.25. When he is united, divide him.

Passage #8.21. Concept for Reaping Advantage by Utilizing Formlessness on One Level to Engineer, Then Profit by, Utter Inflexibility on a Different Level

XI.55. Set the troops to their tasks without imparting your designs; use them to gain advantage without revealing the dangers involved. Throw them into a perilous situation and they survive; put them in death ground and they will live. For when the army is placed in such a situation it can snatch victory from defeat. [Emphasis supplied.]

★★★★★★

E

Strategist Should Make a Situation's Natural Dynamics Work for Her

Theme #9 is about exploiting dynamics already present in a situation to advance one's interests. Many Sun Tzu ideas find a place here, reflecting Sun Tzu's keen appreciation of war's larger context (Passage #1.1) conjoined with the inherently dynamic quality of Sun Tzu's core concept of shi.

Viewed as a complement to Theme #9, Theme #10 stands for being prepared, as cold-bloodedly as necessary, to slice through any normative boundary – ethical, customary, legal, or other – that stands in the way of exploiting favorable dynamics or sidestepping unfavorable ones. Expressed in purest form, Theme #10 is the Sun Tzu of legend, the Sun Tzu who had no compunction about making an example of slaughtered (real) concubines to tighten up the command and control of his (simulated) army. How far the exploits of this legendary Sun Tzu actually align on ideas present in the Sun Tzu text can be debated. Theme #10 explores relevant issues.[1] There is definite potential for a substantially more nuanced understanding of Sun Tzu's thinking as it relates to Theme #10 than is suggested by the often crude summary statements widely encountered in modern Sun Tzu literature.

Both Themes #9 and #10 have important Chinese and, more specifically, Warring States cultural content. It should be noted, however, that the thrust of Theme #10 – even in a comparatively nuanced understanding – is decidedly non-Confucian. Sun Tzu's deception emphases have particularly fostered tension with Confucian mores that emerged as the dominant strand of traditional Chinese thought and culture.[2]

Chinese states and societies of Sun Tzu's time were already complex. They would have harbored a wealth of civilian dynamics providing Theme #9 possibilities for a Sun Tzu–inspired strategist. Those environments would also have harbored many normative constraints, however alien by modern criteria some of them might seem.[3] A basic Sun Tzu insight rooted in Sun Tzu's thinking about

---

[1] As noted in the Introduction (p. 13), the traditional source for the concubines story is a biography compiled centuries after any historical Sun Tzu. Mair, p. 133 labels it a "pseudo-biography."

[2] See, e.g., Raphals's "Sunzi versus Xunzi" article cited in #7 footnote 11.

[3] For a perspective on the evolution of norms in early Chinese society, focusing on the transition between the Spring and Autumn period and the Warring States period that followed, see Lewis's

deception and espionage is that overriding those constraints in quest of greatest efficacy – seemingly an amoral, indeed often immoral act – might actually promote a (relatively) more benign outcome than that likely to result from the horrors, devastation, and waste of bitter, large-scale conventional warfare. See verses XIII.1– XIII.2 (Passage #10.6).

THEME #9 BUILDING ON NATURAL FORCES AND MOMENTUM

Sun Tzu's water images:

Passage #9.1 (verses VI.27–VI.31) (first water image: opportunistic dynamics of water, a standard for an army to emulate)

Passage #9.2 (verses V.5–V.7) (second water image: inexhaustible flow of great rivers as image for dynamics of military creativity)

Passage #9.3 (verses V.13–V.16 + VI.21–V.25) (third water image: torrential flow of water, together with other images for military momentum and timing of its unleashing)

Passage #9.4 (verse IV.20) (fourth water image: release of pent-up waters as image for achieving military momentum)

physical or biological dynamics setting context:

Passage #9.5 (verse I.5) (conforming military operations to dynamics of weather and climate)

Passage #9.6 (verse IX.6) (taking up a suitable position vis-à-vis a river's current)

Passage #9.7 (verses XII.1–XII.14) (dynamics of a conflagration, to use for profit or avoid)

Passage #9.8 (verse XI.39) (from physical to social dynamics: storm eliciting cooperation between foes)

Passage #9.9 (verse IX.13) (dynamics of health and disease)

psychological and biophysical processes affecting an army in the field:

Passage #9.10 (verses VII.20–VII.25) (psychological and biophysical dynamics of soldiers)

Passage #9.11 (verses II.16–II.18) (psychological dynamics of rage and greed in soldiers)

Passage #9.12 (verse XI.44) (superstitious practices undermining morale)

Passage #9.13 (verses XI.2 + XI.11 + XI.15) (problematic *mentalité* of soldiers fighting on home turf)

Passage #9.14 (verses XI.23 + XI.30 + XI.33–XI.37 + XI.49 + XI.55) (death ground psychology as a special psychological dynamic)

Passage #9.15 (verses VII.30–VII.33) (psychological dynamics associated with defeat and retreat)

(footnote 3 cont.)

book cited in Background footnote 11. A central harmonic of the cultural shift he describes (see his p. 246) is the manner in which "the 'matters of highest concern' substantiated through violence changed from the honor of the individual and lineage derived from the service of the ancestral spirits to the hierarchic bonds that linked men together and preserved order in society."

Passage #9.16 (verses IX.32–IX.34 + IX.36–IX.43) (disequilibrium dynamics, pointing to an army in difficulties)

Passage #9.17 (verses X.9–X.16) (dynamics that could portend an army's disintegration)

dynamics on a level of a country's resources and grand strategy:
Passage #9.18 (verses II.5 + II.11–II.14) (adverse socioeconomic dynamics brought about by a major war effort, especially a protracted war)

Passage #9.19 (verses III.18–III.23) (pathologies in civil–military relations that rival polities can exploit)

warning about failure to build on success and the favorable dynamics it produces:
Passage #9.20 (verse XII.15) (failure to exploit momentum born of past successes)

★★★★★★

In a modern parlance, Theme #9 is the dynamic systems face of Sun Tzu. Its core idea, which Sun Tzu often conveys through water images, is one of piggybacking strategic control on more fundamental processes, natural or social, that are not (or at least not fully) of that strategist's making. As Shakespeare (employing his own water image!) put it, "There is a tide in the affairs of men which, taken at the flood, leads on to fortune."[1]

The two-part idea thus expressed – Shakespeare's "tide" and "taken at the flood" – is undeniably appealing: the strategist does not have to do it all by herself. Sun Tzu deserves kudos here, not so much for his specific insights into dynamic processes (where Sun Tzu's concepts tend to be rudimentary, at times simplistic), as for advocating for an underlying idea, whose varied expressions surface in many different ways throughout the text. Framed in general terms, that idea envisions successful strategic or tactical action as a kind of epicycle. Such action builds on and exploits autonomous dynamics, to which it stands in relationships like trigger event; catalyst or intensifier; source of mid-course correction, etc. For example, in the background of verse V.14's image of raptor and prey (Passage #9.3) are effects of gravity enlisted in support of the kill.[2] The insightful strategist rides to victory on advantageous dynamics like that; the maladroit one is unseated by them. This vein of conceptualization has deep roots in traditional Chinese thought."[3] Building on favorable dynamics already present in a situation is also commonly a cost-avoiding step, so that Theme #9 dovetails with Theme #3's focus on extraordinarily cheap gains.[4]

---

[1] William Shakespeare, *The Tragedy of Julius Caesar*, act 4, scene 3.

[2] A natural history source (Ehrlich et al., cited in Appendix 9 footnote 4) observes: "One of the most spectacular sights in the world of birds is a kill of another bird by a stooping Peregrine Falcon. The falcon plunges steeply downward, wings partially closed, at speeds that can exceed 150 miles an hour." A different Theme #9 case involving quickest flight paths is analyzed in Kenneth J. Arrow, "On the use of winds in flight planning," *Journal of Meteorology*, 1949, 6(2), 150–59.

[3] Ames's "Introduction" to his Sun Tzu translation develops pertinent cultural background.

[4] A close relationship between Themes #3 and #9 is underscored by traditional Sun Tzu commentator Du Mu, commenting on verse V.25's image of boulders rolling down a mountain (Passage #9.3):

Meanwhile a Sun Tzu–inspired strategist will be delighted to egg on, and facilitate in any way possible, an existing process or trend that is deleterious to the enemy.

In his efforts to gain purchase on dynamic phenomena, Sun Tzu repeatedly ran up against the severe limits of his analytical toolkit for characterizing and analyzing dynamics he instinctively felt to be important.[5] That is a likely reason for Sun Tzu's frequent recourse to metaphor in profiling relevant processes. Three strands of Sun Tzu's dynamics thinking serve to introduce his basic concepts and emphases:

Focus on individual and group psychological dynamics. For Sun Tzu, a focus on dynamics is never far removed from phenomena of human nature, emotions, and the human group. His frequent use of physicalistic imagery notwithstanding, Sun Tzu is deeply attuned to warfare's human face, subsuming both the psychological dynamics of the commanding general and that of the troops (which appear side by side in Passage #9.10's verse VII.20). Social processes are rarely easy to capture in credible models, a challenge with which analysts and model-builders struggle to this day.[6] Possibly reflecting widespread cultural stereotypes and biases of his time, Sun Tzu seems prepared to envision the behavior of common soldiers (presumably on both sides) in simplistic, even at times inanimate, terms – a standpoint that influences his thinking about leadership (Theme #13).[7] In truly serendipitous fashion this facet of Sun Tzu, analytically limited as it may seem, resonates with modern agent-based computational modeling directions (though it is also plain that Sun Tzu had no

---

Their unstoppable descent is due to the mountain, not to the boulders themselves. *The main drift of this chapter [Sun Tzu V] is the importance in war of relying on the energy of the situation itself, using speed and sudden attack. In this way, a great deal can be achieved with little expenditure of force.*

(Minford, p. 175, emphasis supplied; in Sun Tzu's verse V.25 analogy, boulders = soldiers)

[5] Note, however, Sun Tzu's foray into analyzing wind patterns (Passage #9.7's verse XII.11), which contains a valid physical science insight. See #9 footnote 56 below.

[6] See George C. Homans, *The Human Group* (New York: Harcourt, Brace, 1950). See also David J. Bartholomew, *Stochastic Models for Social Processes* (3rd ed.; Chichester/New York: Wiley, 1982).

[7] See, e.g., verses V.24 and V.25 in Passage #9.3 (soldiers as logs or stones); verse XI.49 in Passage #9.14 (soldiers as sheep). Lewis, Background footnote 11 sums up the perspective:

> As a corollary to assigning sole agency and responsibility to the commander and his officers, the rhetoric of the military treatises tends to identify soldiers with various mindless natural objects or with animals. Sometimes they are compared to logs or stones which are rolled by the commander, and sometimes they assume the guise of flowing water, the storm, or of beasts, objects that have their own dynamism but lack calculation or deliberation.

(pp. 110–11; see also p. 231)

Of course, any view of common soldiers as lacking agency is a cultural one, and (like so many other stereotyped perspectives) need not mesh with the reality. For a remarkable archaeological find of letters home (written on wooden boards) by two ordinary Qin soldiers fighting in the last stages of the Qin campaign against Chu (224–223 BC) at the tail end of the Warring States period, see pp. 362–63 of Robin D. S. Yates, "Soldiers, scribes, and women: literacy among the lower orders in early China," in *Writing and Literacy in Early China* (Li Feng and David Prager Branner, eds.; Seattle: University of Washington Press, 2011), pp. 339–69; Wilkinson, §9.8.3, p. 166. As Yates notes:

> These letters indicate that ordinary Qin soldiers in the field could consign their concerns [e.g., about a mother's health] to writing, that the postal system was sufficiently organized and operational, even in a time of war, and that written communications could be passed back and forth between soldiers at the front and people at home.

(p. 363, also cautioning that it is "quite possible that the two brothers solicited the help of a literate colleague to write these letters")

inkling of how very simple rules may, through their interacting effects, be capable of producing highly non-obvious "emergent" patterns in a population).[8]

Water images. Given Sun Tzu's unabashedly human foci, it is in some ways striking, perhaps even a little paradoxical, that some of his most important and compelling insights about dynamics grow out of images of how water behaves:

- water's ineffable ability to work its way downhill, tunneling through the slightest crack or crevice to do so (Passage #9.1);
- renewable water resources, illustrated by flow of great rivers (Passage #9.2);
- the formidable momentum that a body of water can acquire (Passages #9.3 and #9.4).

One way of interpreting this emphasis is that Sun Tzu's intuition told him that many of the greatest military opportunities grow out of situations involving or approximating fluid dynamics, either literally or metaphorically.[9] Certainly Sun Tzu's water images come in handy in conjuring situations that have, or develop, their own momentum and tendencies not requiring the general's ongoing intervention. Indeed, much as is frequently the case with water, those tendencies may be beyond centralized human control, though Sun Tzu never says as much (possibly since to do so would be too candid an admission of the commanding general's limitations).

Focus on dynamics involving an abrupt qualitative change or "rupture." A leading human behavior example would be the "supranormal" combat reflexes that Sun Tzu expects from troops who find themselves in death ground. Further examples involve fire or water (e.g., fire attack as an agent of panic; or effects of pent-up waters, suddenly released, as a metaphor for momentum of a successful attack sweeping away the enemy).[10] Other types of actual or incipient rupture center on breakdowns in an army's chain of command.[11] Although Sun Tzu's ability to develop any of these promising directions is blunted by his lack of analytical tools,

---

[8] In a kindred vein, Minford, p. 289 describes parts of early Chinese thought, not limited to Sun Tzu, as viewing ordinary people of the Warring States as entities "to be treated like so many automata."

A compelling (if make-believe) account of an amazingly realistic automaton fashioned of leather, wood, glue, and lacquer appears in an early Daoist text, the *Liezi*. For 1912 translation of the pertinent *Liezi* passage by Lionel Giles – not long after appearance of Giles's Sun Tzu translation (1910) – see Joseph Needham (with research assistance of Wang Ling), *Science and Civilisation in China*, Vol. 2: *History of Scientific Thought* (Cambridge, UK: Cambridge University Press, 1956), p. 53. (The *Liezi* text is traditionally attributed to a Warring States thinker, though is likely to have been produced at a much later time. For further *Liezi* background see *Early Chinese Texts* [cited on p. xxii above], pp. 298–308.)

A classic contribution to agent-based modeling is Thomas Schelling's "Dynamic models of segregation," *Journal of Mathematical Sociology*, 1971, 1(2), 143–86. Although his main applications focus is civilian, Schelling's article notably (albeit very briefly) on p. 186 alludes to tipping "say the U.S. Army."

[9] It is worth noting that fire dynamics (Passage #9.7) also has fluid dynamics aspects. See Sheldon R. Tieszen, "On the fluid mechanics of fires," *Annual Review of Fluid Mechanics*, 2001, *33*, 67–92.

[10] Turning attention from metaphorical to literal water, flash flood possibilities are illustrated by verse IX.16's "Heavenly Wells" (Passage #7.5, also Passage #9.24A), the latter "so named because it is a significant depression ... dangerous because the runoff of rainwater from unexpected storms can inundate [the area]." See Sawyer, pp. 323–24 footnote 145. Verse XI.15a (Passages #11.18, #9.24A) is also motivated by flash flood perils.

[11] Verses IX.41–IX.43 (Passage #9.16) and X.11–X.14 (Passage #9.17) point to examples.

there are recognizable gropings here toward what is sometimes described as war's nonlinear level.[12] However, Sun Tzu's basic optimism regarding predictability of dynamics of interest to him – especially notable in his treatment of death ground behavior – suggests limited awareness of just how complicated and non-intuitive nonlinear dynamics can be.[13]

The kinds of dynamics about which Sun Tzu has something to say comprise a kind of map of war as Sun Tzu knew it. Passages #9.1–9.20 offer a more systematic tour of that map than a translation of the Sun Tzu text in its traditional order can provide. Inevitably judgment calls arise as to which parts of the text to include and what to say about them. Any military situation, certainly one involving actual combat, has immensely dynamic qualities by reason of the enemy's "living reaction" of which Clausewitz took note. Yet elements of the pertinent "situation" and changes in it are rarely, if ever, spelled out by Sun Tzu in what a modern analyst would consider adequate detail. In short, we lack a clear sense of the moving parts. That makes it difficult to be clear about specifics of the dynamics Sun Tzu envisioned or sought to emphasize; much detail has to be filled in, and there is rarely just one way of doing this.[14] Faced with such limitations, the present goal is to give a flavor of Sun Tzu's Theme #9 thinking without aspiring to exhaust it, a goal having points of similarity to how Theme #2 addresses Sun Tzu's terrain thinking.

So compelling are Sun Tzu's water images that they are chosen here, rather than any direct Sun Tzu discussion of human psychology, to lead off the present passage selection (see Passages #9.1–#9.4, which in some cases supplement water images with other imagery). In parsing those images it must be recognized that a hefty strand of ideal-type thinking runs throughout. Actual military situations would rarely if ever unfold without a hitch in the way Sun Tzu portrays. Clausewitzian friction as well as the tyranny of logistics applies to Sun Tzu's wars no less than to modern ones.[15]

Attention then shifts (Passages #9.5–#9.9) to a non-metaphorical level involving, respectively, effects of weather or climate on military operations; a river's current;

---

[12] See Alan Beyerchen, "Clausewitz, nonlinearity, and the unpredictability of war," *International Security*, 1992–93, 17(3), 59–90.

[13] A candidate for some awareness by Sun Tzu of the complexity of dynamic systems is verse VI.31 (Passage #9.1). Its context is the early Chinese theory of "five elements" or "five phases," one version of which envisioned a recurring cycle, in which each stage "destroyed" the next one: water > fire > metal > wood > earth > water. See Brooks & Brooks, Introduction footnote 36, pp. 84–85 (for further background see #5 footnote 103). They argue that verse VI.31 signals skepticism, i.e., that "the Five Elements have *no* constant order, and thus no predictive value" (emphasis in original). See also Zhang Dainian, #3 footnote 80, p. 99 ("The *Sunzi* may represent an earlier understanding of the five agents [i.e., five elements] according to which there was no necessary sequence"). If that interpretation is adopted, verse VI.31 shows Sun Tzu tackling an important flip side of the idea that animates Theme #9, namely, the question of what should *not* be treated as a reliable dynamic on which to build.

[14] Theme #4's "unfinished painting" idea (see p. 141 above) again has relevance here.

[15] Sawyer, Introduction footnote 55, p. 61 sounds an important cautionary note:
These analogies [like the onrush of pent-up water tumbling stones along] and the [shi] concept's configurational aspect have prompted a few writers, especially contemporary Westerners, to rapturously ponder its semi-mystical transformational possibilities. However, their suggestion that merely positioning the army to accord with a situation's latent potential will allow it to emerge victorious without any significant effort lacks historical substantiation and is not countenanced by contemporary PRC [People's Republic of China] theoreticians.

fire; a storm on water; spread of disease. Here Theme #9's core concept of piggy-backing on effects of natural dynamics not of the strategist's making (or in the case of fire, not fully so) is quite transparent, requiring little by way of inference or elaboration of assumptions.

Next, Passages #9.10–#9.17 train a spotlight on social or psychological dynamics (all of them exploitable by one side or the other) occurring *within* an army – one's own or the enemy's. Because some sort of human dynamics is present, if sometimes sotto voce, in every situation Sun Tzu discusses, there is particular need for judgment in picking examples here. Passage #9.10 is a clear, albeit not comprehensive, statement of Theme #9's core idea. Passage #9.11 offers examples of dynamics useful to Sun Tzu's general, followed by Passages #9.12 and #9.13 illustrating two types of unfavorable dynamics that an alert foe might turn to profit. Turning to the idea of behavioral discontinuity, Passages #9.14–#9.17 round up cases where Sun Tzu's treatment of army dynamics points toward the possibility of a qualitative change or "rupture" in the situation, in some instances favorable, in others adverse.

Passages #9.18 and #9.19 turn to a macro level of analysis involving the state as a whole. Perhaps reflecting the fact that Sun Tzu is not by instinct a "builder," both passages point to negative or adverse contingencies. Socioeconomic dynamics take center stage in Passage #9.18. Sources other than Sun Tzu suggest what has been called a "remarkable degree of economic understanding" in early China in service of the needs of statecraft and warfare.[16] But Sun Tzu himself is not interested in statecraft of that type. Considerably more in line with Sun Tzu's primary interests is Passage #9.19's focus on tensions in civil–military relations, which (taking a third-party meddler's standpoint) is a fine illustration of Sun Tzu's Theme #9 thinking in action.

Verse XII.15 (Passage #9.20) is a particularly pure expression of the core concept of Theme #9, albeit again focused on a failure rather than a success case. Here Sun Tzu extends the emphasis on momentum found in some of the water passages, chiding a successful general (or ruler) not to rest on laurels, but to exploit an existing string of successes to generate more of the same.

Roads Not Taken by Sun Tzu in Developing Theme #9

Since dynamics of many varieties are ubiquitous in war, Theme #9 "roads not taken" could be explored in many directions; only a sampler of omissions will be noted here. One omission looms especially large, since it marks a major chasm between Sun Tzu's and Mao's thought (to which Sun Tzu's thinking is at times uncritically assimilated). Notwithstanding Sun Tzu's alertness to many kinds of psychological and sociological factors in war, the Sun Tzu text contributes no ideas about how to exploit grassroots social conflict for strategic gain. Put differently, the Sun Tzu text is truly about warfare of, and between, elites. Its interest in dynamics does not extend to larger sociopolitical processes or to the vast social changes that were occurring in the Warring States period.[17]

---

[16] See Brooks & Brooks, Introduction footnote 36, p. 39.
[17] For a snapshot of those changes see Lewis, Background footnote 11, pp. 5–6. A possible glancing recognition of social change appears via verse VIII.21. See p. 311 discussion below.

Sun Tzu makes only the skimpiest of mentions of mobilization for war and undertakes no discussion of, or even allusion to, its innumerable details.[18] In his time as in our own, mobilization processes would involve complex flows of people, objects, and information, which would present numerous opportunities for piggy-backing strategic action on naturally occurring dynamics (e.g., building on the profit motives of arms makers).[19] Yet Sun Tzu shows no interest in this topic, which might require delving into dynamics of cooperation rather than zero-sum conflict alone.

A further road not taken in Sun Tzu's Theme #9 development can be illustrated by a simple, humorous story from the Mencius text, part of the Confucian canon:[20]

> In Song, there was a man who was worried that his rice plants were not growing, so he pulled them longer. After doing this for a long time, he came home and said to his people, I am really tired out today; I have been helping the rice plants to grow. His son hurried out to look, and the rice plants were all withered . . .

An elementary strategic moral here is that wishfully hothousing a genuine trend present in a situation can backfire – perhaps badly. The annals of political–military (and economic!) history are littered with errors of that type; a military example might be premature promotion of a talented young officer. Yet Sun Tzu offers no compa-rably clear statement warning of downsides of hothousing a trend. Here is a case where the Confucian tradition seems to pull ahead of Sun Tzu's teaching.

On a different analytical plane, notwithstanding repeated metaphorical allusions to water and its dynamics, Sun Tzu notably makes no explicit mention of operations actually taking place *on* water. It is reasonable to wonder how Sun Tzu's Theme #9 ideas might have changed, and perhaps been deepened, had his military focus extended to naval operations, a type of warfare familiar in at least some parts of pre-imperial China (including the state of Wu with which Sun Tzu is traditionally associated).[21] Reflecting effects of currents, tides, and winds on both navigation and naval combat, applied fluid dynamics permeates problems encountered in operations afloat, both in riverine contexts and in littoral zones or on the open ocean. As a conjecture, a strategist imbued with Sun Tzu's basic mindset who was also versed in operations taking place on water might take cognizance of a larger range of strategic and tactical Theme #9 possibilities than Sun Tzu's land warfare focus elicited.

Sun Tzu (2) ("Sun Tzu Extended") and Sun Tzu (3) ("Sun Tzu Analogical") Frontiers of Theme #9

As noted in Theme #1 discussion, Sun Tzu is at home in emotional territory, as many writers on strategy (particularly those of a rational choice persuasion) are not. There are modern cases where dramatic strategic success has pivoted on a successful

---

[18] Virtually Sun Tzu's entire coverage of mobilization activities appears in verse VII.1 (Passage #5.8).

[19] For glimpses of a weapons industry in early China see Background footnote 67 (citing sources).

[20] Translation is from Brooks & Brooks, Introduction footnote 36, p. 189.

[21] A naval text *Wu Zixu shuizhan bingfa* ("Wu Zixu's Naval Warfare Methods") attributed to Wu Zixu (died 484 BC, a military colleague of the putative historical Sun Tzu who likewise served the state of Wu) appears to have been lost in antiquity, though fragmentary quotations from it have been assembled. See p. 114 and footnote 41 of Milburn, "*Gai Lu*: a translation and commentary on a yin-yang military text excavated from Tomb M247, Zhangjiashan," *Early China*, 2010–2011, *33/34*, 101–40; also, p. 135 footnote 17 of Andrew Chittick, "The transformation of naval warfare in early medieval China: the role of light fast boats," *Journal of Asian History*, 2010, 44(2), 128–50.

judgment call by innovators who, following in Sun Tzu's footsteps (wittingly or not), have accurately read latent emotional dynamics potent enough to reshape a strategic landscape. Two otherwise utterly dissimilar cases, one twentieth-, one twenty-first-century, come to mind (though both center on dynamics in large populations, not the mind of a single key actor – general or ruler – that is Sun Tzu's basic focus).

The first case is the pioneering recognition by Mao – perhaps alone among early twentieth-century Chinese leaders – of what a firestorm could be unleashed by tapping into the hatred and bitterness of Chinese peasants against the landlord class, a firestorm that Mao's Chinese Communist Party could then manipulate for decisive strategic gain.[22] The second case – far less toxic on its face than Mao's, though certainly capable of generating new species of major trouble – would be social media entrepreneurs harnessing the voyeuristic and narcissistic tendencies of countless yuppies and others, with the unprecedented sharing of personal information thereby unleashed. Both classes of dynamics give much scope for applications of the core Theme #9 idea of piggybacking strategic interventions on naturally occurring processes, in each case prominently involving emotion-fueled social conflicts.

Formal modeling directions also have a part to play in further development of Sun Tzu's Theme #9 thinking. As seems appropriate in view of the fundamentally dynamic nature of Sun Tzu's *shi* (see p. 176 above, citing analysis of Li Ling), a natural framework here is a dynamics systems one.[23] A relevant analytical direction, having both mathematical and computational aspects, comes from the Boorman–Levitt island cascade model.[24]

---

[22] See Griffith, #8 footnote 27, p. 14 (compare the spirit of Passage #9.11's verse II.16):

> The Chinese peasant, in his own expressive idiom, "ate bitterness" from the time he could walk until he was laid to rest in the ancient and crowded family burial plot beneath the cypress trees. This was feudal China. Dormant within this society were the ingredients that were soon to blow it to pieces.

Pertinent emotional dynamics – a better phrase might be raw hatred – is conveyed by a woodcut of artist Li Hua (1907–1994), entitled "Attack." See Westad, #4 footnote 74, facing p. 178 there.

[23] Relatedly, Arthur Waldron has suggested a formal interpretation of Sun Tzu's focus on "pivots" (ji), a term with concrete roots in description of the firing mechanism of a crossbow but also equipped with a generalized meaning referring to pivotal moments portending changes in a strategic situation (see p. 183 above). Waldron's "Foreword" to Mair's translation of Sun Tzu sets out the basic concept in qualitative terms: "Sun Zi is what might be called a 'differential strategist,' concerned with timing, turning points, and the ever-changing play of circumstances and opportunities." See Mair, p. xxv; p. xxiv amplifies this line of thinking with a sketch of a simple, impressionistic formal model.

Compare an observation of war correspondent Frederick Palmer regarding a battle in the Russo-Japanese War he was witnessing: "And little things – little in this great affair – began to speak of tendencies, at least." See Palmer, #5 footnote 92, p. 280.

[24] This model was originally developed in an evolutionary biological context. Biological roots or inspirations are common in computational model-building, as artificial neural networks, genetic algorithms, and other computational areas attest. The island cascade model is presented and analyzed in Scott A. Boorman & Paul R. Levitt, *The Genetics of Altruism* (New York: Academic Press, 1980), pp. 35–163; p. 378 there provides a short description of the concept in that setting.

A summary statement is Scott A. Boorman, "Island models for takeover by a social trait facing a frequency-dependent selection barrier in a Mendelian population," *Proceedings of the National Academy of Sciences, USA*, 1974, 71(5), 2103–107 (communicated by Kenneth J. Arrow).

Analysis in the 1980 book provides additional detail on model parameters and equations (which are tailored to the original biological context, though capable of being generalized).

This model starts by postulating a population where a type of strategic competition is unfolding between what will here be labeled type Ss and type Ds. Invoking categories inspired by Sun Tzu, these might, for example, be interpreted as an army's "sharp" (type S) versus "demoralized" (type D) soldiers (cf. Passage #9.10's verse VII.21; also, Passage #7.13's discussion of verse VII.28). In an elaboration of the story, those "demoralized" troops might be incipient deserters or turncoats.

A signature assumption of the island cascade model is that the relevant population is partitioned into "islands" selectively connected by channels that allow for a limited degree of interaction between pairs of "adjoining" islands: in short, an island network, which the model assumes to be connected.[25] What drives the competition is that the dynamics at each island involves dynamics with a threshold β_{crit} (in the simplest case, the same threshold for all islands, which also have identical internal dynamics). The dynamics is best illustrated by an example, emphasizing the simplest case (many elaborations are possible). If in a given time period n the proportion $\beta_i(n)$ of type Ds at island i exceeds β_{crit}, then (ceteris paribus) local pressures operating within that island will produce a larger $\beta_i(n+1)$ in time period $n+1$, hence favoring the Ds and ultimately, again ceteris paribus, propelling their proportion toward 1 (i.e., complete victory for the Ds, with Ss eliminated) – an army heading for disintegration.[26] But if $\beta_i(n)$ falls short of β_{crit} then the local pressures operate in the opposite direction, giving rise to a smaller $\beta_i(n+1)$ and ultimately driving the proportion of Ds down to zero (i.e., complete replacement of the Ds by the Ss). In the first case, morale plummets and defeat looms. In the second case, morale soars, underwriting victory. Both outcomes may be viewed as expanding on a concept expressed in verse V.19 (Passage #5.9), which reads in relevant part "courage or cowardice [depends] on circumstances [shi]."

What makes the dynamics complicated is, of course, the interaction between islands, some of which may be above, some below, β_{crit} at a given point in time. Although several stories could be told depending on the initial distribution of Ss and Ds, from a strategic perspective one natural starting point involves analyzing a situation where just one island is initially all Ds; all the other islands, which could be many, are initially all Ss. Positing that starting point, a basic cascade insight is that there are circumstances under which (depending on the structure of the island network and the intensity of interactions between adjoining islands) even the single island initially containing all Ds can set in motion an "island cascade" leading to replacement of Ss by Ds everywhere. Importantly (and herein lies a key finding of the

---

[25] For examples of island networks see Boorman & Levitt, previous footnote, pp. 80, 145, 157. In human contexts, "islands" are often defined by organizational boundaries. Military units provide one natural source of examples of islands. Rulers' courts would be another. Moving in a direction of physically enclosed spaces, so too would walled cities in early China, or particular enclosures within them (see #3 footnote 34). Modern rise and proliferation of complex organizations of many types, frequently with sharply etched insider/outsider boundaries (e.g., key-card-enabled building access), also makes an island-structured population concept highly relevant to analyzing human social dynamics across a wide range of twenty-first-century contexts, including ones relevant to formulation and implementation of grand strategy.

[26] Cf. verse X.12 (Passage #9.17). As Cao Cao's commentary observes (Minford, p. 256): "The officers are strong and want to advance, but the rank and file are weak and disintegrate. The end result is defeat."

model) such an island cascade may still be possible when the initial proportion of Ds in the overall population of all the islands falls short, even far short, of β_{crit}.

Depending on context, such an outcome may be situation-transformative, possibly handing strategic victory to some offstage actor like one of the meddlesome neighbors of Passage #9.18's verse II.5 or Passage #9.19's verse III.23.

Viewed through a Sun Tzu lens, an island cascade might be seen as an application of the concept of verse V.21 (Passage #9.3) – "a skilled commander seeks victory from the situation [shi]," etc. – taking the pertinent "situation" to be the prevailing island network, strength of interactions between adjoining islands, and the distribution of Ss and Ds across the islands. Figure 4 illustrates a cascade success case where β_{crit} is slightly greater than .25 (here understanding "success" to be that of the *enemy*, along lines pointed by verse III.23's "A confused army leads to another's victory"). Note that presence of the islands is indispensable to takeover by the Ds. Were the boundaries between islands to be erased, throwing their separate populations together into one large population (with the same dynamics operating there), the proportion of the Ds would dwindle to zero.

Possibility of island cascade success in the circumstances just sketched does not mean that such success should be regarded as "easy" or a "foregone conclusion."[27] Explorations with various parameter choices in the original biological application point to the difficulty: the range of parameters in which island cascade success is possible is typically narrow, suggesting a phenomenon that is (a) hard to make happen yet (b) possibly game-changing when it does happen. A further basic feature of the model is that even seemingly tiny modifications in prevailing parameters, such as changes in the strength of interactions between islands or factors determining the size of the threshold β_{crit}, may spell the difference between complete victory and complete defeat.[28]

The island cascade model offers a concrete mechanism whereby an extreme underdog may be able (if the island network structure and other parameters are right) to turn existence of an island-structured population into a recipe for complete victory – or, in the evocative language of verse III.11 (Passage #4.1), takeover of "All-under-Heaven." The island cascade's central idea of building cumulatively on past island "tippings" to tip yet more islands comports well with the thrust of Sun Tzu's verse XII.15 (Passage #9.20) (now, of course, read through the eyes of a foe, for

---

[27] The island cascade model is richly endowed with threshold effects (in fact of several different types) and, as such, it combines well with the concept of ji in Sun Tzu, of which Mair, p. xliv writes:

> Ji. Pivot, moment of change (functions somewhat like a tipping point), the instant just before a new development or shift occurs . . . It cannot be stressed too heavily that ji by itself does not mean "opportunity" nor does it mean "crisis," although it is closer to the latter than to the former because of the extreme instability of a given situation and the unforeseen consequences that may follow.

[28] Such flip-flops of island cascade outcomes – victory into defeat or defeat into victory – combine well with Lau & Ames, p. 56: "Small, pointed alterations [e.g., of a military situation] can have precipitous and cascading consequences."

Tendency of island cascade dynamics to produce either complete victory or complete defeat (stalemates are possible, but not typical) also combines well with *quan sheng* of verse IV.7 (Passage #5.4) which (as Lau, 1965 article, pp. 334–35 citing this verse, notes) always means "complete victory."

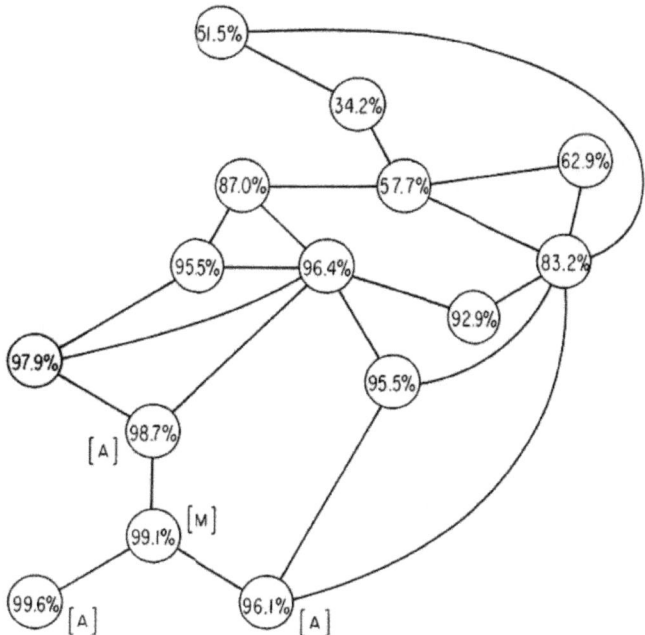

Figure 4. Snapshot of demoralization of an army via island cascade process.
The Ds start at 100% Ds in island M only, with Ss everywhere else.

Proportions of Ds (converted to percentages) are here shown after 300 time periods.* All islands are now above β_{crit} (see main text), which means that local pressures at all islands favor further increases in the proportions of Ds, with the island network being on a road to complete takeover by Ds, eliminating Ss.

Positing equal-sized islands, the overall initial proportion of Ds is $1/15 < \beta_{crit}$, so that in the absence of the island network – i.e., were all island boundaries to be erased – the island cascade would fail.

*This example is derived from Boorman & Levitt (1980) (cited in #9 footnote 24), p. 84, fig. 3.2, where it appears with other examples. Labels [M] and [A] identify the "mother site" [M] where the cascade starts (initially all Ds at that island) and its immediately adjoining islands [A].

whom island cascade success means takeover by the Ds).[29] Best prospects for cascade success commonly arise where the cascade starts at an island situated in a relatively peripheral position in an island network.[30] For this and other reasons, many of the characteristics of island cascade dynamics resonate with the 1927–49

[29] Relatedly, the island cascade model also suggests a path to integrating two apparently disjointed parts of Sun Tzu XII (verses XII.1–XII.14, XII.15–XII.19), an interpretive puzzle that has challenged Sun Tzu commentators and analysts since at least the Song dynasty. See Appendix 9, Part 3, pp. A-180ff.

[30] See Boorman & Levitt, #9 footnote 24, pp. 144–51. For a distinct vein of formal model-building leading to related results see Hassanpour, #6 footnote 47 (citing Boorman & Levitt [1980] on his p. 74).

history of the Chinese revolution, especially operations of the Chinese Communist Eighth Route Army fanning out across North China during the 1937–45 Sino-Japanese War with step-by-step development of guerrilla base areas in tandem with grassroots political mobilization.[31] The resulting consolidation of Communist power in North China proved crucial to Communist victory in the 1945–49 Chinese civil war that followed.[32] As a Chinese student introduced to the island cascade model once observed: "That is the story of our revolution."[33]

★★★★★★

While wars or other strategic competitions between two actors are Sun Tzu's central focus, there are also multi-actor competitions where Sun Tzu has ideas to contribute. A key verse here is verse III.23 (Passage #9.19), calling attention to how third parties may interfere with, and benefit from, a seemingly two-sided struggle (in that case involving discordant views of ruler and general). The present discussion now takes a closer look at multi-actor cases, distinguishing among three species of multi-actor competitions. All involve human psychological predilections for discovering or imposing linear orderings, often invidious ones.[34]

A first species of such competition centers on struggles for prestige. That is a type of social dynamic to which Sun Tzu repeatedly alludes, directly or indirectly.[35] It is a dynamic with autocatalytic tendencies, one that is prone to produce personal and institutional arrogance (among perceived victors), and bitterness, sometimes

---

[31] From a Sun Tzu (3) standpoint, note a tradeoff here, departing from Sun Tzu on one level (since Sun Tzu says nothing about political mobilization; see p. 282 above) in exchange for opening a door to empirically anchored exploration of some of Sun Tzu's dynamics ideas – for example, the meaning and role of sanctuary in cascade success (cf. verse IV.7, which appears in Passages #4.4 and #5.4).

For a brief sketch of how Chinese Communist Eighth Route Army operations might be analyzed from an island cascade standpoint see Scott A. Boorman, "Mathematical and simulation models (with research & model evaluation suggestions)," *Trajectories* (Newsletter of the American Sociological Association Comparative and Historical Sociology Section), Fall 2010, 22(1), pp. 14–20.

[32] For historical analysis see Westad, #4 footnote 74.

[33] The existing island cascade model should be treated as a baseline one. Greater realism would require elaborating it, e.g., to probe impact of local setbacks on prospects for cascade spread. One way that might be done is by adding stochastic effects to the basic model (whose equations are deterministic). For steps in that direction see Boorman & Levitt, #9 footnote 24, pp. 154–55.

[34] See Siegwart Lindenberg, "The direction of ordering and its relation to social phenomena," *Zeitschrift für Soziologie*, 1977, 6(2), 203–21.

[35] See, e.g., verse X.19 (Passage #8.18) on fame as motivator; verse I.23 (Passage #4.10) on arrogance as an exploitable characteristic; verse IX.37 (Passage #9.16) on a general's lack of prestige; verse XII.18 (Passage #8.16) on missteps born of rage and resentment.

Also relevant here is the concept of *wei*, commonly translated as "awesomeness," appearing in verse XI.52 (Passage #4.17A). Regarding *wei* see Rand, Background footnote 95, p. 108; *One Hundred Unorthodox Strategies* (translated, with historical introduction and commentary by Ralph D. Sawyer, with the collaboration of Mei-chün Lee Sawyer; Boulder, CO: Westview Press, 1996), chapter 14, "Awesomeness," pp. 52–56. (This last Chinese work, possibly dating from late Southern Song times, exposits a set of 100 military concepts, heavily rooted in Sun Tzu though also incorporating other influences.)

long-lasting (among perceived losers). It is also a timeless dynamic, unlikely to be erased or reduced very much by technological progress or social change.[36] The modern face of prestige competition is likely to have major organizational aspects. Modern governments and defense establishments, worldwide, are home to innumerable organizations unknown in Sun Tzu's simpler time which are prone to engage in parochial quests for prestige (a version of Martin Shubik's "games within the game").[37] Such organizational prestige aspirations are often pursued with a muscular and sustained vigor exceeding that found in quests for prestige by individuals.

Once quest for prestige, individual or organizational, takes root, emotional factors and perceptual distortions intrude, even ones that cut against self-interest (as verses XII.17–XII.19 [Passage #1.8] warn). A door opens for hostile or opportunistic third parties to induce actors in the ruler's court to take imprudent steps, or to engineer new or deepened conflicts among them – or even to lead some to become disloyal or traitors.[38] Meanwhile the latest prestige-fueled gossip about who's up, who's down may all too easily prove all-consuming, diverting participants' attention from reports of ill-led, ill-trained, ill-nourished armies collapsing on the frontier.

A second species of competition (sometimes co-occuring with prestige battles, though the underlying forces are different) is a struggle for "purity," which again may be individual or organizational. The second dynamic appears in Sun Tzu via possibilities for exploiting an adversary general's sense of "honor" (see verse VIII.21 [Passage #6.10]). Presence of verse VIII.21 in the text might reflect vestiges of a Spring and Autumn period "culture of honor" still present in Chinese society in Sun Tzu's time.[39] In modern times, battles for purity (i.e., being perceived as "pure" in one's own eyes as well as those of whatever audiences are deemed relevant) may reflect influence of organized belief systems, ideological or theocratic; or it may reflect some version of tribalism. Some early Chinese thinkers noted that "purity itself seems to spawn antagonism."[40] As the history of twentieth-century Communism illustrates, purity is fodder for strategic manipulation, sometimes even better than prestige, first, because interpretations of "purity" often clash (cf. Mao's Cultural Revolution!); second, because quest for purity can be tinged with loftier purpose, so that inhibitions go down and clashes escalate naturally. Again, the ultimate beneficiary (and, at times, also the instigator) may be a third party lurking on the sidelines.

---

[36] For a classic statement, starting with a vignette from Tolstoy's *War and Peace*, see C. S. Lewis, "The inner ring," from Lewis's *Transposition, and Other Addresses* (London: Geoffrey Bles, 1949), pp. 55–64.

[37] A source for this Shubik concept is given in #7 footnote 46.

[38] For a relevant exercise in role-playing ("As a Russian, I would attempt . . .") see p. 867 of Eccles, "Allied staffs," US Naval Institute [USNI] *Proceedings*, 1953, 79(8), 859–67. See also positive comment on this Eccles article by the Chief of Naval Operations, Admiral Robert B. Carney, USN, published in USNI *Proceedings*, 1953, 79(10), 1129–30 (noting prestige and political desires of Allied countries).

During World War II, Carney had served on a five-man deception planning team reporting to Admiral William Halsey. See Rothstein & Whaley, #7 footnote 1, p. 395 (Halsey post-war reminiscence).

[39] Regarding a "culture of honor" of the Spring and Autumn period, from which the society of the successor Warring States period diverged, see Lewis, Background footnote 11, chapter 1, "The warrior aristocracy," pp. 15–52, especially pp. 36–43, "Warfare and honor."

[40] Sawyer, #1 footnote 3, p. 578 note 171 (citing Han dynasty sources).

A third species of competition involves yet a different kind of rivalry, this time to have one's product (or that of one's organization) – good, service, capability, etc. – be rated as being of high (or, with an ordinal twist, highest) "quality." In a military context, a basic example would be evaluation of troops as having a high level of operational readiness.[41] There is a deeply dynamic aspect to this process, as any commander who continually struggles to sustain the readiness of his forces is acutely aware. Pertinent evaluations may be formalized, applying explicit criteria of judgment (cf. the calculations in the temple made famous by verse I.28 [Passage #1.2]); others are informal, tacit, even casual. Sometimes the parties doing the evaluations are on one's own side. Sometimes they are drawn from broader populations and audiences. Sometimes evaluations emanate, often indirectly, from rivals or enemies. But whatever their origin, "quality" evaluations always contain major subjective components and at times may be largely in the eye of the beholder. The possibility of diverging civilian vs. military criteria of judgment regarding an army's state of readiness is real (cf. Passage #9.19's verse III.21) and by no means limited to Sun Tzu's time. It is a short step from there to third-party manipulation of those quality judgments, exploiting differences of opinion in ways that serve an outside actor's strategic purpose (Passage #9.19's verse III.23). Compounding such vulnerabilities, psychological stresses on participants commonly associated with endemic quality wrangles create further pressure points, a path to realizing the aspirations of verse I.24 (Passage #4.10), "Keep him under a strain and wear him down."

In modern times, a basic case of this third dynamic arises from the difficulty of accurately gauging the reliability of technological superiority, especially when it is conjoined with pushing the state of the art. In modern military history overconfidence in evaluating such superiority has often led to massive overpromising with regrettable – and exploitable – consequences in both peace and war. Passage #9.19's list of pathologies in the civil–military relationship also has relevance to modern R&D projects where clashing criteria of judgment for evaluating "success" are a magnet for controversy and conflict. Drawing a supporting idea from a different part of Sun Tzu (Passage #9.16), verse IX.36,

> IX.36.
> When at night the enemy's camp is clamorous, he is fearful.

invites an updated reading as a suggestion to monitor an adversary's communications traffic for signs of incipient unease about a high-tech project on the brink of not living up to expectations. Once that dynamic is identified, third-party exploitation might, for example, include steps geared toward increasing to a crescendo the "clamor" to which verse IX.36 alludes, possibly aiming to engineer "spills" of classified information as intramural disputes escalate; or alternatively, finding some new way of feeding the underlying unease – for example, manufacturing an adverse "omen" (verse XI.44 in Passage #9.12), perhaps a rumor of an impending legislative hearing or inspector general probe.

★★★★★★

---

[41] Theme #12 (pp. 398–99) revisits readiness evaluation as a facet of Sun Tzu's "know yourself."

Set of principal passages used to illustrate Sun Tzu Theme #9 (building on natural forces and momentum)

Passage #9.1. First Water Image: Opportunistic Dynamics of Water, a Standard for an Army to Emulate

(Although other images enter too, Passages #9.1–#9.4 feature Sun Tzu's oft-cited water images, each showcasing some aspect of the dynamics of water.[42] Passage #9.1 employs a water image to suggest how the movements of an army should be shaped and guided by exploiting a foe's areas of weakness. Reflecting the fact that he is putting forward this idea without known support or encouragement from Chinese mathematical work of his time, or indeed of much later times, Sun Tzu may be forgiven for not contemplating the pitfalls of becoming trapped in local optima that his water image here should also suggest.)

VI.27. Now an army may be likened to water, for just as flowing water avoids the heights and hastens to the lowlands, so an army avoids strength and strikes weakness.

VI.28. And as water shapes its flow in accordance with the ground, so an army manages its victory in accordance with the situation of the enemy.

VI.29. And as water has no constant form, there are in war no constant conditions.

VI.30. Thus, one able to gain the victory by modifying his tactics in accordance with the enemy situation may be said to be divine.

VI.31. Of the five elements, none is always predominant; of the four seasons, none lasts forever; of the days, some are long and some short, and the moon waxes and wanes.[43]

Passage #9.2. Second Water Image: Inexhaustible Flow of Great Rivers As Image for Dynamics of Military Creativity

(Inexhaustibility of flow of rivers is the water image here and the associated military idea is a skilled general's similarly inexhaustible creativity or ingenuity.[44] Although sometimes overlooked among Sun Tzu's water imageries, verse V.6's concept is especially interesting because it highlights the flow of rivers as a *renewable* resource. Such a focus on renewability harmonizes with "ecological Sun Tzu" previously introduced in discussion of Sun Tzu on deception [Theme #7]. It also harmonizes with a concept of logistics as *sustained* support. See p. 49 above.)

V.5. Generally, in battle, use the normal force [zheng] to engage; use the extraordinary [qi] to win.[45]

---

[42] Ordering of Passages #9.1–#9.4 reflects Brooks's proposed order of Sun Tzu chapter accretions (see Table 12A on p. A-35).

[43] Regarding the "five elements" theory see #5 footnote 103 (basic background) and #9 footnote 13 (further dynamics aspects). One of those five elements is water.

[44] However, limits to Sun Tzu's thinking about military innovation should be noted. See pp. 215–16.

[45] For analysis of the qi/zheng concept pair see Theme #14.

V.6. Now the resources of those skilled in the use of extraordinary forces [qi] are as infinite as the heavens and earth; as inexhaustible as the flow of the great rivers.

V.7. For they end and recommence; cyclical, as are the movements of the sun and moon. They die away and are reborn; recurrent, as are the passing seasons.

Passage #9.3. Third Water Image: Torrential Flow of Water, Together with Other Images for Military Momentum and Timing of Its Unleashing

(Leading off with a further water image, these verses are Sun Tzu's most sustained development of topics pertaining to military momentum and its control.[46] Moving focus away from water images, verse V.22 suggests conceptualizing a "person" as a packet of dynamics in a biophysical wrapping. From that dynamics perspective, a strategist's choice to rely on one individual rather than another for a given purpose amounts to choosing among alternative sources of timing and momentum to translate his concepts into action.[47] The primacy Sun Tzu accords to war's human dimensions is on display here.)

V.13. When torrential water tosses boulders, it is because of its momentum [shi];[48]

---

[46] Illustrating relevant modern research, see Hsue-shen Tsien (Qian Xuesen), "Problems in motion of compressible fluids and reaction propulsion," Ph.D. thesis, California Institute of Technology, Pasadena, CA, 1938.

[47] A principle of traditional Chinese warfare articulated in Needham & Yates, pp. 33–35 – and illustrated by verse V.25's image of troops as boulders rolling down a high mountain – is "independent spontaneous movement":

> All individual persons (the enemy, soldiers, allies, officials, and the like) have their own intentions, psychological characteristics and drive, under any given set of circumstances. If one wants to lead them or stop them, *one should use their own potential, 'might' (shi), to achieve this, not just give orders or prohibit something.* One should understand the mental characteristics of others ("roundness of stones") and then create the necessary circumstances ("put the stones on the high slide"); *only then will the people do what is needed, usually without realising that they are fulfilling the commander's plans.*

> (Needham & Yates, pp. 33–34, emphases supplied; in their accompanying discussion, Sun Tzu verses V.13 and V.16 are cited)

Agent-based modeling (see p. 301 and #9 footnote 8) suggests a path to harmonizing this line of thinking with Warring States tendencies to regard common soldiers as entities that "have their own dynamism but lack calculation or deliberation" (see Mark Lewis observation quoted in #9 footnote 7). Compare p. 146 of Michael W. Macy & Robert Willer, "From factors to actors: computational sociology and agent-based modeling," *Annual Review of Sociology*, 2002, 28, 143–66: "Agents adapt by moving, imitating, replicating, or learning, but not by calculating the most efficient action."

[48] Sawyer, p. 314 note 75 makes a case that the image here is one of pent-up water being released.

V.14. When the strike of a hawk breaks the body of its prey, it is because of timing.[49]

V.15. Thus the momentum [shi] of one skilled in war is overwhelming, and his attack precisely regulated.

V.16. His potential [shi] is that of a fully drawn crossbow; his timing, the release of the trigger.

... (omitted verses [Passage #7.7] have a natural deception interpretation. Such "deception sandwiched in" suggests affinity between deception and Sun Tzu's ideas about dynamic systems and their control.[50])

V.21. Therefore a skilled commander seeks victory from the situation [shi] and does not demand it of his subordinates.

V.22. He selects his men and they exploit the situation [shi].

(Owing to a glitch in Griffith's numbering, there is no verse V.23 in his translation. There is no untranslated Chinese text intervening between verses V.22 and V.24.)

V.24. He who relies on the situation [shi] uses his men in fighting as one rolls logs or stones. Now the nature of logs and stones is that on stable ground they are static; on unstable ground, they move. If square, they stop; if round, they roll.

V.25. Thus, the potential [shi] of troops skilfully commanded in battle may be compared to that [shi] of round boulders which roll down from mountain heights.[51]

Passage #9.4. Fourth Water Image: Release of Pent-up Waters As Image for Achieving Military Momentum

(Having positioned his forces like pent-up waters, the general releases the dam, then natural dynamics does the rest of the job.[52] This verse is a good example of

---

[49] Verse V.14 points to an intersect of physics and biology. For modern analysis see Robin Mills, Hanno Hildenbrandt, Graham K. Taylor, & Charlotte K. Hemelrijk, "Physics-based simulations of aerial attacks by peregrine falcons reveal that stooping at high speeds maximizes catch success against agile prey," PLOS Computational Biology https://doi.org/10.1371/journal.pcbi.1006044 (visited October 6, 2021) (analysis of dynamics of aerial predation "combining guidance and control laws inspired by missile theory with a detailed simulation model of the biology and physics of animal flight"). The peregrine falcon's range extends to North China. See Appendix 9 footnote 4.

[50] See Passage #7.7 (where the omitted verses V.17–V.20 appear). Deception content of verse V.18 is the focus of Appendix 7 anchor footnote 14. Regarding verse V.17 see Appendix 7 footnote 13.

[51] At the end of this verse (which is also the end of Sun Tzu V) Griffith omits two characters that Mair, p. 94 translates "it is all a matter of configuration [shi]" – a summary comment that nicely integrates the preceding verses into Sun Tzu V's overall thrust, where the shi concept takes center stage.

[52] For analysis of relevant physical dynamics (a military angle is injected by noting wartime dam destruction by military action), see Oscar Castro-Orgaz & Hubert Chanson, "Ritter's dry-bed dam-break flows: positive and negative wave dynamics," Environmental Fluid Mechanics, 2017, 17(4), 665–94.

Sun Tzu's knack for using carefully chosen physical dynamics images to anchor and illustrate his military thinking.)

IV.20. It is because of disposition that a victorious general is able to make his people fight with the effect of pent-up waters which, suddenly released, plunge into a bottomless abyss.

Passage #9.5. Conforming Military Operations to Dynamics of Weather and Climate

(Early Chinese tradition ritually dictated autumn and winter as the "killing time of the year," suitable for military campaigning as well as for other official activities like hunting and punishments associated with violence and death.[53] Given Sun Tzu's pragmatic streak it is tempting – though still a conjecture – to understand verse I.5, which is part of Sun Tzu's net assessment material, as signaling a break with that tradition in favor of a less stylized stance vis-à-vis weather and climate, much as verse XIII.4 [Passage #1.19] betokens a break with divination.)

I.5. By weather I mean the interaction of natural forces [yin/yang]; the effects of winter's cold and summer's heat and the conduct of military operations in accordance with the seasons.

Passage #9.6. Taking up a Suitable Position vis-à-vis a River's Current

(Focus here shifts from Passage #9.1–#9.4's water imagery to actual water. Verse IX.6 conveys very simple but practical advice, since all manner of things – some nasty – can be floated your way when you are situated downstream.)

IX.6. If you wish to give battle, do not confront your enemy close to the water. Take position on high ground facing the sunlight. *Do not take position downstream.* [Emphasis supplied.]

Passage #9.7. Dynamics of a Conflagration, to Use for Profit or Avoid

(One way of approaching these verses, guided by verse XII.12, is as an effort at crafting a type of dynamic systems counterpart to Sun Tzu's prescriptive terrain rules. Unsurprisingly, given how volatile fire can be, Passage's #9.7 actionable advice is sketchier than is Sun Tzu's terrain advice. However, at least one verse – verse XII.11 – finds noteworthy support in modern physical science. As with Sun Tzu's terrain prescriptions, it is possible to discern in Passage #9.7 a goal of making sure that commanders think through aspects of situations, advantageous or the opposite, that they might otherwise overlook or neglect.)

---

[53] The traditional prescriptions are discussed in Lewis, Background footnote 11, "The Calendar of Violence," pp. 138–46. At least by Han times there is evidence of "sharp nullification" of those prescriptions when militarily inconvenient. See p. 680 of Derk Bodde, review of Mark Lewis, *Sanctioned Violence in Early China*, published in *Journal of the American Oriental Society*, 1992, 112(4), 679–81.

XII.1. There are five methods of attacking with fire. The first is to burn personnel; the second, to burn stores; the third, to burn equipment; the fourth, to burn arsenals; and the fifth, to use incendiary missiles.

XII.2. To use fire, some medium must be relied upon.

XII.3. Equipment for setting fires must always be at hand.

XII.4. There are suitable times and appropriate days on which to raise fires.

XII.5. "Times" means when the weather is scorching hot; "days" means when the moon is in Sagittarius, Alpharatz, I, or Chen constellations, for these are days of rising winds.[54]

XII.6. Now in fire attacks one must respond to the changing situation.

XII.7. When fire breaks out in the enemy's camp immediately co-ordinate your action from without. But if his troops remain calm bide your time and do not attack.[55]

XII.8. When the fire reaches its height, follow up if you can. If you cannot do so, wait.

XII.9. If you can raise fires outside the enemy camp, it is not necessary to wait until they are started inside. Set fires at suitable times.

XII.10. When fires are raised up-wind do not attack from down-wind.

XII.11. When the wind blows during the day [for a long time] it will die down at night.[56]

XII.12. Now the army must know the five different fire-attack situations and be constantly vigilant.

---

[54] As Mair, p. 160 note 3 observes, the latter part of verse XII.5 – proposing propitious days for launching fire attacks based on position of the moon – lacks scientific support. A trace of astrology is visible in Sun Tzu's thinking here. For additional background see Appendix 1 footnote 54 (noting that astrology and astronomy were not conceptually differentiated in traditional Chinese thinking).

[55] Du Mu's commentary (Minford, p. 307) points to an important connection here between physical and psychological dynamics: "The prime object of fire is to throw the enemy into confusion and then to attack. Fire is not in itself the means for defeating the enemy. So attack as soon as you hear of the outbreak of fire. Once the fire has died down and order is reestablished, an attack will be futile."

[56] There is a nugget of valid scientific information here. For the underlying physics see John M. Wallace & Peter V. Hobbs, Atmospheric Science: an introductory survey (2nd ed.; Amsterdam/Boston, MA: Elsevier Academic Press, 2006), §§9.3–9.4, pp. 391–404. Fig. 9.18 on p. 395, a "Sketch of variation of wind speed with local time on a sunny day over land, as might be measured at different heights," shows diminution of wind speed to a very low level c. 9 p.m.–5 a.m. See also Roland V. Stull, An Introduction to Boundary Layer Meteorology (Dordrecht: Kluwer Academic, 1988), §2.5, pp. 45–47.

Read from this perspective, verse XII.11 conveys actionable information. Night winds are not to be relied on in mounting incendiary attacks; even when wind blows persistently during the day, a continuing night wind should not be expected. This point seems especially relevant as a caution to commanders not to defer fire attacks until nightfall, possibly hoping for a combination of high, gusty wind and darkness to trigger the sought-for confusion to which Du Mu points (see previous footnote).

Appendix 9 anchor footnote 9 calls attention to Chinese text signifying "for a long time," which Griffith's translation omits. A textual variant in verse XII.11 is also discussed and rejected there.

XII.13. Those who use fire to assist their attacks are intelligent; those who use inundations are powerful.

XII.14. Water can isolate an enemy but cannot destroy his supplies or equipment.

Passage #9.8. From Physical to Social Dynamics: Storm Eliciting Cooperation between Foes

(Here seemingly adverse dynamics elicit a potentially beneficial "win–win" outcome. But Sun Tzu does not further develop this promising idea, which in one modern terminology suggests that a potentially catastrophic environment might overcome extreme political polarization, at least temporarily.)

XI.39. Should one ask: "Can troops be made capable of such instantaneous co-ordination?" I reply: "They can." For, although the men of Wu and Yue mutually hate one another, if together in a boat tossed by the wind they would co-operate as the right hand does with the left.

Passage #9.9. Dynamics of Health and Disease

(While Sun Tzu's basic prescription – occupy sunny high ground – could be faulted as simplistic, he at least acknowledges the underlying population health issue and the need to address it, which many works of military theory, some of them far more recent, fail to do.[57] The instinct to relate an army's health prospects to its position on a landscape has some claim to be treated as a remote ancestor of the modern area that flies under a banner of landscape epidemiology.[58] A major further consideration, embedded in the "hundred diseases" figure of speech, is that a general may need to contend at the same time with concurrent outbreaks of multiple diseases having very different dynamics and remedies.[59])

---

[57] Verse IX.13 alludes explicitly to both elements of the yin/yang duality, which over the course of Chinese history has been one of the conceptual cornerstones of Chinese medicine. Its content could be further developed in that direction. However, as Minford, pp. 232–33 cautions in his discussion of this verse, the Sun Tzu text – by contrast to some other early Chinese military treatises – is not particularly invested in the broader reaches or applications of yin/yang thinking. As Minford also notes, a simple sunny/shady understanding of verse IX.13 "seems perfectly natural in this context."

[58] Uriel Kitron, "Landscape ecology and epidemiology of vector-borne diseases: tools for spatial analysis," *Journal of Medical Entomology*, 1998, 35(4), 435–45.

[59] A case in point comes from the American Civil War, where "malaria, typhoid, and dysentery assailed [General McClellan's] army in the swampy encampments of the Chickahominy River, reducing troop strength by one-third" and derailing Union plans for early seizure of Richmond (Sartin, #3 footnote 12, p. 583). Late nineteenth-/early twentieth-century Iranian experience with sequential epidemics of several types – cholera, bubonic plague, flu – is analyzed by Amir Afkhami, *A Modern Contagion: imperialism and public health in Iran's age of cholera* (Baltimore, MD: Johns Hopkins University Press, 2019).

IX.13. An army prefers high ground to low; esteems sunlight and dislikes shade. Thus, while nourishing its health, the army occupies a firm position.[60] An army that does not suffer from countless diseases [*bai ji*, lit. "hundred diseases"] is said to be certain of victory.

Passage #9.10. Psychological and Biophysical Dynamics of Soldiers

(Central to verse VII.21 and its surrounding verses is the core Theme #9 idea of dynamics not of the strategist's making – here ones of a human factors type – that a canny general can proceed to exploit.[61] On an important interpretive aspect, verse VII.21 can be read expansively, applying the same idea to timescales much longer than a single day, as in the several stages of a campaign, war, etc.[62])

VII.20. Now an army may be robbed of its spirit and its commander deprived of his courage.[63]

VII.21. During the early morning spirits are keen, during the day they flag, and in the evening thoughts turn toward home.

VII.22. And therefore those skilled in war avoid the enemy when his spirit is keen and attack him when it is sluggish and his soldiers homesick. This is control of the moral factor.

VII.23. In good order they await a disorderly enemy; in serenity, a clamorous one. This is control of the mental factor.[64]

VII.24. Close to the field of battle, they await an enemy coming from afar; at rest, an exhausted enemy; with well-fed troops, hungry ones. This is control of the physical factor.[65]

VII.25. They do not engage an enemy advancing with well-ordered banners nor one whose formations are in impressive array. This is control of the factor of changing circumstances.

---

[60] Sawyer, p. 323 note 142 portrays the relevant terrain in straightforward terms: "[l]ife-supporting terrain ... ground that has sunlight, grass for the animals, brush and trees for firewood, and especially potable water."

[61] Passage #9.10's emphases seem likely to be influenced by characteristics of Sun Tzu's armies, heavily composed of peasant levies of quite possibly wobbly morale and commitment.

[62] See Preliminaries footnote 4 (citing traditional commentator Mei Yaochen and modern sources).

[63] Griffith's "spirit" or "spirits" in verses VII.20–VII.22 as well as "moral factor" in verse VII.22 is qi 氣, a concept basic to traditional Chinese culture and which in a military context has overtones of fighting spirit or morale. For further background see #3 footnote 80 (citing sources). Qi 氣 is a distinct concept from qi 奇, variously translated as "crafty," "cunning," "unorthodox," etc. (see Theme #14).

[64] Mair, p. 103 renders the first clause as "Confront chaos with control," which has a more abstract flavor that aligns well on dynamic systems thinking.

[65] Regarding the great, even paramount, significance of the fatigue factor in ancient or medieval warfare see #6 footnote 89. A different perspective on fatigue comes from a modern quarter (General Lucian K. Truscott, Jr.), who observed of part of the hard-fought World War II Italian campaign: "There has never been a battle that wasn't won by tired troops." See Rick Atkinson, *The Day of Battle: the war in Sicily and Italy, 1943–1944* (New York: Henry Holt, 2007), p. 547.

Passage #9.11. Psychological Dynamics of Rage and Greed in Soldiers

(These verses pertain to why men fight. They point to tools available to a Sun Tzu general to manipulate his troops' motivations for best performance.[66])

II.16. The reason troops slay the enemy is because they are enraged.

II.17. They take booty from the enemy because they desire wealth.

II.18. Therefore, when in chariot fighting more than ten chariots are captured, reward those who take the first. Replace the enemy's flags and banners with your own, mix the captured chariots with yours, and mount them.

Passage #9.12. Superstitious Practices Undermining Morale

(Verse XI.44 suggests a corrosive dynamic, ripe for exploitation by the enemy. Yet Sun Tzu fails to articulate any corresponding counter-dynamic – e.g., a social process on which a general might build to achieve compliance with his goal.[67])

XI.44. He prohibits superstitious practices and so rids the army of doubts. Then until the moment of death there can be no troubles.[68]

Passage #9.13. Problematic *Mentalité* of Soldiers Fighting on Home Turf

(A common interpretation of the dispersive ground dynamic is that soldiers fighting on home ground will be prone to be preoccupied with their homes and families, with resulting loss of combat potential and risks of going AWOL or desertion.[69] The first half of verse XI.11 suggests that Sun Tzu accepted the limited top-down controllability of social dynamics of that type – a noteworthy concession given Sun Tzu's very high expectations of the general's performance.)

XI.2. When a feudal lord fights in his own territory, he is in dispersive ground.

. . .

XI.11. And therefore, do not fight in dispersive ground; do not stop in the frontier borderlands.

. . .

XI.15. In dispersive ground I would unify the determination of the army.

---

[66] Mair, p. 82 makes this goal explicit: "if one wants the soldiers to kill the enemy, one must enrage them; if one wants the soldiers to derive profits from the enemy, one must induce them with goods." Of course, this type of incentive scheme should be understood in a context of Sun Tzu's time.

[67] This omission is particularly noteworthy because of a larger principle of early Chinese warfare that Needham & Yates call the principle of "independent spontaneous movement." For description see #9 footnote 47 above.

[68] A possibly clearer translation of the second sentence is: "The troops will then fight to the death rather than run away from the battlefield" (which more fully brings out relevant dynamics aspects).

[69] To be clear, verse XI.2's anchoring in this type of psychological dynamic remains inferential and other interpretations are possible. See Appendix 2 anchor footnote 4 (noting two alternatives).

Passage #9.14. Death Ground Psychology As a Special Psychological Dynamic

(Sun Tzu here profiles psychological dynamics of troops in "death ground," whereby extreme circumstances catalyze troops to fight with a ferocity far beyond "all charted limits of morale," to draw a phrase from US Marine Corps historian Lynn Montross.[70] The root principle might be labeled Sun Tzu's "phase transition theory of combat behavior." It involves a kind of behavioral rupture triggered by a sense of having no way out, save by fighting ferociously against seemingly impossible odds. Importantly, such rupture is not inconsistent with fighting smartly, indeed may demand it.[71] Sun Tzu here posits a class of dynamics on which an astute and ruthless general may ride to victory.

Three aspects of Passage #9.14 content should be particularly noted:

(a) the timescale on which a death ground response occurs may be more elongated than that suggested by labels such as "primal ferocity";[72]

(b) relatedly, that response may involve planning and coordination steps, with the lead times those usually require, thereby creating a role for not only the unbridled ferocity of the troops but also the acumen of the general;[73]

---

[70] It is instructive to compare and contrast the kinds of psychological dynamics described in Passage #9.14 with what has been termed the "doctrine of controlled beserkerdom," for which the Czarist-era Russian general and military thinker Mikhail Dragomirov (1830–1905) was an influential advocate. Dragomirov envisioned a Russian common soldier who could "on command, start and stop himself from going berserk." See p. 270 of Jan Plamper, "Fear: soldiers and emotion in early twentieth-century Russian military psychology," *Slavic Review*, 2009, 68(2), 259–83 (also pointing out that this concept depended on an implicit notion of a rational self, hence – paradoxically – was tinged with Enlightenment thinking and modernity). There is no reason to believe that Sun Tzu thought in any such terms. For Sun Tzu, the relevant behavior is catalyzed by absence of any viable alternative, i.e., it is a byproduct of the *structure* of the situation. For examples see verses XI.33 and XI.37 in Passage #9.14.

[71] While also making clear that a "single surge of morale" is part of the package, traditional commentator He Yanxi, discussing a death ground situation, quotes a saying, "He who is in dire straits and can find no stratagem, is lost" (Minford, pp. 274–75). He goes on to describe a relatively complex battle plan that includes an attack with shock troops on the enemy's rear as well as a noisy direct assault.

[72] The phrase "primal ferocity" is from p. 57 of David Graff's article cited on p. 98 above.

[73] For an example involving Cao Cao's generalship see #1 footnote 92. Verse XI.23's reference to desperate troops "following commands implicitly" supports the point, since in the hands of a talented general such troops may be able to pull off ambitious stratagems. Herein is a key difference between Sun Tzu's thinking and madly reckless fighting styles. For a cross-cultural review of the latter (which exhibit substantial internal diversity as a category, ranging from Nordic berserks to Malay and Malabar *amoks* to Aztec wolf warriors to Arapaho Plains Indians "no-retreat men") see Michael P. Speidel, "Berserks: a history of Indo-European 'mad warriors'," *Journal of World History*, 2002, 13(2), 253–90.

However, a contrary strand of opinion should be noted. As Du You describes a death ground situation: "The commander has no further strategies" (Ames, p. 218, translating from the *Tongdian*).

(c) some situations depicted here involve both serious ground [army's deep incursion into hostile territory] and death ground, suggesting that these terrain types may at times overlap in Sun Tzu's thinking.)

XI.23. In death ground I could make it evident that there is no chance of survival. For it is in the nature of soldiers to resist when surrounded; to fight to the death when there is no alternative, and when desperate to follow commands implicitly.

. . .

XI.30. The general principles applicable to an invading force are that when you have penetrated deeply into hostile territory your army is united, and the defender cannot overcome you.

. . .

XI.33. Throw the troops into a *position from which there is no escape* and even when faced with death they will not flee. For if prepared to die, what can they not achieve? Then officers and men together put forth their utmost efforts. In a desperate situation they fear nothing; when there is no way out they stand firm. Deep in a hostile land they are bound together, and there, *where there is no alternative*, they will engage the enemy in hand to hand combat.[74] [Emphases supplied.]

XI.34. Thus, such troops need no encouragement to be vigilant. Without extorting their support the general obtains it; without inviting their affection he gains it; without demanding their trust he wins it.

(Verse XI.44 [Passage #9.12], which Griffith's editing relocates, appears here in the traditional text.[75])

---

[74] (continuing line of analysis in #9 footnote 70) It might be tempting to align such behavior on a modern concept of "weapon morale" rooted in "rigorous discipline, hard training, confidence in one's leaders, one's weapons, one's ability to use them, and above all by pride in one's ability to accept great risk and hardship" (Eccles, Preliminaries footnote 5, p. 247), with a modern contrast case being "soda fountain morale" sustained by flow of creature comforts and other goodies. A basic military insight is that under pressure soda fountain morale crumbles whereas weapon morale endures. Yet Passage #9.14 – reinforced by other parts of Sun Tzu – suggests that Sun Tzu is thinking here along other lines, focusing not on weapon morale but rather on a starkly structural factor: how a desperate situation may leave soldiers *no choice* but to fight or die, thereby eliciting the courage of desperation. This point is bluntly made, in an after-action analysis of a battle, in Shiji 92 (Watson translation cited on p. xxiii above), Han Dynasty I, pp. 170–71 (quoted on p. 32 above).

[75] As a conjecture, superstitious beliefs (or at least certain types of them) may have been one of the few social forces in Warring States China capable of nullifying troops' supranormal level of performance profiled in verses XI.33 and XI.55. Cf. Kierman observations quoted in #7 footnote 101.

Indirect support for such a conjecture comes from the Mozi text, advising the general "to exercise strict control of experts in the esoteric arts, such as ether-watchers . . . and shamans . . . to have them close at hand, and *not to permit them to relay their prognostications to the soldiers and officers at large for fear that their mantic utterances might scare them.*" See p. 25 of Robin D. S. Yates, "The history of military divination in China," *East Asian Science, Technology, and Medicine (EASTM)*, 2005, No. 24, 15–43 (emphasis supplied).

XI.35. My officers have no surplus of wealth but not because they disdain worldly goods; they have no expectation of long life but not because they dislike longevity.

XI.36. On the day the army is ordered to march the tears of those seated soak their lapels; the tears of those reclining course down their cheeks.

XI.37. But throw them into *a situation where there is no escape* and they will display the immortal courage of Zhuan Zhu and Cao Gui. [Emphasis supplied.]

. . .

XI.49. He burns his boats and smashes his cooking pots; he urges the army on as if driving a flock of sheep, now in one direction, now in another, and none knows where he is going.

. . .

XI.55. Set the troops to their tasks without imparting your designs; use them to gain advantage without revealing the dangers involved. Throw them into a perilous situation and they survive; put them in death ground and they will live. For when the army is placed in such a situation it can snatch victory from defeat.

Passage #9.15. Psychological Dynamics Associated with Defeat and Retreat

(Certain military situations may appear overwhelmingly advantageous, yet might trigger an outsized fighting response by the enemy troops, turning their impending defeat into victory – and for that reason call for judgment and possible restraint by the side that appears to have a winning hand. That type of situation, seen here in several variants, involves the same death ground psychology spelled out in Passage #9.14, though now it is the foe who may exhibit that mindset.[76])

VII.30. Do not thwart an enemy returning homewards.

VII.31. To a surrounded enemy you must leave a way of escape.[77]

VII.32. Do not press an enemy at bay.

VII.33. This is the method of employing troops.

Passage #9.16. Disequilibrium Dynamics, Pointing to an Army in Difficulties

(Emphasis here is on observable signs of trouble. Because the chain of command is the backbone of any army, especially noteworthy are verses IX.37–IX.38 and IX.41–IX.43 pointing to incipient instability in that chain.)

---

[76] Traditional commentator Du Mu makes a related observation. See Giles, p. 137. Also expressing a similar concept, see account of the aftermath of the battle of Boju (noted in Introduction footnote 33) in Durrant, Li, & Schaberg, *Zuo Tradition (Zuozhuan)* (cited on p. xxiii above), Vol. 3, p. 1755.

[77] Another way of reading this verse would understand the advice to involve creating only the *appearance* of a way out, precisely to avoid eliciting an awesome "death ground" response from trapped troops. See Passage #7.21A discussion (quoting Du Mu interpretation).

IX.32. When his troops lean on their weapons, they are famished.

IX.33. When drawers of water drink before carrying it to camp, his troops are suffering from thirst.

IX.34. When the enemy sees an advantage but does not advance to seize it, he is fatigued.

. . .

IX.36. When at night the enemy's camp is clamorous, he is fearful.

IX.37. When his troops are disorderly, the general has no prestige.

IX.38. When his flags and banners move about constantly he is in disarray.

IX.39. If the officers are short-tempered they are exhausted.

IX.40. When the enemy feeds grain to the horses and his men meat and when his troops neither hang up their cooking pots nor return to their shelters, the enemy is desperate.

IX.41. When the troops continually gather in small groups and whisper together the general has lost the confidence of the army.[78]

IX.42. Too frequent rewards indicate that the general is at the end of his resources; too frequent punishments that he is in acute distress.

IX.43. If the officers at first treat the men violently and later are fearful of them, the limit of indiscipline has been reached.[79]

Passage #9.17. Dynamics That Could Portend an Army's Disintegration

X.9. Now when troops flee, are insubordinate, distressed, collapse in disorder or are routed, it is the fault of the general. None of these disasters can be attributed to natural causes.

X.10. Other conditions being equal, if a force attacks one ten times its size, the result is flight.

X.11. When troops are strong and officers weak the army is insubordinate.

X.12. When the officers are valiant and the troops ineffective the army is in distress.

X.13. When senior officers are angry and insubordinate, and on encountering the enemy rush into battle with no understanding of the feasibility of engaging and without awaiting orders from the commander, the army is in a state of collapse.

X.14. When the general is morally weak and his discipline not strict, when his instructions and guidance are not enlightened, when there are no consistent rules to guide the officers and men and when the formations are slovenly the army is in disorder.

X.15. When a commander unable to estimate his enemy uses a small force to engage a large one, or weak troops to strike the strong, or when he fails to select shock troops for the van, the result is rout.

---

[78] For an alternative reading of this verse see Appendix 4 anchor footnote 16. The military upshot seems likely to be about the same.

[79] Lit. "is extremely unwise" or, as Goldin, Introduction footnote 7, p. 17 puts it, "is the epitome of incompetence." For a military rationale for Griffith's choice of translation see Appendix 9 footnote 19.

X.16. When any of these six conditions prevails the army is on the road to defeat. It is the highest responsibility of the general that he examine them carefully.

Passage #9.18. Adverse Socioeconomic Dynamics Brought about by a Major War Effort, Especially a Protracted War

(Although Sun Tzu does not say so, the severe economic repercussions he describes may well set in motion further dynamics involving rural instability: abandonment of land, migration, banditry and other grassroots violence, widespread malnutrition with resulting decreased labor supply, and more.[80])

II.5. When your weapons are dulled and ardour damped, your strength exhausted and treasure spent, neighbouring rulers will take advantage of your distress to act. And even though you have wise counsellors, none will be able to lay good plans for the future.

...

II.11. When a country is impoverished by military operations it is due to distant transportation; carriage of supplies for great distances renders the people destitute.

II.12. Where the army is, prices are high;[81] when prices rise the wealth of the people is exhausted. When wealth is exhausted the peasantry will be afflicted with urgent exactions.

II.13. With strength thus depleted and wealth consumed the households in the central plains will be utterly impoverished and seven-tenths of their wealth dissipated.

II.14. As to government expenditures, those due to broken-down chariots, worn-out horses, armour and helmets, arrows and crossbows, lances, hand and body shields, draft animals and supply wagons will amount to sixty per cent of the total.

Passage #9.19. Pathologies in Civil–Military Relations That Rival Polities Can Exploit

III.18. Now the general is the protector of the state. If this protection is all-embracing, the state will surely be strong; if defective, the state will certainly be weak.

III.19. Now there are three ways in which a ruler can bring misfortune upon his army.

---

[80] Griffith, p. 21 paints a brief word picture. On the extremes to which hunger can drive people see Wilkinson, §13.2.2, p. 208; on peasant uprisings see Wilkinson, §24.2, pp. 342–43. Related analysis spanning some two millennia of Chinese history, picking up just after the Warring States era (see table 1 on their p. 361), is C. Y. Cyrus Chu & Ronald D. Lee, "Famine, revolt, and the dynastic cycle: population dynamics in historic China," *Journal of Population Economics*, 1994, 7(4), 351–78.

[81] Rising prices in the vicinity of an army is an age-old and ubiquitous phenomenon. For examples from Qing dynasty military operations see Perdue, Introduction footnote 52, pp. 243, 253.

III.20. When ignorant that the army should not advance, to order an advance or ignorant that it should not retire, to order a retirement. This is described as "hobbling the army".

III.21. When ignorant of military affairs, to participate in their administration. This causes the officers to be perplexed.

III.22. When ignorant of command problems to share in the exercise of responsibilities. This engenders doubts in the minds of the officers.

III.23. If the army is confused and suspicious, neighbouring rulers will cause trouble. This is what is meant by the saying: "A confused army leads to another's victory."

Passage #9.20. Failure to Exploit Momentum Born of Past Successes

(Failure to exploit momentum – even snatching defeat from the jaws of victory – takes center stage in verse XII.15.[82] Dynamics of several kinds come together here. Beyond military dynamics, which have both psychological and logistics aspects,[83] political dynamics involving alliances formed and broken may also enter the picture, especially when success is great enough. So too may a ruler's growing mistrust of an overly successful general, one of the great recurring patterns in Chinese history.)

XII.15. Now to win battles and take your objectives, but to fail to exploit these achievements is ominous and may be described as "wasteful delay".

★★★★★★

THEME #10 NO SCRUPLES AND CONTROLLED VICIOUSNESS

---

roots of a no-scruples mindset:

Passage #10.1 (verse I.1) (catalyst to dispensing with scruples: war as existential conflict)

Passage #10.2 (verses XII.18–XII.19) (war as existential conflict, cont.)

Passage #10.3 (verse II.3) (further catalyst to dispensing with scruples: focus on a quick win)

deception and betrayal at the core:

Passage #10.4 (verse I.17) (license to deceive? Doctrinal foundation for absence of scruples, including profitable betrayals)

Passage #10.5 (verses VII.12–VII.13) (license to deceive, cont.)

Passage #10.6 (verses XIII.1–XIII.2) (endorsement of espionage métier)

---

[82] Starting with the traditional commentators, verse XII.15 has seen a number of interpretations, though there is broadly based modern support for Griffith's understanding (which builds on that of traditional commentator Mei Yaochen). For analysis see Appendix 5 anchor footnote 21.

[83] Factors favoring exploitation of success could include the infusion of fresh logistics and human resources from captured enemy materiel and troops of which Sun Tzu speaks in verse II.20 (Passage #3.5), describing it as "winning a battle and becoming stronger."

Passage #10.7 (verse XIII.10 + XIII.19) (betrayals in espionage: expendable agents)

Passage #10.8 (verse XIII.22) (betrayals at the highest levels of grand strategy and statecraft)

no scruples on a boutique scale, illustrated by passages that allow such an interpretation:

Passage #10.9 (verse XI.28) (seizing something the enemy cherishes)

Passage #10.10 (verses VIII.17–VIII.24) (exploiting personality traits, possibly in vicious ways)

Passage #10.11 (verse XIII.15) (enforcing intelligence security with death, possibly including death of innocents)

Passage #10.12 (verse XIII.16) (systematic planning for targeted killing)

no scruples on a larger or mass scale:

Passage #10.13 (verses XI.33–XI.50) (throwing one's own people into extraordinary peril, without knowledge of your plan to do so, as scheme to elicit supranormal performance)

Passage #10.14 (verses XI.54–XI.55) (related statement, alongside encouragement to disregard customary norms)

Passage #10.15 (verses XI.7 + XI.13 + XI.20 + XI.30–XI.31) (advice to plunder)

five codicils to Theme #10, each suggesting a potential source of limitation or constraint:

Passage #10.16 (verse I.4) (first codicil: harmonious relation between ruler and subjects)

Passage #10.17 (verses III.1–III.3 + III.10–III.11) (second codicil: taking the enemy state intact, in preference to destroying it, and comparable accomplishments vis-à-vis enemy military units)

Passage #10.18 (verses II.18–II.20) (third codicil: caring for captives)

Passage #10.19 (verse X.20) (fourth codicil: accompanying one's troops on high-risk missions, and sharing their fate, as factor mitigating leader deception by a general?)

Passage #10.20 (verse XIII.13) (fifth codicil: implicit constraints on interrogation methods?)

★★★★★★

Warring States China was not a gentle or a merciful place. The temper of those times sets a stage for Theme #10:

> By the fifth century B.C., it had become clear to all contenders that the only alternative to winning was to perish. And as these rivals for the throne of a unified China grew fewer, the stakes and the brutality of warfare increased exponentially. During this period, warfare was transformed from a gentlemanly art to an industry, and lives lost on the killing fields climbed to numbers in the hundreds of thousands. Itinerant philosophers toured the central states of China, offering their advice and services to the contesting ruling families. Along with the Confucian, Mohist, and the Legalist philosophers who joined this tour was a new breed of military specialists schooled in the concrete tactics and strategies of waging effective warfare. Of these military experts, history has remembered best a man named Sun Wu from the state of Wu, known honorifically as "Sun-tzu" or "Master Sun."

(Ames, pp. 3–4)

Using Sun Tzu's own language, this type of warfare, and the societies that gave rise to it, itself constitutes a kind of death ground, "terrain in which you will survive only if you fight with all your might but perish if you fail to do so."[1]

Under these circumstances it should come as no surprise that the methods of warfare espoused by many of the itinerant experts should be very open to extreme measures, possibly of more than one kind – indeed, might actively embrace such measures and strive to dream up new ones. That mindset was reflected in military texts of the era (the genre of work known as bingfa, or "military methods"), spawning a secondary competition among works of that ilk. As Li Ling, a modern Sun Tzu authority, puts it, "The more immoral, the better the bingfa"![2]

Against this backdrop, it should be little wonder that Sun Tzu – a bingfa text – shows traces of engagement with toxic facets of strategy for which suitable descriptive labels might be "no scruples" or "controlled viciousness."[3] Of course, such labels should be equipped with criteria of judgment for their application. Here complications intrude. Warring States China presents a complex and moving analytical target, by no means historically or doctrinally homogeneous. It is also a long-ago world regarding many of whose pertinent aspects, among them beliefs and attitudes of common soldiers, our sources are inherently very limited. Faced with those challenges, a first Theme #10 analytical step is to inventory Sun Tzu passages advocating for practices that might reasonably be regarded as transgressing modern norms pertaining to armed conflict. That initial Sun Tzu harvest plainly overfishes, in the sense that many (perhaps most) of such modern norm-violating practices would have been widely accepted in Warring States China, as they have been in numerous other times and places in world history (though Warring States China also harbored voices that set very high ethical bars, questioning the legitimacy of many practices pertaining to warfare and even of war itself).[4] Yet this initial textual harvest is actually strikingly sparse. It reveals a Sun Tzu whose statements germane to Theme #10 are more limited, more focused, and in key ways also more nuanced than Sun Tzu's classic image, certainly in the West, suggests.[5]

---

[1] Verse XI.10 (Passage #2.1), translation from Lau's 1965 article, p. 328.

[2] Li Ling, #1 footnote 84, p. 67. In a larger discussion (pp. 66–68 there), he also observes (p. 66): The emergence of bingfa [with Sun Tzu as a major example] is directly related to the idea that war is based on deception. The weak may use whatever tactics – including the most unscrupulous ones – to achieve strategic advantage. This has nothing to do with morality. But what is bingfa? . . . It is an art, it is first and foremost to break the gentlemanly rules that previously governed the rule of the battlefield: use whatever means available, regardless of morality.

[3] The more basic concept is "no scruples," with "controlled viciousness" amounting to a certain no-scruples style. Viciousness refers to willingness to inflict harm in a deliberately cruel or violent way, for maximum effect. "Controlled" adds a harmonic of calculated action. Albeit from outside the Sun Tzu text, the story of Sun Tzu and the concubines (p. 329 below) establishes a relevant tonality.

[4] Quoting Mohist writings on point, Brooks & Brooks, Introduction footnote 36, pp. 94–97 note that "Antiwar sentiment appears in an already developed form at the beginning of the [fourth century BC]."

[5] In point of fact, Sun Tzu's sense of nuance did not always gain traction in Warring States conditions, including the ones that ultimately mattered most. Often overlooked by aficionados of Sun Tzu's ideas is a key point of historical context relevant to Sun Tzu's counsel to take the enemy state, army, etc. intact (Passage #10.17). Appealing as those aspirations may be in the abstract, they

To a first approximation, Sun Tzu's posture with respect to Theme #10 might be described as encouraging a strategist to cast off self-imposed normative constraints – or at least any that would hobble that strategist's attainment of his ultimate core purpose.[6] In two words, no scruples. Two modern scholars put it almost as succinctly: "the [Sun Tzu] book is concerned with attaining mastery by any means, fair or foul."[7] This is the Sun Tzu of Sima Qian's famous biographical sketch. As told by Sima Qian, Sun Tzu slays two concubines of the ruler in order to shock a larger group of them to stop giggling at him and submit to military drill discipline (and, as the intended larger lesson, to instruct the ruler in a general's proper handling of his army).[8] The norm-transgressing character of Sun Tzu's action (at least from the appalled ruler's standpoint) is underscored by the ruler's response to Sun Tzu's announcement that "the troops have now been properly disciplined": "The Commander [meaning Sun Tzu] may return to his chambers to rest. I have no desire to descend and review the troops."[9] The ruler also complains about his stomach.

---

fail to align on the actual course of Chinese history as the Warring States period drew to a close, where devastation of adversaries by the conquering Qin became the order of the day. As Brooks & Brooks (Introduction footnote 36, p. 120) put it, "as the wars went on, Qin preferred to devastate conquered cities and massacre surrendered armies … *This retaliation against resistance to Qin seems to have assisted, not retarded, the progress of Qin to final victory.*" (Emphasis supplied.) See also Wilkinson, §59, p. 786 ("with the Shang Yang reforms of 356 BC, Qin began to prepare for all-out war").

Illustrating how tricky historical meaning-making can be, parts of the modern Sun Tzu literature display images of the terracotta soldiers archaeologically recovered from the tomb of the founding emperor of Qin who died 210 BC. Yet his unification of China, by the above account, came about, not by virtue of his adherence to Sun Tzu's wise counsel, but from outright flouting of it!

[6] Because normative constraints impose a type of form, there is continuity here with Sun Tzu's Theme #8 formlessness (*wu xing*) thinking, especially *wu xing* 5 and 6 (see Figure 3 on p. 270 above).

[7] See Gawlikowski & Loewe, Background footnote 79, p. 446.

[8] See Ames, pp. 32–34; Mair, pp. 133–35; Nienhauser, Vol. VII (1994), pp. 37–38; Nienhauser, Vol. VII (2021), pp. 69–71 (both cited on pp. xxiii–xxiv) (further information relevant to the concubines story is contained in footnotes there). See also pp. 12–13 above (noting challenges to that story's authenticity). It should again be emphasized that this story is not found in the thirteen-chapter Sun Tzu text.

[9] Ames, p. 34. It is, however, worth noting a less widely known version of the concubines story, from the same archaeological find that recovered the Han strips. Translated by Ames, pp. 193–96, it appears to point to a slightly kinder, gentler version of the Sun Tzu persona. An excerpt illustrates:

"Master Sun [addressing the ruler, the King of Wu]:	"It is only important that it be what Your Majesty wants to do. We can use [for purposes of the exercise] noble persons, we can use common folk, we can use your court ladies. We will train the men on the right and the ladies on the left. … [missing material].
[King Helü] said:	'I want to use my court ladies.'
Master Sun replied:	'Many of the court ladies lack the stamina. I would rather use … [missing material]." (Ames, p. 193; underlined comments in brackets added here for clarification)

This alternative version lacks some of the pathos of the later Sima Qian version, since the two (simulated) officers (the ones who may wind up being executed on Sun Tzu's orders, though missing material in the archaeologically recovered text makes that upshot inferential) are identified there as Sun Tzu's chariot driver and his arms bearer – possibly themselves grizzled old

The mindset illustrated in this story suggests one basic reason Sun Tzu's military theory is so streamlined, its exposition so compact, and (as long as one accepts the premises!) its applicability broad indeed. As Clausewitz observed: "Theory becomes infinitely more difficult as soon as it touches the realm of moral values."[10]

Apocryphal as it almost certainly is, the story of the concubines also captures a key element of finely titrated control that sets Sun Tzu's thinking apart from sheer wanton ruthlessness or brutality. For example, in that story, following Sun Tzu's first (unheeded) order to the concubines but before any concubine has been slain, Sun Tzu offers a patient, didactic explanation to the ruler that a commander must allow for the possibility that his order has not been clear. Sun Tzu then repeats the order (in fact, several times), to exclude that contingency, thus harnessing communications redundancy in service of command-and-control – a strikingly modern harmonic![11] Only after the orders have been repeatedly given and disobeyed are two concubines slaughtered. A larger principle is that vicious measures are not to be taken for their own sake and must allow for context (albeit not necessarily in ways that would align on modern norms!).

The Sun Tzu text is rich in tonalities conducive to its being read as a tacit encouragement of controlled viciousness. Yet only in the barest handful of instances are arguably norm-transgressing expedients spelled out by Sun Tzu (even applying the standards of our own age).[12] The unkindly temper of Sun Tzu's time suggests that such rather oblique handling of the meat of Theme #10 does not stem from any wish to mask or "sanitize" measures that nowadays would be repugnant. Much more likely, it is a byproduct of the text's brevity and abstract level, in which many military and strategic topics clamor for attention and not all receive it. But planned or not, Theme #10's sparse textual anchoring compels a modern analyst to do some sleuthing to piece together what we can of Sun Tzu's relevant mentality.

As a starting point for developing a textually anchored lens on the no-scruples Sun Tzu, one need search no further than the first verse of the text:

I.1.

War is a matter of vital importance to the State; the province of life or death; the road to survival or ruin. It is mandatory that it be thoroughly studied.

(from Passage #1.1)

Passage #10.2 hammers home the same existential warning. From these starting points to jettisoning scruples is a short, albeit not inevitable, step. Taking that step is encouraged by Sun Tzu's Theme #3 emphases, notably his proccupation with the "quick win" that sidesteps many of the costs of war (Passage #10.3).

---

soldiers, not the king's two favorite concubines who are made unit commanders in the better-known version.

[10] Clausewitz, On War, Preliminaries footnote 1, p. 136.

[11] Further illustrating the controlled viciousness exemplified by the concubines story, note that it is two concubines – the simulated unit commanders – who are killed on Sun Tzu's order. In this context 2 > 1: two object lessons might be more effective than a single object lesson (which might be discounted by the audience for a variety of reasons, e.g., because interpreted as personal animus).

[12] Illustrating the point, a modern analyst has pungently described verses I.17 et seq. (Passage #7.1) as "the book's most infamous passage" (Sawyer, Introduction footnote 55, p. 58). Yet read literally, the nuggets of advice offered there are, on their face, little more than a collection of ruses de guerre familiar from many places in the history of warfare, none of them particularly eyebrow-raising.

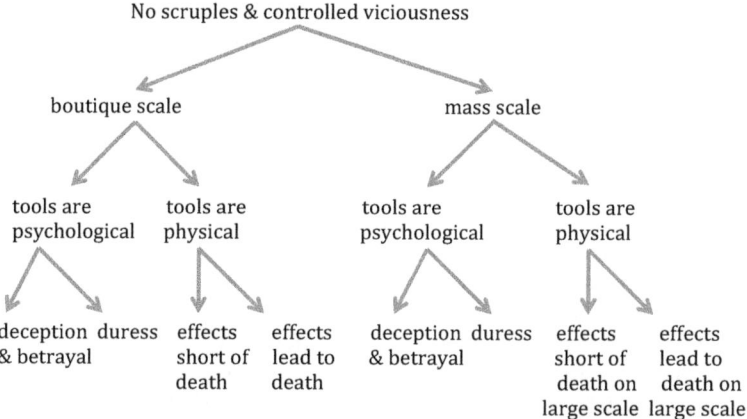

Figure 5. Heuristic typology of potentially norm-transgressing actions in Sun Tzu.
The present structure is designed to help organize and clarify Sun Tzu's Theme #10 thinking. To that end, detail is kept to a minimum. Many refinements or elaborations are plainly possible.*
    Order of listing ("boutique" first, then "mass"; "psychological" first, then "physical"; "deception and betrayal" first, then "duress") broadly aligns on emphases found in Sun Tzu.

*Omitted in Figure 5, because there is so little in the Sun Tzu text about them, are norm-transgressing actions that confer on the object of the action not a cost but a benefit. A candidate for an example is verse XI.54 (Passage #10.14), encouraging a general to "bestow rewards without respect to customary practice." However, the interpretation of verse XI.54 and its place in Sun Tzu's thinking remains unclear. For further analysis of this verse see p. 275 above.
    Likewise omitted, since Sun Tzu does not directly touch on this topic at all, are psychological or information warfare steps of the sort that aim to spread disinformation, fake news, etc. on a mass scale. These may shade over into persuasion campaigns directed against human groups or populations. See William J. McGuire, *Constructing Social Psychology: creative and critical processes* (New York: Cambridge University Press, 1999), "Developing effective persuasion campaigns," pp. 136–79.

Figure 5's typology is a way of structuring Sun Tzu's Theme #10 thinking. Often (though not always) "no scruples" operates on a "boutique scale" – opting for small-scale measures that suffice to get the job done (a concept that brings in Theme #3's emphasis on being cheap as well as Theme #4's focus on winning by non-military ways and means). Sun Tzu's predilection for such measures is coded into Figure 5 via the first distinction it draws, between "boutique scale" measures (whose immediate target is individuals or small groups) and ones on a "mass scale" (whose target is larger groups or populations).[13] Passages #10.4–#10.12 collect Sun Tzu no-scruples pronouncements that allow for an interpretation in boutique terms (often because the action involved targets just one person, namely, the enemy general or ruler). The broader effects of such an action may, of course, be far from boutique,

[13] This ideal-type binary classification could clearly be elaborated as endpoints on a continuum.
    Sun Tzu's calculating mindset (Theme #1) helps explain a preference for actions having "boutique" rather than mass effects. The latter may be less controllable, at least with available means of control.

especially if it is successful in its intended purpose. Passages #10.4–#10.12 are then followed by a noticeably smaller set of passages (Passages #10.13–#10.15) that, in one way or another, point to no scruples on a larger, possibly mass scale.

Combing the Sun Tzu text for explicit allusions to boutique physical measures that might be characterized as norm-transgressing (at least in our own time) yields only scattered candidates, none likely to transgress prevailing Warring States norms.

Verse XI.28 (Passage #10.9) might cover hostage-taking, though that is inferential. Hostage-taking as a practice was well known in Chinese history by Sun Tzu's time, and long persisted in the face of criticism because it was useful.[14] Verse XIII.15 (Passage #10.11) counsels willingness to take draconian steps against not only a loose-mouthed spy but also against anyone he talked to about secret matters (evidentiary standards are not mentioned, and the advice might extend to kill-on-suspicion). A door may then open to killing innocents (conceivably children) to seal off the security breach. Yet this type of step, indeed much more far-reaching forms of guilt by association, was a commonplace in early China, so Sun Tzu is here doing little more than signing on to conventional thinking in his own time.[15] If there is anything noteworthy about the operational advice of verse XIII.15, it is that the targeted killing set is defined using something like social network criteria (i.e., people to whom the operative talked or might have talked), as contrasted with criteria based on cultural categories alone (e.g., kinship).[16]

Continuing with a physical measures focus, verse XIII.16 (Passage #10.12) confirms, en passant, an acceptance of a targeted killing agenda (its scope is not specified). Ranging beyond the espionage chapter, verse XI.37 (Passage #10.13) gives implied plaudits, via its historical allusions, to assassinating rulers. In early Chinese context, as in many other eras of world history, those foci too are unremarkable.[17]

Verse XI.44 (Passage #10.13) is advice to stamp out superstitious practices.[18] Given everything else we know about Warring States armies and culture, it is hardly implausible that implementing that advice might involve putting to death individuals

---

[14] See Yang, #6 footnote 27, p. 519. On a point of clarification, verse XI.28's advocacy for "seize something he cherishes" need not be confined to boutique (small scale) targets (e.g., the target might be a capital city). However, suitable boutique targets may be a longer list (and more feasible to seize).

[15] For background on early Chinese practices involving vengeance and collective punishments, see Lewis, Background footnote 11, pp. 55–56 and 80–94. See also Crispin Williams, "Early references to collective punishment in an excavated Chinese text: analysis and discussion of an imprecation from the Wenxian covenant texts," *Bulletin of the School of Oriental and African Studies, University of London*, 2011, 74(3), 437–62 (also citing Susan Roosevelt Weld, 1990 Harvard Ph.D. dissertation).

[16] In the Western Han dynasty, a later though still early time, the stipulated punishment for rebellion – a crime considered heinous – included, not only the execution of the chief culprit (sometimes cut in two at the waist), but also beheading of his "'three groups of relatives" (parents, wives and children, brothers and sisters, and theoretically also his grandparents and grandchildren). See p. 317 of A. F. P. Hulsewé, "Royal rebels," *Bulletin de l'École française d'Extrême-Orient*, 1981, 69, 315–25.

[17] See Sawyer, #1 footnote 3, pp. 116–20, noting that (p. 116) "During the Warring States period assassination continued to be actively employed by generals and rulers to swiftly resolve difficult situations, especially when confronted by disadvantageous battlefield odds." The "swiftly resolve" aspect harmonizes with Sun Tzu's Theme #3 (cheap expedients) thinking.

[18] Mair, p. 120 is closer than Griffith to the Chinese text: "prohibit talk of omens and banish doubts."

deemed ringleaders in spreading such practices or the rumors to which they give rise. But that is again a conjecture. Sun Tzu says nothing specific to corroborate it.

Still focusing on physical measures, but now shifting attention to the righthand part of the Figure 5 tree, there is even less Sun Tzu content to generate a set of relevant passages. Nothing in Sun Tzu explicitly endorses killings on a mass scale, apart, of course, from battlefield events (and even there Sun Tzu's Theme #4 emphasis on winning without fighting points strongly in a different direction, casting a witheringly skeptical eye on the need for any bloody battles at all). The nearest approach to an exception is Sun Tzu's advice to plunder fertile country (Passage #10.15), which could lead to large-scale starvation deaths of civilian men, women, and children (though it should be noted that such outcomes have at times been taken in stride in modern warfare).[19]

To this point in the overview, Sun Tzu's Figure 5 scorecard has few entries. But as soon as focus shifts from physical to psychological measures that smack, or might smack, of no scruples or norm-transgressing viciousness, the picture changes. Prominently positioned in Sun Tzu I (and reinforced, were any reinforcement necessary, by verse VII.12 [Passage #10.5], expressing a like sentiment), verse I.17 (Passage #10.4) should be considered basic to understanding Sun Tzu's distinctive slant on Theme #10 issues. Its bald observation about centrality of deception in war notably comes without limitations or scope conditions. Conjoined with Sun Tzu's emphasis on winning without fighting (Theme #4), verse I.17 is a recipe for norm-transgressing ruthlessness of many varieties, possibly physical but definitely psychological.

Here a distinction should be drawn. "Deception" is a broad and versatile concept. In many times and places in world military history, Western as well as Chinese, deception – even used extensively – has been a mainstay of successful military action, often reducing war's human costs (sometimes for both sides).[20] In a context of warfare, there is nothing about such reliance that by itself necessarily connotes exceptional viciousness or lack of scruples. What may enter Theme #10 norm-transgressing territory is when a deception carries with it a further element of betrayal – of deliberately violating someone's trust in you, with devastating consequences for that person (who may be thoroughly unsuspecting up to the moment of betrayal or, in some cases, even after it). A key to grasping the toxic element of the Sun Tzu mentalité is precisely Sun Tzu's apparent, if implicit, willingness to have recourse, without scruples, to profitable betrayals.[21]

---

[19] Traditional commentator Du You puts the situation starkly: "the countryside will have nothing left for [the army] to pillage." See Ames, p. 200 (translating from Du You's *Tongdian* encyclopedic work).

[20] On deception as a cost-saving measure see pp. 240–41 above and #7 footnote 9 there.

[21] There is, of course, much cultural content surrounding application of the concept of "betrayal," which builds on the yet more fundamental concept of trust. By way of entrée to the topic, more useful than any dictionary definition is Charles Tilly's treatment of "trust networks" in his *Trust and Rule* (Cambridge, UK: Cambridge University Press, 2005), which mobilizes many pre-modern examples (less so, non-Western ones). The concept of betrayal is noted at the outset (see Tilly's Preface, p. xii).

Albeit phrased in somewhat antiseptic language (e.g., "threat to well-being," not bloody death), a definition of betrayal found in modern business literature does what a useful definition should do, namely, to provide a structural framework for analyzing applications of the concept in a disciplined way (possibly incorporating culture-specific content). Betrayal is defined as "a voluntary violation of mutually known pivotal expectations of the trustor by the trusted party

Such willingness dovetails with strands of Sun Tzu that predispose to favor a "logic of short-term advantage."[22] Sun Tzu's major focus on inducing estrangements (see p. 149 above) also suggests that a strategic actor imbued with Sun Tzu's thinking might exhibit considerable sangfroid in shrugging off yet one more estrangement from an erstwhile friend or ally. Sun Tzu's penchant for envisioning all conflict in zero-sum terms (see pp. 66, 145–46, 152) might tend to encourage a blunting of perceptions, placing disagreement or divergence of interests with an ally on a par with war with an enemy, and thereby making quite conceivable turning on that ally when the time seems right. These considerations suggest that, for a Sun Tzu–inspired actor, betrayals may be a routine, business-as-usual, matter.[23] There is a potent "splitter" dynamic at work here, and Sun Tzu leaves wide open where it might lead.

One reason a "friction-free betrayal" mindset is so revealing a facet of Sun Tzu, far more so than any endorsement of mass mayhem would be, is that willingness to betray raises deep issues, not only of strategic and tactical control, but also of identity in its political, social, and personal senses. Without checks and balances (which are not apparent in Sun Tzu) verse I.17 points toward steps that could destabilize identity, possibly many identities – perhaps including that of the deceiver.

Two additional parts of Sun Tzu shed further light on this facet of his thinking.

(trustee), which has the potential to threaten the well-being of the trustor." See p. 548 of A. R. Elangovan & Debra L. Shapiro, "Betrayal of trust in organizations," *Academy of Management Review*, 1998, 23(3), 547–66.

[22] "Logic of short-term advantage" is a phrase used by Mark Elvin. See p. 119 above. Illustrating the point, Sun Tzu sheds no light on advantages of long-term alliances or approaches to building them.

[23] Elite politics in early China brims with examples of wild fluctuations in the fortunes of prominent actors, often brought about by a betrayal. Reflecting on the case of Wu Zixu, a contemporary of the putative historical Sun Tzu – a distinguished general done in by a fabricated allegation (see #6 footnote 9) – Sawyer, p. 107 observes that "the ease with which men shifted allegiances and were willingly embraced by other factions or even states is astounding."

From a few centuries on, a further vivid sense of turn-on-a-dime loyalties comes from the fate of the talented early Han general Han Xin (died 196 BC). Han Xin's military talent was spotted and brought to the attention of the future Han emperor by Xiao He, a key early supporter. Han Xin was duly appointed general-in-chief and went on to win many military victories (one of which is recounted on pp. 30–32 above); but was then said by the Han ruler to be plotting a revolt. As he was being led away, bound by the emperor's guards, Han Xin lamented:

> It is just as men say. ... When the cunning hares are dead, the good dog is boiled; when the soaring birds are gone, men put away the good bow; when the enemy states have been defeated, the ministers who plotted their downfall are doomed. The world is now at peace, and so it is fitting that I be boiled!"
>
> (Shiji 92 in Watson translation [cited on p. xxiii above], *Han Dynasty I*, p. 181)

Han Xin narrowly escaped execution then; was demoted; then again came under suspicion of fomenting rebellion, was arrested again (this time at the instigation of the empress but with connivance from Xiao He in the emperor's temporary absence, and was executed. So Xiao He was both Han Xin's talent discoverer and his executioner.

When the emperor upon his return learned of this development, he reacted with Sun-Tzu-esque ambiguity, being "both pleased and saddened by the news" (Watson translation, *Han Dynasty I*, p. 182). For further background on these and other dizzying twists and turns see Aihe Wang, "Creators of an emperor: the political group behind the founding of the Han empire," *Asia Major*, Third Series, 2001, 14(1), 19–50, especially pp. 40ff. there chronicling the self-destruction of this group.

One is, of course, the espionage chapter which is permeated with betrayal harmonics over and above those associated with garden-variety spying, in particular assigning a starring role to double agents:[24]

> XIII.9.
>
> Doubled agents [*fanjian*] are enemy spies whom we employ.

Passage #10.7 also makes clear Sun Tzu's willingness to treat some spies as "expendable," setting them up to be caught. Read literally, Sun Tzu's definition of "expendable agent" (verse XIII.10 in Passage #10.7) might actually cover volunteers who have made a deliberate (and heroic) decision to sacrifice their life for their country. However, widespread understanding of the concept, as it crystallized in later Chinese history, clearly gravitated toward applying the "expendable" label to agents unaware that they were being sacrificed. The latter practice, involving an element of betrayal, is a candidate for "vicious Sun Tzu."[25] It certainly raises issues that have troubled thoughtful intelligence professionals in our own time.[26]

But potentially even stronger stuff (since espionage is well known to be a dirty business!) is a Sun Tzu general's behavior as profiled in Passages #10.13 and #10.14. One natural reading of this material is as advocacy (especially clear in the latter passage's verse XI.55) for a general's making a conscious decision to insert his people into a desperate situation, to extract "supranormal" combat performance from them as a steppingstone to some larger military end – meanwhile deliberately keeping them in the dark, poor dumb bastards that they are, about the extremity of their impending peril:[27]

> XI.55.
>
> Set the troops to their tasks without imparting your designs; use them to gain advantage *without revealing the dangers involved*. Throw them into a perilous situation and they survive; put them in death ground and they will live. For when the army is placed in such a situation it can snatch victory from defeat.
>
> (from Passage #10.14, emphasis supplied)

---

[24] For more on *fanjian*, along with complications in understanding the concept, see pp. 368–71 below.

[25] A flavor of some of the duplicity that may be involved here comes from traditional commentator Du You (Giles, p. 167) (a plausible but unstated premise is that steps are also taken by one's own side to ensure that these unfortunate spies are in fact captured by the enemy):

> We ostentatiously do things calculated to deceive our own spies, who must be led to believe that they have been unwittingly disclosed. Then, when these spies are captured in the enemy's lines, they will make an entirely false report, and the enemy will take measures accordingly, only to find that we do something quite different. The spies will thereupon be put to death.

[26] See R. V. Jones, *Reflections on Intelligence* (London: Heinemann, 1989), "Intelligence Ethics," pp. 35–56, in particular pp. 41–43, "Can Agents (and Others) Be Expendable?"

For background on usage of the concept of "expendable agent" in the history of Chinese espionage, see Sawyer, #1 footnote 3, pp. 156–67 ("Expendable and Double Agents"), cautioning (p. 156) that the "expendable agent" label came to be expansively applied to agents who simply died in the course of performing their mission rather than being deliberately sacrificed by their own side.

[27] Lewis, Background footnote 11, p. 106: "various accounts of battles in the late Warring States and the Qin–Han interregnum show that the commanders consciously employed stratagems, including maneuvering the army into positions where retreat was impossible, to elicit a psychology of desperation in which the troops had no choice but to fight." Such a pattern suggests (though, of course, does not prove) that ethical or other inhibitions created scant barriers to such Warring States practices.

335

A rationale for such a course of action is not hard to find: the armies Sun Tzu knew were heavily, probably predominantly, large conscript armies, ever prone to melt away or flee if given half a chance.[28] Responding to that challenge, Sun Tzu's advice to the general is blunt: give your troops a Hobson's choice. Do so by plunging them without warning into a situation where their only option is to fight bitterly and prevail, or die. Note that this advice follows hard on the heels of verse XI.54's encouragement to a general to transgress norms or conventions in control of his army.

A timeless military fundamental is that social trust (by whatever name known), importantly including trust in leaders, is a crucial factor in combat morale of soldiers undergirding military success.[29] There is reason to think that Warring States armies were no exception to that pattern.[30] Indeed Sun Tzu himself calls attention to the importance of social trust between the general and his troops.[31] The sort of conduct by the general profiled in verse XI.55 would likely undermine any trust previously present, and with it the reliability of his troops – unless, of course, that general was somehow able to engineer a death ground situation without being suspected of having done so.

Since Sun Tzu is plainly aiming to render sound military advice for the military conditions he knew, these basic considerations suggest that Sun Tzu's general should be supremely effective at concealing the existence of his premeditated intent to insert his own people into a death ground situation, both before and after the fact. Sun Tzu's call for a general to be inscrutable and maintain absolute secrecy about his intentions (Passage #10.13's verses XI.42–XI.46) comports well with such a goal. But purely defensive information security measures may not suffice to achieve the necessary information denial.[32] Soldiers can be extremely adept at piecing together clues where their own necks are at stake. More active deceptions of the general's troops may hence be needed in order to deliver on verse XI.43's advice and cloak a general's true plan effectively. An example might be deployment of a "double" of the

---

[28] Cf. a common explanation for Sun Tzu's dispersive ground advice. See #2 footnote 52.

[29] Pointing to morale as "the greatest single factor in successful war," Eisenhower went on to observe that "in any long and bitter campaign morale will suffer unless all ranks thoroughly believe that their commanders are concerned first and always with the welfare of the troops who do the fighting." See Dwight D. Eisenhower, *Crusade in Europe* (Garden City, NY: Doubleday, 1948), p. 210.

[30] A "new style of social trust on which (in Mencius's view) the mass army depended," and which was taking hold, albeit unevenly, in the Warring States period, is noted on p. 81 of A. Taeko Brooks, "Heaven, li, and the formation of the Zuozhuan 左傳," *Oriens Extremus*, 2003–4, 44, 51–100 (p. 81 there also describes Sun Tzu as the "earliest tactical manual of that army"). Mencius was a Confucian, hence not of Sun Tzu's philosophical persuasion, but there is no reason a Confucian could not have accurately spotted the crucial part played by social trust – so easy to shatter, so hard to rebuild – in effective workings of a mass-infantry army. (Regarding Mencius [Mengzi] see Brooks & Brooks, Introduction footnote 36, pp. 54, 135.)

[31] See verse IX.50 (Passage #13.31), where Mair's translation (Mair, p. 112) brings out the social trust idea with great clarity (more so than does Griffith's translation): "The reason I can ensure that my orders are carried out all along is that there is mutual trust between me and my multitudes [of soldiers]." See also #13 footnote 93 (traditional commentator Zhang Yu on verse IX.50).

[32] Sometimes, of course, defensive measures might suffice. For example, a closely held spy report might tell the general that enemy troops are atop a ravine, making it death ground for unwitting troops he is about to order to enter it.

commanding general to accompany his troops on their unwitting death ground mission, thereby quieting possible suspicions of leader deception.[33] But this line of thinking also suggests that squeezing reliable social-trust-preserving military gain from verse XI.55's death ground advice may embroil Sun Tzu's general in a web of deception of his own people, which may have to be sustained indefinitely or at least for a long time.[34] Navigating the ensuing complexities and contradictions could be (or become) a formidable task indeed ("Oh what a tangled web we weave when first we practice to deceive").[35]

There are, to be sure, other possible interpretations that might mute a trust-destroying interpretation of verse XI.55. Tonality of an ancient poem, perhaps based on an early marching song, conveys something of the mood of conscripted Warring States common soldiers worn down by campaigning.[36] That tonality is suggestive of an army held together more by coercion and inertia (e.g., pressures to stay in formation when arrayed for battle) than by social trust in a leader. Simply put, a general's verse XI.55 behavior may not produce a collapse of trust because that trust may not exist in the first place, making any issue of that general's "betrayal" of his troops substantially moot. Of course, a situation of non-existent trust does not bode well for the the army's morale under the intense stresses of combat operations – but possibly the enemy army will be in the same no-trust situation and the two deficits of trust might cancel each other out (cf. Sun Tzu on net assessment!).

That Sun Tzu might have been thinking along "no trust" lines finds support from the fact that Passage #10.13's verses XI.40 and XI.49 both point to various ways, all

---

[33] Chinese (and other) military history is no stranger to use of doubles of the commanding general for a variety of deception purposes. For a Zhuge Liang ruse of that type see Romance of the Three Kingdoms (Moss Roberts translation cited on p. xxiv above), chapter 101, pp. 779–80. Such a ruse might come in handy in crafting a death ground stratagem, because it gives the troops a (false) impression that the general is prepared to share their fate (which might help avert a collapse of trust in the general). See Passage #10.19 analysis of verse X.20 (pp. 355–56 below).

[34] In a comment relating to verse XI.46, traditional commentator Zhang Yu notes: "In war, deception is to be prized; and it is not just the enemy that must be deceived, but one's own men too. They must be enabled to follow but not to understand" (Minford, p. 288, emphasis supplied). Giles, p. 132 picks up on Zhang Yu's observation, which he translates: "The axiom, that war is based on deception, does not apply only to deception of the enemy. You must deceive even your own soldiers. Make them follow you, but without letting them know why." Cao Cao's commentary on verse XI.43 can be also read as pointing to a general's deception of his troops that goes beyond simply keeping them in ignorance of his plans (though that reading is not definitive). See Appendix 10 footnote 13.

[35] For a sense of the narrow path Sun Tzu's general may have had to tread see Lewis, Background footnote 11, p. 125: "Of course the military treatises also insisted that the commander had to be totally trustworthy in applying rewards and punishments to his own men, but the maneuvers, manipulations, and deceptions that constituted this art in the field made any claims to complete reliability impossible." Cf. also Fleming & Zyglidopoulos, #7 footnote 48 ("snowballing" of deception over time).

Further complicating the general's task, the thrust of verses IX.42–IX.43 – insisting on a need for the general to be opaque (see Theme #8 analysis, pp. 273–77) – might trigger some of the same trust-destroying pitfalls as more affirmative leader deception steps. Keeping subordinates in the dark could spur burgeoning of rumors and suspicions (well founded or not), possibly of the darkest sort.

[36] Quoted in #2 footnote 21 above, along with observations of E. Bruce and A. Taeko Brooks.

highly visible, of foreclosing possibilities of retreat and thereby artifically engineering a death ground situation. Because the general's role as a prime mover of that engineering would be obvious, this material suggests Sun Tzu's general's indifference to any leader-trust-destroying effects such measures might have.

Drawing traction from within Warring States culture, another interpretation (not inconsistent but different in emphasis) is suggested by tendencies to view soldiers as akin to inanimate objects, lacking volition or agency, or as dumb beasts:[37]

> XI.49.
> ... [the general] urges the army on as if driving a flock of sheep.

Read from that standpoint Sun Tzu's verse XI.55 counsel may not actually represent an endorsement of betrayal of one's men. One does not betray inanimate objects, or for that matter sheep either.[38]

In seeking to clarify whether Sun Tzu is in fact advocating for a form of betrayal by the general of his troops, the problem is a familar one. Sun Tzu simply does not provide anything like enough context, including information about crucial intangibles like the state of play of social trust or other expectations, to provide a clearcut resolution. Relatedly, not all armies were the same, or perceived as such by their general.[39] For example, the highly competent Han dynasty general Han Xin distinguished an army of "troops that I had trained and led from past times" from an army of "men rounded up from the market place" – i.e., riff-raff – that he had put in death ground to elicit a combat response born of desperation (as in fact it did).[40]

In sum, a general's willingness to deceive/betray his own people needs to be recognized as consistent with Sun Tzu's Passages #10.13 and #10.14 advice (though less grossly manipulative interpretations are possible).[41] What seems unarguable is that some who fall under Sun Tzu's sway will definitely think in that way, and proceed to act upon that thinking.

---

[37] For further background see #9 footnotes 7 and 8.

[38] Furthermore, by cultural criteria of Sun Tzu's time what nowadays might seem like a betrayal may have fallen well within the scope of a general's admissible conduct legitimated by the army's "bonds" (yue), which by Warring States times had come to form the basis of military regulations. For a view of the role played by such bonds and their tendency to become increasingly one-sided (binding servants to masters), see Lewis, Background footnote 11, pp. 67–78, 95.

[39] For example, it is quite possible that elite troops, committed to battle by a general inspired by verse XI.55, might view their desperate situation as a challenge, not a betrayal. It may indeed not be possible to define "death ground" in purely objective terms. Social constructions of reality enter too.

[40] Han Xin's comment (from a Shiji battle description) is quoted on p. 32 above. It is worth noting that Han Xin is unfazed by the fact that his troops drawn up with their backs to a river would surely have caught onto the fact that he had intentionally placed them in a death ground situation there, which suggests that in his estimate preserving social trust with these riff-raff troops was irrelevant.

[41] It is always conceivable, of course, that Sun Tzu's views pertaining to leader deception were simply not fully worked out, or that the text reflects conflicting inputs. Brooks & Brooks, Introduction footnote 36, p. 98 suggests a tension between the more accommodating emphases of Sun Tzu IX's Passage #13.31 (as they put it, "the new-style leader had to consider the feelings of the recruits") and Sun Tzu XI's stark affirmations regarding thrusting troops into death ground. In their summary, "Later on, the element of persuasion vanishes, and orders are followed."

Roads Not Taken by Sun Tzu in Developing Theme #10

Among the most important Theme #10 omissions is lack of any clear recognition by Sun Tzu of how lack of scruples can sometimes be counter-productive from a broader self-interested perspective.[42] Pursuing that line of thinking might hold seeds of a theory incorporating aspects of cooperation as well as conflict. Several parts of the text could serve as jumping off points for such an inquiry (e.g., Passages #10.16–#10.20, especially verse I.4 [Passage #10.16] regarding harmony of ruler and people). But Sun Tzu never follows up on those leads.

Once no-scruples and controlled viciousness are accepted as basic premises of Sun Tzu's thinking, a striking deficit is that – even using a low bar – so few verses explicitly spell out particularly nasty or underhanded expedients. Perhaps reflecting Sun Tzu's inattention to alliance issues, there is no mention of a state's double-crossing an allied state. In addition, there is nothing in Sun Tzu suggesting steps to build up trust with an eye to making its subsequent betrayal more profitable.

Notably, there is also no Sun Tzu content alluding to any number of relevant capabilities that would certainly have existed in Sun Tzu's time (among them, physical and mental torture of captives). Unlike the Arthaśāstra among early Indian strategic writings, there is no allusion to poisons for targeted attacks on individuals.[43] Sun Tzu makes no explicit mention of hostage-taking. Holding aside Passage #10.11's endorsement of a death penalty, to be inflicted not only on a loose-mouthed spy but also on anyone he talked to about secrets, Sun Tzu makes no mention of the draconian punishments that were a staple of early Chinese law and society and that figure prominently in some early Chinese military works other than Sun Tzu.[44]

On a plane of statecraft, Sun Tzu offers no advice to a ruler on how to eliminate rivals, by duplicity or force, in the bloody succession struggles dotted throughout Chinese history.[45]

---

[42] From a formalist perspective, this omission is in keeping with Sun Tzu's failure to distinguish between local and global optima in Passage #9.1's otherwise highly insightful water image.

[43] See, e.g., Olivelle, #1 footnote 39, Arthaśāstra, Book Fourteen, "On Esoteric Practices," pp. 421–34 (recipes for attacks aimed at inducing instant death, death within a month, blindness, etc.).

[44] For example, a discussion of "orders for severe punishments" in the Wei Liaozi, another of the Seven Military Classics (cited earlier in a Theme #3 connection, see #3 footnote 7), begins:

> If a general commanding one thousand men or more retreats from battle, surrenders his defenses, or abandons his terrain and deserts his troops, he is termed a "state brigand." He should be executed, his family exterminated, his name expunged from the registers, his ancestral graves broken open, his bones exposed in the marketplace [etc.].
>
> (Sawyer, Seven Military Classics, p. 263)

One is rather reminded here of a wistful comment of the great historian of science, Otto Neugebauer, "Of all the civilizations of antiquity, the Egyptian seems to me to have been the most pleasant." See p. 71 of his The Exact Sciences in Antiquity (2nd ed.; Providence, RI: Brown University Press, 1957).

For further discussion of Sun Tzu's inferred posture vis-à-vis punishments see #13 footnote 95.

[45] For historical overview see Wilkinson, §17.1.1, pp. 270–72 ("Succeeding to the throne"). In verse XI.37 (Passage #10.13) Sun Tzu does make laudatory mention of Zhuan Zhu, whose assassination of a ruler of Wu cleared the way for Helü, patron of the putative historical Sun Tzu, to become king. But verse XI.37's focus lies elsewhere and its mention of Zhuan Zhu seems largely rhetorical.

On a larger scale of mayhem tinged with deception, there is no Sun Tzu content endorsing or pointing to practices of persuading besieged cities to surrender on promise of leniency, only to betray that promise as their populations were tortured, slaughtered, or enslaved. Sun Tzu also makes no mention of "scorched earth" steps by a defender; or of reprisals or similar punitive measures. One arena where Sun Tzu might possibly be expected to have weighed in in favor of mass killings (e.g., on grounds of expediency) would be killing captives.[46] But here, for reasons anchored in Theme #3 (cheap gains), Passage #10.18 actually takes a stance urging good treatment of captives (presumably with a goal of getting positive future use out of them). Here is a check on certain kinds of vicious measures, albeit one premised on efficacy, not humanitarian impulses.

This is not to say that Sun Tzu would have had clean hands with respect to these or comparable measures. Especially when set alongside Sun Tzu's larger historical context, the evidence suggests otherwise.[47] However, Sun Tzu's omissions just surveyed signal that Sun Tzu's fundamental interests lie elsewhere. Once again, Sun Tzu emerges as a fundamentally military thinker preoccupied with army-versus-army encounters. Relatedly, there is no Sun Tzu coverage of domestic internal security measures or civilian population control (thus setting Sun Tzu apart from the Arthaśāstra or writings of Niẓām al-Mulk in Persian tradition).[48]

Separately, there is also nothing in Sun Tzu touching on self-defilement, a stratagem discussed under Theme #8 as a possible expression of formlessness (see p. 280 above, noting Han dynasty case) but which might also be construed as a form of norm-transgressing controlled viciousness toward self.

Sun Tzu (2) ("Sun Tzu Extended") and Sun Tzu (3) ("Sun Tzu Analogical") Frontiers of Theme #10

The possibility and sometime-utility of no-scruples, norm-transgressing measures is a topic whose importance all strategists, theorists and practitioners alike, must be prepared to acknowledge and in some manner address. Depending on the specific strategist, where that response fits into a larger canvas of strategic theory and

---

[46] From the latter part of the Warring States period a prominent example would be Qin's reputed mass slaughter of Zhao soldiers who had surrendered in the Changping campaign of 260 BC (the "Chinese Cannae"; see pp. 265–66 above). Illustrating both strengths and limitations of modern archaeology, at least one mass grave associated with Changping has been identified, but analysis of its skeletal contents fails to corroborate the Shiji's famous claim that some 400,000 Zhao troops who had surrendered were buried alive en masse. See Shi Jinming 石金鳴 and Song Jianzhong 宋建忠 (Shanxi Provincial Institute of Archaeology), "Changping zhi zhan yizhi yong lu 1 hao shigu keng fajue jianbao" 長平之戰遺址永錄1號屍骨坑發掘簡報, Wen Wu 文物 (Cultural Relics), 1996 (6), pp. 33–40 and p. 98 (p. 40 provides an English-language title and abstract). For the Shiji claim (also noting some discrepancies across Shiji chapters in details provided) see Wilkinson, §24.5.3, pp. 346–47.

[47] It is worth noting that the patron of the putative historical Sun Tzu, King Helü of Wu (who had come to power by the assassination route through the knife work of Zhuan Zhu, see #10 footnote 45 and further discussion in Passage #10.8), apparently found it necessary to "hunt down and kill as many junior members of the royal house of Wu as he could" (Milburn, #1 footnote 41, p. 12).

[48] A very limited exception is fleeting treatment of related topics in verse XI.58 (Passage #10.21A).

practice varies enormously. Choice points abound. Is the enemy ruler or command-ing general exempt, or might he (or she) personally be on the receiving end of vicious treatment? Through Passages #10.9 and #10.10, inter alia, Sun Tzu implies there is no such exemption in his way of war. Is no-scruples action geared to an individual or a population level (or both)? Sun Tzu implies a predilection for the former. Also, once controlled vicious measures start to be taken, what, exactly, does "control" mean? What, if any, residual limitations or checks-and-balances are envi-sioned? Scattered passages in Sun Tzu where some form of limitation is at least implied (Passages #10.16–#10.20) merit attention here.

Further analysis should start by highlighting an important constructive strand present in Sun Tzu's thinking and reflected in Theme #10 substance and emphases. In the nineteenth and early twentieth centuries when war first entered the machine age (to use Bernard Brodie's framing), it brought with it unprecedented possibilities for viciousness on a truly mass scale, some of it inflicted on armies, but often far more on civilian bystanders. Holding aside endorsement of plundering (Passage #10.15), geared in the first instance to meeting pressing needs of an army that would otherwise go hungry, it should be clearly noted that *no encouragement of mass viciousness against civilians is found in Sun Tzu*. Very importantly, there are basic parts of Sun Tzu's thought – notably Passage #10.17's emphasis on the value of capturing the enemy intact, and, more broadly, a focus on fighting cheaply – that strongly suggest that Sun Tzu would take a negative stance toward tendencies to equate "strategy" with destruction.[49] There is harmony here between Sun Tzu's thinking and the strategy-as-control concept analyzed under Theme #5 "frontiers" (see pp. 187–94 above). The latter concept took shape in the 1950s, partly as a rejection of strategy-as-destruction motifs associated with the early nuclear age ("massive retaliation," etc.). In scaled-back forms, strategy-as-destruction thinking persists in some quarters to the present day. Versions of it surface in struggles against terrorism, certainly if progress in such struggles is measured by body count of dead terrorists.[50] *Judiciously deployed, Sun Tzu's thinking can be a valuable antidote to strategy-as-destruction distortions,* presenting a major case of "*What did Master Sun know that we still don't (or have yet to absorb adequately)?*"

Moving beyond this point, evaluating Sun Tzu's Theme #10 legacy becomes murkier. Doing so becomes especially complicated once betrayal implications of Sun Tzu's deception thinking start to enter the picture.[51] Relevant analysis and critique could plainly move in many directions, most of them having limited Sun

---

[49] Here mention should again be made of the chasm between Sun Tzu's thinking and practices of the state of Qin, the ultimate survivor of the Warring States. See #10 footnote 5 above.

[50] Even more pervasive strategy-as-destruction fallacies permeate much terrorist thinking. See, e.g., Boorman, Introduction footnote 4, p. 114 note 61 regarding the "strategy" of al-Qa'ida.

[51] A testbed for developing Sun Tzu–inspired thinking relating to betrayal issues is M. Hausner, J. Nash, L. Shapley, & M. Shubik, "'So long sucker' – a four-person game," pp. 359–61 in Martin Shubik (ed.), *Game Theory and Related Approaches to Social Behavior* (New York: Wiley, 1964). Built into the structure of that game is the requirement that "individuals form coalitions but, in order to win, someone in a coalition [has] to double-cross his partner." Perhaps understandably, playing this game from time to time triggered emotional outbursts on the part of humans playing it – which is precisely a point that might have engaged Sun Tzu's interest. For more on the game, with a relevant anecdote, see Martin Shubik, "John Forbes Nash, Jr. (1928–2015)," *Science*, June 19, 2015, 348(6241), p. 1324 (from which the summary statement just quoted is taken).

Tzu textual grounding beyond very general statements like Passage #10.4's verse I.17. As a way of keeping analysis anchored in Sun Tzu – and in that sense intellectually disciplined – Passages #10.13 and #10.14, with their extended development of Sun Tzu's "death ground" thinking, provide a natural focus. The following discussion revisits this part of Sun Tzu from Sun Tzu (2) and (3) standpoints.

First, it should be stressed that verses XI.33 (Passage #10.13) and XI.55 (Passage #10.14) do in fact build on a timeless military and indeed broader behavioral insight, illustrated in Western military lore by an observation attributed to Napoleon: "How many things apparently impossible have nevertheless been performed by resolute men who had no alternative but death."[52] While refinements and qualifications could be added, on this descriptive level Sun Tzu's thinking stands the test of time well.[53]

However, as underscored by previous Sun Tzu (1) analysis, Sun Tzu's Passages #10.13 and #10.14 contain other ingredients too. The most important Theme #10 ingredients center on the notion of withholding vital information from one's own people for reasons of tactical expediency.[54]

In modern times, instances of successful cryptanalysis give a new lease on life to a leader deception interpretation of Sun Tzu's advice, as when a commander knows through closely held cryptologic means where a death ground situation is, and proceeds to commit forces there (or allows them to remain there) without apprising them in advance of what overwhelming odds they face.[55] In a departure from Sun Tzu's Passages #10.13 and #10.14 scenario, that modern commander's motivation is commonly not to extract supranormal performance but to prevent cryptanalysis triumphs from becoming known to the enemy so that they can continue to bear fruit. But the upshot for the forces thus committed seems much the same as in Sun

---

[52] See *Operational Level of War – Its Art*, Preliminaries footnote 8, p. 1–19, "Military Maxims of Napoleon," Maxim No. 67. Perhaps particularly striking, since his thinking on many matters is so often contrasted with Sun Tzu's, is an observation of Clausewitz, *On War*, Preliminaries footnote 1, p. 284:

> The hard-pressed army, not expecting help where none can be forthcoming, can only trust to the high morale that despair breeds in all courageous men. At that point the greatest daring, possibly allied with a bold stratagem, will seem to be the greatest wisdom. Where success is out of reach, an honorable defeat will at least grant one the right to rise again in days to come.

Clausewitz's first sentence aligns closely on Sun Tzu's core death ground insight; the second sentence, on the observation that death ground may harbor a place for military behavior more complex than primal ferocity or kindred responses (see Passage #9.14 analysis). Clausewitz's third sentence goes beyond Sun Tzu, both in its realistic appreciation of the possibility of failure which contrasts with verse XI.55's unqualified optimism; and also by suggesting that, phoenix-like, even a tactical defeat, possibly martyrdom, may lay groundwork for later strategic gain – a sophisticated idea. There is kinship here between Clausewitz and Sun Tzu's fourth path to *shi* (pp. 178, 183 above).

[53] For pertinent insights from behavioral and neuroscience quarters see #1 footnote 91.

[54] For modern background on leader deception, noting that the phenomenon has not received as much analytical attention as it merits, see Jennifer A. Griffith, Shane Connelly, & Chase E. Thiel, "Leader deception influences on leader-member exchange and subordinate organizational commitment," *Journal of Leadership & Organizational Studies*, 2011, 18(4), 508–21.

[55] For discussion of relevant decision problems see Jones, #10 footnote 26, pp. 42–43 (also noting a variant scenario, more immediately relevant to Sun Tzu's playbook, where crucial information comes, not from cryptanalysis, but from a well-placed secret agent).

Tzu's death ground scenario, indeed quite possibly worse (since the situation they are entering may be a truly hopeless one for them).

Yet cryptologic successes, intangible and abstract as they commonly are, have unusual potential for being kept robustly secret, which at least means that the commander's information denial actions, whether or not ultimately justified ethically or strategically, may work as intended (a fact that contributes to their strategic and tactical coherence). Other sorts of secrets may be harder to keep, especially in military and civilian environments with large administrative staffs. Then it becomes important that the opportunistic military calculus that animates verse XI.55 (at least in one natural interpretation) makes no allowance for what might be labeled the "day after" problem. Widespread suspicion may emerge that the general has engaged in a form of leader deception, permitting his unwitting troops to enter a "death ground" situation (or even engineering it) as steppingstone to other goals (one of which, of course, might be combat performance from them that makes him look good in the eyes of his superiors!). An observation of Admiral Eccles expresses an eternal verity that applies here: "The men are not deceived."[56]

Troops who see themselves thrown into a questionable death ground situation might, perhaps not unreasonably, respond by deserting[57] – or by going over to the enemy and serving as active soldiers in the enemy's army, precisely as did many erstwhile Chinese Nationalist troops who, often feeling abandoned by their leaders, went over to the Communist side during the 1945–49 Chinese civil war.[58] A foe's profiting from contingencies like those harmonizes excellently with Sun Tzu's encouragement of cheap successes (Theme #3) – but now those cheap successes would be successes of an *enemy* who has read Sun Tzu's playbook (see Passage #10.18)! Regarding any turncoat possibilities Passages #10.13 and #10.14 are notably silent (which is in many ways striking, given conditions in Warring States China as well as Sun Tzu's willingness to contemplate possibilities of far-less-than-perfect loyalty on the part of spies).[59]

For these reasons, which modern information and communication conditions (among them, social media) may easily reinforce by promoting circulation of gossip and rumors, well founded or not, loose ends in Sun Tzu's Passages #10.13 and #10.14 thinking start to appear. They should serve as a caution to modern readers prone to find military or strategic gems in everything Sun Tzu has to say.[60]

---

[56] See Eccles, Background footnote 62, p. 233. Eccles's preceding sentences are also pertinent: Throughout history, men have cheerfully accepted danger and hardship when a positive, professionally competent command is exercised with courage and integrity. Throughout history military men in the ranks of our services have recognized this element of integrity in their military seniors and have responded in kind – whether for good or for bad, depends on the seniors.

[57] Griffith, p. 122 footnote 1 notes a suggestion of traditional commentator Mei Yaochen that the soldiers described in verse IX.41 (Passage #9.16) are planning to desert.

[58] Mao to Chen Yi et al. (3 September 1947), quoted by Westad, #4 footnote 74, p. 174: "Whenever you capture enemy soldiers, just fill them into your own forces." As Sun Tzu remarks in verse II.20 (Passage #10.18), "this is called 'winning a battle and becoming stronger.'"

[59] Mair, p. 142 note 7: "Changing colors is a common theme in accounts of warfare of the Warring States period."

[60] A further source of basic limitation on Sun Tzu's death ground thinking, elementary but fundamental, is that it involves a high-risk scenario unlikely to work successfully over and over

Yet despite these criticisms, Sun Tzu's thinking retains great potential to stimulate productive lines of strategic analysis. It is worth posing a Sun Tzu (3) challenge question as to what non-combat scenario might, on some level, capture the distilled essence, even if not the letter, of Passages #10.13 and #10.14 and, at the same time, possibly align better than this part of Sun Tzu on modern social and information conditions. Framed in abstract terms, the following are suggested basic elements of such a scenario:

(1) deliberate engineering of a situation,

(2) to maneuver one's own people,

(3) without their foreknowledge,

(4) into an extraordinarily difficult situation for them,

(5) to elicit a behavioral response otherwise not normal,

(6) that can be harnessed to the great advantage of the situation's engineer,

(7) at least so long as that engineer's role is not recognized (or only dimly so).

A concept for such a scenario – as manipulative as any conjured by Sun Tzu – is suggested by accounts of the former Soviet Union, where living *nalevo* ("on the left," meaning breaking or bending the rules) was a fact of life for individuals high and low. According to this scenario, a well-placed strategic actor ("rulemaker"), who is in a position to set the rules for many people, deliberately goes about creating rules that are vastly complex and intractably burdensome. Initially, so runs the scenario, most people will try to comply, but frustrations mount as rules continue to proliferate and they discover that it is becoming impossible to get anything done, including essential tasks. At that point – and here is a rupture analogous to onset of troops' "nonlinear" behavior depicted in verses XI.33 and XI.55 – they will start to flout those rules out of necessity and in many cases out of sheer frustration too. A cunning mentality then takes root. That will in turn give rise to many instances of what has been felicitously termed "stable reciprocating blackmail," whereby actors X and Y each have knowledge of the other's rule violations, thereby laying foundations for a newfound cooperation between them.[61] Positing that the rulemaker has surveillance capabilities to spot the rulebreakers, or at least enough of them (who, at the rulemaker's discretion, may or may not be significantly penalized for their noncompliance),[62] the resulting situation harbors immense opportunities for advantageous situation control by an unscrupulous rulemaker. That could include pressuring known rulebreakers to carry out tasks they would not otherwise do.

Meanwhile veiling the existence of the manipulative scheme serves to buttress the rulemaker's legitimacy and longevity in that role, even preserving a modicum of

---

again. No one – not even Sun Tzu's ideal general – can be that good at parsing risk! A military history example is the case of a Roman commander, Gaius Flaminius, who by design placed his army in death ground and went on to win a victory. Analyzing factors underlying Flaminius' decision, see article by Bell (cited in Table 2 note i, p. 98 above), p. 413, observing that "Flaminius' expedient was a great deal too dangerous to be repeated." (Bell's analysis also cites Sun Tzu, though in support of a different point.)

[61] The phrase "stable reciprocating blackmail" describing this outcome comes from Professor Leon Lipson of the Yale Law School (personal communication). The concept of cooperation born of a difficult or hostile environment evokes the type of cooperation elicited by the storm in verse XI.39 (Passage #10.13), albeit that in this Sun Tzu scenario cooperation is for overt, not sub rosa, reasons.

[62] Drawing on the spirit of verse XI.54 (Passage #10.14) some malefactors might even be rewarded!

social trust between leaders and followers (if not necessarily among followers!). Under modern information conditions, concealment of the rulemaker's manipulative design may also be easier than veiling the Sun Tzu general's scheme in Passages #10.13 and #10.14, for all the reasons that complex rulemaking tends to be an opaque activity and may often be credibly passed off as born of enlightened, indeed altruistic motives.[63]

The specifics of this rulemaking scenario are, of course, not exactly the same as those in Passages #10.13 and #10.14. In particular, the rulemaking may have straightforward goals quite apart from entrapping the populace in a web of non-compliance. Furthermore, life-and-death outcomes on a large scale are not nearly as salient in this "complex rules" scenario as they are in the bitter combat one in verses XI.33 and XI.55 (though life-and-death consequences for individuals might yet arise).[64] Even so, the rulemaking scenario just sketched has a whiff of Sun Tzu about it. Ever greater twenty-first-century reliance on complex software and algorithms (the ultimate "general's garden" for makers of complex rules![65]) creates unprecedented opportunities for a never-ending stream of new rulemaking and resulting countermoves on the part of those affected by the rules, possibly involving "creative compliance."[66] Here is the germ of not one but many scenarios having overtones of Sun Tzu.[67]

✶✶✶✶✶✶

The fact remains that Sun Tzu's aura of intellectual respectability means that Sun Tzu's Theme #10 content could be used to justify actions repugnant in our own time, if not necessarily in that of the Warring States. Given the casual ethics so often found in civilian life, verse I.17 (Passage #10.4), treated as a license to deceive (also to betray), holds major potential being for a doctrinal slippery slope, creating a frankly volatile part of Sun Tzu's broader legacy. Yet there are many other parts of Sun Tzu – among them Theme #1 (calculation), Theme #3 (cheap expedients), Theme #4 (winning without fighting), and Theme #8 (formlessness) – that do not seem inherently toxic.

In thinking clearly about these issues, it is unfortunate that many of Sun Tzu's audiences, especially non-military ones, take an equivocal stance, plainly intrigued by

---

[63] Obfuscatory pronouncements by the rulemaker plainly have a part to play here, suggesting opportunities for giving play to Sun Tzu's deception thinking (Theme #7).

[64] Death on a vast scale did, of course, occur in the Soviet era "gulag archipelago" (though the present Theme #10 analysis finds no natural place for such a concept in Sun Tzu's own thinking).

[65] Image is from strategy work "The General's Garden," attributed to Zhuge Liang of the Three Kingdoms era (though commonly regarded as being of later origin). See Preliminaries footnote 13 above.

[66] See #4 footnote 72 (citing sources).

[67] One direction for followup comes from combinatorial game theory, of which it has been observed that "some of the most exciting games are obtained by breaking some of the rules." That comment, of course, evokes Theme #10, especially when conjoined with the further comment that "your opponent in a game can be very mean!" See pp. 494–95 of Aviezri S. Fraenkel, "Combinatorial games: selected bibliography with a succinct gourmet introduction," pp. 493–537 in Nowakowski, #5 footnote 13.

Sun Tzu's devious or nasty undercurrents while at the same time cloaking themselves in the lofty ambiance and aspirations of an acknowledged world classic.[68]

By way of response (and invoking the memory of nuclear strategist Albert Wohlstetter, who by first training was a mathematical logician), it may be useful to conceptualize Sun Tzu's overall strategic thought as erected on a set of "axioms" expressing the fourteen themes (some involving more than one axiom) and then to explore the different Sun Tzus obtained by picking and choosing among those axioms, accepting some, rejecting or modifying others. One possibility would be to weaken Theme #10, e.g., ruling out deaths of innocents like the concubines or at least some of the people verse XIII.15's talkative leaker may have spoken with; or murder of an individual acting in a diplomatic capacity;[69] or practices of "plunder" (including its modern update, major financial fraud, especially fraud large enough to create systemic risk). Tailoring of Sun Tzu might also reflect more than ethical considerations alone, making other adjustments to adapt Sun Tzu's counsel to strategic and tactical conditions in other times, places, and contexts.[70] Precedent for a "tailored Sun Tzu" in fact already exists in one Inner Asian context from as early as Song times.[71]

It is worth noting that ethically motivated tailoring of Sun Tzu might in some cases actually boost the usefulness and efficacy of the remaining parts of Sun Tzu's

---

[68] This is neither a new nor a specifically Western phenomenon. Comparable ambivalence is evidenced by the author of a late Ming work of stories about various swindles, the title of whose first edition (c. 1617) may be translated as "A New Book for Foiling Swindlers, Based on Worldly Experience." As its English-language translators point out, the book's author, Zhang Yingyu (fl. 1612–17), "appends to the stories comments that vacillate between stern disapproval of the crooks' predations and connoisseurial appreciation of their ingenuity. Such equivocality permeates the book ..." See Zhang Yingyu, *The Book of Swindles: selections from a late Ming collection* (Christopher Rea and Bruce Rusk, trans.; New York: Columbia University Press, 2017), p. xiv.

  Unsurprisingly, Rea and Rusk's Translators' Introduction also points to Sun Tzu (see their p. xvi), noting a mention of Sun Tzu in the book's 1617 preface.

[69] See anecdote of Han dynasty vintage described on p. 209 above. The concept of diplomatic immunity is an ancient one, already found in ancient Indian epic poetry. See Sarva Daman Singh, *Ancient Indian Warfare with Special Reference to the Vedic Period* (foreword by Sir Mortimer Wheeler; Leiden: E.J. Brill, 1965), p. 150.

[70] For example, for a solo (or small) strategic actor – commonly civilian – in a defensive position Sun Tzu's formlessness thinking is among the most practically useful parts of Sun Tzu. Cf. *Nüjie* analysis, pp. 153–57 above.

  Meanwhile much of Theme #13 (whose focus is leadership topics) may be substantially irrelevant to many aspiring civilian appliers of Sun Tzu, who may not be in charge of any formal organization at all, indeed might not even be a member of such an organization.

  A contender for being jettisoned in many civilian uses of Sun Tzu would be verse VIII.16. Its advice, if interpreted as taking a hard-line position against intentions-based planning, seems out of place (indeed often counter-productive) in many civilian contexts. See Passage #4.28A analysis (critiquing verse VIII.16 from this standpoint).

[71] Pioneering steps in a direction of tailoring Sun Tzu content to meet a strategic actor's current needs were taken a millennium ago by an unknown translator of Sun Tzu into Tangut, the language of an Inner Asian state that was then contesting Song dynasty power. See analytical introduction to Ksenia Kepping, *Sun' TSzy v tangutskom perevode. Faksimile ksilografa* (Moscow: Nauka, 1979) as well as briefer treatment in Kepping & Gong, Preliminaries footnote 13, pp. 14–16. For geopolitical background on the Song/Tangut struggle see F. W. Mote, *Imperial China 900–1800* (Cambridge, MA: Harvard University Press, 1999), chapter 8, "Origins of the Xi Xia [Tangut] state," pp. 168–90.

advice. A case in point might be verse XIII.15, the soundness of whose death-dealing advice is challenged in Passage #10.11 analysis (where verse XIII.15 appears).[72]

The "three Sun Tzus" of pp. 1–2 above then becomes the "many Sun Tzus."[73] The upshot is a kind of doctrinal lattice of Sun Tzu (3) strategies of conflict, of varying degrees of ethical acceptability and practical utility. Exploration of Sun Tzu (3) frontiers would be the healthier for recognizing and analyzing this lattice, forcing aspiring modern appliers of Sun Tzu to scrutinize their own thinking, explicitly clarifying which parts of Sun Tzu's counsel they are comfortable with, which not.[74]

★★★★★★

Set of principal passages used to illustrate Sun Tzu Theme #10 (no scruples and controlled viciousness)

Passage #10.1. Catalyst to Dispensing with Scruples: War As Existential Conflict

(The existential perspective of this passage makes it a catalyst, and on some levels a justification, for Sun Tzu's "no scruples" thinking. Such a perspective accurately describes some kinds of wars, notably all-out wars; other kinds of wars, less well.)

I.1. War is a matter of vital importance to the State; the province of life or death; the road to survival or ruin. It is mandatory that it be thoroughly studied.

Passage #10.2. War As Existential Conflict, cont.

(Positing Brooks's proposed order of Sun Tzu chapter accretions, for a human generation or more these verses would have been located at the end of the text.[75])

XII.18. A sovereign cannot raise an army because he is enraged, nor can a general fight because he is resentful. For while an angered man may again be happy, and a resentful man again be pleased, *a state that has perished cannot be restored, nor can the dead be brought back to life.*

---

[72] See also #11 footnote 73. For a different analytical plane where ethical and strategic considerations may stand to reinforce one another rather than clashing, see p. 140 above (quoting Liddell Hart on ethical pressure as an instrument of grand strategy).

[73] The concept of a prominent figure as having many faces, sometimes inconsistent, is by no means foreign to Chinese tradition. See, e.g., p. 126 of Michael Nylan, "The many Dukes of Zhou in early sources" (cited in #6 footnote 13 above), noting "the protean character of the monumental figures within story-telling traditions in China."

[74] Compare Minford, pp. xxxi–xxxii, "The Art of Reading The Art of War," suggesting that
   The balance for each reader (soldier, trader, corporate manager, marriage partner, tennis player, cook, driver) lies somewhere between the extraordinary wisdom embedded in its pages and its less attractive or acceptable implications. Each reader must negotiate this fascinating but treacherous terrain . . .

(p. xxxi)

and also quoting a Ming dynasty thinker to the effect that one should "Emerge from the mud untainted; understand cunning but do not use it" (p. xxxii).

[75] See Table 12A on p. A-35. Making a similar observation see Brooks, p. 59.

XII.19. Therefore, the enlightened ruler is prudent and the good general is warned against rash action. Thus the state is kept secure and the army preserved. [Emphasis in verse XII.18 supplied.]

Passage #10.3. Further Catalyst to Dispensing with Scruples: Focus on a Quick Win

II.3. Victory is the main object in war. If this is long delayed, weapons are blunted and morale depressed. When troops attack cities, their strength will be exhausted.

Passage #10.4. License to Deceive? Doctrinal Foundation for Absence of Scruples, Including Profitable Betrayals

(Although this verse speaks in general terms of deception, not of betrayal – a narrower analytical category that presumes a pre-existing level of trust – lack of any qualifications or sense of limits accompanying verse I.17, conjoined with Sun Tzu's broad understanding of "war," suggests that it could cover either possibility. Verse I.17's implied endorsement of profitable betrayals might extend to almost any human conflict situation. Carrying this line of thought to an extreme – and there is nothing in the Sun Tzu text that would preclude doing so – verse I.17 could be read as encouragement actively to seek out opportunities for profit-by-betrayal in all human relationships, no matter how close or longstanding or how imbued with attributes of mutual trust and confidence.)

I.17. All warfare is based on deception.

Passage #10.5. License to Deceive, cont.

(These verses invite thinking about deception in a context of strategic style. Verse VII.13's allusion to fire, which has its treacherous aspects, might be given a civilian dual use reading as tacit encouragement of opportunistic betrayals.)

VII.12. Now war is based on deception. Move when it is advantageous and create changes in the situation by dispersal and concentration of forces.
VII.13. When campaigning, be as swift as the wind; in leisurely march, majestic as the forest; in raiding and plundering, like fire; in standing, firm as the mountains. As unfathomable as the clouds, move like a thunderbolt.

Passage #10.6. Endorsement of Espionage Métier

(In a marvelously backhanded move, Passage #10.6 transmogrifies unbridled reliance on espionage, an activity rich in betrayals, into what might be viewed as a positive moral commandment – failure to make fullest use of espionage as a crime against humanity![76])

---

[76] See suggestion of Giles, p. 162 (previously mentioned in #4 footnote 2 above; also quoted by Minford, p. 316). Of course, a "crime against humanity" framing points to a modern category that might have puzzled Sun Tzu (whose focus here as elsewhere is on achieving highest efficacy).

XIII.1. Now when an army of one hundred thousand is raised and dispatched on a distant campaign the expenses borne by the people together with the disbursements of the treasury will amount to a thousand pieces of gold daily. There will be continuous commotion both at home and abroad, people will be exhausted by the requirements of transport, and the affairs of seven hundred thousand households will be disrupted.

XIII.2. One who confronts his enemy for many years in order to struggle for victory in a decisive battle yet who, because he begrudges rank, honours and a few hundred pieces of gold, remains ignorant of his enemy's situation, is completely devoid of humanity. Such a man is no general; no support to his sovereign; no master of victory.

Passage #10.7. Betrayals in Espionage: Expendable Agents

(Observations of traditional commentators translated by Griffith, p. 146 support understanding Sun Tzu's expendable agents as unwitting pawns.[77])

XIII.10. Expendable agents are those of our own spies who are deliberately given fabricated information.

. . .

XIII.19. And it is by this means [the double agent] that the expendable agent, armed with false information, can be sent to convey it to the enemy.

Passage #10.8. Betrayals at the Highest Levels of Grand Strategy and Statecraft

(An implied claim here is that exploiting duplicity at the highest level may be both acceptable and efficacious as a tool of strategy, though the cases Sun Tzu invokes might be better described as involving defectors, not spies.[78] The time periods involved are so ancient that distinguishing history from myth is not really possible. Behavior tinged with betrayal of trust is further illustrated by the cases of Zhuan Zhu and Cao Gui, both of whom receive kudos in Passage #10.13's verse XI.37. The specifics are somewhat different.[79] By premeditated arrangement, Zhuan Zhu, a retainer of an aspirant to the throne of Wu, assassinated Wu's reigning king whom the aspirant had invited to an ostensibly friendly social occasion – a banquet. The means of assassination was a knife hidden in the belly of a braised fish. Cao Gui, on the other hand, was a commander of the state of Lu who took the opportunity of the swearing of a covenant between rulers of Qi and Lu, a solemn diplomatic occasion at which both men were present, to hold the Qi ruler at knife point in order to persuade him to relinquish territory captured from Lu. Common to these two cases is that the assailant exploited the cover of a social or diplomatic event, using it to give him access to his target at close quarters.)

---

[77] For broader understanding of this category of spy see p. 335 and #10 footnote 26 above.

[78] See p. 192 and Passage #5.19 (p. 205) above, where verse XIII.22's historical content is critiqued.

[79] For further background on these two cases (citing sources) see Appendix 6 footnote 2.

XIII.22. Of old, the rise of Yin [Shang dynasty] was due to Yi Zhi, who formerly served the Xia [dynasty]; the Zhou [dynasty] came to power through Lü Ya, a servant of the Yin.

Passage #10.9. Seizing Something the Enemy Cherishes

(While not by itself inherently norm-transgressing – for example, Sun Tzu's advice here might apply to seizing a logistics base on which an enemy supply line depends – this verse invites play of strategic imagination along "nasty" lines, possibly including hostage-taking.[80])

XI.28. Should one ask: "How do I cope with a well-ordered enemy host about to attack me?" I reply: "Seize something he cherishes and he will conform to your desires."

Passage #10.10. Exploiting Personality Traits, Possibly in Vicious Ways

(In many strictly military contexts – where, for example, "compassion" might refer to an enemy general's concern about the fate of hard-pressed contingents of his army – Sun Tzu's proposals here do not seem norm-transgressing. If transposed to civilian conflict situations, however, they may take on a different hue. For example, verse VIII.22, applied without scruples, could point to getting strategic traction from manipulating a parent's worries about the safety of a child or an adult offspring's worries about an aging parent's welfare.[81] Various recipes found here invite other civilian generalizations, some of them nasty – say, substituting health worries for cowardice, or fondness for gambling for recklessness, possibly with financial ruin replacing death as the sought-for outcome.)

VIII.17. There are five qualities which are dangerous in the character of a general.
VIII.18. If reckless, he can be killed;
VIII.19. If cowardly, captured;
VIII.20. If quick-tempered you can make a fool of him;
VIII.21. If he has too delicate a sense of honor you can calumniate him;
VIII.22. If he is of a compassionate nature you can harass him.
VIII.23. Now these five traits of character are serious faults in a general and in military operations are calamitous.
VIII.24. The ruin of the army and the death of the general are inevitable results of these shortcomings. They must be deeply pondered.

---

[80] Sawyer, p. 220, translates "I would say, seize something that they love for then they will listen to you," which makes Sun Tzu's point even a little more sharp-edged (especially in civilian applications).

[81] Cf. M. W. Parker, Vaughn R. A. Call, Ruth Dunkle, & Mark Vaitkus, "'Out of sight' but not 'out of mind': parent contact and worry among senior ranking male officers in the military who live long distances from parents," *Military Psychology*, 2002, 14(4), 257–77.

Passage #10.11. Enforcing Intelligence Security with Death, Possibly Including Death of Innocents)

(Verse XIII.15 mentions no standard of proof. It is also worth noting that there could be perils in following verse XIII.15's advice too mechanistically, since executing the agent might foreclose getting to the bottom of the surrounding circumstances, possibly hobbling one's own side's espionage or counterespionage operations.)

XIII.15. If plans relating to secret operations are prematurely divulged the agent and all those to whom he spoke of them shall be put to death.[82]

Passage #10.12. Controlled Viciousness: Systematic Planning for Targeted Killing

(The matter-of-fact way in which planning for targeted killing is mentioned here suggests that, for Sun Tzu, such measures were thoroughly unremarkable, part and parcel of how warfare was conducted in his time.)

XIII.16. Generally in the case of armies you wish to strike, cities you wish to attack, *and people you wish to assassinate*, you must know the names of the garrison commander, the staff officers, the ushers, gate keepers, and the bodyguards. You must instruct your agents to inquire into these matters in minute detail. [Emphasis supplied.]

Passage #10.13. Throwing One's Own People into Extraordinary Peril, without Knowledge of Your Plan to Do So, As Scheme to Elicit Supranormal Performance

(Parts of this material also appear in Passages #8.9 and #9.14. The core question in Passage #8.9 is whether to understand Sun Tzu's general as keeping all of his officers and men in the dark about his plans, as verse IX.43 appears to counsel, or whether an exception might be made for a select few high-level subordinates. The core question of Passage #9.14 centers on the nature and circumstances of the "supranormal" combat behavior Sun Tzu expects.

The focus of Passages #10.13 and #10.14 differs from both of these. Now the core question is the existence and implications of leader deception in engineering a death ground scenario. That question is analyzed in detail on pp. 335–38 above.

Undoing Griffith's rearrangement, Passage #10.13 presents Sun Tzu's content in the traditional order of the Chinese text, a step that helps to clarify Sun Tzu's logic flow here. Appendix 10, where Passage #10.13 again appears, further analyzes that logic flow together with assorted textual and interpretive issues.)

XI.33. Throw the troops into a position from which there is no escape and even when faced with death they will not flee. For if prepared to die, what can they not achieve? Then officers and men together put forth their utmost

---

[82] For a textual variant, possibly enlarging the circle of targets, see Appendix 10 footnote 3.

efforts. In a desperate situation they fear nothing; when there is no way out they stand firm. Deep in a hostile land they are bound together, and there, where there is no alternative, they will engage the enemy in hand to hand combat.

XI.34. Thus, such troops need no encouragement to be vigilant. Without extorting their support the general obtains it; without inviting their affection he gains it; without demanding their trust he wins it.

XI.44. He prohibits superstitious practices and so rids the army of doubts. Then until the moment of death there can be no troubles.

XI.35. My officers have no surplus of wealth but not because they disdain worldly goods; they have no expectation of long life but not because they dislike longevity.

XI.36. On the day the army is ordered to march the tears of those seated soak their lapels; the tears of those reclining course down their cheeks.

XI.37. But throw them into a situation where there is no escape and they will display the immortal courage of Zhuan Zhu and Cao Gui.

XI.38. Now the troops of those adept in war are used like the "Simultaneously Responding" snake of Mount Chang. When struck on the head its tail attacks; when struck on the tail, its head attacks, when struck in the centre both head and tail attack.

XI.39. Should one ask: "Can troops be made capable of such instantaneous co-ordination?" I reply: "They can." For, although the men of Wu and Yue mutually hate one another, if together in a boat tossed by the wind they would co-operate as the right hand does with the left.

XI.40. It is thus not sufficient to place one's reliance on hobbled horses or buried chariot wheels.

XI.41. To cultivate a uniform level of valour is the object of military administration. And it is by proper use of the ground that both shock and flexible forces are used to the best advantage.

XI.41a. (Ames, p. 159; segment omitted by Griffith) "Thus the expert in using the military leads his legions as through he were leading one person by the hand. The person cannot but follow."

XI.42. It is the business of a general to be serene and inscrutable, impartial and self-controlled.

XI.43. He should be capable of keeping his officers and men in ignorance of his plans.

... *(Conforming to the Chinese text order, verse XI.44 is now located immediately following verse XI.34.)*

XI.45. He changes his methods and alters his plans so that people have no knowledge of what he is doing.

XI.46. He alters his camp-sites and marches by devious routes, and thus makes it impossible for others to anticipate his purpose.

XI.50. He fixes a date for rendezvous and after the troops have met, cuts off their return route just as if he were removing a ladder from beneath them.

XI.48. He leads the army deep into hostile territory and there releases the trigger.

XI.49. He burns his boats and smashes his cooking pots; he urges the army on as if driving a flock of sheep, now in one direction, now in another, and none knows where he is going.

XI.47. To assemble the army and throw it into a desperate position is the business of the general.

Passage #10.14. Related Statement, alongside Encouragement to Disregard Customary Norms

(If verses XI.51–XI.53 are an interpolation, which seems possible, and are deleted accordingly, then verses XI.54–XI.55 immediately follow verse XI.23 [Passage #2.1], which likewise focuses on the behavior of troops in death ground.[83])

XI.54. Bestow rewards without respect to customary practice; publish orders without respect to precedent. Thus you may employ the entire army as you would one man.

XI.55. Set the troops to their tasks without imparting your designs; use them to gain advantage without revealing the dangers involved. Throw them into a perilous situation and they survive; put them in death ground and they will live. For when the army is placed in such a situation it can snatch victory from defeat.

Passage #10.15. Advice to Plunder

(References to "plunder" in the Sun Tzu text should be understood broadly. Plunder may have included seizing desired items other than grain or other food supplies.[84] There is also nothing in Sun Tzu to suggest a distinction between plundering the supplies of an enemy army versus plundering from a defenseless civilian population, which may be left starving as a result. Certainly modern attempts at drawing a neat conceptual distinction between permissible foraging or capture of enemy supplies and impermissible pillaging or looting directed against noncombatants should not be imposed on Sun Tzu's early time.[85]

Although it is clear from context that Sun Tzu's basic focus here is on predatory behavior by an army in enemy territory, these verses leave unresolved whether Sun Tzu's advice would change when that territory becomes, or is about to become,

---

[83] Making a case that verses XI.51–XI.53 are an interpolation, see Appendix 8 anchor footnote 19.

[84] Verse II.17 (Passage #10.26A) is suggestive on point, though details are lacking. Further context comes from the observation that "[i]n Chinese history it is rarely possible to make a clear-cut distinction between soldiers, militia and bandits as different social groups." See p. 350 of Barend J. ter Haar's review of Herbert Franke, *Studien und Texte zur Kriegsgeschichte der südlichen Sungzeit* (Wiesbaden: Otto Harrassowitz, 1987), published in *T'oung Pao*, 1991, 77(4/5), 348–52.

[85] Even in modern times drawing a practical distinction between starvation measures directed against an enemy army vs. a civilian population has proved elusive. See p. 234 of George Alfred Mudge, "Starvation as a means of warfare," *The International Lawyer* (American Bar Association), 1970, 4(2), 228–68.

part of one's own state. For further textual background, notably limited, on what Sun Tzu's position, if any, might have been on that issue see pp. 445–48 below.)

XI.7. When the army has penetrated deep into hostile territory, leaving far behind many enemy cities and towns, it is in serious ground.

. . .

XI.13. In focal ground, ally with neighbouring states; in deep [i.e., serious] ground, plunder.[86]

. . .

XI.20. In serious ground I would ensure a continuous flow of provisions.[87]

. . .

XI.30. The general principles applicable to an invading force are that when you have penetrated deeply into hostile territory your army is united, and the defender cannot overcome you.

XI.31. Plunder fertile country to supply the army with plentiful provisions.

Five codicils to Theme #10, pointing to potential sources of limitation

The following five passages identify Sun Tzu content that, in various ways, may be read as pointing away from maximum feasible unscrupulousness, though in each case for reasons predicated on efficacy, not altruism.

Passage #10.16. First Codicil: Harmonious Relation between Ruler and Subjects

(Verse I.4 points to a basic source of limitation on unbridled lack of scruples. Unfortunately Sun Tzu never returns to elaborate on this principle, undeniably crucial as it is. There is no theory of good government in Sun Tzu.)

I.4. By moral influence I mean that which causes the people to be in harmony with their leaders, so that they will accompany them in life and unto death without fear of mortal peril.

Passage #10.17. Second Codicil: Taking the Enemy State Intact, in Preference to Destroying It (and Comparable Accomplishments vis-à-vis Enemy Military Units)

III.1. Generally in war the best policy is to take a state intact; to ruin it is inferior to this.

III.2. To capture the enemy's army is better than to destroy it; to take intact a battalion, a company or a five-man squad is better than to destroy them.

III.3. For to win one hundred victories in one hundred battles is not the acme of skill. To subdue the enemy without fighting is the acme of skill.

. . .

---

[86] For translation consistency, "deep ground" here should read "serious ground." See #2 footnote 61.

[87] Perhaps counterintuitively, most of the traditional commentators interpret verse XI.20 as also being advice to plunder, as contrasted with putting other logistics arrangements in place (e.g., involving some sort of supply line). For further analysis see #2 footnote 64.

III.10. Thus, those skilled in war subdue the enemy's army without battle. They capture his cities without assaulting them and overthrow his state without protracted operations.

III.11. Your aim must be to take All-under-Heaven intact. Thus your troops are not worn out and your gains will be complete. This is the art of offensive strategy.

Passage #10.18. Third Codicil: Caring for Captives

(Here again is a textual basis for a limitation on certain extreme measures, now pertaining to maltreating captives or simply slaughtering them, as the Qin ruler was said to have done to captured soldiers after his great Changping victory.[88])

II.18. Therefore, when in chariot fighting more than ten chariots are captured, reward those who take the first. Replace the enemy's flags and banners with your own, mix the captured chariots with yours, and mount them.

II.19. Treat the captives well, and care for them.

II.20. This is called "winning a battle and becoming stronger".

Passage #10.19. Fourth Codicil: Accompanying One's Troops on High Risk Missions, and Sharing Their Fate, As a Factor Mitigating Leader Deception by a General?

(Lest the modern reader understand verse X.20 in an overly modern way, the fate of straw dogs noted in early Daoist thought should be mentioned. As D. C. Lau describes the practice, building on the Zhuangzi text, "straw dogs were treated with the greatest deference before they were to be used as an offering, only to be discarded and trampled upon as soon as they had served their purpose."[89]

Yet presence of verse X.20 might still operate to soften the betrayal-of-one's-own harmonics plausibly present in Sun Tzu's Passages #10.13 and #10.14.[90]

In particular, verse X.20 appears to cast the general as taking much the same risk as his underlings, as contrasted with sitting in a location of personal safety while his troops are battling against frightful odds and – so he calculates – still prevailing for death ground psychology reasons.[91] However, Sun Tzu leaves open

---

[88] See #10 footnote 46.

[89] See footnote accompanying Lau's translation of Laozi V, which begins "Heaven and earth are ruthless, and treat the myriad creatures as straw dogs; the sage is ruthless, and treats the people as straw dogs." See Lau, #1 footnote 34, p. 61 (also referring to Zhuangzi text, on which see #1 footnote 59).

[90] General support for this interpretive point comes from remarks on verse X.20 of assorted traditional commentators, emphasizing that a good general shares hardships with his troops. See Griffith, p. 128; Mair, pp. 153–54 note 3.

[91] A case in point might be the early Han general Han Xin's presence with troops he had deliberately placed in death ground. See battle account from his Shiji biography summarized on p. 32 above.

A contrast case noted by traditional commentator Du Mu involved a general of Eastern Jin dynasty times (317–420 AD) who "fearing the consequences of defeat, had a light getaway skiff constantly moored at the ready beside his warship, which greatly undermined the morale of his men." See Minford, pp. 223–24.

whether the scenario envisioned by Passages #10.13 and #10.14 always involves the entire army or might involve only an included part of it. In the latter case, part of that army might be placed in death ground, while the general remains with a different part, quite likely the main force. Then content of X.20 seems irrelevant as a factor mitigating leader deception/betrayal on the general's part.)

X.20. Because such a general regards his men as infants they will march with him into the deepest valleys. He treats them as his own beloved sons and they will die with him.

Passage #10.20. Fifth Codicil: Implicit Constraints on Interrogation Methods?

(Verse XIII.13 might be read as implying a pragmatic warning about unbridled recourse to harsh but unreliable interrogation methods. Of course, we do not know if Sun Tzu intended for any such implication to be drawn.)

XIII.13. He who is not sage and wise, humane and just, cannot use secret agents. And he who is not delicate and subtle cannot get the truth out of them. [Emphasis supplied.]

★★★★★★

$$\boxed{F}$$

Strategist Should Have an Accurate Grasp of the Significant Information

At the heart of the versatility of Sun Tzu's thinking – and a basic reason it is so extraordinarily conducive to digital age applications – stands its unswerving emphasis on the pivotal importance of information as a resource for strategic actors.

Building on the present thematic approach to Sun Tzu, three analytical distinctions help structure and sharpen Sun Tzu's information foci for modern audiences.[1]

The first involves a division of labor between Themes #11–#12, on the one hand, and Themes #5–#10, on the other. The former themes focus on how a Sun Tzu-inspired strategist would go about *observing and making sense of the world* (self included!) as a prelude to action. By contrast, Themes #5–#10 emphasize *changing that world*, shaping it in advantageous directions.[2] Yet as is often the case with Sun Tzu, his thinking defies so neat a partition. A closer approximation to the relevant division of labor would focus on where pertinent storytelling starts. In Themes #11 and #12 that storytelling picks up in a realm of steps taken vis-à-vis information; action consequences follow, but come after. In the case of Themes #5–#10 the story starts at a later point in the flow of events, in a realm of military operations. In stories of the latter type, the play – so to speak – begins in the second act.[3] But that flow of events inevitably modifies the information landscape, so that (drawing on Sun Tzu's own image) the actual dynamic is "as a ring is without a beginning or an end."[4]

A second distinction separates Themes #11 and #12 (whose focus is on acquiring information about the world) from Theme #1 (whose focus is on calculations building on that information). As verse XIII.4 (Passage #1.19, also appearing in Passage #11.6) illustrates, this boundary is again not always sharp. It may be further blurred by linguistic tendencies in the classical Chinese language. However, there is a gain in conceptual clarity from recognizing this distinction.

---

[1] For further perspective, building on the present study of the fourteen themes, see Conclusion section entitled "Sun Tzu As Prophet and Pioneering Theorist of Information Warfare" (pp. 504–13 below).

[2] Cf. Marx's Eleventh Thesis on Feuerbach – "Philosophers have only *interpreted* the world, in various ways; the point, however, is to *change* it." (Emphases in original.) See Robert C. Tucker (ed.), *The Marx–Engels Reader* (2nd ed.; New York: Norton, 1978), p. 145.

[3] Image from Padgett & Powell, #2 footnote 43, p. 434.

[4] Ames, p. 119, translating verse V.12. For verse V.12 translation issues see Appendix 14 footnote 3.

A third distinction pertains to how Themes #11 and #12 relate. Here it is hard to improve on Sun Tzu's succinct word picture "know the enemy and know yourself" (verse III.31), one of the most compelling ever offered by any writer on strategy. Theme #11 is about "knowing the enemy." Theme #12 about "knowing yourself" (a concept extended there to cover knowing allies, neutrals, bystanders, etc.).

★★★★★★

THEME #11 KNOW THE ENEMY

role of intelligence:
Passage #11.1 (verses III.31–III.33) (know the enemy and know yourself)
Passage #11.2 (verses X.22–X.26) (know the enemy, know yourself, adding environmental context)
Passage #11.3 (verses IV.1–IV.4) (role for intelligence in spotting enemy vulnerabilities; sense of the limits of intelligence)
Passage #11.4 (verse III.3) (broader vistas for intelligence work: winning without fighting)

intelligence needs and sources of intelligence:
Passage #11.5 (verses I.1–I.14) (agenda for strategic intelligence)
Passage #11.6 (verses XIII.1–XIII.23) (Sun Tzu on espionage)
Passage #11.7 (verses VI.13–VI.14 + VI.17 + VI.20–VI.23) (need for field intelligence; aggressive reconnaissance)

scope for imagination and flexibility in intelligence work:
Passage #11.8 (verses V.5–V.7) (imagination and resourcefulness in intelligence activities)
Passage #11.9 (verses VI.26–VI.31) (flexibility and adaptiveness in intelligence activities)

cover for spies and covert operators:
Passage #11.10 (verse IX.18) (affinity of spies and covert operators for complex environments and the opportunities for good cover they afford)

psychological operations:
Passage #11.11 (verse VIII.17–VIII.24) (knowing the enemy general; psychological operations targeting him)
Passage #11.12 (verses I.22–I.25) (further psychological operations)

intelligence interpretation and evaluation:
Passage #11.13 (verse XI.51) (classificatory and all-source instincts in intelligence work)
Passage #11.14 (verses IX.45–IX.46) (overconfidence and related psychological pitfalls)
Passage #11.15 (verses V.17–V.19) (signal vs. noise)
Passage #11.16 (verse VIII.16) (focus on enemy capabilities, not enemy intentions)

Passage #11.17 (verse IX.44) (incongruity testing)
Passage #11.18 (verse IX.15a) (creative early warning methodology)
Passage #11.19 (verse I.28) (quantitative intelligence analysis)

★★★★★★

A hallmark of the Sun Tzu text is its rich intelligence content. All thirteen chapters of the text make useful contributions to it. In order to avoid duplication of ideas addressed under other themes, the present Theme #11 development primarily focuses on Sun Tzu XIII, Sun Tzu's espionage chapter. That chapter is a genuine gem – forceful, thought-provoking, and well structured in ways that other parts of the Sun Tzu text are sometimes not. It is a manifesto for spies everywhere – above all, for the existence and coherence of the spy business as well as for its larger role. Certainly the existence of Sun Tzu XIII adds a significant additional dimension to the text, offering a way of reading through fresh eyes much of what Sun Tzu says in other chapters. Sun Tzu XIII's contributions, which have undoubtedly added much to the Sun Tzu text's towering overall reputation as a strategy classic, open paths from Sun Tzu (1) to Sun Tzu (2) and (3) material (indeed many more such paths than Sun Tzu's otherwise rather sparse treatment of "winning by non-military ways and means" [Theme #4] would do in the absence of Sun Tzu XIII). For purposes of this impact assessment it matters little, if at all, that there is a strand of modern scholarly opinion that regards Sun Tzu XIII as a latecomer to Sun Tzu's thirteen chapters, quite possibly the work of another hand; or that some of Sun Tzu XIII's enthusiasm for double agents (*fanjian*) might be faulted for overreaching.[1]

In tackling Theme #11, a productive initial focus is what might be labeled Sun Tzu's "intelligence style," representing an included part of his fourteen themes strategic style. Attention then shifts to a set of more concrete Sun Tzu–inspired principles of intelligence work (Table 4). *As a streamlining step, because Theme #11 analysis is structured around Sun Tzu XIII, Sun Tzu XIII verses will generally be cited in Theme #11 development without mention of Passage #11.6 (where all of them appear).*

Sun Tzu's intelligence style is analyzed here in terms of four characteristics.

The first centers on Sun Tzu's unequivocal position (often absent from Western strategy writing until well into the twentieth century) that secret intelligence work must permeate – indeed must be the foundation for – any major strategic undertaking. A basis for that stance comes from Sun Tzu's spotlighting of the pivotal role of foreknowledge, specifically excluding supernatural sources (verses XIII.3–XIII.4 in Passage #11.6). Such foreknowledge is a sine qua non for putting into practice Sun Tzu's overall style of warfare, much of which indeed makes little sense in its absence.

A second ingredient of the style is Sun Tzu's notably systematic and calculating approach to intelligence work. A no-nonsense tonality is set by Sun Tzu's very first verse (verse I.1 in Passage #11.5) and is further developed by emphases running throughout Sun Tzu XIII. It harmonizes with Sun Tzu's view of strategist-as-inconspicuous-elite, privileging formlessness (*wu xing*) of those in that

---

[1] Brooks's proposed order of Sun Tzu chapter accretions (see Table 12A on p. A-35) places Sun Tzu XIII at the extreme tail end of his conjectured timeline – *c.* 272 BC, slightly more than a human generation after the next-most-recent chapter (identified by him as Sun Tzu I, *c.* 309 BC). A "late arrival" view of Sun Tzu XIII is also suggested by Sawyer, Introduction footnote 37, p. 100.

role.[2] That mindset is by no means shared by all spies or spymasters, or the rulers they have served.[3] Throughout recorded history, the personalities and operating styles of those in this line of work, even some of the more effective (or lucky) ones, have often run toward the colorful, at times even when that tendency (and the attention it attracts) interferes with doing the job. Down to modern times, such a tendency has not infrequently been associated with flashes of outright amateurishness, or at least a tolerance for it.

Sun Tzu's intelligence mindset is devoid of any hint of such aberrations. His methodical approach – the phrase "essence of professionalism" applies well here – is epitomized by the cost–benefit observations at the beginning of Sun Tzu XIII. The same mindset also appears via a spirit of digging for potentially invaluable minutiae that is woven into the fabric of verse XIII.16. Above all, a methodical mindset is conveyed by the sheer existence of Sun Tzu XIII as a repository of systematic intelligence doctrine, possibly the first such doctrine statement in human history (or at least the earliest that has survived).[4] Perhaps the most striking feature of that doctrine statement is the manner in which it implicitly accepts, and indeed treats as an article of faith, the possibility of imposing order at the edge of chaos – achieving a disciplined and dispassionate engagement with a world of lying and betrayals.[5]

A third feature of Sun Tzu's intelligence style – amply anchored in Sun Tzu XIII but also supported by other parts of the text – is an overriding focus on the human dimensions of intelligence work. For Sun Tzu, a focus on the human face of intelligence starts with a paramount emphasis on spies. Even though Sun Tzu also contemplates intelligence-gathering by human beings who may not be spies (e.g., the observations producing Passage #11.26A's field intelligence), Sun Tzu's time faced a paucity of alternatives to spies. A case in point involves verses VI.21–VI.23 (Passage #11.7). Although these point to aggressive use of military reconnaissance methods, absence of cavalry in the armies Sun Tzu knew would have greatly limited this type of intelligence-gathering.[6] Meanwhile, the absence of paper records and the awkwardness and limited portability of bamboo strips "books" sets the information environment of Sun Tzu's era worlds apart from the plethora of information sources so conducive to "open source" intelligence collection in a modern society.

But Sun Tzu's focus on human factors in intelligence involves more than simply a predilection for a certain type of intelligence tool, the spy. It also involves an affinity for intelligence work directed toward human intelligence targets, which brings to the fore topics like human emotions and personality, as contrasted with military-technical realms involving weapons, order of battle, etc.

---

[2] See Passage #8.10 and Passage #8.1's verse VI.25, and further observations on pp. 64 and 71.

[3] The story of Sun Tzu and the concubines, albeit almost certainly apocryphal, is a story about a kind of professionalism (itself a type of style), which Sun Tzu epitomizes and the ruler does not.

[4] A serious competitor for this appellation is the Arthaśāstra, though modern scholarship suggests a somewhat later origin. For Arthaśāstra dating issues see Olivelle, #1 footnote 39, pp. 25–31.

[5] A contrast case – precisely the sort of pitfall Sun Tzu's intelligence style demands that a spymaster avoid – is evocatively captured by a comment of one French general on another general who also served in the first Indochina war, "remembering the [professional] deformation of the man who served in intelligence and whose reasoning finally becomes crooked because he must deal with so many crooked people ..." See René Cogny, "La libre confession du Général Cogny," L'Express (Paris), December 6, 1963 (translated and quoted by Fall, #4 footnote 51, p. 27).

[6] For discussion of dating of the advent of cavalry in Warring States China see pp. 29–30 above.

A fourth feature of Sun Tzu's intelligence style – again a source of much connectivity between Sun Tzu XIII and other Sun Tzu ideas – concerns the distinctively "activist" flavor of Sun Tzu's approach to intelligence work.[7] On the one hand, military operations of the sort favored by Sun Tzu find in intelligence operations a rich resource. Verse XIII.23 says as much. Conversely, the style in which Sun Tzu envisions military operations as being conducted, if transposed by analogy to intelligence realms, harbors many ideas for how to plan and carry out intelligence operations. For example, Theme #6 – attacking the adversary's strategy – holds lessons for non-military intelligence warfare of a spy-vs.-spy type. Those lessons may be even more useful there than in guiding purely military operations since intelligence operations plans may often be more vulnerable to upset than their conventional warfare counterparts (cf. verse XIII.14's "Delicate indeed! Truly delicate!"). Further transposable insights, some particularly relevant to counterintelligence work, appear in Sun Tzu's coverage of "formlessness" (*wu xing*, see Theme #8).

What has just been said still doesn't quite adequately capture Sun Tzu's specific slant on intelligence operations. Cowboy (to use one modern terminology) Sun Tzu may be, but at the same time he is a singularly careful cowboy. It must never be forgotten that Sun Tzu likes to bet only on sure things, a predilection visible in verse III.31 (Passage #11.1) and overwhelmingly corroborated throughout the text. It is hard to imagine Sun Tzu whiling away the hours before the culmination of a major intelligence operation playing "Luck Be a Lady Tonight" over and over again on a record-player (or whatever its musical counterpart might have been in Warring States China!).[8] Sun Tzu's penchant for active measures hence always needs to be situated in a larger context of Sun Tzu's assumptions shaped by Warring States conditions and culture. Examples would include feasibility of the necessary calculations, including delicate psychological ones (Theme #1), as well as belief that any determinate situation holds seeds of advantage (Theme #5). *If there is a soft spot in Sun Tzu's intelligence thinking, it largely lies in some of those assumptions.* Those premises must be analyzed if Sun Tzu's intelligence style is to be accurately understood and its implications, involving both strengths and limitations, mastered.

Much as strategic style may be usefully viewed as a more fluid generalization of doctrine, so too doctrine is a way of pinning down style, making it easier to study, teach, and practice. The doctrinal content of Sun Tzu XIII provides a vehicle for clarifying many particulars of Sun Tzu's intelligence thinking (which often receives positive press; less commonly, a specific point-by-point analysis). Unusually in the Sun Tzu text, Sun Tzu XIII is an entire chapter whose ideas unfold from beginning to end in a taut, crisply logical way. Building on that unfolding, Sun Tzu XIII's content is disaggregated in Table 4 into a baker's dozen of thirteen ideas supported by an analytical outline. In keeping with Sun Tzu's lack of a strategy/tactics distinction (see pp. 47–48 above), most of the ideas in Table 4's inventory could find application on both strategic and tactical intelligence levels, though a few of them (e.g., Idea #12) plainly align on strategic intelligence. One would certainly like to know more

---

[7] A pertinent Russian phrase is *aktivnye meropriyatiya* ("active measures"), described by one veteran KGB officer as "the heart and soul of Soviet intelligence."

[8] See June 10, 2000 *Washington Post* obituary of Kermit Roosevelt, Jr.

Table 4 *Overview of Sun Tzu's espionage chapter*

Part 1: Menu of Sun Tzu's espionage ideas, with textual anchoring

Reflecting Theme #5's central position in Sun Tzu's thought, all these ideas have a
  Theme #5 aspect. By definition of spying, all involve deception (Theme #7). The
  qi/zheng ("crafty/straightforward") concept pair (Theme #14) also shows its
  influence throughout.

	Griffith verses
Idea #1: Cost-effectiveness of spying as compared with conventional military operations (relates to Theme #3, cheapness)	XIII.1–XIII.2, reinforced by XIII.6
Idea #2: Foreknowledge as key to political–military success; centrality of spying to achieving it (relates to Theme #1, calculation)	XIII.3–XIII.4, reinforced by XIII.23
Idea #3: Botanizing spies: five-fold typology of kinds of spies	XIII.5–XIII.11
Idea #4: Concept of espionage network and its far-reaching tentacles (relates to Theme #8, formlessness)	XIII.6 and XIII.21, reinforced by XIII.14 and XIII.16
Idea #5: Seamless/intimate relationship between general (or ruler) and spies (this idea is a kind of espionage cousin of Theme #13, leadership and harmony in the chain of command)	XIII.12, reinforced by XIII.21 and XIII.23
Idea #6: Spymaster as "humanistic" role (relates to Theme #12, know yourself; also, Theme #13, leadership and harmony)	XIII.13, reinforced by XIII.23
Idea #7: Spying as a most delicate endeavor (relates to Theme #1, calculation)	XIII.12–XIII.14, reinforced by XIII.19 and XIII.23
Idea #8: Spying as ubiquitous – universal in scope and utility (relates to Theme #4, winning by non-military ways and means; also to Theme #10, no scruples and controlled viciousness)	XIII.14 and XIII.23; XIII.21 can be interpreted as spying on one's own spies using double agents (fanjian)
Idea #9: Imperative of effective command control of secret information, ensuring both authorized access and denial of unauthorized access (the former relates to Theme #12; the latter to Theme #8, formlessness)	XIII.12 and XIII.21 (access); XIII.6 and XIII.12 (denial of access in general); XIII.15 (steps to seal off a leak)

Table 4 (*cont.*)

Idea #10: Affinity between spying and data, both accurate and false (relates to Theme #4, winning by non-military ways and means; also to Theme #2, terrain)	XIII.4 and XIII.16 (need for accurate and detailed data); XIII.19 (planting false data)
Idea #11: Central role of *fanjian* (commonly translated as "double agents") in successful espionage (relates to Theme #6, attacking the adversary's strategy)	XIII.9 (definition of *fanjian*) and XIII.17–XIII.21 (*fanjian* recruitment and use)
Idea #12: Potential of well-placed figures (closely associated with a regime) "who, having inside information, could effectively topple the power at the center" if their loyalties shift (Ames, p. 297; relates to Theme #9, building on natural forces and momentum)	XIII.22, reinforced by XIII.6 and XIII.21 (spy network in place to identify potential side switchers on either side)
Idea #13: Affinity between successful espionage and talents of spies	XIII.23

Table 4, Part 2: Analytical outline of espionage chapter

Unlike most of Sun Tzu's chapters, the espionage chapter is a structured essay developing a coherent, indeed in many ways elegant, argument. Part 2 of Table 4 provides a coup d'oeil of that argument, presented in an analytical outline format. Seen through modern eyes, the most distinctive – and problematic – part of that argument is the central role Sun Tzu accords to *fanjian* (see pp. 368–71 below).

1. Case for spying: spying (cheap and effective) vs. conventional warfare (expensive and risky) (verses XIII.1–XIII.2)
2. Spying contrasted with alternative information sources (verses XIII.3–XIII.4)
3. Anatomy of a spy network
   3.1. Profile of a well-functioning spy network (verses XIII.5–XIII.6)
   3.2. Description of five types of spies (verses XIII.7–XIII.11), two of which are subverted insiders, possibly some of a third type of spies as well
4. Unleashing the potential of spies
   4.1. Harmony (1): seamless, thoroughly confidential, relationship of spy apparatus to command level of authority, sustained by generous rewards (verse XIII.12)
   4.2. Harmony (2): characteristics of ideal spymaster (verse XIII.13)
   4.3. Characteristics of the capability: spying as a fine-grained instrument (verse XIII.14, first part)
   4.4. Playing offense: opportunities for spying as ubiquitous (verse XIII.14, second part)
   4.5. Playing defense: response in the event of a leak (verse XIII.15)
   4.6. Reaching deep: uses of spying to obtain detailed information on individuals of interest (verse XIII.16)

Table 4 (cont.)

5. Double agents (fanjian) as key to successful espionage
   5.1. Recruiting fanjian (verse XIII.17)
   5.2. Building and using a spy network based on fanjian (verses XIII.18–XIII.20)
       5.21. Using expendable spies to pass false information to enemy high command (verse XIII.19)
   5.3. Fanjian as foundation for effective command control of a spy network (verse XIII.21); crucial importance of fanjian incentives
6. Spies (or defectors) as catalysts to regime transition (verse XIII.22, pointing to historical cases)
7. Summation on spying: spying and secret operations as foundation of all military or strategic effort (verse XIII.23)
   7.1. Need for the "most intelligent" (shang zhi) people as spies (verse XIII.23)

about specifics of the Warring States intelligence activities that inspired Sun Tzu's thinking: their scale, their organization, the daily problems they faced, their faults, stress points, and organizational failures. Unfortunately, much as with Warring States logistics apparatus (though for somewhat different reasons), our knowledge of those specifics seems destined to remain scant.

These thirteen ideas are now analyzed one by one, from a perspective oriented to Warring States conditions.[9] How those ideas and their implementation fare in the modern world will be taken up in later Sun Tzu (2) and (3) discussion (pp. 374ff.).

Intelligence idea #1 – spying as cheap! – is very effectively introduced in verses XIII.1–XIII.2 and reinforced by verse XIII.6 (via its allusion to a well-functioning spy network as a ruler's "treasure" [bao]). As noted in analyzing Theme #3 (cheap measures), a core selling point of espionage by human means is its utter cheapness by comparison with conventional military operations, on which Cicero famously opined: "Nervos belli, pecuniam infinitam" ("The sinews of war is endless money").

Intelligence idea #2 – centered on sources and methods – is the subject of verses XIII.3–XIII.4. The latter verse offers a menu of sources (mainly what would nowadays be called supernatural ones) whose efficacy Sun Tzu summarily rejects in preference to "men who know the enemy situation," by which he means spies.[10]

Intelligence idea #3 is a classificatory exercise: botanizing types of spy (verses XIII.5–XIII.11). At the root of this classification, which should not be construed too rigidly, is where the spies hail from:

---

[9] Table 4's list aligns recognizably, though not exactly, on a disassembly of Sun Tzu XIII into component segments by Shi Zimei (fl. twelfth–thirteenth century AD), a Southern Song commentator on the Seven Military Classics (he is not one of the Sun Tzu commentators on the traditional list of eleven, pp. 42–44 above). Shi Zimei's Sun Tzu XIII analysis is translated by Sawyer, #1 footnote 3, pp. 143–56.

[10] For further analysis see Passage #1.19 in main text and Appendix 1 (where verses XIII.3–XIII.4 also appear) as well as Table 5 on pp. 405–406 below.

- one's own state (two types, one deemed "expendable,"[11] one "living"[12]);
- the enemy state (three types).

The latter three types are differentiated by the populations they draw on:

- the enemy country's general populace;
- its officialdom;
- its spies ("enemy spies whom we employ" [Chinese *fanjian*, see verse XIII.9]).

The *fanjian* category receives outsized attention – certainly by the standards of a short text – in verses XIII.17–XIII.21. Other ways of categorizing spies would plainly be possible.[13] Given psychological emphases that permeate Sun Tzu, it is plausible (though unstated) that the rationale for Sun Tzu's choice of typology is based on an assessment that these five types (in Sun Tzu's time) presented different agent recruitment and handling challenges.[14] Sun Tzu's attention to types of spies drawn from the enemy's country harmonizes with a more basic concept – common to several Sun Tzu themes – of conjuring ways of "using the enemy against himself" (Ames, p. 84).

Intelligence idea #4 (see verses XIII.6 and XIII.21) is Sun Tzu's version of the spy network concept. In modern terminology, verse XIII.6's focus is on network functioning, not network structure.[15] In Sun Tzu's ideal-type case, the network is functioning harmoniously. A sine qua non of such functioning is invisibility, for which Ames's translation's elegant image is the "imperceptible web."[16] Formlessness (Theme #8) enters here. Via the enemy's *fanjian* (double agents on the foe's side), possibilities of network failure are also contemplated (see Idea #11).

Intelligence idea #5 – verse XIII.12, fleshed out by verses XIII.21 and XIII.23 – underscores the need for a tight, indeed intimate, relationship between the intelligence apparatus and ultimate intelligence consumers (commanding general or

[11] For traditional commentator Du You's vivid illustration of this category see #10 footnote 25 above.

[12] Giles, p. 167: "This is the ordinary class of spies, properly so-called, forming a regular part of the army." This category could also include individuals serving in a diplomatic capacity. Sawyer, p. 135 notes that living spies "were often talented individuals of exceptional perspicacity who could be dispatched to foreign states, sometimes in diplomatic guise, to observe and then report back."

[13] A quite different, occupation-based, typology of spies comes from the *Arthaśāstra*, for example noting spies "disguised as persons engaged in harlotry, different kinds of ascetics, of bards, of fortellers, cripples, of workers, of those who prepared food and drinks, and physicians." See Ludwik Sternbach, "Legal position of prostitutes according to Kauṭilya's Arthaśāstra," *Journal of the American Oriental Society*, 1951, 71(1), 25–60 (quotation is from p. 54, Sanskrit terms omitted). Regarding medical practitioners as secret agents see Patrick Olivelle, "The medical profession in ancient India: its social, religious, and legal status," *Journal of Indian Medicine*, 2017, 9, 1–21.

[14] However (and lest Sun Tzu's intelligence thinking be uncritically credited with overmuch modernity), it is worth noticing that Sun Tzu's classification of spies involves five types, and that the number five had numerological significance in early Chinese thought. See p. A-50f. (comment on Passage #1.21).

[15] A fleeting glimpse of cognizance of a special case of network structure appears in verse XIII.16's reference to *ye zhe* (Mair's "appointment secretary") as an espionage target. See p. 108 above (analyzing Passage #2.2).

[16] See #2 footnote 72 (sourcing inspiration for his translation to a concept in the *Laozi* Daoist classic).

ruler).[17] A likely implication (in Sun Tzu's time) is face-to-face contact between general or ruler and spies in their employ; or, short of that, at least existence of significant interpersonal trust (e.g., involving credible personal assurances). Sun Tzu leaves unresolved whether every general was expected to be his own case officer; or whether someone else in an intelligence officer role might be interposed between spies and the commanding general or ruler.[18] However, Sun Tzu's thrust (and what little else we know about early Chinese espionage practices) suggests a very simple intelligence organization by modern standards, probably a largely ad hoc one.[19]

Intelligence idea #6 – dissecting the spymaster role in "humanistic" terms (to use one modern framing) – is set forth in verse XIII.13. Important Chinese cultural content enters here. A three-pronged interpretation of verse XIII.13, building on observations of commentator Shi Zimei, would identify a competent spymaster as having:[20]

(a) the wisdom needed to "know men";

(b) the personal qualities needed to motivate spies (a distinct task); and

(c) the subtlety and delicacy of perception to interpret and evaluate spy reports, a further distinct task.

These criteria suggest an ideal spymaster as one who, far from being a narrow specialist, possesses genuine depth and breadth of knowledge of, and appreciation for, the human condition writ large. Relevant here is a flair for empathy: intuitive understanding of a very broad spectrum of human mentalities, motivations, and life circumstances. A corollary, indirectly supported by verse XIII.23 (see Idea #13), is that such understanding needs to range beyond the baser types thereof (albeit that these have always tended to be the coin of spy recruitment and spy dealings).[21]

---

[17] **Note on present usage of term "spymaster":** There is no counterpart term in the Sun Tzu text. The advantage of the term "spymaster" is its flexibility. It could cover case officers (in Sun Tzu's time possibly the commanding general or ruler himself, though for practical reasons of limited time and attention the handling of some, possibly most, spies seems likely to have been farmed out to others). The term spymaster could also apply to assorted other persons having some significant role, official or unofficial, in recruiting or running spies or human intelligence assets. Here and elsewhere in engaging with Sun Tzu's espionage thinking, it seems best not to be too rigid about definitions, trying to force Sun Tzu's thinking into modern conceptual boxes.

[18] Traditional commentator Mei Yaochen's commentary suggests the former possibility: "Secret agents receive their instructions within the tent of the general, and are intimate and close to him." See Griffith, p. 147.

Illustrating an instance of a general serving as his own operative, Sawyer, Introduction footnote 55, p. 144 notes an instance where the great general (and Sun Tzu commentator) Cao Cao took a personal hand in an estrangement plot "[r]ather than employing spies and double agents."

[19] Cf. G. H. Donaldson, "Modern idiom in an ancient context: another look at the strategy of the Second Punic War," *Greece & Rome*, 1962, 9(2), 134–41, cautioning (p. 136) that "intelligence service" as a terminology "conjures up visions of a complex organization" not suited for conditions of that time.

[20] See Sawyer, #1 footnote 3, pp. 150–51. Regarding Shi Zimei see #11 footnote 9 above.

[21] A vivid image of what a spymaster should *not* be comes from a rhetorical question posed by Shi Zimei (translation from Sawyer, #1 footnote 3, p. 151): "How can a drunken and sated general who must contend with force, yet lightly engages in battle, manage to discern the affairs of spies?"

It is instructive to compare verse XIII.13's enumeration of human qualities of the ideal spymaster with Sun Tzu's list of traits of the ideal commanding general found in verse I.7 (Passage #1.1, the net assessment passage, see p. 75 above):[22]

> (Griffith: sage & wise, humane & just, delicate & subtle)
> spymaster: *sheng zhi,*   *ren yi,*[23]   *wei miao*
>   ↓   ↓
> commanding general: *zhi, xin, ren, yong, yan*
> (Griffith: wisdom, sincerity, humanity, courage, strictness)

For a spymaster, "delicate & subtle" might be a plausible substitute for a general's "sincerity" (not a spymaster job requirement!). Omission of "strictness" as a spymaster attribute could well reflect the fact that, unlike a general commanding a mass-infantry army, a spymaster (at least in Sun Tzu's early time) may have had no need to ride herd on a large organization. Omission of "courage" as a spymaster attribute may signal belief that such a trait (in particular in its physical sense) is also irrelevant to spymaster job performance, albeit that a strong stomach might be essential.[24]

Intelligence idea #7 – spying as a most delicate endeavor, a fine-grained activity extraordinarily different in kind from the coarse-grained activity that conventional warfare represents – is expressed by verses XIII.12–XIII.14 with further aspects conveyed by verses XIII.19 and XIII.23.[25] The delicacy of interpersonal relationships in spying may be treated as a source of opportunity (because therein may lie possibilities for subtle strategic combinations and manipulations foreign to conventional warfare or indeed any form of warfare in a military or tools-of-violence sense); and as a source of risk (especially because a single false move may destroy interpersonal trust; and that trust, once lost, may prove irretrievable).[26]

Intelligence idea #8 – opportunities for spying as ubiquitous – is conveyed by verses XIII.14 and XIII.23. It plainly undergirds Sun Tzu's focus on winning by non-military ways and means (Theme #4). Carried to a logical limit (as Sun Tzu's abstractness and absence of scope conditions so often encourages doing!), Idea #8 sets a stage for internal spying operations no less than ones directed against foreign adversaries. Indeed a ruler's spies themselves need to be spied on: the starring role Sun Tzu accords to *fanjian* (Griffith's "doubled agents") guarantees that need. Sun Tzu's closest approach to a specific statement is verse XIII.21, if read as implying that it is through one's double agents that one's own spies may themselves be spied on.

---

[22] Chinese characters for these concepts are provided by the Glossary in the online annex (pp. A-3ff.)

[23] Common translations for *ren* and *yi* (Griffith's "humane" and "just") are "benevolence" and "righteousness," respectively. The archaeologically recovered Han strips text omits *yi*.

[24] In the half dozen times *yong* (courage) appears elsewhere in Sun Tzu, an obvious context is a battlefield one or, in the case of verse XI.37 (Passage #6.5), that of a knife-wielding assailant.

[25] Here is a major illustration of Theme #5's second path to advantage. See pp. 178, 180–81 above.

[26] Denma, p. 269 offers an insightful impressionistic word picture:

> The spy is an authentic source of knowledge that must be cherished, nurtured and rewarded. In this relationship everything is intensified: the kinship, the potential for new knowledge, the consequences. This weapon is so sharp that it must be kept concealed. Otherwise it may unpredictably injure itself or you. This has its own rules. One must be a sage to use spies, to feel their vulnerability, to keep oneself and them from jeopardy, to manage the delicacy of their operations amid the enemy, to interpret and apply their knowledge, to handle reversals and false information. Things can go quickly and disastrously wrong.

This line of thinking bolsters Sun Tzu's credentials as one who intuitively grasps the "trap door" potential of double agentry. At the same time, Sun Tzu may well expect too much of *fanjian* (a topic to which the present analysis will shortly turn, see Idea #11 below).

Intelligence idea #9 – centered on the timeless problem of preserving secrecy of espionage information, even while giving unfettered access to those who legitimately need it – is addressed, from different angles, in XIII.12 and XIII.21 (access); XIII.6 and XIII.12 (access denial); and XIII.15 (dealing with leakers). That level of attention, in a short document notable for its taut structuring, may be treated as a marker of emphasis. The material on access denial has resonance with Sun Tzu's "formless-less" thinking (Theme #8), starting with Passage #8.9's observation that

XI.42.
It is the business of a general to be serene and inscrutable, impartial and self-controlled.

Intelligence idea #10 has two aspects, both placing in center stage what would nowadays be labeled "data." First, Idea #10 stands for collecting accurate, timely intelligence data to support military or intelligence operations. Verse XIII.4 is the centerpiece here. It is supplemented by verse XIII.16 with its overtones of collecting meticulously detailed intelligence, in some ways a notably modern harmonic. Beyond this, matters having to do with double agents – a natural Sun Tzu emphasis in light of Idea #11 (see below) – are inherently a data-intensive activity, one where the tiniest factual details may be all-important.[27]

Second, Idea #10 points to a need to fabricate information compelling and (up to a point!) accurate enough to exert a sought-for effect on the enemy (verse XIII.19). One would definitely like to know any Sun Tzu tips on how to do such fabrication most effectively. Unfortunately he vouchsafes us no details.

Intelligence idea #11 trains a spotlight on uses of *fanjian* (lit. "turned spy," an agent category Sun Tzu defines in verse XIII.9 to mean "enemy spies whom we employ"), the subject of verses XIII.17–XIII.21.[28] Centrality of the *fanjian* role is a prominent, even defining, feature of Sun Tzu XIII.[29] Two basic assumptions regarding importance and use of *fanjian* are set forth by a traditional commentator:

---

[27] "These are the little things upon which the very life of the agent depends." See Stanley P. Lovell, *Of Spies & Stratagems* (Englewood Cliffs, NJ: Prentice-Hall, 1963), p. 23, quoted (with further analysis) on p. 23 of Erving Goffman's "Expression games: an analysis of doubts at play," in Goffman's *Strategic Interaction* (Philadelphia: University of Pennsylvania Press, 1969), pp. 3–81.

[28] In the following discussion, *fanjian* – which Griffith renders "doubled agents" – is often left untranslated, to provide interpretive flexibility and minimize possible modern conceptual baggage.

[29] At the outset it should be noted that verse XIII.9's definition of *fanjian*, if read alone, could be interpreted in two ways. It could mean "double agent" (and traditional commentator Li Quan so reads it, see Griffith, p. 146). It could also mean enemy agents who are duped by false information fed to them, and in that sense are "used" (traditional commentator Du Mu reads verse XIII.9 in that way; see also discussion in Nienhauser, Vol. VII [2021 revised edition cited on p. xxiv above], p. 505 footnote 12).

However, as Giles, p. 166 – resoundingly rejecting the second interpretation – points out, "that it is not what Sun Tzu meant is conclusively proved by his subsequent remarks about treating the converted spy generously" (see verses XIII.17 and XIII.21). Also noting the high level of analytical integration of Sun Tzu XIII (see Table 4), the present study adopts the first (double agent) interpretation.

insider knowledge regarding recruitment targets: "the doubled agent [*fanjian*] knows those of his own countrymen who are covetous as well as those officials who have been remiss in office. These we can tempt into our service."[30]

insider knowledge pertaining to the enemy's knowledge, perceptions, and plans: "doubled agents know in what respects the enemy can be deceived."

A third assumption (a strong one!) is also implicit in Sun Tzu's treatment:

fanjian bona fides can be ascertained with confidence.

Therein lies a rub – and a puzzle. The annals of intelligence history, worldwide, strongly suggest that double agents are hard to handle under the best of circumstances and the information they provide is seldom the master key to major intelligence success (which is exactly what Sun Tzu XIII appears to claim). Even the most basic issues presented by a double agent can defy crisp answers. Is the agent an actual turncoat – or a fake one (in which case at least some of that agent's information is tainted, and relying on it may prove deadly)? Is the agent, even if genuinely working for your side at the moment (whatever that may mean!), likely to revert back to serving the other side (for example, once released back to the enemy, as one reading of verse XIII.17 contemplates)?[31] When a situation involves multiple double agents (a possibility that Sun Tzu's thinking does nothing to exclude and may in fact invite) the ensuing complexities and uncertainties may well spiral out of control.[32]

Yet verses XIII.17–XIII.21 ignore these problems entirely and make the task of turning, then running, a double agent seem almost routine, and double agentry a near-panacea for intelligence problems. Importantly, there is no allusion to a need to put special strings in place to back up double agent reliability.[33] Something seems amiss here, either with Sun Tzu or with our understanding of the *fanjian* category. Given the prominence of *fanjian* in Sun Tzu XIII's scheme of things, and of espionage in Sun Tzu's way of war, this is no minor matter. Perhaps the meaning of *fanjian* needs rethinking?[34] Or perhaps a key piece of context is being overlooked?[35]

[30] This quotation and the next are from traditional commentator Zhang Yu, as translated by Griffith, pp. 148, 149. Giles, p. 172 also provides translations of the same Zhang Yu material.

[31] Regarding interpretations of verse XIII.17 see #11 footnote 75 and Appendix 11 footnote 8.

[32] The closest Sun Tzu comes to a concrete prescription for dealing with double agent reliability issues is found in verse XIII.21, which could be read as implying use of double agents to ascertain the reliability of other double agents. Yet it seems eminently reasonable to be a little queasy about the soundness of this "solution." A game theory article sums up pitfalls: "The issue with which Sun Tzu seems to have the greatest difficulty . . . is the possibility that both sides will use agents to feed each other false information." See Niou & Ordeshook, #5 footnote 7, p. 173.

[33] Contrast the *Arthaśāstra* (Olivelle, #1 footnote 39, p. 79): "He [the king] should appoint double agents after taking their [i.e., the double agents'] wives and children into custody. . ."

[34] Although the taut analytical structure of Sun Tzu XIII suggests a precise usage there, the label *fanjian* was often loosely applied in early (and later) Chinese usage. For example, after reviewing a number of *fanjian* references in the Shiji, Kierman finds no clear reason to regard *any* of them as double agents in the modern sense. Instead, a common denominator of those *fanjian* was that each was in some way involved in perpetrating a deception, so that Kierman proposes to call them "deception agents." See Frank A. Kierman, Jr., *Ssu-ma Ch'ien's Historiographical Attitude As Reflected in Four Late Warring States Biographies* (Wiesbaden: Otto Harrassowitz, 1962), pp. 68–69 note 31. A related but distinct interpretive possibility is noted by Sawyer, #1 footnote 3, p. 89: "in common use from the Warring States on *fanjian* rarely meant 'doubled agent,' but instead someone recruited from the other side to act as an agent in place." As Sawyer also points out, that would make *fanjian* fall under a heading of "inside agents" (verse XIII.8), thereby conflating Sun Tzu secret agent categories.

[35] A modern case of usefulness of double agents pivoting on special circumstances (albeit ones very different from conditions prevailing in Warring States China) is the famous British "Double-Cross

One possibility is that conditions in Sun Tzu's early time and place were somehow distinctively conducive to existence and advantageous use of *fanjian*. Widespread affirmations of the virtue of loyalty to the contrary, there is reason to believe that Warring States China was in fact a very low-loyalty environment heavily driven by personal ambition (and, of course, greed), suggesting that opportunities to recruit double agents may have continually presented themselves. Paradoxically, it *also* seems to have been an environment where turncoats were often accorded what to modern eyes seems a remarkably high level of trust.[36]

As a conjecture, a tendency to place trust in turncoats – quite possibly more than they merit – may in part have been a byproduct of Sun Tzu's information-sparse age, which meant that key decision makers involved in evaluating *fanjian* credibility had few information channels on which to draw, so that discrepant information was often not available.[37] Furthermore, much of the secret information provided by *fanjian* may itself have been simple and short-term, of a type that could be divulged and acted on without further adieu (thus reducing opportunities for complex wrinkles to creep into a drawn-out developing double agent situation with influences and counter-influences galore). Meanwhile, weighing Sun Tzu's *fanjian* thinking from a vantage anchored in what we know of Warring States culture, some strands of early Chinese thought gave grounds for optimism that human facades could be penetrated to reveal underlying truths.[38] Such a belief (however problematic by modern standards) might have gone a considerable way to buttress an assumption that a *fanjian*'s bona fides could be reliably ascertained.

System" of World War II, whereby "agents trusted by the [German] enemy were used to feed carefully orchestrated misinformation into his intelligence system." Michael Howard, *Strategic Deception in the Second World War* (London: Pimlico, 1992), p. x observes that the efficacy of the system rested on advantages "so extraordinary indeed that one would be rash to assume that we or indeed anyone else could ever possess them in comparable measure again." One of those advantages was cryptologic access to German capabilities and intentions. A second was the high level of security within the UK itself, leading to "apprehension and control of all active enemy agents in Britain." (Howard, p. x)

[36] Regarding both aspects of Warring States China see Sawyer, p. 107 (quoted in part in #10 footnote 23). As Sawyer, p. 107 goes on to observe:

> A general who suddenly lost favor, whether through battlefield defeat or political machination, might flee to the enemy and – contrary to expectation – not only avoid execution but actually be enfeoffed, his troops and followers being equally welcomed. A political figure who became entangled in court intrigues might also find refuge in either friendly or enemy states, and be granted significant power as a minister, advisor, or even a strategist for military campaigns against his former state.

For further background see Yuri Pines, "Friends or foes: changing concepts of ruler–minister relations and the notion of loyalty in pre-imperial China," *Monumenta Serica*, 2002, 50, 35–74.

[37] See #7 footnote 13 (quoting Richards J. Heuer, Jr.). The level of confidentiality implied by verse XII.12 may have meant that only one person had any involvement in evaluating a *fanjian*'s credibility, reducing possibilities for discrepant information to intrude. Overconfidence biases may also operate to reinforce judgment errors (see #11 footnote 88, citing cognitive science work of Griffin and Tversky).

[38] This points to a body of early Chinese thought that Sawyer has labeled the "problematic theory of transparency." See #8 footnote 26. However, the Sun Tzu text makes no allusion to such a theory.

Severe limitations on long-distance communications and transportation, and especially on written communications (in an era antedating writing on paper, so that "files" would not exist as a path to storing and retrieving political–military secrets and perhaps also planting false or misleading information), may also have worked to limit some of the more complex double agent shenanigans.

There are still other possible avenues of explanation, but we are unlikely ever to be able to confirm them.[39] Perhaps the puzzle of Sun Tzu's faith in *fanjian* is best turned on its head. Accepting that the centrality Sun Tzu accorded to *fanjian* – along with acceptance of their trustworthiness – was in fact a sound judgment call, what insight can be extracted regarding the war, politics, and logistics Sun Tzu knew?

Turning to the next idea, Intelligence idea #12 is anchored, if very briefly, in verse XIII.22, alluding to purported circumstances associated with two different dynastic transitions that in Sun Tzu's time would already have been ancient history.[40] In parsing verse XIII.22 it seems clear that the detailed history (or imagined history) is beside the point. Standing back from antiquarian allusions, a robust message is that a well-placed insider's switch of loyalties may suffice to produce major regime change, especially if the existing regime is already in decline. By implication, inducing such a switch of loyalties would be a Holy Grail for an adversary's intelligence activities. Reading verses XIII.21 and XIII.22 together also suggests that Sun Tzu may see *fanjian* as a major line of defense against such an attack, since in Sun Tzu's view they are well positioned, among parties serving one's own side, to spot impending treachery and nip it in the bud.

Intelligence idea #13 poses the intriguing thought (via the first part of verse XIII.23) that the very best espionage results are attained by making use of the most intelligent people. That notion is glossed by Shi Zimei's commentary, opining that "Spies must successfully undertake actions beyond the capability of other men."[41] This point could be challenged, since crucial information can always come from a spy (or intelligence asset) who simply happens to see something or overhear a bit of unguarded conversation and passes it on. But if coordinated with Sun Tzu's larger emphases and priorities, verse XIII.23 comes into sharper focus. A basic reason the "most intelligent" (*shang zhi*) people make especially valuable spies is the fact that it is they who are most likely to be able to worm their way into the inner circle and trust and confidence of Sun Tzu's highest priority intelligence target, namely, the enemy general (or ruler) – even one who takes to heart Sun Tzu's counsel of surrounding himself with a protective shroud of *wu xing* "formlessness" (Passage #8.9). In particular, it may require *shang zhi* individuals to report insightfully, not just on facts

---

[39] One possibility is suggested by a variant "cumulative" interpretation of the *fanjian*'s contributions as set forth by Sun Tzu in verses XIII.17–XIII.20 (Passage #7.10). Somewhat reminiscent of the "pyramid" thinking of verses IV.16–IV.18 in Passage #1.7, it is reflected in Mair's translation (Mair, pp. 130–31). See #7 footnote 85 for further specifics. By scaling back the direct contributions expected of *fanjian*, this variant interpretation helps modulate Sun Tzu's seemingly exorbitant faith in *fanjian* (though the basic puzzle of their perceived efficacy persists). In any case, the variant interpretation goes beyond what the Sun Tzu Chinese text literally says.

A different possibility, which should not be overlooked, is that Sun Tzu XIII, gem though it is in many ways, reflects thinking of a theorist not attuned to the messy realities of espionage practice.

[40] For background see p. 205 above, discussing Passage #5.19 (where verse XIII.22 also appears).

[41] Sawyer, #1 footnote 3, p. 155 (translating from Shi Zimei's commentary).

learned from participation in an inner circle, but on matters requiring nuanced evaluative judgment (e.g., signs of latent conflict in the enemy's high command).[42]

★★★★★★

By its nature intelligence work must navigate an extraordinarily rich web of interconnections. Unusual situations and offbeat interpretations abound and can at times be game-changers. Sun Tzu's Theme #3 emphasis on "outliers" fits well here. The result is many intellectual trap doors, far more than is the case with ordinary military knowledge. With a weather eye on the possibilities for analytical complexity thus arising, organization of Theme #11 passages has been kept simple.

Coverage starts with a first set of passages (Passages #11.1–#11.4) situating Sun Tzu's intelligence thinking in context of his broader military and strategic thought.

A second set of passages (Passages #11.5–#11.7) then presents the heart of Sun Tzu's substantive intelligence thinking. While for many modern audiences the net assessment verses (Passage #11.5) and the espionage chapter (Passage #11.6) take center stage here, also basic to Sun Tzu's way of war are military intelligence emphases found in Passage #11.7. These combine a call for good field intelligence with endorsement of aggressive scouting or reconnaissance measures.[43]

A third set of passages (Passages #11.8–#11.12) is culled from parts of Sun Tzu outside Sun Tzu XIII. Read from an intelligence standpoint, a strand running throughout this material is emphasis on resourcefulness in intelligence work, drawing leverage from physical (Passage #11.10) or psychological (Passages #11.11, #11.12) features of the environment in which intelligence activities are taking place.

---

[42] See also #4 footnote 86 (amplifying this point). It is instructive to compare and contrast verse XIII.23's emphasis on *shang zhi* with an observation of a 1790 French commentator who wrote:

> The ablest ambassador can do nothing without spies and he would achieve even less if he chose them from the gutter. Taken from the higher ranks of society they are necessarily more expensive. To fulfill his mission worthily, an ambassador must be ready to buy anyone from the secretary to the valet, from the serving-maid of the favorite mistress to the lady-in-waiting of the Queen.

See Herman, #1 footnote 35, p. 11. Verse XIII.23 is pointing to something rather different, namely the need for spies who possess exceptional intelligence (cognitive ability, emotional intelligence, social observation skills, etc.). Social access and exceptional intelligence may be correlated, depending on the makeup of a particular society's "higher ranks"; but they are definitely not identical.

[43] **Note on present usage of term "field intelligence":** "Field intelligence" is here used as a convenient umbrella term to describe matters pertaining to the intelligence needs and activities of a Warring States Chinese army in the field. A contrast case would be strategic or political intelligence (though here, as always, intelligence categories often blend and blur, and many important developments may occur at the boundaries of categories). "Field intelligence" is the title of Sawyer, #1 footnote 3, chapter 15, pp. 427–66. The concept is usefully illustrated in an early modern Western context by Lee Kennett, "French military intelligence, 1756–1763," *Military Affairs*, 1965–66, 29(4), 201–204 (contrasting field intelligence with higher-level intelligence networks run out of Versailles).

A fourth set of passages (Passages #11.13–#11.19) begins with material touching on timeless challenges of intelligence analysis:

- classification of sources, with a nascent "all-source" concept (Passage #11.13)
- pitfalls of overconfidence and related psychological factors (Passage #11.14)
- distinguishing signal from noise, in an intelligence meaning of those concepts (Passage #11.15)
- focus on enemy capabilities, not enemy intentions (Passage #11.16)

This fourth set of passages concludes with three specific intelligence analysis topics where Sun Tzu has particularly noteworthy – albeit telegraphically expressed – things to say (certainly if one keeps in mind how ancient a text this is):

- incongruity testing (Passage #11.17)
- creative early warning methodology (Passage #11.18)
- quantitative methods in intelligence analysis (Passage #11.19)

Roads Not Taken by Sun Tzu in Developing Theme #11

While Sun Tzu's espionage chapter (Sun Tzu XIII) has been lauded in many circles, military and civilian alike, it is less commonly dissected for what it does *not* say. Here, of course, it is essential to keep squarely in view Sun Tzu's simpler time, and to avoid faulting him for failing to anticipate and delve into the complexities of modern intelligence establishments, with their vast bureaucratic structures, multibillion dollar budgets, and surfeit of high technology.

One resounding omission in Sun Tzu is deficit of attention to espionage tradecraft – a striking deficit for one as interested in spies and spying as Sun Tzu is. Much as is the case with military weapons and training topics, this is a level of specifics in which Sun Tzu is simply not interested. Also, incisive as it is, Sun Tzu XIII offers few tips on how spies might be actually employed – i.e., espionage counterparts to the many military operational tips Sun Tzu offers for employment of armies.

A second omission – from a modern perspective a comparably basic one – is absence of any concept or indeed credible foreshadowing of an intelligence *service*, a type of durable complex organization as contrasted with an ad hoc one.[44]

As regards intelligence analysis – making sense of information obtained from spies or others, and gauging its credibility – Sun Tzu's coverage is limited, heavily centered on characteristics of the good spymaster who must carry out that task (verse XIII.13 in Passage #11.6). One likely reason for that limitation is that much intelligence analysis in Sun Tzu's time probably amounted to exercise of intuitive judgment by one individual – namely, the spymaster (who may often have been the same person as the commanding general or ruler) – perhaps conjoined with informal conversations with that person's trusted associates.[45] Verses I.1–I.14 (Passage #11.5),

---

[44] A contrast case would be the *Arthaśāstra*, which touches on workings of a multi-tier intelligence organization. Olivelle, #1 footnote 39, p. 479 gives a flavor: "the roving spies report to those running the spy establishments, and they in turn transmit the information to the central authorities."

[45] A military staff apparatus described in a different Warring States text, the *Liutao*, might have supported a more formalized intelligence analysis process. Sawyer, *Seven Military Classics*, pp. 60–62 translates the relevant passage, which notably mentions staff "responsible for discussing divergent views." However, the extent to which the *Liutao*'s staff description depicted

supplemented by verse I.28 (Passage #11.19), might point to a more elaborate, systematic analysis process, quite possibly involving multiple people – but such a process may have taken place just once, prior to launching a war.

Sun Tzu's rather scanty coverage of intelligence interpretation and evaluation may also in part be sourced to Sun Tzu's sidelining of problems of judgment under uncertainty (see pp. 64–66). Even in the absence of probability calculus (or any other favored systematic approach, divination included) it might still be possible to offer some pointers pertinent to coping with uncertainty, obviously a critical issue in exercise of intelligence judgment.[46] But Sun Tzu does not do this.[47] A corroborator of this de-emphasis is Passage #11.26A, which brims with field intelligence advice, but allocates no attention to how to handle uncertain or conflicting observations.

As discussed on pp. 368–71 above, Sun Tzu's treatment of *fanjian* sidesteps some genuinely hard problems, showing little if any sign of engagement with rarified higher realms of counterintelligence work.

On a distinct (though related) topic, Sun Tzu says nothing about the important problem of intelligence-sharing in alliances, where a mix of cooperation and conflict (not Sun Tzu's forte) is often at the heart of what is shared and what is not.

Sun Tzu (2) ("Sun Tzu Extended") and Sun Tzu (3) ("Sun Tzu Analogical") Frontiers of Theme #11

A structured standpoint on frontiers of Theme #11 can be developed around the same analytical template introduced above: How have the four features of Sun Tzu's intelligence style and thirteen intelligence ideas from the espionage chapter fared in our own time? Which strands are enduring? Which are contested? What updates or modifications might be called for? What is missing – i.e., where are the holes?

Responses to these questions must be put in a context of a modern intelligence community as a major emergent political arena with its own political and bureaucratic dynamics. The Sun Tzu text here shows its age, harboring no hint of this fact.

Of Sun Tzu's intelligence style's four signature features (pp. 359–61 above), the first (intelligence as foundational for any major strategic undertaking) and the second (intelligence professionalism) seem, with some wobbles, to be more or less securely ensconced on the strategic landscape, worldwide. Of course, a limiting factor here is the sheer scale of modern intelligence effort by great powers, vastly outstripping, by orders of magnitude in head counts of personnel involved, anything Sun Tzu could have imagined. For that reason alone, the "most intelligent people" for which verse XIII.23 (Passage #11.6) calls are often in short supply.

---

actual military practice is not clear. See #1 footnote 2 above. (For *Liutao* dating see Background footnote 50.)

[46] There are strands of early Chinese thought that contribute relevant insights (e.g., pertaining to sources of perceptual distortion, stance vis-à-vis dissonant information, etc.), but philosophical works are more in evidence here than are Chinese military classics. See Sawyer, #1 footnote 3, Part Four, "Theories of Evaluating and Intelligence," chapters 10–14, pp. 286–423, in particular his chapter 10, "Basic theory and issues," pp. 289–308.

[47] See Theme #1 analysis (pp. 65–66 and #1 footnote 35 above).

The third ingredient of Sun Tzu's intelligence style – a focus on human intelligence (HUMINT)[48] – finds a far more ambivalent reception in the contemporary intelligence world. In many ways the position of HUMINT today is precarious, as intelligence organizations worldwide – perhaps lacking the ideal spymasters of verse XIII.13! – succumb ever more to the siren lure of high-tech collection means. These commonly mesh better than HUMINT with the operating styles of Weberian modern bureaucracies (which tend to find HUMINT disorderly and just a bit annoying).

Evaluating the pros and cons of such a stance is beyond the present study's scope, even were systematic data available to do so. Of course, Sun Tzu's early time – when even written paper records did not exist – provided scant opportunities for species of intelligence other than HUMINT. The result is to give Sun Tzu's HUMINT focus something of a Hobson's choice flavor (and even there limitations of repertoire are easy to spot).[49] But the more important point is that Sun Tzu's intuition told him, in no uncertain terms, that effective use of HUMINT (and engagement with the broader human dimensions of intelligence work to which it is a point of entry) can be a potent path to achieving major, even situation-transformative results.[50] That instinct retains vitality. Updating it to twenty-first-century conflicts should challenge a dichotomy between HUMINT and high-tech methods and mindsets. In particular, no amount of high tech is likely to remove, any time soon, the need for human intervention in the design and operation (as well as maintenance and updating) of high-tech systems, thus creating innumerable access points for hostile actors.[51] Indeed the more high tech there is, and as contractors and supply chains proliferate, the more such access points there will be (see also p. 382 below for further digital-era discussion).

The fourth ingredient of Sun Tzu's intelligence style is its "activist" focus. On its face, such a focus, updated to modern times, might suggest that Sun Tzu would wholeheartedly embrace the covert operations/active measures strand of intelligence work (as contrasted with secret intelligence-gathering alone). In fact, a more nuanced position seems called for. Particularly under modern communications conditions (now including social media) many types of covert operations, especially large or long-running ones, are prone to come to light – and to incur steep political costs, then or later, costs that can far outstrip any benefits gained. Such an upshot clashes with both Sun Tzu Theme #8 (formlessness) and Theme #3 (cheapness). For that reason, the modern legacy of Sun Tzu's activist intelligence style aligns best on what are sometimes termed "boutique" intelligence operations – small-scale (in resources needed), customized, nimble, truly secret (both now and later).

One place where the "activist" strand of Sun Tzu's intelligence style unequivocally finds a modern home is in cyber warfare (see Table 3, pp. 151–52 above on cyber warfare as a potentiator of political warfare). But the possibilities do not stop with

---

[48] Sometimes defined as "any information that can be gathered from human sources."

[49] For example, most orders and plans were probably oral (aside from those delivered to a body of troops using the mass-communications tools of Sun Tzu's time like flags or drums and gongs), so there would have been no files or written records for a spy to steal or sneak a peek at.

[50] An example would be Theme #6's archetypical concept of attacking the enemy general's mind.

[51] Here it matters little whether or not the subverted individual is formally an employee of an adversary secret service, hence a *fanjian* in the sense of double agent.

the Internet (which is commonly treated as cyber warfare's realm). Many kinds of digital environment, some never connected to the Internet (often for security reasons), are conducive to clandestine operations of many sorts that combine high efficacy with cheapness and low visibility (e.g., because their effects may not be visible for a long time, if ever). Such operations constitute a close-to-ideal-type contemporary case of the fourth stylistic feature of the "intelligence world according to Sun Tzu."

Attention now turns to how the Sun Tzu XIII ideas inventory summarized in Table 4, Part 1 (pp. 362–63) fares in the modern era. Coverage here is intended to be illustrative, not exhaustive, placing emphasis on structural issues and trends.

Intelligence idea #1: Cost-effectiveness of spying as compared with conventional military operations

Weighing contemporary applicability of Idea #1 involves a choice of metrics. If the benchmark, following Sun Tzu, is the money cost of fighting a major conventional war, almost any level of money spent on spying seems cheap by comparison. In peacetime, however – while Idea #1 retains a kernel of truth there too – the multiple-orders-of-magnitude disparity expressed by verses XIII.1–XIII.2 is shrinking, once the full costs of a high-tech intelligence apparatus (from satellites to gadgets for spies to supporting R&D, etc.) are taken into account. Simply put, spying on a great-power scale is no longer as dirt cheap as it was in Warring States China. Yet seen through a more customized lens Idea #1 retains vitality, even taking on intensified force. Compelling modern examples abound in digital high tech – e.g., how much sensitive information can leave supposedly secure premises on a portable hard drive (and how much does such a drive cost)? What sort of drastic effect might a certain tiny modification in software code have (and how much time and effort is required to make that modification as compared with the costs it may impose, now or later, especially if its effects are of low visibility and go undetected)?[52]

Also relevant to modern evaluations of Idea #1 are other yardsticks (e.g., moral costs to a society, depending on what type of spying is envisioned and its methods). However, gauging that type of cost enters conceptual territory foreign to Sun Tzu.

Intelligence idea #2: Foreknowledge as key to political–military success, and centrality of "men who know the enemy situation" to achieving it

This idea is timeless, though in the modern era what Griffith here translates as "men" should be updated to "people and machines," the latter being shorthand for computing and related high-tech capabilities and all the data they produce and analyze (including troves of open-source intelligence). Idea #2 nowadays also finds a home in economic realms no less than in political–military ones (which is, of course, why industrial espionage thrives). Read broadly and with a dash of poetic license – which is often the best way of getting the most out of Sun Tzu – Sun Tzu's verse XIII.4 subsumes a warning against basing decisions on a priori reasoning having slender (if any) relevant empirical basis. In modern times, such a warning might, for example, apply to computation-intensive simulations involving many

---

[52] This line of thinking ties back to Theme #3 "frontiers" discussion. See pp. 122–23 above.

assumptions and scant data (see p. 72 above; Table 5 on pp. 405–406 below). It could also be a caution against substituting reliance on cultural or psychological theories of an adversary's mentality or behavior for Sun Tzu's people "who know the enemy situation."[53]

Narrowing the interpretation of what verse XIII.4 stands for, it is also a heads-up that old-fashioned espionage may sometimes get results that advanced algorithms, even conjoined with big data, cannot. In particular, the still-primitive state of "general" artificial intelligence (as contrasted with machine learning, a more limited capability) suggests a continuing niche for the talented spies contemplated by verse XIII.23 – people who can observe a complex situation, read it insightfully, and then communicate an accurate intuitive grasp of what is actually going on there.

One way in which Sun Tzu's otherwise healthy and robust Idea #2 may require modern editing is an unplanned side effect of the bureaucratic phenomenon itself. A heavily bureaucratized opponent – perhaps nowadays comprising a whole land-scape of complex organizations – may *itself* not know what it is going to do, or even what it wants to do.[54] In that sense such an organization may indeed be "without ascertainable shape" (verse VI.24 in Passage #8.1). Then even Sun Tzu's best-placed, most perspicacious spy may be at a loss to provide actionable foreknowledge.

Intelligence idea #3: Botanizing spies

For twenty-first-century social structures, with their intricate and often counter-intuitive connectivities (small world networks, etc.), cyber as well as human, a search for fixed categories to which to assign spies, generalizing the spirit if not the letter of verses XIII.5–XIII.11 (Passage #11.6), does not seem particularly promising. Of course, some types of classification of spies still have important uses (say, categoriz-ing them by language skills). Some typologies of agents may also serve orientation or training purposes.[55]

---

[53] See pp. 20–21 above (noting risks of stereotyped understanding of "the" Chinese strategic culture or tradition). An insightful lens on some of the pitfalls of approaches to analyzing strategic behavior that pivot on oversimplified theories is Daniel Bell, "Ten theories in search of reality: the prediction of Soviet behavior in the social sciences," *World Politics*, 1958, 10(3), 327–65. His first section ("Enter Pirandello") leads off with a memorable challenge: "Hegel once said that what was reasonable was real. Each of the theories to be discussed seems reasonable, yet not wholly real. Something may be wrong with Hegel, the theories, or both. The reader will have to be the judge."

[54] Related analysis is pioneered in a seminal paper of Michael D. Cohen, James G. March, & Johan P. Olsen, "Garbage can organizational choice," *Administrative Science Quarterly*, 1972, 17(1), 1–25. For exploratory military applications of this line of thinking see James G. March & Roger Weissinger-Baylon (eds.), *Ambiguity and Command: organizational perspectives on military decision making* (Marshfield, MA: Pitman, 1986), also cited in Boorman, Introduction footnote 4, p. 110 note 28.

[55] For one schematic classification of HUMINT sources (not closely aligned on Sun Tzu's five categories of spy) see Herman, #1 footnote 35, pp. 61–63 and his Figure 7 (p. 63), "Humint's pyramid of source sensitivity, quantity and value," with "relatively non-sensitive, bread-and-butter [sources] at its base [Herman's 'casual travellers, experts'] and increasingly sensitive ones toward the apex," culminating in agents/informers in place (Herman, p. 61). In-between would be sources such as defectors; political opponents, exiles, alternative governments; business contacts; refugees, etc.

Classificatory niceties aside, it is worth noting that several of Sun Tzu's five types of spy (inside agents and double agents; depending on specifics, also some native agents) are all versions of the subverted insider, suggesting a level of attention to subversion by Sun Tzu that is worth keeping in mind today.

Intelligence idea #4: Concept of espionage network and its far-reaching tentacles

Possibly Sun Tzu's most useful contemporary contribution here – certainly a growing point for it – involves his terrain thinking rather than the espionage chapter (once again illustrating that Sun Tzu's thirteen chapters are often most productively approached as a unified and interconnected body of thought). For illustrative leads see Theme #2 "frontiers" (directions for social network analysis derived from Passage #2.1). Verse XIII.6, whose focus is a well-functioning spy network, points by implication to a need to analyze why some clandestine networks achieve success while others fail. While that problem is never directly addressed by Sun Tzu, some of Sun Tzu's military ideas may be repurposed to suggest durable tips into how to take an adversary's espionage network to pieces – i.e., how to make it fail. A case in point would be taking cues from Theme #6's "soft counter-strategist" mode to sideline careers of the enemy's best available intelligence or counterintelligence talent.

Intelligence idea #5: Seamless/intimate relationship between general (or ruler) and his spies

The need for such a relationship remains as great today as in Sun Tzu's time. Yet translating Sun Tzu's Idea #5 to modern contexts is not entirely straightforward.

To start with, the sheer scale of modern intelligence establishments creates barriers even to assigning a clear meaning to a "seamless," "intimate," etc. relationship between a chief of state or other top-level decision maker and an intelligence organization or (more likely nowadays) set of such organizations.[56] A modern updating of Idea #5 would also bring in a quite different cast of characters than Idea #5 as distilled from Sun Tzu. A natural primary focus of such a modern update would be intelligence officers on one's own side, nowadays a large and heterogeneous as well as bureaucratically structured population, most of whom are not "spies" in a traditional sense. Sun Tzu does not even hint at existence of such a population, placing emphasis on spies and spies (or human assets) alone. There are trust-building challenges here – big ones! – without precedent in a simpler time.

A distinct pitfall in updating Idea #5 to modern conditions involves the highly (and diversely) technical nature of modern intelligence work. In large part for that reason, risks arise when a non-professional in intelligence matters (which describes many chiefs of state, worldwide) starts to engage with raw intelligence and draw action implications from it (a practice that the Sun Tzu verses informing Idea #5 might be read as encouraging). At the same time, political considerations unknown in Sun Tzu's age (e.g., ones born of need for plausible deniability in a world of many social and other media) often operate to encourage nominal, and perhaps on some levels real, distancing between top-level decision makers and their intelligence

---

[56] The closeness of the pertinent relationship is expressed by traditional commentator Du Mu on verse XIII.12: "These are 'mouth to ear' matters." See Griffith, p. 147; Minford, p. 321.

apparatus. For both reasons, verse XIII.21, as a call for the ruler "to have full knowledge of the activities of the five sorts of agents," may not fit modern conditions.[57]

In sum, Idea #5 is a case where structural constraints of the modern era have diverged so far from the situations Sun Tzu envisioned that Sun Tzu's precepts afford limited useful guidance and could prove counter-productive. The most durable part of Idea #5 largely boils down to advice to the modern prince to invest in building mutual trust and confidence with his intelligence apparatus, whose performance – if strong – he should indeed appreciate as truly a "treasure" (as verse XIII.6 suggests).

Intelligence idea #6: Spymaster as "humanistic" role

This is an idea, which finds textual expression in verse XIII.13,

> XIII.13.
> He who is not sage and wise, humane and just, cannot use secret agents. And he who is not delicate and subtle cannot get the truth out of them.

whose perceptiveness and depth is not eroded by the passage of millennia. Here Sun Tzu still has things to teach contemporary practitioners, presenting a case of *"What did Master Sun know that we still don't (or have yet to absorb adequately)?"*[58] Idea #6 is also a somewhat subtle concept could that could be unpacked and developed in many ways, which may vary depending on whether the focus is a case officer or someone in a higher management role in a modern intelligence hierarchy.

Idea #6, however, is also one that is decidedly swimming against the modern tide, indeed several such tides. One is, of course, the bureaucratization of intelligence, which affords little scope for humanist persuasions of any kind. A second is the ever-more-high-tech nature of intelligence work. A third factor, not to be discounted, is the effect of modern Western educational systems which de-emphasize (often flatly devalue) "Great Books" learning geared toward clarification of fundamental social and civic values, in favor of structuring knowledge along expert system lines as a raft of interlocking technical specialties.

Against this backdrop, Idea #6 carries live implications for the education of future spymasters, especially ones on career tracks destined to carry them to high levels. Here steps to inculcate a broad intuitive grasp of the human condition (among them, education in areas like grand strategy) are a much-needed accompaniment to acquisition of bureaucratic and technocratic skills.[59] How to do this effectively under modern conditions is best approached as a Sun Tzu educational research frontier.

On a distinct (though related) plane, the thrust of the latter part of verse XIII.13 merits being read as weighing in against interrogation that seeks its results from

---

[57] Mair's translation of this first part of verse XIII.21 has a somewhat more circumspect flavor and for that reason may mesh better with twenty-first-century intelligence needs: "The lord must be kept well informed about the affairs of the five different types of agents" (Mair, p. 131).

[58] In a modern framing, something of the spirit of Idea #6 is conveyed by an observation of Sir Richard Livingstone, a classicist by background and World War II–era Vice-Chancellor of Oxford: "by a mere technician I mean a man who understands everything about his job except its ultimate purpose and its place in the order of the universe. They are a very common type." See his *Some Tasks for Education* (London: Oxford University Press, 1946), p. 13. Sun Tzu's spymaster is no mere technician.

[59] See Linda Kulman, *Teaching Common Sense: the Grand Strategy Program at Yale University* (Foreword by Henry Kissinger; Westport, CT: Prospecta Press, 2016).

physical abuse and harsh measures more generally.[60] Modern experience suggests that there are still lessons to be learned here in aligning the interrogation practices of twenty-first-century security services on this interpretation of Sun Tzu verse XIII.13.

Intelligence idea #7: Spying as a most delicate endeavor

This Sun Tzu idea retains full force. Indeed the delicacy of managing the human relations at the heart of HUMINT – which is Sun Tzu's focus here and in Idea #6 – is now joined by further species of delicacy required by the management and manipulation of the digital systems and structures that permeate twenty-first-century spy work. The latter type of delicacy has many facets. Among them are the vagaries of memory preserved in digital devices, which may (often more readily than pre-digital forms of memory) be manipulated in nuanced ways for strategic purposes; possibilities for outsized effects from small software or algorithm modifications (previously discussed in Theme #3 context, see pp. 122–23 above); further possibilities for low-visibility tinkering with hardware; and, of course, the challenges of positioning vis-à-vis the volatile dynamics of social media.

All these demands mean that, for spymasters, a phenomenal number of things can go wrong, requiring exquisite titration of risk, intellectual suppleness, and (ideally) ability to learn constructive lessons from failure. One stumbling block is a tendency of many very real effects of digital technology to seem just a little "unreal" to those who did not grow up with that generation of the technology. Given the fast pace of the technology's evolution, that statement might apply to many senior intelligence officials, worldwide (perhaps even more so, to leaders and policymakers they serve).

Viewed from this perspective, Idea #7 suggests a compelling case for continuing education in relevant technology issues and skill-sets that allows no exemptions for seniority or high rank. Indeed the same continuing education model might usefully be applied to the modern prince (e.g., national leader) and his immediate circle.

Intelligence idea #8: Opportunities for spying as ubiquitous – spying as universal in scope and utility

At base, this is a philosophical stance, viewing the world as a network of possibilities for spying, great and small. Unchanging basics of human nature give an "evergreen" quality to Sun Tzu's Idea #8, making its core insight as pertinent now as in the Warring States era. At the same time, Idea #8 is reinforced and extended by digital trends that vastly enlarge the cast of characters who can enter the spy game in a significant way. Those trends are placing in a host of hands significant clandestine and covert operations capabilities that, as recently as mid-twentieth century, would have been out of reach to all but a handful of statesmen, soldiers, and intelligence officials, along with a few trusted associates, worldwide.

A taste of Idea #8, though certainly not its sum total, would treat *any person as a running log (cumulative record) of personal vulnerabilities, including errors that person has ever made* (compare Table 3 on pp. 151–52 above). Digital capabilities are now increasingly making that record legible to all who care to look.

---

[60] However, as cautioned in Passage #10.20 analysis (p. 356 above), such a reading – while not inconsistent with what Sun Tzu says in verse XIII.13 – may not reflect Sun Tzu's intended meaning.

Intelligence idea #9: Imperative of effective command control of secret information, ensuring both authorized access and denial of unauthorized access

While the basic concept is timeless, methods for attempting to implement it have shifted drastically since Sun Tzu's time in a direction of complex legal, organizational, and technological measures used to create and support a classified information system. This is conceptual territory – at whose heart stands design of a structure – unexplored by Sun Tzu who, as previously noted (pp. 66–67, 146, 193, etc.), is not at all a builder by instinct. Yet far from being ideal structures, classified information systems commonly harbor deep flaws that often accumulate and ramify as those systems age. The approaches to information security they embody tend to get out of control – too much classified information, often vastly overclassified, yet also too many people with authorized access to it, creating logistics and security problems galore. Of course, over-complexity of a classified information system is a modern phenomenon about which Sun Tzu has nothing directly to say. However, such a system offers a new, rich and complex, arena for applying Sun Tzu's principles, which point to creative manipulations that turn to advantage the system's failures. Extrapolating Sun Tzu's thinking into Sun Tzu (3) territory, here is a candidate for *"What did Master Sun know that we still don't (or have yet to absorb adequately)?"*

Intelligence idea #10: Affinity between spying and data, both accurate and false

This is an outstandingly modern harmonic, one where Sun Tzu's thinking aligns well on modern conditions. Building on Sun Tzu's HUMINT emphases, one major Sun Tzu–inspired direction for developing the accurate data facet of Idea #10 would involve collecting and exploiting biographical intelligence (see verse XIII.16 as well as Passages #11.11 and #11.12).[61] Modern data resources, many of them publicly available (e.g., social media footprints), superbly lend themselves to developing such intelligence (possibly in combination with social network analysis, itself a kind of offshoot of the biographical intelligence concept). Indeed the point could be put more strongly. Taking to heart the "know the enemy and know yourself" counsel of verse III.31 (Passage #11.1), and with a weather eye on tendencies in game theory and other rational choice literature to approach strategic problems in a depersonalized, disembodied way, biographical intelligence – including its collective biography extensions – is a further candidate for *"What did Master Sun know that we still don't (or have yet to absorb adequately)?"*

Turn next to false "data." In light of acute twenty-first-century concerns about "fake news," the starring role Sun Tzu accords to disseminating false information (verse XIII.19) – an "active measure" that could be regarded as the culmination of Sun Tzu XIII – is squarely on target for modern conditions, even prescient. Some update of verse XIII.19 is called for since Sun Tzu attributes to double agents the knowledge needed to craft such fabricated information, which expendable agents (a different Sun Tzu category) then spread.[62] Under twenty-first-century conditions

---

[61] This type of intelligence has been defined, in antiseptic governmental prose, as "that component of intelligence that deals with individual foreign personalities of actual or potential importance."

[62] Here, however, a distinction should be kept in mind. The target of the fabricated information to which verse XIII.19 alludes is presumably the only decision maker who matters to Sun Tzu,

that focus seems far too narrow. Nowadays open-source intelligence may be able to provide all the insights needed to create credible false information, which the Internet can then proceed to circulate without further strategic intervention being needed.

Intelligence idea #11: Central role of *fanjian* in successful espionage

Taken at face value, Sun Tzu's starring role for *fanjian* – "enemy spies whom we employ" (verse XIII.9) – in his firmament of spies is at the very least controversial, possibly downright wrong-headed. Under modern conditions even a resolutely HUMINT focus need not by any means pivot on running double agents. Modern HUMINT – often in combination with information from non-HUMINT quarters, much of it from specialized high tech, much also publicly available on the web – harbors innumerable intelligence opportunities unknown to Sun Tzu. They include ways of secret agent recruitment (verse XIII.18), deception steps (verse XIII.19), and uses of agents in place (verse XIII.20) that, contrary to Sun Tzu XIII's affirmations, do not in any way depend on double agents. Meanwhile the historical record relating to double agent uses suggests that this highly specialized line of intelligence work should be approached in a caveat emptor spirit, definitely not expecting too much by way of tidy or even actionable results. Today as in Sun Tzu's time, central here is the often poorly tractable challenge of verifying where a double agent's actual loyalties, if any, reside.

Yet even in early Chinese usage the *fanjian* concept showed signs of overflowing the boundaries set for it by Sun Tzu's verse XIII.9 definition.[63] Taking cues from the "deception agent" understanding of *fanjian* proposed by Frank Kierman,[64] a modern candidate for a *fanjian* would be a subverted computer system or network administrator, who may be in a position to carry out modern versions of the kind of deception steps that Sun Tzu aspired to accomplish with help from *fanjian* (indeed very possibly, with support from contemporary technology, a lot better!).[65]

A more radical updating of the *fanjian* concept – perhaps increasingly pertinent in a world of machine intelligence – would pose the question of where the "loyalties" of a machine (or computer network) lie, entertaining the possibility that those loyalties might stay "turned" long after the human being who turned them is out of the picture – retired, deceased, even unmasked, etc. In a society of algorithms, the strategic implications run deep. They remain neglected in most modern treatments of strategy, even ones that combine well with espionage motifs (like Liddell Hart's *Strategy*, which took shape too soon in the long twentieth century to take account of impacts of digital technology).

---

namely, the enemy general (or ruler). By contrast, the target of modern "fake news" is commonly a much larger and commonly more diffuse audience or set of audiences.

[63] Sawyer, #1 footnote 3, pp. 89, 135, 166 notes a range of possibilities. See also #11 footnote 29.

[64] See #11 footnote 34 (citing *Shiji* study by Kierman).

[65] Beyond this, a modern benefit of early Chinese conceptual confusion – or flexibility! – regarding the meaning and scope of application of *fanjian* is to encourage more and deeper thinking about what, exactly, constitutes a human intelligence "asset" and what such an asset may realistically be expected to accomplish within the "iron cage" of constraints imposed by a highly developed social structure. The Eccles–Rosinski framework (pp. 187ff.) offers a starting point for such an inquiry.

Intelligence idea #12: Potential of well-placed figures having inside information to bring about destabilization of political power at the center, even regime change, should their loyalties shift

The thrust of Idea #12, whose political sensitivity is obvious, has as many active implications for our own time as for Sun Tzu's. In a pattern not unknown in pioneering doctrine statements, a major offshoot of Idea #12 is left untouched by Sun Tzu: namely, how a political authority's reliance on secret intelligence – a reliance that Sun Tzu's thinking strongly encourages – might foster a situation where the intelligence apparatus, or part of it, starts to pursue agendas of its own. While those agendas might involve toppling the visible center of power (the focus of verse XIII.22, involving famous dynastic transitions in early Chinese history), there could be other agendas comparably corrosive in their long-term effects (e.g., low-visibility displacement of the ostensible power center by a rival one; or exploitation of intelligence secrecy as cover for unauthorized side projects, conceivably major enough to create systemic risk). In a modern context, concerns like these point to unending need to monitor the structure and dynamics of a complex intelligence apparatus. Reflecting a simpler time, Sun Tzu, via verse XIII.21, has just one idea to contribute here – namely, that *fanjian* may play a role in unmasking unauthorized activities by a leader's own intelligence people.

Idea #13: Affinity between successful espionage and talents of human beings used as spies (or in other observer roles)

Particularly as regards intelligence deemed "strategic," there is a timeless quality to this insight. It applies to more than spies or spying in a narrow sense. The core insight – playing up the immense importance of the talents of an observer, possibly a participant one – has relevance to other kinds of observers (journalists, for example) who gather open-source information on the "atmospherics," as well as more specific details, of tangled conflict-fraught situations.[66] Putting the point with a bow to order-of-magnitude thinking (see Theme #3), a highly talented observer may be able to glean, not just slightly more but an order of magnitude more useful insight into a pertinent situation than one less talented would produce.

There is no question that modern intelligence establishments have a voracious need for personnel of very high quality, including personnel with high observational skills. While that principle is widely endorsed, there remains important room for improvement in putting it into practice. An important modern implication of verse XIII.23 is to challenge tendencies – not unknown in modern security services, including those of Western democracies – to reject or marginalize otherwise highly

---

[66] An example comes from Hong Kong of the 1950s, of which it has been written that

Hong Kong offers a research base where it is possible, given discipline and discrimination on the part of the observer, to acquire this 'feel' of Communist China's living problems, its tensions, aspirations, and frustrations. The colony offers practical advantages as well, for it provides regular contact with an experienced community of observers directly concerned with China (Asian, British, European, and American government officers, journalists, and businessmen) as well as occasional contact with diplomatic and other travelers from China.

See p. 598 of Howard L. Boorman, "The study of contemporary Chinese politics: some remarks on retarded development," *World Politics*, 1960, 12(4), 585–99.

qualified agents or potential agents, including personnel of their own services who have a dark skin color; a "wrong" religious faith; or who otherwise fall in some societally or politically disfavored category. In twenty-first-century security services such a fallacy can greatly impair the operational effectiveness toward which Sun Tzu's teaching is relentlessly directed. Here is a major example of "*What did Master Sun know that we still don't (or have yet to absorb adequately)?*"

★★★★★★

Set of principal passages used to illustrate Sun Tzu Theme #11 (know the enemy)

Passage #11.1. Know the Enemy and Know Yourself

III.31. Therefore I say: "Know the enemy and know yourself; in a hundred battles you will never be in peril.

III.32. When you are ignorant of the enemy but know yourself, your chances of winning or losing are equal.[67]

III.33. If ignorant both of your enemy and of yourself, you are certain in every battle to be in peril."

Passage #11.2. Know the Enemy, Know Yourself, Adding Environmental Context

(Value-added beyond Passage #11.1 comes from Passage #11.2's emphasis on knowing the physical environment – and, if "terrain" and "weather" are understood metaphorically, the sociopolitical environment too – in which conflict operations are unfolding.)

X.22. If I know that my troops are capable of striking the enemy, but do not know that he is invulnerable to attack, my chance of victory is but half.[68]

X.23. If I know that the enemy is vulnerable to attack, but do not know that my troops are incapable of striking him, my chance of victory is but half.

X.24. If I know that the enemy can be attacked and that my troops are capable of attacking him, but do not realize that because of the conformation of the ground I should not attack, my chance of victory is but half.

X.25. Therefore when those experienced in war move they make no mistakes; when they act, their resources are limitless.

X.26. And therefore I say: "Know the enemy, know yourself; your victory will never be endangered. Know the ground, know the weather; your victory will then be total."

---

[67] Here and in Passage 11.2, no probabilistic thinking on Sun Tzu's part should be inferred. Roughly speaking, verse III.32 points to a medium-risk "win some, lose some" situation, to be contrasted with that described in verse III.33, which is high-risk.

[68] Giles, p. 111 translates the ending of verse X.22 "... we have gone only halfway towards victory" (and similarly for verses X.23 and X.24). That translation accords with the Chinese text's thrice repeated *sheng zhi ban ye* 勝之半也, lit. "victory's half."

Passage #11.3. Role for Intelligence in Spotting Enemy Vulnerabilities; Sense of the Limits of Intelligence

(Through their emphasis on enemy vulnerabilities, and knowing what they are, these verses are a window on Sun Tzu's intelligence-oriented style of war.)

IV.1. Anciently the skilful warriors first made themselves invincible and awaited the enemy's moment of vulnerability.

IV.2. Invincibility depends on one's self; the enemy's vulnerability on him.

IV.3. It follows that those skilled in war can make themselves invincible but cannot cause an enemy to be certainly vulnerable.

IV.4. Therefore it is said that one may know how to win, but cannot necessarily do so.

Passage #11.4. Broader Vistas for Intelligence Work: Winning without Fighting

III.3. For to win one hundred victories in one hundred battles is not the acme of skill. *To subdue the enemy without fighting is the acme of skill.* [Emphasis supplied.]

Passage #11.5. Agenda for Strategic Intelligence

(Passage #11.5 may be read as profiling what knowledge of the enemy side one needs to have as part of a larger, inherently comparative task of net assessment. Viewed through that lens, verses I.11–I.14 set forth a strategic intelligence agenda with specific intelligence targets. That agenda is notably broad, including political intelligence [verses I.4 and I.11], biographical intelligence [verses I.7 and I.11], and organizational and logistics intelligence [verses I.8 and I.11–13].)

(Passage #11.5 = verses I.1–I.14 = Passage #1.1 on pp. 74–76 above)

Passage #11.6. Sun Tzu on Espionage

(This is the entirety of Sun Tzu XIII. It may be a late addition to the Sun Tzu text.)

XIII.1. Now when an army of one hundred thousand is raised and dispatched on a distant campaign the expenses borne by the people together with the disbursements of the treasury will amount to a thousand pieces of gold daily. There will be continuous commotion both at home and abroad, people will be exhausted by the requirements of transport, and the affairs of seven hundred thousand households will be disrupted.

XIII.2. One who confronts his enemy for many years in order to struggle for victory in a decisive battle yet who, because he begrudges rank, honours and a few hundred pieces of gold, remains ignorant of his enemy's situation, is completely devoid of humanity. Such a man is no general; no support to his sovereign; no master of victory.

XIII.3. Now the reason the enlightened prince and the wise general conquer the enemy whenever they move and their achievements surpass those of ordinary men is foreknowledge.[69]

XIII.4. What is called "foreknowledge" cannot be elicited from spirits, nor from gods, nor by analogy with past events, nor from calculations.[70] It must be obtained from men who know the enemy situation.

XIII.5. Now there are five sorts of secret agents to be employed. These are native, inside, doubled, expendable, and living.

XIII.6. When these five types of agents are all working simultaneously and none knows their method [dao] of operation, they are called "The Divine Skein" and are the treasure of a sovereign.

XIII.7. Native agents are those of the enemy's country people whom we employ.

XIII.8. Inside agents are enemy officials whom we employ.

XIII.9. Doubled agents [fanjian] are enemy spies whom we employ.

XIII.10. Expendable agents are those of our own spies who are deliberately given fabricated information.

XIII.11. Living agents are those who return with information.

XIII.12. Of all those in the army close to the commander none is more intimate than the secret agent;[71] of all rewards none more liberal than those given to secret agents; of all matters none is more confidential than those relating to secret operations.

XIII.13. He who is not sage and wise, humane and just, cannot use secret agents. And he who is not delicate and subtle cannot get the truth out of them.[72]

XIII.14. Delicate indeed! Truly delicate! There is no place where espionage is not used.

XIII.15. If plans relating to secret operations are prematurely divulged the agent and all those to whom he spoke of them shall be put to death.[73]

---

[69] The attention that Sun Tzu XIII gives to the ruler is worth noticing. The ruler appears alone (or is mentioned before the general) in verses XIII.3, XIII.6, XIII.21, and XIII.23. Much of the rest of the Sun Tzu text (with the exception of the latter part of Sun Tzu XII, also a late addition to the text in Brooks's order of chapter accretions, see Table 12A on p. A-35) aligns much more readily on the perspective of the general than on that of the ruler. If the ruler appears at all, he is often cast as a source of interference with the general's task. Perhaps here is a clue that Sun Tzu XIII is work of another hand?

Drawing a contrast between Sun Tzu's focus on the general and focus on the ruler in the "Methods of War" chapter of a Legalist classic, see work of Yuri Pines cited in Appendix 8 anchor footnote 19.

[70] Understandings vary as regards the content of these several categories, particularly the segment Griffith translates as "analogy with past events." See Appendix 1 anchor footnote 25. Table 5 on pp. 405–406 below revisits relevant issues from a generalized standpoint.

[71] The specifics are open to varying interpretations. Griffith comes close to casting the commanding general as a case officer (which, in the armies of Sun Tzu's time, may well often have been the case). Ames, p. 170 formulates the relationship in a more open-textured way: "Thus, of those close to the army command, no one should have more direct access than spies ..."

[72] Sawyer, p. 232: "... he cannot perceive the substance in intelligence reports."

[73] Expanding on a comment in Passage #10.11 discussion of verse XIII.15 (p. 351 above), it is worth challenging Sun Tzu's rather blunt-edged advice here. Unless it is thoroughly clear what

XIII.16. Generally in the case of armies you wish to strike, cities you wish to attack, and people you wish to assassinate, you must know the names of the garrison commander, the staff officers, the ushers, gate keepers, and the bodyguards.[74] You must instruct your agents to inquire into these matters in minute detail.

XIII.17. It is essential to seek out enemy agents who have come to conduct espionage against you and to bribe them to serve you. Give them instructions and care for them. Thus doubled agents [fanjian] are recruited and used.[75]

XIII.18. It is by means of the doubled agent [fanjian] that native and inside agents can be recruited and employed.

XIII.19. And it is by this means that the expendable agent, armed with false information, can be sent to convey it to the enemy.

XIII.20. It is by this means also that living agents can be used at appropriate times.[76]

XIII.21. The sovereign must have full knowledge of the activities of the five sorts of agents. This knowledge must come from the doubled agents [fanjian], and therefore it is mandatory that they be treated with the utmost liberality.[77]

---

the agent's game was (which implies more effective interrogation methods than probably existed then, or even today), imposing a death penalty would foreclose gleaning any future insights from that quarter into that game's nature, scope, and extent of success – not necessarily the best advice in Sun Tzu's world of double agentry and betrayal!

[74] The specifics here are, of course, geared to relevant roles as they existed in Sun Tzu's time. For discussion of those roles see Passage #2.2 (pp. 107–109 above), where verse XIII.16 also appears.

[75] Reflecting an interpretive ambiguity in a key Chinese character here (see Appendix 11 footnote 8) there is a split of opinion on how to understand the specifics. Mair's translation (Mair, p. 130) reads:
XIII.17.
I must further seek out the enemy agents who come to spy on us, ply them with benefits, induce them to come over to our side, and *then release them back to the enemy.*
[Emphasis supplied.]
This alternative interpretation appears to advise forgoing the strings on double agents – big ones! – that come with having physical custody of those individuals (who, upon release back to the enemy, might easily prove fickle or otherwise unreliable). Absent some other consideration that Sun Tzu does not supply, the logic of the situation appears to favor Griffith's understanding.

[76] Sun Tzu appears to be pointing here to timing factors in spy work (which is often the key to effective integration with uses of other forms of power, hence the efficacy of spies). A passage from the *Romance of the Three Kingdoms* (Moss Roberts translation cited on p. xxiv above), p. 662, has Zhuge Liang upbraiding a spy he has caught: "Why have you missed the date? A good spy can't afford to be so careless!"

[77] Inferentially (though other interpretations exist), a key reason "this knowledge must come from the double agents [fanjian]" is that they know what the other side knows, hence are in a good position to know about any unauthorized activities by the agents on one's own side. Of course, that line of thinking idealizes what *fanjian* can be expected to know (compartmentalization of intelligence information is not just a modern invention; cf. also the thrust of verse XIII.12). It also presumes that the reliability of *fanjian* is known, a point discussed in detail on pp. 368–71 above.

XIII.22. Of old, the rise of Yin [Shang dynasty] was due to Yi Zhi, who formerly served the Xia [dynasty]; the Zhou [dynasty] came to power through Lü Ya, a servant of the Yin.[78]

XIII.23. And therefore only the enlightened sovereign and the worthy general who are able to use the most intelligent people as agents are certain to achieve great things. Secret operations are essential in war; upon them the army relies to make its every move.

Passage #11.7. Need for Field Intelligence; Aggressive Reconnaissance

(These verses showcase Sun Tzu the practical land warfare soldier, alert to the importance of reconnaissance and scouting, along with anti-scouting measures to frustrate the enemy's scouting efforts.)

VI.13. If I am able to determine the enemy's dispositions while at the same time I conceal my own then I can concentrate and he must divide. And if I concentrate while he divides, I can use my entire strength to attack a fraction of his. There, I will be numerically superior. Then, if I am able to use many to strike few at the selected point, those I deal with will be in dire straits.

VI.14. The enemy must not know where I intend to give battle. For if he does not know where I intend to give battle he must prepare in a great many places. And when he prepares in a great many places, those I have to fight in any one place will be few.

... (knowledge or information conditions are not explicit in the elided verses VI.15–VI.16)

VI.17. If one knows where and when a battle will be fought his troops can march a thousand li and meet on the field. But if one knows neither the battleground nor the day of battle, the left will be unable to aid the right, or the right, the left; the van to support the rear, or the rear, the van. How much more is this so when separated by several tens of li, or, indeed, by even a few!

... (knowledge or information conditions are not explicit in the elided verses VI.18–VI.19)

VI.20. Therefore, determine the enemy's plans and you will know which strategy will be successful and which will not;

VI.21. Agitate him and ascertain the pattern of his movement.[79]

---

[78] For historical and other background on verse XIII.22 see Passage #5.19 (pp. 205–206 above).

[79] The *Wuzi* text (see Background footnote 27) goes substantially further than Sun Tzu in describing a reconnaissance operation carefully crafted to probe the enemy general's deception abilities:

> Order bravos in command of some *élite* troops to try him [the enemy] out. Their sole purpose is to flee, *not to gain anything but to observe how the enemy reacts*. If his actions are in unison and his discipline good, and when he pursues and pretends to be unable to catch up, when he sees an advantage but pretends to be unaware of it, then the [enemy] general is wise and you should not engage him.

> (Griffith, p. 163, emphasis supplied)

VI.22. Determine his dispositions and so ascertain the field of battle.

VI.23. Probe him and learn where his strength is abundant and where deficient.

Passage #11.8. Imagination and Resourcefulness in Intelligence Activities

(Generalizing beyond a military context, verse V.6 can be read as encouragement from Sun Tzu's thinking about qi and zheng [Theme #14] for a strategic actor to explore creative but untested intelligence capers provided that they are suitably "cheap" [Theme #3].[80])

V.5. Generally, in battle, use the normal force [zheng] to engage; use the extraordinary [qi] to win.

V.6. Now the resources of those skilled in the use of qi are as infinite as the heavens and earth; as inexhaustible as the flow of the great rivers.[81]

V.7. For they end and recommence; cyclical, as are the movements of the sun and moon. They die away and are reborn; recurrent, as are the passing seasons.

Passage #11.9. Flexibility and Adaptiveness in Intelligence Activities

(Read in an analogical spirit – a reading possible in Sun Tzu's time no less than in our own – these verses have important intelligence content, suggesting that intelligence work no less than warfare itself should emulate the opportunism of water, ever seeking out weakness. Verse VI.26 is particularly good guidance for intelligence work in realms of psychology and information.)

VI.26. Therefore, when I have won a victory I do not repeat my tactics but respond to circumstances in an infinite variety of ways.

VI.27. Now an army may be likened to water, for just as flowing water avoids the heights and hastens to the lowlands, so an army avoids strength and strikes weakness.

VI.28. And as water shapes its flow in accordance with the ground, so an army manages its victory in accordance with the situation of the enemy.

VI.29. And as water has no constant form, there are in war no constant conditions.

VI.30. Thus, one able to gain the victory by modifying his tactics in accordance with the enemy situation may be said to be divine [shen].[82]

VI.31. Of the five elements, none is always predominant; of the four seasons, none lasts forever; of the days, some are long and some short, and the moon waxes and wanes.

---

[80] Going beyond the letter of the text, a possible implication of verse V.6's river image is that many (most?) qi steps will yield little intelligence return, even as much of the flow of great rivers has historically gone to waste from a standpoint of advancing human purposes. For water control projects as "one of the great themes of Chinese history" (language is Wilkinson's) see #3 footnote 34.

[81] Mair, p. 92 uses the expressive idea of "repertoire" to capture a facet of verse V.6 that also applies to intelligence work writ broadly: "Therefore, he who is good at devising unconventional tactics has a repertoire that is as unlimited as heaven and earth, as inexhaustible as the rivers and streams."

[82] Ames, p. 127 translates shen as "inscrutable," which works better here. See #5 footnote 97.

Passage #11.10: Affinity of Spies and Covert Operators for Complex Environments and the Opportunities for Good Cover They Afford

(Read both narrowly and in a broader analogical spirit, verse IX.18 is an ecological vantage on spying. A generalizable idea is that environments characterized by high complexity and irregularity provide natural cover for spying and active measures of many sorts. See pp. 242, 243–45, and pp. 253–55 above. In the background is the larger point that establishing and maintaining good cover is often the greatest challenge of secret intelligence work.)

IX.18. When on the flanks of the army there are dangerous defiles or ponds covered with aquatic grasses where reeds and rushes grow, or forested mountains with dense tangled undergrowth you must carefully search them out, for these are places where ambushes are laid and spies are hidden.[83]

Passage #11.11. Knowing the Enemy General; Psychological Operations Targeting Him

(Sun Tzu's focus on human factors is much in evidence here. Read broadly, these verses are a placeholder for psychological operations targeting any aspect of an adversary's personality deemed exploitable. Viewed through a lens of human development and the life course, they also broadly encourage gathering, analyzing, and actively exploiting biographical intelligence.[84])

VIII.17. There are five qualities which are dangerous in the character of a general.

VIII.18. If reckless, he can be killed;

VIII.19. If cowardly, captured;

VIII.20. If quick-tempered you can make a fool of him;

VIII.21. If he has too delicate a sense of honor you can calumniate him;

VIII.22. If he is of a compassionate nature you can harass him.

VIII.23. Now these five traits of character are serious faults in a general and in military operations are calamitous.

VIII.24. The ruin of the army and the death of the general are inevitable results of these shortcomings. They must be deeply pondered.

Passage #11.12. Further Psychological Operations

(Putting verse I.25 into practice may call for influencing multiple actors in a coordinated way to induce desired estrangements, a task that may call for the kind of delicate judgment highlighted in Passage #11.6's verse XIII.14.)

---

[83] Mair, p. 110 renders the end of verse IX.18 as "places where ambushes and snipers are found." For further discussion of the "snipers" possibility, which points in a direction of "active measures" in an intelligence sense, see #6 footnote 5.

[84] Although the bulk of that tradition postdates Sun Tzu, the multistranded Chinese biographical tradition provides further relevant context. For overviews see Wilkinson, §9, "Biography," pp. 155–80 and D. C. Twitchett, "Chinese biographical writing," in W. G. Beasley & E. G. Pulleyblank (eds.), *Historians of China and Japan* (London: Oxford University Press, 1961), pp. 95–114.

I.22. Anger his general and confuse him.

I.23. Pretend inferiority and encourage his arrogance.

I.24. Keep him under a strain and wear him down.

I.25. When he is united, divide him.

Passage #11.13. Classificatory and All-Source Instincts in Intelligence Work

(Verse XI.51 points to intelligence needs spanning disparate types and sources of intelligence, thereby staking a claim to be a remote ancestor of classificatory as well as all-source instincts in modern intelligence work.[85] The latter's signature emphasis on analytical integration of information from diverse sources is not explicit in verse XI.51, though might be inferred from its warning about substandard generalship.[86])

XI.51. One ignorant of the plans of neighbouring states cannot prepare alliances in good time; if ignorant of the conditions of mountains, forests, dangerous defiles, swamps and marshes he cannot conduct the march of an army; if he fails to make use of native guides he cannot gain the advantages of the ground. *A general ignorant of even one of these three matters is unfit to command the armies of a Hegemonic King.*[87] [Emphasis supplied.]

Passage #11.14. Overconfidence and Related Psychological Pitfalls

(A basic harmonic here is avoiding overconfidence, possibly born of reliance on overly simplistic planning factors like troop counts or other preconceived ideas.[88]

---

[85] See Herman, #1 footnote 35, chapter 6, "All-source analysis and assessment," pp. 100–112, describing (p. 100) all-source intelligence work as involving analysis of "information from a wide variety of sources, both overt and covert." He also emphasizes that "[u]nlike collection, there is nothing esoteric about all-source analysis itself."

[86] A somewhat different genre of analytical integration, showing by example that the instinct was not foreign to early Chinese warfare, is the archaeologically recovered Mawangdui military map of c. 181–168 BC, a time period within living memory of the Warring States era. See pp. 35–36 above.

[87] Sun Tzu's emphasis on local guides, here and in verse VII.11 (Passage #2.9A), should serve as a salutary reminder of Sun Tzu's early time. Relevant context is provided by Robin Waterfield, *Xenophon's Retreat: Greece, Persia, and the end of the Golden Age* (Cambridge, MA: Belknap Press of Harvard University Press, 2006), p. 133:

Familiar as we are with maps, it is hard for us to project our minds back to an era when local knowledge was the only reliable resource – and when without it whole armies could get lost, pass close by sources of food without knowing it, or suffer the morale-draining terrors of uncertainty. Every day of the retreat, the Greeks [in Xenophon's army] sent out scouts . . . But they also needed local guides, and preferably more than one at a time, so that they could cross-check, even when the guides had been required to give them sacred pledges of their honesty and sometimes hostages too.

[88] Regarding overconfidence as a widespread cognitive phenomenon, see p. 432 of Dale Griffin and Amos Tversky, "The weighing of evidence and the determinants of confidence," *Cognitive Psychology*, 1992, 24(3), 411–35 (references omitted):

The significance of overconfidence to the conduct of human affairs can hardly be overstated. Although overconfidence is not universal, it is prevalent, often massive, and difficult to eliminate . . . This phenomenon is significant not only because it demonstrates the

Verse XI.45 amounts to a forerunner of modern critiques of "bean-counting" versions of net assessment that rely on counts of military resources or assets.[89])

IX.45. In war, numbers alone confer no advantage. Do not advance relying on sheer military power.

IX.46. It is sufficient to estimate the enemy situation correctly and to concentrate your strength to capture him. There is no more to it than this. *He who lacks foresight and underestimates his enemy will surely be captured by him.* [Emphasis supplied.]

Passage #11.15. Signal vs. Noise

(Verse V.17 points to a noise-rich environment, with verse V.18 suggesting the further idea that such noise may be artificially created to confuse the enemy.[90] Because no intelligence analyst role is specified, the challenge of distinguishing "signal" from "noise" is seemingly left up to the general himself, part of his "expedient assessment" of battlefield developments discussed on pp. 58–59 above.)

V.17. In the tumult and uproar the battle seems chaotic, but there is no disorder; the troops appear to be milling about in circles but cannot be defeated.

V.18. Apparent confusion is a product of good order; apparent cowardice, of courage; apparent weakness, of strength.

V.19. Order or disorder depends on organization; courage or cowardice on circumstances; strength or weakness on dispositions.

Passage #11.16. Focus on Enemy Capabilities, Not Enemy Intentions

(With an eye on civilian applications of Sun Tzu's ideas, it is worth noting that verse VIII.16's counsel, if interpreted as a crisp rejection of basing actions on an estimate of another strategic actor's intentions, fits military intelligence situations better than most civilian ones. In the latter, social context makes some sort of reliance on an estimate of others' intentions virtually unavoidable.[91])

discrepancy between intuitive judgments and the laws of chance, but primarily because confidence controls action . . .

[89] For brief critique of "bean counting" see p. 131 of Joshua M. Epstein, *Measuring Military Power: the Soviet air threat to Europe* (London: Taylor & Francis, 1984), chapter V, "The art of war and the craft of threat assessment," pp. 131–72.

[90] Verse V.18 need not involve manufactured "noise," though it can be read in that way (as can verse V.17). See Appendix 7 anchor footnote 14 (verse V.18) and footnote 13 (verse V.17). Plainly pointing to artfully created noise is verse IX.21 (Passage #7.5; also Passage #11.26A), whose focus is a foe's planting "many obstacles in the undergrowth" to achieve a disorienting effect. See #7 footnote 68. An example involving a dust cloud, from a *Zuozhuan* passage, is quoted in Appendix 11 footnote 30.

[91] See Passage #4.28A (where verse VIII.16 is critiqued from a "dual use" standpoint). See also pp. 212–14, weighing alternative ways of interpreting verse VIII.16.

VIII.16. It is a doctrine of war not to assume the enemy will not come, but rather to rely on one's readiness to meet him; not to presume that he will not attack, but rather to make one's self invincible.

Passage #11.17. Incongruity Testing

(This Sun Tzu verse, which also appears as Passage #7.15, points to spotting and analyzing incongruities in a situation, a task close to the heart of intelligence analysis.[92] It is worth noting that verse IX.44 immediately precedes the two verses in Passage #11.14, which also have an intelligence analysis focus. As Cao Cao's commentary suggests, the enemy antics profiled in verse IX.44 might be "a ruse to gain time for some unexpected flank attack or the laying of an ambush" [Minford, p. 245]. The onus is then on the observer – the opposing general or his adviser – to size up potential trap doors lurking in this situation, rather than getting mesmerized by an attention-grabbing performance. In a variant scenario, military-historical analysis of ancient warfare suggests that soldiers' instincts of fear and self-preservation, dampening eagerness to engage in close combat, might in and of themselves have been enough to produce the kind of noisy stand-off depicted in verse IX.44, conceivably lasting for hours.[93] The evaluation for which verse IX.44 calls then needs to address the following nuanced question: is the enemy's observed behavior attributable to a natural hesitancy of his soldiers to engage in close combat, hence not particularly incongruous under the circumstances, or on scrutiny of the situation does incongruity persist – e.g., might there be an enemy stratagem in the making?)

IX.44. When the enemy troops are in high spirits,[94] and, although facing you, do not join battle for a long time, nor leave, you must thoroughly investigate the situation.

Passage #11.18. Creative Early Warning Methodology

(Long before big data and machine intelligence to analyze it, Sun Tzu is here seen doing the best he can to craft a type of early warning methodology, geared to avoiding a scenario where an army gets caught in a flash flood situation.[95] More broadly, presence of verse IX.15a – from what Brooks's analysis indicates is the earliest layer of the Sun Tzu text – helps establish a basic place for weather intelligence in Sun Tzu's way of war. Discerning a comparably basic Sun Tzu (2) or (3) role for climate and other environmental intelligence is an extrapolation, but one that finds some support in Passage #11.2's verse X.26. From a still

---

[92] Compare Harris, #7 footnote 89, p. 558 ("incongruity testing is the guts of intelligence analysis").

[93] See, e.g., Sabin, Background footnote 30, p. 72 (also citing classic study of du Picq, #6 footnote 89).

[94] The term Griffith renders as "high spirits" is nu, more standardly rendered "angry," "fury," etc.

[95] Verse IX.15a is excluded by Griffith from the main text of his translation. See Appendix 11 footnote 13, critiquing Griffith's decision to do so.

broader standpoint, this material shows Sun Tzu in sync with the Duke of Wellington's famous agenda: "All the business of war, and indeed all the business of life, is to endeavour to find out what you don't know by what you do; that's what I called 'guessing what was at the other side of the hill.'"[96])

IX.15a. When rain falls in the upper reaches of a river and *foaming water descends* those who wish to ford must wait until the waters subside.[97] [Emphasis supplied.]

Passage #11.19. Quantitative Intelligence Analysis

(This verse establishes Sun Tzu's place in history as a pioneering advocate for bringing quantitative analysis to bear on intelligence needs – a concept unrivaled in writings found in most ages of war down to modern times.)

I.28. Now if the estimates made in the temple before hostilities indicate victory it is because calculations show one's strength to be superior to that of his enemy; if they indicate defeat, it is because calculations show that one is inferior. With many calculations, one can win; with few one cannot. How much less chance of victory has one who makes none at all! By this means I examine the situation and the outcome will be clearly apparent.

(Passage #8.1 contains content relevant to counterintelligence. See also pp. 284–87.)

★★★★★★

---

[96] *The Croker Papers. The Correspondence and Diaries of the Late Right Honourable John Wilson Croker, LL.D., F.R.S, Secretary to the Admiralty from 1809 to 1830* (Louis J. Jennings ed.; 2nd ed., revised; London: John Murray, 1885), Vol. III, pp. 276–77.

[97] There is a credible scientific nugget here. An aquatic ecology source notes that "[i]ncreases in foam abundance will often follow rainstorms that transport the surfactants to the stream or along lake shores on windy days." See Jeffrey C. Davis, "What causes foam in streams and lakes?" (The Aquatic Restoration and Research Institute [ARRI], Alaska Clean Water Action Grant No. 05-02), available at www.arrialaska.org/foam-in-streams.html (visited June 23, 2021). More specifically, modern science lends some support to the appearance of foam as an early warning indicator of intense rains upstream that Sun Tzu's general would not directly see (Professor Mary Power, Department of Integrative Biology, University of California, Berkeley, personal communications, January 28, 2020 and February 5, 2021).

Flash flood risks are made vivid by a modern observer's comment (which, of course, must be read in context of a China far more widely deforested than in Sun Tzu's era) that "in many parts of China . . . absence of forest cover allows almost all precipitation to run into the streams as soon as it falls. Thus a stream bed may be dry one day, and the next, after a heavy rain, become a raging torrent." See David D. Barrett (Colonel, USA, Ret.), *Dixie Mission: the United States Army observer group in Yenan, 1944* (Berkeley, CA: Center for Chinese Studies, University of California, Berkeley, 1970), p. 39.

For a textual issue pertaining to verse IX.15a's interpretation see Appendix 11 anchor footnote 14.

THEME #12 KNOW YOURSELF (AND OTHER NON-HOSTILE ACTORS AND AUDIENCES)

basics:

Passage #12.1 (verses III.31–III.33) (knowing self as well as adversary)

Passage #12.2 (verses X.22–X.26) (knowing self and adversary, adding environmental context)

Passage #12.3 (verses I.1–I.14) (agenda for self-knowledge)

Passage #12.4 (verses XIII.3–XIII.4) (relevant empirical data as cornerstone of knowing oneself)

Passage #12.5 (verses V.17–V.19) (challenge of observing chaotic situations)

Passage #12.6 (verses XI.42–XI.43) (spirit of unobtrusive observation)

Passage #12.7 (verse IX.15a) (spirit of imaginative use of available data)

knowing operational readiness:

Passage #12.8 (verse XI.32) (knowing operational readiness)

Passage #12.9 (verses VII.20–VII.25) (more about knowing operational readiness)

Passage #12.10 (verse IX.13) (knowing sanitary and public health conditions affecting readiness)

Passage #12.11 (verses IX.47–IX.50) (knowing the loyalty and reliability of your troops)

Passage #12.12 (verse X.9–X.16) (knowing factors producing failures of readiness or effectiveness)

Passage #12.13 (verses VII.2–VII.9) (more about factors producing readiness or effectiveness failures)

knowing your country's economic situation:

Passage #12.14 (verses II.1–II.5 + II.9–II.14) (knowing your country's economic situation and the costs of war)

Passage #12.15 (verse XIII.1) (keeping track of war's devastating effects on civilian economic life)

knowing key strategic actors and the state of civil–military relations on your side (perspective is that of the ruler or, in Passages #12.19 and #12.20, also that of the commanding general):

Passage #12.16 (verse III.18) (know your commanding general, and pick him carefully!)

Passage #12.17 (verse X.19) (know your commanding general: importance of assessing motivation)

Passage #12.18 (verse I.15) (know your commanding general's qualities of judgment)

Passage #12.19 (verse XIII.13 + XIII.23) (know your spymaster: qualities to be sought)

Passage #12.20 (verses III.19–III.23) (know the state of civil–military relations)

knowing persons of interest (similar steps could apply to anyone):

Passage #12.21 (verses V.21–V.22) (knowing people, the better to pick them)

Passage #12.22 (verses VIII.17–VIII.24) (knowing persons of interest by their personality flaws)

Passage #12.23 (verses I.22–I.25) (actively probing persons of interest, the better to know them)

Passage #12.24 (verses VI.20–VI.23) (further probing moves)

Passage #12.25 (verses XIII.15–XIII.16) (gathering detailed information on persons of interest)

general social and organizational observation tips:

Passage #12.26 (verses IX.19–IX.44) (general observer tips, based on analogical reading of Sun Tzu)

✶✶✶✶✶✶

Like "the fleet the gods forgot" – an image born of the disaster that befell the US Asiatic fleet at the outset of World War II – Theme #12 is at risk of being Sun Tzu's lost theme. Certainly Sun Tzu's sharp-edged espionage content siphons attention away from Theme #12, for all the reasons that many people find espionage fascinating – much more so than, say, logistics (which, via the task of evaluating the operational readiness of one's own army, is a basic part of Theme #12). "Intelligence" is also a roomy rubric and it is possible to stretch it to fold some (though not all) Theme #12 content into Theme #11, thereby scaling back any perceived need to tackle Theme #12 as a separate category.

But trusting Sun Tzu's instincts – embodied in his illustrious two-part "know the enemy and know yourself" dictum (verse III.31 in Passage #12.1) – suggests that greater insight may be had by providing Theme #12 with a robust identity on a parity with, yet distinct from, that of Theme #11, the inevitable overlaps notwithstanding. That is the analytical path pursued here.

To a first approximation, to be refined below, the Theme #12 "know yourself" imperative places three demands on Sun Tzu's general (and, by extension, the ruler):

(a) Knowing one's army

(b) Knowing conditions in one's own state

(c) Knowing oneself as an individual person and decision maker.

The third task, about which Sun Tzu says little directly, is the most analytically elusive. It certainly involves major Chinese cultural aspects.[1] There is overlap between task (c) and the topmost rung of Theme #8's heuristic "formlessness ladder," which calls for introspection by the general or ruler to clarify priorities and prune extraneous or incidental goals or preferences (see *wu xing* 6 discussion, pp. 277–78 above).

In the present Theme #12 development, a choice has been made to include with Theme #12 Sun Tzu content relevant, not only to "knowing yourself," but also to "knowing" assorted non-hostile actors of interest. Those other actors, many of whom enter the picture via tasks (a) and (b) above, are a diverse cast of characters.

---

[1] Background for this third task is Lisa Raphals, "Thirteen ways of looking at the self in early China," *History of Philosophy Quarterly*, 2009, 26(4), 315–36 (p. 336 points to a "need to roam beyond the usual sources" in developing this topic).

They may be individuals or organizations. They may be political allies, neutrals, bystanders, or others (for example, weapons makers in one's home country).[2]

A key to Theme #12 passage analysis is a spirit of analogical thinking. It enables assorted Sun Tzu advice – framed on its face in terms of observing and sizing up the enemy – to do double duty, with mostly minor modifications, as advice about assessing non-hostile actors, including people or organizations on one's own side.[3]

Passages #12.1 and #12.2 launch Theme #12 development. They involve the same Sun Tzu content with which Theme #11 began, but now viewed through a different lens – "know yourself" rather than "know the enemy." While Passage #12.1 is more famous, Passage #12.2 actually makes more of Theme #12 contribution since its mentions of "terrain" and "weather" can be read as placeholders for environmental factors relevant to knowing yourself.

Passage #12.3 likewise repeats verses (verses I.1–I.14) present in Theme #11. Focus now, however, should be on observation and evaluation of one's *own* side's capabilities and weaknesses, rather than (as with Theme #11) those of the enemy.[4] Such inward-directed scrutiny may in some ways be easier to carry out than is scrutiny of a foe, where sources of information are often scant or unreliable. Yet observation of one's own side still requires data, in Sun Tzu's time probably mainly gleaned from knowledgeable observers (see verse XIII.4 in Passage #12.4, now broadly read to include knowledgeable observers who are not necessarily in clandestine roles). Evaluating that data runs up against the inherent complexity of situations involving many actors (even if non-hostile). In the Warring States era few analytical tools were available to cope with that complexity, nor does Sun Tzu offer much by way of guidance. Read in an analogical spirit, verse V.17 (Passage #12.5) presents a perspective on some of the observational challenges, which good organization (e.g., logistics organization) may help to surmount (verse V.19). The next two passages each contribute a useful concrete idea, centering respectively on unobtrusive observation (Passage #12.6) and imaginative use of available data (Passage #12.7).

Inward-directed scrutiny may also be very difficult for other basic reasons. Some types of introspection – a ruler's scrutinizing his own polity, his own regime,

---

[2] The Chinese-language text of verse III.31 refers to *zhibi zhiji*, lit. "know the other, know yourself." In the perceptions of a Sun Tzu–inspired strategist, assorted people, organizations, polities, etc. might shift back and forth and back again between categorization as *ji* (self) and as enemy or "other" (*bi*). A fine example of dizzying shifts in alliance patterns comes from the history of the state of Zheng, which occupied an exposed strategic position on the Central Plain (cf. Sun Tzu's "focal ground"):

> Zheng was allied with Jin in 632–612 [BC]. In 608 it concluded an alliance with Chu; in 606 it re-established friendly ties with Jin only to abandon it for Chu in 603; next year another alliance with Jin was concluded; in 599 Zheng shifted its alliance to Chu and then immediately back to Jin; in 598 an alliance with Chu was concluded, betrayed and established once more in 597.

See p. 288 footnote 19 of Yuri Pines, "'The One That Pervades the All' in ancient Chinese political thought: the origins of 'The Great Unity' paradigm," *T'oung Pao*, 2000, 86(4/5), 280–324. For further background on Zheng's exposed situation see Appendix 2 footnote 9.

[3] Here is another facet of the important "dual use" idea undergirding Theme #4 development (pp. 142ff. above).

[4] A third aspect of these same verses is self/enemy comparison – i.e., net assessment (Passage #1.1).

himself – may well disclose inconvenient truths. Sun Tzu's verse XII.18 (Passage #1.8) suggests recognition of how easily distortions born of emotional dynamics can creep in here. Traditional Chinese thought is well versed in such possibilities. An attempt at remedial measures is represented by the remonstrance tradition in Chinese history, where already in pre-Confucian times "records abound in examples of ministers who doggedly remonstrated with rulers, often at great cost to themselves" where an issue of principle was perceived to be at stake.[5]

When canvasing for Sun Tzu "know yourself" passages with which to populate Theme #12 – thus passing from the general agenda set out in (a)–(c) above (p. 396) to specifics of the Sun Tzu text – it is helpful to group passages around four strands of "knowing oneself," all having roots in Sun Tzu on net assessment (verses I.1–I.14, see Passage #1.1 = #12.3):

- readiness self-knowledge, centering on the readiness of one's army (includes "consumer logistics" issues, which overlap with personnel matters);[6]
- producer logistics and economic self-knowledge (focus is relevant conditions in one's home state and "sinews of war potential" available there);
- political and related self-knowledge of a population-level or mass-psychology type;
- command self-knowledge, pertaining to knowing the strengths and weaknesses of high command on one's own side as well as self-scrutiny by a leader.

For Sun Tzu, much readiness self-knowledge takes the form of answers to a few basic, highly concrete questions.[7] For example: Is the army hungry? Is it tired? Is it

---

[5] See Hucker, #4 footnote 8, pp. 6–7. See also David Schaberg, "Remonstrance in Eastern Zhou historiography," *Early China*, 1997, 22, 133–79, observing (p. 142) that "remonstrances assert inherited truths when they are most endangered, as when a ruler indulges destructive passions or takes foolish military risks." In the imperial period of Chinese history, later than Sun Tzu's time, remonstrance functions were institutionalized via the censorate, which had surveillance functions too. In different ways both functions were pertinent to "knowing yourself." In the Song dynasty a widely known name for the censors was "the eyes and ears of the emperor." See Wilkinson §17.1.3.3, p. 274.

[6] The producer/consumer logistics distinction in military analysis is introduced on p. 36 above. For coordinated definitions of operational readiness and, separately, effectiveness see p. 53 of Admiral Eccles's "Suez 1956 – some military lessons," *Naval War College Review*, March 1969, pp. 28–56 (essay reprinted in B. Mitchell Simpson III, ed., *War, Strategy, and Maritime Power* [New Brunswick, NJ: Rutgers University Press, 1977], pp. 221–50):

> Readiness = Degree of ability of a unit/ship to perform its designed mission. It includes status of personnel, equipment, supplies, maintenance, facilities, intelligence, and training. It also incorporates "performance," "endurance," and "preparedness."
> Effectiveness = Performance × Availability × Utilization.

Note that "knowing yourself" might include retrospective analysis of combat or other effectiveness. Suggesting that such analysis was in the repertoire of early Chinese warfare, see Han Xin's after-action conference (reciting principles found in Sun Tzu) recounted in the Shiji (see p. 32 above).

[7] Traditional commentator's Zhang Yu's gloss on verse I.11 also captures a spirit of readiness evaluation: "In which army are the chariots most solid, the horses fittest, the officers bravest, the weapons keenest ...?" See Minford, p. 110. Insight into Sun Tzu's grasp of the concept of readiness also comes from verses II.1–II.2 (Passage #12.14), specifically the allusion there to "materials such as glue and lacquer" (which underscores that readiness may crucially depend

fearful? Is it homesick? Are essential stores and equipment readily at hand? A more integrative (holistic) perspective on operational readiness is suggested by verses I.11–I.14 in Passage #12.3.[8] Reinforcing need for such an integrative approach to readiness is the sheer size of armies in Sun Tzu's time. Size suggests a major logistics apparatus to support readiness, about whose workings we unfortunately know very little.[9] Envisioning logistics in a broad sense as embracing a full range of "means of war" extending beyond supply alone, and running the Sun Tzu text through that conceptual filter, turns up quite a lot of relevant material. Pertinent passages include Passages #12.8–#12.13, with verse VII.21 (Passage #12.9) being particularly noteworthy for its explicit recognition that readiness is a "wasting asset," ever prone to dwindle as time passes, so that its evaluation and restoration calls for constant vigilance and effort. Also noteworthy is Passage #12.9's further emphasis on human aspects of readiness as embodied in the Chinese concept of qi (for which there is no exact English-language counterpart, but which roughly translates in this context as "fighting spirit" or "morale").[10] That emphasis lays groundwork for integrating personnel issues with logistics ones, from which they are often divorced in modern military organization and thinking.[11]

Likewise bearing on knowing readiness and effectiveness is the planning factors material, some of it quantitative, appearing in Passage #12.13. Further relevant Sun Tzu content is assigned to Theme #1 (calculation). Particular note should be taken of Passage #1.7 there, which emphasizes how military calculations build on one another. Read through an operational readiness lens, the core idea of Passage #1.7 is highly relevant to readiness evaluation and reporting, since fudging at lower echelons or somewhere in a supply chain may cumulate to serve up to a commanding general a largely imaginary picture of his army's readiness. Relatedly, verse IX.45 (Passage #1.18) is a warning that there is much more to operational effectiveness – the payoff of readiness! – than the size of one's army (a point missed by many potentates in history, Eastern and Western alike). Although Sun Tzu does not spell out the point in so many words, an easy inference from Passages #12.8 and #12.9 is that a large army's voracious supply needs may actually operate to reduce, not enhance, the overall readiness and effectiveness of the force as army size increases.

A second type of self-knowledge swings the spotlight to producer logistics and the economic roots of military power. Here Sun Tzu says relatively little (in particular Sun Tzu's producer logistics coverage is next to nil) but what he does say (Passages #12.14 and #12.15) – even if the specific statistics he cites are taken to be only

---

on small or specialized items like these). See, e.g., Sawyer, Background footnote 19, p. 317: "From the late Shang [dynasty] onward lacquer was applied to protect the finished bow against moisture."

[8] Supporting a view of readiness as an amalgam of multiple factors see verse VII.9 (Passage #12.13), when read as Ames does. See also Appendix 5 footnote 10 (noting that Ames's translation accurately captures the idea of multiple essential ingredients expressed in the Chinese text, as Griffith's translation does not).

[9] For a snapshot of Warring States military logistics see pp. 33ff. above.

[10] Regarding qi 氣 (not the same concept as qi 奇 of the qi/zheng concept pair) see #3 footnote 80.

[11] There is kinship here between Sun Tzu and Admiral Eccles's perspective on personnel matters, approaching them as an integral part of the logistics process, albeit one with unique features. See Eccles, Background footnote 52, p. 52; Eccles, #2 footnote 6, chapter 14, "Personnel," pp. 117–29.

figurative – suggests an instinct for quantitative economic observation with particular attention to a protracted war's devastating impact on a civilian economy.

A third type of self-knowledge is political in a broad sense. Taking cues from verses I.4 and I.11 (Passage #12.3), a core part of this third type is knowing the extent and depth of the people's loyalty to the ruler and factors affecting it – e.g., the general state of mind of the populace of the home state, sources of satisfaction, grievances, etc. (among the latter might be war's economic devastation depicted in Passages #12.14 and #12.15). Unfortunately, Sun Tzu does not elaborate on how a relevant analysis should be carried out (and for allied states too!). However, before becoming too critical of Sun Tzu, it should not be forgotten that as recently as World War I the belligerent powers had at best spotty ability to gauge public opinion in their own country, even when signs of war-weariness of their populace began to surface as the war dragged on and casualties mounted into the millions.

A fourth type of self-knowledge pertains to "knowing" one's own side's high command. In the first instance, this means that the ruler should know traits of his own commanding general (Passages #12.16–#12.18) (and pick him carefully!). It would also mean knowing one's spymaster (Passage #12.19), if that is in fact a different individual (a point on which Sun Tzu is not clear).[12] A further strand of the relevant self-knowledge, which both ruler and commanding general need, is knowing the state of civil–military relations on one's own side, since problems there will inevitably ramify to undercut all facets of military endeavor (Passage #12.20).

This fourth bundle of self-evaluation tasks has immense potential for being politically sensitive. Note that Passage #12.3 (and also the related verse I.28 in Passage #1.2) leaves unresolved the all-important issue of who is supposed to undertake (or have input into) the relevant assessment and how to deal with the likely differences of opinion that then arise. For reasons that align on the centrality Sun Tzu accords to the commanding general's role (Theme #13), Sun Tzu has far more to say about evaluating that general than the ruler. In practice, the general also needs to "know" (evaluate) the ruler too, a task likely to call for considerable fancy footwork.

A steppingstone to all four types of self-knowledge is the assessment of men, to borrow the gendered title of an Office of Strategic Services study published shortly after World War II.[13] Although terse, Passage #12.21 serves as an anchor, focusing on personnel selection tasks (and, by implication, the "knowing people" analysis on

---

[12] Given our scant knowledge of specifics of the intelligence organization Sun Tzu had in mind (see pp. 364, 366 above) it is not clear whether the commanding general (or, for that matter, the ruler) is expected to serve as his own spymaster or even his own pro se operative (see #11 footnotes 17 and 18). The answer to that question clearly bears on the latitude of choice available in picking a spymaster. If a spymaster is the commanding general (or yet more so, the ruler) that latitude may be nil.

[13] See Office of Strategic Services (OSS) Assessment Staff, *Assessment of Men; Selection of Personnel for the Office of Strategic Services* (New York: Rinehart, 1948). In Sun Tzu's era the implied gender limitation was probably largely in effect. There may have been a few enlightened exceptions. From a period immediately following the Warring States, a plausible female candidate for being a "person of interest" to a Sun Tzu–inspired strategist was a wealthy widow-entrepreneur named Qing who took over the family cinnabar mining business, became very wealthy in consequence, and was honored by the First Emperor of Qin (r. 221–210 BC). See *Shiji* 129 (Watson translation cited on p. xxiii above), *Han Dynasty* II, p. 440. Yates, #9 footnote 7, p. 367 also recounts her story

which such selection builds). Sun Tzu's key insight here centers on verse V.21's concept of "seeking victory from the situation." In application to personnel selection this points to the possibility that an individual of limited or specialized capabilities may still be able to contribute in an important way to a winning situation if placed in exactly the right role or position.[14]

Starting with Passage #12.22, the arc of Theme #12 development moves on to knowledge – here to be interpreted as knowledge pertaining to key non-hostile actors – regarding specific personality traits and other characteristics of such individuals. That knowledge will inevitably become relevant if a self-knowledge project is pursued seriously. With minor adjustments in framing, Passages #12.22–#12.25 need little reworking to suggest types of knowledge pertinent to knowing a wide range of non-enemy "persons of interest." Passage #12.22, in particular, is a fine start at setting forth a catalog of personality flaws worth knowing because they are raw material for manipulations by self, the enemy, or third parties.

Sun Tzu's psychological emphases and instincts are on full parade here. They harmonize with an important further body of early Chinese thought pertaining to "knowing people," one that gives credence to the possibility that astute observation will be able to pierce attempts at dissembling, thereby laying bare specific individuals' inner natures.[15] Relevant here is a dual use reading of Passages #12.23 and #12.24, treating them (to adapt William James!) as a "moral equivalent of reconnaissance" – probing persons of interest with an eye to eliciting reactions and obtaining information about their behavioral repertoires and response patterns. While Sun Tzu's immediate focus in these passages is clearly interactions with a foe, the kinds of probings he suggests – some notably aggressive – may be even more useful as vehicles for acquiring deeper knowledge of key personnel on one's own side, in particular because those probings may be unexpected (think "soft targets"!) and so stand to shake loose a greater volume of useful information.[16] That may be especially true when the source of those probings, and possibly even the existence of any probing at all, can be kept veiled (which ties back to the unobtrusive observation harmonic of Passage #12.6). Here is the nucleus of a Sun Tzu–inspired methodology for better knowing one's own side, avoiding faulty personnel choices and improving intramural negotiation outcomes (from a Sun Tzu general's perspective, of course!).

Redirected to actors on one's own side and divested of their targeted killing aspects, the Passage #12.25 verses suggest how information crucial to sizing up people and situations may pivot on what may appear at first to be minutiae (possibly including casual social contacts, cf. verse XIII.15's "all those to whom he spoke").

(noting that her success in a highly regulated economic activity almost certainly meant that she was literate).

[14] See Passage #4.6 (where verses V.21–V.22 appear) and the related Appendix 4 anchor footnote 4.

[15] See #8 footnote 26. See also Appendix 11 footnote 25 (Chinese physiognomy thinking).

[16] Aggressive probings as a tool of personnel evaluation are a practice that finds support in other early Chinese sources. See, e.g., *The Annals of Lü Buwei* (*Lüshi chunqiu*), #2 footnote 71, p. 109:

> [in evaluating men] the six tests are: make them elated in order to test their self-control; make them happy in order to test their tendency to wantonness; make them angry in order to test their self-restraint; make them apprehensive in order to test their unique features; make them sad in order to test their empathy for others; and oppress them in order to test their resolve.

On its face, the last main text Theme #12 passage (Passage #12.26) concerns military field intelligence topics (mostly antiquarian), hence is not a naural part of Theme #12 development at all. However, reading it through an analogical lens suggests a reinterpretation along Theme #12 lines. That reinterpretation is far enough removed from a surface reading so that discussion of Passage #12.26 is assigned to Theme #12 frontiers (see Table 6 and related discussion, pp. 404ff.).

Roads Not Taken by Sun Tzu in Developing Theme #12

A major hole is lack of explicit attention by Sun Tzu to systematic analysis of one's own side's strategic objectives – certainly a core aspect of "knowing oneself." To be sure, the ever-serviceable net assessment passage (Passage #12.3) makes some contribution here, mostly related to avoiding launching a losing war; as do scattered other Sun Tzu observations such as verse XI.28 (Passage #8.17) if parsed as implicit advice to prune objectives to shed extraneous concerns. But for the most part Sun Tzu treats strategic goals as a given – if to be analyzed at all, then by someone other than Sun Tzu (or Sun Tzu's general). On a grand strategy plane it is quite possible that one is seeing here a facet of Warring States China mentalité, where a zero-sum perspective prevailed and conquest objectives – swallowing up one's neighbors – were plain to all. Even here, however, there remains substantial scope for clarifying national interest (i.e., that of a given state actor), challenging assumptions and setting priorities. The Sun Tzu text offers little help here.[17] The same is true for analysis of objectives on a lower plane of military strategy or operations.[18]

On a yet more fundamental level, Sun Tzu does not qualify as a broad-ranging student of human nature or the human condition. The Sun Tzu text is narrower by far in actual focus than the canon of Chinese classics perpetuated by the Confucian tradition and the imperial examination system. The same narrowness also sets Sun Tzu apart from the Daoist classic, the Laozi, with which Sun Tzu's thinking has affinities.[19] Sun Tzu is not an administrator, nor a jurist, nor a wide-ranging social or political philosopher. In consequence, the "know yourself" exhortation in Passages #12.1 and #12.2 is by no means developed as broadly or deeply as it might be, including by standards set by other Warring States schools of thought.

Meanwhile, Sun Tzu's treatment of logistics – a crucial part of "knowing your-self" in military terms – remains fragmentary, even allowing for the simpler condi-tions of Sun Tzu's time. Sun Tzu takes no real cognizance of the true complexity of a major logistics effort, certainly an effort on a scale adequate to support an army of

---

[17] Illustrating the point, verse III.1 (Passage #4.1),

   III.1.

   Generally in war the best policy is to take a state intact; to ruin it is inferior to this.

   provides guidance pertinent to selecting objectives. Realistically, however, its counsel is far too generic to do more than scratch the surface of analysis-of-objectives tasks faced by working strategists.

[18] Verse XII.16 (Passage #1.23) touches on what could be regarded as analysis of objectives on a level of both grand and military strategy. However, focus there is on spelling out a division of labor between general and ruler, not giving guidance as to how each should approach their respective tasks.

[19] For one modern authority's perspective see Brooks, p. 66 (quoted in part in #4 footnote 47).

100,000 or more in the field (see Passages #12.14 and #12.15). In many military contexts, logistics constraints exert an enormous, at times decisive, influence on campaigns (a point of which Griffith as a military professional was thoroughly aware, but which many civilians, ancient and modern alike, woefully underappreciate).[20] Logistics involves supply; transportation; weapons logistics; maintenance and repair; personnel; medical and veterinary services;[21] base development; and more. All of these come with information needs which, if not met on a timely basis, mean that the general is flying blind on crucial matters. Sun Tzu points once in broad terms to an overall supply system (verse I.8 in Passage #12.3); takes cognizance of long-distance transportation burdens (Passages #12.14 and #12.15) with further observations relevant to transportation in Sun Tzu VII (Passage #12.13); indirectly alludes to repair and maintenance issues (verse II.14 in Passage #12.14); and touches on parts of medical logistics (Passages #12.10 and #12.13). Personnel issues get more attention, but there are holes here too, notably including a lack of interest in training topics.[22] Base development activities – a major and complex topic especially where offensive operations are concerned (and such operations are a core Sun Tzu emphasis) – are not mentioned at all. Nor, apart from glancing mentions, does Sun Tzu analyze civilian roots of logistics effort. The quality of one's army's logistics system as a whole is a major type of military self-knowledge that Sun Tzu the soldier might have applied his deep military acumen to analyzing, yet did not do so.

Instead of engaging with logistics complexities, Sun Tzu is intent on finding simple solutions that lift logistics constraints, a central example being his focus on "living off the enemy" (Passage #3.5's verse II.15).[23] Yet however militarily laudable

---

[20] See review of Engels's study of the logistics of Alexander the Great (cited in Background footnote 56) by G. L. Cawkwell, published in The Classical Review, New Series, 1980, 30(2), 244–46:
> This book deserves the attention of all ancient historians. It is all too easy in the mind to move armies like pencil-points across maps, and forget the factors which limit and delay. The larger the army the slower its progress and the more formidable the problem of providing it with food and drink; and the more one's nose is rubbed in the facts of logistics the better.
>
> (p. 244)

[21] Dental services too, if available. For dental issues identified in skeletal remains from what may have been a military base of the state of Zhao to defend its northern border, see Evan Hardy, "A bioarchaeological examination of the skeletal remains of Warring States period Tuchengzi, Inner Mongolia, China," Master's Thesis, Department of Archaeology, Simon Fraser University, Fall 2015, cited in Evan Hardy, Deborah C. Merrett, Hua Zhang, Quanchao Zhang, Hong Zhu, & Dongya Y. Yang, International Journal of Paleopathology, March 2019, 24, 1–6.

[22] As Mair notes, the Sun Tzu text "is concerned with the theory and tactics of war, not such mundane matters as training and practice. ... nor does it utter a word about the drilling of soldiers." See Mair, "Soldierly Methods" (cited on p. xxii above), p. 1. A contrast case is the Wuzi, another early Chinese military work (see Background footnote 27), which does touch on training matters. See Brooks, p. 62; Griffith, p. 160; Sawyer, Seven Military Classics, p. 204.

Scouring the Sun Tzu text for coverage, state of training does figure in Passage #1.1's net assessment exercise – see verse I.12 there – but that strand remains undeveloped by Sun Tzu. A further possible reference to training is verse IX.49 (Passage #12.11). Giles, p. 98 and Ames, p. 144 so interpret it. However, the orders mentioned in verse IX.49 might also be operational, not training, ones.

[23] One hint that Sun Tzu was well aware of the complexity of logistics issues (even though he proceeded to shy away from full engagement with them) is his alertness (see above) to specifically long-distance transportation issues. In modern times such problems have often been a catalyst to

such an aim is, it is is unlikely to be fully feasible for long, positing that the basic instrument of power remains the mass-infantry army. Sun Tzu's astuteness in pursuing and making use of psychological knowledge cannot fully compensate for his lack of comparable attention to war's equally fundamental logistics aspects.

An intriguing counterfactual question is whether seriously developing logistics issues might have led Sun Tzu into such a thicket of detail as to produce a fundamentally different work, its strategic and psychological insights overshadowed by a mass of logistics details. Giving logistics its due might then have generated a military manual having much the same fate as did Sun Tzu's advice against assault on cities – possibly sound enough at a certain historical moment but antiquarian only a little later in Chinese history (to say nothing of more recent times!).

There are the makings of a paradox here. Taking full cognizance of the logistics implications of "knowing yourself" could have wound up producing a military treatise which (if it survived at all) might have joined the second or third tiers of surviving early Chinese military works, rather than towering above all the rest.

Sun Tzu (2) ("Sun Tzu Extended") and Sun Tzu (3) ("Sun Tzu Analogical") Frontiers of Theme #12

Exploring in its natural generality Sun Tzu's thinking about sources of information – as pertinent to knowing yourself as to knowing the enemy – Table 5 assembles twenty-first-century interpretations of four categories of information source listed in verse XIII.4 (Passage #12.4). Sun Tzu's fourth category (the one Sun Tzu endorses) is generalized as pointing to the crucial importance of insightful human observers.

Table 6 (Passage #12.26) is a further vehicle for exploring Sun Tzu (2) and (3) frontiers of Theme #12 in a way that stays grounded in the Sun Tzu text. Nominally about field intelligence or surveillance of an enemy army (with a few further pointers on interpreting a foe's verbal signals) these verses from Sun Txu IX have attracted little interest from modern Sun Tzu audiences.[24] Besides being antiquarian, some of their content seems seriously vulnerable to deceptions by an even slightly sophisti-cated foe who has access to a copy of Sun Tzu's playbook.[25] But this material takes a new lease on life when given an analogical reading, one that shifts focus away from an enemy army and instead applies the spirit if not the letter of Passage #12.26's numerous tips to observation of assorted friends, allies, neutrals, bystanders, and other non-enemy "persons of interest," along with interpretation of their behaviors and communications. Motivation for doing so might come from a ruler or general intent on probing areas of strength and weakness in his own complex war machine;

---

logistics creativity. A case in point would be the Qing war against the Dzungars which led the Qing to create supply routes spanning vast distances, a logistics accomplishment with few (if any) parallels prior to modern times. See Perdue, Introduction footnote 52, chapters 5–7, pp. 174–299 (his pp. 522–23 and map 7 on p. 273 provide an overview). See also #2 footnote 6 and #6 footnote 48 (World War II experience).

[24] One exception is Giles, p. 88, who observes of this part of Sun Tzu that "much of [it] is so good that it could almost be included in a modern manual like Gen. Baden-Powell's 'Aids to Scouting.'"

[25] All of the Table 6 verses except verse IX.44 appear in their original guise as military field intelligence tips in Passage #11.26A, where they are further analyzed in that military context.

Table 5 *Sun Tzu's favored and disfavored information sources, with selected modern updates*

Table 5 explores generalizations of verse XIII.4 (Passage #12.4), which expresses Sun Tzu's views regarding merits and demerits of four kinds of information sources.

Verse XIII.4 category of source	Universalistic interpretation	Candidates for examples
First category ("spirits or gods") *rejected by Sun Tzu*	information derived from non-empirical belief systems	theologically or ideologically inspired decision making; cult leader pronouncements
Second category ("analogy with past events") *rejected by Sun Tzu*	information derived from interpreting "symbols of things" (note a)	history formulaically invoked as guide to action, with little or no analysis (note b)
Third category ("astrological calculations") (note c) *rejected by Sun Tzu*	calculations that outrun the available empirical data or are based on irrelevant data (note d)	computer models based on slender, if any, relevant empirical data (note e)
Fourth category ("men who know the situation") (note f) *endorsed by Sun Tzu*	observations of astute, knowledgeable, unbiased observers	some spies, political analysts or commentators, journalists, business people, travelers

Note a. The phrase "symbols of things" is Mair, p. 129's translation of Sun Tzu's second information category. Mair's "symbols of things" is more flexible than Griffith's "analogies with past events."

Note b. For analysis of pitfalls of historical analogies in decision making see Richard E. Neustadt & Ernest R. May, *Thinking in Time: the uses of history for decision-makers* (New York/London: The Free Press/Collier Macmillan, 1986). Particularly relevant to Sun Tzu's warning is their notion of "allure" that may render certain analogies "irresistible" (Neustadt & May, pp. 48, 56). A twenty-first-century candidate might be the concept of a "digital Pearl Harbor." Short of irresistible, but still possibly major in effects, is what they term a "seductive" analogy:

> A choice has to be made in a hurry. An obvious analogy surfaces. It is seductive, because it helps supply a decision rule when there seems to be no time for analysis to "bubble up." Because it has that attraction, it does not get close inspection. The decision rule then holds even though initial *Likenesses* fade or even disappear.

> (Neustadt & May, p. 66, emphasis in original)

Of course, some analogies have uses other than as prescriptive decision guides. See, e.g., Lau, #2 footnote 48, p. 262, suggesting how breakdowns of analogies may at times yield analytical insight. For further observations about analogies and their limits see Howarth, Introduction footnote 13, pp. vii–viii.

Table 5 (cont.)

Note c. Ames's "astrological calculations" catches relevant tonality better than Griffith's "calculations."

Note d. An early Chinese example would be divination exercises which, from some standpoints, were awash in data, but data of a type nowadays deemed irrelevant. See Appendix 1 footnote 23.

Note e. E.g. agent-based models whose behavioral rules lack empirical support.

A prescient warning comes from traditional commentator Mei Yaochen's commentary, which Giles, p. 163 imbues with a quasi-Newtonian thrust: "the laws of the universe can be verified by mathematical calculation [more literally, 'the law of heaven and earth can be verified from the calculation of numbers']: but the dispositions of an enemy are ascertainable through spies and spies alone."

Note f. In the interest of generality, "enemy situation" has been here replaced by "situation." The Han strips also abbreviate Sun Tzu's thought to "men who know" (omitting reference to "enemy").

---

or it might involve strictly civilian agendas, possibly non-state actor ones, still further afield from Passage #12.26's avowed context. To be clear, there is no evidence that Sun Tzu himself envisioned this kind of analogical reading of Passage #12.26, though some civilian readers of Sun Tzu in Warring States times might have extracted similar sustenance from it.[26]

Turning to specifics, Table 6 sorts Passage #12.26 content into five clusters. Read in an analogical spirit, each cluster centers on one class of clues that social or organizational observers might draw on as a "way of accessing and making sense of their mundane surroundings."[27] To that end, a common denominator of Cluster 1's tips is using non-verbal *visible* indicators to shed light on *non-visible* (or not yet visible) phenomena of interest.[28] Of particular interest is the line of thinking found in verse IX.44 and parts of verse XI.19, giving a heads-up that "something is wrong with this picture" – a type of warning that could find uses in many civilian settings. Read in a generalized way, verses IX.23–IX.24 (pertaining to reading patterns in dust clouds) suggest a niche for communications traffic analysis of patterns in the "chatter" that social or political conflict situations elicit (even when the specific content of those conversations is not knowable from such analysis).[29]

A further placeholder for using proxy indicators to learn about non-visible (or not yet visible) phenomena of interest is verse IX.20 ("When trees are seen to move, the

---

[26] Cf. Han Feizi's comment quoted in #4 footnote 6.

[27] This framing draws on p. 121 of Gagan D. S. Sood, "Circulation and exchange in Islamicate Eurasia: a regional approach to the early modern world," *Past & Present*, No. 212, 113–62 (August 2011).

[28] Very much on point here is the observation of Lord Wellington previously quoted in connection with Passage #11.18 (see p. 394 above). In the twenty-first century, the challenge thus laid down is, of course, what animates many machine learning agendas.

[29] For a snapshot of traffic analysis see p. 140 of David Kahn, "The intelligence failure of Pearl Harbor," *Foreign Affairs*, 1991, 70(5), 138–52.

Table 6 *(Passage #12.26). Sun Tzu the observer: general tips, based on analogical reading of Sun Tzu*

The basic concept here involves rethinking these verses as tips for observing a non-enemy organization of interest (which, of course, does not preclude a neutral, "arm's-length," or even adversarial relationship to the observer, as illustrated by, say, cases of defense contractors and government procurement agencies).

Viewed through that lens, some of these field intelligence tips plainly transpose better, or in a more literal or specific way, to the observational challenge than do other tips. Disciplined imagination is a basic resource in getting the most out of the reinterpretation exercise.

With four exceptions, the present clustering aligns on the order of the traditional Chinese text. For a list of those exceptions, with further examples of possible civilian generalizations of this part of Sun Tzu, see Appendix 12.

1. Non-verbal clues shedding light on observee intent, non-visible actions, situations, etc. (a primary focus of these tips is the environment, which in Sun Tzu's original context is physical or biological, that provides context for strategic action)

IX.44. When the enemy troops are in high spirits, and, although facing you, do not join battle for a long time, nor leave, you must thoroughly investigate the situation. (note a)

IX.19. When the enemy is near by but lying low he is depending on a favourable position. When he challenges to battle from afar he wishes to lure you to advance, for when he is in easy ground he is in an advantageous position.

IX.20. When the trees are seen to move the enemy is advancing.

IX.21. When many obstacles have been placed in the undergrowth, it is for the purpose of deception. (note b)

IX.22. Birds rising in flight is a sign that the enemy is lying in ambush; when the wild animals are startled and flee he is trying to take you unaware.

IX.23. Dust spurting up in high straight columns indicates the approach of chariots. When it hangs low and is widespread infantry is approaching.

IX.24. When dust rises in scattered areas the enemy is bringing in firewood; when there are numerous small patches which seem to come and go he is encamping the army.

IX.35. When birds gather above his camp sites, they are empty.

2. Verbal clues from observee's communications as a window into observee's plans or goals (note c)

IX.25. When the enemy's envoys speak in humble terms, but he continues his preparations, he will advance.

IX.26. When their language is deceptive but the enemy pretentiously advances, he will retreat.

IX.27. When the envoys speak in apologetic terms, he wishes a respite.

IX.28. When without a previous understanding the enemy asks for a truce, he is plotting.

Table 6 (cont.)

3. Observed resource commitments that give insight into the observee's intent, plans, etc. (note d)

IX.29. When light chariots first go out and take position on the flanks the enemy is forming for battle.

IX.30. When his troops march speedily and he parades his battle chariots he is expecting to rendezvous with reinforcements. (Mair, p. 111 translates differently: "... he seeks to engage in battle with me.")

IX.31. When half his force advances and half withdraws he is attempting to decoy you.

4. Actions indicative of stress or disequilibrium in observee organization

IX.32. When his troops lean on their weapons, they are famished.

IX.33. When drawers of water drink before carrying it to camp, his troops are suffering from thirst.

IX.34. When the enemy sees an advantage but does not advance to seize it, he is fatigued.

IX.36. When at night the enemy's camp is clamorous, he is fearful.

IX.37. When his troops are disorderly, the general has no prestige.

IX.38. When his flags and banners move about constantly he is in disarray.

IX.39. If the officers are short-tempered they are exhausted.

5. Clues to more extreme organizational disequilibria, possibly portending drastic organizational failure

IX.40. When the enemy feeds grain to the horses and his men meat and when his troops neither hang up their cooking pots nor return to their shelters, the enemy is desperate. (note e)

IX.41. When the troops continually gather in small groups and whisper together the general has lost the confidence of the army. (note f)

IX.42. Too frequent rewards indicate that the general is at the end of his resources; too frequent punishments that he is in acute distress.

IX.43. If the officers at first treat the men violently and later are fearful of them, the limit of indiscipline has been reached. (note g)

Note a. "Angry" or "fierce" is closer to the Chinese text than "high spirits." See #7 footnote 96. A generalizable idea is that the observee organization may be "putting on a performance" to distract attention from a planned surprise move. Such performances are found in civilian environments as well as in battlefield ones. The observer challenge is not to become fixated on watching such "theater" but to probe for its motivation.

Note b. In non-military settings verse IX.21 generalizes as a warning to the observer not to allow high-quality attention – ever a scarce resource – to be siphoned off by attention to complex but extraneous features of a social or organizational environment, placed there for just that purpose.

Table 6 (cont.)

Note c. For further specifics see Passage #7.16 where verses IX.25–IX.28 also appear. Although the stratagems depicted do not seem especially sophisticated, there is an important generalizable concept here: verbal behavior can be a cheap way of obfuscating actual intent or plans.

Note d. The key generalizable idea here is an early warning one: spotting signs of major resource commitments, whatever they may be, that are irreversible or very difficult or costly for an organization to reverse.

Note e. A generalizable idea is observed steps (say, consuming irreplaceable assets) that suggest desperation. In modern organizational contexts such steps might, for example, be financial ones.

Note f. Reading it as Griffith does (Mair, p. 111 translates similarly), verse IX.41 is a placeholder for insight from observing aberrant human group dynamics. That can be a far more demanding task than observing one focal individual (in Sun Tzu's world, usually the general or the ruler). On that point social network observers – faced with the daunting task of distilling 0–1 matrix information from observing interactions in a large and lively human group – might wearily concur! (For a variant understanding of verse IX.41 that would revert focus to observing just the general – one key individual – see Appendix 4 anchor footnote 16.)

Note g. Appendix 9 footnote 19 suggests that Griffith's translation – which is not a literal one – is a way of emphasizing that the top man cannot do all this by himself. Other actors play an important part too. That in turn calls for observing *not one but two distinct, interacting sets of social dynamics*: one involving upper and middle echelons meting out the punitive sanctions, the other involving the rank and file that are those sanctions' objects/victims and audience. Labor movement history suggests a vein of non-military examples.

The resulting trajectory of interactions may lead on to organizational collapse.

Estimating if (and when) such a collapse will occur can outrun the capabilities of even a sharp observer. Model-building has a part to play here. It can leverage capabilities of that observer, even suggesting possibilities that might not otherwise occur. As a case in point, coupled dynamic systems may behave in ways that neither system taken in isolation would do. See S. Smale, "A mathematical model of two cells via Turing's equation," in *The Collected Papers of Stephen Smale* (F. Cucker & R. Wong, eds.; Singapore/River Edge, NJ: Singapore University Press/World Scientific, 2000), Vol. 2, pp. 969–78.

---

enemy is advancing"). An organizational example might be many midnight takeout deliveries to headquarters as a sign that an organization is in crisis mode.

Cluster 2 – one where Sun Tzu's specific advice needs little editing for civilian applications – contains tips for reading between the lines of verbal communications received from a third party. Experience suggests that this is an invaluable skill-set in countless social and organizational environments – as useful in reading what one's own people are professing as what allies, neutrals, bystanders, and, of course, enemies are saying. True to the salience of deception in Sun Tzu's thinking, tips in the second cluster center around puncturing some sort of deceptive façade. Admittedly, such tips could invite more tricky deceptions precisely geared toward pulling the rug out from under someone who naively relies on them. However, deception efforts in many ordinary social or business settings (undoubtably including some in Warring States China) may not run so deep, indeed may be amateurish. As long as the Cluster 2 tips (and others of similar ilk) are treated as factors to consider, not as rules to be mechanically applied, their heuristic value is defensible. Additions to Sun Tzu's set of Cluster 2 tips are clearly possible, creating growing points for doctrine.[30]

Cluster 3's thrust is one of inferring plans or goals, particularly short-term ones, from observed organizational resource commitments. A generalized message of verses IX.29 and IX.30 is that the often ponderous, highly time-phased actions of a complex organization (of which a mass-infantry army in Sun Tzu's time would be one example, though certainly not the only one) may provide early warning of a plan-in-progress and insight into the timing of that plan's implementation.[31]

Table 6's last two clusters center on using visible signs of weakness to infer greater underlying weakness. Cluster 4 focuses on serious, but possibly containable, organizational stresses; Cluster 5, on major, even catastrophic, impending failures.

Cluster 4's verse XI.33 suggests the value of watching for logistics problems manifested by corner-cutting or cheating behavior – in a nutshell, an organization's underground economy level.[32] Verse IX.38 ("When his flags and banners move constantly he is in disarray") invites a generalized reading pointing to how continual reorganizations may be a symptom of deeper unresolved organizational or leadership problems. Elaborating this line of thought, symptoms of such disarray might include frequent turnover of personnel in key jobs; a high incidence of vacancies or of personnel in "acting" roles; continual policy flip-flops evidenced by rules and regulations that repeatedly change, then change back again.

Cluster 5's verses involve early warning signs of deeper organizational failure.[33] Each expresses a generalizable insight that could find applications in many situations, civilian as well as military, involving non-hostile as well as adversary organizations.[34]

---

[30] Illustrating potential for further development, a candidate for an addition to Sun Tzu's Cluster 2 is Shakespeare's "The lady doth protest too much, methinks" (*Hamlet*, act 3, scene 2).

[31] Sawyer, p. 209 renders verse IX.30 in a way that plays up a scheduling aspect: "One whose troops race off but [who] deploys his army into formation is implementing a predetermined schedule."

[32] Or, in a Soviet-era Russian terminology, living *nalevo*. See p. 344 above.

[33] Cluster 5 verses are not the only place Sun Tzu touches on major organizational failure possibilities. Other examples include verses X.9–X.16 (Passage #12.12); also, verse XI.25 (Passage #6.16).

[34] Expressed in general terms not limited to Sun Tzu's time (or to warfare), Cluster 5's ideas are:

Verse IX.40 (focus = logistics): Deliberate failure to conserve critical organizational resources as sign of desperation (by implication, also of impending desperate – and dangerous – steps);

Verse XI.41 (focus = informal organization): Aberrant rank-and-file communications as sign of lost confidence in leadership;

Verse IX.42 (focus = formal organization): Overuse of special incentives, positive or negative, as sign of authority breakdown;

Verse IX.43 (focus = leadership): Tyrannical punitive sanctions directed against rank-and-file personnel, followed by fear of them, as sign of atrocious, unstable leadership.

Standing back from details, a harmonic running throughout the verses in Passage #12.26 is imagination in making the most of available data. In modern times, many opportunities of that type – foreshadowed in Sun Tzu's still-paperless world by Sun Tzu's reading patterns in the dust – involve exploiting to the fullest possible extent data generated as a byproduct of routine social and organizational processes. Even today possibilities of that type are often unutilized or underutilized.[35] In the Internet age, here is major grist for *"What did Master Sun know that we still don't (or have yet to absorb adequately)?"*

★★★★★★

Many high-level military officers, worldwide, now spend more time in national capitals or otherwise immersed in deeply civilianized social structures than they do on any battlefield, actual or simulated. The upshot is a sea-change in the menu of necessary skill-sets, albeit that old-fashioned combat ones remain vital. In the twenty-first century a core part of Sun Tzu's "know yourself" involves and requires acute social and organizational observation skills. Honing these is as important for the present time as sharpening Passage #12.26 (Table 6) expertise was for Sun Tzu's.[36]

A particularly crucial strand of such observational capabilities involves what might be labeled bureaucracy-watching. A basic reason for this need is the sheer organizational complexity of the modern national security landscape, embodying not just dozens but hundreds of major organizational players and actors, some with strong institutional identities and all with their own local histories and institutional cultures and legacies of past dealings with other organizational actors. In Sun Tzu's time, of course, anything like that extent of organizational complexity was unknown.

Yet the spirit of Sun Tzu retains vitality here. A Passage #12.2 verse reads:

> X.24.
> If I know that the enemy can be attacked and that my troops are capable of attacking him, but do not realize that because of the conformation of the ground I should not attack, my chance of victory is but half.
>
> [Emphasis supplied.]

In the modern era, such "conformation of the ground" might be given an extended meaning to include features of one's own side's bureaucratic terrain. Recognizing

---

[35] Some of the possibilities are illustrated by Harrison White's vacancy chain analysis, using garden-variety organizational records to identify and collect a previously unrecognized species of data from which significant insights may be derived. See White's book cited in Part D footnote 5 (p. 236 above). For a lawyer's perspective on the issue of unrecognized or underrecognized data see BakerHostetler's *Discovery Advocate* blog post by James A. Scherer (December 16, 2013), distinguishing
  – "dark data" (data organizations keep unknowingly) and
  – "dusty data" (described as "data the organization – or someone within it – kind of knows about, but is still cloaked with mystery and obscured by time" (emphasis in original).
  See www.discoveryadvocate.com (visited October 6, 2021).

[36] Replacing "agents" (i.e., spies) by "observers," the first sentence of verse XIII.23 (Passage #11.6) conveys a relevant perspective:
  XIII.23.
  And therefore only the enlightened sovereign and the worthy general who are able to use the most intelligent people as agents are certain to achieve great things. . . .

the pitfalls that lurk there, before they trip you up whilst all your attention is focused on a capable external foe, is a challenge worthy of a modern Sun Tzu.[37]

★★★★★★

Set of principal passages used to illustrate Sun Tzu Theme #12 (know yourself and other non-hostile actors and audiences)

Passage #12.1. Knowing Self As Well As Adversary

(Passages #12.1 and #12.2 take a large step on Sun Tzu's part toward placing analysis of self on parity with analysis of the adversary.)

(Passage #12.1 = verses III.31–III.33 = Passage #11.1 on p. 384 above)

Passage #12.2. Knowing Self and Adversary, Adding Environmental Context

(Passage #12.2 = verses X.22–X.26 = Passage #11.2 on p. 384 above)

Passage #12.3. Agenda for Self-Knowledge

(For Theme #12 purposes verses I.1–I.14 should be read for the requirements for self-knowledge they impose – self-knowledge that, in conjunction with intelligence information on the adversary, constitutes essential input to a net assessment. This is no easy task. In Warring States China, as in much more recent times, potent psychological and political forces operate to distort carrying out – or block altogether – the self-assessment part of a rigorous net assessment.[38])

(Passage #12.3 = Passage #11.5 = verses I.1–I.14 = Passage #1.1 on pp. 74–76 above)

---

[37] An offshoot of this line of thinking, and a candidate for a new set of failures to "know yourself," arises from widespread lack of transparency in the workings of the machine learning algorithms on which bureaucracies increasingly rely to make key decisions. There is a bundle of issues here. Some are technical, e.g., involving "unexplainable algorithms" (deep learning neural networks are a rich vein of examples). Some are psychological or social (e.g., loss of trust resulting from lack of transparency). For a structured approach to relevant artificial intelligence (AI) issues see Ron Schmelzer, "Towards a more transparent AI" (May 23, 2020), available at www.forbes.com/sites/cognitiveworld/2020/05/23/towards-a-more-transparent-ai (visited October 6, 2021). Cf. also Neville-Neil, "Know your algorithms" cited in #6 footnote 56.

[38] Ardant du Picq, a thoughtful nineteenth-century French army officer with service in Crimea and Syria who died in the Franco-Prussian War, wrote:

> When one occupies a high command there are many things which he does not see. The general in-chief, even a division commander, can only escape this failing by great activity, moved by strict conscientiousness and aided by clairvoyance. This failing extends to those about him, to his heads of services. These men live well, sleep well; the same must be true of all! They have picked, well-conditioned horses; the roads are excellent! They are never sick; the doctors must be exaggerating sickness! They have attendants and doctors; everybody must be well looked after! Something happens which shows abominable negligence, common enough in war. With a good heart and a full belly they say, "But this is infamous, unheard of! It could not have happened! It is impossible! etc."

See du Picq, #6 footnote 89, chapter on "Command, general staff, and administration," pp. 209–10.

Passage #12.4. Relevant Empirical Data As Cornerstone of Knowing Oneself

(These verses from the espionage chapter are included here for their relevance to self-knowledge, as contrasted with "knowing the enemy." Pertinent self-knowledge would certainly include a general's knowledge of his own army from men, including staff and subordinate commanders, who know its situation.[39] Building on the spirit of Passages #12.1 and especially #12.2, Passage #12.4 also points to need for well-founded empirical data pertaining to the allies, neutrals, bystanders, and others who form part of the environment of conflict.)

XIII.3. Now the reason the enlightened prince and the wise general conquer the enemy whenever they move and their achievements surpass those of ordinary men is foreknowledge.

XIII.4. What is called "foreknowledge" cannot be elicited from spirits, nor from gods, nor by analogy with past events, nor from calculations. It must be obtained from men who know the enemy situation.[40]

Passage #12.5. Challenge of Observing Chaotic Situations

(The overwhelming multiplicity of actors and motives in civilian environments can create arenas as chaotic as battlefields, albeit by utterly different paths. Verses V.17–V.19 underscore the challenge faced by an observer, in both military and civilian contexts, in recognizing underlying patterns and regularities amid that chaos. A Chinese term for "chaos" appearing in verse V.17 is *fenyun* – which the text evocatively reduplicates as *fenfen yunyun* – referring to the "helterskelter nature of threads that have come undone, going 'every which way.'"[41])

V.17. In the tumult and uproar the battle seems chaotic, but there is no disorder; the troops appear to be milling about in circles but cannot be defeated.

V.18. Apparent confusion is a product of good order; apparent cowardice, of courage; apparent weakness, of strength.

V.19. Order or disorder depends on organization; courage or cowardice on circumstances; strength or weakness on dispositions.

Passage #12.6. Spirit of Unobtrusive Observation

(Although more commonly interpreted from a standpoint of preventing information leaks, the qualities here demanded of the general also set a stage

---

[39] For a Warring States example of a general's acquiring such self-knowledge see Passage #12.8.

[40] Interpretations of the categories listed in the first sentence of verse XIII.4 vary, particularly the category that Griffith renders as "analogy with past events." See pp. A-48f. and Appendix 1 anchor footnote 25. For further development in Sun Tzu (2) and (3) directions see Table 5 (pp. 405–406 above).

[41] See Shaughnessy, Background footnote 5, p. 174. Verse V.17 actually deploys two phrases – *fenyun*, *hundun* – both in this context signifying "chaos." The former phrase seems particularly apropos as a description of strategic environments where diverse actors are pursuing disparate goals, with their sequences of steps taken in pursuit of those goals often criss-crossing (like multiple small-world search chains simultaneously unfolding in quest of different search targets).

for the general's unobtrusive observation of conditions in his army and its environment.)

XI.42. It is the business of a general to be serene and inscrutable, impartial and self-controlled.

XI.43. He should be capable of keeping his officers and men in ignorance of his plans.

Passage #12.7. Spirit of Imaginative Use of Available Data

(Interpreted as an exhortation to watch for foaming waters as early signs of a coming flood, verse IX.15a comes as close as anything in Sun Tzu to the spirit of elementary-yet-razor-sharp observation epitomized in our own time by the Sherlock Holmes stories. A modern analyst states the underlying idea in general terms: "Sun Tzu's original purpose is to teach people to observe those subtle signs, predict their future development and take measures accordingly."[42])

IX.15a. When rain falls in the upper reaches of a river and foaming water descends those who wish to ford must wait until the waters subside."[43]

Passage #12.8. Knowing Operational Readiness

(This and the next five passages have relevance to evaluation of operational readiness, a basic military self-knowledge task in any age of war. A late Warring States example of a general's gauging operational readiness by informal observation of his own army involved Wang Jian, a general who played a key role in the unification of China under Qin.[44] In a 224–223 BC campaign against the state of Chu marked by much waiting and undoubtedly boredom and frustration on the part of his troops – some of whose letters home have, remarkably, survived[45] – and desiring to ascertain their state of readiness, Wang instructed underlings to find out how his men were amusing themselves. Upon learning that they were engaged in demanding athletic competitions, Wang Jian sized up his troops as ready to fight, engaged the enemy, and triumphed. An army's recreational pursuits thus did collateral duty as a readiness indicator!)

XI.32. Pay heed to nourishing the troops; do not unnecessarily fatigue them. Unite them in spirit; conserve their strength. Make unfathomable plans for the movements of the army.

---

[42] See Zhu Jun 朱軍, Sunzi bingfa shiyi 孫子兵法釋義 (Beijing: Hai chao chubanshe, 1992), p. 137 (earlier edition is cited by Sawyer, p. 302 and p. 323 note 144).

[43] For a textual issue in verse IX.15a see Appendix 11 anchor footnote 14.

[44] The story appears in Wang Jian's Shiji 73 biography (Watson translation cited on p. xxiii above), Qin Dynasty, p. 129; Nienhauser, Vol. VII (1994), p. 175; Nienhauser, Vol. VII (2021), p. 320–21 (both cited on pp. xxiii–xxiv). It also appears in traditional commentator Chen Hao's commentary on verse XI.32. See Giles, p. 124; Minford, p. 280.

[45] See #9 footnote 7 (noting archaeological findings described by Yates).

Passage #12.9. More about Knowing Operational Readiness

(These verses recognize that readiness is a *dynamic* phenomenon, ever at risk of eroding, with resulting need for never-ending updates on the state of readiness.)

VII.20. Now an army may be robbed of its spirit and its commander deprived of his courage.[46]

VII.21. During the early morning spirits are keen, during the day they flag, and in the evening thoughts turn toward home.[47]

VII.22. And therefore those skilled in war avoid the enemy when his spirit is keen and attack him when it is sluggish and his soldiers homesick. This is control of the moral factor.[48]

VII.23. In good order they await a disorderly enemy; in serenity, a clamorous one. This is control of the mental factor.

VII.24. Close to the field of battle, they await an enemy coming from afar; at rest, an exhausted enemy; with well-fed troops, hungry ones. This is control of the physical factor

VII.25. They do not engage an enemy advancing with well-ordered banners nor one whose formations are in impressive array. This is control of the factor of changing circumstances.

Passage #12.10. Knowing Sanitary and Public Health Conditions Affecting Readiness

(Focus here is on recognizing and avoiding circumstances conducive to disease, a basic factor in promoting military readiness.)

IX.13. An army prefers high ground to low; esteems sunlight and dislikes shade. Thus, while nourishing its health, the army occupies a firm position. An army that does not suffer from countless diseases is said to be certain of victory.[49]

Passage #12.11. Knowing the Loyalty and Reliability of Your Troops

(The start of verse IX.47 points to a need for the general to probe beneath the surface of his army's demeanor and the hubbub and ritual that attends army life, to gauge the true loyalty of his troops. That is a major self-knowledge challenge calling for a finely honed blend of observation and intuition. These verses also point to need for an additional type of self-knowledge, centering on quality of

---

[46] Regarding the Chinese concept of *qi* that appears here – which Griffith here and elsewhere in Passage #12.9 variously translates as "spirit" or "spirits" or "moral factor" – see #3 footnote 80.

[47] The timeframe in verse VII.21 need not be limited to a single day, a point that has implications for readiness evaluation tasks. See Preliminaries footnote 4 (citing sources).

[48] Giles, p. 67 plays up an observer aspect: "A clever general, therefore, avoids an army when its spirit is keen, but attacks it when it is sluggish and inclined to return. This is the art of studying moods."

[49] Lit. "the hundred diseases." See p. 114 above. For further discussion see Passage #9.9.

implementation of orders.[50] Accurate organizational knowledge of that type is often in short supply in highly authoritarian organizations, which Sun Tzu's armies were, since underlings there do not relish being bearers of bad news.)

IX.47. If troops are punished before their loyalty is secured they will be disobedient. If not obedient, it is difficult to employ them. If troops are loyal, but punishments are not enforced, you cannot employ them.

IX.48. Thus, command them with civility [wen] and imbue them uniformly with martial ardour [wu] and it may be said that victory is certain.[51]

IX.49. If orders which are consistently effective are used in instructing the troops, they will be obedient. If orders which are not consistently effective are used in instructing them, they will be disobedient.

IX.50. When orders are consistently trustworthy and observed, the relationship of a commander with his troops is satisfactory.

Passage #12.12. Knowing Factors Producing Failures of Readiness or Effectiveness

(These verses catalog assorted failures of which a general should take heed.)

X.9. Now when troops flee, are insubordinate, distressed, collapse in disorder or are routed, it is the fault of the general. None of these disasters can be attributed to natural causes.

X.10. Other conditions being equal, if a force attacks one ten times its size, the result is flight.[52]

X.11. When troops are strong and officers weak the army is insubordinate.

X.12. When the officers are valiant and the troops ineffective the army is in distress.

X.13. When senior officers are angry and insubordinate, and on encountering the enemy rush into battle with no understanding of the feasibility of engaging and without awaiting orders from the commander, the army is in a state of collapse.

X.14. When the general is morally weak and his discipline not strict, when his instructions and guidance are not enlightened, when there are no consistent rules to guide the officers and men and when the formations are slovenly the army is in disorder.

X.15. When a commander unable to estimate his enemy uses a small force to engage a large one, or weak troops to strike the strong, or when he fails to select shock troops for the van, the result is rout.

---

[50] Implementation issues are also noted in verse I.11 in the net assessment passage (Passage #12.3).

[51] Verse IX.48 incorporates traditional Chinese cultural content via the wen/wu concept pair. For background on these twin concepts see Passage #13.31 discussion.

[52] Applying concepts from #12 footnote 6 ("Effectiveness = Performance × Availability × Utilization"), verse X.10 illustrates low (or zero) effectiveness resulting from misutilization of a force.

X.16. When any of these six conditions prevails the army is on the road to defeat. It is the highest responsibility of the general that he examine them carefully.

Passage #12.13. More about Factors Producing Readiness or Effectiveness Failures

(Since Warring States generals must have often used forced marches to spring surprises – verse XI.29 [Passage #3.7] says as much – these verses are a warning to be aware that such marches can greatly degrade an army's readiness and effectiveness when it arrives at its destination.)

VII.2. Nothing is more difficult than the art of manoeuvre. What is difficult about manoeuvre is to make the devious route the most direct and to turn misfortune to advantage.

VII.3. Thus, march by an indirect route and divert the enemy by enticing him with a bait. So doing, you may set out after he does and arrive before him. One able to do this understands the strategy of the direct and the indirect.

VII.4. Now both advantage and danger are inherent in manoeuvre.

VII.5. One who sets the entire army in motion to chase an advantage will not attain it.

VII.6. If he abandons the camp to contend for advantage the stores will be lost.

VII.7. It follows that when one rolls up the armour and sets out speedily, stopping neither day nor night and marching at double time for a hundred li, the three commanders will be captured. For the vigorous troops will arrive first and the feeble straggle along behind, so that if this method is used only one-tenth of the army will arrive.

VII.8. In a forced march of fifty li the commander of the van will fall, and using this method but half the army will arrive. In a forced march of thirty li, but two-thirds will arrive.

VII.9. It follows that an army which lacks heavy equipment, fodder, food and stores will be lost.

Passage #12.14. Knowing Your Country's Economic Situation and the Costs of War

(Focus here shifts to a different major facet of "knowing oneself," pertaining to the economic and logistics foundations on which all military effort builds.[53] While the numbers stated in verses II.13 and II.14 could be figurative, their mention opens a door to acquiring self-knowledge of a country-level statistical type. A corollary is that ruler and general need to be able to track, if possible on a quantitative level, the impact of sustained warfare on the economic foundations of the state along with its warmaking capacities – human, material, financial. Albeit that verses II.9 and II.10 are at some pains to sidestep relevant constraints,

[53] See pp. 33–38 above for overview of Warring States logistics and military economics.

verse II.11's reference to "distant transportation"[54] suggests cognizance of what has been called the "logistics bridge" connecting civilian economic activity to the logistics needs of an army in the field.[55] In many societies such a logistics bridge requires meshing two very different, often conflicting cultures, civilian and military.[56] In the conditions of Warring States China – where the state, its people, its economy, and its army were fused as far as possible into one warmaking machine – such cultural conflicts were held to a minimum.[57] However, they stand in the background of civil–military relations issues identified in Passage #12.20.)

II.1–II.2. (Translation from Lau's 1965 article, p. 324, which starts with a caption supplied by Sun Tzu: "*The method of employing troops.*"[58]) "When one thousand fast four-horse chariots, one thousand four-horse wagons covered in leather, and one hundred thousand armour-clad troops are used, if provisions have to be transported over a distance of a thousand li, then what with expenditure at home and in the field, on the guest advisers, on materials such as glue and lacquer, and on the supply of chariots and armour, it will cost one thousand pieces of gold every day before one hundred thousand troops can be raised."

II.3. Victory is the main object in war. If this is long delayed, weapons are blunted and morale depressed.[59] When troops attack cities, their strength will be exhausted.

II.4. When the army engages in protracted campaigns the resources of the state will not suffice.

II.5. When your weapons are dulled and ardour damped, your strength exhausted and treasure spent, neighbouring rulers will take advantage of your distress to act. And even though you have wise counsellors, none will be able to lay good plans for the future.

. . .

---

[54] The Chinese phrase is *yuan shu* and another translation would be "long-distance transportation."

[55] Eccles, Background footnote 52, pp. 53–57 (three charts developing Eccles's "logistics bridge" concept). For relevant World War II background, in a context where long-distance transportation and advanced base issues held center stage, see also Boorman, Introduction footnote 4, p. 99.

[56] Eccles, Preliminaries footnote 5, pp. 71–76 (inherently dual nature of the logistics process).

[57] That certainly did not eliminate all imperfections in the logistics bridge (e.g., fraud) and need for measures to try to spot them (a form of "knowing oneself"), then stamp them out. See, e.g., the *Mozi* text of ~242 BC as translated by Brooks & Brooks, Introduction footnote 36, p. 119: "Let the wounded return home to heal their wound and be cared for. Provide a doctor who will give medicines … Have an officer go regularly to the village to see if the wound has healed … In the case of those who falsely wound themselves to avoid service, put the whole family to death." For background and scholarship on the *Mozi* text see Background footnotes 12 and 80.

[58] Rationale for using Lau's translation rather than Griffith's is given in Appendix 1 footnote 18.

[59] Applied to the home front, verse II.3 points to need for observation seeking symptoms of war-weariness. For pioneering modern analysis see Lewis F. Richardson, "War moods," *Psychometrika*, 1948, I: 13(3), 147–74, II: 13(4), 197–219.

II.9. Those adept in waging war do not require a second levy of troops nor more than one provisioning.

II.10. They carry equipment from the homeland; they rely for provisions on the enemy. Thus the army is plentifully provided with food.

II.11. When a country is impoverished by military operations it is due to distant transportation; carriage of supplies for great distances renders the people destitute.

II.12. Where the army is, prices are high; when prices rise the wealth of the people is exhausted. When wealth is exhausted the peasantry will be afflicted with urgent exactions.

II.13. With strength thus depleted and wealth consumed the households in the central plains will be utterly impoverished and seven-tenths of their wealth dissipated.

II.14. As to government expenditures, those due to broken-down chariots, worn-out horses, armour and helmets, arrows and crossbows, lances, hand and body shields, draft animals and supply wagons will amount to sixty per cent of the total.

Passage #12.15. Keeping Track of War's Devastating Effects on Civilian Economic Life

XIII.1. Now when an army of one hundred thousand is raised and dispatched on a distant campaign the expenses borne by the people together with the disbursements of the treasury will amount to a thousand pieces of gold daily. There will be continuous commotion both at home and abroad, people will be exhausted by the requirements of transport, and the affairs of seven hundred thousand households will be disrupted.[60]

Passage #12.16. Know Your Commanding General, and Pick Him Carefully!

III.18. Now the general is the protector of the state. If this protection is all-embracing, the state will surely be strong; if defective, the state will certainly be weak.

Passage #12.17. Know Your Commanding General: Importance of Assessing Motivation

(This verse points to a crucial need, part of the larger "know yourself" problem faced by rulers, for a ruler to ascertain the loyalty of the general to his chief. Over the long course of Chinese history many a Chinese ruler got that assessment wrong – one way or the other – greatly to his cost. The difficulty of the challenge may be gauged from the fact that Sun Tzu's general, as a core job qualification, is *supposed* to be a consummate deceiver – and deception need not always stop with deceiving the enemy in another state.)

---

[60] As traditional commentator Mei Yaochen puts the point succinctly, in terms geared to an agricultural society: "There will be a shortage of men at the plow." (Minford, p. 315.)

X.19. And therefore the general who in advancing does not seek personal fame, and in withdrawing is not concerned with avoiding punishment, but whose only purpose is to protect the people and promote the best interests of his sovereign, is the precious jewel of the state.

Passage #12.18. Know Your Commanding General's Qualities of Judgment

(This verse – as here understood – points to a need to know what sort of military and strategic judgment a ruler's choice of general will exhibit.[61] That assessment involves a judgment about compatibility with Sun Tzu's strategic style.)

I.15. If a general who heeds my strategy is employed he is certain to win. Retain him! When one who refuses to listen to my strategy is employed, he is certain to be defeated. Dismiss him!

Passage #12.19. Know Your Spymaster: Qualities to Be Sought

(This is Sun Tzu's spymaster counterpart to a "know your general" imperative.)

XIII.13. He who is not sage and wise, humane and just, cannot use secret agents. And he who is not delicate and subtle cannot get the truth out of them.

. . .

XIII.23. And therefore only the enlightened sovereign and the worthy general who are able to use the most intelligent people as agents are certain to achieve great things. Secret operations are essential in war; upon them the army relies to make its every move.

Passage #12.20. Know the State of Civil–Military Relations

(Dispassionate observation and evaluation on the part of both ruler and general is essential here. That does not make the task at all easy. Potent countervailing forces – personal, political, organizational – are certain to intrude, operating to cloud clarity of observation and making the observer tasks implicit in this passage among the hardest of any on which Sun Tzu touches.)

III.19. Now there are three ways in which a ruler can bring misfortune upon his army.
III.20. When ignorant that the army should not advance, to order an advance or ignorant that it should not retire, to order a retirement. This is described as "hobbling the army".
III.21. When ignorant of military affairs, to participate in their administration. This causes the officers to be perplexed.
III.22. When ignorant of command problems to share in the exercise of responsibilities. This engenders doubts in the minds of the officers.
III.23. If the army is confused and suspicious, neighbouring rulers will cause trouble. This is what is meant by the saying: "A confused army leads to another's victory."

---

[61] Other readings of verse I.15 are possible. See #13 footnote 10 and Appendix 6 anchor footnote 3.

Passage #12.21. Knowing People, the Better to Pick Them

(Focus in Passages #12.21–#12.26 now shifts to more generic observation targets, which could be any person or organization relevant to Sun Tzu's general or ruler. At the heart of Passage #12.21 stands a need to be a shrewd judge of people.[62])

V.21. Therefore a skilled commander seeks victory from the situation [shi] and does not demand it of his subordinates.

V.22. He selects his men and they exploit the situation [shi].[63]

Passage #12.22. Knowing Persons of Interest by Their Personality Flaws

(Read in an analogical spirit, the content of these verses suggests similar scrutiny of non-enemy persons of interest, civilian as well as military. Notice, however, that while Sun Tzu's ideas here and in the next two passages lay groundwork, the thinking represented there has not yet developed – at least not explicitly – to a point where it contemplates crafting and exploiting a customized psychological profile of a *specific* person of interest.[64])

VIII.17. There are five qualities which are dangerous in the character of a general.

VIII.18. If reckless, he can be killed;

VIII.19. If cowardly, captured;

VIII.20. If quick-tempered you can make a fool of him;

VIII.21. If he has too delicate a sense of honor you can calumniate him;

VIII.22. If he is of a compassionate nature you can harass him.

VIII.23. Now these five traits of character are serious faults in a general and in military operations are calamitous.

---

[62] Ames, p. 87 notes that Sun Tzu's emphasis "on the effective selection of military personnel reflects a fundamental assumption in the tradition." Early Chinese approaches to that task were diverse. Some involved "physiognomizing" a person to ascertain their qualities and future from visible bodily features – a pseudoscience that had its devotees as well as detractors over much of Chinese history. See Sawyer, #1 footnote 3, pp. 346–47. Other approaches involved versions of what would now be called ego network analysis, evaluating a person of interest through that individual's network ties, scrutinizing their friends and associates. See, e.g., Lüshi chunqiu text, quoted under title of *The Annals of Lü Buwei* in #2 footnote 71. Verses V.21–V.22 further broaden focus, now relating personnel selection to sizing up a larger ongoing situation and a focal individual's contribution to it. See Appendix 4 anchor footnote 4. Such sizing up might, for example, involve an estimate of likely effects if that individual were removed; or a key social network tie involving that person ceased to exist; or that tie's content changed (as when a formerly favored general runs afoul of the ruler, as illustrated by the story of Wu Zixu, a colleague of the putative historical Sun Tzu; see #6 footnote 9).

[63] A textual variant pointing elsewhere is analyzed and rejected in Appendix 4 anchor footnote 4.

[64] See Raphals, Introduction footnote 42, p. 137:
A crucial aspect of both [Sun Tzu and the methods of Zhuge Liang as portrayed in the *Romance of the Three Kingdoms*] is psychological knowledge, which enables both the prediction of an enemy's likely actions and a knowledge of how to mislead him effectively. The *Romance* differs from the *Sunzi*, however, in its emphasis on detailed psychological understanding of an individual enemy. This strand in the *Romance* seems to represent an evolution in military thought over that of the *Sunzi*.
(Emphasis in original; Zhuge Liang's "empty city" stratagem – see pp. 249–50 above – is cited.)

VIII.24. The ruin of the army and the death of the general are inevitable results of these shortcomings. They must be deeply pondered.

Passage #12.23. Actively Probing of Persons of Interest, the Better to Know Them

(Thinking analogically, the steps noted in Passages #12.23 and #12.24 may be recast as approaches to probing non-enemy persons of interest.)

I.22. Anger his general and confuse him.
I.23. Pretend inferiority and encourage his arrogance.
I.24. Keep him under a strain and wear him down.
I.25. When he is united, divide him.

Passage #12.24. Further Probing Moves

VI.20. Therefore, determine the enemy's plans and you will know which strategy will be successful and which will not;
VI.21. Agitate him and ascertain the pattern of his movement.
VI.22. Determine his dispositions and so ascertain the field of battle.
VI.23. Probe him and learn where his strength is abundant and where deficient.

Passage #12.25. Gathering Detailed Information on Persons of Interest

(An important generalizable focus of these verses is collecting detailed information about persons of interest, including social or contact network information.)

XIII.15. If plans relating to secret operations are prematurely divulged the agent and all those to whom he spoke of them shall be put to death.
XIII.16. Generally in the case of armies you wish to strike, cities you wish to attack, and people you wish to assassinate, you must know the names of the garrison commander, the staff officers, the ushers, gate keepers, and the bodyguards. You must instruct your agents to inquire into these matters in minute detail.

Passage #12.26. General Observer Tips, Based on Analogical Reading of Sun Tzu

(Passage #12.26 = verses IX.19–IX.44, which Table 6 analyzes in five clusters:

[1] non-verbal clues – many of them broadly ecological or environmental – to observee organization intent, non-visible actions, situations, etc.;
[2] verbal clues;
[3] clues from observed resource commitments;
[4] clues to observee disequilibrium;
[5] clues to extreme disequilibrium, possibly foreshadowing catastrophic failure of the organization being observed.)

★★★★★★

G

Strategist Should Manage the Interfaces

"Interface" in present usage is a modern concept whose roots lie in physical science and engineering disciplines. For serendipitous reasons it happens to work well as a conceptual tool for structuring Sun Tzu's approach to leadership topics, starting with the political level (the ruler) and extending in the military realm down to the level of common soldiers. That is the focus of Theme #13.

For a slightly different reason – summarized by the observation that "[i]n military operations, the basic interface is with the enemy" – the concept of an interface is also a serviceable backdrop for Theme #14 development.[1] Theme #14 addresses the military concepts of qi and zheng, a concept pair for which there is no truly adequate English-language translation.[2] Sun Tzu assigns a basic role to each of these concepts in managing the all-important interface with the enemy, while also setting a stage for analyzing "switches" between the two, zheng → qi, qi → zheng. Adopting a switching perspective helps avoid a common pitfall in the modern Western Sun Tzu literature. Possibly animated in part by a desire to "sell Sun Tzu" by playing up Sun Tzu's crafty aspects, this pitfall involves outsized attention to the qi (crafty, surprise, unconventional, etc.) strands of Sun Tzu's thinking, crowding out its more mundane-appearing zheng features. The resulting imbalance distorts the essential place of both qi and zheng as basic ingredients of Sun Tzu's way of war.

Running throughout both Themes #13 and #14 is a further, and in some ways distinctively Chinese, cultural concept of harmony, which is an apt rubric for succinctly summarizing much of Sun Tzu's thinking about both themes.[3] As Lau and Ames, p. 85 describe it, "Harmony is achieved through the art of productively contextualizing the available ingredients, whether they be foodstuffs, farmers, or infantry. ... It is the capacity to make the most of any situation, even when that

---

[1] See Harrison C. White, *Identity and Control* (revised ed.; Princeton, NJ: Princeton University Press, 2008), p. 80, making the further Clausewitzian observation that in war "the classic though not sole way of dealing across the interface is the engagement."

[2] Griffith renders qi and zheng as "extraordinary" and "normal," respectively. Lau's 1965 article favors "crafty" and "straightforward," while Sawyer opts for "unorthodox" and "orthodox" and Mair for "unconventional" and "conventional." There are other translation possibilities too.

[3] To be clear, "harmony" in present usage does not correspond to any one specific concept in Sun Tzu.

situation is war." Such a concept, finding manifestations of "harmony" even amid seemingly profound disorder, encourages thinking in an integrated way about many of the major intellectual currents in Sun Tzu. It reflects deep emphases in traditional Chinese civilization – including but by no means limited to the Confucian tradition – on concepts of harmony and balance, centered in ideal-typical human relations but also extending to harmony between humankind and nature.[4]

★★★★★★

THEME #13 LEADERSHIP AND HARMONY IN THE CHAIN OF COMMAND (AND AVOIDING ITS PATHOLOGIES)

the command role:
Passage #13.1 (verses V.15–V.16) (essence of the command role)

people/ruler interface:
Passage #13.2 (verses I.4 + I.11–I.14) (moral influence or *dao*)
Passage #13.3 (verses II.10–II.13) (unraveling harmony between ruler and people; warning signs of incipient regime failure?)
Passage #13.4 (verse XIII.1) (unraveling harmony, cont.)
Passage #13.5 (verses III.1 + III.3 + III.10–III.11) (basis for harmony between ruler and people of a conquered state?)
Passage #13.6 (verses XI.7 + XI.13 + XI.20 + XI.30–XI.31) (ruler and new subjects: plunder as Sun Tzu blind spot?)

ruler/general interface:
Passage #13.7 (verse XII.16) (division of labor between ruler and general)
Passage #13.8 (verses III.18 + III.29) (central importance of the general)
Passage #13.9 (verse II.21) (central importance of the general, cont.)
Passage #13.10 (verses VII.1–VII.2) (overview of general's role; multiple interfaces are pertinent here, but all involve the general)
Passage #13.11 (verse I.7) (qualities of an ideal general)
Passage #13.12 (verses XI.42–XI.43) (further attributes of ideal general)
Passage #13.13 (verses VIII.17–VIII.24) (disabling character flaws of a general)
Passage #13.14 (verse I.15) (hiring and firing generals)
Passage #13.15 (verse X.19) (general's contribution to harmonious civil–military relations)
Passage #13.16 (verses III.19–III.23) (pathologies of civil–military relations)
Passage #13.17 (verses VIII.7–VIII.8) (general need not always obey commands of the ruler)
Passage #13.18 (verse X.18) (more on same idea)

---

[4] This a good point at which to underscore that Sun Tzu was no Confucian, and that the Sun Tzu work took shape in an intellectually and politically decentralized pre-imperial era of Chinese history, well before Confucian ideas and institutions attained their later near-hegemonic sway.

general/army interface:
Passage #13.19 (verse III.27) (successful leadership)
Passage #13.20 (verse X.20) (Sun Tzu general's posture vis-à-vis his troops – soldiers as infants)
Passage #13.21 (verse XI.49) (Sun Tzu general's posture vis-à-vis his troops – soldiers as sheep)
Passage #13.22 (verses IV.19–IV.20) (troops well organized and led – an illustration)
Passage #13.23 (verses V.17–V.19) (troops well organized and led – a further illustration)
Passage #13.24 (verses XI.41–XI.41a) (morale as a quality of troops requiring management)
Passage #13.25 (verse XI.44) (morale as a quality of troops requiring management, cont. – prohibiting superstitious practices)
Passage #13.26 (verses VII.20–VII.25) (managing morale as a dynamic task, unfolding in real time)

tools of leadership:
Passage #13.27 (verse IV.15) (tools of leadership: cultivating the *dao*)
Passage #13.28 (verse XI.24) (skill-sets that generals need: overview of core elements)
Passage #13.29 (verses V.21–V.22) (tools of leadership: choice of personnel)
Passage #13.30 (verse I.8) (tools of leadership: organizational and logistics doctrine and methods)
Passage #13.31 (verses IX.47–IX.50) (tools of leadership: applied group psychology, emphasizing timing and consistency, with a *wen/wu* Chinese cultural dimension)
Passage #13.32 (verse XI.54) (tools of leadership: imagination and flexibility in use of rewards and orders)

pathologies of leadership:
Passage #13.33 (verses X.9–X.16 + X.21) (pathologies of leadership)
Passage #13.34 (verses IX.37 + IX.41–IX.43) (pathologies of leadership, cont.)
Passage #13.35 (verses II.3–II.5) (pathologies of leadership, cont.: protracted war and attacks on cities as sources of flagging morale)
Passage #13.36 (verses VII.7–VII.8) (pathology of another kind – decapitation of high command)

interface with spies:
Passage #13.37 (verses XIII.12–XIII.13 + XIII.15–XIII.17 + XIII.21 + XIII.23) (relationships with secret agents)

★★★★★★

Sun Tzu's treatment of war's leadership aspects is best approached as a hybrid. On the one hand, it expresses certain universals not limited by time or place. On the other, it is also permeated by ideas and standpoints reflecting Warring States Chinese society, a society profoundly different from any modern one, Chinese included.[1] Given Sun Tzu's focus on the commanding general's role, present coverage commences with two verses from Sun Tzu V (Passage #13.1), providing a sharp and in many ways universal depiction of successful exercise of military command:[2]

V.15.

Thus the momentum [shi] of one skilled in war is overwhelming, and his attack precisely regulated.

V.16.

His potential [shi] is that of a fully drawn crossbow; his timing, the release of the trigger.

Via this crossbow image Sun Tzu puts his finger on several defining aspects of the command role, in effect providing a structural analysis of it. Boiled down to its essentials, the crux of that role is the ability of a commander – which becomes progressively greater the higher the level of command – to make decisions that:

(a) have major consequences, often – as the crossbow does – affecting individuals and populations at a distance;

(b) require little effort and are capable of being swiftly taken (like pulling on a crossbow trigger);[3]

(c) have consequences that are also irreversible (including loss of life, possibly on a large scale).[4]

The Sun Tzu text is oriented toward the general rather than the ruler, an emphasis that does much to differentiate Sun Tzu's core analytical interests from those of a political theorist or philosopher. Indeed it might not be too much to see Sun Tzu's entire perspective on war as pivoting on one role occupied by one individual – the commanding general. All else in Sun Tzu's purview derives its importance from its relevance to that general and tasks he must do or the obstacles he faces. Subordinate officers at any level are scarcely mentioned.[5] *On a point of terminology, Theme #13*

---

[1] A concrete anchoring for this observation comes from Wilkinson, §36.1, p. 488: "To somebody brought up on early twenty-first century Chinese cuisine, Ming dishes would still seem familiar, and possibly, as some have claimed, also those of Song, but anything before the Song [founded in 960 AD] would probably be difficult to recognize as Chinese food as we know it today."

[2] For further analysis of *shi*, sometimes identified as Sun Tzu's central concept, see Theme #5.

[3] An illuminating modern comment here is an observation of Eisenhower's chief of staff Walter Bedell Smith, to the effect that a major command decision may involve little more than a nod of the commander's head. See his *Eisenhower's Six Great Decisions: Europe, 1944–1945* (New York: Longmans, Green, 1956), p. 60 (also quoted by Eccles, Background footnote 52, p. 39).

[4] Sun Tzu has a well-honed sense of irreversibility of choices. Allusions to irreversibility or equivalents appear in numerous places in the text (thereby also bolstering Theme #1's focus on accurate calculation before making a decision). A non-exhaustive sampler would include:

(1) verses V.16 (Passage #13.1) and XI.48 (Passage #10.13), invoking a metaphor of pulling a trigger (*faji*), as of a crossbow (Griffith, p. 137 footnote 3, following commentator Wang Xi, explicitly associates this idea with taking an irreversible action; Giles, p. 133 translates verse XI.48 using the card-playing image of "show[ing] his hand," which suggests taking an irreversible *informational* action);

(2) verse X.3 (Passage #2.4A), on terrain easy to get out of but difficult to return to;

(3) verse XII.18 (Passage #1.8), focusing on the irreversible consequences that can arise from a decision taken by a ruler or general in the heat of passion propelled by uncontrolled emotions;

(4) verses XII.1–XII.13 (Passage #9.7), on fire attacks (which, of course, have irreversible aspects).

[5] Verses X.11–X.14 (Passage #13.33) draw a distinction between "officers" li (verses X.11 and X.12) and high-ranking or "big" officers da li (verse X.13), cautioning (by easy implication) that even the latter need to remember that they are only subordinates. Verses VII.7 and VII.8 (Passage #13.36) also take note of high-ranking subordinate officers but shed no particular light on command relations issues.

references to the "general" will therefore refer (unless otherwise indicated) to the commanding general and to him alone, never to lesser generals.

For Sun Tzu the general is a single human being, albeit one with idealized attributes. Sun Tzu accords no attention to any institutional structures that support the individual exercise of high command and that, in modern times, have often come to overshadow it (see Theme #1 analysis, pp. 56–57). In particular, the Sun Tzu text contains no allusion to a general staff, an institution that in post-Napoleonic times has come to be inseparable, in military establishments worldwide, from the exercise of command. While such a staff may have existed in Sun Tzu's time – and there is some evidence from other early Chinese military writing that it did – we know nothing about how it worked or the division of labor between staff and general.[6] What is clear is that – whatever the role such a staff might have played in Sun Tzu's armies – that role did not attain such salience for Sun Tzu as to call for explicit mention of it. That elision of staff roles imbues Sun Tzu's thinking about command with an essential simplicity, setting it apart from modern analyses of high command as a set of complex institutions.[7]

As a byproduct (surely unintended) this same simplicity smooths a way for Sun Tzu's thinking to be transposed to civilian life (Theme #4).

Sun Tzu's overall theory of command may be usefully structured from a standpoint inspired in part by sociological role theory, in part by ideas from systems analysis (both deployed here as modern analyst tools). The overall system to be analyzed might be described as command and its human environment. By contrast to three "subsystems" commonly associated with the famous Clausewitzian "trinity" (respectively involving the people; the commander and his army; the government), a similar, slightly richer five subsystem formulation may be distilled from Sun Tzu:[8]

- the people
- the ruler
- the commanding general
- the army
- the espionage apparatus (found only in Sun Tzu XIII, possibly a late addition to the Sun Tzu text)

To these should be added allies and neutrals – and, of course, the enemy – as "associated systems."

---

[6] Noting a military staff description in the Liutao text see p. 34, #1 footnote 2, and #11 footnote 45. Like Sun Tzu, the Liutao is one of the Seven Military Classics (see p. ix footnote 7). Referring to the early Chinese states of Yue and Wu, the latter of which the putative historical Sun Tzu served, Milburn, #1 footnote 41, p. 8 observes: "Even less is known about the military and naval command structures" than about the "civil administration ... [which seems] to have gone largely unrecorded."

[7] See, for example, Boorman, Introduction footnote 4, p. 108 note 14 (US as having some ten general staffs [!]). For further background on the modern evolution and structure of general staffs see also Otto L. Nelson, Jr., National Security and the General Staff (Washington, DC: Infantry Journal Press, 1946), reviewed by Henry Eccles in US Naval Institute Proceedings, 1947, 73(6), 720–21.

[8] This capsule summary of the Clausewitzian "trinity" bypasses some issues in understanding Clausewitz, not necessary to revisit here. For discussion see Edward J. Villacres & Christopher Bassford, "Reclaiming the Clausewitzian trinity," Parameters, Autumn 1995 issue, pp. 9–19.

Using systems terminology Sun Tzu's five subsystems give rise to four "interfaces":

(1) people/ruler
(2) ruler/general (this includes Sun Tzu's thinking about civil–military relations)
(3) general/army
(4) general (or ruler; Sun Tzu is not quite clear about this)/espionage network.

The following diagram offers a visual presentation:

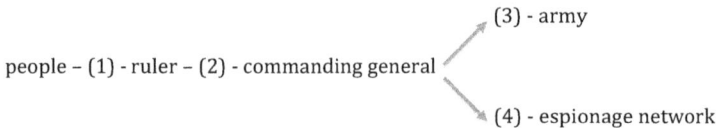

people – (1) - ruler – (2) - commanding general
(3) - army
(4) - espionage network

The bulk of Sun Tzu's extensive Theme #13 coverage is readily recognizable as a description, usually framed in ideal-type normative terms, of one or more of these four interfaces. With a bow to a longstanding Chinese cultural penchant for numbered lists these interfaces might be labeled the Four Harmonies. Moving from left to right in the diagram, the elements of Sun Tzu's theory of leadership, including its problematic patches and limitations, can be placed in a clear perspective.

Commencing the Four Harmonies description with the people/ruler interface, Passage #13.2 profiles an ideal-type harmony in that relationship. Such harmony is by no means guaranteed; Passages #13.3 and #13.4 are word pictures implying that the exorbitant economic and human costs of war could unravel it. Such concern is consistent with Sun Tzu's call for taking an enemy state intact rather than ruining it (verse III.1 in Passage #13.5). The warfare of Sun Tzu's time was primarily warfare between elites. The lion's share of a newly conquered population was therefore most likely little affected by change at the top and – if conquest meant "taking intact," as verse III.1 (Passage #13.5) advises – might even have stood to benefit from it. That would in turn lay groundwork for achieving harmony between the conquered populace and its new ruler in the early aftermath of a successful war, facilitating consolidation of territorial gains.[9] However, Sun Tzu's unqualified endorsement of plunder (Passage #13.6) muddies the picture. Plausible scenarios involving plunder, during a war or after it, that harms civilian lives and livelihoods are at odds with verse I.4 (Passage #13.2) applied to a post-war populace in a conquered state.

[9] A pleasing image of a beneficent conqueror, one that has no counterpart in Sun Tzu, is provided by another early text, the Yi Zhou shu (for Yi Zhou shu background see Appendix 1 anchor footnote 8):

Therefore, conquest over an enemy state does not extend to its subjects. Execute only the executor, and no more. Promote talented local officers and enfeoff them; select worthy men and pay respect to them. Seek out the widows and orphans among them, and give compassion and relief to them. Give audience to their elders and treat them with proper ritual reverence. Raise salaries and increase ranks across the board. Assess the guilt or innocence of those accused of crimes, and release those jailed wrongly. Distribute funds from the treasury and hand out grain from the granaries, in order to relieve their masses. Do not treat their resources as your private booty. [etc.]

(McNeal, Background footnote 58, p. 60; see also his p. 108)

Shifting to the ruler/general interface, Passage #13.7 sets forth an ideal-type division of labor between commanding general and ruler. Passages #13.8 and #13.9, which take on particular force in a context of Warring States China, underscore just how crucial is that general's role. Fleshing out that role, Passage #13.10 gives a snapshot of the general's tasks as soon as war breaks out. Passages #13.11 and #13.12 zoom in on the general-qua-person, profiling ideal-type attributes of the human being who fills that role. True to Sun Tzu's alertness to human frailties, Passage #13.13 explores the flip side – disabling character flaws a general may exhibit.

Passage #13.14 is a noteworthy nugget of Sun Tzu advice (of potential interest to flag rank officers the world over!) on hiring and firing generals:

> I.15.
> If a general who heeds my strategy is employed he is certain to win. Retain him! When one who refuses to listen to my strategy is employed, he is certain to be defeated. Dismiss him!

A corollary of verse I.15 is the basic importance of compatible "styles" on a level of top command (here meaning strategic styles of the ruler and his general).[10]

Against this backdrop, Passages #13.15–#13.18 turn to the ever-delicate issue of relations between general and ruler (the people enter here, but decidedly sotto voce).[11] A key harmonic (inevitably a source of concern to even the most secure ruler) is the power inherent in the general's role. Certainly if its incumbent is talented, that power could easily have potential to unseat the ruler himself. This

---

[10] A variant reading of verse I.15 (see Appendix 6 anchor footnote 3) would cast Sun Tzu as leaving if his counsel is not heeded (in particular, by the ruler). That substitution of dramatis personae does not affect the basic idea that "strategic styles" of key players need to be compatible.

[11] Lewis, Background footnote 11, p. 122 sketches basic terms of that relationship, centering on the concept of ruler and commanding general as occupying what might be described as "separate spheres," certainly once a war has begun:

> [S]ince the commander was to be the unchallenged and absolute leader of a force in the field, he was obliged to make frequent decisions on matters of life and death without any recourse to the distant ruler. The successful performance of his role required immediate and unhesitating reactions to situations in constant flux, and any factor that blocked his freedom of decision or introduced an element of doubt or hesitation into his mind could prove disastrous. While other officials could be simple agents of the ruler's will, his organs of sense or his limbs, the commander had to be ruler in his own sphere, a lesser sage in service of a greater. *The reconciliation of these two authorities, however, was problematic, because the virtues of the commander and the principles that guided his actions potentially contradicted and undercut the virtues and principles of the ruler.*
>
> [Emphasis supplied.]

Here, however, as with so much historical context relevant to Sun Tzu, it is necessary to tread somewhat carefully. Describing a somewhat earlier period (the period in which the putative historical Sun Tzu lived), a comment of Olivia Milburn, *Cherishing Antiquity: the cultural construction of an ancient Chinese kingdom* (Cambridge, MA/London: Harvard University Asia Center, distributed by Harvard University Press, 2013), p. 159 paints a picture where the ruler is the commanding general:

> The historical significance of these campaigns (for example, the campaign against Chu during the reign of King Helü, which led to the sack of their capital [a campaign in which the putative historical Sun Tzu is said to have played a part], or the victory over Qi at the battle of Ailing in the reign of King Fuchai) has been much discussed. *What has been less frequently mentioned is that the kings of Wu directed these triumphant campaigns personally.* Although these monarchs could have called upon the services of such distinguished generals as Sun Wu [i.e., Sun Tzu] and Wu Zixu, it is clear that, in fact, the kings took personal control of their armies.
>
> [Emphasis supplied.]

issue is too sensitive for Sun Tzu to tackle more than obliquely; the closest he comes to doing so is verse X.19 (Passage #13.15), which alludes in blandly favorable terms to "promot[ing] the best interests of his sovereign." Tellingly, when Sun Tzu does tackle pathologies in the ruler/general relationship those pathologies pertain to a different problem, that of misguided encroachment by the ruler on the commanding general's military sphere (verse III.29 in Passage #13.8; Passages #13.16–#13.18).

Moving down the chain of command, the next focus is the general/army interface. Two aspects of it can be distinguished. One pertains to the commanding general's relation to his senior subordinate commanders. A second pertains to that general's relation to the army at large, including rank-and-file troops.

Regarding the first, it is tempting to treat the commanding general's authority as absolute, or virtually so. However, note should be taken of a possible limitation on that authority in Sun Tzu's time. That limitation is sketched in a comment of Mair regarding Sun Tzu's "dispersive ground" category, suggesting that it is[12]

> [s]o called because the feudal lords do not send their troops to serve under the coordinated command of the leading general of the state to which they owe fealty. Instead, the feudal lords retain control of their troops, even though they may be fighting for the cause of a king to whom they owe allegiance.

Although the Sun Tzu text does little to enlighten us on this key aspect of command relations[13] – which might, of course, have been in flux in the period when the text took shape – it clearly would have basic consequences for exercise of leadership and command, hobbling the more soaring ambitions of Sun Tzu's art of generalship.[14]

Shifting to the army-at-large part of the general/army interface, Passage #13.19 is an insightful, if clearly idealized, summary statement:[15]

> III.27.
> He whose ranks are united in purpose will be victorious.

Less universal in thrust, and certainly more likely to elicit controversy in Sun Tzu's modern audiences, are the next two passages (Passages #13.20 and #13.21). These have Sun Tzu's general presiding over troops cast as children or sheep, images that

---

[12] Mair, p. 155 note 2, glossing verse XI.2 (Passage #2.1) which defines dispersive ground. For alternative bases for Sun Tzu's dispersive ground thinking see Appendix 2 anchor footnote 4.

[13] A hint of the type of loosely cobbled command relations that Mair's interpretation suggests is verse X.13 (Passage #13.33), pointing to insubordinate steps by high-level subordinates who "do battle on their own initiative" (Mair, pp. 114–15).

[14] Note should be taken here of comparative-historical analysis of John Levi Martin, *Social Structures* (Princeton, NJ: Princeton University Press, 2009), distinguishing *anti-transitivity* and *transitivity* in military command relations. A transitive command relation is one where a superior can issue binding orders directly to a subordinate's subordinate. An anti-transitive relation requires that orders be routed via the immediate subordinate, with all the possibilities for delay, distortion, or outright disobedience that entails. See Martin's book, pp. 210ff. Somewhat oversimplifying Martin's findings, Martin's analysis identifies and analyzes significant gains in military effectiveness arising from replacing the anti-transitivity characteristic of patronage pyramids with transitive military command relations. See Martin's chapter 7, in particular his pp. 270ff.

For highlights and critique of the argument see also the review of Martin's book by Gagan D. S. Sood, "Asking about origins," *Science*, November 27, 2009, 326(5957), 1190–91.

[15] Emphasizing Warring States aspirations to total unity of effort, see Lewis, Background footnote 11, "The Commander and the Army," pp. 104–14, especially p. 105 (referring to Sun Tzu).

suggest diminished independent "agency."[16] Characteristics of Warring States society and its belief systems are here on display and, from many modern standpoints, show a clear limitation on Sun Tzu's leadership concepts.[17]

While still focusing on ideal types, Passages #13.22–#13.26 become more concrete. Passages #13.22 and #13.23 provide two different snapshots of troops well organized and led. Passage #13.24 points to an army's fighting spirit as not monolithic – any army is an amalgam of different degrees of valor – hence as an attribute of armies needing affirmative management. Passage #13.25 takes a stand on a related issue, calling for stamping out superstitious beliefs. Passage #13.26 contributes the important further point that management of fighting spirit is a dynamic task, unfolding in real time.

Passages #13.27–#13.32 inventory a repertoire of tools of leadership available to Sun Tzu's general to carry out these tasks successfully. Chinese cultural content is visible here, though (as should be expected) many of the tools align on leadership universals familiar from all ages of warfare. Culled from five of Sun Tzu's thirteen chapters, those tools make for a somewhat eclectic but informative list:[18]

- cultivating the *dao* (Passage #13.27) (a concept with many facets, some highly abstract, some more concrete; for Sun Tzu, the *dao* is generally practical)[19]
- knowledge of human character (or emotions) as one of the core skill-sets a general needs as part of the *dao* of generalship (Passage #13.28)
- choice of personnel (Passage #13.29)
- organizational and logistics methods (Passage #13.30)
- implementation steps, including training (Passage #13.2's verses I.11–I.13; Passage #13.31)[20]

---

[16] For further discussion see p. 301 and #9 footnotes 7 and 8. For agency as a sociological concept see Mustafa Emirbayer & Ann Mische, "What is agency?," *American Journal of Sociology*, 1998, 103(4), 962–1023. In defense of sheep, it should perhaps be noted – Warring States Chinese stereotypes to the contrary – that modern animal behavior research suggests that sheep are not actually so stupid.

[17] However, comparable imageries appear sporadically down into modern times. See, e.g., Peter Kenez, "The ideology of the white movement," *Soviet Studies*, 1980, 32(1), 58–83, e.g., p. 78: "The officers [of the anti-Bolshevik Volunteer Army, from the time of the Russian Civil War] looked on their peasant soldiers with condescension. Indeed, they thought of the soldiers as children, who both needed and liked to be told what to do."

[18] For analysis of capabilities and limitations of command, control, and communications methods available to armies of Sun Tzu's time, and their implications for Sun Tzu's way of war, see Appendix 15.

[19] Cf. Mair, p. 138 note 7 (glossing verse I.4 content in the Sun Tzu text): "This is not the metaphysical entity of early Taoist discourse nor the ethical and cosmological construct of Confucian thought. Rather, in the Sun Zi, it is a more practical, methodological notion." For basic background on *dao* as a Chinese philosophical concept see Appendix 1 footnote 3. For comments on occurrence of *dao* in verse IV.15 (Passage #13.27), which has some exceptional features, see Appendix 13 footnote 13. For broader background on the cultural context of translation see also Appendix 1 anchor footnote 2.

[20] The Chinese text of verse I.11 refers to the carrying out or implementation (*xing* 行) of *fa ling* (Griffith's "regulations and instructions"). Sun Tzu's attention to the importance of implementation is noteworthy, especially given the oft-neglected status of implementation issues into recent times. Regarding the latter see Jeffrey L. Pressman & Aaron Wildavsky, *Implementation*

- applied group psychology emphasizing timing and consistency (this tool has a *wen/wu* or civil/martial Chinese cultural dimension) (Passage #13.31)[21]
- imagination and flexibility in use of rewards and orders (Passage #13.32) (may be read as a Sun Tzu version of "thinking outside the box" in leadership practices)
- exploitation of terrain features to stabilize possibly wobbly morale (Passage #13.24)

This last tool is an important point of contact with Sun Tzu's well-developed terrain ideas (Theme #2) and, like them, has potential to be generalized in various ways, possibly to social network settings.

Making use of these tools also calls for cognizance of dynamics – some favorable, some not – already present in the situation (see Theme #9), including:

- dynamics associated with biophysical processes (hunger, fatigue, etc.) (Passage #9.10 = Passage #13.26)
- soldiers' natural motivations, in particular rage and greed (Passage #9.11)
- spread of superstitious practices and rumors (Passage #9.12 = Passage #13.25)
- lure of hearth and home when fighting on home turf, with its tendencies to sap fighting spirit and operational readiness and effectiveness) (Passage #9.13)
- behavior of soldiers in extreme ("death ground") circumstances (Passage #9.14)

Still focusing on the general/army interface, Passages #13.31 and #13.33–#13.35 delve (in what, for Sun Tzu, is considerable detail) into leadership failures. In an important affirmation of a principle of command responsibility, the most developed of those treatments finds Sun Tzu at pains to spell out (verse X.9 in Passage #13.33) that the failures he enumerates are the fault of the general, *not* of "natural causes" (a category that would naturally cover the *sua sponte* behavior of the general's troops).

Failures of a different sort, pointing to what would nowadays be called a decapitation strike against an army's high command, are noted by verses VIII.18 and VIII.19 (Passage #13.13) (focus is commanding general) and verses VII.7 and VII.8 (Passage #13.36) (focus is other senior officers).[22] Sun Tzu is silent as to what damage control steps should be taken to recover from such eventualities.

Turning to the remaining interface listed on p. 428 – that between general (or ruler) and espionage network – Passage #13.37 rounds up relevant substance

---

(Berkeley, CA: University of California Press, 1973). Commenting (as of 1973 publication date) on a deficit of a body of modern analytical work addressing policy implementation issues, they memorably observe (p. 166): "It must be there; it should be there; but in fact it is not." Especially with more attention to logistics issues in many quarters that deficit is gradually being rectified, but often unevenly.

   Xing 行 should noted as being a different character from xing 形 analyzed on pp. 175–76 above.

[21] Notwithstanding traditional Chinese cultural content present there, some of the basic human relations ideas found in Passage #13.31 have counterparts in modern military experience. See #13 footnote 92 (comments of George Homans on baseball competitions between warship crews).

[22] Read literally, verse XI.57 (Passage #6.21A) also points to a decapitation strike, though what Sun Tzu says there may be simply a figure of speech. A further candidate for decapitation thinking in Sun Tzu is Mair's "sniper interpretation" of verse IX.18 (Passage #7.5). See #6 footnote 5.

gleaned from Sun Tzu's espionage chapter. What Sun Tzu says here mostly pertains to describing ideal-type relations between a commanding general (or ruler) and his spies (though further interfaces would exist between spies and espionage targets, as verse XIII.16 implies). Apart from verse XIII.15 (recommending execution of a loose-mouthed spy and those to whom he talked), failures in those relations (including the ever-present possibility of double agents working for a foe) are not explored.

★★★★★★

As perhaps befits an infantry warfare treatise in an age of mass armies, Theme #13 contains almost twice as many main text passages as most of the other Sun Tzu themes. Sun Tzu's Theme #13 emphases may be compactly summarized as a call for "harmony" throughout the chain of command (and also extending beyond it to harmony between ruler and people). A "principle of inversion" may give insight here. It involves taking Sun Tzu's "harmonies" and standing each of them on its head, seeking to spot where the human relations Sun Tzu lauds were most likely to fall apart in armies of his time.[23] Contenders include verses XI.41a (Passage #13.24) and XI.54 (Passage #13.32) alluding to an army so perfectly fused that it appears to the general to be like one man responding to his will, when set alongside "the difficulty of securing the loyalty of troops who, unlike the chariot elites, had not been trained to war all their lives" (Brooks, p. 60). The key point here is that the armies Sun Tzu knew – heavily made up of peasant conscripts in massive numbers – seem to have been rather brittle social formations. If properly handled they were potent enough by the standards of their day, yet they were ever prone to shatter and disintegrate absent the general's strong controlling hand.[24]

Read in a similar spirit, the list of qualities of an ideal spymaster in verse XIII.13 (Passage #13.37) invites a contrarian reading as a list of Sun Tzu concerns about ways in which that principal's relations with his agents may fall apart!

Yet to do Sun Tzu justice as a thinker, he does not leave it entirely up to his audience to figure out where his most basic concerns lie. He does not merely play the role of cheerleader for harmony when it does exist, but also points out some of

---

[23] This analytical move has some similarities with deconstructing an organizational mission statement, using its often lofty verbiage as a source of pointers into actual unresolved organizational tensions and conflicts.

[24] Such brittleness – reminiscent of that of the famous swords of Wu (see p. 84 above) – gives a plausible reason why Sun Tzu shows such interest in creating situations where troops have no *alternative* but to fight fiercely because, put simply, they have literally no other option for survival (as contrasted with an emphasis on steps geared to building high and positive morale, about which Sun Tzu says little). Examples of "no way out" verses would include verses XI.33 and XI.55 (Passage #9.14) depicting death ground situations. A separate example of "no way out" control comes from traditional commentator Du Mu, who starkly observes: "The Military Law states: 'Those who when they should advance do not do so and those who when they should retire do not do so are beheaded.'" See Griffith, pp. 106–107 (context of Du Mu's remark is verse VII.18 [Table 15A in Appendix 15]).

Standing back from details, the key point is that Sun Tzu's leadership ideas are a product of Warring States China and they fit that context well. However, some of those ideas do not generalize well. See also #9 footnote 74 (contrasting modern concepts of morale with Sun Tzu's thinking).

the most crucial ways it may fail and (in some cases) to remedial steps.[25] By far the most interesting material here concerns pathologies at the heart of civil–military relations, hence (certainly in Warring States times) at the heart of the state itself.

That is the focus of Passages #13.16–#13.18 and of verse III.29 (Passage #13.8).

Here we encounter Sun Tzu's most important (and controversial!) legacy to the theory of command: his explicit endorsement of the possibility that the general may, under certain circumstances, disregard commands of the ruler.[26] Sun Tzu's most prominent concern here involves the army in the field and the crucial importance of avoiding interference in its combat operations by the ruler – a remote figure, quite possibly an unversed civilian. However, verse III.21 (Passage #13.16) also alludes to pernicious effects of a ruler's ignorant interference with military administration, which might occur at any time. Circumstances where a break in the chain of command is permissible are explicitly stated in Passage #13.18. Candidates for further examples are implicit in Passages #13.16 and #13.17. Framed in modern terminology, a basic factor here is *asymmetric information*: the general has essential information the ruler lacks. Verse III.22 (Passage #13.16) notably warns of the pitfalls of command-sharing by a ruler ignorant of command problems. That might extend to instances where a ruler lacks essential understanding of the nature of the military command role itself (a possibility well captured by the ruler's disturbed reaction to Sun Tzu's execution of the concubines in the famous story).[27]

Reading between the lines, it is possible to detect a certain empathy on Sun Tzu's part for a general who must constantly cope with the bumbling intrusions of an ill-informed, unversed, yet activist ruler. But it is essential to be clear what Sun Tzu is *not* advocating for here. No implication should be drawn that Sun Tzu endorses a *political* challenge by the general to the ruler's authority, even a challenge short of raising a flag of outright rebellion (the latter possibility a perennial nemesis of Chinese rulers across much of Chinese history).[28] Any challenge to the ruler's authority that does take place is to be confined to the military realm and centers on military judgments of a military-operational (or military-administrative) nature: that is, judgments falling within the commanding general's natural sphere.

Figure 6 is a snapshot of Sun Tzu's relevant thinking. Notably, Sun Tzu makes no allusion in this civil/military relations context to the deception thinking for which he is so famous (Theme #7). In that sense, Sun Tzu's openness to a general's defying

---

[25] As noted above (p. 432) such steps might, for example, include use of physical terrain features to stabilize morale. See Passage #1.41A analysis of verse XI.41 (which also appears in Passage #13.24).

[26] Continuing analysis in #13 footnote 11 above, that advice needs to be set in a context of Warring States times and that era's civil–military relations. Lewis, Background footnote 11, p. 127 amplifies:

> This separation of the army from civil society did not exist solely in theory, for the historical records indicate that the military camp and the army in the field were in fact legally and ritually separated from the state, and that in the army the orders of the state's ruler often yielded to those of the commander.

[27] See p. 329 above. For clarity, it should again be noted that this story, which is almost certainly apocryphal, does not appear anywhere in the thirteen chapters of the Sun Tzu text.

[28] As Needham & Yates, p. 52 caution: "the army's right to independence was formulated always with its subordinate position in mind; only within its limited instrumental function was it given freedom."

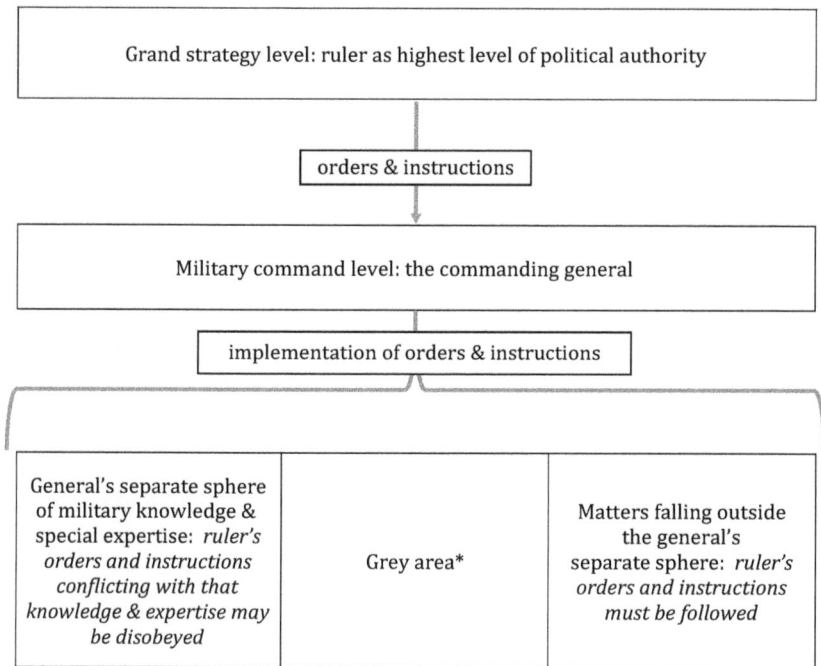

Figure 6. <u>Ruler and his commanding general: refusal to obey orders and its limits.</u>

This figure is a representation of concepts distilled from Sun Tzu's thinking pertaining to the relationship between ruler and general, focusing on circumstances where the latter may legitimately disobey the former's orders or instructions.

*Needham & Yates, p. 52 importantly note: "In reality the army was not fully independent of the State bureaucracy, and was never completely separated from it."

That observation evokes basic structural issues expressed by Admiral Eccles's concept of the "logistics bridge," one of whose termini is in the civilian economy of a country and the other is in the combat operations of its armed forces. Drawing on modern experience, Eccles emphasized that the duality of that bridge, with one end in civilian life, one in warfare – a duality inherent in the nature of the military logistics process – is a source of never-ending tension and debate, starting with all the reasons that civilian and military concepts and criteria of judgment often differ. See Eccles, Background footnote 62, pp. 64–65. See also Passage #12.14 discussion (where the logistics bridge concept also appears).

orders from the highest level (the ruler) is less Sun-Tzu-esque than might appear. Nonetheless the seriousness of the issues and possibilities Sun Tzu raises, even in hindsight of more than two millennia, imbues his laconic statements with a timeless electricity.

Roads Not Taken by Sun Tzu in Developing Theme #13

In the case of Theme #13 such roads need to be delineated with particular care, since it is easy to impute to Sun Tzu's treatment of command "omissions" that, on scrutiny, turn out to involve issues and concerns foreign to Sun Tzu's society, culture, and historical context. For example, Sun Tzu's thinking conveys scant recognition of the surpassing value of the teamwork that is a mainstay of successful modern militaries.[29] But quite possibly that is because a Warring States mindset would have considered such a concept (with its overtones of cooperation among peers) alien, incompatible with the intensely hierarchical nature of the "bonds" (yue, a term of art) on which the coherence of Warring States armies as social formations was built.[30]

A first "road not taken" is a corollary of the fact that Sun Tzu's standpoint remains strictly instrumental and, above all, fundamentally military. Sun Tzu never engages with the kinds of searching questions regarding the purpose and organization of the state that captured major attention from leading exponents of several early Chinese schools of thought – not least among them the founders of the Confucian tradition that would shape China for the next two millennia and more. For example, an overriding concern of the Huainanzi, a wide-ranging collection of essays (including considerable military content) dating from the second century BC, has been described as an "attempt to define the essential conditions of a perfect socio-political order."[31] That sort of question concerns Sun Tzu not one whit.

Relatedly, it bears underscoring (see also pp. 224, 251, 279 above) that – by dint of ignoring it – Sun Tzu treats the idea of sociopolitical identity as unproblematic or (as Thomas Schelling might have put it) as "parametric."[32] That stance places limits on the analytical depth of Sun Tzu's leadership thinking. Dilemmas of sociopolitical identity (who are we? what side am I on?) simply do not concern Sun Tzu – strikingly so in light of Sun Tzu XIII's avowed interest in "double-cross" in the form of double agentry. At most such dilemmas hover in the wings of Sun Tzu's thinking, e.g., via its emphasis on creating estrangements.[33]

Here is a basic way in which Sun Tzu falls short of emerging as a major political theorist in his own right. There is, of course, still a sense in which Sun Tzu's often inspired thinking about strategy and tactics is indeed relevant to the crafting of a healthy, robust sociopolitical identity, for all the reasons that actual identities are the cumulative upshot of battles lost and won (some of them in the drawing rooms of

---

[29] Verse XI.39 (Passage #9.8) might be read as an exception. But there it is a storm on water that galvanizes cooperation between archenemies, each concerned with their own short-term survival – a specialized context not paralleled in most modern cases of successful teamwork, military or other.

[30] Regarding those "bonds" see #10 footnote 38 above (citing analysis of Mark Lewis).

[31] See p. 189 of Huainanzi entry in Early Chinese Texts (cited in #5 footnote 9 above).

[32] Illustrating the point, verse VII.30 (Passage #9.15), "Do not thwart an enemy returning homeward" might evoke the story recounted in the Anabasis of Xenophon, a contemporary of the Warring States period. But that initial impression is misleading. Xenophon's army, which had marched into Mesopotamia originally in a mercenary capacity, found itself no longer in that role as a result of a military defeat. Suddenly having no status, "they had become a vagabond army of freebooters, no longer mercenaries, with no purpose beyond self-preservation" (Waterfield, #11 footnote 87, p. 113). The challenge facing Xenophon and his men was then not only a strategic and logistics challenge – i.e., how to extricate themselves from a precarious situation deep in unfriendly territory – but also a challenge involving sociopolitical identity, which Xenophon's men constantly struggled to discover in their circumstances. Sun Tzu's verse VII.30 lacks any hint of a counterpart to such quest for identity.

[33] See p. 149 above (smorgasbord of Sun Tzu passages having an estrangement focus).

polite society). But developing that relevance remains a task for the Sun Tzu (2) or (3) analyst. Sun Tzu himself did not attempt it.[34]

Consistent with these Sun Tzu omissions (though the political sensitivity of the issue is also a likely contributing factor) is Sun Tzu's lack of engagement with the possibility that a too-successful general may pose a threat to his nominal chief.

A second road not taken by Sun Tzu is much more specifically military. There is nothing in Sun Tzu that foreshadows modern concepts of "mission command" (sometimes known as *Auftragstaktik*), whereby senior commanders set goals and junior ones implement them with a high degree of independent initiative.[35] There are, to be sure, a handful of places in Sun Tzu where Sun Tzu contemplates an army being deployed by design in some version of extended form, which would perforce entail some degree of decentralized decision making.[36] More broadly, the day-to-day running of a large military force, not to mention its combat employment, makes some degree of delegation unavoidable. *What is absent in the Sun Tzu text, however, is any endorsement of empowering junior (or any subordinate) officers to undertake decentralized execution of the general's intent in ways that draw on their individual initiative and imagination.* Sun Tzu is perfectly capable of making clear military values that he likes. Decentralization of that stripe is not one of them.[37] Tellingly, the only times that

---

[34] As a conjecture, Sun Tzu's lack of attention to identity issues may go hand in hand with a relentless emphasis on control conjoined with equally relentless denial of the impact of chance. For relevant thinking see Harrison C. White, "Control to deny chance, but thereby muffling identity" (review of James S. Coleman, *Foundations of Social Theory*, 1990), *Contemporary Sociology*, 1990, 19(6), 783–88.

[35] For modern versions (whose specifics vary across different national militaries) see Eitan Shamir, *Transforming Command: the pursuit of mission command in the U.S., British, and Israeli Armies* (Foreword by H. R. McMaster; Stanford, CA: Stanford University Press, 2011). A thumbnail description of the core idea is provided on Shamir's p. 3, with a key ingredient being "trust in the subordinate's ability to act wisely and creatively without supervision when faced with unexpected situations."

[36] Candidates include verses XI.24 (Passage #13.28) and VII.12 (Passage #1.6). By way of military historical context, the Changping battlefront is said to have extended for hundreds of li (see #7 footnote 91), which would inevitably have required some degree of decentralization (indeed on both sides). There is also verse V.22 (Passage #13.29), which (taken in isolation and read through modern eyes) might conceivably be given a mission command reading. However, such a reading finds scant support in what we know of dominant patterns of Warring States military culture.

Traditional commentator Du Mu (commenting on another part of Sun Tzu) recounts an anecdote where an enterprising officer in a Warring States Chinese army operated independently and conducted what amounted to a small, successful – but unauthorized – raid on the enemy. For his pains, he was beheaded. See Griffith, pp. 106–107 (Giles, pp. 64–65 renders "officer" as "soldier"; the Chinese text does not resolve the issue of whether the man was an officer). The temper of the times thus revealed would have scarcely encouraged subordinates to take independent initiatives! Reinforcing reluctance to do so would be the practical difficulty, in an era when orders were almost certainly oral (or communicated en masse by flags, drums, etc.), of getting prior authorization to take unscripted independent actions in a way that a subordinate could genuinely trust and rely on.

[37] Shamir's study cited in #13 footnote 35 makes plain the extent to which mission command presupposes and builds on a compatible military culture – one where "[s]ubordinates were not relegated to the status of robots, simply following orders, but rather were regarded as individuals capable of making independent judgments" (Shamir, p. 15). That is a perspective worlds apart from Sun Tzu's paternalistic and controlling emphases (Passages #13.20, #13.21, #13.24, #13.32, etc.).

Sun Tzu mentions officers at levels below the commanding general is when they are giving him trouble – e.g., by being insubordinate or by falling prey to death or capture or simply by being weak.[38]

A third road not followed by Sun Tzu pertains to what Sun Tzu does not have to say about logistics as a function of command or the exercise of command in logistics matters.[39] In large-scale conventional warfare this is one of the most basic arenas where harmony can greatly facilitate military success and lack of harmony can precipitate military failure.[40] Logistics command issues must have been a basic concern to generals of Sun Tzu's time. As Brooks, p. 60 puts it succinctly, "a mass-infantry force requires large stocks of food and arms." With at most a few limited exceptions,[41] logistics harmony is the "missing harmony" in Sun Tzu's otherwise notably thorough (for his time) inventory of the harmonies needed for effective exercise of military command and prosecution of a war. Addressing logistics harmony in a comparably thorough way would have required much more discussion of organizational details such as supply and transport arrangements. The closest Sun Tzu comes to acknowledging need for systematic approaches to addressing matters of that type is verse I.8 (Passage #13.30). But the focus there remains limited. It is part of the backdrop to a net assessment exercise to be conducted before the war (or possibly a campaign) starts. Verse I.8 says nothing about how to tackle the dynamic challenges involved in running a logistics organization under the immense stresses of a campaign and sustained major combat.

Sun Tzu's neglect of logistics generalship is perhaps partially understandable given the relative simplicity of the logistics of ancient warfare, certainly when compared with that of modern times. However, the effect of this neglect is to foreclose any significant treatment by Sun Tzu of the many logistics complications and complexities that are a natural major part of military analysis. Because in all armies logistics woes, left unattended, can very easily come back to smother the commander, a corollary of Sun Tzu's "logistically sanitized" perspective is to divorce Sun Tzu's otherwise incisive treatment of the command role from many day-to-day realities of a commander's life – for example, how he spends his time; which tasks he does himself and which he delegates (a topic that points back to Sun Tzu's omission of any discussion of general staff functions; see p. 427 above). The omission of "logistics generalship" is one of the few points where Sun Tzu's otherwise hard-bitten realistic treatment of military issues could legitimately be faulted for unreality.

---

[38] E.g., verses X.13 (Passage #13.33) (insubordination by senior officers), VII.7–VII.8 (Passage #13.36) (death or capture of senior officers); verse X.11 (Passage #13.33) (weak officers).

[39] For further overview and critique of Sun Tzu's somewhat fragmentary coverage of logistics topics, of which his neglect of logistics generalship is one aspect, see pp. 402–404 above.

[40] See Henry E. Eccles, "Logistics and strategy," *Naval War College Review*, March 1962, pp. 15–31, stating (p. 23) a principle of "logistics harmony" based on his World War II Pacific logistics experience: "An effective logistic system must be in harmony on the one hand with the economic system of the nation, and on the other hand with the tactical concepts and environment of the combat forces."

[41] The most prominent is the notion that an army should live off the enemy (Passage #3.5), which could be seen as promoting a type of logistics harmony, obviating need to maintain a lengthy and vulnerable supply line. Yet this prescription is scarcely a panacea (and is also likely to promote anything but harmonious relations with a civilian populace devastated by effects of plunder).

Sun Tzu (2) ("Sun Tzu Extended") and Sun Tzu (3) ("Sun Tzu Analogical") Frontiers of Theme #13

Any theory of leadership and civil–military relations is unavoidably rooted in the cognitive patterns, practices, and institutions of its time and place. Those roots range far and wide across the larger society and polity, shaped by concepts and worldviews not merely of elites but of other people too. For these reasons unanchored efforts at generalization of such a theory are at best risky, often misleading.[42] To get a sense of just how alien to modern sensibilities was the society of Warring States China, Mark Lewis's *Sanctioned Violence in Early China* merits reading in tandem with Roger Gould's research on nineteenth-century Corsican vendettas.[43] While socially accepted bloody vengeance by non-state actors was a feature of both societies, Warring States society carried the concept vastly further. In Warring States China, as Mark Lewis has argued in *Sanctioned Violence*, "all men engaged in licit violence."[44] By comparison, Gould's Corsica analysis reveals a relatively far tamer social order. Vengeance in Corsica – while it certainly existed as a social practice – was rare and rarely extended beyond the nuclear family, with a key determinant being whether or not the act avenged was perceived as an attack on the solidarity of the familial group.[45] Vengeance was typically carried out by an avenger standing in a close kinship relation to the original victim.[46] Semi-professional third parties were not involved. By contrast, in Warring

---

[42] An example of the need to take account of social context comes from research of Peter S. Bearman, "Desertion as localism: army unit solidarity and group norms in the U.S. Civil War," *Social Forces*, 1991, 70(2), 321–42, finding that (quoting summary by Charles Tilly, #10 footnote 21, pp. 21–22):

> Early in the war, commitment to a locality and commitment to the Confederate cause as a whole aligned neatly. In these circumstances, locally recruited companies that had kept their members stuck together with determination. As the war proceeded, however, overall losses introduced increasing discrepancies between national and local solidarity; collective connection to the same locality simultaneously activated commitments to people at home and facilitated collective defection from the national military effort. Collectively, members of defecting military units withdrew their trust ... from faltering national authorities.

Discussing practices in the Qin army, Yates observes that men serving in the same small unit came from the same locale and would therefore "know each other intimately, would probably also be related to each other by blood and/or marriage, and would therefore be prepared to fight to the death to save the other members of their unit." See p. 31 of Robin D. S. Yates "Law and the military in early China," in Nicola Di Cosmo (ed.), *Military Culture in Imperial China* (Cambridge, MA: Harvard University Press, 2009), pp. 23–44. Illustrating a marked difference from the Confederate army, such practices did not encourage Qin soldiers to desert because, simply put, Qin had draconian sanctions against desertion on both the battlefield and the home front that the Confederacy lacked. See Lewis, Background footnote 7, pp. 612 (home front), 615 (battlefield: "if one member of a squad of five fled in battle, the other four members would be punished, unless they captured their fleeing comrade or were able to offer the head of an enemy to redeem their guilt," etc.).

[43] See Roger V. Gould, "Revenge as sanction and solidarity display: an analysis of vendettas in nineteenth-century Corsica," *American Sociological Review*, 2000, 65(5), 682–704.

[44] Lewis, Background footnote 11, p. 53.

[45] Gould, #13 footnote 43, p. 699: "kin groups in Corsica used revenge to demonstrate solidarity after actions by a rival group had placed that solidarity in doubt."

[46] Gould, #13 footnote 43, p. 697: "nuclear kin were not merely the main focus of retaliation (after offenders themselves); they were likewise the primary recruitment pool for avengers."

States China "wandering swordsmen" (you xia) – another translation would be "wandering vigilantes" – routinely helped people seek revenge (and also commonly mobilized assistants to facilitate that task).[47]

Holding these caveats in mind, it is still worth probing Sun Tzu Theme #13 material for insights potentially useful in other times and places, including our own. Often the actual use of Sun Tzu's thinking is to encourage and sharpen certain lines of inquiry even if the conceptual transfer to modern times is far from exact.[48]

Running throughout Theme #13 is Sun Tzu's unwavering belief in the centrality of the commander in all military effort, the linchpin of its success or failure. Possibly the single most important point Sun Tzu makes about that central role is the command responsibility principle articulated in Passage #13.33:

> X.9.
> Now when troops flee, are insubordinate, distressed, collapse in disorder or are routed, *it is the fault of the general*. None of these disasters can be attributed to natural causes.
>
> [Emphasis supplied.]

and given teeth by verse I.15 (Passage #13.14):

> I.15.
> If a general who heeds my strategy is employed he is certain to win. Retain him! When one who refuses to listen to my strategy is employed, he is certain to be defeated. Dismiss him!

Divesting what Sun Tzu says of its immediate Warring States context, Sun Tzu here puts his finger on a class of fundamental problems with which contemporary political–military institutions still struggle. In a twenty-first-century world where modern high command is becoming ever more heavily bureaucratized and institutionalized, worldwide, there are signs that Sun Tzu's command responsibility principle is becoming increasingly watered down – or flatly honored in the breach. Even though Sun Tzu does not have (or indeed claim to have) all the answers, his clear-sighted cognizance of a type of command responsibility concept – in fact more clear-sighted than is the case in some modern treatments – is a contender for "*What did Master Sun know that we still don't (or have yet to absorb adequately)?*"

An exceptionally important cluster of Sun Tzu ideas – certainly among the most debate-provoking in the entire treatise – centers on Sun Tzu's identification of circumstances where orders from a ruler need not (or even *should not*[49]) be obeyed by the commanding general who is that ruler's subordinate. Given the brevity of the

---

[47] Lewis, Background footnote 11, pp. 8off.

[48] For example, verse XI.44 (Passage #13.25) offers a lens through which to scrutinize George Patton's World War II effort to squelch cartoonist Bill Mauldin, whose work – which Patton regarded as "set[ting] such a damned bad example with his unsoldierly Willie and Joe" cartoon characters – was regularly appearing in the newspaper of the US 45th Division. Responding to Patton's directive to "get rid of Mauldin and his cartoons," the commanding general of the 45th, Troy Middleton – who supported Mauldin's work – asked that Patton's order be put in writing. Patton dropped the matter. Middleton's rationale was that "he believed [Mauldin's cartoons] boosted morale and also attracted readers to the division newspaper, which he used to kill unhelpful rumors." See Ricks, *The Generals* (cited in Table 2 note h on p. 97 above), p. 114, drawing on Frank James Price, *Troy H. Middleton: a biography* (Baton Rouge: Louisiana State University Press, 1974), p. 160. From a Sun Tzu perspective, both commanders were trying, each in his own way, to follow the spirit of verse XI.44's advice to "rid the army of doubts," but came to loggerheads over how to do so!

[49] See verse X.18 (Passage #13.18), following Minford's translation (Minford, pp. 259–60). The grammar of the Chinese text does not clarify whether "may not" or "should not" is intended here.

Sun Tzu text, and its tendency to avoid repetition, the number of times spread over several chapters that Sun Tzu touches on this topic may be treated as a marker of emphasis (Passages #13.16–#13.18; see also Passage #13.8's verse III.29).[50] In contemporary political–military settings, worldwide, Sun Tzu's position endorsing at least some types of military disobedience can hardly escape controversy.

Abstracting from a Warring States milieu and belief systems, two broad considerations may be identified as motivating Passages #13.16–#13.18 and Sun Tzu's counsel regarding justifiable disobedience there. The first pivots on the highly distinctive nature of military command expertise, where qualification for high command involves a blend of technical knowledge, practical experience, and intuition. Even talented civilians rarely measure up. A second consideration, pertinent to many ages of war, Sun Tzu's included, centers on communication barriers to informed exercise of command by a geographically remote ruler (or other higher authority).[51]

Further analysis appropriately begins by analyzing the nature and role of the expertise relevant to a given strategic situation. Much of the reason for Sun Tzu's modern popularity stems, of course, from perceived opportunities for transposing Sun Tzu's military ideas to civilian arenas where military power (and the expertise needed to wield it effectively) may not be involved at all, or be only a bit player. In such primarily non-military settings, if Sun Tzu's disobedience thinking is to retain validity three questions need to be answered in the affirmative:

(a) (existence) Is there a body of "special expertise" (a counterpart to military expertise) that a subordinate charged with implementation of orders or policy has mastered but the ruler (or other highest political authority level) has not?

(b) (relevance) Is that special expertise so crucial to the problem at hand that the judgment there of the subordinate is in fact likely to be superior to that of the higher authority?

(c) (larger context) Taking account of all relevant factors – a broad canvas that includes values of political legitimacy, stability, and integrity of the chain of command – is the subordinate who possesses that special expertise justified in an act of principled disobedience, flouting orders from higher authority?

Starting with Questions (a) and (b), and by contrast to the military case that is Sun Tzu's focus and forte, it is by no means self-evident that "special expertise," certainly expertise

---

[50] It is worth noting that these passages hail from three separate Sun Tzu chapters (Sun Tzu III, VIII, and X). In Brooks's theory of the genesis of the text (see Table 12A on p. A-35), Sun Tzu X is assigned an early date (c. 342 BC), Sun Tzu III a much later one (c. 312 BC) near the end of Brooks's conjectural sequence. Supplying a date for Sun Tzu VIII (the only chapter left undated in Brooks's Sun Tzu chapter chronology), Mair, p. 29 provisionally assigns it a relatively early date (c. 336 BC). Sun Tzu content relevant to flouting the ruler's orders hence spans both "early" and "late" Sun Tzu (though Sun Tzu III stops short of expressly endorsing disobedience, which Sun Tzu VIII and X explicitly do).

[51] A third factor – one that hovers in the background of Passage #13.16 – could enter when the general evaluates the superior civilian authority as exhibiting what nowadays might be labeled "diminished competence" (a concept which, of course, could have many meanings and where pertinent criteria of judgment could also vary greatly). However, by contrast to Sun Tzu's repeated allusions to diminished competence *down* the chain of command (e.g., soldiers as children, sheep, logs, etc.), Sun Tzu says not a word about a comparable situation that might exist *up* that chain.

of a type crucial to a particular crisis or problem, exists in every civilian case. The point might be put more strongly. In many civilian contexts claims of "special expertise" not shared by the highest political level may be little more than a veneer for differing judgment about how a strategic problem should be handled. Invoking Sun Tzu's disobedience verses to justify flouting orders of a political superior may then be little more than window dressing for an illegitimate challenge to legitimate superior authority – emphatically *not* a result countenanced by anything Sun Tzu says or implies. No flag of revolution is present in Sun Tzu.

A direction worth probing is whether *some types* of civilian expertise share with military expertise the attribute of being a domain of knowledge where Questions (a) and (b) have affirmative answers. Candidates for modern examples might include expertise relating to certain digital realms; or biological or biomedical expertise (e.g., disease-related). Another source of examples might be country- or region-specific knowledge, particularly when anchored in a subordinate's personal acquaintance with relevant leaders, perhaps also their families and entourages – the sort of knowledge that an experienced diplomat who has had years of dealings with key "players" in a region might be expected to possess (and few other people do).

Yet modern responses to Questions (a) and (b) must also allow for the magnificent capabilities of contemporary communications systems. Those systems give the top political level access to vastly better and more timely information (including expert advice) than capabilities available to higher political echelons in ages past, thus weakening one of the most compelling factual supports for Sun Tzu's disobedience verses (namely, a remote Warring States ruler's lack of timely data).

But here further nuances need to get their due. Information available to the top political level may *appear* to be more comprehensive than it actually is. For example (as Sun Tzu might have appreciated), human relationships crucial to the situation "on the ground" may be missing from the databases and masked or muted by even the best communications technology. Moreover, even with the best advice the top level may lack the intuition crucial for synthesizing all available information to reach a sound decision involving an alien technical area.[52] Meanwhile the same technology that gives higher-ups timely access to superb information can also serve up undigested raw data to top-level leaders unversed in its careful evaluation and interpretation. The upshot, all too easily, may be catastrophic judgment errors.

In weighing pros and cons of disobedience, the timescale on which a crisis or other situation unfolds is also relevant. Bracketing some ultra-fast-moving crises (e.g., cyber, financial, biological) there may exist adequate time for consultation between superior and subordinate levels, including of written types unknown in Sun Tzu's time (which, it should be recalled, antedated invention of paper as a medium of written communication). In nearly all cases consultation (backed, if necessary, by the subordinate expert's option to resign[53]) seems a modus operandi preferable to the subordinate (the counterpart to Sun Tzu's general) taking matters into that

---

[52] Cf. Jerome Bruner, Introduction footnote 2, p. 58 on need for both intuitive and analytical thinking, noting that "Usually intuitive thinking rests on familiarity with the domain of knowledge involved and with its structure, which makes it possible for the thinker to leap about, skipping steps and employing shortcuts ..." (which analytical thinking should later recheck as time and opportunity permit).

[53] A contingency anticipated by some readings of verse I.15 (Passage #13.14). See #13 footnote 72.

subordinate's own hands and acting disobediently. Even if the political leadership cannot truly become an instant expert, Question (c) is answered in the negative.

Yet there remains a nagging feeling that Sun Tzu's spirit of principled defiance of superior political authority by an expert subordinate might sometimes reach a better outcome. Consultative outcomes – much like committee ones – often predispose to "split the difference" compromises, precluding the diamond-like brilliance of Sun Tzu's way of war (or civilian counterparts thereto) at its apex of efficacy.[54]

In the disobedience-centered strand of his thinking, as is often the case with Sun Tzu, Sun Tzu's enduring constructive contribution is an educational one, prompting better awareness of structural issues and encouraging probing questions.

******

An outlier in Sun Tzu's Theme #13 development – but deeply relevant to twenty-first-century problems – is Sun Tzu's nascent theory of "intelligence command." Although his explicit development of that theory comes entirely from Sun Tzu XIII, there is important grist for its further development that draws on other Theme #13 content, applying it to produce advice on running an intelligence establishment, possibly including a modern one. For example, verse XI.54 (Passage #13.32) can be read as encouraging unconventional uses of rewards like medals or honors in intelligence contexts.

Echoing Passage #13.29, the nuanced nature of much intelligence work means that often the single most crucial judgment call is assigning the right person to tackle a given problem. An inspired choice here may improve results, not by, say, 50 per cent, but by a factor of ten or more – a possibility that resonates with Sun Tzu's quest for order-of-magnitude (or even greater) gains that animates Theme #3. Perhaps even more importantly than in warfare, choice of the right person for a particular intelligence task depends, not just on innate attributes and talents, but on the sources of shi present in the situation into which that individual is being inserted.[55]

But in transposing Sun Tzu's Theme #13 thinking from military to intelligence realms it is again Passages #13.16–#13.18 – hovering around the issue of legitimate disobedience – that call for the most careful scrutiny. Applying this material in an intelligence rather than a military context, Sun Tzu is here broaching the possibility that *under certain conditions* an intelligence chief may (in some sense, legitimately) disobey an order from lawful political authority. The phrase "nervous making" that was a favorite of Yale Law School professor Arthur Leff applies with full force here. A complicating factor (here more than in many cases of potential disobedience involving a general fighting a conventional war) is the role played by extreme secrecy

---

[54] From the Great Depression era of US economic history, critique of the head of the Federal Reserve Bank of New York (1928–40), a lawyer named George Harrison, is worth noting here:

> As time went on, however, he reverted to his natural character, that of an extremely competent lawyer and excellent administrator, who wanted to see all sides of an issue and placed great value on conciliating opposing points of view and achieving harmony. He was persuasive yet too reasonable to be truly single minded and dominant.

See Milton Friedman & Anna Schwartz, *A Monetary History of the United States, 1867–1960* (Princeton, NJ: Princeton University Press, 1963), p. 414.

[55] For relevant analysis of verses V.21–V.22 (Passage #13.29) see Appendix 4 anchor footnote 4.

and attempts, commonly by multiple actors and interests, to exploit such secrecy to the hilt. A second complicating factor is the deception-oriented mindset of many intelligence professionals which may run deeper by far than counterpart mindsets found in most military commanders, even ones who self-define as aficionados of deception. A further overlay of complication is the layered bureaucracy of modern intelligence organizations, presenting the specter of cascading breakdowns in integrity of command in a complex and quite possibly already tangled intelligence chain of command (a pattern unimagined in Sun Tzu's relatively simple intelligence world).

Sun Tzu does not tell us enough about relevant intelligence matters to clarify his specific position, if any, on the intelligence disobedience issue just raised.[56] However, the full range of relevant Sun Tzu principles – including not only Passages #13.16–#13.18 but also what Sun Tzu has to say about deception (Theme #7), calculation (Theme #1), and, of course, double agents (fanjian) (Passage #11.6) – suggests that Sun Tzu's legacy in this area could be a double-edged sword.

On the one hand, careful study of Sun Tzu can serve a constructive educational purpose. It can alert a broad non-specialist audience – some of whose members may occupy, or rise to, high decision-making positions – concerning the depths of manipulation and duplicity possible in intelligence matters, some of them corrosive of the integrity of the intelligence chain of command over which incumbents in those positions nominally preside. Such study can help provide a visceral sense of intelligence deception possibilities when everyone who matters thinks and acts as Sun Tzu might have done. Modern thinking about complex organizations and public policy commonly misses or marginalizes Sun-Tzu-esque possibilities – perhaps in part reflecting the influence of economists, who commonly (and on one level appealingly!) gravitate to a straightforward mindset, one that is prone to take a plain vanilla view of deception if deception is considered at all.[57]

On the other hand, in a twenty-first-century world where intelligence establishments worldwide have increasingly come to rival, or to exceed, in usable forms of power the capabilities of traditional military establishments, potential intelligence applications of Sun Tzu's Passages #13.16–#13.18 enter – as the court-martial passage in The Caine Mutiny puts it well – "the most dangerous possible ground."[58]

The potential for an intentional break in the chain of command at the interface between legitimate political authority and the intelligence apparatus, or parts of it, should be recognized as a fundamental soft spot of contemporary world politics. Sun Tzu's Passages #13.16–#13.18 point to the possibility of deep failures at that interface and can also, albeit unwittingly, help to legitimate or at least to justify them.

---

[56] For context, here it is important to reiterate that, in Sun Tzu's time, intelligence organization seems likely to have been ad hoc, rather than being vested in a durable formal organization. Indeed the ruler or general might have been his own case officer, at least on some occasions. See p. 366 above. See also #11 footnote 44, noting that the Arthaśāstra depicts a multi-tier intelligence organization in a way the Sun Tzu text does not.

[57] Cf. a relevant observation of Mark Granovetter, pointing out "a peculiar assumption of modern economic theory, that one's economic interest is pursued only by comparatively gentlemanly means." See p. 488 of Granovetter, "Economic action and social structure: the problem of embeddedness," American Journal of Sociology, 1985, 91(3), 481–510.

[58] Herman Wouk, The Caine Mutiny, a novel of World War II (Garden City, NY: Doubleday, 1954), p. 449.

Sharper analysis and theory, building on Sun Tzu but also going fundamentally beyond him, is needed here. Such theory will not resolve this class of problems, but it can shed valuable light on them and thereby better prepare future political leaders for the intelligence command and oversight challenges they must face.

✶✶✶✶✶✶

Set of principal passages used to illustrate Sun Tzu Theme #13 (leadership and harmony in the chain of command)

Passage #13.1. Essence of the Command Role

(Sun Tzu's aspirations for the command role, conveyed here by a crossbow image, lay groundwork for analysis of what commanders do. At the same time, those aspirations may well have outrun the capabilities of military command control systems of Sun Tzu's era, indeed of all later eras prior to the digital age.)

V.15. Thus the momentum [shi] of one skilled in war is overwhelming, and his attack precisely regulated.

V.16. His potential [shi] is that of a fully drawn crossbow; his timing, the release of the trigger.

Passage #13.2. Moral Influence or *Dao* (Focus = People/Ruler Interface)

(*Dao* as used here seems best imbued with a pragmatic focus, rather than being treated as standing for a highbrow philosophical concept. See p. 431 above.)

I.4. By moral influence [dao] I mean that which causes the people [min] to be in harmony with their leaders, so that they will accompany them in life and death without fear of mortal peril.

. . .

I.11. If you say *which ruler possesses moral influence* [dao], which commander is the more able, which army obtains the advantages of nature and the terrain, in which regulations and instructions are better carried out, which troops are the stronger;

I.12. Which has the better trained officers and men;

I.13. And which administers rewards and punishments in a more enlightened manner;

I.14. I will be able to forecast which side will be victorious and which defeated. [Emphasis supplied.]

Passage #13.3. Unraveling Harmony between Ruler and People; Warning Signs of Incipient Regime Failure? (Focus = People/Ruler Interface)

(A natural reading of Passage #13.3 is that certain types of foreign wars, especially protracted ones, destroy the livelihood and economic prospects of a state's population. From that unfortunate outcome to regime failure from internal strife or foreign conquest may well be a short step. At the very least, it becomes easier for an enemy to recruit well-placed spies from a state's disgruntled populace! Verse II.12 may also be taken as a warning about possible social instability in conquered or occupied territory arising from draconian exactions levied there.[59])

II.10. They carry equipment from the homeland; they rely for provisions on the enemy. Thus the army is plentifully provided with food.

II.11. When a country is impoverished by military operations it is due to distant transportation; carriage of supplies for great distances renders the people destitute.

II.12. Where the army is, prices are high; when prices rise the wealth of the people is exhausted. When wealth is exhausted the peasantry will be afflicted with urgent exactions.

II.13. With strength thus depleted and wealth consumed the households in the central plains will be utterly impoverished and seven-tenths of their wealth dissipated.

Passage #13.4. Unraveling Harmony, cont. (Focus = People/Ruler Interface)

(Verse XIII.1 suggests Sun Tzu's recognition that the costs of war may include adverse effects on the populace of newly conquered territory. Grassroots instability there could undercut a victor's capacity to wage war after next.[60])

XIII.1. Now when an army of one hundred thousand is raised and dispatched on a distant campaign the expenses borne by the people together with the disbursements of the treasury will amount to a thousand pieces of gold daily. There will be continuous commotion both at home *and abroad*, people will be exhausted by the requirements of transport, and the affairs of seven hundred thousand households will be disrupted. [Emphasis supplied.]

Passage #13.5. Basis for Harmony between Ruler and People of a Conquered State?

(The "preservation" prioritized in verse III.1 may reduce the need to cope with a resentful, recalcitrant population of the conquered.[61] Notice, however, that it is perfectly possible to take a state intact – say by guile or subversion – only to proceed to plunder it, a sequence of steps not unknown in modern finance.)

---

[59] In the background here stands the "whose peasants are they?" question posed in #3 footnote 64.

[60] It is notable that Qin, the sole ultimate victor of the Warring States era which was famed for its draconian methods, was "never able to repress opposition entirely" (e.g., what they labeled "gang robbers and bandits"), a problem logistically reflected in their military's voracious consumption of supplies of feathers for arrows. See Yates, Background footnote 53, p. 304.

[61] Sawyer, p. 310 note 40. Discussing fourth-century BC war and peace in China, and quoting Sun Tzu on plundering, Brooks & Brooks, Introduction footnote 36, p. 106 also quote a passage from the Mozi text: "They [the conquerors] cut down grainfields and fell trees; they tear down the inner and outer walls and fill in ditches and ponds; they seize and kill sacrificial animals and burn down ancestral temples; they kill the people and exterminate the aged and weak . . ." As Brooks & Brooks dryly note, "It may be doubted that this will win the affections of the conquered." (For Mozi background see Background footnotes 12 and 80.)

III.1. Generally in war the best policy is to *take a state intact; to ruin it is inferior to this.*

. . .

III.3. For to win one hundred victories in one hundred battles is not the acme of skill. *To subdue the enemy without fighting is the acme of skill.*

. . .

III.10. Thus, those skilled in war subdue the enemy's army without battle. They *capture his cities without assaulting them* and overthrow his state without protracted operations.

III.11. Your aim must be to *take All-under-Heaven intact.* Thus your troops will not be worn out and your gains will be complete. This is the art of offensive strategy. [Emphases supplied.]

Passage #13.6. Ruler and New Subjects: Plunder As Sun Tzu Blind Spot?

(A question arises here as to whether Sun Tzu's advice to plunder while in enemy territory, which might soon become one's own territory if the war is won, could bring about social instability there, depriving the victor of much of the conqueror's prize. Sun Tzu's comments favoring plunder – spread across several chapters, see p. A-224 for a roundup – seem to betoken a fixation on military success at the expense of the quality of the peace that follows.[62]

Even as guidance for the active conduct of military operations there are reasons to be cautious about Sun Tzu's overly facile plundering advice. A maxim attributed to Napoleon warns that "Nothing can be designed better to disorganize and destroy an army than pillage."[63] This warning raises a further major class of issues on which Sun Tzu is again notably silent.[64])

XI.7. When the army has penetrated deep into hostile territory, leaving far behind many enemy cities and towns, it is in serious ground.

. . .

---

[62] Presenting a notable contrast to Sun Tzu on point, the Yi Zhou shu, a different early text, faces up to challenges confronting a conqueror who seeks to turn military conquest into lasting political gain. See #13 footnote 9 above. Recapitulating advice found in the Yi Zhou shu, see also McNeal, Background footnote 58, p. 108: "The territory is not to be looted, behavior fitting a marauding band that has no intention of occupying and governing its defeated enemies."

Cf. a counsellor's famous response to a challenge question posed by the Han ruler: "Your Majesty may have won the empire on horseback, but can you rule it on horseback?" See Wilkinson, §59.5.7, p. 800. For the story see Shiji 97 (Watson translation cited on p. xxiii above), Han Dynasty I, p. 226.

[63] See *Operational Level of War – Its Art*, Preliminaries footnote 8, p. 1–26, "Military Maxims of Napoleon," Maxim No. 107. See also van Creveld, Background footnote 48, pp. 7–8, describing European armies of c. 1560–1660 as "marauding bands of armed ruffians, devastating the countryside they crossed" whose "commanders were also incapable of keeping them under control and of preventing desertion." There is little reason to regard Warring States Chinese armies as immune from that type of disciplinary breakdown.

[64] Possibly somewhat ameliorating Sun Tzu's endorsement of plunder is the fact that early Chinese armies also bartered for, or purchased, some of the provisions needed to sustain them in the field. See Sawyer, Introduction footnote 55, p. 17. Although Sun Tzu says nothing about this directly,

XI.13. In focal ground, ally with neighbouring states; in deep [i.e., serious] ground, plunder.[65]

. . .

XI.20. In serious ground I would ensure a continuous flow of provisions.[66]

. . .

XI.30. The general principles applicable to an invading force are that when you have penetrated deeply into hostile territory your army is united, and the defender cannot overcome you.

XI.31. Plunder fertile country to supply the army with plentiful provisions.

Passage #13.7. Division of Labor between Ruler and General (Focus = Ruler/ General Interface)

(Verse XII.16's reference to the "good general" [*liang jiang*] as executing the ruler's plans points by implication to one thing that good generals do *not* do, which is to raise a flag of rebellion. That sets important context for Sun Tzu's endorsement of the principle that, under certain circumstances, a general may disobey the ruler's orders. See Passages #13.17 and #13.18 below.)

XII.16. And therefore it is said that enlightened rulers deliberate upon the plans, and good generals execute them.

Passage #13.8. Central Importance of the General (Focus = Ruler/General Interface)

(This and the following passage document the towering importance Sun Tzu attaches to the role of the commanding general.[67] See also Passage #13.15 below.)

III.18. Now the general is the protector of the state. If this protection is all-embracing, the state will surely be strong; if defective, the state will certainly be weak.

. . .

III.29. He whose generals are able and not interfered with by the sovereign will be victorious.

---

such practices may stand in the background of Passage #13.3's verse II.12 ("Where the army is, prices are high," etc.). Here, however, note should be taken of van Creveld's important observation (van Creveld, Background footnote 48, p. 7) that "[e]stablishing markets takes time" – and time is precisely what Sun Tzu's way of war is exceedingly loath to set aside, certainly while active warfighting operations are ongoing. See verse XI.29 (Passage #3.7), "Speed is the essence of war. . . ."

[65] For translation consistency, "deep ground" here should read "serious ground." See #2 footnote 61.

[66] The weight of traditional commentator opinion points to reading verse XI.20 as a call for an invader to live off the enemy (or the enemy's population) – i.e., as further encouragement to plunder. For discussion see #2 footnote 64.

[67] It is worth underscoring again that, for Sun Tzu, the spotlight is always on the commanding general, and on him alone. A vivid contrast case is the War Gallery of 1812 in the Hermitage Museum in St. Petersburg, Russia, created by Carlo Rossi in 1826 in commemoration of Russian arms in the Napoleonic wars. In that gallery hang portraits of over 300 generals who were heroes of the 1812 War.

Passage #13.9. Central Importance of the General, cont.

II.21. Hence what is essential in war is victory, not prolonged operations. And therefore the general who understands war is the Minister of the people's fate and arbiter of the nation's destiny.

Passage #13.10. Overview of General's Role (Multiple Interfaces Are Pertinent Here, but All Involve the General)

(Although Sun Tzu vouchsafes us no details, a full range of mobilization steps stands in the background of verse VII.1 – among them, ones pertaining to putting in place a command structure; production and distribution of weapons and heavy equipment; setting up supply arrangements; etc.[68] Such tasks would inevitably come to involve the commanding general, directly or indirectly through staff, along with a raft of home front actors – i.e., putting in play further interfaces that must be successfully navigated.[69])

VII.1. Normally, when the army is employed, the general first receives his commands from the sovereign. He assembles the troops and mobilizes the people. He blends the army into a harmonious entity and encamps it.[70]

VII.2. Nothing is more difficult than the art of manoeuvre. What is difficult about manoeuvre is to make the devious route the most direct and to turn misfortune to advantage.

Passage #13.11. Qualities of an Ideal General

(Verse I.7 invites reflection on what military virtues *fail* to make the list. Human relations qualities are prominently represented here, but even in that sphere Sun Tzu's list has some notable omissions. As Lau & Ames observe: "*De* (excellence/

---

[68] In a common reading, verse II.10 (Passage #13.3) calls for equipment to be brought along with the army from the homeland (where it would have to be manufactured or otherwise procured). For further background on this part of verse II.10, noting its lack of specificity, see Appendix 3 footnote 6.

Also pointing to a range of procurement needs are verses II.1–II.2 (Passage #3.4). From a far more recent time, where the sources are much better, a sense of the immense complexity of mobilizing for a major campaign with a large army is found in Peter C. Perdue, "Military mobilization in seventeenth and eighteenth-century China, Russia, and Mongolia," *Modern Asian Studies*, 1996, 30(4), 757–93.

[69] Yet further interfaces are implicit in verse XI.58 (Passage #4.20A), setting forth a battery of tasks involving the home front and border control that must be addressed without delay once war starts. Some of the latter interfaces are noted by Lewis, Background footnote 7, p. 629: "barriers [at the borders of states] also allowed states to check those who entered and exited their borders and to collect transit taxes from merchants who carried goods from state to state."

[70] See Appendix 13 footnote 8 for discussion of a split of opinion among traditional commentators and modern translators on how to understand the segment of the text that Griffith renders as "blends the army into a harmonious entity." However, the implications of that split of opinion for the military substance of verse VII.1 do not seem major, since such harmonizing would be a necessary task whether or not Sun Tzu mentions it explicitly.

character) does not occur in the *Sun Tzu* at all . . ."[71] Meanwhile a general's innovative qualities like technological or logistics inventiveness or ingenuity are likewise not visible in verse I.7. "Wisdom" [*zhi*] might be elastic enough to cover such qualities, though there is scant reason to think that Sun Tzu understood it in that way.)

I.7. By command I mean the general's qualities of wisdom, sincerity, humanity, courage, and strictness.

Passage #13.12. Further Attributes of Ideal General

(These qualities, analyzed on pp. 273–77, play up skill in information denial.)

XI.42. It is the business of a general to be serene and inscrutable, impartial and self-controlled.

XI.43. He should be capable of keeping his officers and men in ignorance of his plans.

Passage #13.13. Disabling Character Flaws of a General

VIII.17. There are five qualities which are dangerous in the character of a general.
VIII.18. If reckless, he can be killed;
VIII.19. If cowardly, captured;
VIII.20. If quick-tempered you can make a fool of him;
VIII.21. If he has too delicate a sense of honor you can calumniate him;
VIII.22. If he is of a compassionate nature you can harass him.
VIII.23. Now these five traits of character are serious faults in a general and in military operations are calamitous.
VIII.24. The ruin of the army and the death of the general are inevitable results of these shortcomings. They must be deeply pondered.

Passage #13.14. Hiring and Firing Generals

(Griffith's understanding of verse I.15 puts a spotlight on personnel selection at the top, advocating for unhesitating relief of underperforming generals.[72])

I.15. If a general who heeds my strategy is employed he is certain to win. Retain him! When one who refuses to listen to my strategy is employed, he is certain to be defeated. Dismiss him!

---

[71] Lau & Ames, p. 199 note 45.
[72] An alternative reading of the Chinese text shifts focus to Sun Tzu's relationship to the ruler (e.g., whether to continue in an advisory role or depart, presumably in quest of a different patron). Regarding that alternative, which like Griffith's version pertains to stability vs. turnover at the top but now focuses on a different type of turnover, see Appendix 6 anchor footnote 3 discussing verse I.15 interpretation issues.

Passage #13.15. General's Contribution to Harmonious Civil–Military Relations

X.19. And therefore the general who in advancing does not seek personal fame, and in withdrawing is not concerned with avoiding punishment, but whose only purpose is to protect the people and promote the best interests of his sovereign, is the precious jewel of the state.

Passage #13.16. Pathologies of Civil–Military Relations (Focus = Ruler/ General Interface)

(Verses III.19–III.23 suggest that disharmony in relations between ruler and commanding general may induce further disharmony within the army and its officer corps. Such a pattern is reminiscent of Roger Gould's concept of "conflict waves," whereby disruptions in particular social network ties ramify to produce further disruptions in adjoining network ties, setting off a social network version of a chain reaction.[73])

III.19. Now there are three ways in which a ruler can bring misfortune upon his army.
III.20. When ignorant that the army should not advance, to order an advance or ignorant that it should not retire, to order a retirement. This is described as "hobbling the army".
III.21. When ignorant of military affairs, to participate in their administration. This causes the officers to be perplexed.
III.22. When ignorant of command problems to share in the exercise of responsibilities. This engenders doubts in the minds of the officers.
III.23. If the army is confused and suspicious, neighbouring rulers will cause trouble. This is what is meant by the saying: "A confused army leads to another's victory."

Passage #13.17. General Need Not Always Obey Commands of the Ruler (Focus = Ruler/General Interface)

VIII.7. There are some roads not to follow; some troops not to strike; some cities not to assault; and some ground which should not be contested.
VIII.8. There are occasions when the commands of the sovereign need not be obeyed.

Passage #13.18. More on Same Idea (Focus = Ruler/General Interface)

(Verse X.18 delineates circumstances in which the ruler's commands need not be obeyed by the general. Read in tandem, verse X.19 – which has already appeared in Passage #13.15 and is therefore not repeated here – makes two crucial further points. The first is that the general must not refrain from making an appropriate

---

[73] See Roger V. Gould, *Collision of Wills: how ambiguity about social rank breeds conflict* (foreword by Peter Bearman; Chicago: University of Chicago Press, 2003), p. 155.

choice through fear of later punishment. The second is that purity of motives must attend a disobedient general's conduct.)

X.18. If the situation is one of victory but the sovereign has issued orders not to engage, the general may decide to fight. If the situation is such that he cannot win, but the sovereign has issued orders to engage, he need not do so.

Passage #13.19. Successful Leadership (Focus = General/Army Interface)

(In Western military literature the idea succinctly expressed by verse III.27 is sometimes labeled "conceptual unity." Verse III.27 leaves unresolved how that conceptual unity is to be attained, especially in large organizations like Sun Tzu's mass-infantry armies whose ranks are chock full of conscripts, many of them probably miserable.)

III.27. He whose ranks are united in purpose will be victorious.

Passage #13.20. Sun Tzu General's Posture vis-à-vis His Troops – Soldiers As Infants (Focus = General/Army Interface)

(There should be no illusion that verse X.20 signals kindly motives or a humanitarian mindset. Illustrating the point, the Sun Tzu text entertains the concept of deliberately inserting one's troops into desperate circumstances to induce them to fight more fiercely. See Passage #9.14 as well as further Theme #10 analysis. The point is rammed home by the later Warring States military theorist Sun Bin, who opines that "The commander should look on his troops as he would an infant ... [but] he should use them as he would so much dirt or straw."[74])

X.20. Because such a general regards his men as infants they will march with him into the deepest valleys. He treats them as his own beloved sons and they will die with him.

Passage #13.21. Sun Tzu General's Posture vis-à-vis His Troops – Soldiers As Sheep (Focus = General/Army Interface)

(The underlying idea is that troops – or at least the common soldiers – have limited agency or, as Minford puts it, were "to be treated like so many automata."[75])

XI.49. He burns his boats and smashes his cooking pots; *he urges the army on as if driving a flock of sheep*, now in one direction, now in another, and none knows where he is going. [Emphasis supplied.]

---

[74] Lau & Ames, p. 162. Verse X.20 translation appears in accompanying endnote (p. 230 note 350).
[75] See #9 footnote 8 (where Minford's summary statement is also quoted) as well as #9 footnote 7.

Passage #13.22. Troops Well Organized and Led – An Illustration

IV.19. Thus a victorious army is as a hundredweight balanced against a grain; a defeated army as a grain balanced against a hundredweight.

IV.20. It is because of disposition [xing] that a victorious general is able to make his people fight with the effect of pent-up waters which, suddenly released, plunge into a bottomless abyss.

Passage #13.23. Troops Well Organized and Led – A Further Illustration

V.17. In the tumult and uproar the battle seems chaotic, but there is no disorder; the troops appear to be milling about in circles but cannot be defeated.

V.18. Apparent confusion is a product of good order; apparent cowardice, of courage; apparent weakness, of strength.

V.19. Order or disorder depends on organization [shu]; courage or cowardice on circumstances [shi]; strength or weakness on dispositions [xing].[76]

Passage #13.24. Morale As a Quality of Troops Requiring Management

XI.41. *To cultivate a uniform level of valour is the object of military administration.* And it is by proper use of the ground that both shock and flexible forces are used to the best advantage. [Emphasis supplied.]

XI.41a. (Ames, p. 159; segment omitted by Griffith) "Thus the expert in using the military leads his legions as though he were leading one person by the hand. The person cannot but follow."

Passage #13.25. Morale As a Quality of Troops Requiring Management, cont. – Prohibiting Superstitious Practices

(One modern scholar has suggested that verse XI.44 aims "not so much to exclude from the military sphere the use of divination as to preserve the categorical authority of commanders by preventing the troops from turning to other voices that may contradict them or sow doubt about orders issued by their superiors."[77])

XI.44. He prohibits superstitious practices and so rids the army of doubts. Then until the moment of death there can be no troubles.

---

[76] For discussion of these three concepts see pp. 175–76 (shi and xing) and p. 200 (shu, shi, and xing).

[77] See p. 155 footnote 12 of Albert Galvany, "Signs, clues and traces: anticipation in ancient Chinese political and military texts," *Early China*, 2015, 38, 151–93.

Passage #13.26. Managing Morale As a Dynamic Task, Unfolding in Real Time

VII.20. Now an army may be robbed of its spirit [qi] and its commander deprived of his courage.[78]

VII.21. During the early morning spirits [qi] are keen, during the day they flag, and in the evening thoughts turn toward home.[79]

VII.22. And therefore those skilled in war avoid the enemy when his spirit [qi] is keen and attack him when it is sluggish and his soldiers homesick. This is control of the moral factor [qi].

VII.23. In good order they await a disorderly enemy; in serenity, a clamorous one. This is control of the mental factor.

VII.24. Close to the field of battle, they await an enemy coming from afar; at rest, an exhausted enemy; with well-fed troops, hungry ones. This is control of the physical factor.

VII.25. They do not engage an enemy advancing with well-ordered banners nor one whose formations are in impressive array. This is control of the factor of changing circumstances.

Passage #13.27. Tools of Leadership: Cultivating the Dao (Focus = General/Army Interface)

(This and the next several passages round up leadership tools discussed by Sun Tzu for molding and running an army. Verse IV.15 sets a baseline, expressing a very broad, holistic concept that might also find application to the people/ruler interface. Given its abstractness, along with Han strips variation in the text, it is not surprising that verse IV.15 has been imbued with a variety of emphases.)

IV.15. Those skilled in war cultivate the Dao and preserve the laws [fa] and are therefore able to formulate victorious policies.

Passage #13.28. Skill-Sets That Generals Need: Overview of Core Elements

(This verse spells out three major branches of knowledge required of a land warfare general.[80] But it leaves a basic question unresolved: Must Sun Tzu's general know logistics too?[81] For further analysis see p. 438 above.)

---

[78] For background on the important Chinese concept of qi 氣 appearing here and elsewhere in this passage (where it is variously translated as "spirit[s]" or "moral factor") see #3 footnote 80. Repeating a point of clarification there, qi 氣 is a distinct concept from qi 奇 ("crafty," "unconventional," etc.) appearing in the qi/zheng concept pair (Theme #14).

[79] Importantly, verse VII.21 may be understood in a generalized way, envisioning a timescale that could be much longer than a single day. See Preliminaries footnote 4.

[80] From a Sun Tzu (2) and (3) vantage there are education, training, and research implications here.

[81] This is the title theme of a 1953 essay by Lieutenant General (later General) W. B. Palmer, USA, "Commanders must know logistics," Operational Level of War – Its Art, Preliminaries footnote 8, pp. 5–6 to 5–14 (article originally published in Army Information Digest, April 1953, pp. 2–14).

XI.24. The tactical variations appropriate to the nine types of ground, the advantages of close or extended deployment, and the principles of human nature are matters the general must examine with the greatest care.[82]

Passage #13.29. Tools of Leadership: Choice of Personnel (Focus = General/Army Interface)

(Read literally, the Chinese text of verse V.22 says no more than that the commander should select the right person. Beyond that, various interpretations can be given.[83] The one favored here envisions personnel as selected based on the "fit" between their talents and the shi [strategic advantage] of the situation.[84] In effect, specific people are to be used as a kind of transmission belt to amplify shi.[85] See also Passages #4.6, #9.3, and #12.21, discussing verses V.21–V.22 from perspectives of other themes.)

V.21. Therefore a skilled commander seeks victory from the situation [shi] and does not demand it of his subordinates.
V.22. He selects his men and they exploit the situation [shi].

Passage #13.30. Tools of Leadership: Organizational and Logistics Doctrine and Methods (Focus = General/Army Interface)

(This verse from Sun Tzu's net assessment material involves technical terminologies whose meaning and translation could vary depending on how one interprets Sun Tzu's words.[86] Three sets of ideas may be discerned:

---

[82] The Chinese phrase Griffith translates as "human nature" is ren qing. As a stand-alone concept qing can be translated as "emotions" (clearly a Sun Tzu emphasis). For background on the concept of qing in Chinese philosophy see Zhang Dainian, #3 footnote 80, pp. 383–87.

[83] For example, Griffith, p. 93 footnote 1 reads verse V.22 as support for military personnel choices devoid of factors like nepotism or favoritism (i.e., guided by military professionalism).

As Griffith implies, though does not say explicitly, verse V.22 may pack the most punch in a context where potent, if possibly weakening, social forces cut *against* a commander's flexibility in selecting key personnel. In the case of Warring States China, such social forces are noted by Yates, Introduction footnote 28, p. 224. They would have most likely included lingering influence of a hereditary nobility and officer assignments based on ascribed, not achieved, status.

[84] The shi concept, which is often regarded as central to Sun Tzu's ideas, is introduced and analyzed on pp. 175ff. above.

[85] See Appendix 4 anchor footnote 4 analysis (which also takes a position on a verse V.22 textual issue). A modern illustration of the basic idea would be Eisenhower's decision to place Patton in charge of pursuing the retreating German army across France, dubbing him (in a post-war memoir) "the finest leader in military pursuit that the United States Army has known." That personnel choice looked beyond Patton's substantial personality flaws as a leader to recognize Patton's superlative if specialized command skills. The former had earned him a reprimand from Eisenhower, his superior officer, for repeated incidents where Patton had mistreated hospitalized US soldiers (e.g., accusing them of cowardice). See Ricks, The Generals (cited in Table 2 note h on p. 97 above), pp. 59–64.

[86] For a snapshot of some of the issues, citing analysis of Li Ling, see Kidder Smith's discussion of this segment of Sun Tzu I in http://learn.bowdoin.edu/suntzu/content/chapter_01/section_11.html (visited October 6, 2021).

- one pertains to military organization
- one pertains to office/rank
- one pertains to provision/logistics.[87]

Griffith's choice of "doctrine" as a translation of *fa* is not a standard one, though it is not wrong. It may reflect his attempt to find English words that convey Sun Tzu's military substance to modern audiences.[88] The term "doctrine" could risk overly modern connotations – suggestive of detailed, written statements – though that is not an invariable feature of all doctrine and the sometimes startling sophistication of Warring States Chinese warfare also needs to be kept in mind.[89] Other leading translators imbue *fa* with less of a modern military ring – e.g., "regulation" [Ames], "method" [Mair], "method and discipline" [Giles]. Sawyer renders *fa* "the laws [for military organization and discipline]," with the bracketed words being his.[90])

I.8. By doctrine [*fa*] I mean organization, control, assignment of appropriate ranks to officers, regulation of supply routes, and the provision of principal items used by the army.

Passage #13.31. Tools of Leadership: Applied Group Psychology, Emphasizing Timing and Consistency, with a *Wen/Wu* Chinese Cultural Dimension (Focus = General/Army Interface)

(These verses set forth practical leadership insights in which timing and consistency are core ingredients. Also basic, and imbued with distinctively Chinese cultural content, is the *wen/wu* – civil/martial – concept pair appearing in verse IX.48. In present Sun Tzu context, the *wen/wu* idea might be roughly summarized as describing leadership that induces those being led simultaneously to fear the leader [*wu* aspect] and to draw close to him [*wen* aspect].[91])

IX.47. If troops are punished before their loyalty is secured they will be disobedient.[92] If not obedient, it is difficult to employ them. If troops are loyal, but punishments are not enforced, you cannot employ them.

---

[87] Here Griffith's "regulation of supply routes" is Sawyer's "management of logistics." See Sawyer, p. 304 note 8. By conveying a more general concept, one that is also notably flexible, the latter translation has some advantages, though Sun Tzu's thinking falls well short of a comprehensive approach to the logistics process.

[88] Boorman & Boorman, Introduction footnote 39, p. 131.

[89] In modern military literature the term "doctrine" finds many definitions. It is commonly associated with prescriptive rules that are in some sense authoritative (which in Sun Tzu context calls to mind Sun Tzu's terrain rules), though modern usage is often looser.

[90] For further analysis of *fa* and various possible translations see Appendix 1 anchor footnote 5.

[91] This description of *wen* and *wu* draws on a first-century BC Han work translated by McNeal, Background footnote 58, p. 15. See Appendix 13 footnote 15. The *wen/wu* concept pair – centered on the complementarity of civilian (*wen*) and military (*wu*) realms – had a basic place in early Chinese thought, though verse IX.48 is Sun Tzu's only explicit mention of it.

[92] A kindred idea in modern garb is suggested by an observation of George Homans, reflecting on his personal experience commanding a warship during World War II: "A ship's party, when all

IX.48. Thus, command them with civility [*wen*] and imbue them uniformly with martial ardour [*wu*] and it may be said that victory is certain.

IX.49. If orders which are consistently effective are used in instructing the troops, they will be obedient. If orders which are not consistently effective are used in instructing them, they will be disobedient.

IX.50. When orders are consistently trustworthy and observed, the relationship of a commander with his troops is satisfactory.[93]

Passage #13.32. Tools of Leadership: Imagination and Flexibility in Rewards and Orders (Focus = General/Army Interface)

(There is a tension, even seeming contradiction, between the first part of verse XI.54 and other parts of Sun Tzu – for example, verse X.14 in the next passage [Passage #13.33] below. Yet there are possible ways around this difficulty, in particular if verse XI.54 is given a relatively restricted realm of application.[94])

XI.54. Bestow rewards without respect for customary practice; publish orders without respect to precedent.[95] Thus you may employ the entire army as you would one man.

Passage #13.33. Pathologies of Leadership (Focus = General/Army Interface)

(In verse X.9 Sun Tzu articulates a version of a command responsibility principle, placing onus for the adverse outcomes listed here squarely on the commanding general, the result of poor leadership or in some cases flawed tactical judgment. Unusually for Sun Tzu, there is also cognizance here of a middle level of the direction of war, distinct from both the general and the rank-and-file soldiers.)

---

hands, officers and men, go ashore to play a game of baseball against another ship, the loser providing the beer, reinforces in the ship's company a sense of unity, *but only if this unity exists to begin with.*" See Homans, #6 footnote 72, p. 296 (emphasis supplied).

[93] Traditional commentator Zhang Yu plays up the importance of reciprocal trust and confidence here: "The general has confidence in the men under his command, and the men have confidence in obeying him. Thus the trust is mutual." See Minford, p. 247 (adapting Giles, p. 99).

[94] See, e.g., Appendix 8 anchor footnote 19 (suggesting a verse XI.54 interpretation that pivots on treating the preceding verses XI.51–XI.53 as an interpolation).

[95] Interestingly, verse XI.54 explicitly mentions only rewards, not punishments (an apparent emphasis which behavioral psychologist B. F. Skinner might have applauded on basic operant conditioning grounds), though not enough context is provided to clarify Sun Tzu's thinking on point. It is worth noting that the term *xing* 刑, signifying "punishment" (commonly in a corporal punishment sense), is notable by its absence from the Sun Tzu text. Mair, p. xlv comments that "one might well expect to find [this character 刑] in a work of strategy such as the *Sun Zi*, but it does not occur even once." Another term for punishments (*fa*) does appear in Sun Tzu via the phrase *shang fa*, "rewards and punishments," in verse I.13 (Passage #13.2), as well as in verses IX.42 (Passage #13.34, also referring to *shang*) and IX.47 (Passage #13.31). However, that *fa* is more generic than *xing* and has less harsh connotations. In something of that spirit, traditional commentator Du Mu offers a comment on verse I.13's *shang fa*: "Neither should be excessive." See Griffith, p. 66.

To be clear, *fa* 罰 in *shang fa* is a different concept from *fa* 法 ("law," "method," etc.) appearing in verse I.8 (Passage #13.30).

X.9. Now when troops flee, are insubordinate, distressed, collapse in disorder or are routed, it is the fault of the general. None of these disasters can be attributed to natural causes.

X.10. Other conditions being equal, if a force attacks one ten times its size, the result is flight.

X.11. When troops are strong and officers weak the army is insubordinate.

X.12. When the officers are valiant and the troops ineffective the army is in distress.

X.13. When senior officers are angry and insubordinate, and on encountering the enemy rush into battle with no understanding of the feasibility of engaging and without awaiting orders from the commander, the army is in a state of collapse.

X.14. When the general is morally weak and his discipline not strict, when his instructions and guidance are not enlightened, when there are no consistent rules to guide the officers and men and when the formations are slovenly the army is in disorder.

X.15. When a commander unable to estimate his enemy uses a small force to engage a large one, or weak troops to strike the strong, or when he fails to select shock troops for the van, the result is rout.

X.16. When any of these six conditions prevails the army is on the road to defeat. It is the highest responsibility of the general that he examine them carefully.

. . . (the intervening verses mostly center on success, not failure, cases, intermixed with material on permissible disobedience to the ruler, see verse X.18 in Passage #13.18 above.)

X.21. If a general indulges his troops but is unable to employ them; if he loves them but cannot enforce his commands; if the troops are disorderly and he is unable to control them, they may be compared to spoiled children, and are useless.

Passage #13.34. Pathologies of Leadership, cont. (Focus = General/Army Interface)

(While Sun Tzu IX context can be read as pointing to the enemy, Sun Tzu's comments here could equally apply to a general's relationship with his own troops.)

IX.37. When his troops are disorderly, the general has no prestige.[96]

. . .

IX.41. When the troops continually gather in small groups and whisper together the general has lost the confidence of the army.[97]

---

[96] Giles, p. 93 renders verse IX.37 in a blunt way that cuts to the heart of effective control of an army: "If there is a disturbance in the camp the general's authority is weak."

[97] For textual and interpretive issues arising in this verse see Appendix 4 anchor footnote 16. The military gist, which points to erosion of top-down authority, is robust.

IX.42. Too frequent rewards indicate that the general is at the end of his resources; too frequent punishments that he is in acute distress.

IX.43. If the officers at first treat the men violently and later are fearful of them, the limit of indiscipline has been reached.[98]

Passage #13.35. Pathologies of Leadership, cont.: Protracted War and Attacks on Cities As Sources of Flagging Morale (Focus = General/Army Interface)

(These verses can also be read as a warning of impending trouble at the interface of ruler and people.)

II.3. Victory is the main object in war. If this is long delayed, weapons are blunted and morale depressed. When troops attack cities, their strength will be exhausted.

II.4. When the army engages in protracted campaigns the resources of the state will not suffice.

II.5. When your weapons are dulled and ardour damped, your strength exhausted and treasure spent, neighbouring rulers will take advantage of your distress to act. And even though you have wise counsellors, none will be able to lay good plans for the future.[99]

Passage #13.36. Pathology of Another Kind – Decapitation of High Command

(For good leadership practices to exist that leadership must itself be extant. These verses are one instance where Sun Tzu qualifies his usual enthusiasm for speed in executing military operations.)

VII.7. It follows that when one rolls up the armour and sets out speedily, stopping neither day nor night and marching at double time for a hundred li, *the three commanders will be captured*. For the vigorous troops will arrive first and the feeble straggle along behind, so that if this method is used only one-tenth of the army will arrive.[100]

VII.8. In a forced march of fifty li *the commander of the van will fall*, and using this method but half the army will arrive. In a forced march of thirty li, but two-thirds will arrive. [Emphases supplied.]

Passage #13.37. Relationships with Secret Agents (Focus = Interface between General or Ruler and Espionage Network)

(Sun Tzu here points to a much more complex and nuanced relationship between spymaster and spies than that which suffices in a general's dealings

---

[98] Griffith's translation outruns the Chinese text, which simply says "it is extremely unwise." For a suggested military rationale for Griffith's choice of translation see Appendix 9 footnote 19.

[99] There is some ambiguity here as to who it is that is "wise," though the central idea of "no good options" is robust. For further specifics see Appendix 5 footnote 7.

[100] For clarification of "the three commanders" phrase see Appendix 13 footnote 20.

with officers and men in a mass-infantry army, a far larger, less specialized organization.[101] Verse XIII.16 points to a need for Sun Tzu's spies to deal across unspecified further interfaces, probably some civilian, to obtain the required information.)

XIII.12. Of all those in the army close to the commander none is more intimate than the secret agent; of all rewards none more liberal than those given to secret agents; of all matters none is more confidential than those relating to secret operations.

XIII.13. He who is not sage and wise, humane and just, cannot use secret agents.[102] And he who is not delicate and subtle cannot get the truth out of them.

. . .

XIII.15. If plans relating to secret operations are prematurely divulged the agent and all those to whom he spoke of them shall be put to death.

XIII.16. Generally in the case of armies you wish to strike, cities you wish to attack, and people you wish to assassinate, you must know the names of the garrison commander, the staff officers, the ushers, gate keepers, and the bodyguards. You must instruct your agents to inquire into these matters in minute detail.

XIII.17. It is essential to seek out enemy agents who have come to conduct espionage against you and to bribe them to serve you. Give them instructions and care for them. Thus doubled agents are recruited and used.

. . .

XIII.21. The sovereign must have full knowledge of the activities of the five sorts of agents. This knowledge must come from the doubled agents, and therefore it is mandatory that they be treated with the utmost liberality.

. . .

XIII.23. And therefore only the enlightened sovereign and the worthy general who are able to use the most intelligent people as agents are certain to achieve great things. Secret operations are essential in war; upon them the army relies to make its every move.

★★★★★★

---

[101] E.g., as profiled in verse IX.50 (Passage #13.31). However, it is also worth noting that there is a clear harmonic in Sun Tzu XIII that plays up material rewards for spies (see verses XIII.12, XIII.17, and XIII.21) – a relatively unnuanced type of incentive – rather than, say, playing on less tangible and possibly more nuanced motivations for spying (e.g., personal revenge, belief systems, etc.).

[102] For analysis of these attributes of an ideal spymaster, see pp. 366–67 above.

THEME #14 HARMONIOUS INTEGRATION OF *QI* AND *ZHENG* APPROACHES IN STRATEGY AND TACTICS

qi/zheng concept pair:
Passage #14.1 (verses V.3–V.12) (Sun Tzu's qi/zheng concept pair)

first interpretation of qi/zheng pair: qi as unexpected action, zheng as expected action:
Passage #14.2 (verses I.18–I.20 + I.26) (capitalizing on false enemy expectations and unpreparedness)
Passage #14.3 (verse XI.61) (switch of style to exploit unpreparedness)
Passage #14.4 (verse XI.29) (speed as tool for upending enemy expectations)

second interpretation of qi/zheng pair: focus on "spiking," then "tilting" or "toppling" the enemy:
Passage #14.5 (verse XI.9) (terrain roots of Sun Tzu's focus on "spiking," then "toppling")
Passage #14.6 (verses IX.16–IX.17) (spiking an adversary by exploiting extreme terrain features)
Passage #14.7 (verse IX.5) (spiking, then toppling, with reference to a water crossing)
Passage #14.8 (verses X.3–X.4) (spiking, then toppling, by exploiting awkward terrain situations)
Passage #14.9 (verses IX.31 + IX.44) (battlefield cases of Sun Tzu's focus on spiking, then toppling)
Passage #14.10 (verses VII.20–VII.22) (psychological paths to spiking, then toppling, an army or its general)
Passage #14.11 (verses VIII.17–VIII.24) (spiking, then tilting or toppling, a general by exploiting character traits)
Passage #14.12 (verses I.22–I.25) (further psychological paths to spiking the enemy)
Passage #14.13 (verses VIII.14–VIII.15) (spiking by burdening a foe or rival with a task, against a coercive backdrop)
Passage #14.14 (verses XIII.10 + XIII.16 + XIII.19) (false information fed to the enemy as versatile tool for effective spiking, as prelude to toppling)

inept self-spiking:
Passage #14.15 (verse X.21) (self-spiking, administratively inept variety, by reason of excessive kindliness to subordinates)
Passage #14.16 (verses IX.47 + IX.49) (self-spiking, administratively inept variety, cont.)
Passage #14.17 (verses X.9–X.16) (self-spiking by reason of a general's professional incompetence: ineptness of administrative or tactical varieties)
Passage #14.18 (verses III.7–III.9) (self-spiking, tactically inept variety, by reason of undertaking a protracted siege operation or impetuously ordering a costly assault)
Passage #14.19 (verse XII.18) (self-spiking, strategically inept variety, by reason of a ruler's or general's uncontrolled emotions)

Passage #14.20 (verse XIII.2) (self-spiking, grand strategically inept variety, by reason of failure to cultivate espionage capabilities)

Passage #14.21 (verses III.18–III.23) (self-spiking, grand strategically inept variety, by reason of failure to keep civil–military relations in good repair)

reality checks:

Passage #14.22 (verses VII.30–VII.33) (reality check: pitfalls in passing from spiking to toppling)

Passage #14.23 (verses XII.7–XII.8) (reality check: pitfalls in *zheng* → *qi* transition in fire attack)

more complex agenda:

Passage #14.24 (verses VI.13–VI.16) (more complex agenda: spiking the enemy by forcing sprawling commitments, then toppling the foe by exploiting local superiority)

★★★★★★

Theme #14 is an outlier among the fourteen themes because Sun Tzu's explicit attention to it is limited to one part of one chapter (verses V.3–V.12, see Passage #14.1). The focus of that passage is a pair of concepts which Griffith translates as "extraordinary" (*qi*) and "normal" (*zheng*) and treats as pertaining to different kinds of military forces.[1]

The *qi/zheng* contrast is a basic strand of Chinese military thinking for which no close counterpart exists in Western military or strategic thought.[2] Sun Tzu is the earliest traditionally transmitted text to invoke it in a military context. The *qi/zheng* distinction finds uses on what would nowadays be described as both a tactical and a strategic level.[3] It also appears in early Chinese usage in contexts that range beyond warfare.[4]

Important groundwork for clarification of the *qi/zheng* disinction comes from work of contemporary Sun Tzu authority Li Ling, analyzing eight different traditional and modern interpretations he has identified.[5] Sifting these (along with some other modern sources), with an eye to spotting militarily or strategically enduring

---

[1] Taking cues from word order in Sun Tzu, Theme #14 analysis will refer to *qi/zheng*, not *zheng/qi*.

[2] Liddell Hart's master theme of the "indirect approach" to strategy or tactics is a potential contender for an analog to Sun Tzu's *qi* in Western military thought. But for Liddell Hart the contrast case of "direct approach" serves mostly as a conceptual foil for conveying the merits of an indirect approach or, as he himself puts it, "More and more clearly has the lesson emerged that a direct approach to one's mental object, or physical objective, along the 'line of natural expectation' for the opponent, tends to produce negative results" (Liddell Hart, #4 footnote 2, p. 25). That is a far cry from Sun Tzu's thinking calling for productive *combinations* of *qi* and *zheng*, each playing an essential role.

[3] Griffith, p. 43: "Qi and zheng operations may be launched . . . on strategic [as well as tactical] levels."

[4] *Laozi* LVII reads in relevant part: "Govern the state by being straightforward [*zheng*]; wage war by being crafty [*qi*]." (Lau, #1 footnote 34, p. 118.) Note that this formulation mentions *zheng* before *qi*.

[5] See Li Ling, #1 footnote 84, pp. 177–91.

substance, the present analysis concentrates on two clusters of meanings.[6] One cluster aligns on the state of play of *expectations*, the enemy's and one's own.

A second aligns on *types of military operations* (or, in a variant, of military forces). Both clusters find support in a long-lost bamboo strips text known as "*Qi and zheng*" (cited below as *QZ Lost Book*). That work, archaeologically recovered in the same find that recovered the Han strips text of Sun Tzu, has been analyzed in detail by Li Ling.[7]

Meaning cluster 1. A first cluster of meanings of qi and *zheng* receives a traditional exposition in one of the *Seven Military Classics*. It should be noted that this traditional exposition postdates Sun Tzu by a thousand years or more.[8] Lau's 1965 article is a modern exposition of the first cluster's central ideas. The basic point, which is highly elementary (at least so long as one brackets nuances of human psychology!), is that if the foe foresees your strategy or tactic, then that strategy or tactic is *zheng*; a strategy or tactic not foreseen (i.e., "surprise") is qi.[9]

An offshoot (amounting to an intensification) of the concept of "qi as surprise" is a qi move whose possibility has not even crossed the foe's mind.[10] Edgar Allan Poe's famous "purloined letter" story provides a fine example of "hiding in plain sight" as a qi expedient.

Importantly, a qi step need not involve deliberate deception. Even if you did not deliberately create qi, but by chance your enemy is taken off-guard, that would be qi too.[11] *A strategy or tactic may hence qualify as qi even though it has none of the trappings of guile.*

---

[6] Both of these clusters subsume a variety of shades of meaning, depending on details. With two exceptions (one associated with the Chinese Communist general Lin Biao, not principally known as a military theorist; another with François Wildt, see p. 465), the other qi/zheng interpretations considered by Li Ling center on traditional Chinese mantic practices and/or battlefield troop formation theories. Those topics are recondite and antiquarian. Especially given that the Sun Tzu text says next to nothing about either, their analysis is beyond the present study's scope.

[7] Li Ling, #1 footnote 84, pp. 178–80, comments positively on the intellectual quality of the *QZ Lost Book*, which has survived relatively complete. It shows much influence of the "five elements" (*wu xing* 五行) theory, traces of which also appear in Sun Tzu (see #5 footnote 103 and #9 footnote 13).

For clarity, it should be noted that the *QZ Lost Book* is *not* part of a corpus of Sun Tzu–related material from the same archaeological find that Ames describes as standing in a "commentarial" relationship to Sun Tzu's thirteen chapters (see Background footnote 82). Positing that the five elements theory is a relatively late arrival (a view advocated by Brooks & Brooks, Introduction footnote 36, p. 85), the *QZ Lost Book*'s perspective on qi and *zheng* might represent a time later than Sun Tzu.

[8] The relevant work is the *Tang Taizong Li Weigong wendui* ("Questions and replies between Tang Taizong and Li Weigong" or *Questions and Replies*, for short). Although attributed to Li Jing, a prominent early Tang general (571–649 AD) who is the Li Weigong of the title, *Questions and Replies* is widely viewed as being of late Tang or Song dynasty vintage (see Sawyer, *Seven Military Classics*, p. 313 and pp. 488–90 note 4; Wilkinson, §28.4.3, p. 351). For pertinent qi/zheng content see Sawyer, *Seven Military Classics*, p. 337. Sawyer, Introduction footnote 55, p. 243 refers to it as a "variant of the indefinitely extendible tactical conundrum 'I know that you know that I know.'"

[9] Elaborating on this interpretation and, relatedly, proposing to translate qi as "crafty" and *zheng* as "straightforward," see Lau, 1965 article, pp. 330–31 (quoted at length on pp. A-236f.).

[10] See Lau, 1965 article, p. 321 footnote 9 (previously quoted in relevant part in #5 footnote 78).

[11] Li Ling, #1 footnote 84, pp. 178–80. Interpreting the *QZ Lost Book*, Li Ling observes "To strike is like using *zheng*; to *strike unexpectedly* (qi) and the enemy cannot respond back, then you win." (Quoted from Li Ling, #1 footnote 84, p. 180, emphasis supplied.)

Meaning cluster 2. A second cluster of meanings of qi/zheng brings in more specifically military context, aligning the distinction on specific types or functions of military operations (or, in a variant favored by Griffith, on different kinds of military forces).[12] This second cluster finds anchoring in Sun Tzu's verse V.5 (Passage #14.1), aligning zheng on "engaging" the enemy (in Mair's translation, "joining battle"); qi, on steps geared to clinching a victory. The second cluster is further developed by two statements from Cao Cao's commentary, which have served as an anchor for much traditional Chinese thinking about qi and zheng:[13]

> to attack first is zheng, to attack later is qi,
> frontal attack is zheng, attacking from the side is qi.

Both these ideas broadly align on verse V.5. The QZ Lost Book points in a similar direction but on a notably more abstract plane, thereby opening a door for applications ranging beyond land warfare (or indeed any specific medium of conflict):[14]

> To use xing ["form"] to react to xing, this is zheng; to use wu xing [Griffith's "formlessness," see Theme #8] to react to xing, this is qi.[15]
>
> ... If you are the same as your enemy [i.e., you have the same xing as your enemy, or you employ the same tactic as your enemy], you cannot win. Therefore, to create a [winning] difference is qi.

Symmetry-breaking steps stand at the heart of both these last two formulations.

A relevant perspective on qi and zheng has been proposed by Benjamin Wallacker in a brief study published shortly after Lau's 1965 article.[16] Wallacker's work is worth attention, not for its etymological arguments (which can be challenged), but for the way in which it fleshes out Sun Tzu's verse V.5 concept, staking out a middle ground

---

[12] Lau's 1965 article (pp. 330–31) roundly criticizes Griffith's alignment of qi and zheng on different types of military forces, which Griffith labels "extraordinary" (qi) and "normal" (zheng). Yet Griffith's understanding seems actually not that far off Lau's. See, e.g., Griffith, pp. 42–43 and p. 91 footnote 1: "Should the enemy perceive and respond to a qi manoeuvre in such a manner as to neutralize it, the manoeuvre would automatically become zheng." References to qi forces (qi bing) appear in the Shiji. Cf. Shiji 92 battle account summarized on pp. 30–32 above, where Watson translates qi bing as "surprise troops." Importantly, as this Shiji passage also suggests by implication, a force might qualify as qi bing because of the use to which that force is put as contrasted with its intrinsic attributes (weapons, training, etc.), thereby bringing the focus back to military operations.

To the extent that Griffith's understanding of qi aligns on military forces with distinctive attributes (as contrasted with distinctive uses), that understanding may in part reflect influence of Griffith's World War II military background with the First Marine Raiders on Guadalcanal, a period when the modern special forces concept was still early in its development. In that era the Raiders were among US military's closest approximations to a qi force with distinctive characteristics and capabilities (even though, reflecting conventional military attitudes of that time, higher command authorities left the Raider's qi potential largely untapped; see #14 footnote 51 below).

[13] Li Ling, #1 footnote 84, p. 180.

[14] Li Ling, #1 footnote 84, p. 179.

[15] Li Ling, #1 footnote 84, p. 180 glosses this thought by suggesting that using wu xing to react to xing means using an intentionally created (and in that sense "customized") xing to react to the enemy's xing, as contrasted to relying on a ready-made ("off the shelf") xing. Drawing on Theme #8's "ladder of formlessness" (see Figure 3 on p. 270 above), wu xing 5 (p. 277) is particularly relevant here.

[16] See Benjamin E. Wallacker, "Two concepts in early Chinese military thought," Language, 1966, 42(2), 295–99. Highlights of his conclusions are briefly noted by Lau & Ames, p. 58. Wallacker's etymological methodology is controversial. It should therefore be clearly emphasized that the present use of Wallacker's terminology does not rely on that methodology (on which see pp. A-245ff. below).

between Cao Cao's commentary interpretations (overly narrow, at least when taken literally) and that of the *QZ Lost Book* (overly abstract). The key idea is simple but compelling. As Wallacker states it succinctly (using a clever imagery), after first "'spiking' the foe, keeping him fixed vulnerably to his position, we bring in our 'tilting' forces to knock him off balance and even to topple him over."[17]

Combing the Sun Tzu text for verses and passages that align on this understanding of *qi* and *zheng* (even though, apart from Passage #14.1, not using *qi/zheng* terminology) turns up a goodly harvest, identifying a vein of Sun Tzu content otherwise easily missed. The extent of that harvest is one clue that Wallacker's spiking/tilting imagery captures something useful about Sun Tzu's *qi/zheng* thinking.[18] A further clue that this analytical direction is productive comes from the natural land warfare connection between difficult terrain and military flexibility loss (Wallacker's "spiking"). Such anchoring in terrain issues, a major Sun Tzu focus, is notably lacking in both Cao Cao's and the *QZ Lost Book*'s explanations of *qi* and *zheng*.

Standing back from the details, there is no need to see a contradiction between the first and second cluster of meanings just discussed. Many applications of the *qi/zheng* distinction contain traces of both, and in some situations they coalesce (as when, for example, use of *zheng* creates a "stalemate" situation typified by both low flexibility and high predictability, in fact for both sides, which a surprise *qi* move then breaks in a way favorable to one side, tilting or toppling the other).[19] Neither meaning cluster captures everything of interest about *qi* and *zheng*; other interpretations have been proposed and also have their uses.[20] Each may be viewed as

---

[17] For vigorous challenge to that imagery's etymological origins see Mair, "Soldierly methods" (cited on p. xxii above), pp. 79–80. From within the Sun Tzu text, conceptual support for a tilting/toppling distinction (though without using that terminology) is found in Passage #14.11, whose verses VIII.18 and VIII.19 point to two different "toppling" outcomes, verses VIII.20–VIII.22 to three "tilting" ones.

[18] Wallacker's proposed English-language terminology for talking about *qi* and *zheng* is helpful in another way as well, sidestepping semantic pitfalls lurking in some widely used translations of those concepts. A case in point would be translating *qi* as "unorthodox," *zheng* as "orthodox." That terminology runs up against the military reality that both are well represented in influential theory and doctrine statements (not least the Sun Tzu text itself!) and in that sense might both lay valid claim to being "orthodox." It also has overtones of casting *zheng* ("orthodox") as the more basic category, with *qi* ("unorthodox") as a derived category – a casting that could be at odds with the *qi/zheng* order of mention in the Sun Tzu Chinese text if treated as a marker of emphasis.

[19] This "stalemate" conceptualization is Li Ling's. See Li Ling, #1 footnote 84, p. 177. However, as a concept "stalemate" seems too narrow. It would not, for example, fit Passage #14.7's river-crossing scenario which puts the army undertaking the crossing in an unenviably awkward position that Wallacker's "spiking" terminology captures very effectively.

[20] Particular mention should be made here of an essay by a French scholar, François Wildt, of which Li Ling, #1 footnote 84, pp. 182–84 writes very favorably (even treating it as a preferred interpretation). That essay is available in Chinese as François Wildt 魏立德, "Guanyu Sunzi bingfa zhong de shuli luoji" 關於孫子兵法中的數理邏輯 (On the arithmetic logic in Sun Tzu), in *Sunzi xintan: zhong-wai xuezhe lun Sunzi* 孫子新探：中外學者論孫子 (*New Research on Sun Tzu: Chinese and foreign scholars' studies of Sun Tzu*) (Beijing: Jiefangjun chubanshe, 1990), pp. 122–30.

Noting that in the early Chinese language the concept of *qi* 奇 did double duty, signifying both "odd number" (a meaning in which it is now pronounced *ji*) and "extraordinary," Wildt develops a case for anchoring a military/strategic interpretation of *qi* in Sun Tzu on *yuji* 余奇 ("remainder" in a mathematical sense). His analysis provides a fresh way of thinking about *qi/ji* in

valid – up to a point. Then the creative spirit of Sun Tzu steps in, and encourages us to entertain competing interpretations which cross-fertilize one another, catalyzing fresh ideas. To avoid overcomplicating the following discussion, Lau's analysis will be taken as placeholder for the first cluster of *qi/zheng* meanings; Wallacker's analysis (divested of its etymological origins), as placeholder for the second cluster.

The most basic problem with Lau's analysis of *qi* and *zheng* in his 1965 article is that it is devoid of substantive context, thus failing to draw traction from the sensitivity to context that is one of Sun Tzu's strongest suits. Relevant here is an action dilemma posed by Thomas Schelling in his essay "The Reciprocal Fear of Surprise Attack."[21] Describing a scenario where an armed homeowner surprises an armed burglar (incipiently leading to a shoot-out that neither party wants), Schelling notes that a basic difficulty facing an attempt to solve this problem in "he thinks I think" mode is that "nothing generates the series." Much as in Schelling's scenario, Lau's version of a "calculus" of *qi* and *zheng* lacks a mechanism for generating the pertinent expectations – and is therefore too ungrounded to give practical guidance. For example, the adversary's relevant expectations may be unknown; may fluctuate, perhaps too wildly to make the *qi/zheng* distinction useful (cf. the phrase "does not know his own mind" as applied to a vacillating enemy decision maker[22]); or may be so all-embracing as to span all possibilities, excluding nothing (a pattern by no means foreign to how bureaucracies think!).

To get further, it is essential (as Schelling's scenario suggests) to identify some source of constraint that imposes more structure on the problem at hand. *It is precisely here that Wallacker's analysis stands to make a basic contribution. In a nutshell, that contribution is to train a spotlight on sources of inflexibility in the enemy's posture; to recognize that inflexibility when it exists (an observer's or analyst's task), possibly also taking steps to induce or intensify it (a strategist or tactician's task); then to exploit it to the hilt.*[23] Sources of inflexibility, which are the stuff of *zheng*, are diverse. They may be thoroughly tangible and physical (Max Weber's image of the "iron cage" has roots here!). Via the role of

---

its its military sense. For further analytical directions inspired by Wildt's work, see Appendix 14, pp. A-247ff.

[21] See Schelling, #5 footnote 8, pp. 207–209.

[22] See p. 123 above (*weiqi*-inspired example from the *Zuozhuan*; see #3 footnote 46 for citation).

[23] For a relevant line of thinking, citing practices found in Russian chess strategy and ancient Parthian warfare, see Karl E. Deutsch, *The Nerves of Government: models of political communication and control* (with new introduction; New York: Free Press of Glencoe, 1966), p. 274 note 8:

> This strategy [geared to encouraging loss of flexibility by the foe] differs significantly from the familiar one of keeping one's opponent "off balance." To keep an adversary off balance may mean, among other things, to prevent him from committing himself thoroughly to any course of action. The Russian chess strategy, like the warfare of the ancient Parthians, would on the contrary encourage him to make such a commitment, in the hope of turning this commitment later to his (opponent's) disadvantage. The latter strategy, unlike the former, can employ deliberate pauses of activity, as well as positive action. The difference between the two strategies resembles thus, in some respects, the difference between boxing and jujitsu.

> [This Deutsch footnote also appears on p. 50 of Admiral Eccles' "Strategy: the essence of professionalism," *Naval War College Review*, December 1971, pp. 43–51.]

A focus on major inflexibility – or, put differently in terms suggested by Deutsch's analysis, a commitment that forecloses other options – also complements the *wu xing* ("formlessness") concept central to Sun Tzu (Theme #8), which emphasizes the antithesis, namely, extremes of flexibility.

terrain, physical inflexibility is perhaps the most basic case of inflexibility in the conventional warfare Sun Tzu knew. Sources of inflexibility may also be intangible – informational, cognitive, emotional, organizational, etc. (or a mix thereof). From whatever source derived, inflexibilities may be an outcome of clever stratagems that maneuver an adversary into a low-flexibility posture. They may also – and very importantly – be brought about, at least in part, by a general's or ruler's ineptness, a contingency to which Sun Tzu pays much attention. In any case, exploiting the enemy's inflexibility to "tilt" or "topple" that foe is a task for qi.[24] Certainly "surprise" is one major path to that end. However, actually achieving tilting or toppling calls for more than surprise alone. Clearly relevant too is how much muscle backs up that surprise;[25] also, how long the surprise action can escape detection.[26] Some qi steps may be efficacious precisely because they never become known to the foe at all, which means that unthinkingly equating qi to "surprise" (a common English translation of qi) risks overlooking some potent possibilities (e.g., invisible or low-visibility algorithm sabotage, see Theme #6, pp. 220ff. above).

Yet Lau's expectations-oriented vein of thinking also offers value-added. It is particularly well suited to analyzing qi and zheng in elite civilian contexts where physical realities are often little more than stage props for dueling expectations. Those expectations are frequently far more fine-grained and nuanced than those associated with military situations. They are also deeply embedded in social structure and social context. Civilian social structure is itself enormously complicated and messy (a reality that military professionals sometimes find disconcerting!). In such a setting, the notion that "under irregular circumstances, what is regular becomes

---

[24] Of course, certain types of inflexibility may create advantage for the side that lacks it – an area of common ground between parts of Schelling and Sun Tzu's "death ground" thinking (Passage #8.21). But that is a specialized situation. More usually, flexibility is to be prized. As Padgett & Ansell, #2 footnote 43, p. 1264 observe, "Victory, in Florence, in chess, or in go [weiqi] means locking in others, but not yourself, to goal-oriented sequences of strategic play that become predictable thereby."

[25] In the annals of US military history, a classic example of a qi step not accompanied by sufficient muscle to throw the adversary off balance for more than a fleeting time is the July 30, 1864 Battle of the Crater in the American Civil War, which the Union forces commenced by detonating a large mine under fixed Confederate defenses, blowing a hole in their line. Failure promptly to exploit that Union success made it short-lived. The Confederates quickly recovered, launching counterattacks that resulted in heavy Union casualties. The upshot was a return to many more months of trench warfare of a zheng operations type. As Grant later wrote: "It was the saddest effort I have witnessed in the war." For an account of the episode see Alfred P. James, "The Battle of the Crater," Journal of the American Military Foundation, 1938, 2(1), 2–25.

[26] A case in point is a failed qi maneuver attempted by Griffith himself, then commanding the First Marine Raiders on Guadalcanal. See Griffith, Introduction footnote 20, p. 136:
> Leaving a reinforced company to contain the enemy blocking force [zheng], Griffith sought room to maneuver, and led the remainder of the battalion up a precipitous jungle-covered spine from which he could move down on the rear of the Japanese [qi]. The slow and exhausting development of a battalion in Indian file was not completed until midafternoon, and was discerned by Japanese on an adjacent ridge. Another fire fight ensued in which the battalion commander [Griffith] was wounded. Attempts to push along the low ground or the ridge brought instant reaction from mortars, machine guns and automatics. The Raiders were stalled.

itself irregular" – a Lau & Ames gloss on qi and zheng[27] – launches a productive line of inquiry, probing questions pertaining to discrepancies between "myth system" and "operational code";[28] who is likely to violate sociocultural norms; and what forces might lead those norms themselves to shift.[29] All those issues are intertwined with prevailing expectations. All would have arisen in Warring States civilian applications of Sun Tzu, even if our sources shed scant light on how they were addressed.

Further developing the analysis of qi and zheng calls for a more searching look at their relationship. A hint of asymmetry is signaled by Passage #14.1's consistent mention of qi before zheng.[30] Lauding with vivid metaphor – Griffith's verse V.7 reads well as poetry – qi's creative potential and the extent of capability that goes with skill in its use, verses V.6 and V.7 suggest that qi offers near-boundless scope for military or strategic creativity. Sun Tzu mentions no zheng counterpart.[31]

But here it is important to avoid being swept away by the flow of Sun Tzu's qi rhetoric. The thrust of Passage #14.1 needs to be absorbed as a whole. That thrust plainly accords a fundamental role for zheng (Lau's "straightforward"), pedestrian and downright dull as it may often be. To put the point more sharply: "The compleat Sun Tzu" (to adapt the title theme of an early exposition of game theory[32]) is *not only* the highly manipulative qi Sun Tzu, the Sun Tzu of Theme #7 (deception) fame for whom, carried to its logical limit, all strategy devolves into "games within the game."[33] It is *also and equally* the zheng Sun Tzu whose straightforward actions are rooted in meticulous planning and calculating (Theme #1), adroit use of terrain (Theme #2), and the critical scrutiny of operational readiness that is a basic part of

---

[27] See Lau & Ames, pp. 232–33 note 366 (citing further early Chinese sources).

[28] "Myth systems present the official picture – the norms that are supposed to apply. . . . Operational codes refer to the norms that are actually applied – the way of doing business." See p. 400 footnote 35 of Patrick J. DeSouza, "Regulating fraud in military procurement: a legal process model," *Yale Law Journal*, 1985, 95(2), 390–413 (citing scholarship of the Yale Law School's W. Michael Reisman).

[29] Here mention should be made of a further interpretation of qi and zheng proposed by Lewis, Background footnote 11, pp. 122ff., rendering qi as "extraordinary" and zheng as "normative." See his p. 125: "I have translated zheng as 'normative' because, in addition to the specifically military usage, it had the broader meaning of 'correct' or 'properly regulated' and could be synonymous with its homophone zheng (政), which meant 'to regulate' or 'govern.'" See also his p. 293 note 89: "The army was constituted as a social body through the *normative rules*, but victory in the field was won through *extraordinary maneuvers*." (Emphasis supplied.) By suggesting a way of breaking out of a narrowly military interpretation of the qi/zheng concept pair, Lewis's "normative" encourages finding applications of the qi/zheng distinction in civilian strategic contexts otherwise beyond its reach.

[30] Wallacker's conceptualization of qi and zheng conveys this asymmetry better than Lau's. The point is illustrated by a "guessing game" example Lau uses to convey the difference between qi and zheng (1965 article, pp. 330–31, quoted on p. A-236f.). That game centers on two options, one expected (zheng), one unexpected (qi), that are otherwise entirely interchangeable and in that sense are structurally equivalent and symmetric. Departure from symmetry – not always initially obvious – is a feature of other paired concepts in early Chinese thought. See Lau, #5 footnote 29, p. 353.

[31] Relatedly, Mair, p. xlvi observes that "[o]f these two terms, the more difficult to grasp is qi."

[32] *The Compleat Strategyst, being a primer on the theory of games of strategy* (New York: McGraw-Hill, 1954) by John D. Williams (who headed the Mathematics Division of the RAND Corporation).

[33] The "games within the game" image is Martin Shubik's. See p. 251 above.

knowing oneself (Theme #12). It is clear from the Sun Tzu text that Sun Tzu's way of war stands for a harmonious coordination of qi and zheng.[34] Not all of life, military or civilian, reduces to a succession of qi stunts![35] Such a standpoint might impress some modern audiences as un-Sun-Tzu-esque. But to harp incessantly on the qi face of Sun Tzu to the exclusion or marginalization of zheng is to misread Sun Tzu.[36]

Developing the point a step further, for reasons that Sun Tzu's Passage #14.1 conveys through images like the flow of great rivers or the passing of the seasons, there is also an inescapably dynamic aspect to the interplay of qi and zheng that no static pair of conceptual boxes (or static formal models) captures. Although both Lau's and Wallacker's interpretations accommodate this dynamic aspect, Lau's is particularly effective in doing so. A shift in an adversary's expectations may occur abruptly, possibly as a result of new information or a flash of suspicion, thereby converting a qi action into a zheng one.[37] Sun Tzu's emphasis on warfare's information level sets basic context here. Information is a quicksilver entity. It can travel swiftly; can spread by unpredictable zigzag paths (cf. Theme #2 discussion of small world networks, pp. 99–101 above); and can also mutate in the course of transmission, possibly several times over. Conversely, a zheng action may turn into a qi one – for example, when the adversary comes to take a static situation too much for granted and lets down his guard, thereby setting the stage for a direct assault by the foe that qualifies as qi for exactly that reason.

The dynamism and mutability of qi and zheng places extraordinary demands on a general's ability to toggle back and forth between qi versus zheng aspects or levels of

---

[34] Compare *Questions and Replies* (see #14 footnote 8 above), which takes a position that if there is no zheng, qi will have no use at all. See summary by Li Ling, #1 footnote 84, pp. 181–82.

[35] This perspective in turn presents the question of how to calibrate the qi/zheng mix. A pioneering step toward a quantitative answer comes from an observation in a lost work attributed to Cao Cao that survives in the *Questions and Replies*. It reads (see Li Ling, #1 footnote 84, p. 180): "If you are two to the enemy's one, then one tactic [shu] is zheng, one tactic is qi. If you are five to the enemy's one, then three tactics are zheng, two tactics are qi."

By way of clarification of this terse, somewhat obscure statement, Cao Cao's reference to 2 (or 5) tactics to the enemy's one might be interpreted as referring, not as a direct reference to numerical superiority, but rather to how many qi or zheng tactics could be implemented – in the simplest scenario, simultaneously – making use of different parts or units of one's army (any one of which could be employed either in a qi way or a zheng way, but not both). Generalizing from what seems best taken as an example, the idea appears to be that the representation of qi in that qi/zheng mix should diminish – perhaps gradually – as the number of different available qi-or-zheng choices grows larger. Cao Cao's stipulation that the enemy has just one available tactic might be interpreted as a simplifying assumption (e.g., in one modern framing, as a way of preventing a decision theory problem from burgeoning into a full-fledged game-theoretic one, derailing Cao Cao's simple rules).

[36] Some of this failing may perhaps be laid at Sun Tzu's feet. Lau & Ames, p. 59 suggest that fuller clarification of the interdependence and complementarity of qi and zheng may have had to await Sun Bin's later military treatise (lost until a partial text was archaeologically recovered in 1972). As they translate from the Sun Bin text, "An excess of surprise operations [qi] . . . will overshoot the mark of victory" (Lau & Ames, p. 175; interpretation issues are discussed in their pp. 234–35 note 377).

[37] Such a shift may also be gradual, rather than discontinuous. For further analysis see Appendix 1 anchor footnote 29.

a military situation.[38] Rising to that challenge is surely one of the most difficult tasks facing any commander, particularly since the nature of both problems and solutions tends to be very different as between qi and zheng. Extending earlier discussions of style switches in Themes #5 and #7 contexts, zheng-to-qi or qi-to-zheng switches are often where mistakes get made and the mask of command drops a little, so that information leaks.[39] In Sun Tzu's version of land warfare, the latter switches may actually be harder for commanders to pull off than many of the style switches implicit in verse VII.13 (Passage #5.14), precisely because zheng-to-qi or qi-to-zheng switches commonly occur when there is direct contact with the enemy and depend in no small part on what the enemy general and army do (hence are not fully under one's own general's control). Also, zheng-to-qi or qi-to-zheng switches call for a general to control not only the posture of his army but also his own frame of mind. One reading of Sun Tzu's emphasis on the need for a commander to be opaque or inscrutable (Theme #8, pp. 273–77 above; a relevant verse is Passage #8.9's verse XI.42) is precisely that others should not see such mindset-switching while it is taking place – thereby divulging possibly actionable information to enemy spies.

One of Sun Tzu's important original contributions to the art of war is to set a stage for better understanding of switches of military posture and a general's frame of mind, zheng $\rightarrow$ qi, qi $\rightarrow$ zheng.

✦✦✦✦✦✦

Because the qi/zheng concept pair is explicitly mentioned only in Passage #14.1, judgment is needed in combing the Sun Tzu text for further passages illustrating qi, zheng, or (preferably) both concepts working together. A case can be made that this task is particularly important for zheng because qi finds major anchoring in several other themes (among them, Themes #5–#8), whereas zheng operations lack a comparable alternative home. Here Wallacker's interpretation shows its practical worth.

Following Passage #14.1, passage roundup continues with Passages #14.2–#14.4. These further anchor Lau's interpretation of qi by identifying Sun Tzu statements that focus on upending enemy expectations (which, especially in the case of Passage #14.2, may also be specifically engineered with that goal in mind).

Attention then shifts to Wallacker's interpretation. In keeping with terrain roots of Sun Tzu's ideas, treatment starts with inflexibility born of challenging terrains (Passages #14.5–#14.8). A sizable suite of further passages (Passages #14.9–#14.21) covers other scenarios that lend themselves to interpretation as involving some version of "spiking." The "tilting" or "toppling" to follow is sometimes spelled out, but Sun Tzu often leaves it up to his audience to infer the obvious.

A noteworthy feature of this material is the amount of attention Sun Tzu pays to "self-spiking" cases, where major flexibility loss occurs with no visible nudge from any adversary (see Passages #14.15–#14.21). One basic reason that spiking/toppling works as well as Sun Tzu expects it to is that people are remarkably good at self-spiking – and in Warring States China, as elsewhere in world history, there have

---

[38] For observations on qi/zheng role-switching see also Christopher C. Rand, *Military Thought in Early China* (Albany, NY: State University of New York Press, 2017), pp. 46–47.

[39] See pp. 181–82 (Theme #5), pp. 247–48 (Theme #7). To be clear, it should be noted that not all style switches considered there fall under a rubric of zheng $\rightarrow$ qi or qi $\rightarrow$ zheng.

often been scant safeguards in place to bar inept actions taken by persons in high places.[40] As this observation suggests, many of the sources of self-spiking are emotional. Some reflect poor tactical judgment. Others are logistical (understanding logistics in its comprehensive sense). In many ages of war faulty logistics concepts are a major source of self-spiking.[41] A self-inflicted loss of flexibility can, of course, be a source of cheap advantage to a foe, providing another example of how Sun Tzu's ideas interlock (here Themes #14 and #3). Particularly worth noting – because now the "self-spiking" has a flavor of larger organizational or systemic failure – are Passages #14.20 and #14.21, pointing respectively to failure to cultivate an espionage apparatus and failure to keep civil–military relations in good repair.[42]

Lest mastery of qi and zheng seem too easy, Passages #14.22 and #14.23 offer reality checks, warning of scenarios in which an intended spiking/toppling sequence may go awry (as it did for Griffith's Marine Raiders on Guadalcanal).[43]

Passage #14.24 is suggestive of more complex qi/zheng scenarios. However, Sun Tzu vouchsafes us no details that would further develop relevant thinking.

Roads Not Taken by Sun Tzu in Developing Theme #14

Interpretive caution is needed here, given that we have limited information to flesh out the qi/zheng dichotomy as actually envisioned by Sun Tzu (or whoever compiled the text). With that caveat in mind, two Theme #14 candidates for analytical paths that might have been followed by Sun Tzu, but were not followed, center respectively on two military universals, namely, intelligence and logistics.

First, intelligence: Implicit in existence of a Sun Tzu chapter on espionage is a notion that intelligence operations may be the qi that complements the zheng of conventional military operations. Whetting appetite, Wallacker's analysis readily extends to suggest a Sun Tzu–inspired blueprint for successful subversion: An army-in-being pins down the enemy's army (zheng), though without engaging it in costly major combat; the spy network then topples the enemy regime by clandestine measures (qi). Yet Sun Tzu himself says nothing about this most Sun-Tzu-esque scenario.

Second, logistics: Military operations comprise not only tactical operations but also, just as importantly, logistics ones. Logistics operations harbor a natural place for a qi/zheng distinction. "Normal" logistical operations tend to be highly routinized and in that sense are straightforward (as well as "expected," in fact by both sides).

---

[40] Cf. William J. Goode, "The protection of the inept," *American Sociological Review*, 1967, 32(1), 5–19. A contrast case to inept self-spiking would be "artful" self-spiking. See verse XI.55 (Passage #8.21).

[41] Eccles, Background footnote 52, p. 321 (perils of self-deception in logistics matters). There is an interesting divergence here between Sun Tzu's emphases and Barton Whaley's. The latter's masterful study of stratagem explicitly excludes cases of self-deception. See Whaley, #7 footnote 4, p. 83.

[42] In Warring States contexts, care needs to be exercised in pointing to "institutions," a term that could import extraneous modern connotations. Cf. #6 footnote 4 (pertinent strands of traditional Chinese thought); #11 footnote 19. Relatedly, Lewis, Background footnote 7, p. 603 observes: "In reality, the Warring States have no histories of institutions, but rather biographies of individuals."

[43] For description of the circumstances see #14 footnote 26 above.

Major logistics operations of that type are needed to support *zheng* tactical operations by a mass-infantry army engaging the enemy (see Passage #14.1's verse V.5). What's more, normal logistics operations also tend very effectively to pin down the troops assigned to do them! Those attributes combine to make such logistics operations a natural candidate for being considered *zheng*. Improvised logistics operations, departing from routine in ways that (conjoined with tactical operations) can throw the enemy off balance, are the stuff of qi.[44] There is a whiff of qi logistics in Sun Tzu's counsel for an invading army to live off the enemy (see Passages #3.5 and #8.2–#8.3). But Sun Tzu does not develop either his qi/zheng or his logistics thinking far enough to explore what the qi/zheng distinction might mean in a logistics context in the warfare of his time.

Shifting to a wholly different analytical plane, there is also nothing in Sun Tzu about impaling a rival or foe on the horns of an ethical dilemma – a potentially outstanding way of freezing an adversary's decision-making processes (a type of "spiking" preparatory to tilting or toppling; cf. the fate of the unfortunate police inspector Javert in Victor Hugo's *Les Misérables!*). Here may be yet one more bit of evidence that Sun Tzu's focus was overridingly military, with civilian applications of the text being a later, presumably civilian, invention.

Sun Tzu (2) ("Sun Tzu Extended") and Sun Tzu (3) ("Sun Tzu Analogical") Frontiers of Theme #14

Pieces of the qi/zheng distinction are well known, if not by that name, in the doctrines of major modern militaries.[45] Yet there remains much scope for fresh,

---

[44] A combination of "normal" and improvised logistics is exemplified by practices of the Mongol army in the time of Chinggis Khan. See May, Introduction footnote 52, chapter 4, "The care of the army: logistics, supply, and medical care," pp. 58–68. A famous example of Mongol logistics improvisation was described by Marco Polo (1254–1324):

> And in case of great urgency they will ride ten days on end without lighting a fire or taking a meal. On such occasion they will sustain themselves on the blood of their horses, opening a vein and letting the blood jet into their mouths, drinking until they have had enough, and then staunching it.
>
> (quoted in May, Introduction footnote 52, p. 61)

For relevant planning factors (based on blood supply and physical needs of horse, caloric needs of rider, etc.) see John Masson Smith, Jr., "Mongol campaign rations: milk, marmots, and blood?," *Journal of Turkish Studies*, 1984, 8, 223–28 (also cited in Wilkinson, §24.13, p. 357).

[45] For example, there is a first-cousin resemblance between the two halves of a military operation as envisioned by Wallacker – the zheng part that "spikes" the enemy (a fixing action idea, see Rand, #14 footnote 38, p. 45), the qi part that tilts or topples him – and modern military thinking envisioning a "fixing" force and a "striking" force. One formulation refers to using "the fixing force to hold enemy forces in position, to help channel attacking enemy forces into ambush areas, and to retain areas from which to launch the striking force." See p. 17 of LTC (Retired) Michael T. Chychota and LTC (Retired) Edwin L. Kennedy, Jr., "Who you gonna call? Deciphering the difference between reserve, quick reaction, striking, and tactical combat forces," *Infantry*, July–September 2014, 103(3), 16–19.

As with many such distinctions, apparent simplicity can veil debates, sometimes long-running. For a flavor of relevant issues see Major Gregory J. Borden, USA, "'True' tactical reserves in

creative applications of qi/zheng ideas. The distinction is a versatile conceptual tool that can shed light on strategic debates and deficiencies across many contexts, some involving no clearly defined adversary. Both Lau's and Wallacker's interpretations can find uses here, as can other interpretations of the qi/zheng pair. To exploit the qi/zheng distinction to full potential it is best not to play interpretation favorites.[46]

Taking a standpoint that focuses on failure cases – often a sound perspective in talking about strategic issues! – a simple typology serves to organize a suite of contexts where the qi/zheng distinction has applications (the arrows are used to suggest that addressing one failure frequently illuminates existence of another failure, possibly harder to fix):

(a) existence of a *qi/zheng* distinction needs clearer recognition

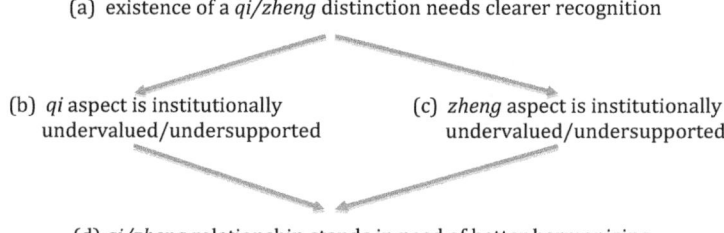

(b) *qi* aspect is institutionally undervalued/undersupported

(c) *zheng* aspect is institutionally undervalued/undersupported

(d) *qi/zheng* relationship stands in need of better harmonizing

Table 7 assembles examples of (a)–(d), on which the following discussion provides running commentary. There is extensive grist here for "*What did Master Sun know that we still don't (or have yet to absorb adequately)?*" There is also considerable resonance between this part of "applied Sun Tzu" and the analysis of innovation, with particular reference to the obstacles innovators often face.[47]

Emerging technology provides many examples of (a). Technological advances are often pioneered with a thoroughly straightforward purpose in mind (zheng in the sense of Lau), though also harboring major, sometimes low-visibility, potential for crafty (qi) uses (possibly of more than one type). In the early days of a technology even the existence of qi uses of it for nefarious purposes – for example, the possibility that the technology could be weaponized, even devastatingly so – may be only hazily recognized, if recognized at all, being overshadowed by the zeal (and, to be fair,

---

striking force operations: pilfery of combat power at the line of contact?" (monograph, School of Advanced Military Studies, United States Army Command and General Staff College, Fort Leavenworth, Kansas, 1995).

[46] Flexibility is also desirable on the issue of whether the qi/zheng distinction should be aligned (as Griffith's translation can be read as suggesting) on different kinds of forces, resources, etc.; or, alternatively, on different kinds of operations or ways of using assets (Wallacker). Both can be valid perspectives. For related analysis of qi as a military concept see pp. A-247ff.

[47] A natural division of labor should be noted between Theme #1 and Theme #14 "frontiers." Theme #1 frontiers analysis points to how Sun Tzu's ideas might be applied to suggest ways to *stymie* innovation by an adversary (pp. 72–74 above). Various of the Theme #14 frontiers topics in Table 7 point to how Sun Tzu's qi/zheng thinking can also be harnessed to *facilitate* innovation by one's own side.

Table 7 Qi/zheng *distinction: further areas of application, with focus on strategic error avoidance*

Case (a): *Qi/zheng* distinction is unrecognized (or stands in need of clarification)
    Example 1: Failure to grasp qi aspects of a new or emerging technology

Case (b): *Qi/zheng* distinction is recognized, but qi ("surprise") aspect is undervalued
    Example 2: Failure on the part of a military establishment dominated by zheng forces and associated vested interests to understand and appreciate the need for, and nature of, qi forces (such failure can take various forms, depending on the state of the art in warfare and supporting institutions and technology)
    Example 3: Failure of a military-industrial establishment invested in zheng logistics capabilities to understand and appreciate the need for, and contribution of, qi logistics (e.g., logistics improvisation and capabilities that support it)
    Example 4: Failure of a military establishment invested in existing capabilities (zheng) to understand and appreciate the need for strong R&D to improve them (qi)
    Example 5: Failure of a military-industrial R&D establishment, focused on development of improved versions of existing systems (zheng), to understand and appreciate a further need to pioneer systems of radically new kinds (qi)
    Example 6: Focus on denial-of-access attacks on digital systems (zheng) with deficit of attention to low-visibility algorithm sabotage vulnerabilities, particularly as carried out by knowledgeable subverted insiders with extensive access (qi)

Case (c): *Qi/zheng* distinction is recognized, but zheng ("straightforward") aspect is undervalued
    Example 7: Focus on imaginative tactical ("warfighting") operations (qi) – the kinds of successes highly visible in battles and campaigns of "great captains" – to the detriment of attention to need for sustained logistics support (zheng)
    Example 8: In the law (which has many modern tangencies to logistics as well as to warfare), focus on litigation and appellate decisions (qi) to the detriment of transactional and legislative law activities whose focus is designing and building legal structures and devices in much the same sense that engineers design and build physical ones ("law as engineering" concept) (zheng)

Case (d): *Qi/zheng* distinction is recognized, but its two parts stand in need of better harmonization
    Example 9: Failure to inculcate seamless switching between qi and zheng activities and foci of attention, as sine qua non of effective performance in high command roles (and by extension, also an essential task for education for such roles)
    Example 10: Failure to foster an adequate supply of personnel cross-trained to a high level of proficiency in both military affairs (qi) and their continuations by non-military means (also qi) and in public administration and state-building (zheng)
    Example 11: Failure to devise paths to "getting action" (qi) that move a prevailing social order – local, national, or transnational – in a desired direction, but in a sustainable way (zheng) that does not shortly revert to a pre-existing status quo

often also the prosocial impulses) of inventors and their backers and support networks. Digital technologies are a major case in point, though scarcely the only one.[48] Sometimes even a novel abstract concept that does not require a new technology to implement it (say, a financial innovation) may harbor qi perils.[49]

Cases of (b) – involving a deficit of institutional support for qi – are common byproducts of the internal politics of a military establishment. Historically, there are many cases where the constituencies supporting the kinds of operations, forces, and weapons that sustain Sun Tzu's qi are very different from those that sustain zheng (stuff of the "big battalions"). A classic example from early in the twentieth century was the struggle, in which Liddell Hart played a part, to create, support, and institutionalize armored warfare capabilities (a modern contender for the qi force role played by cavalry in many ages of land warfare).[50] Harking back to the era of the US Army Indian scouts of the nineteenth century, there have been many qi vs. zheng struggles between advocates of special forces capabilities, variously conceived, and advocates of conventional military forces.[51] As with any political question, those struggles combine aspects of a competition for resources, a factional contest, a battle over ideas, and a crisis of identity.[52] While details are beyond the scope of the present study, such struggles involve attempts to find a legitimate and stable niche for a certain type of qi military capability, in the teeth of often bitter opposition

---

[48] As internet pioneer Leonard Kleinrock observed: "One other aspect of today's Internet we did not foresee was the emergence of the dark side (in all its manifestations) that plagues us today." See p. 35 of George Varghese, "An interview with Leonard Kleinrock," *Communications of the ACM* (Association for Computing Machinery), 2019, 62(11), 31–36. For a different example from the dawn of the nuclear age see Nuel Pharr Davis, *Lawrence and Oppenheimer* (New York: Simon & Schuster, 1968), pp. 52–53.

An observation of University of Chicago physicist David Awschalom, referring to quantum Internet initiatives, has broader applicability: "You can be sure that we haven't yet thought of some of the most important things this technology will do," he said, "It would take extraordinary arrogance to believe you've done that" (*Science*, June 4, 2021, 372(6546), 1029). Cf. verse I.23 (Passage #14.12)!

[49] Cf. Enron's thousands of Special Purpose Entities (SPEs). See C. William Thomas, "The rise and fall of Enron," *Journal of Accountancy*, 2002, 193(4), 41–48. For further examples see also Howarth, Introduction footnote 13, chapter 4, "Implications (1) – Professional Ethics," pp. 97–147, especially pp. 106–108 (on topic of "Lawyers and the Crash" of 2008).

[50] Wallacker, #14 footnote 16, p. 299 goes further, labeling cavalry as "qi par excellence." However, Chinese armies in Sun Tzu's time may well have lacked cavalry. See pp. 29–30 above.

[51] Illustrating some intramural conflicts, in 1944, with World War II still in full swing, the Marine Raiders (whose First Raider Battalion Griffith had commanded on Guadalcanal) were disbanded and their personnel folded into other parts of the US Marine Corps. Regarding the motivations underlying that step, a historical study observes: "Senior Marine officers had never really taken to the concept of separate 'elite of the elite' units, and as the requirement for such units came into question [with the changing nature of the Pacific War], this opposition became more effective." See Charles L. Updegraph, Jr., *U.S. Marine Corps Special Units of World War II* (Washington, DC: History and Museum Division, Headquarters, US Marine Corps, 1972), p. 34. See also pp. 708–709 of David Thomas, "The importance of commando operations in modern warfare 1939–82," *Journal of Contemporary History*, 1983, 18(4), 689–717.

[52] This list of four concurrent dynamics is inspired by a similar list on p. 9 of David Howarth, "Is Law a Humanity (or is it more like Engineering)?," *Arts & Humanities in Higher Education*, 2004, 3(1), 9–28.

from much larger, deeply entrenched, and (in an institutional sense) highly conservative military interests.[53] Those internecine struggles are also commonly tinged with a generational conflict, with some younger individuals challenging older and far more senior ones in an intensely authoritarian setting where military rank is the basic currency of credibility.[54] Here, as Liddell Hart intuitively recognized in his *Strategy* and other writings, Sun Tzu's emphasis on qi, as articulated in verses V.5–V.7 (Passage #14.1), provides a conceptual resource.[55] Backed by Sun Tzu's prestige and durability over many centuries, this strand of Sun Tzu stands to contribute intellectual muscle to an often much-needed leveling of the playing field of policy debate.

Other case (b) struggles, usually less colorful and therefore of lower visibility, play out within the logistics sphere of warfare. As previously noted (pp. 471–72), "normal" logistics actions (usually comprising the lion's share of logistics) are zheng, both in a sense of being expected (that's what logistics planning is all about!) and also "spiking" the person or organization charged with doing the logistics work (that work is often burdensome!). The upshot is a tendency of major logistics systems to be enormously routinized, with all the attendant implications for bureaucratization and its inflexibilities (zheng mindset). Yet there remains an often poorly recognized and poorly supported need for a very different (qi) variety of logistics behavior that is prepared to break pattern and unhesitatingly improvise imaginative, unscripted logistics initiatives when a situation calls for it. In China of the Three Kingdoms, a spirit of qi logistics shows itself in the famous "borrowed arrows" ruse perpetrated by Zhuge Liang on Cao Cao.[56] Modern advances in logistics, especially digital ones where software allows preplanning for rare as well as common contingencies, can narrow, but never eliminate, the niche of qi logistics.

Sun Tzu's qi thinking also lends moral support to unceasing investment in military and other strategic R&D activities, for whose fruits Sun Tzu's allusion to "as inexhaustible as the flow of the great rivers" in verse V.6 (Passage #14.1) provides outstanding imagery. Although a need for military R&D is well accepted today, that has not always been so in relatively recent historical times. A case in point is Napoleon's failure to take military advantage of the seeds of many modern

---

[53] A flavor of clashing standpoints, one that Sun Tzu might have appreciated, is found in an observation attributed to a nineteenth-century chief of scouts (Stanton G. Fisher) about the US Army regulars of his time, summing up a scout's perspective: "Uncle Sam's boys are too slow for this business."

[54] For relevant background see Norman B. Ryder, "The cohort as a concept in the study of social change," *American Sociological Review*, 1965, *30(6)*, 843–61.

[55] The Liddell Hart/Sun Tzu connection dates to 1927 when Liddell Hart received a letter mentioning Sun Tzu from Sir John Duncan (then commanding the Defence Force that the War Office had dispatched to Shanghai to protect British interests there during Chiang Kai-shek's Northern Expedition). It continued in Liddell Hart's later career, evidenced by his selection of a sampler of Sun Tzu passages – among them, verse V.5 (Passage #14.1) in Giles's translation, Griffith's not yet being extant – to lead off the front material for Liddell Hart's 1954 *Strategy: The Indirect Approach* (which proselytizes relentlessly for the indirect approach concept). The same Sun Tzu sampler appears in *Strategy*'s 1967 revised edition (cited in #4 footnote 2). See also Liddell Hart's "Foreword," lauding Sun Tzu, located at the beginning of Griffith's 1963 Sun Tzu translation (quoted in part on p. 19 above).

[56] For this colorful (if probably apocryphal or exaggerated) tale see #8 footnote 9 above.

military innovations whose precursors were already in evidence at the beginning of the 1800s.[57]

Even within the R&D sphere the *qi/zheng* distinction may also find applications. Much R&D is geared to improving existing weapons and other systems (*zheng*). Other R&D focuses on creating fundamentally new kinds of weapons or systems (*qi*). These often aspire to create a war-winning edge and, confounding skeptics, sometimes deliver on promise (thereby infusing research content into verse V.5's "use the *zheng* to engage; use the *qi* to win").[58] The institutional conservatism of armies the world over betokens a need to keep a constant weather eye on ways of preserving a niche for R&D of a *qi* variety.[59] Perhaps paradoxically, when those ways are given bureaucratic institutional embodiments the result may amount to a version of *zheng* in service of *qi*. In the modern US defense establishment the Defense Advanced Research Projects Agency (DARPA) would be a major example. So too in historical hindsight would be the work of early RAND;[60] that of the old (pre-Ma-Bell-breakup) Bell Laboratories, Murray Hill, NJ and other locations;[61] and, on a level of innovative concepts, the early post–World War II US Naval War College.[62]

---

[57] See Andrew Roberts, *Napoleon and Wellington* (London: Weidenfeld & Nicolson, 2001), p. 41, observing of Napoleon that "Submarines, rockets, steamships, observation balloons and what later became known as shrapnel were all in their infancy in his time, but their potential military applications were more or less ignored by the emperor." He quotes Major General J. F. C. Fuller, writing in 1936, to the effect that "had Napoleon, in 1805, placed down a challenge to the mechanical intellect of France to produce a weapon 100% more efficient than the [British] Brown Bess [flintlock musket], it is almost a certainty that, in 1815, he would have got it; that he would have won Waterloo, and that the whole course of history would have been changed."

[58] Worth again noting here is François Wildt's interpretation of *qi* as "remainder," which represents a departure from both Lau's and Wallacker's lines of interpretation of the *qi/zheng* distinction. See #14 footnote 20 above. In R&D contexts, qi-as-remainder points to taking a hard second look at abandoned projects or discarded technologies left over as a residue of earlier R&D efforts (which, in that sense, constitute a "remainder"), with an eye to giving them a new lease on life in novel, possibly "recombinant," uses. Noting that the "evolution of technology is not always linear but has lots of 'reuses' or 'rising from the ashes,'" see Liu et al., Preliminaries footnote 17, p. 69. More broadly, this article advocates for historical thinking as a perspective too often neglected by computer scientists, noting (p. 70) the same Warring States era computational tool mentioned in #1 footnote 82.

[59] Here Mair's translation of *qi* as "unconventional" (Mair, p. 91) works well. For related reasons, associated with propensities of many research establishments to favor scientific and technological orthodoxy, so too does Sawyer's translation of *qi* as "unorthodox" (Sawyer, p. 187).

[60] For perspective on the early years of RAND see Augier, March, & Marshall, #3 footnote 48, whose title theme ("The flaring of intellectual outliers") conveys a significant message.

[61] For operations research contributions of the old Bell Labs see Cree S. Dawson, Charles J. McCallum, Jr., R. Bradford Murphy, & Eric Wolman, "Operations research at Bell Labs through the 1970s," *Operations Research*, 2000, 48(2), 205–15 (Part I); 48(3), 351–61 (Part II); 48(4), 517–26 (Part III).

[62] See Eccles, Background footnote 52, pp. 319–20, emphasizing need for not only technological research but also what he terms "idea research" in strategy, logistics, and warfare. Given that Sun Tzu is the earliest traditionally transmitted military text to invoke it, Passage #14.1's *qi/zheng* distinction is itself an early Chinese contender for work of that type. See #6 footnote 36. For further background on idea-level contributions of the early post–World War II Naval War College in which concepts of Admirals Eccles and Wylie, among others, took shape, see Boorman, Introduction footnote 4.

Although many cyber warfare attacks are an obvious contender for being billed as *qi*, there is also value to be had from exploring a *qi/zheng* fault line within that capacious realm. The well-known genre of "denial of access" attacks directed against computer systems and networks lends itself to being conceptualized as a species of *zheng* operation. Supporting such characterization from a Wallacker perspective, denial-of-access stops dead in its tracks the otherwise dramatic mobility of information in digital form, thereby very effectively "spiking" those who depend on it (with consequences spilling over to the biophysical world, often giving rise to major loss of flexibility there as well). Algorithm sabotage can be far more insidious, with even its occurrence often not being known to its victim (at least for a long time) with concomitant potential for both "tilting" and "toppling" (possibly because a non-sabotaged rival can move much more quickly). That profile, conjoined with the wealth of targets born of a "society of algorithms," aligns well on Sun Tzu's verses V.6 and V.7 (Passage #14.1), making algorithm sabotage a candidate for qi-within-qi.[63]

Referring back to the p. 473 diagram, cases of (c) swing the spotlight from qi to *zheng*. Contrary to those qi advocates who discern in Sun Tzu a friend only to their half of the *qi/zheng* divide, it is definitely wrong to treat Sun Tzu as making a case for qi alone. Equally important is the intellectual support that Sun Tzu's thinking can offer in situations where it is a *zheng* capability that needs advocacy and bolstering. Over much of the history of warfare (and in more subtle forms sometimes persisting to the present day), a leading military example would be systematic undervaluing of logistics, a quintessentially *zheng* military activity (even though, as noted above, logistics improvisation has aspects of qi). There is now near-universal recognition that US military logistics capabilities, no less than US high-tech intelligence ones, are areas of great contemporary US military competence and often comparative advantage. Yet there remains need for vigilance to avoid slipping back to an era when logistics assignments for an officer betokened a career derailed.

A knight's-move away from logistics in its military sense, deep in the heart of civilian life stands another major human activity – the law.[64] The *qi/zheng* distinction has vitality here too. Specifically, the law's legislative and transactional strands – heavily concerned with the design of durable structures (statutes, treaties, stable business arrangements, etc.) – occupy a *zheng* niche. That niche contrasts with the qi niche of litigation epitomized by courtroom drama and surprises associated with it.[65] There is a family resemblance between litigation and combat engineering.[66] Both those activities involve creating structures and devices *just resilient enough* to last out the course of the battle, creating a place for "bricolage" and improvisation. Both popular perceptions and Anglo-American legal education have long tended to focus on litigation, often to a point of crowding out attention to legal planning work

---

[63] For analysis of algorithm sabotage as a military concept see Boorman & Levitt, #3 footnote 43.

[64] In twenty-first-century contexts the law is, of course, heavily intertwined with the profession of arms, in no small part as a byproduct of the vast defense contractor establishment and its supply chains, conjoined with the complexity of logistics activities.

[65] This distinction between legislative or transactional and litigation-focused strands of the law builds on law-as-engineering concept of David Howarth. See pp. 193–94 above.

[66] Howarth, Introduction footnote 13, p. 93.

geared to less glamorous tasks of building legal structures and devices resilient enough to serve needs of clients, public or private, for many years to come. Far jazzier is trial advocacy and the high-profile appellate litigation (sometimes also a source of surprises!) to which it gives rise when great issues of the day are at stake. The zheng half of Sun Tzu's qi/zheng thinking encourages more and deeper analytical and educational attention to legislative and transactional strands of the law. Many challenges of twenty-first-century grand strategy, among them ones associated with climate change responses and other environmental issues, have roots there.

Whereas (b) and (c) are about giving legitimacy and adequate support to each of qi and zheng, (d) showcases a sometimes harder problem: managing and harmonizing the often intensely dynamic relationship between qi and zheng. Sun Tzu's thinking calls for a strategic actor to be able to switch back and forth between use of these two different modalities of power, which – as Sun Tzu's ring image in verse V.12 (Passage #14.1) elegantly suggests – also continually mutate into one another (complicating the task of surefooted switching). Today's qi form of military power may become established and its advocates establishmentarian, no longer pioneers and prophets but vested interests, converting that form of power into tomorrow's zheng. The "living reaction" aspect of warfare that Clausewitz emphasized comes into its own here. A clandestine operation epitomizing qi may become public, with its highly publicized fallout giving grist for a zheng operation by the adversary, geared to reputationally spiking the side that initiated it. Or a zheng logistics system, apparently working so smoothly, may be thrown into abrupt confusion by a foe's qi sabotage operation, triggering an urgent need for qi logistics on the part of the discombobulated side. Zigzags like these create psychological and organizational demands that few strategic actors ever meet. Under twenty-first-century conditions, a particular stumbling block arises from the fact that qi capabilities are difficult (even perhaps impossible!) to bureaucratize fully or to reduce to programmable rules.[67]

Digital technology multiplies both qi and zheng options and opportunities, creating formidable challenges for qi/zheng harmonization. The nature of software is conducive to straightforward extension of existing software to routinize new or previously unaddressed contingencies, giving rise to an unprecedentedly rich and ever-expansible repertoire of straightforward (zheng) moves. Yet that same software is commonly also rife with opportunities for unexpected modifications that "tilt" (or topple!) the performance of one who relies on it. Meanwhile, on a different plane of the technology, durability of digital memories creates unprecedented opportunities for inept "self-spiking" (as noted earlier, a Sun Tzu emphasis), one of whose active ingredients is incautious use of email or social media. Compounding the difficulty of qi/zheng harmonization is, of course, the sheer tempo of digital life.

---

[67] The difficulty of bureaucratizing qi stands in the background of a question rhetorically posed by Secretary of Defense Robert M. Gates in a September 29, 2008 speech to a National Defense University audience: "Why did we have to go outside the normal bureaucratic process to develop counter-IED [improvised explosive device] technologies, to build MRAPs [mine resistant ambush protected vehicles], and to quickly expand our ISR [intelligence, surveillance, and reconnaissance] capability?" (Gates's question is quoted in Boorman, Introduction footnote 4, p. 102.)

At the highest grand strategic level, cases of (d) – centering on need for better harmonization of qi and zheng – point to a Daoistic strand of traditional Chinese thought that is broader than Sun Tzu's (though often regarded as standing in close kinship to it). The *Laozi* Daoist classic envisions war as the province of qi by contrast to governance of a state, which is zheng.[68] In an ever more networked twenty-first-century world, that interpretation of qi and zheng underscores a need for personnel whose education and career experience makes them surefooted on both sides of this divide. Even today grand strategy in its traditional foreign policy sense remains a substantially distinct arena from domestic socioeconomic policy: the actors, tools, institutions, histories, and norms are different, calling for different intellectual, leadership, and problem-solving styles. If integration of qi and zheng capabilities is to be realized in a modern policy environment, there is need for a deep bench of individuals comfortable with both arenas and styles, along with switching back and forth between them. As ever more issues come to straddle the foreign policy/domestic policy divide (abetted by the extreme mobility of digital information) that need will only increase. Therein lies a task for the education of a rising generation of grand strategists to become versed in balancing zheng and qi in the *Laozi*'s sense.

Applying the qi/zheng distinction on a yet broader canvas, social organization has a way of serving up what Harrison White has felicitously termed a "Sargasso Sea" of social obligation and context.[69] Most goal-directed action in such a setting must perforce respect a wealth of normative or conventional constraints, making such action zheng (in fact, in both Lau's and Wallacker's senses).[70] Taken one at a time those constraints are often not major. But taken together and cumulatively, they commonly make next-to-impossible achieving truly major change in a prevailing social order. Occasionally, some actor – a few successful revolutionaries come to mind, as do some religious leaders, as do a few entrepreneurs – actually manages to break free of this "Sargasso Sea" to achieve what White calls "getting action": changing in a basic way how things are done. That is an accomplishment – whether for good or ill depends on context – worthy of being labeled qi in Sun Tzu's sense. Yet very commonly, whatever change is temporarily brought about by qi steps proves unsustainable in the longer run. Powerful social forces not under the control of any strategic actor propel the system back to its former equilibrium (a pattern encountered time and again in the aftermath of shakeups in Communist societies throughout the twentieth century). Zheng and qi have failed to work together to lock in the change.

Taking cues from the spirit of Sun Tzu's qi/zheng thinking, there remains room for understanding much better than we do today what strategic actions, a blend of qi and zheng, may be potent enough to displace a complex sociopolitical order to a new, stable equilibrium fundamentally different from the old (ideally also superior to it).

******

---

[68] See #14 footnote 4 (quoting *Laozi* LVII in D. C. Lau's translation).

[69] See p. 230 of Harrison C. White, *Identity and Control* (Princeton, NJ: Princeton University Press, 1992), chapter 6, "Getting Action," pp. 230–86.

[70] This perspective on zheng moves in a direction of Lewis's proposed "extraordinary" (qi) and "normative" (zheng) interpretation of the qi/zheng distinction. See #14 footnote 29.

Set of principal passages used to illustrate Sun Tzu Theme #14 (harmonious integration of *qi* and *zheng* approaches)

Passage #14.1. Sun Tzu's Qi/Zheng Concept Pair

(This is the only part of Sun Tzu in which the qi/zheng distinction explicitly appears.[71] Note that Sun Tzu identifies a place for both qi and zheng in both defensive [verse V.3] and offensive [verse V.5] modes of warfare.)

V.3. That the army is certain to sustain the enemy's attack without suffering defeat is due to qi and zheng operations.

V.4. Troops thrown against the enemy as a grindstone against eggs is an example of a solid acting upon a void.[72]

V.5. Generally, in battle, use the zheng to engage [he]; use the qi to win.

V.6. Now the resources of those skilled in the use of qi are as infinite as the heavens and earth; as inexhaustible as the flow of the great rivers.

V.7. For they end and recommence; cyclical, as are the movements of the sun and moon. They die away and are reborn; recurrent, as are the passing seasons.

V.8. The musical notes are only five in number but their melodies are so numerous that one cannot hear them all.

V.9. The primary colours are only five in number but their combinations are so infinite that one cannot visualize them all.

V.10. The flavours are only five in number but their blends are so various that one cannot taste them all.

V.11. In battle there are only qi and zheng, but their combinations are limitless; none can comprehend them all.[73]

V.12. (Translation following Lau, 1965 article, p. 331.[74]) "The qi and the zheng produce each other endlessly like a ring and who is there that can exhaust the possibilities?"

Passage #14.2. Capitalizing on False Enemy Expectations and Unpreparedness

(This passage and the next two serve as anchors for Lau's 1965 article's interpretation of qi as involving unexpected action, zheng as expected action.[75]

---

[71] In a departure from Griffith's translation, the concepts qi and zheng are deliberately left untranslated in Passage #14.1, with a goal of preserving a full measure of interpretive flexibility for purposes of Theme #14 analysis. Also, in Passage #14.1's verse V.11 the order of mention of qi and zheng (qi first, zheng second) reinstates the Chinese word order that the translation by Griffith, p. 92 has reversed (a reversal that could be regarded as a translation flaw). See #7 footnote 64.

[72] Verse V.4 involves the further traditional Chinese military concept of xu/shi ("solid/void"), anchoring the principle that what is solid should be used to strike what is empty (see p. 113 above). Li Ling, #1 footnote 84, p. 178 regards the xu/shi concept as an extension of qi/zheng thinking.

[73] Weighing combinatorial vs. other perspectives on Passage #14.1 see Appendix 1 anchor footnote 29. Appendix 1 footnote 30 compares and contrasts Sun Tzu's relevant emphases with those of early Indian mathematical/computational thinkers (Pāṇini, Piṅgala).

[74] Rationale for using Lau's translation rather than Griffith's is given in Appendix 14 footnote 3.

[75] See Lau, 1965 article, pp. 330–31 (quoted in Appendix 14, where Passage #14.2 again appears).

In verses I.18–I.20 deception is involved: false expectations are created by one's own actions. Verse I.26 and the latter part of verse XI.29 [Passage #4.4] covers not only that possibility but also simply catching the enemy off-guard – *also* a case of qi.[76] The latter possibility may often be cheaper [Theme #3]: no prior investment in deception is needed. The natural target set may also be larger, starting with the many ways in which the enemy may be logistically unprepared.)

I.18. Therefore, when capable, feign incapacity; when active, inactivity.
I.19. When near, make it appear that you are far away; when far away, that you are near.
I.20. Offer the enemy a bait to lure him; feign disorder and strike him.

. . .

I.26. Attack where he is unprepared; sally out when he does not expect you.[77]

Passage #14.3. Switch of Style to Exploit Unpreparedness

(If one isn't too much of a purist – Giles, p. 149 calls attention to timidity of hares – Sun Tzu's imagery here is quite effective in bringing to life a distinction between expected action [*zheng*] and unexpected/surprise action [*qi*].)

XI.61. Therefore at first be shy as a maiden. When the enemy gives you an opening be swift as a hare and he will be unable to withstand you.

Passage #14.4. Speed As Tool for Upending Enemy Expectations

(This verse points to sheer speed as a path to creating opportunities to exploit unpreparedness. It underscores how nimble actors, possibly civilian, may especially stand to profit from qi.)

XI.29. Speed is the essence of war. Take advantage of the enemy's unpreparedness; travel by unexpected routes and strike him where he has taken no precautions.

Passage #14.5. Terrain Roots of Sun Tzu's Focus on "Spiking," then "Toppling"

(Focus now shifts to the alternative interpretation of qi/zheng to which Wallacker's work points. "Spiking," "toppling," and "tilting" are all Wallacker terminologies, introduced on pp. 464–65 above. To be clear, in verse XI.9 it is the *foe* who stands to benefit from a terrain-based spiking and is positioned to do a toppling.)

---

[76] See p. 463 above (citing analysis of Li Ling).

[77] Verse I.26 can also be given a reading as advice to attack objectives where "it never occurred to the enemy that one would attack there" or to "make for places where it never occurred to him you would make for." Words in quotes are from Lau's 1965 article, p. 321 footnote 9, advocating for similar interpretations of (a) verse I.26; (b) the second sentence of verse XI.29 (Passage #14.4); and (c) the second part of verse VI.5 (Passage #14.25A). Such a reading has overtones of hypergame thinking. See #5 footnote 15 as well as Niou & Ordeshook, #5 footnote 7, p. 172.

XI.9. Ground to which access is constricted, where the way out is tortuous, and where a small enemy force can strike my larger one is called "encircled".

Passage #14.6. Spiking an Adversary by Exploiting Extreme Terrain Features

(Sawyer sums up the common denominator here, which squarely focuses on low-flexibility situations: "in every case the basic principle is simply that configurations of terrain that constrict an army's movement are to be avoided."[78])

IX.16. Where there are precipitous torrents, "Heavenly Wells", "Heavenly Prisons", "Heavenly Nets", "Heavenly Traps", and "Heavenly Cracks", you must march speedily away from them. Do not approach them.[79]

IX.17. *I keep a distance from these and draw the enemy toward them. I face them and cause him to put his back to them.* [Emphasis supplied.]

Passage #14.7. Spiking, Then Toppling, with Reference to a Water Crossing

(An army straddling two sides of a river is in a highly vulnerable position, effectively "spiked." In the next round it is at high risk of being "toppled."[80])

IX.5. When an advancing enemy crosses water do not meet him at the water's edge. It is advantageous to allow half his force to cross and then strike.[81]

Passage #14.8. Spiking, Then Toppling, by Exploiting Awkward Terrain Situations

(These verses provide further support for treating Sun Tzu's interest in "spiking" – i.e., in low-flexibility situations – as having roots in terrain issues. The perspective here, somewhat confusingly, alternates between self and enemy. In verse X.3 it is you who are at risk of getting spiked, as Sun Tzu's "difficult to return to" language suggests. In verse X.4 it is the enemy.)

X.3. Ground easy to get out of but difficult to return to is entrapping.[82] The nature of this ground is such that if the enemy is unprepared and you sally

---

[78] See *Sun Pin Military Methods* (translated, with introduction and commentary by Ralph D. Sawyer, with the collaboration of Mei-chün Lee Sawyer; Boulder, CO: Westview Press, 1995), p. 115.

[79] For thumbnail interpretations of these several extreme terrain types see #7 footnote 66.

[80] As de Crespigny, *Background* footnote 95, p. 278 starkly observes: "many battles turned into slaughter when one side caught the other half-way [in a river-crossing operation]."

[81] From Western military history, a fine example of the basic tactical concept here is the battle of Stirling Bridge (1297) during the First War of Scottish Independence. Exploiting the vulnerability of an English force that was slowly crossing the narrow bridge, a drawn-out process, the Scots waited until "as many of the enemy had come over as they believed they could overcome" (observation from the *Chronicle of Walter of Guisborough*, a fourteenth-century source), then crushed the force that had crossed. See Michael Prestwich, "The Battle of Stirling Bridge: an English perspective," in Edward J. Cowan (ed.), *The Wallace Book* (Edinburgh: John Donald, 2007), pp. 64–76. For further discussion, focusing on the military calculation involved on the Scottish side, see also Appendix 1 footnote 50.

[82] Ames, p. 147 translates verse X.3 content in a way that brings out the "spiking" aspect with clarity: Terrain that allows your advance but hampers your return is entangling. On entangling ground, if you go out and engage the enemy when he is not prepared, you might defeat him.

out you may defeat him. If the enemy is prepared and you go out and engage, but do not win, it is difficult to return. This is unprofitable.

X.4. Ground equally disadvantageous for both the enemy and ourselves to enter is indecisive. The nature of this ground is such that although the enemy holds out a bait I do not go forth but entice him by marching off. When I have drawn out half his force I can strike him advantageously.

Passage #14.9. Battlefield Cases of Sun Tzu's Focus on Spiking, Then Toppling

(When armies face off in the run-up to a probable battle, verse IX.44 warns against becoming mesmerized by the enemy's pre-battle theatrics – a form of cognitive spiking! – neglecting to be alert to the very real possibility of a stratagem [qi step] in the making.[83] Replacing combat by pre-combat theatrics, verse IX.31 warns of falling for a feigned retreat that induces a misguided commitment of forces, thereby serving the enemy's spiking/toppling purpose.)

IX.31. When half his force advances and half withdraws he is attempting to decoy you.

. . .

IX.44. When the enemy troops are in high spirits, and, although facing you, do not join battle for a long time, nor leave, you must thoroughly investigate the situation.

Passage #14.10. Psychological Paths to Spiking, Then Toppling, an Army or Its General

(While terrain- or battlefield-related loss of flexibility illustrates the "spiking" idea with particular clarity, Passage #14.10 suggests that sheer passage of time may itself erode flexibility, as Griffith's term "sluggish" suggests.[84])

VII.20. Now an army may be robbed of its spirit and its commander deprived of his courage.

VII.21. During the early morning spirits are keen, during the day they flag, and in the evening thoughts turn toward home.

VII.22. And therefore those skilled in war avoid the enemy when his spirit is keen and attack him when it is sluggish and his soldiers homesick. This is control of the moral factor.

But when the enemy is prepared, *if you go out and engage him and fail to defeat him, you will be hard-pressed to get out, and will be in trouble.*

[Emphasis supplied.]

[83] Goldin, Introduction footnote 7, p. 17: "Most commentators agree that the [enemy] commander in [verse IX.44] is planning a spectacular surprise-attack."

[84] Importantly, that passage of time may be longer than a single day. See Preliminaries footnote 4.

Passage #14.11. Spiking, Then Tilting or Toppling, a General by Exploiting Character Traits

(This passage serves up a menu of "If X then Y" stratagems, where X is a psychological trait that predisposes to low flexibility, and Y is a tilting or toppling effect that results from exploiting it. A "toppling" upshot is illustrated by verses VIII.18 and VIII.19; a "tilting" one by verses VIII.20–VIII.22 where the likely outcome is a loss of poise and resulting errors of judgment.)

VIII.17. There are five qualities which are dangerous in the character of a general.
VIII.18. If reckless, he can be killed;
VIII.19. If cowardly, captured;
VIII.20. If quick-tempered you can make a fool of him;
VIII.21. If he has too delicate a sense of honor you can calumniate him;
VIII.22. If he is of a compassionate nature you can harass him.
VIII.23. Now these five traits of character are serious faults in a general and in military operations are calamitous.
VIII.24. The ruin of the army and the death of the general are inevitable results of these shortcomings. They must be deeply pondered.

Passage #14.12. Further Psychological Paths to Spiking the Enemy

(Inducing loss of flexibility is a unifying idea here. In the case of verse I.25, flexibility loss arises from rifts in the enemy camp that may lead to decision-making paralysis there – a type of spiking on a human group dynamics level. Verse I.26, which has already appeared in Passage #14.2 and is therefore not repeated here, points to followup steps that tilt or topple a foe thus spiked.)

I.22. Anger his general and confuse him.
I.23. Pretend inferiority and encourage his arrogance.
I.24. Keep him under a strain and wear him down.
I.25. When he is united, divide him.

Passage #14.13. Spiking by Burdening a Foe or Rival with a Task, against a Coercive Backdrop

VIII.14. He who intimidates his neighbours does so by inflicting injury upon them.
VIII.15. He wearies them by keeping them constantly occupied, and makes them rush about by offering them ostensible advantages.

Passage #14.14. False Information Fed to the Enemy As Versatile Tool for Effective Spiking, As Prelude to Toppling

(Sun Tzu's *qi/zheng* thinking suggests that the goal of verse XIII.19's fabricated information, possibly building on verse XIII.16's program of detailed information-gathering, may be to induce a misguided commitment or irreversible estrangement in the enemy camp – both exemplifying a "spiking" idea.)

XIII.10. Expendable agents are those of our own spies who are deliberately given fabricated information.

. . .

XIII.16. Generally in the case of armies you wish to strike, cities you wish to attack, and people you wish to assassinate, you must know the names of the garrison commander, the staff officers, the ushers, gate keepers, and the bodyguards. You must instruct your agents to inquire into these matters in minute detail.

. . .

XIII.19. And it is by this means [the double agent] that the expendable agent, armed with false information, can be sent to convey it to the enemy.

Passage #14.15. Self-Spiking (Administratively Inept Variety, by Reason of Excessive Kindliness to Subordinates)

(A harmonic shared by Passage #14.15 and several of the passages that follow is never to allow oneself to be spiked by unfettered emotions – your own or your subordinates'. If so, you are fair game for being toppled!)

X.21. If a general indulges his troops but is unable to employ them; if he loves them but cannot enforce his commands; if the troops are disorderly and he is unable to control them, they may be compared to spoiled children, and are useless.

Passage #14.16. Self-Spiking (Administratively Inept Variety, cont.)

IX.47. If troops are punished before their loyalty is secured they will be disobedient. If not obedient, it is difficult to employ them. If troops are loyal, but punishments are not enforced, you cannot employ them.

. . .

IX.49. If orders which are consistently effective are used in instructing the troops, they will be obedient. *If orders which are not consistently effective are used in instructing them, they will be disobedient.* [Emphasis supplied.]

Passage #14.17. Self-Spiking by Reason of a General's Professional Incompetence (Ineptness of Administrative or Tactical Varieties)

(A common denominator of the scenarios here is that all lead to major loss of flexibility. Verse X.9 pins blame squarely on the general for these adverse outcomes, thereby pointing to a self-spiking idea.)

X.9. Now when troops flee, are insubordinate, distressed, collapse in disorder or are routed, it is the fault of the general. None of these disasters can be attributed to natural causes.

X.10. Other conditions being equal, if a force attacks one ten times its size, the result is flight.

X.11. When troops are strong and officers weak the army is insubordinate.

X.12. When the officers are valiant and the troops ineffective the army is in distress.[85]

X.13. When senior officers are angry and insubordinate, and on encountering the enemy rush into battle with no understanding of the feasibility of engaging and without awaiting orders from the commander, the army is in a state of collapse.

X.14. When the general is morally weak and his discipline not strict, when his instructions and guidance are not enlightened, when there are no consistent rules to guide the officers and men and when the formations are slovenly the army is in disorder.

X.15. When a commander unable to estimate his enemy uses a small force to engage a large one, or weak troops to strike the strong, or when he fails to select shock troops for the van, the result is rout.

X.16. When any of these six conditions prevails the army is on the road to defeat. It is the highest responsibility of the general that he examine them carefully.

Passage #14.18. Self-Spiking (Tactically Inept Variety), by Reason of Undertaking a Protracted Siege Operation or Impetuously Ordering a Costly Assault

(Two different types of self-spiking are present here. One involves settling down to a protracted siege, with its inevitable self-inflicted loss of flexibility. A second involves running out of patience and committing to an ineffective, costly assault.[86])

III.7. The worst policy is to attack cities. Attack cities only when there is no alternative.

III.8. To prepare the shielded wagons and ready the necessary arms and equipment requires at least three months; to pile up earthen ramps against the walls an additional three months will be needed.

---

[85] Griffith, p. 126 footnote 3, notes that the literal meaning of the Chinese term *xian* he translates as "in distress" is "[b]logged down, or sinking, as in a morass. The idea is that if the troops are weak the efforts of the officers are as vain as if the troops were in a bog." A bog or morass is an alternative imagery conveying much the same no-flexibility idea that "spiking" does. The same term *xian* also appears in verse IX.16 (Passage #14.6), describing the type of ground that Griffith translates as "Heavenly Traps" and Ames (#7 footnote 66) as "quagmires."

[86] The enormously costly Japanese front assaults during the siege of Port Arthur in the 1904–1905 Russo-Japanese War are noted by Giles, p. 19 as a case in point. For a perspective on those assaults see annotation of verse III.9 in Ochiai Toyosaburō 落合 豊三郎, *Sonshi reikai* 孫子例解 (*Sun Tzu by Examples*) (Tokyo: Gunji Kyōiku Kai [Military Education Society], 1917), pp. 88ff. (book by an Imperial Japanese Army Lieutenant General who had served in a different part of the Russo-Japanese War). Criticism of the Japanese Port Arthur operation is clearly implied, though expressed in a carefully phrased way in general terms and conditional sentences. (Peter Kornicki, Emeritus Professor of Japanese Studies, University of Cambridge, personal communication, April 1, 2021.)

III.9. If the general is unable to control his impatience and orders his troops to swarm up the wall like ants, one-third of them will be killed without taking the city. Such is the calamity of these attacks.

Passage #14.19. Self-Spiking (Strategically Inept Variety), by Reason of a Ruler's or General's Uncontrolled Emotions

XII.18. A sovereign cannot raise an army because he is enraged, nor can a general fight because he is resentful. For while an angered man may again be happy, and a resentful man again be pleased, a state that has perished cannot be restored, nor can the dead be brought back to life.

Passage #14.20. Self-Spiking (Grand Strategically Inept Variety), by Reason of Failure to Cultivate Espionage Capabilities

(A pertinent sense of self-spiking has never been better conveyed than by Henry Stimson's observation – famous or infamous, take your pick! – that "Gentlemen do not read each other's mail."[87])

XIII.2. One who confronts his enemy for many years in order to struggle for victory in a decisive battle yet who, because he begrudges rank, honours and a few hundred pieces of gold, remains ignorant of his enemy's situation, is completely devoid of humanity. Such a man is no general; no support to his sovereign; no master of victory.

Passage #14.21. Self-Spiking (Grand Strategically Inept Variety) by Reason of Failure to Keep Civil–Military Relations in Good Repair

(The pathologies Sun Tzu here describes point to major loss of flexibility, most likely in war and diplomacy alike.)

III.18. Now the general is the protector of the state. If this protection is all-embracing, the state will surely be strong; if defective, the state will certainly be weak.
III.19. Now there are three ways in which a ruler can bring misfortune upon his army.
III.20. When ignorant that the army should not advance, to order an advance or ignorant that it should not retire, to order a retirement. This is described as "hobbling the army".
III.21. When ignorant of military affairs, to participate in their administration. This causes the officers to be perplexed.
III.22. When ignorant of command problems to share in the exercise of responsibilities. This engenders doubts in the minds of the officers.

---

[87] Henry L. Stimson & McGeorge Bundy, *On Active Service in Peace and War* (New York: Harper, 1948), p. 188.

III.23. If the army is confused and suspicious, neighbouring rulers will cause trouble. This is what is meant by the saying: "A confused army leads to another's victory."

Passage #14.22. Reality Check: Pitfalls in Passing from Spiking to Toppling

(An implied warning here is that pinning an enemy unduly, leaving no exit option – while on its face a textbook case of spiking – may elicit a death ground response, whose primal fury could lead to tilting or toppling of one's own army.[88])

VII.30. Do not thwart an enemy returning homewards.
VII.31. To a surrounded enemy you must leave a way of escape.[89]
VII.32. Do not press an enemy at bay.
VII.33. This is the method of employing troops.

Passage #14.23. Reality Check: Pitfalls in *Zheng* → *Qi* Transition in Fire Attack

(A fire is set in the enemy's camp, possibly originating by qi means,[90] aiming to produce wild chaos there.[91] Applying the spirit of the first part of Passage #14.1's verse V.5, that chaos amounts to a type of spiking. If, however, enemy forces remain calm the sought-for spiking has failed.[92] Then mindlessly unleashing a followup military assault on the enemy's camp is most unlikely to catch the enemy off-guard – i.e., no qi effect. Indeed committing to such an attack might easily wind up "spiking" one's own troops!)

XII.7. When fire breaks out in the enemy's camp immediately co-ordinate your action from without. But if his troops remain calm bide your time and do not attack.
XII.8. When the fire reaches its height, follow up if you can. If you cannot do so, wait.

---

[88] See Griffith, p. 110; Giles, pp. 69–70; Minford, pp. 211–13 (assorted Chinese military history examples provided by the traditional commentators).

[89] Another way of understanding this verse is that an apparent escape route is left open as only as a deception. See #9 footnote 77.

[90] Perhaps set by a traitor, as Cao Cao's commentary suggests. See p. 43 above.

[91] An important clarifying comment of traditional commentator Du Mu is that "the prime object of fire is to throw the enemy into confusion." See Minford, p. 307, previously quoted in #9 footnote 55.

[92] The deeply hierarchical nature of Warring States armies may have often given level-headed officers the top-down control needed to avert fire-induced panics, making such panics elusive for would-be attackers. For analytical background see James S. Coleman, *Foundations of Social Theory* (Cambridge, MA: Belknap Press of Harvard University Press, 1990), pp. 203–15 ("Escape Panics"). His analysis clarifies that "the chance of such a panic can be reduced by providing a natural focal point for attention, *to which persons will naturally look for direction, that is, to which they will transfer control.*" (See Coleman, p. 215, emphasis supplied.) In the case of Sun Tzu's armies, an obvious candidate for such a focal point would be a relevant officer, conceivably even the general.

Passage #14.24. More Complex Agenda: Spiking the Enemy by Forcing Sprawling Commitments, Then Toppling the Foe by Exploiting Local Superiority

(Although Sun Tzu does not use the qi/zheng terminology, these ideas are again recognizably in play, now on a wider canvas. The foe is compelled by lack of information to commit forces to the defense of many places [zheng]. Those forces can then be picked off at will under circumstances unfavorable to them [qi]. Interpreted on a grand strategy plane, this part of Sun Tzu has intellectual kinship with Paul Kennedy's "imperial overstretch" concept.[93])

VI.13. If I am able to determine the enemy's dispositions [xing] while at the same time I conceal my own then I can concentrate and he must divide. And if I concentrate while he divides, I can use my entire strength to attack a fraction of his. There, I will be numerically superior. Then, if I am able to use many to strike few at the selected point, those I deal with will be in dire straits.[94]

VI.14. The enemy must not know where I intend to give battle. For if he does not know where I intend to give battle he must prepare in a great many places. And when he prepares in a great many places, those I have to fight in any one place will be few.

VI.15. For if he prepares to the front his rear will be weak, and if to the rear, his front will be fragile. If he prepares to the left, his right will be vulnerable and if to the right, there will be few on his left. And when he prepares everywhere he will be weak everywhere.

VI.16. (Translation from Lau, 1965 article, p. 330.[95]) "It is the one who has to prepare against his enemy who is few and the one who makes his enemy prepare against him who is many."

★★★★★★

---

[93] See Paul M. Kennedy, *The Rise and Fall of the Great Powers: economic change and military conflict from 1500 to 2000* (New York: Random House, 1987), p. 515. See also Passage #4.25A.

[94] Mair, p. 97 translates "Being able to strike his few forces with my numerous forces imposes constraints on those who would engage in battle with me." This mutes Griffith's "in dire straits" but helps underscore the enemy's predicament as a product of a straightjacket of constraints.

[95] Rationale for using Lau's translation rather than Griffith's is given in Appendix 3 footnote 9.

Conclusion

Demystifying Sun Tzu and Future Directions

There is a story in the *Zhuangzi* text, one of the great classics of Daoism, which (in summary by *Zhuangzi*'s translator, Burton Watson) concerns

> a man named Nanrong Zhu who went to visit the Daoist sage Laozi in hopes of finding some solution to his worries. When he appeared, Laozi promptly inquired, "Why did you come with all this crowd of people?" The man whirled around to see if there was someone standing behind him. Needless to say, there was not; the "crowd of people" that he came with was the baggage of old ideas, the conventional concepts of right and wrong, good and bad, life and death, that he lugged about with him wherever he went.[1]

The moral of this story extends to Sun Tzu. Thinking about Sun Tzu needs to challenge, and in many instances to jettison, a host of pre-existing beliefs and assumptions accumulated from diverse quarters, Chinese and non-Chinese alike, in some cases over centuries. Ignoring countless imaginary Sun Tzus, the goal should be to identify and analyze, through the lens of each of the "three Sun Tzus" (pp. 1–2 above), the thought patterns and intellectual harmonics actually present in the Sun Tzu text, along the way demystifying them.[2] With the ever-greater penetration of Sun Tzu into world cultures, including twenty-first-century global culture from *Star Trek* to marriage manuals, that task is becoming ever more timely.[3]

In the present study the vehicle of choice has been a certain type of "reverse engineering" of Sun Tzu, whose immediate product is the fourteen themes along with textual anchoring and substantive analysis of each. That textual immersion reveals that the Sun Tzu text's distribution of attention – perhaps the best measure we have of what Sun Tzu regarded as important – is rather different from widely held conceptions of it. The real Sun Tzu is more straightforward and less "crafty" than commonly envisioned.[4] There are shards of credible physical science insight. In other respects Sun Tzu shows himself as a thinker to be more structural, less

---

[1] See Watson, #1 footnote 59, pp. 3–4 (summary) and pp. 251–52 (translation of *Zhuangzi* passage).

[2] Sun Tzu's need for demystification is palpable. Swathes of modern literature invoking Sun Tzu's name have a tonality that calls to mind a literary critic's observation, noting how an "aura of ultimate secrets is felt like a powerful magnetic field whose source lies just beyond the reach of our present means of inquiry." See p. 77 of Peter Barry, "Citizens of a lost country: Kawabata's *The Master of Go* and James's 'The Lesson of the Master'," *Comparative Literature Studies*, 1983, 20(1), 77–93. Cf. also Schwartz, Introduction footnote 42, p. 237 ("mystical metaknowledge that lies beyond all language").

[3] A perceptive overview of the state of Sun Tzu studies, cast in the form of a July 5, 2018 book review, is John F. Sullivan, "Reviewing a new Sun Tzu translation: is there any blood left in this old stone?," available at https://thestrategybridge.org/thebridge/2018/7/5/reviewing-a-new-sun-tzu-translation-is-there-any-blood-left-in-this-old-stone (visited October 6, 2021).

[4] Given scholarly doubts as to whether Sun Tzu was an actual historical person (see pp. 12–13 above) it should be reiterated that this anthropomorphic framing is merely a convenient shorthand.

aphoristic than the popular image. In still other ways, he is a lot more military. In fact, time and again the key to reading Sun Tzu is to understand that his counsel reflects timeless forces and constraints with which warriors have struggled since the dawn of history.

A useful conceptual aid to thinking about strategy and strategists is that of a "level" of strategic analysis. An analytical level may be described as a class of problems and scenarios whose exploration a particular brand of strategic analysis finds congenial and on which it achieves intellectual traction. Modern examples of "levels" are illustrated by work of Thomas Schelling and, separately, of Erving Goffman. Importantly, an analytical level has no aspirations to be comprehensive, thus setting it apart from any grand system – philosophical, ideological, or other. Mastery of an analytical level gives insight into certain kinds of problems only; it may have little or no relevance to other problems equally pressing and worthy of analysis.

Sun Tzu represents an analytical level in the study of strategy, one associated with a unique mix of insights and observations that has no close counterpart in any other single surviving work on strategy before or since.[5] One hallmark of a level of strategic analysis is that its presence is easy to spot even in contexts where names of its pioneer or pioneers, their foundational writings and favorite terms of art, nowhere appear. In Sun Tzu's case, for example, many of the *Romance of the Three Kingdoms* stories illustrate ideas that could have come straight out of Sun Tzu's thinking (though Sun Tzu is not invoked by name); as do some (though not all) scenarios found in Mao's revolutionary warfare writings.

Yet these and other examples also illustrate (and modern cases like Schelling and Goffman corroborate) the limitations of a "level." For example, Maoist thinking cannot be reduced to "applied Sun Tzu"; still less can "Chinese strategic tradition" writ large. Nor should Sun Tzu be regarded as a comprehensive fountain of cyber warfare wisdom. Much like any weapon system, an analytical level in strategic thinking comes equipped with both capabilities and limitations. In Sun Tzu's case, this point of view returns us to the twin questions with which the present study began:[6]

What did Master Sun know that we still don't (or have yet to absorb adequately)?

What are Sun Tzu's limitations or blind spots?

A Structure for Sun Tzu: The Fourteen Themes

The fourteen themes approach that the present book has implemented builds on some simple premises. If one wishes to engage with Sun Tzu's ideas, one first has to find out what Sun Tzu said. Read the text! What's more, reading Sun Tzu is a precision task. The Sun Tzu text should be read in its entirety, not just plucking from it a few isolated

---

[5] Sun-Tzu-qua-analytical-level is well conveyed by Lau & Ames, p. 43: "... the *Sunzi* presents a profound way of thinking which has broad application, and which cannot be reduced to techniques or tactics." A compatible perspective comes from McNeal, Background footnote 58, p. 18: "it is likely that the genre of [early Chinese] military texts as a whole created an intellectual space in which a *timely* discourse continued to unfold long after the competing states of the Eastern Zhou period were demolished" (emphasis in original). McNeal's "intellectual space" concept succinctly expresses the vistas opened by Sun Tzu (2) and Sun Tzu (3) analytical directions.

[6] See Introduction, pp. 21–22 for overview of how the present study addresses those two questions.

ideas or one-liners.[7] The text repays reading through a Sun Tzu (2) lens of generaliz-able military content and a Sun Tzu (3) one of broader strategic insights, while never losing sight of Sun Tzu (1) conditions operating in Sun Tzu's own time.

Applying the fourteen themes approach to Sun Tzu suggests that the Sun Tzu text is coherent, indeed for the most part remarkably lucid.[8] A useful modern analogy comes from the law: like a well-crafted law review article Sun Tzu alerts you to the issues its subject matter presents. Also like many such articles he rarely shows you how to "solve" resulting puzzles. As regards its expository aspects, a fitting label for the Sun Tzu text would be "semi-organized." Shortfalls in intellectual organiza-tion, while real enough, seem no worse than those encountered in many modern writings on strategy (where the effects of poor intellectual organization may indeed be more pernicious, since those writings are often verbose, whereas the Sun Tzu text is admirably sparse). As regards Sun Tzu's substance, the "Axiom of Conceptual Unity" for Sun Tzu (p. 15 above) fits well. Unsurprisingly (given what we know of the life and times of the text), there are scattered intellectual tensions and inconsistencies in Sun Tzu's thinking.[9] But those tensions are far outweighed by powerful consistencies in substance anchored by many interconnections among Sun Tzu's ideas.[10]

Against this backdrop, the fourteen themes structure developed in the present study offers a tool for exploring Sun Tzu's thought in a disciplined and analytically integrated way.[11] Viewed as intellectual archaeology, that structure offers a way of

---

[7] Important to note here are strands of Sun Tzu's thinking that would-be appliers (including some self-identified Sun Tzu enthusiasts!) may be prone to misuse or distort, to their cost. An example (see p. 278 above) would be taking Sun Tzu's deception counsel so much to heart that one starts to prize deception *for its own sake*. That is a recipe for alienating friends and allies, even to a point of creating a weapon directed, however inadvertently, at oneself. Conjoined with Sun Tzu's encouragement of plunder, that same prizing of deception may also foster underestimation of plunder's systemic risks in a digitally enabled, tightly interconnected world.

A culprit in such cases is a mechanistic understanding of Sun Tzu, in application if not in theory.

[8] In particular, the fourteen themes analysis reveals a work far less opaque than a comment of Lau, 1965 article, p. 329 footnote 21 – referring to the Sun Tzu text as "extremely cryptic" – suggests. To be sure, sporadic obscurities and traces of textual corruption are present, probably of multiple authorship too; but these are limited in their effects on the substance and are fundamentally minor.

Some of Sun Tzu's interpretive ambiguities arise because we lack context that Sun Tzu (or whoever compiled the text) assumed, perhaps correctly, would be familiar to the intended audience.

[9] Possibly the most basic case in point is the tension between the Sun Tzu text's fundamental identity as a military treatise, focused on conventional warfare, and the fact that much of Sun Tzu's winning edge derives from comparative advantage in spheres of grand strategy and intelligence. See Part C footnote 2 (p. 171) and pp. 178ff. Another example is the tension between the prescriptive rules in Sun Tzu's terrain material and apparent rejections of any such rules in other parts of Sun Tzu (see pp. 185–86). In both cases, however, the identified tension is also a stimulus to creative thought.

[10] A similar view is voiced by Griffith, p. 12. Reinforcing a point made in the Introduction (pp. 15–16), it should be reiterated that conceptual unity in Sun Tzu's substance need not require sole authorship.

[11] A contrast case to the structured standpoint represented by the fourteen themes is the stance of Ming dynasty Sun Tzu commentator Liu Yin, who wrote that Sun Tzu should be read "in a lively manner, like pearls rattling around in a dish, with no prescribed order" (Minford, p. xxxi). While Liu Yin's way of reading Sun Tzu might sometimes catalyze drawing fresh substance from the text, in favor of the fourteen themes approach is that is provides a range of natural controls and

retrieving something like the "whole Sun Tzu" – Sun Tzu's thought as it might have appeared to educated Chinese in many epochs of Chinese history, some of whom had invested in the text to a point of memorizing it. Here the fourteen themes approach offers four specific contributions.

A first contribution – of basic importance for all three Sun Tzus, but especially for Sun Tzu (3) purposes – is to recognize and address head-on the challenge of Sun Tzu's "multivocality," which is much more of a stumbling block to mastering Sun Tzu's ideas than are any textual or translation ambiguities.[12] "Multivocality" is a way of saying that the same snippet of Sun Tzu text commonly sends different messages to multiple audiences, civilian as well as military. In many ways multivocality is an appealing feature of Sun Tzu, enabling a remarkable amount to be said with Sun Tzu's signature terseness. Yet that same characteristic has potential to disorient all but the most dedicated readers. Compounding that complication, Sun Tzu is also prone to "switch horses in mid-stream": initially appearing to address one set of audiences, then (commonly without any signal recognizable to modern readers) abruptly switching a sentence or two later to a different set of audiences whose concepts and interests often overlap with the first set yet are not identical.[13] Moving on through the text another such switch follows, and another, and another. The upshot is a "garden of forking paths" in reading Sun Tzu – instances where interpretive pathways (each with its associated audiences) diverge and those divergences lead on to further divergences. It is little wonder that tendencies like these can convey an aura of jumbled thinking, even though the multivocality that underlies the pattern can sometimes also reveal buried interconnections among Sun Tzu's ideas.[14]

Here the fourteen themes analysis, thoroughly grounded in the entire text, pays its way. Via the passage identifications it establishes a basic intellectual control on the formidable multivocality present in the text. A key to that control is taking in stride the possibility that the same Sun Tzu verse may appear in several passages assigned to different themes. Indeed such multiplicity is the norm, not the exception.[15] A further

crosschecks on understanding Sun Tzu's ideas, reducing chances that Sun Tzu's words are invoked to advocate for something foreign to Sun Tzu's thinking (possibly, on further scrutiny, at odds with Sun Tzu's own likely position).

[12] "Multivocality" is a sociological and strategic concept pioneered by John Padgett & coworkers and applied by them to characterize the strategic and communications style of the great Florentine banker and political actor Cosimo de' Medici (1389–1484) and its implications for robust action. It is introduced in a different context on pp. 103–105 above with #2 footnote 43 providing references.

[13] Many examples congregate around the dual use (civilian plus military applications) passages assembled to illustrate Theme #4 (winning without fighting). But there is also much internal diversity in military audiences, and here too Sun Tzu often speaks at one and the same time to audiences with widely differing military occupations, professional persuasions, and problems of interest.

[14] Recognizing multivocality in Sun Tzu helps to clarify why efforts at summarizing Sun Tzu are often problematic, since they attempt (usually without comment) to ignore or gloss over it.

[15] An example already appears at the beginning of the text, in verses I.1–I.14. From this one set of verses Passages #1.1, #11.5, and #12.3 distill, respectively, net assessment, adversary strategic intelligence, and self-knowledge agendas. In the present study, Sun Tzu verses have a median of three main text thematic assignments; some have eight or in one case nine. See pp. A-28f. for further details.

key is the three-way distinction among Sun Tzu (1), (2), and (3), which serves as an additional tool for differentiating Sun Tzu messages sent to different audiences. The resulting intellectual product – a structure, not just a list – substantially reduces risk of disorientation born of multivocality (though not fully eliminating it).

A second contribution of the fourteen themes structure is to clarify holes or omissions in Sun Tzu's thinking, for reasons well captured by an observation of E-tu Zen Sun: "when individual pieces fall into place within a general framework, the lacunae become more obvious."[16] Those holes are the focus of the next section (pp. 497ff.)

A third contribution is to serve as a kind of conceptual sieve to catch a type of intellectual "remainder" that might otherwise escape attention: a variety of interesting, possibly important, Sun Tzu conceptual directions or leads that slip through the mesh of the fourteen themes and are *not* caught by it. "Orphan ideas" is a serviceable label to describe such Sun Tzu content. Even though orphan ideas perforce "hide in plain sight" since the Sun Tzu text is so short, the vast Sun Tzu secondary literature commonly ignores them. In the present study, identifying such byways of Sun Tzu's thinking is a collateral benefit of the close reading of Sun Tzu elicited by the primary goal of developing a detailed census of Sun Tzu passages to illustrate the fourteen themes.

Orphan Sun Tzu ideas enrich the primary fourteen-themes-centered harvest from Sun Tzu, sometimes suggesting fresh directions of Sun Tzu (2) and (3) type. A case in point would be a strategic or tactical "style" interpretation of verse VII.13 (Passage #5.14), noted in Theme #5 analysis (pp. 181–82 above), and the broader challenges of analyzing styles and style-switching it suggests.

In rounding up orphan ideas a cautionary question should be kept in mind: When is such an enrichment of Sun Tzu an overreach? To give that question a Chinese twist: if a journey of 10,000 miles begins with a single step, how much credit should be given to Sun Tzu for taking that first step? While no useful general answer to that question seems likely, a suggested heuristic is a "rule of two." Specifically, if Sun Tzu textual support (possibly implicit) for a given orphan idea *can be found in at least two different parts of the text* that idea is a candidate for being treated as a basic part of Sun Tzu's legacy, and its broader Sun Tzu (2) and (3) implications are worth attention. Switching is one example, since relevant Sun Tzu content appears not only in verse VII.13 but also in Sun Tzu's material on qi and zheng (Passage #14.1), which points (if implicitly) to the challenge of switching between qi and zheng (pp. 469–70 above). A second example comes from verses III.23 (Passage #4.12) and II.5 (Passage #4.13), both of which point to possibilities for meddling by opportunistic third parties in a fraught domestic political situation. A third example is Sun Tzu's imaginative methodology for flash flood prediction (Passage #11.18's verse IX.15a) set alongside Sun Tzu's cognizance of weather and climate planning issues (Passage #1.1's verse I.5).[17] There is a launching point here for an

---

[16] See p. 90 of Sun's review of Needham's physics volume, *Technology and Culture*, 1964, 5(1), 89–92.

[17] That Sun Tzu methodology shares with Zhang Heng's pioneering seismoscope of the second century AD (see #4 footnote 50) an ambition to develop a capability to detect remote physical events.

"environmental Sun Tzu" that puts Sun Tzu's credibility behind other searches for early warning of environmental changes or "pivot" points.[18]

A fourth contribution of the fourteen themes puts Sun Tzu (2) and (3) in center stage. Reflecting on World War II naval logistics experience, historian (and later World Bank official) Duncan Ballantine observed with great insight that the function of logistics is "to bridge the gap between two normally alien spheres of activity," thereby being at the same time "the military element in the nation's economy and the economic element in its military operations."[19] Generalizing this image of a bridge suggests a way of thinking about the Sun Tzu (2) and (3) "frontiers" discussions accompanying each of the fourteen themes.

Those discussions can be likened to bridge-building exercises – now of an abstract intellectual variety – connecting Sun Tzu's ancient text with a wide range of arenas where Sun Tzu's ideas can inspire and at times guide modern thinking and research. Like other kinds of bridges, intellectual bridges are diverse. Some are, as is sometimes said, "bridges to nowhere." Some are long (building them involves quite a few conceptual steps!) and perhaps also a little prone to sway. A case in point might be the "bridge" leading from Sun Tzu's terrain thinking to modern computing's rule-based expert systems. Others are short, sturdy, and can bear much traffic (say, the bridge from Sun Tzu's focus on attacking the adversary's strategy to algorithm sabotage concepts). Some connect territories with similar "ecologies" (say, Sun Tzu's lessons for modern mobile warfare); others, seemingly very different ones (say, transposition of Sun Tzu's terrain thinking to social network analysis; or of his formlessness ideas to terrorist/counterterrorist hider–seeker games). Some bridges connect two large territories (say, Sun Tzu Theme #6 and attacks on a modern army's logistics); others, important but narrower ones (say, Sun Tzu's death ground thinking and psychological manipulations of human groups on the cusp of catastrophe). Some (like Bailey bridges in combat engineering!) may be relevant at just one crucial juncture (e.g., Passage #11.6's verses XIII.1–XIII.2 as a nudge to a wavering decision maker to take the plunge of creating his own spy service!).

No less than its physical counterpart, intellectual bridge-building requires materials. A core contribution of the passages assembled to develop the fourteen themes is to clarify what Sun Tzu intellectual materials are available for a Sun Tzu (2) or (3) bridge-builder to work with. Those textual materials are richer and more diverse than the oft-quoted parts of Sun Tzu to which so much of the modern Sun Tzu literature limits its attention.[20] Knowing the available materials with specificity sheds light on what type of Sun Tzu (2) or (3) bridges can reasonably be expected, where their other ends might be most usefully placed, and what sort of traffic they are likely to be able to bear. Here is much scope for disciplined analogical thinking.

---

[18] An inferred Sun Tzu focus on tipping points (cf. Arthur Waldron's insightful Sun Tzu interpretation noted in #9 footnote 23) might lead "environmental Sun Tzu" (p. 244 above) to take an interest in the state of the Antarctic Thwaites Glacier, whose erosion has been identified as a predictor of rising sea levels associated with climate change (*Science*, February 7, 2020, 367(6478), 608).

[19] Duncan Ballantine, U.S. *Naval Logistics in the Second World War* (Princeton, NJ: Princeton University Press, 1947), p. 3 (quoted by Eccles, Preliminaries footnote 5, p. 72).

[20] A case in point would be Sun Tzu material assembled in Table 3 (pp. 151–152) in support of developing the concept of cyber warfare as a potentiator of political warfare. That material mobilizes ingredients culled from all thirteen of Sun Tzu's chapters.

Reality Check: Sun Tzu's Limitations

Perspicacious as Sun Tzu is, there is a risk of attributing too much omniscience to the text, tendencies that the modern Sun Tzu literature often reinforces, albeit unwittingly.[21] One set of Sun Tzu limitations has already been addressed in the "roads not taken" sections that accompany each of Themes #1–#14. The focus there is on holes or omissions in Sun Tzu involving matters that, *under Warring States military, intellectual, social, and technological conditions,* might plausibly have been addressed in some way in the text (yet were not).

The present analysis now lifts the restriction just stated, approaching Sun Tzu's limitations and blind spots from a much broader perspective informed by universal military and strategic theory.[22] As with the "roads not taken" material, Sun Tzu's deficits analyzed below may sometimes suggest paths to advantage in a strategic competition with an adversary whose thinking reflects Sun Tzu's influence.

Table 8 is a menu of major omissions or blind spots in Sun Tzu. That list, which is intended to open up questions, not to be exhaustive, corroborates that Sun Tzu did not have all the good ideas about strategy (an observation possibly reassuring to aspiring twenty-first-century recruits to strategy or grand strategy careers, ever on the lookout for new dragons to slay!). The following discussion tracks Table 8's topic order.[23]

First and in some ways foremost, Sun Tzu explicitly addresses one kind of war only, namely, *conventional land warfare between state actors of comparable motivations and capabilities.*[24] The closest approach to an exception is the espionage chapter which could be interpreted as a roadmap for a spy-vs.-spy war, possibly taking place in

---

[21] A contributing factor here is that Sun Tzu's ideas are almost always presented without qualifications or scope conditions that might alert us to limitations. Cf. Sawyer, pp. 319–20 note 120.

[22] As noted on pp. 17–18 above, the possibility and utility of such a universal theory is a premise of the present study. Although there is no single work one can point to as definitively setting forth the elements of such a theory, Eccles's 1965 book (Preliminaries footnote 5) is a major early post–World War II synthesis anchored in a strategy-as-control and logistics perspective. A further far-reaching synthesis is Lawrence Freedman's 2013 *Strategy: a history* (#4 footnote 41). Karl Deutsch's *The Nerves of Government* (cited in #14 footnote 23) remains important as a conceptual foundation for melding military theory with modern technological advances and perspectives. Further basic ideas about control, developed in a much more technical vein, are found in H. S. Tsien (Qian Xuesen), *Engineering Cybernetics* (New York: McGraw-Hill, 1954). Note especially Tsien's chapter 18, "The control of error," pp. 268–84 (building on ideas of von Neumann – not von Neumann's game theory work but von Neumann's research that pioneered foundations of reliable computing).

A list of thinkers whose writings lay many of the foundations for a universal theory is set forth by RADM J. C. Wylie, USN, *Military Strategy: a general theory of power control* (New Brunswick, NJ: Rutgers University Press, 1967), p. 7, mentioning Machiavelli, Clausewitz, Mahan, Corbett, Douhet, Liddell Hart, and Mao. The Western-centric tilt of that list is noteworthy and is a limitation. There is no reason why contributors to universal military theory should be Western any more than are, say, contributors to universal algorithmic knowledge (see Appendix 1 footnote 30 on Pāṇini and Piṅgala).

[23] Here mention should again be made of the Eccles–Rosinski strategy-as-control framework as a tool for clarifying Sun Tzu's limitations. See Theme #5 "frontiers" analysis (pp. 187ff. above).

[24] By contrast to Sun Tzu, the Zuozhuan takes cognizance of at least two kinds of war, distinguishing intrastate warfare from warfare between states. See Durrant, Li, & Schaberg, *Zuo Tradition (Zuozhuan)* (cited on p. xxiii above), Vol. 1, p. 505 ("When arms are taken up within a domain, it is unrest; when without, it is an incursion").

Table 8 *Important holes or limitations in Sun Tzu's treatment of strategy and warfare*

1. Sun Tzu considers no kind of war other than conventional land warfare between state actors of comparable motivations and capabilities. Sun Tzu takes no account of asymmetric warfare of any sort. With very limited exceptions, Sun Tzu's focus is on two-sided, not multi-sided, conflicts.

    Accordingly, Sun Tzu shows no attention to, or cognizance of, Clausewitz's all-important "what kind of war" question, which is in essence a diagnostic question.

2. There is no attention to the task of analyzing and choosing strategic objectives (a task that blends with statecraft and analysis of the core sociopolitical identity of the strategist and the cause she serves).

    Importantly, Sun Tzu's focus is on complete victory; limited war objectives are not contemplated. Relatedly, Sun Tzu has difficulty extending his ideas to analyzing and planning strategic action on a long-run timescale. Sun Tzu's emphasis on rapid action and rapid results is a limiting factor here.

3. More broadly, Sun Tzu undertakes no visible engagement with the kind of soul-searching challenge of assumptions exercise that is a hallmark of thorough strategic analysis. As a major case in point, Sun Tzu's fundamental focus is always zero-sum conflict. Cooperative, "win–win," etc. possibilities are not entertained.

    Unalloyed pursuit of efficacy drives Sun Tzu's thinking throughout. There is no trace of any softening of a core agenda of "attaining mastery by any means, fair or foul" (Gawlikowski & Loewe, quoted on p. 329).

4. Sun Tzu's concept of human nature and motivation is narrow. Common soldiers are likened to children, sheep, or inanimate objects. Relatedly, Sun Tzu pays no heed to the potent role in warfare often played by belief systems (e.g., religion, ideology, patriotism). Emphasis is on exploiting human frailties and foibles, by contrast to encouraging positive aspects of human potential and flourishing. Also indicative of the kind of society Warring States China represented, Sun Tzu's thinking accords a basic place to *fanjian* (double agents) and their greed-driven loyalty.

5. There is no engagement with vistas opened by decentralized command and control of military forces or operations. In particular, the Sun Tzu text contains no traces of the concept of mission command, with its signature encouragement of initiative and imagination on the part of subordinate officers.

6. The civilian population is offstage and mostly invisible – as are its communications systems (a particularly important omission in contemplating Sun Tzu applications to a twenty-first-century world of email, social media, and digital devices). Sun Tzu lacks a concept of the civilian audiences of war.

7. There is no cognizance of guerrilla nor, save possibly on a highly abstract level, of irregular warfare.

8. There is no cognizance of naval warfare or of operations of any kind on water. In particular, there is no reference to water transport, in many ways the major pre-digital form of good logistics.

9. Attention to non-military means and tools of conflict (including diplomacy) is limited, the one major exception being spies. Material on alliances, especially alliance-building, is undeveloped.

Table 8 (cont.)

10. There is no apparent recognition or analysis of the fundamental distinction between strategy and tactics (a distinction that Sun Tzu might indeed have rejected).
11. Logistics coverage is limited. In particular, Sun Tzu lacks a comprehensive concept of the nature and sustained functioning of a logistics system comprising a full range of the means of war.
12. Ideas for guiding judgment under uncertainty (a type of judgment needed in strategy, logistics, and tactics alike) are lacking.
13. Also lacking is explicit attention to military innovation and the challenges it presents.

peacetime. But there is not enough detail there to make it a full-fledged alternative focus. From a twenty-first-century standpoint a particularly important Sun Tzu omission is lack of coverage of asymmetric warfare of any type. Sun Tzu's focus, limited to symmetric versions of conventional warfare (and for the most part involving two sides only), should be noted as greatly simplifying the analytical problems Sun Tzu's advice manual needs to tackle.

Second (and expanding on Theme #12's "roads not taken" discussion) Sun Tzu offers scant guidance on how to analyze objectives, certainly not on a plane of grand strategy where both actions taken and their effects commonly unfold only over the long run – even that "long run" on which John Maynard Keynes famously pronounced.[25] One symptom of Sun Tzu's deficit is verse IV.7's unnuanced concept of "complete victory" (quan sheng). Such a concept might have been adequate in a context of internecine warfare among the Warring States, where the basic strategic goal was conquest of one state by another. But an exclusive focus on "complete victory" is plainly inadequate for many other kinds of war and conflict.[26] For example, Sun Tzu has little useful to contribute to analyzing the strategic objectives of Chinese rulers facing the chronic challenge of defending against incursions by Inner Asian peoples; or, in some eras of Chinese history, projecting Chinese power across Central Asia. Sun Tzu's "complete victory" rhetoric could prove misleading to those whose expectations are pumped up by his counsel, especially on a grand

---

[25] "In the long run we are all dead." See J. M. Keynes, *A Tract on Monetary Reform* (London: Macmillan, 1923), p. 80. Illustrating that Sun Tzu runs into difficulty in getting analytical purchase on strategic issues where a truly long-run perspective is needed, note the rather unsatisfying quality of Sun Tzu's comments on alliance-building and, more generally, Sun Tzu's lack of insight into how to navigate strategic challenges on the cusp of cooperation and conflict. Lofty as are its ambitions, the open-ended ambiguity of verse III.11 (Passage #3.3) is also a case in point. See #3 footnote 61.

[26] An "incomplete victory" idea has deep roots in parts of Chinese strategic thought (e.g., Needham & Yates, p. 37). It is epitomized by outcomes in the game of *weiqi* where the game routinely ends with each player controlling at least some territory on the board (see Boorman, Introduction footnote 13, pp. 22–25, 112–13). No counterpart exists in Sun Tzu's treatment of strategy or grand strategy.

strategic plane. Taken at face value, such rhetoric has potential to derail delicate strategic calculations where they are most needed. That in turn may encourage overreaching that an astute foe (who may have also read Sun Tzu!) can exploit.

Amplifying this deficit, there is also a shortage of Sun Tzu material calling for a strategist regularly to examine – and to challenge – that strategist's own most basic, most cherished assumptions. Sun Tzu remains wedded to a zero-sum view of conflict. There is no concept of "win–win" cooperation with implications for reshaping strategic goals on both sides, tackling the question of "what am I willing to settle for?" In traditional Chinese theory and practice of governance, a time-honored source of challenge of assumptions was a loyal minister's (or a censor's) act of remonstrating with a ruler about what was seen to be a misguided choice or policy – a step requiring courage on the underling's part, for all the reasons that the powerful commonly do not receive criticisms kindly.[27] Nothing in Sun Tzu envisions the general remonstrating with the ruler with an eye to correcting faulty strategic assumptions.[28]

An unmet need to challenge assumptions spills over into, and in some respects weakens, Sun Tzu's otherwise incisive thinking about shi (strategic advantage). While such thinking may indeed elicit heights of strategic creativity (see Theme #5), unless accompanied by a searching challenge of assumptions it also harbors a risk of self-deception, one that sees more possibilities for success than are in fact realistically present in a given situation at a given time and may prompt misguided actions as a result (possibly including failure to compromise with a foe).

A related omission (though one that certainly streamlines Sun Tzu's theory of war[29]) is lack of concern for any issues of an ethical type – much to the consternation of Confucians who came after Sun Tzu.[30] In the short run, disregard of ethics or related prosocial values may often be a strategic and tactical asset. In the longer run, it can easily prove a recipe for grand strategic failure.[31]

Failure to challenge assumptions also sets a tone for Sun Tzu's concept of human nature and motivation. That concept is fundamentally narrow – narrower by far than that found in some of the major Chinese belief systems that were taking shape in Sun

---

[27] On the remonstrance function and its risks for remonstrators see Hucker, #4 footnote 8, pp. 6–7.

[28] Sun Tzu's delineation of circumstances where the general may flout orders from the ruler addresses a different problem. In many ways it is a simpler one.

[29] See again Clausewitz's observation, "Theory becomes infinitely more difficult as soon as it touches the realm of moral values," previously quoted on p. 330 above.

[30] Groping for exceptions to Sun Tzu's ethical void, it might be possible to discern ethical content in verse I.4 pertaining to a ruler's dao (Griffith's "moral influence," see Passage #13.2). But Sun Tzu's development of that concept is so scanty, and also so infused with considerations of advantage (for the ruler or his general), that inferring ethical content must perforce invoke some source outside the Sun Tzu text. For a flavor of the Confucian debate with Sun Tzu see Raphals, #7 footnote 11.

[31] Compare Liddell Hart, #4 footnote 2, p. 336:

> while the horizon of strategy is bounded by the war, grand strategy looks beyond the war to the subsequent peace. It should not only combine the various instruments [of conflict], but so regulate their use as to avoid damage to the future state of peace – for its security and prosperity. The sorry state of peace, for both sides, that has followed most wars can be traced to the fact that, unlike strategy, the realm of grand strategy is for the most part terra incognita – still awaiting exploration, and understanding.

[Italics in original.]

Tzu's time. Apart from general, ruler, and spymaster, the human beings appearing in the Sun Tzu text are profiled in stylized, even perfunctory terms, consistent with a low-dimensional, indeed mechanistic, understanding of human behavior and interactions. Sun Tzu is chiefly concerned with exploiting human frailties; most of the emotions he contemplates have an antisocial or at least a negative quality. An army's troops are to be manipulated by exploiting rage or greed. Motivating spies centers on the greed factor: cash or cash-equivalent payoffs. Although Sun Tzu is perfectly willing to take advantage of existing social trust, there is scant discussion of ways of creating it (apart, of course, from the sort of situation-driven temporary "trust" that arises when troops are plunged into a desperate situation!). Altogether unmentioned is the role of deeply held beliefs (e.g., religious, ideological, patriotic, etc.) that in many ages of war have been a wellspring of morale and dedication to duty for soldiers and civilians alike. Concepts of sociopolitical identity, and the loyalty that often attends it, are likewise left offstage.[32] As a corollary of these emphases and omissions, it seems entirely possible that a Sun Tzu–inspired strategic actor might underestimate the motivating power of ideas and ideals, with resulting errors of strategic judgment.

As an unsurprising corollary of such premises, Sun Tzu's vision of command and control appears to be very much a top-down one, consistent with Sun Tzu being a true believer in centralization. Positive outcomes are always scripted as emanating from the commanding general, never from imaginative independent actions taken by underlings.[33] Sun Tzu offers no word picture of how the day might be won by an inspired subordinate taking an initiative sua sponte. The resulting perspective is worlds apart from important strands of modern military, as well as civilian management, thinking that play up gains from decentralized initiatives by junior personnel.

Meanwhile, the civilian populations of the Warring States make scant appearance in Sun Tzu; the non-Sinitic peoples on China's northern and western frontiers (long a looming threat to China's polities), none at all. There is, to be sure, an isolated reference to "the people" in Passage #1.1's verse I.4 ("By moral influence I mean that which causes the people to be in harmony with their leaders ...") and further scattered references appear elsewhere in the text. But for Sun Tzu, "the people" (or any major stratum or social formation thereof) never emerges as an active force in the situation. Nor, save glancingly (e.g., Passage #3.5's warning that a protracted war can render the people destitute), do "the people" enter as a source of major constraint. These omissions may help explain Sun Tzu's uncritical stance regarding plunder. Sun Tzu's wars unfold on a sociologically desolate landscape.[34]

---

[32] There is nothing in Sun Tzu that suggests Sun Tzu's anticipation of a deep truth stated in modern language by Padgett & Powell, #2 footnote 43, p. 312 (discussing, inter alia, the China of Mao and Deng Xiaoping): "In the short run, actors make relations. In the long run, relations make actors."

[33] Regarding lack of mission command thinking in Sun Tzu see pp. 437–38 above.

[34] A sharply contrasting modern perspective is conveyed by Strachan, #2 footnote 12, pp. 278–79:

> Much more challenging for strategy is another of [Rupert Smith's] insights, that contemporary warfare is a form of theatre, played out by a small, separate group (i.e., a professional and not a conscript army), orchestrated by a team of unseen directors, stage managers and lighting engineers, but watched by many more. *The people are the audience for war."*

[Emphasis supplied.]

Amplifying this perspective see Rupert Smith, #4 footnote 90, pp. 286–87.

That in turn creates basic barriers to attempts to align "people's war" thinking (Maoist or other) on Sun Tzu.

This deficit points to another. There is no cognizance by Sun Tzu of guerrilla or irregular warfare, some types of which have a prominent place in Chinese history in the form of peasant revolts. That deficit of attention is again consistent with Sun-Tzu-the-centralizer, for all the reasons that classic guerrilla warfare tends to be decentralized, often heavily so (see pp. 280, 282 above).

Separately, Sun Tzu also takes no cognizance of naval warfare or of operations on the sea or rivers.[35] This is a particularly significant omission in a work on strategy and warfare, since the sea is a sphere of human action that lends itself to ambitious, at times subtle, strategic and grand strategic thinking (in part because, prior to the advent of the cyber domain, water transport was the leading form of good logistics).

With the prominent exception of espionage, Sun Tzu also pays only very limited attention to specific non-military tools of conflict, starting with diplomacy (which was a well-developed and flourishing activity in Chinese context from times earlier than the Warring States era). Here again Sun Tzu's zero-sum standpoint makes its presence felt (for example, underwriting Sun Tzu's lack of interest in methods of negotiation or in persuasion campaigns).

As noted in the Preliminaries chapter (pp. 47–48 above), the Sun Tzu text makes no effort to distinguish conceptually between strategy and tactics. Any attempt to spot even the germ of such a distinction in the text encounters difficulties from the outset. In particular, the Sun Tzu text frequently commingles and blurs what would nowadays be considered strategic, operational, and tactical military contexts.

Sun Tzu's omission of the strategy/tactics distinction fails to take advantage of one of the most useful pieces of military theory. People who make good strategists and those who make good tacticians are often different people (as is their relevant education and training, to touch on two further topics in which Sun Tzu is also not interested). Some of Sun Tzu's modern disciples may have successfully absorbed Sun Tzu's insights *tactically* even while neglecting them *strategically*. That is an almost anti–Sun Tzu posture, for all the reasons that Sun Tzu stands for a conceptually integrated, holistic perspective on war and conflict. But it is also a type of blind spot worth watching for, when confronting a Sun Tzu–inspired antagonist.[36]

Turning to logistics – war's third fundamental ingredient, interlocking with strategy and tactics but having an identity distinct from both – deficits in Sun Tzu's thinking again appear. Sun Tzu has no clearly articulated general concept of logistics, certainly in its comprehensive modern sense as comprising all the "means of war."[37] Bits and pieces of logistics and logistics analysis (or lore) do appear in the Sun Tzu text in scattered locations, but there is no foreshadowing there of any attempt to integrate those diverse considerations into a framework that might encourage and facilitate coherent engagement with logistics problems as a unified

---

[35] By contrast to this naval deficit, Sun Tzu verse V.14 (Passage #9.3), with its image of a raptor striking its prey, holds seeds of air warfare thinking. Cf. #9 footnote 49 (relevant missile theory application).

[36] Griffith, pp. 177–78 makes a related argument concerning the nature and limitations of Sun Tzu's influence on Japanese military strategy and tactics in World War II. For the general issue of strategic vs. tactical understanding of Sun Tzu see also Conclusion footnote 47 below.

[37] See Eccles, Background footnote 52; Boorman, Introduction footnote 4.

whole, with resulting gains in efficacy of military effort. That crucial conceptual advance would have to await much more recent times. Of course, logistics problems were far simpler in Sun Tzu's era than in modern warfare, thus reducing (though not eliminating) the need for, and value of, an integrated perspective on the logistics process. Limits of Sun Tzu's logistics thinking particularly show when considering applications of Sun Tzu to warfare in Central Asia, where in many contexts it is simply impossible for an army to live off the land or its people as Sun Tzu counsels.

A common feature of strategy, logistics, and tactics is that all require judgment under uncertainty. Sun Tzu's thinking grew out of a culture that had not developed probability theory or kindred quantitative approaches to judgment under uncertainty – a workhorse of modern analytical thinking about military and other decision problems.[38] If one accepts common interpretations of verse XIII.4 (Passage #1.19), Sun Tzu explicitly distances himself from divination or related practices that, in his time, would have represented available approaches to navigating uncertainty, however fallacious they may seem by modern standards. Yet he offers no serviceable substitute. Therein lies another glaring deficit in Sun Tzu's thinking.[39]

From a contemporary perspective, among the most important omissions in Sun Tzu is lack of explicit attention to military innovation and the asymmetries in capabilities to which such innovation may lead. While Sun Tzu does take note of some of the more advanced military technology of his day (notably the crossbow), there is no attempt to take into account the challenges of ongoing military innovation (with ensuing need for coordinated development of doctrine).[40] Some of Sun Tzu's ideas can in fact be repurposed to tackle certain innovation-related issues (see, e.g., Theme #1 "frontiers" analysis focusing on attacks on R&D, pp. 72–74 above). But steps of that type require analogical thinking that goes beyond the letter of the text.

An oblique cognizance of military innovation possibilities might be distilled from a pair of verses in Sun Tzu V (V.6 and V.7, see Passage #14.1) lauding the creative potential of the qi mode of warfare. But there is little reason to think that Sun Tzu here had in mind anything more than imaginative stratagems crafted from familiar ingredients, as contrasted with more fundamental forms of innovation ranging into previously unexplored forms of warfare (see pp. 215–16 above on the

---

[38] There remains room for sharpening our understanding of how early Chinese military theory and practice coped with issues of judgment under uncertainty. For a larger context see Elvin, #1 footnote 32. See also Wilkinson §37.1, p. 516 (citing further work of Elvin on styles of scientific thinking).

[39] Possibly related to Sun Tzu's lack of engagement with judgment under uncertainty is absence of any consideration by Sun Tzu of how a general might exploit his own poorly controlled emotions, turning them to his advantage. That line of thought might have enriched Sun Tzu's thinking about the role of emotions in warfare (cf. Schelling's ideas about the rationality of appearing irrational, etc.).

[40] The very existence of the Sun Tzu text may, of course, itself represent a type of innovation spurred by the growing scale and intensity of early Chinese warfare. But that did not translate into explicit cognizance by Sun Tzu of issues like weapons reliability vs. pushing the state of the art; evolution of force structure; ongoing impact of social or technological change on personnel issues; etc. In that sense, Sun Tzu's perspective on war is a static one. Changing military conditions remain offstage.

Mongols). As a conjecture, it seems possible that a Sun Tzu–inspired actor, lacking relevant guidance from the text and galvanized by Sun Tzu's dictate (Themes #1 and #5) to think creatively and ever-successfully – no failures allowed! – might fall into a trap of pursuing dubious "exotic projects" in quest of outsized positive results obtained at a cheap cost (Theme #3).

In fairness to Sun Tzu, on this and several other points it needs to be recognized that in Sun Tzu's time some key Chinese strategic concepts (and terminology used to talk about them) were still early in their development, limiting Sun Tzu's quiver of available ideas for engaging with some of the deficits just discussed.[41] In reading Sun Tzu, it is important never to lose sight of the fact that this truly is an early text – and that distinguished pioneering thinkers do not always "get it right."

For the type of warfare he knew, and given the extreme conciseness of his text, Sun Tzu is indeed remarkably thorough.[42] One way of approaching the Sun Tzu text, consistent with its provenance in a time of great societal and military change, would envision it as a result or cumulative residue of "trying out" many different analytical directions and angles within the broad parameters of war as Sun Tzu knew it. Unsurprisingly, those directions were not always entirely consistent and some of them would prove more productive and enduring than others.[43] Although such a "testbed" reading of Sun Tzu needs to be carefully distinguished from a project of experimental science (a concept foreign to Sun Tzu's thinking), it harmonizes with the larger intellectual ferment of Warring States China. It is a useful lens on the text.

Sun Tzu As Prophet and Pioneering Theorist of Information Warfare

Given the catalog of limitations just reviewed, a puzzle arises: how can Sun Tzu be in some ways so narrow and yet at the same time so remarkably generalizable?

Grist for an answer grows out of the close and fertile relationship between Sun Tzu and ideas about the role of information in human conflict. Sun Tzu has credible claim to be regarded as the first major pioneer of warfare's information level, certainly the first whose work survives. Of course, some of the relevant ideas and certainly some of the ambiance of information warfare may be traced to times antedating Sun Tzu.[44] Yet it would not be overstrong to assert that Sun Tzu's short tract takes a decisive intellectual step toward transposing (elevating) war and its analysis from a tangible realm – the physical and kinetic one that has preoccupied warriors from the beginning of recorded history (indeed earlier) – to an intangible one, as qualitatively different from kinetic aspects of warfare as bits are from atoms. It is that all-important, in key ways liberating, step that enables Sun Tzu's ideas to

---

[41] See, e.g., p. 176 and #5 footnote 3 above (shi and xing as evolving military concepts).

[42] A statement from a biography of Balzac transposes well to Sun Tzu: "what matters to a novelist is not to know *all things* but to know *thoroughly* – to be the master of his material, and to penetrate to the heart of the world which actually surrounds him." See André Maurois, *Prometheus: the life of Balzac* (Norman Denny, trans.; New York: Harper & Row, 1965), p. 159 (emphases in original).

[43] An example would be verse III.7's unqualified warning against attacking cities (Passage #3.3), advice that would be at least partly out of date at most a few generations later in Chinese warfare.

[44] Sun Tzu VII's discussion of command control technologies of his time notably refers to an earlier military treatise, now lost. See verse VII.17 (Passage #15.2A) and Griffith, p. 106 footnote 5.

have so much vitality in so many contexts remote from Warring States China.[45] In a nutshell, the most survivable and broadly applicable parts of Sun Tzu are the parts capable of being interpreted as information-realm advice. For just that reason, the twenty-first-century ascendancy of the digital realm unlocks Sun Tzu's relevance and applicability to an extent without parallel in earlier human history.[46]

Illustrating this observation, a majority of the growing points of Sun Tzu's thinking identified in the present book (see Table 9) center on information issues.[47]

Drawing on the fourteen themes analysis, a profile of Sun Tzu as information warrior will now be sketched, with particular attention to Sun Tzu's digital-age reincarnation. From the outset it should be clearly recognized that the emphases of Sun Tzu's information warfare ideas differ in some basic ways from those of the modern area that flies under that banner. In developing those emphases, it is therefore important not to succumb to the "gravitational attraction" of modern information warfare thinking. Doing so would risk reducing the exercise to rediscovering, in ancient Chinese garb, what we already know.[48]

In matters pertaining to strategy and information Sun Tzu stands for an emphasis on human minds and emotions, never on technology per se. The relevant spirit of Sun Tzu is evocatively captured by a pre-battle episode from the *Zuozhuan's* account of the state of Jin's encounter with Chu at Chengpu (632 BC), a consequential major battle of its time: "The Prince of Jin dreamed that he was wrestling with the Master of Chu. The Master of Chu was bending over him and was sucking out his brains."[49]

Such emphasis on the human dimension of war amounts to an intellectual style choice, one that is entirely understandable given Sun Tzu's early time. It is plainly a

---

[45] This is not to gainsay the value of many of Sun Tzu's ideas as they pertain to physical aspects of war or conflict. However, there is a level of insight in, and also a unity to, Sun Tzu's information-realm thinking that somewhat diminishes when Sun Tzu ideas are applied to physical or kinetic levels of war or other conflict (where they also have more competition from other thinkers, ancient and modern).

[46] Even some of Sun Tzu's limitations identified in Table 8 (pp. 498–99) seem less acute in digital realms than in physical ones. For example, Sun Tzu's zero-sum proclivities (discussed in various contexts, see pp. 66, 145–46, 152, 279, 305, 334, etc.) harmonize with tendencies in computer science, especially in cryptography but also more broadly, "to think as adversarially as possible." See p. 14 of Don Monroe, "Pure randomness extracted from two poor sources," *Communications of the ACM* (Association for Computing Machinery), 2017, 60(1), 13–15.

[47] The Table 9 list rounds up ideas flagged in earlier chapters under a rubric of "*What did Master Sun know that we still don't (or have yet to absorb adequately)?*" (a question that should not be construed in too narrow or literal-minded a way).

On a point of clarification regarding Table 9's last entry, even though a strategy/tactics distinction is not found in Sun Tzu, the text erects few if any obstacles to applications of Sun Tzu's ideas in both strategy and tactics. Indeed the fluid texture of Sun Tzu's thinking in many ways encourages doing so. For a relevant insight of Ralph Sawyer see Preliminaries footnote 3.

[48] Cf. Roger Ames's warning, quoted on p. 176 above, regarding pitfalls facing a translator.

[49] See Durrant, Li, & Schaberg, *Zuo Tradition (Zuozhuan)* (cited on p. xxiii above), Vol. 1, p. 417. The translation continues "and because of this the prince was afraid." The punchline of the story is actually a little more complicated, because a creative adviser to the prince was able to give this seemingly ominous dream an auspicious spin, adding the Sun-Tzu-esque observation: "What is more, we are softening them [Chu] into submission." For background on Chengpu, see Wilkinson, §24.1.2, p. 340.

Table 9 *Inventory of selected growing points for modern applications of* Sun Tzu's *thinking*

This list is not exhaustive. Indeed one of the most important and appealing attributes of Sun Tzu's thinking is that fresh applications of it continually turn up, sometimes in unexpected places.

Example and Sun Tzu theme	Brief statement of concept
pp. 68–69 (Theme #1)	never deviating from "calculating mode" (contemporary case in point: don't click on that link!)
p. 69 (Theme #1)	recognizing that calculation comes in multiple species
pp. 69–70 (Theme #1)	pushing the state-of-the-art in net assessment
pp. 72–74 (Themes #1 and #6)	exploiting sabotage vulnerabilities of R&D projects
p. 121 (Theme #3)	recognizing and accepting limits of counterterrorism
pp. 148–49 (Theme #4)	using subversion to achieve strategic advantage (shi)
pp. 177ff. (Theme #5)	mastering four paths to advantage (shi), building on an expansive "definition of the situation"
pp. 193–94 (Theme #5)	drawing shi from rule system design and manipulation (modern examples: legislative and transactional law)
pp. 220–23 (Theme #6)	exploiting algorithm sabotage, often of low visibility
pp. 252–53 (Theme #7)	making fullest appropriate use of deception
pp. 253–55 (Theme #7)	taking advantage of complex environments, including new or emerging ones, as cover for deception
pp. 283–84 (Theme #8)	exploiting formlessness of undersea warfare
p. 341 (Theme #10)	rejecting strategy-as-destruction in favor of strategy-as-control
pp. 379–80 (Theme #11)	understanding humanistic aspects of spymaster role
p. 381 (Theme #11)	exploiting a classified information system (a modern concept) as arena for strategic manipulations
p. 381 (Theme #11)	using biographical intelligence to full potential
pp. 383–84 (Theme #11)	making fullest use of the "most intelligent" (shang zhi) people as spies (or more broadly, observers)
pp. 406–11 (Theme #12)	making the most use of available data sources
p. 440 (Theme #13)	adhering to command responsibility principle
p. 474 (Table 7) (Theme #14)	applying qi/zheng distinction and finding new applications for it
p. 502 (relevant to all 14 themes)	recognizing pitfalls of applying Sun Tzu's thinking on a tactical but not on a strategic level

different style, vastly more "low tech," than that of twenty-first-century information warfare preoccupied with higher mathematics, electronics, and advanced digital technology applications.[50] Yet it is in the nature of styles to invite creative recombinations.[51] Contemporary digital creations are advancing to a point where some of them are assuming credible human-like characteristics. That opens a direct path for transposing Sun Tzu's human-beings-centered thinking to digital applications areas such as robotics (where some humanoid robots have demonstrated an ability to manipulate people emotionally).[52] Meanwhile machine learning technology is well launched on a path to creating machines that make very human decisions.[53]

A more traditional path to Sun-Tzu-as-twenty-first-century-information-warrior would focus, not on anthropomorphic aspects of machines, but on ever more tightly interlocking human + machine combinations, with the machines increasingly holding sway over the things that matter most to the humans. Here Sun Tzu–inspired manipulations would characteristically begin with the human side, exploiting human frailties and foibles. But the goal would be achieving results beneficial to the manipulator that only digital capabilities provide (or that they can potentiate by orders of magnitude). Table 3 (pp. 151–52 above) opens up this territory, drawing textual anchoring from all thirteen chapters of Sun Tzu.[54]

Against this backdrop, the sharp edge of Sun Tzu's way of information warfare is now profiled in terms of four "active ingredients."[55] They are by no means mutually exclusive and often lend themselves to being deployed in combinations, sometimes of still higher efficacy. The most important analytical payoff here has less to do with specific ideas than with conveying a Sun Tzu–inspired mindset – in a nutshell, the kinds of angles and approaches that "thinking like Sun Tzu" should conjure.[56]

Horizon broadening (and with it opportunities for novel deceptions). A first active ingredient centers on an expansive, creative understanding of the strategic situation facing a conflict actor, conjoined with uses of deception to exploit it (Themes #5

---

[50] There are some similarities between this Sun Tzu style of information warfare and the lowbrow (if ingenious) style of deception that a modern analyst (Amrom Katz) has labeled Type B deception, contrasting it with highbrow Type A deception that he profiles as "very modern, very technological, very elegant." See Amrom Katz, "Styles in deception" in Rothstein & Whaley, #7 footnote 1, pp. 145–46.

[51] Boorman, #5 footnote 24; White, #14 footnote 69, pp. 283–86.

[52] Not altogether a new idea. For reference in an early Daoist text to a manmade automaton that exhibits some capabilities of that type see #9 footnote 8.

[53] One relevant analytical direction is Joshua C. Peterson, David D. Bourgin, Mayank Agrawal, Daniel Reichman, & Thomas L. Griffiths, "Using large-scale experiments and machine learning to discover theories of human decision-making," *Science*, June 11, 2021, 372(6547), 1209–14.

[54] If one were to pick *one* area of modern cyber warfare that best captures such Sun Tzu emphases, a leading contender would be phishing – especially "spear phishing" whose target is a specific human actor in a strategically important job or role. The concept of using phishing to gain surreptitious access to an adversary's email system, lurking there (possibly for a long time) to study patterns in that adversary's behavior, then building on the "know the enemy" knowledge thus obtained to pick exactly the right time and way to strike, is close to vintage Sun Tzu, putting into practice themes of calculation, deception, and knowing the enemy (Themes #1, #7, and #11).

[55] Order of presentation of these four ingredients should be treated as flexible.

[56] A further probe into that Sun Tzu mindset comes from scrutinizing "ultra-versatile" verses that receive an especially high number of thematic assignments. For a list of those verses, in which psychologically oriented thinking about information issues figures prominently, see p. A-29.

and #7). The digital realm, with its mirror-imaging of vast swathes of the analog world and ever-proliferating points of connectivity to physical and biological entities, provides a wealth of opportunities for Sun Tzu's horizon-broadening instincts at the heart of Theme #5 to flourish.[57] The point might be put more sharply: almost any major use of computing has potential to enlarge the definition of the situation in exploitable ways, if only because that application itself can become a focus of manipulation. Supply chains present a particularly tempting target for horizon broadening. The vistas on multi-tiered biological organization that modern biology is opening hold a cornucopia of additional possibilities. A shared feature of these and other examples is that conceptualizing the "situation" more expansively than is routinely done (or is convenient to do) can turn up many unheralded sources of leverage or advantage.[58]

Expanding relevant horizons is first and foremost an act of imagination, hence in the first instance falls in an information realm, not a physical one.[59] A fine comment of a *weiqi* player, defeated by an artificial intelligence capability, captures a pertinent spirit: "We have certain patterns in our minds when we play, so this is the kind of move we would never think about."[60] It is important here that Sun Tzu approaches "deception" in an exceptionally broad way. Sun Tzu's winning edge draws leverage from two information fundamentals:

- expanded horizons go hand in hand with new complexities, creating a richer space for deceptions;
- expanded horizons also tend to make all but the most resourceful foe more vulnerable to deception, since operations there (a) often fall outside that foe's "comfort zone" and (b) may also seem just a little unreal (until, of course, their game-changing effects are felt!).[61]

Extraordinary and normal measures, working in combination. A second active ingredient is an information warfare adaptation of Sun Tzu's *qi/zheng* concept (Theme #14). It centers on ensuring that the adversary's scarce high-quality attention

---

[57] Cf. T. S. Eliot's haunting concept of a "wilderness of mirrors," here applied to the ways in which the digital world contains innumerable partial, often somewhat distorted, mirror images of the biophysical world, and vice versa.

[58] Here the internet-of-things concept also makes its presence felt.

[59] There is intellectual kinship here with modern strategist Albert Wohlstetter's quest for the "dominant solution," meaning a "radical reformulation of an issue so that it became amenable to solutions that were not within the parameters of the conventional solution, but that were clearly superior to them." See Rosen, #1 footnote 54, p. 283 (noting examples from command, control, and communications networks as well as pathbreaking uses of improved accuracies of unmanned weapon systems).

[60] See https://qz.com/639952/googles-ai-won-the-game-go-by-defying-millennia-of-basic-human-instinct (visited October 6, 2021).

[61] A relevant warning about "comfort zones" comes from Amrom Katz in his contrast between two styles of deception (see Conclusion footnote 50 above for citation). His central point is that a strategic actor who finds complex high-tech deception ("Type A") competitions congenial, and excels at winning them, may be vulnerable to deception by a foe who adopts a more low-tech deception style ("Type B"), even though that low-tech competition is in many respects far less "sophisticated." As he pithily sums up the resulting out-of-comfort-zone vulnerability: "In short, it's not our style."

is engaged in all-consuming, possibly pressing, but still ultimately routine information-handling tasks (*zheng*) – in an age of information overload an easy goal to realize! – then using "methods outside the regular methods" (to draw on Cao Cao's inimitable framing) to pull the rug out from under the foe (*qi*). Theme #9 also has a part to play here. The adversary's preoccupation with *zheng* steps present and pending – which take on the role of a "fixing force" in conventional warfare – often grows out of, indeed is demanded by, other dynamics present in the situation (themselves commonly mundane, such as a need to meet a never-ending onslaught of administrative deadlines or to do routine maintenance of existing social network ties).[62]

In pulling off this type of caper, the sheer complexity of modern life is an immense asset to a Sun Tzu–inspired actor. Not only does it serve to waterlog the cognitive and observer capabilities of most people (so that they become uncurious about unusual goings-on in their vicinity), but it also provides ready-to-hand camouflage for unusual steps of a nefarious nature because those steps can commonly be lost amid an endlessly churning surfeit of mundane and often artificial details.[63]

Weaponized formlessness. A third active ingredient centers on attacking the adversary's strategy (Theme #6), here placing emphasis on manipulation of the information that informs and shapes it – the more low-visibility that manipulation is, the better.[64] Ideally such an attack or manipulation should remain "formless" to a point where its target's suspicions are not aroused even after the event (Theme #8).[65] Formless manipulation is one of Sun Tzu's most potent ideas, often underappreciated because deception culminating in surprise attack captures the lion's share of attention from Sun Tzu's modern audiences. In a distinctively Sun Tzu half-twist, when presence of any manipulation at all goes unperceived, even its possibility (and its seriousness) may become itself an object of heated debate, creating fissures in the enemy's camp – and thereby realizing Sun Tzu's ambition to create estrangements there.

Although attacks on strategy may take diverse forms in information realms no less than in physical ones, in a contemporary world of algorithms a particularly deep quiver of possibilities involves algorithm sabotage (see pp. 220ff. above). Because so much consequential human decision making is increasingly being delegated, at least in major part, to complex mathematical-logical algorithms – whose workings are

---

[62] This application of Sun Tzu gives new vitality to Roberta Wohlstetter's classic admonition "In short, we failed to anticipate Pearl Harbor not for want of the relevant materials, but because of a plethora of irrelevant ones." See her *Pearl Harbor; Warning and Decision* (Stanford, CA: Stanford University Press, 1962), p. 387.

[63] See discussion of Barton Whaley's "Principle of Naturalness" on p. 255 above.

[64] Compare Minford, p. 174, juxtaposing verse V.20 (Passage #6.15) with an observation of Sun Tzu analyst François Jullien:

> To manipulate the "other" [the enemy], is to make him want to do "of his own accord and will" the very thing that I want him to do, and which I predict will do him harm (but which he thinks will be to his profit). He thinks he is making his own decisions, but in fact I am the one indirectly leading him on.

[65] Strategic problems tend to be more abstract than tactical ones, and for that reason may particularly lend themselves to "formless" measures – say, steps geared toward unobtrusively eroding an existing consensus, modifying what issues are deemed "salient" by an adversary's leadership, or manipulating that adversary's perceptions of risk, self-interest, preference, feasibility, etc.

commonly opaque save to a handful of specialists (sometimes to them too!) – algorithm sabotage stands to be a major growing point for applications of Sun Tzu.

A contrast arises here between the everyday physical world, on the one hand, and the digital world, on the other. In the former, there are powerful practical constraints – physical, social, bureaucratic – on the elaborateness of the effective deception it is feasible to practice there. By contrast, in complex software and algorithm realms it is difficult to set a priori bounds on the number of nested layers of deception or manipulation that are possible. *Indeed there may be no such bounds.*

Algorithm sabotage is a deep and versatile military concept, a core part of Sun Tzu's living legacy to information warfare. Unleashing this aspect of Sun Tzu draws traction from the fact that vetting human beings with access to information systems remains a soft spot of existing computer and network security measures.[66] Indeed that very fact may offer further ways of cheaply disrupting an adversary's poise by communicating to that adversary a sense, true or false, that subverted insiders are "everywhere yet nowhere" – a classic Sun-Tzu-esque exploitation of formlessness.

The human touch. A fourth active ingredient centers on using human spies, a category best broadly conceived (Theme #11, supported by Theme #4 which could involve any of the $2^7 - 1 = 127$ combinations of Max Weber's seven value-spheres).[67] Sun Tzu's rallying cry is verse XIII.14 (Passage #11.6),

XIII.14.

Delicate indeed! Truly delicate! There is no place where espionage is not used.

Here note should also be taken of Sun Tzu's emphasis, not just on any human spy but on exceptionally talented spies (verse XIII.23), an aspect of Sun Tzu's fascination with outliers (Theme #3).

Even while the frontiers of algorithmic capability continue to advance, there will always be information warfare tasks beyond the reach of the state of the computing art, whatever it happens to be. Among those tasks are certain types of sophisticated sabotage of human relations, inaccessible to computers but feasible for talented spies. Those tasks commonly involve making false information – or more broadly, false or misleading stories and storylines – believable to flesh-and-blood human beings, at least to those who matter.[68] Although digital tools unimagined by Sun Tzu have a part to play here (deepfakes, for one), a larger strategic challenge is making sure that a falsehood is not merely spread far and wide (often an easy task in an internet-enabled world!) but also takes deep root in social or political structures and cognitions, ensuring that its effects persist there for a long time, or at least as long as needed. Psychological, social, and bureaucratic adroitness form a basic part of the human spy's relevant skill-set, for example extracting

---

[66] Underscoring the salience of subverted insiders in Sun Tzu's thinking, two of Sun Tzu's five types of spy are subverted insiders ("inside agents," "double agents"). That label may also fit some spies of a third type (Sun Tzu's "native agents"). See pp. 363, 378 above.

[67] See Appendix 4 footnote 45 (list of Weber's value-spheres, with sources).

[68] See verse XIII.19 (Passage #11.6). Here is a good place to note once more that Sun Tzu's focus is unabashedly elitist, overridingly focused on influencing a *small* set of elite strategic actors (ruler, general, etc.). Such emphasis sets Sun Tzu's thinking apart from strands of modern information warfare geared to influencing the thinking of people en masse (e.g., by disseminating "fake news" to mass audiences).

leverage from the fact that bureaucracies have elephantine memories that often defy attempts to "reset" what the organization "knows." Sun Tzu's estrangement thinking suggests that a major part of wielding that spy's skill-set involves creation and exploitation of breaches of social trust, which once broken can be irretrievable (at least for many years).[69]

The truly sharp edge here comes, not from high tech alone, but from joint manipulation of high tech plus human social structure and processes of types long antedating the digital age, with an age-old and very human tool of conflict – the spy – serving as catalyst.[70]

★★★★★★

There is a further level on which Sun Tzu's ideas about strategy and information, *deployed as an integrated package*, are a resource whose value transcends that of the sum of those ideas treated separately. Much of Sun Tzu's "magic potion" resides, somewhat as does that of an artificial neural network, not in any inventory of discrete parts or propositions (where nowadays Sun Tzu's observations could, of course, be replicated from many unrelated sources, often developing them further than Sun Tzu ever did), but on a pattern level of interlocking ideas. Put in a slightly different way, the power of Sun Tzu's thinking resides in a *network* of concepts and principles, along with the distribution of attention they signal.[71]

Because networks rarely lend themselves to capsule summaries, this perspective on Sun Tzu also suggests why Sun Tzu's contributions cannot be adequately conveyed through aphorisms or reduced to a general ambiance.

A network perspective on the idea level of Sun Tzu is a worthy frontier for future Sun Tzu studies. One way of exploring that frontier would start with a twenty-first-century agenda of topics for Sun Tzu (3) research on issues having a major information warfare dimension.[72] In pursuing that agenda a point worth keeping in mind is that Sun Tzu's value-added often resides in what are sometimes called "cross-cutting concepts" – concepts that cut across boundaries born of modern compartmentalizations of knowledge, helping to jolt existing thinking out of long-established ruts and to connect areas and questions commonly left separate.

---

[69] For relevant larger Sun Tzu context see #13 footnote 4, rounding up examples illustrating Sun Tzu's alertness to irreversibilities.

[70] A line of critique not pursued often enough is whether Sun Tzu accords too much weight to the leverage from psychological paths to advantage. Without particularly singling out Sun Tzu, Needham & Yates, p. 66 (previously cited in #5 footnote 51) tender a similar criticism of pre-modern Chinese military thought in general.

Modern digital capabilities offer a strong riposte to this challenge: While such criticism may in fact be valid for some eras in Chinese and other military history, its force dwindles, perhaps even vanishes, in twenty-first-century environments where psychological and digital dynamics are inextricably intertwined and control over physical assets of many kinds is increasingly digital.

[71] To pick just one example, Li Ling, #1 footnote 84, p. 178 envisions Sun Tzu's qi/zheng concept pair (Theme #14) as the core of Sun Tzu's concept of shi, "strategic advantage" (Theme #5), with the further concept pair xu/shi, "emptiness/fullness" (Theme #3), being an extension of qi/zheng.

[72] Inspiration for such an agenda comes from work of Thomas Schelling. See Schelling, #5 footnote 8, pp. 14–15, listing some half-dozen areas of theory by way of responding to a question he poses (p. 13): "What would 'theory' in this field of strategy consist of?"

Topics for such a Sun Tzu (3) agenda might include:[73]

Core active ingredients:[74]

- theory of turning to advantage the latent potential of a situation by spotting a new, exploitable dimension or level or degree of freedom in it (Theme #5)[75]
- theory of deception in complex environments (Theme #7)
- theory of integrated use of crafty and straightforward information measures and mastering switches between them (Theme #14, supported by Theme #9)
- theory of invisible or low-visibility human social networks and their uses in manipulating social and information structures and processes (Themes #11 and #4)
- theory of algorithm sabotage (Themes #6 and 8)

Further conceptual resources:

- theory of social and organizational observation (Theme #12, supported by parts of Theme #13)
- theory of net assessment and other forms of strategic calculation (Theme #1)
- theory of spotting outsized leverage opportunities, including horizons opened by newfound possibilities for digitally enabled lightning speed; also, by more effective identification and use of "outliers" of many kinds (Theme #3);
- doctrine for operating in complex social network situations building on, while extending, Sun Tzu's thinking about complex terrain (Theme #2)[76]
- theory of thinking about the unthinkable (Theme #10).

A useful perspective on these topics is that the evolution of information technology applications is like a circular staircase: as one ascends one continually returns to the same place, but one level up! That image suggests why experiences may tend to repeat with ever more advanced technology – and why even with the most advanced technology of a given era there remain insights to be gleaned from Sun Tzu.

At the beginning of his great treatise on probability theory, William Feller observed that there, as in other areas of human accomplishment, "intuition" can be trained and developed – even the intuition of collectivities.[77] From that perspective one of Sun Tzu's greatest achievements, perhaps the greatest, consists in advancing

---

[73] Alignment on Sun Tzu themes (shown in parentheses) should be treated as a rough guide only.

[74] Themes #5 and #7 are listed separately. Otherwise the thematic pairings shown here align on the four active ingredients of Sun Tzu's way of information warfare as set forth on pp. 507–11 above.

[75] Sun Tzu's concept of shi captures this idea with admirable, if nearly untranslatable, succinctness. See pp. 175–76 above. As previously noted (p. 178 above), the basic move here should be carefully distinguished from escalation, which (as usually understood) does not involve reimagining a conflict situation in a particularly creative or unorthodox way. Indeed, as the start of Sun Tzu's espionage chapter (verses XIII.1–XIII.2 in Passage #11.6) intimates, an enlightened broadening of how a conflict is approached may facilitate de-escalation – i.e., moving downward on the spectrum of conflict.

[76] For illustrative examples see pp. 99ff. above, supplemented by pp. A-8off. in Appendix 2.

[77] See William Feller, An Introduction to Probability Theory and Its Applications, Vol. 1 (3rd ed.; New York: Wiley, 1968), p. 2.

our individual and collective intuitions about how to think strategically about the many roles of information in human conflict, along the way giving us a sharper sense of good information's contribution to the efficacy of good strategy.

A Sun Tzu for the Twenty-first Century

The key to Sun Tzu's living legacy is to be found, not in things Chinese per se, nor in ever more careful study of the details of the text, but in clarifying the extraordinary harmony between Sun Tzu's way of war and twenty-first-century conditions. Sun Tzu's affinity for strategy's information aspects captures part of this harmony, but there is more to it than that. Contemporary conditions have put firmly in place certain structures – a Sun Tzu–inspired name would be "terrain" – characterized by ambient complexity, worldwide in effects, new in degree if not in kind. The resulting sociopolitical environments harbor unprecedented opportunities to realize Sun Tzu's thinking on an ambitious level. There is true serendipity here. Sun Tzu's ideas transpose uncannily well to a set of conditions that did not exist, even in dim outline, in Sun Tzu's time or indeed for millennia thereafter (or even over most of the twentieth century). It is remarkable that a body of military ideas that emerged, of all places, from mass-infantry warfare should be so relevant to a twenty-first-century world.

The doors thus opened might be profiled as "a different kind of war for the twenty-first century." This twenty-first-century "engine" of Sun Tzu applications has no necessary close connection with twenty-first-century China (though that is certainly one natural setting for it). It may also have little to do with aspirations or agendas of nation-states, save to the extent to which those aspirations set context and constraint. Rather, this different kind of war takes its character from how Sun Tzu's ideas resonate with capabilities and actions of assorted agile strategic actors (many of them not state actors) to whom "operating like Sun Tzu" comes naturally under twenty-first-century conditions.

This observation suggests reorienting a basic question posed at the start of the Introduction (p. 1 above). It is not quite as simple as saying that Sun Tzu knew something we don't. A sharper intellectual target would focus on analyzing evolutionary forces Sun Tzu could not possibly have foreseen or imagined and that are now bringing into a realm of feasibility, in some cases for the first time, concepts that Sun Tzu pioneered under far less propitious circumstances.

The Figure 7 analytical diagram profiles the new world. It is a framework for probing Sun Tzu's efficacy in a twenty-first-century environment defined by the diagram's four vertices.[78] Despite the prominence now widely accorded to internet capabilities (an included part of the lower righthand corner's substance), all four vertices should be treated as comparably important. They also reinforce one another, as conveyed by Figure 7's pattern of arrows suggesting what engineers call a truss structure. In terms of Sun Tzu's own concepts, the four vertices of Figure 7 may be envisioned as a modern counterpart to verse IX.16's list of "extreme terrains."[79]

---

[78] A simplified version of Figure 7 was published in Boorman, Introduction footnote 4, p. 97.
[79] See Passage #7.5, which contains both verses IX.16 and IX.17. No correspondence between Fig. 7's vertices and Sun Tzu's specific extreme terrains, of which verse IX.16 lists six, is implied.

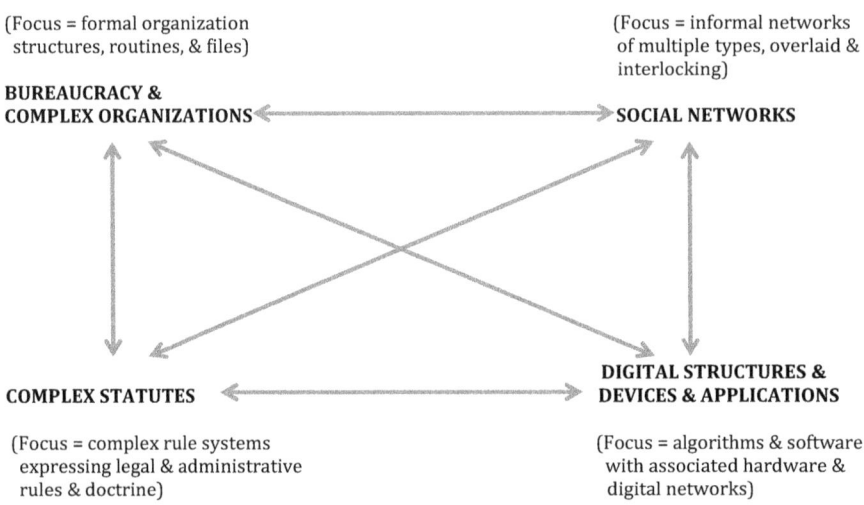

(Focus = formal organization
structures, routines, & files)

(Focus = informal networks
of multiple types, overlaid &
interlocking)

**BUREAUCRACY &
COMPLEX ORGANIZATIONS** ⟷ **SOCIAL NETWORKS**

DIGITAL STRUCTURES &
COMPLEX STATUTES ⟷ **DEVICES & APPLICATIONS**

(Focus = complex rule systems
expressing legal & administrative
rules & doctrine)

(Focus = algorithms & software
with associated hardware &
digital networks)

Figure 7. Four complex environments: a twenty-first-century ecology for Sun Tzu's ideas to flourish.

These four vertices correspond to four types of structures. Each of the four types of structure is:

(1) A major component of the landscape of twenty-first-century social structure and social processes;
(2) A site of massive, and growing, complexity – thereby presenting formidable control challenges for twenty-first-century societies and their leaderships;
(3) Distinct from each of the other three, possessing its own languages, concepts, and principles (in particular, deleting any vertex, or coalescing any two, risks significant analytical distortion because the operational characteristics of each type of structure are different from those of the other ones);
(4) Ever more tightly interlocking with the other three in combined practical effects (the arrows in the figure symbolize that interlock);
(5) Recent in significant aspects (some of those aspects crystallized in the early post–World War II period, still well within living memory; others reflect twenty-first-century developments);
(6) Often low-visibility in impact;
(7) In need of more – and deeper – analysis, singly and in coordination with analyses of the other three.

In the twenty-first century, however, entanglements with social and digital structure make it impossible to follow the wise advice of the first part of verse IX.17, "I keep a distance from these . . ."

The subject matter associated with the four vertices is next briefly sketched from a Sun Tzu standpoint.

Information technology applications: For reasons suggested by the preceding section ("Sun Tzu As Prophet and Pioneering Theorist of Information Warfare"), a natural starting point for a guided tour of Figure 7 is the lower righthand vertex. Sun Tzu's fecundity as a source of strategic inspiration on a twenty-first-century land-scape owes much to the mesh between his ideas and the manipulative potential of information technologies controlled by algorithms and software. Much of that technology involves the Internet – but some does not, since hostile manipulations can exert their influence through computer systems and networks that have never been connected to the Internet (see p. 376 above). Acts of algorithm sabotage may also take place at a software design or testing stage, even before an algorithm has been deployed in the software (or physical) environment that will be its ultimate home. Sun Tzu's thinking has relevance to those cases too. Further rich digital developments grow out of twenty-first-century intersects of computing and biology, setting a stage for transposing Sun Tzu's strategic thinking into biological and biomedical realms. The latter class of Sun Tzu applications is likely to become ever more important as both biology and information technology continue to advance and cross-fertilize.

Social networks: Building on what has just been said about digital environments, an obvious social network arena for operations by a twenty-first-century Sun Tzu is social media – and there is no question that this arena has become, in a very few years, a ubiquitous and as yet relatively uncharted frontier for Sun Tzu–inspired manipulations. Yet social media technological platforms are but one projection of a complex web of interpersonal relationships between flesh-and-blood human beings with whom those platforms coexist, at times uneasily. Sun Tzu's insistence on human psychology as a wellspring of strategic advantage meshes closely with issues of identity and control presented by old-fashioned person-to-person networks, sometimes mediated by advanced technology, other times not.

Complex organizations: The twenty-first century is also a realm of complex organizations and the yet more complex institutional landscapes – involving congeries of many such organizations – to which they give rise.[80] In recent history (well within living memory) many of those complex organizations, worldwide, have passed from being sometimes fragile newcomers to becoming deeply entrenched institutions, some governmental, some for-profit, some private nonprofit. Over the course of their maturation period those institutions have developed an elaborate repertoire of mechanisms (starting with the files they all maintain) for controlling people, both insiders and outsiders. As complex organizations increasingly come to define the basic arena in which human conflict unfolds, those organizational control projects and assorted counter-projects create yet another arena for twenty-first-century Sun Tzus steeped in Sun Tzu's thinking about complex terrain and creating estrangements.

Complex statutes: Behind the world of complex organizations – the stuff of acronyms and news stories – lurks a further, less visible world of complex legal and other rule systems that increasingly enable, empower, shape, and constrain the more visible organizational forms. One name for this phenomenon is the "world of complex statutes."[81] It is increasingly a bedrock of human social structure and

---

[80] An example, the more compelling because visually presented, is a set of c. 150 Executive Branch organizations, large and small, involved in the reorganization that led to the 2002 creation of the US Department of Homeland Security. See "Breaking up is hard. Merging is harder," *New York Times*, June 23, 2002. Relatedly, a common refrain heard from young Washington, DC staffers points to the cumbersome nature of interagency work, often identified as the most frustrating problem they have faced.

[81] "Complex statutes" is used here as a convenient shorthand. Rule systems of high complexity may emanate from many sources of authority, private as well as public.

governance institutions, worldwide – yet at the same time it is a rarified world, terra incognita for most people, even many savvy strategic actors. That world's relevant mechanisms of control (e.g., as channeled through the skill-sets of the statutory or other rule system draftsman) are more abstract, as well as at times more nuanced, than the more visible tools of command control in the repertoire of bureaucracies or other complex organizations. Complex statutes as vehicles of strategic control commonly exert their influence incrementally, often in cumulative ways spanning multiple human generations. Considerable evidence suggests that, short of the rupture created by apocalyptic events (massive war, social revolution, etc.), the ability of human actors to reshape complex statutes in a large-scale way for rational human purposes, once such statutes have taken root on a legal landscape and in the larger social structure of which it forms a part, is distinctly limited. However, that limited controllability does not at all preclude Sun Tzu's thinking from being applied to a complex statute by strategic actors whose parochial goals are unrelated to the human objectives the statute is supposed to serve. Indeed Sun Tzu has an utterly natural place here, for all the reasons that a tiny wording change (or reinterpretation) affecting a single rule – often one of thousands in a particular complex statutory context – may open untold possibilities for artful strategic exploitation.[82]

The four environments just profiled, taken together with all their interactive complications, constitute a natural home for twenty-first-century Sun Tzus. To borrow Sun Tzu's terminology, Figure 7 gives rise to an endless variety of forms (xing) and, along with them, a bottomless quiver of deception or manipulation possibilities. The complexity underlying Figure 7 is a modern counterpart to the complex physical terrain that did much to catalyze Sun Tzu's thought in Warring States China (Theme #2). Absent adequate maps (which Sun Tzu's generals did not possess and which we also lack for much of Figure 7's substance), operating in Figure 7's "terrain" necessitates guides. In his time Sun Tzu repeatedly called for such guides. Yet modern trends toward specialization mean that reliable guides versed in all four corners of Figure 7's "terrain" are few in number, if indeed they can be found at all. That deficit, of course, intensifies the opacity of the Figure 7 environment and compounds deception opportunities present there.

A hallmark of a twenty-first-century Sun Tzu is ability to achieve comparative advantage in the Figure 7 environment, building on better information about it than is possessed (or at least is actively exploited) by peers, rivals, and foes. Such information translates into a never-ending churn of opportunities for gaining or creating still further advantage or, in Sun Tzu's terminology, shi (Theme #5).

What is new about such dynamics – and what sets it apart from the endless ebb and flow of factional politics and conflicts throughout recorded human history, well known to Thucydides and Sima Qian alike – is the truly unprecedented, mind-numbing complexity of the Figure 7 environment. That manmade complexity, greater by orders of magnitude than any present in earlier human history, affords unprecedented opportunities for cover and,

---

[82] Income tax law, including in its international aspects, is an extremely rich arena for Sun Tzu–inspired thinking. See Charles P. Rettig, "Tax enforcement: reading tea leaves in a tax gap environment," *Tax Notes*, March 8, 2010, 126, 1263–70 (p. 1266 quotes Sun Tzu verse III.3 in the Giles translation, "supreme excellence consists in breaking the enemy's resistance without fighting"). I am indebted to Katsiaryna Zinavenka for discussion on point.

especially in light of digital capabilities, for aggressive, speedy, nuanced, yet low-visibility deceptions and manipulations, thereby unleashing Sun Tzu's true potential for our own time.

That new world gives rise to a series of conjectures.

A first conjecture is that increasing the complexity of Figure 7's four environments may *up to a point* stimulate deeper, more creative strategic thinking in many quarters; but then, as that complexity continues to grow, the necessary calculations start to appear so truly formidable as to deter most actors from pursuing them at all. Digital environments are an obvious venue for instances of such "demoralized" behavior, though the same pattern is encountered in all four.

A second conjecture is that discouraged behavior like this will come to describe many, indeed most, "large" strategic actors (among them, many leaders and ruling circles of great powers; also, major bureaucratic actors, public or private). These "big battalions" will become tolerant of unimaginative, often sloppy, solutions, even ones that by no means draw fully on available cognitive and computing resources. They will walk away from Sun Tzu's counsel, though some may give it lip service.[83]

Pushing further, a third conjecture is that the complexity jointly produced by the four environments of Figure 7 will contribute to an erosion, at times outright breakdown, of many established lines of authority, opening a door to sprawling, decentralized pursuit of human agendas of many different kinds.[84] Theme #13 ("Leadership and harmony in the chain of command") makes its presence felt here – but through its failures, not its successes. One way of framing this development is as a kind of phase transition: what previously had been a set of instrumentalities, albeit sometimes clumsy ones, of major, very large players dissolves into a soup of fragments of structure lacking larger coherence. Warlord China of a century ago comes to mind as a twentieth-century partial prototype, as in a different way does Warring States China. The second part of the haunting opening line of the *Romance of the Three Kingdoms* applies well here: "The empire, long divided, must unite; long united, must divide."

A fourth conjecture follows at once. Strategic opportunity of this magnitude is guaranteed to set in motion (in some cases, to bring into being for the first time) opportunistic, aggressive lesser strategic actors capable of exploiting it – indeed a whole population ecology of such actors, many of low visibility at least in some of their activities (though often not fully clandestine).[85] Some of those opportunistic

---

[83] Cf. Mao Zedong's critique of "leaders [who] have no head for strategy and are confused by complicated circumstances; hence, they are at the mercy of these circumstances, lose their initiative and have recourse to passive response." See Mao, #4 footnote 92, pp. 129–30.

[84] These tendencies are often veiled by obfuscatory rhetorics – unsurprisingly so, since all parties (including those whose authority has eroded) may have a stake in keeping up appearances.

[85] Something of the relevant type of strategic opportunity is captured by Stanislaw Ulam, perhaps best known as one of the fathers of the hydrogen bomb, who presciently wrote:

Speaking of the fascination of surprises ... it may be remarked how often it happens that in the game of chess one may observe weak players or even rank beginners getting into deep and fascinating positions. I have often watched amateurs or non-talented beginners, looked at their game after some fifteen moves, observed that their position arrived at perhaps by chance, certainly not by design, was *full of marvelous possibilities for both sides.* And I wonder how it is that the game itself produces these positions of great appeal and art without these simple fellows being even aware of it. I do not know whether an analogous experience is possible in the game

lesser actors will start out as part of, or in service of, state actors, but then start reading Sun Tzu for themselves (actually or metaphorically!), taking to heart his ideas and moving toward de facto (usually low-visibility) autonomy. Such a pattern is partially foreshadowed in Chinese context by the Bo Xilai case.

But in other cases a small group or network, owing allegiance to no particular larger actor, simply sets up shop in the bowels of a complex organizational milieu and commences operating for its own purposes in that environment.[86] Based in nooks and crannies of huge, unwieldy institutional structures such actors will aggressively move to exploit the strategic opportunities afforded by Figure 7's four vertices and their runaway complexity.[87] Doing so with high efficacy will be precisely those actors' core capability. While some of them may have military ties, their métier is likely to center on "winning by non-military ways and means" (Theme #4). Sun Tzu's "active ingredients" profiled in the previous section furnish a basic toolkit and repertoire for achieving that goal. Crucial logistics support comes from ubiquitous availability of modern communications and computing technology; in no earlier era, even a few decades ago, have such rapid, secure, low-visibility tools been available on a global stage, with the opportunities for sophisticated calculating, planning, and coordinating they make possible (Theme #1).

These Sun-Tzu-esque actors will be decidedly agile and are likely to be *relatively* small. Small size tends to favor realization of *wu xing* (formlessness) and all the advantages it affords in strategic competitions with far bigger actors (Theme #8).[88] If their alignment on Sun Tzu is strong, those smaller actors' operations will almost surely be partly illicit, a pattern encouraged by the combined thrust of Themes #3 (cheap expedients), #7 (deception), #10 (no scruples), and #11 (know the enemy). Depending on the legal environment (e.g., applicable privacy laws), giving fullest effect to Themes #1 and #12 may also induce lawbreaking.

What most defines these smaller actors is that they profit from and, wittingly or not, often further develop Sun Tzu's insights in ways that the larger actors cannot or

---

of Go. I cannot myself judge, not knowing much about the intricacies of that beautiful game, but I wonder whether a master looking at a position can tell whether it was arrived at by chance or by a logically developed correct and thoughtful play.

[Emphasis supplied.]

See Stanislaw M. Ulam, *Adventures of a Mathematician* (New York: Scribner, 1976), pp. 279–80.

[86] Relevant here is Charles Tilly's concept of a "trust network" developed in his *Trust and Rule* (cited in #10 footnote 21) and the various "bottom up" strategies that such a network may pursue – among them, concealment, dissimulation, clientage, and predation (Tilly, p. 104). Particularly worth noting here is a "dissimulation strategy," which Tilly characterizes as "feigning conformity by adopting some available public identity, but minimizing both compliance and visibility of internal operations and resources" (p. 104).

[87] Unlike the large, established players, these Sun-Tzu-esque actors commonly have no stake in the smooth running of the system as a whole, hence are less likely to be daunted by the "intellectual and social burdens of a dense complexity" – a challenge about which, candidly, they couldn't care less. (Phrase in quotes is from the concluding sentence on p. 1813 of Richard B. Stewart, "The reformation of American administrative law," *Harvard Law Review*, 1975, 88(8), 1667–1813.)

[88] "Huge fish have no room to turn their bodies around in a ditch that is a few yards wide, but mud loaches can cut circles through it." See *Wandering on the Way: early Taoist tales and parables of Chuang Tzu* (translated with an introduction and commentary by Victor H. Mair; Honolulu: University of Hawai'i Press, 1994), p. 226.

do not try to emulate.[89] Here it is again important to stress that there is little in Sun Tzu that requires a mass of institutional machinery to put it into effect. Much of *Sun Tzu's way of war* actually works better in the absence of such machinery, which tends to retard, not boost. the strategic creativity and flexibility so central to Sun Tzu's thought.[90]

A fifth conjecture – a facet Sun Tzu's "know the enemy and know yourself" – is that the Sun-Tzu-esque actors thus profiled will start to become known to one another as well as to various often sluggish major players. The resulting dynamics are worthy of a full-fledged study. Only a few possibilities will be suggested here.

One class of possible dynamics involves the attempts of a large actor to control (perhaps if feasible to stamp out!) smaller Sun-Tzu-esque actors operating in its own backyard. Lest that dynamic be too quickly or uncritically assimilated to counterinsurgency, it should be noted that those smaller actors will usually have goals very different from a classic revolutionary war agenda. They may well not be at all interested in "taking over" (whatever than means!), instead setting their sights on economic gain or (often relatedly) on certain types of political influence.[91] But a wise observation of David Galula, shaped by his time as a junior French Army officer fighting Front de libération nationale (FLN) rebels in Algeria, transposes well here. A basic asymmetry Galula noted was that insurgents could draw for guidance on a coherent body of people's war doctrine, whereas the counterinsurgent (certainly in that era) had no comparable body of doctrine.[92] A related asymmetry today may plague the large player trying to eradicate or contain a smaller Sun-Tzu-esque actor. The latter has a time-tested doctrine provided by Sun Tzu; the former, maybe not.[93]

In the resulting strategic competitions, formlessness may be a more formidable resource for a Sun-Tzu-esque small actor than are deception capabilities per se. Formlessness of that small actor compels a large actor to take sometimes complicated affirmative steps, struggling to pierce a veil of opacity even to ascertain basic parameters of that small actor, along the way acknowledging that the small actor actually exists and is hard to eliminate – which for a sluggish large actor may be inconvenient truths.

---

[89] However (by contrast to Mao in the caves of Yan'an in the 1930s), few, if any, of these twenty-first-century Sun Tzus are likely to vouchsafe to us candid written descriptions of the story, or perceived active ingredients, of their successes.

[90] Here it is worth noting again that Sun Tzu's general customarily appears on stage alone, seemingly a lone sentient actor like a Daoist hermit-sage, rather than being surrounded by any visible apparatus of aides and staff. Regarding lack of mention of military general staffs by Sun Tzu see p. 427 above.

[91] Here is a departure from "classic" Sun Tzu (1), where conquest is the unalloyed goal.

[92] David Galula, *Pacification in Algeria, 1956–1958* (new Foreword by Bruce Hoffman; Santa Monica, CA: RAND Corporation, 2006), "No doctrine for the counterinsurgent," pp. 64–68.

[93] Relatedly, referring to "building a network [that connects] people or institutions in specific ways for specific purposes," see Anne-Marie Slaughter, *The Chessboard and the Web: strategies of connection in a networked world* (New Haven, CT/London: Yale University Press, 2017), p. 13: "We have no playbook for strategies of connection, or for crafting the tools we need to implement them. To create it, we must turn to network theory as Schelling once turned to game theory. Different networks have different structures and properties for different purposes."

Taking a cue from the idea of the lever scales (Sun Tzu's *quan*) one strategy that Sun-Tzu-esque actors may often use to survive and prosper, capitalizing on low profile, nimbleness, and lack of scruples – all classic facets of Sun Tzu's way of war – involves assuming a formless posture (Theme #8) with respect to long-running conflicts between much larger established actors. Rather than espousing neutrality, such a posture may involve rendering assistance to both sides, possibly of asymmetric and time-varying types.[94] Theme #9, in the form of a keen nose for smelling out trends in a situation, has a place in a Sun-Tzu-esque actor's calculations here.

A distinct dynamic arises when Sun-Tzu-esque actors, operating in the same territory or environment, meet one another. There is clearly potential here for both conflicts and alliances. A relatively unexplored face of Sun Tzu centers on strategy and counter-strategy likely to ensue when one such Sun-Tzu-esque actor clashes with another. In the run-up to such a clash, which could be protracted, one can imagine each of the actors probing again and again to try to feel out the "form" (*xing*) of the other.[95] That feeling-out process, and the tentative encounters to which it gives rise, has potential to generate bizarre outcomes (which might actually provide a useful clue for an outside observer that a low-visibility strategic competition between Sun-Tzu-esque actors is taking place).

There are also alliance possibilities, where one Sun-Tzu-esque actor starts to cooperate with another. Despite possible motives for alliance formation, however, implications of Sun Tzu's focus on deception, lack of scruples, and general tilt toward a logic of short-term gain suggests openness to profitable betrayals and associated major potential for alliance instability. It is also much easier to follow Sun Tzu's counsel when operating in the shadows, or at least in twilight, than under the glare of the publicity that increased size, possibly born of alliance combinations, may trigger. Those considerations also work to limit the extent to which initially small Sun-Tzu-esque actors are likely to emerge as major players on a world stage, either through alliances or autocatalytic successes (albeit that some may quietly come to have the ear of major players there).

Yet even some small Sun-Tzu-esque actors stand to play an important role in the making of twenty-first-century grand strategy through effects they exert as "wild cards." Whether in planned or inadvertent ways their activities have potential to destabilize strategic competitions between far larger great power actors, possibly by dint of sowing internal distrust and estrangements (Theme #6), always a Sun Tzu forte and one where a manipulator's small size may not matter much.[96] Impact of such small actors stands to be enhanced by effects of a further source of

---

[94] This type of stratagem might be seen as a case of "robust action" (see p. 103 above). There are also traces here of an updated (and more abstract) counterpart to the Maoist revolutionary warfare concept of border area bases, artfully situated on the boundary between spheres of influence of large actors unfriendly to one another, giving a small, nimble actor positioned there an opportunity to play off that hostility for survival and possible gain. See Boorman, Introduction footnote 13, pp. 72–73.

[95] For a thumbnail sketch of a potential "Sun Tzu vs. Sun Tzu" scenario see p. 273 above.

[96] Sun Tzu's concepts of *zheng* and *qi* (Theme #14) find uses here. The big players are fully committed, in fact overcommitted, to difficult but straightforward problem-solving (*zheng*). Sun-Tzu-esque small actors proceed to take *qi* ("crafty") steps that, intentionally or not, upset the apple cart.

twenty-first-century instability born of a wide range of immensely potent twenty-first-century weapons, great and small, lethal and non-lethal, having capabilities unimagined in Sun Tzu's time (or even in most quarters throughout much of the twentieth century). Here Sun Tzu's openness to special weapons thinking, via Sun Tzu XII on fire and water, sets a baseline conducive to aggressive special weapons uses by Sun Tzu's modern disciples. Some of those special weapons will be digital. Some may be kinetic or have kinetic effects.

There remains need for clarification of pathways – a version of dynamic net assessment for asymmetric conflict – by which Sun-Tzu-esque strategic actors, at first perhaps locally dangerous (but no more than that), come to present far more serious systemic risks. As Joseph Needham put it, "that is another story, and, as the Szechuanese tea-garden story-tellers say, 'If you want to know how this ended, you should come back at the same time tomorrow evening.'"[97]

★★★★★★

---

[97] Joseph Needham (with others), *Clerks and Craftsmen in China and the West: lectures and addresses on the history of science and technology* (Cambridge, UK: Cambridge University Press, 1970), "The unity of science; Asia's indispensable contribution," p. 29 (published version of UNESCO Month Lecture, Beirut, 1948).

Index A

Index of Chinese Texts

Note: In scholarship on early China it is common practice to refer to a text using the name of its traditionally attributed author. For example, "Laozi" (lit. "Master Lao") may refer either to the text or to its putative author. In such cases, index entries below gather all cases where the reference may be to the text, excluding only those cases – usually few in number – where the reference is clearly to the person, not the text.

Index B

For EU product safety concerns, contact us at Calle de José Abascal, 56–1°,
28003 Madrid, Spain or eugpsr@cambridge.org.

www.ingramcontent.com/pod-product-compliance
Ingram Content Group UK Ltd.
Pitfield, Milton Keynes, MK11 3LW, UK
UKHW042217061225
465726UK00003B/88